Microbes: Diversity and Biotechnology

Microbes: Diversity and Biotechnology

Editors
Prof. S.C. Sati
Dr. M. Belwal
Department of Botany,
Kumaun University,
Nainital – 263 002 (UK)

2012
DAYA PUBLISHING HOUSE®
New Delhi – 110 002

ISBN 978-81-7035-794-0

Published by **Daya Publishing House®**
4760-61/23, Ansari Road, Darya Ganj,
New Delhi - 110 002
Phone: 23245578, 23244987
Fax: (011) 23260116
E-mail: dayabooks@vsnl.com
website: www.dayabooks.com

Laser Typesetting : **Classic Computer Services**
Delhi - 110 035

Printed at : **Chawla Offset Printers**
Delhi - 110 052

PRINTED IN INDIA

Preface

Mycology is one of the oldest known subjects of science dealing with fungi, a very different kind of organisms which are neither plant nor animal but belong to a separate kingdom Mycetae. Fungi have such a fascinating world of microorganism that creates curiosity and attraction among the scientific community. Fungal world embodies diversified groups of organisms which colonize, multiply and survive in nature on many macro and micro-ecological niches and substrates. Fungi are eukaryotic, achlorophyllous, filamentous or unicellular heterotrophic living organisms and are ubiquitous in distribution. Of the estimated 1.5 million species of fungi, more than 98,000 fungal species have now well been described. The utilization and application of fungi by mankind has a long and varied history. They play an important role in biodegradation, recycling of organic matter, pharmaceutical industries, agriculture, medicine, industry, bioremediation, waste management and other activities. It is true that hardly any day passes when we are either harmed or benefited by these organisms.

Microbes, including fungi, constitute an important component of biodiversity and comprise one of the biggest kingdoms in the living world. As there are more than 1 lac species of fungi known which grow and multiply in all diverse habitats in nature. But unfortunately, a little attention has been paid to these organisms. Uses of fungi for mankind has broadened and grown considerably during in the last century. A well known example of this includes the discovery and production of antibiotics and wider utilization of fungi in the food industry, where they are used in the production of various compounds, hormones and enzymes for food processing. The association of many fungi with plants has resulted in their use in agriculture as both biological control agents against plant disease and as plant growth stimulants such as mycorrhizal inoculants. This has now also become a part of biotechnology as well as biochemistry. Therefore, their management, conservation and exploitation for the betterment of mankind is must.

Though there are many national and international Journals that bring out information about fungi and microbes, but almost in scattered state. It is quite desirable to keep all these in a book/ monograph at one place to provide more information about mycological progress especially being done in different corners of India. Therefore, the recent mycological work into a monographic shape is a time demand for the benefit of various persons engaged in the field of mycology in India and abroad.

In this endeavor we have tried to accommodate the articles devoted to different aspects of mycological progress.

It is our humble attempt to bring out an edited monograph titled Contribution to the Mycological Progress. It is an excellent collection of 30 original research/review articles of eminent mycologists and active researchers from different parts of India including Nepal. We hope that this monograph would be very useful to all scientists, teachers, students and readers whosoever is interested on mycology and mycological literature. A line of critical comment on this book is always welcome from the readers/ users of this book.

We wish to express our sincere gratitude to all the contributors of the articles for this book to achieve our objectives. We are thankful to our colleagues Prof. Uma Palni, Prof. Neerja Pande, Dr. Y. S. Rawat, Dr. Lalit Tewari, Dr. Kiran Bargali and Dr. Susma Tamta. We are also grateful to Miss Savita Joshi, Dr. Pratibha Arya, Dr. Kapil Khulbe and many other well wishers for their active assistance time to time. We would also like to thank the publisher, *Daya Publishing House*, New Delhi, for bringing out the monograph timely.

S.C. Sati

M. Belwal

Contents

List of Contributors

Adholeya, Alok
Centre for Mycorrhizal Research, Biotechnology and Management of Bioresources Division, The Energy and Resources Institute (TERI), Indian Habitat Centre, Lodhi Road, New Delhi – 110 003

Aggarwal, Ashok
Botany Department, Kurukshetra University, Kurukshetra – 136 119, Haryana

Arya, Arun
Department of Botany, Faculty of Science, The Maharaja Sayajirao University of Baroda, Vadodara – 390 002

Arya, Pratibha
Department of Botany, D.S.B. Campus, Kumaun University, Nainital – 263 001

Belwal, M.
Botany Department, D.S.B. Campus, Kumaun University, Nainital – 263 001

Bhandari, B.S.
Department of Botany, Birla College Campus, H N B Garhwal University, Srinagar – 246 174, Uttarakhand

Bisht, Saraswati
Department of Botany, D.S.B. Campus, Kumaun University, Nainital – 263 001

Budhathoki, U.
Central Department of Botany, Tribhuvan University, Kathmandu, Nepal

Chandra, Bipin
Department of Plant Pathology (MRTC) College of Agriculture, G.B. Pant University of Agriculture and Technology, Pantnagar – 263 145, US Nagar

Chandra, Bipin
Department of Plant Pathology, College of Agriculture, G.B. Pant University of Agriculture and Technology, Pantnagar – 263 145, US Nagar, Uttarakhand

Chhetri, B.K.
Central Department of Botany, Tribhuvan University, Kathmandu, Nepal

Choksi, Shirali K.
Department of Botany, Faculty of Science, The Maharaja Sayajirao University of Baroda, Vadodara – 390 002

Dhondiyal, Prabha (nee Bisht)
Department of Botany, D.S.B. Campus, Kumaun University, Nainital – 263 001

Dubey, N.K
Centre for Advanced Study, Banaras Hindu University, Varanasi – 221 005, Uttar Pradesh

Guleri, S.
Department of Botany, H N B Garhwal University, Srinagar – 246 174, Uttarakhand

Gupta, R.C.
Department of Botany, Kumaun University, S.S.J. Campus, Almora – 263 601, Uttarakhand

Hosagoudar, V.B.
Tropical Botanic Garden and Research Institute, Palode – 587 116, Thiruvananthapuram, Kerala

Jain, Rachana
Department of Bioscience and Biotechnology, Banasthali University, Banasthali – 304 022, Rajasthan

John, Jinu
Biotechnology Lab, Department of Botany, Faculty of Biological and Chemical Sciences, Dr. H.S.G. University, Sagar – 470 003

Joshi, Savita
Department of Botany, D.S.B. Campus, Kumaun University, Nainital – 263 001

Khulbe, Kapil
D.S.B. Campus, Kumaun University, Nainital – 263 001

Kumar, A.
CAS Botany, Banaras Hindu University, Varanasi – 221 005, Uttar Pradesh

Majumder, M.
Department of Botany, Rajiv Gandhi University, Rono Hills, Itanagar – 791 112

Marihal, Avitha K.
Department of Botany, Karnatak University, Dharwad – 580 003, Karnataka

Mastan, S.A.
Post Graduate Department of Biotechnology, D.N.R. College, P.G. Courses and Research Centre, Bhimavaram – 534 202, Distt. W.G., Andhra Pradesh

Mehrotra, V.S.
Agriculture Division, PSS Central Institute of Vocational Education, 131 Zone II, M.P. Nagar, Bhopal – 462 011, M.P.

Mehta, Archana
Biotechnology Lab, Department of Botany, Faculty of Biological and Chemical Sciences, Dr. H.S.G. University, Sagar – 470 003

Mehta, Pradeep
Biotechnology Lab, Department of Botany, Faculty of Biological and Chemical Sciences, Dr. H.S.G. University, Sagar – 470 003

Mishra, Meenakshi
Lab of Microbial Technology and Plant Pathology, Department of Botany, Dr. H.S. Gour University, Sagar – 470 003, M.P.

Negi, P.S.
Defence Agricultural Research Laboratory, Pithoragarh – 262 501, Uttaranchal

Pande, Neerja
Department of Botany, D.S.B. Campus, Kumaun University, Nainital – 263 001

Pargaien, N.
Department of Botany, D.S.B. Campus, Kumaun University, Nainital – 263 001

Parkash, Vipin
Rain Forest Research Institute (ICFRE), Autonomous Council of Ministry of Environment and Forests, Govt. of India, Deovan, Jorhat – 785 001, Assam

Pokhrel, C.P.
Central Department of Botany, Tribhuvan University, Kathmandu, Nepal

Prabhuji, S.K.
Biotechnology and Molecular Biology Centre, M.G. Post Graduate College, Gorakhpur – 273 001, U.P.

Prakash, B.
CAS Botany, Banaras Hindu University, Varanasi – 221 005, Uttar Pradesh

Radhika, Nair
Department of Botany, Goa University, Goa – 403 206

Ramesh, Ch.
Department of Botany, Karnatak University, Dharwad – 580 003, Karnataka

Rekha
Department of Microbiology, College of Basic Sciences and Humanities, G.B. Pant University of Agriculture and Technology, Pantnagar – 263 145, US Nagar, Uttarakhand

Rodrigues, B.F.
Department of Botany, Goa University, Goa – 403 206

Ruby
Department of Microbiology, College of Basic Sciences and Humanities, G.B. Pant University of Agriculture and Technology, Pantnagar – 263 145, US Nagar, Uttarakhand

Saini, Jitendra
Department of Microbiology, CB SH, GB. Pant University of Agriculture and Technology, Pantnagar – 263 145, US Nagar

Sati, S.C.
Department of Botany, D.S.B. Campus, Kumaun University, Nainital – 263 001

Saxena, Jyoti
Department of Bioscience and Biotechnology, Banasthali University, Banasthali – 304 022, Rajasthan

Saxena, S.
Department of Botany, Shri Guru Ram Rai Post Graduate College, Dehradun – 248 001, Uttarakhand

Sharma, Vinay
Department of Bioscience and Biotechnology, Banasthali University, Banasthali – 304 022, Rajasthan

Shukla, A.K.
Department of Botany, Rajiv Gandhi University, Rono Hills, Itanagar – 791 112

Shukla, R.
CAS Botany, Banaras Hindu University, Varanasi – 221 005, Uttar Pradesh

Singh, Pradeep Kumar
Goverment Naveen College, Janakpur, Korea, C.G.

Singh, Reena
Centre for Mycorrhizal Research, Biotechnology and Management of Bioresources Division, The Energy and Resources Institute (TERI), Indian Habitat Centre, Lodhi Road, New Delhi – 110 003

Singh, S.
Allahabad Agricultural Institute, Deemed University, Allahabad – 211 007, Uttar Pradesh

Singh, Priyanka
CAS Botany, Banaras Hindu University, Varanasi – 221 005, Uttar Pradesh

Sinha, S.K.
Fisheries Department, U.P., Gorakhnath, Gorakhpur – 273 005, U.P.

Sridhar, K.R.
Department of Biosciences, Mangalore University, Mangalagangotri, Mangalore – 574 199, Karnataka

Srivastava, Deepanjali
Department of Home Science, St. Joseph College for Women, Gorakhpur – 273 009, U.P.

Swain, M.R.
Department of Biotechnology, College of Engineering and Technology, Techno Campus, Ghatkia, Bhubaneswar – 751 003

Tewari, Lakshmi
Department of Microbiology, College of Basic Sciences and Humanities, G.B. Pant University of Agriculture and Technology, Pantnagar – 263 145, US Nagar, Uttarakhand

Thatoi, H.N.
Department of Biotechnology, College of Engineering and Technology, Techno Campus, Ghatkia, Bhubaneswar – 751 003

Tripathi, P.
Department of Botany, DAV-PG College, Kanpur – 208 001

Upadhyaya, Manju Lata
Department of Botany, Kumaun University, S.S.J. Campus, Almora – 263 601, Uttarakhand

Vyas, Deepak
Lab of Microbial Technology and Plant Pathology, Deparment of Botany, Dr. H.S. Gour University, Sagar – 470 003, M.P.

Yongam, Y.
Department of Botany, Rajiv Gandhi University, Rono Hills, Itanagar – 791 112

Microbes: Diversity and Biotechnology (2012) *Pages* **1–36**
Editors: **Prof. S.C. Sati & Dr. M. Belwal**
Published by: **DAYA PUBLISHING HOUSE, NEW DELHI**

Chapter 1
The Black Mildews

V.B. Hosagoudar

Tropical Botanic Garden and Research Institute,
Palode – 587 116, Thiruvananthapuram, Kerala

ABSTRACT

The present paper deals with a exhausting account on the Black or dark mildews, foliicolous fungi distributed over two orders namely. Microthyriales and Meliolales with a number of families. These fungi flourish well in tropics and have extended their distribution to subtemperate to temperate regions of the world. Since these fungi do not cause any appreciable pathogenicity on the staple food producing crop plants, attention has not been paid much to this group. Economically, like other biotrophs, these fungi increase the temperature in the areas covered by the black colonies, initiate more respiration, reduce the efficiency of the chlorophyll, reduce total sugar, etc. The whole description is categories on their hosts and complete list of black mildews is enumerated.

Keywords: Foliicolous fungi, Meliolales, Black mildews.

Introduction

Fungi are heterotrophs and absorb prepared food from different sources. Depending upon the source of absorption and association with its partner or associated substratum, they are classified as saprophytes, parasites and symbionts. Parasites obtain nutrients partly or fully from the living tissues of another organism. The parasites which are totally dependent on living organisms are called obligate parasites. The parasites which feed both on living and non-living organisms are called facultative or opportunistic parasites or saprophytes. Hence, the obligate parasites have to adjust and modify themselves with their partners for their survival. Certain obligate parasites kill the host tissues either by producing toxins or enzymes and then absorb nutrients from the hosts. These are called necrotrophs (necros-death, trophy-feeding). Certain parasites absorb nutrients from the living tissues, without killing them, by the specialised organs like appressoria, haustoria or nutritive hyphae. These are called biotrophs (bio-life, trophy- feeding). Fungi differ in their morphology, ecology, life history strategies, etc. The extent of fungal diversity is so high that, it is very difficult to get expertise in all the fungal groups. Hence the present study is restricted here to foliicolous fungi with reference to black mildews. Taxonomically, black mildews or black colony forming parasitic fungi belong to several taxonomic groups, namely, Hyphomycetes, Meliolales, *Schiffnerula* and its anamorphic forms, Asterinales, Meliolinaceae, etc.

In contrast to Powdery mildews, Black or dark mildews are obligate mostly ecto-parasites produce black colonies on the surface of the host plants. The term "sooty moulds" was loosely applied all the black colony forming fungi. Hughes (1976) clearly made a distinction between "sooty moulds' and "black mildews". Sooty moulds are totally distinct from these in their nutritional habit, grow on insect secretion or on nectar produced by the plants and spread on entire surface (irrespective of leaf, petiole, stem or dead bark) of plants. When handled, such colonies stick-on to hands and clothes. Close observation of these uniformly spread dense colonies reveal their association with ants, thrips or nectar glands of the plants. In contrast to these, black mildews are obligate parasites and are specific in infecting their compatible hosts. In short, black mildews are obligate parasites, while, sooty moulds are saprophytes.

Defination

Earlier, the term "Mildew" was referred to Erysiphaceae, Perenosporaceae, Uredinales (Knight, 1818; Murray, 1908), insects (Forsyth, 1824), the fungal growth on old cloth, leather, paper,' or wood (Cooke, 1878). According to Large (1940), it might have originated from the German *mehltau* or Gothic *mealdeau mili* for honey and *tau* for dew (Yarwood, 1978). The other spellings are meledeaw, myldewe, myldeawe, mildewe, mieldew and mealydew (Murray, 1908). By its continuous usage, the term "Mildew" is parallel to the term "fungus". Based on this parallel term, Powdery mildews, downy mildews, sooty moulds, black mildew, etc. came into practice.

These fungi flourish well in tropics and have extended their distribution to subtemperate to temperate regions of the world. Since these fungi do not cause any appreciable pathogenicity on the staple food producing crop plants, attention has not been paid much to this group. Economically, like other biotrophs, these fungi increase the temperature in the areas covered by the black colonies, initiate more respiration, reduce the efficiency of the chlorophyll, reduce total sugars, etc. (Hosag. *et al.*, 1997). Hence, Wellman (1972) stated that "nowhere are these black mildews being made a subject of major pathological study, although agriculturists who observe their crops well, know that at times these fungi are very damaging in their effects."

Methodology

Foliicolous fungi mainly infect leaves, soft stems and tender shoots of herbaceous plants to trees of 30-40 metres in height. Some trees possess the crown only at the upper portion and pose difficulty in noticing the disease and also in the collection. In such cases, recently shed leaves were examined, collected, concerned tree was traced, twig or the reproductive parts were collected for the host identity. For each collection, a separate field number was given, field notes were made regarding the pathogenicity, nature of colonies, nature of infection, locality, altitude, etc. In the field, such infected plants were collected separately in polythene bags with the host twig (preferably with the reproductive parts to facilitate the corresponding host identity). These infected plant parts were pressed neatly between the wire presses and dried in-between blotting papers. Regular transfer of the collections to the fresh and dry blotters ensured the dryness of the collections and such materials were used for the microscopic studies. Plants were indentified with the help of field notes and the subject experts. Later the identity of the host plants were confirmed by matching them with the authentic materials placed in the herbarium.

Preparation of Permanent Slides

Just by the external appearance, it is difficult to state its identity. Hence, the colonies were mounted to reveal their identity. In case of ectophytic or superficial fungi, scrapes were made directly from the infected host parts and mounted in Lactophenol (prepared according to Rangaswamy, 1975). A tinge of Cotton blue was added to Lactophenol to stain hyaline fungal parts. Dematiaceous fungi were first mounted in 10 per cent KOH solution and later transferred to Lactophenol. Both mountants work efficiently and make the septa visible by clearing melanin content.

To study the entire colony in its natural condition, a drop of high quality natural coloured or well transparent nail polish, devoid of gilt, was applied to the selected colonies and carefully thinned with the help of a fine brush without disturbing the colonies. Colonies with hyperparasites showed wooly nature and were avoided. The treated colonies along with their host plants were kept in dust free chamber for half an hour. When the nail polish on the colonies dried fully, a thin, colourless or slightly apple rose coloured (depending upon the colour tint in the nail polish) "film or flip" was formed with the colonies firmly embedded in it. In case of soft host parts, a slight pressure on the opposite side of the leaves and just below the colonies lifts the flip up. In case of hard host parts, the flip was eased off from the edge with the help of a razor orscalpel. A drop of DPX was spread on a clean slide and the flip was spread properly on it. One or two more drops of DPX were again added on the flip and a clean cover glass was placed over it. A gentle pressure on the cover glass brought out excess DPX and it was easily removed after drying. Care was taken to avoid air bubbles. These slides were labeled and placed in a dust free chamber for 1-2 days for drying. These permanent slides were then used for further studies (Hosagoudar and Kapoor, 1984).

For the innate fungi, a rectangular piece of 1-2 mm broad and 2-4 mm long, by keeping the infected spots at the centre was cut from the infected plant part, placed on a clean slide, one to two drops of sterile water dropped on the cut portion, allowed it to soak for 2-3 minutes, sectioned by holding it firmly between the forefinger of the left hand and the slide by using a sharp blade. Thin sections with the desired fungal and the host parts were sorted out, mounted in lactophenol with cotton blue mountant for the microscopic observation.

Preparation of Herbarium

After confirming the identity of the fungus and the corresponding host plant, they were kept in polythene covers. Later, these folders were placed in thick paper envelopes of convenient size with the name of the host, locality, date of collection, place of collection, name of the collector with the field number written on the top corner. These envelopes were serially arranged in a closed rack, based on their collection numbers. Friction between the envelope and the material was avoided to keep the fungal parts intact. Care was taken to place them in a closed system so as to avoid silver fishes and other insects.

Key to the Components of Black Mildews

Most of the time, it will be surprised to know that the colonies mounted are belonging to different genera of related or unrelated groups. Hence, to confirm the identity of the fungi of these black mildews requires more than one subject expert or a seasoned mycologist.

After mounting the colonies, we come across the materials which can be classified as:

1.	Hyphae with green content	....	Blue green algae
1.	Not so	....	2
2.	Produce sporangium like structures with brick-red content	....	Red algae
2.	Producing typical fungal mycelium	....	3
3.	Mycelium constricted at the septa, colonies peeled-off easily	....	Sooty mould
3.	Mycelium not so	....	4
4.	Ascocarp lacking, produce conidia on conidiophores	....	Hyphomycetes
4.	Produce ascocarp	....	5
5.	Produce thyriothecia	....	6
5.	Produce perithecia	....	9
6.	Ostiole present	....	Microthyriales
6.	Ostiole lacking	....	7
7.	Mycelium lacking or very much scanty	....	Parmulariaceae
7.	Mycelium present	....	8
8.	Thyriothecia orbicular, dehisce stellately at the centre	....	Asterinaceae
8.	Thyriothecia oval to elongated, dehisce vertically at the centre	....	Lembosiaceae
9.	Asci unitunicatae	....	Meliolales
9.	Asci bitunicatae	....	10
10.	Thyriothecia/perithecia with radiating cells, central portion dissolved, often 3-4-septate ellipsoidal or sarciniform conidia produced	....	*Schiffnerula*
10.	Not so	....	11
11.	Mycelium non-appressoriate	....	14
11.	Mycelium appressoriate	....	12
12.	Perithecia setose	....	*Balladynopsis*
12.	Perithecia glabrus	....	13
13.	Mycelial setae present	....	*Balladyna*
13.	Mycelial setae absent	....	*Balladynocallia*
14.	Conidia present	....	*Alina*
14.	Conidia absent	....	*Dysrhynchis*

Meliolales

Meliolales Gaumann ex Hawksworth and O. Eriksson, Systema Ascomycetum 5: 142, 1986.

Parasites on vascular plants. Mycelium mostly superficial, appressoriate. Appressoria mostly two celled, rarely many celled. Phialides unicellular. Ascomata flattened-globose to globose, ± ostiolate, peridium smooth, surface cells protruded, often supplemented with setae, appendages; asci on basal hymenium, unitunicate, 2-8 spored, clavate to cylindrical, evanescent; ascospores 1-4 septate, brown at maturity.

Type family: Meliolaceae

Key to the Families of Meliolales

1. Hyphae with phialides; asci clavate; ascospores 3-4 septate ....*Meliolaceae*
1. Hyphae without phialides; asci cylindrical to sub cylindrical, ascospores 1-2 septate ....*Armatellaceae*

Armatellaceae

Armatellaceae Hosag., Sydowia 55: 162, 2003.

Leaf parasites, ectophytes, mycelium appressoriate, phialides and mycelial setae absent. Perithecia on superficial hyphae, globose, verrucose, without appendages and setae; asci 4-8-spored; ascospores 1-2-septate, brown at maturity.

Type genus: *Armatella* Theiss. and Sydow

Armatella and *Basavamyces* are the two genera included in the family Armatellaceae.

Key to the Genera of Armatellaceae

1. Ascospores uniseptate, septum almost median ....*Armatella*
1. Ascospores 2-septate, septa at the distal ends ....*Basavamyces*

Meliolaceae

Meliolaceae Martin ex Hansf., Mycol. Pap. 15: 23, 1946.

Parasitic on vascular plants, mycelium mostly superficial, appressoriate, phialidic. Ascomata flattened-globose to globose, ± ostiolate, peridium with conoid cells, larviform and striated appendages, or with repent or strong setae. Asci unitunicate, 2-4-spored, clavate to cylindrical, evanescent; ascospores 3-4-septate, brown at maturity.

Type genus: *Meliola* Fries

This family includes genera: *Amazonia, Asteridiella, Appendiculella, Irenopsis* and *Meliola* (Hansford, 1961). Hosagoudar (1996) and Hu *et al.* (1996) have followed Hansford (1961). Later, the genera *viz. Cryptomeliola, Endomeliola, Ectendomeliola, Pauhia* and *Prataprajella* have been added to this family (Mibey and Hawksworth, 1997; Hosagoudar, 1992, 1996, 2008; Hughes, 1995; Hughes and Pirozynski, 1994; Hosagoudar and Agarwal, 2008).

Key to the Genera of Meliolaceae

1. Hyphae and appressoria partly endophytic .... *Ectendomeliola*
1. Hyphae and appressoria superficial only.... 2
2. Perithecia flattened-globose, hidden in the radiating mycelium .... *Amazonia*
2. Perithecia globose, discrete, not hidden in the radiating mycelium 3
3. Mycelial setae present .... *Meliola*
3. Mycelial setae absent.... 4
4. Perithecial setae and or larviform appendages present.... 5

4. Both perithecial setae and larviform appendages absent .... *Asteridiella*
5. Both perithecial setae and or larviform appendages present.... 6
5. Only perithecial setae present .... *Irenopsis*
6. Only larviform appendages present .... *Appendiculella*
6. Both perithecial setae and larviform appendages present .... *Prataprajella*

It is evidenced by the Indian Monographs (Hosagoudar, 1996, 2008) that the Indian subcontinent is rich in its meliolaceous fungal flora. Beeli's digital fomula is the prime tool in the identification of meliolaceous taxa and is being modified from time to time.

Digital Formula

Beeli formula consists of 8 digits. The first 4 digits before the stop (left side to the stop) represent the morphological characters like ascospore septation, presence or absence and the nature of the perithecial setae or appendages, presence or absence and the nature of the mycelial setae and the arrangements of appressoria, respectively. The second 4 digits, after the stop, represent the measurements such as length and breadth of ascospores, diameter of perithecia and length of mycelial setae, respectively. The species having both simple and dentate setae is denoted by 1/3, while species having straight and uncinate setae are designated as 1/2. The Beeli formula is modified here to accommodate the genus *Armatella* having 1-2 septate ascospores. Further, for *Prataprajella*, the second digit becomes ¾ or so.

Earlier Used Digital Formula in all the Monographs

I. Morphology (First four digits from left)

1. Normal septation of ascospores
    1. 1-septate
    2. 3-septate
    3. 4-septate
2. Perithecia
    1. Without setae or appendages
    2. With larviform, horizontally striated appendages
    3. With uncinate or coiled setae
    4. With straight setae
3. Mycelial setae (often on perithecia and from subiculam)
    0. Absent
    1. Simple
    2. Simple, entire, uncinate or coiled
    3. Dentate or shortly furcate (up to 30 μm)
    4. Branched (branches more than 30 μm)
4. Appressoria
    1. Alternate or unilateral (less than 1 per cent opposite)
    2. Regularly opposite
    3. Both opposite and alternate

II. Measurements (Second four digits from the full stop)

5. Maximum ascospore length
   1. Below 20 µm long
   2. Between 21-30 µm long
   3. Between 31-40 µm long
   4. Between 41-50 µm long
   5. Between 51-60 µm long
   6. More than 60 µm long
6. Maximum ascospore width
   1. Up to 10 µm broad
   2. Between 11-20 µm broad
   3. Between 21-30 µm broad
   4. More than 30 µm broad
7. Maximum diameter of perithecia
   1. Up to 100 µm
   2. Between 101-200 µm
   3. Between 201-300 µm
   4. More than 301 µm
8. Maximum length of mycelial setae
   1. Up to 300 µm long
   2. Between 301-500 µm long
   3. Between 501-1000 µm long
   4. More than 1000 µm long
   0. Absent.

Modified Digital Formula (Hosagoudar, 2003)

Beeli (1920) proposed a numerical code to categorize the members of the Meliolaceae. Stevens (1927, 1928) modified and extensively used this code in his work. Hosagoudar (1996), Hosagoudar *et al.* (1997), Hu *et al.* (1996, 1999) and Mibey and Hawksworth (1997) have followed Hansford (1961) in using this code as key character and also supplemented the code with other characters to distinguish the individual taxa. However, the Beeli's code or formula has certain limitations, such as:

1. The code does not distinguish all the genera, namely, *Amazonia* Theissen, *Asteridiella* McAlpine, *Pauhia* Stevens and *Prataprajella* Hosag.
2. The code does not distinguish morphology of the head cells of the appressoria
3. The code does not distinguish position of phialides
4. In short, Beeli formula is useful in assembling like species, but it does not separate the taxa with clear demarcation

To overcome these difficulties and to incorporate the newly described genera: *Endomeliola* Hughes and Pirozynski, *Pauhia* Stevens and *Prataprajella* Hosag., the Beeli's digital formula has been modified (Hosagoudar, 2003).

The modified digital formula consists of three groups of digits. The first group of four digits distinguishes all the genera using eleven characters. The second group of five digits uses nineteen characters to denote arrangement of appressoria, the morphology of the head cells of appressoria, location of the phialides in the colony, morphology of the mycelial setae and septation of ascospores. The third group of five digits uses twenty three characters. Appressoria are mostly borne just below the septum and can be distinguished based on the distance between appressoria. This shows whether the appressoria are crowded or sparsely arranged. Then, length and breadth of ascospores, diameter of perithecia and length of the mycelial setae are important. Taxa can be further distinguished by noting the position of the colony on the host, nature of the colony, morphology of mycelium, branching pattern, shape of the cells of appressoria, pattern of distribution of setae, ascospore shape, etc. can be used as an additional characters to distinguish the taxa.

I – Group (Generic characters)

(1) Mycelium

1..........endophytic (*Endomeliola*)

2..........ectophytic

(2) Ascomata

1..........stromatic (*Pauahia*)

2..........flattened-globose and in or with radiating mycelium (*Amazonia*)

3..........globose

(3) Mycelial setae

0..........absent

1..........present (*Meliola*)

(4) Setae and appendages of Ascomata

0..........absent (*Asteridiella*)

1..........with larviform appendages (*Appendiculella*)

2..........with setae (*Irenopsis*)

3..........with setae and appendages (*Prataprajella*)

II – Group (Morphology of appressoria, mycelial setae and position of phialides)

(1) Appressoria (Capitate hyphopodia)

1............alternate and/or unilateral

2............opposite

3............alternate and opposite

(2) Head cells of appressoria

1............entire

2............angulose to slightly lobate

3............sublobate to deeply lobate

(3) Phialides (Mucronate hyphopodia)

1............mixed with appressoria

2............borne on a separate mycelial branch

(4) Mycelial setae

0............absent

1............simple, straight, acute to obtuse at the tip

2............simple, uncinate to coiled

3............dentate or shortly furcate (>30 µm)

4............branched and the branchlets diverged

5............simple and dentate

6............simple and branched or furcate

(5) Ascospores

1............1-septate

2............2-septate

3............3-septate

4............4-septate

III – Group (Measurements)

(1) Length of mycelial cells

1............ up to 10 µm

2............11-20 µm

3............21-30 µm

4............31-40 µm

5............41 µm and above

Length of ascospores

1............up to 20 µm

2............21-30 µm

3............31-40 µm

4............41-50 µm

5............51-60 µm

6............61 µm and above

(2) Width of ascospores

1............up to 10 µm

2............11-20 µm

3............21-30 µm

4............31 and above

(3) Diameter of ascomata

1............up to 100 µm

2............101-200 µm

3............201-300 µm

4............301 µm and above

(4) Length of setae

0............. setae absent
1............ up to 300 μm
2............301-500 μm
3............501-1000μm
4............1000 μm and above.

The first four digits make generic distinction as:

- ☆ 1300 – *Endomeliola*
- ☆ 2100 – *Pauhia*
- ☆ 2200 – *Amazonia*
- ☆ 2300 – *Asteridiella*
- ☆ 2301 – *Appendiculella*
- ☆ 2302 – *Irenopsis*
- ☆ 2303 – *Prataprajella*
- ☆ 2310 – *Meliola*

The first species in the Hansford's Monograph is:

Asteridiella crustacea (Speg.) Hansf., Sydowia 10: 47, 1957. Hansf., Sydowia, Beih. 2: 26, 1961.

Colonies epiphyllous, up to 1.5 mm in diameter. Hyphae substraight to undulate, branching opposite, acute, densely reticulate and subsolid, cells 15-20 x 10-12 μm. Appressoria alternate, antrorse, straight or slightly bent, mostly 30-35 μm long; stalk cells cuneate, 6-14 μm long; head cells clavate with crenate to sublobate margin, 20-25 x 12-18 μm. Phialides numerous in some colonies, rare in others, mixed with appressoria, opposite or alternate, ampulliform, 22-30 x 8-10 μm, neck elongated. Perithecia scattered, verrucose, up to 290 μm in diameter, perithecial wall cells rounded to obtusely conoid, up to 15 μm high; ascospores ellipsoidal, obtuse, 4-septate, constricted at the septa, 60-75 x 30-34 μm.

On leaves of *Drimys* sp., Brazil, Puiggari, type (SPEG).

This description converted into the digital formula as:

2300. 12104. 26430 Colonies epiphyllous, dense; hyphae substraight to undulate, closely reticulate and subsolid; appressoria antrorse, 30-35 μm long ....*Asteridiella crustacea*

In most of the cases, colonies are mixed with other ectophytes and it is also not uncommon that two fungi belonging to the same genus are also inter mixed by making them difficult to write their description correctly, specially incase of spores. In such cases, germinating spores will be traced; the morphology of the appressoria produced by the spores will be compared with that in the colony so as to confirm the identity of the ascospores.

The Order Asterinales

Leaf parasites. Mycelium ectophytic, with or without appressoria, nutrient mycelium and leaf permating stroma present. Ascomata ectophytic, dimidiate, orbicular with radiating cells, astomatous, dehisce stellately at the center; asci globose, spherical, octosporous, bitunicate; ascospores two to many septate, conglobate, hyaline to brown.

Type family: Asterinaceae

Key to the Families of Asterinales

1. Thyriothecia orbicular, dehisce stellately at the center ....*Asterinaceae*
1. Thyriothecia oval to elongated, X or Y shaped, dehisce longitudinally at the center ....*Lembosiaceae*

The Family Asterinaceae

Luttrell in Ainsworth *et al.* (eds.). The Fungi. An advanced Treatise 4: 207, 1973; Arx and Muller, Stud. Mycol. 9: 40, 1975; Hosag., Abraham and C.K. Biju, J. Mycopathol. Res. 39: 62, 2001. Asterinaceae Hansf., Mycol. Pap. 15: 189, 1946.

Leaf parasites. Mycelium ectophytic, with or without appressoria, nutrient mycelium and leaf permating stroma present. Ascomata ectophytic, dimidiate, orbicular with radiating cells, astomatous, dehisce stellately at the center; asci globose, spherical, octosporous, bitunicate; ascospores two to many septate, conglobate, hyaline to brown.

Type genus: *Asterina* Lev.

Key to the Genera of Asterinaceae

1. Appressoria present 2
1. Appressoria absent or very rarely observed 7
2. Appressoria intercalary ....*Asterolibertia*
2. Appressoria lateral 3
3. Appressoria in clusters ....*Ishwaramyces*
3. Appressoria not so 4
4. Mycelium/Thyriothecia setose ....*Trichasterina*
4. Mycelium/Thyriothecia devoid of setae 5
5. Ascospores uniseptate 6
5. Ascospores biseptate and taper towards one end ....*Meliolaster*
6. Septa laid at the middle of ascospore ....*Asterina*
6. Septa laid at the extreme end forming one pinch- off cell ....*Vishnumyces*
7. Appressoria formed only around the stomata ....*Symphaster*
7. Appressoria not formed ....*Prillieuxina*

Key to the Species of Asterinaceae

ACANTHACEAE

Asterina

1. Appressoria alternate only 2
1. Appressoria alternate and opposite ....*A. betonicae*

2. Ascospores more than 20 μm long ....*A. tertia*
2. Ascospores less than 20 μm long ....*A. tertia* var. *africana*

ANACARDIACEAE

Asterina

1. Appressoria alternate and opposite on *Nothopegiae* ....*A. nothopegiae*
1. Appressoria opposite, 10 per cent solitary on *Lannea* ....*A. lanneae*

Asterolibertia

1. On *Mangifera* ... *A. mangiferae*
1. On *Nothopegia* .... *A. nothopegiae*

ANNONACEAE

Asterina

Single species ....*A. miliusae*

Prillieuxina

Single species ....*P. polyalthiae*

Trichasterina

Single species ....*T. goniothalami*

APIACEAE

Asterina

Single species ....*A. hydrocotyles*

APOCYNACEAE

Asterina

Single species ....*A. wrightiae*

ARECACEAE

Asterina

Single species ....*A. arecacearum*

ARISTOLOCHIACEAE

Asterina

Single species ....*A. thotteae*

ASCLEPIADACEAE

Asterina

1. Appressoria unicellular ....*A. tylophorae-indicae*
1. Appressoria 2- celled 2
2. Basal cell gibbous ....*A. cynanchi*
2. Basal cell not so ....*A. toxocarpi*

AVERRHOACEAE

Asterina

Single species ....*A. averrhoae*

BERBERIDACEAE

Asterina

Single species ....*A. goosii*

CAESALPINIACEAE

Asterina

Single species ....*A. saracae*

CELASTRACEAE

Asterina

1. On *Glyptopetalum* ....*A. glyptopetali*
1. On *Microtropis* 2
2. Appressoria alternate ....*A. microtropidis*
2. Appressoria alternate and opposite ....*A. microtropidicola*

CHLORANTHACEAE

Asterina

Single species ....*A. sarcandrae*

CLUSIACEAE

Asterina

1. Appressoria unicellular ....*A. morellae*
1. Appressoria 1- 2 celled ....*A. garciniae*

Prillieuxina

Single species ....*P. garciniae*

COMBRETACEAE

Asterina

1. On *Calycopteris* ....*A. combreti*
1. On *Quisqualis* ....*A. escharoides*

DILLENIACEAE

Asterina

Single species ....*A. acrotremae*

DIPTEROCARPACEAE

Asterina

1. Appressoria ovate, conoid, broadly rounded at the apex ....*A. hopeae*
1. Appressoria conoid, attenuated at the apex .... *A. hopiicola*

Asterolibertia

Single species ....*A. vateriae*

EBENACEAE

Asterina

Single species ....*A. diospyri*

Prillieuxina

Single species ....*P. diospyri*

ELAEGNACEAE

Prillieuxina

Single species ....*P. elaegni*

ELAEOCARPACEAE

Asterina

1. Appressoria cylindrical, uncinate, forked ....*A. elaeocarpicola*
1. Appressoria not so 2
2. Appressoria ovate, oblong ....*A. elaeocarpi* var. *ovalis*
2. Appressoria ovate, conoid, rounded at the apex ....*A. gamsii*

ERICACEAE

Asterina

Single species ....*A. hakgalensis*

ERYTHROPALACEAE

Asterina

Single species ....*A. erythropalicola*

EUPHORBIACEAE

Asterina

1. Appressoria unicellular ....*A. aporusae*
1. Appressoria 2- celled 2
2. On *Glochidion* 3
2. On the other hosts 4
3. Appressoria alternate only ....*A. lobulifera*
3. Appressoria alternate and opposite ....*A. lobulifera* var. *indica*
4. Appressoria alternate only ....*A. phyllanthigena*
4. Appressoria alternate, 1 per cent opposite ....*A. malloticola*

Meliolaster

Single species ....*M. aporusae*

FABACEAE

Asterina

Single species ....*A. millettiae*

FLACOURTIACEAE

Asterina

1. Appressoria unicellular ....*A. granulosa*
1. Appressoria 2- celled ....*A. hydnocarpi*

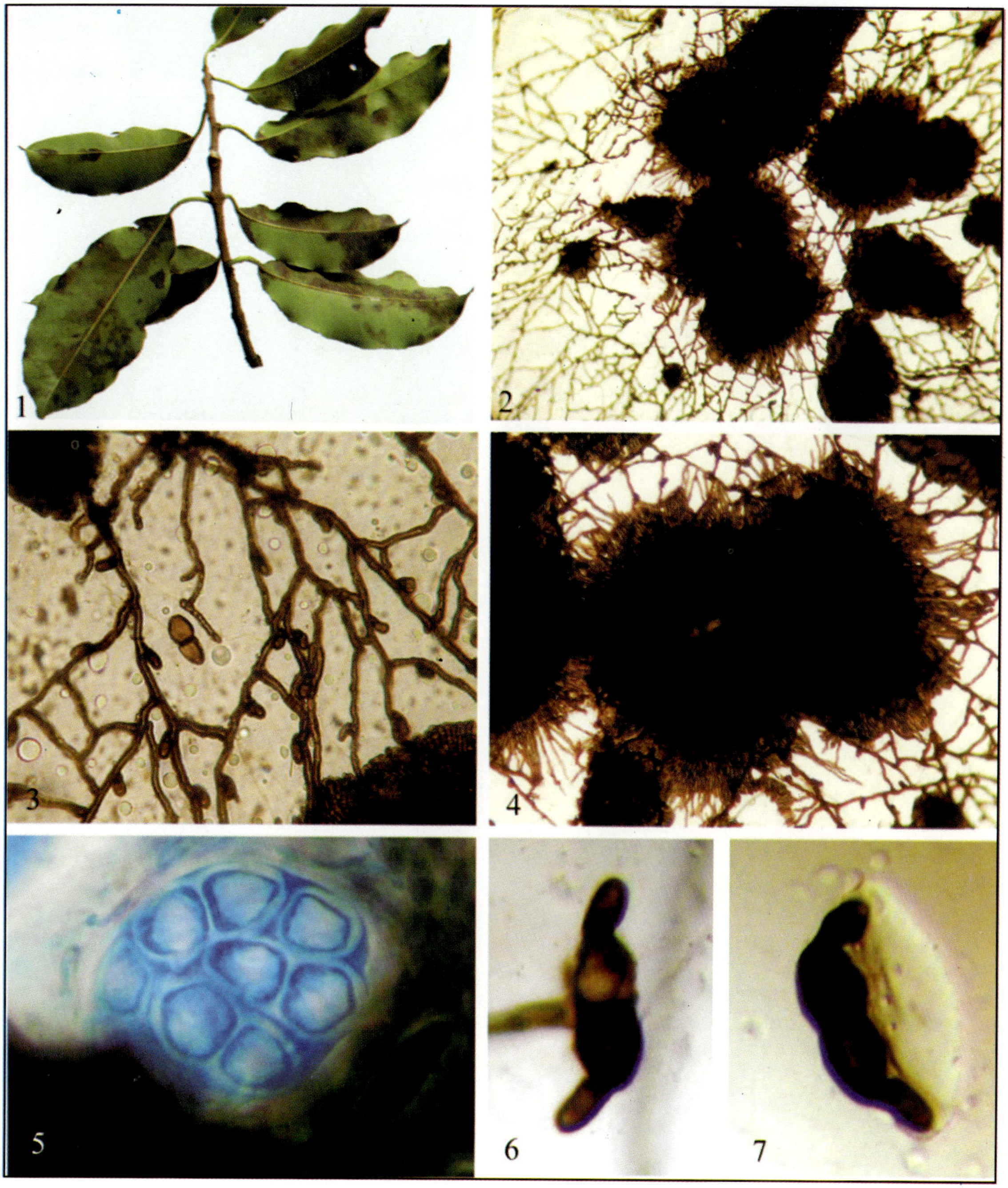

Plate 1.1: ***Asterina*** **sp.**

1: Infected leaves; 2: Colony with thyriothecia; 3: Closely appressed appressoria; 4: Thyriothecia; 5: Ascus; 6 and 7: Germinating ascospores

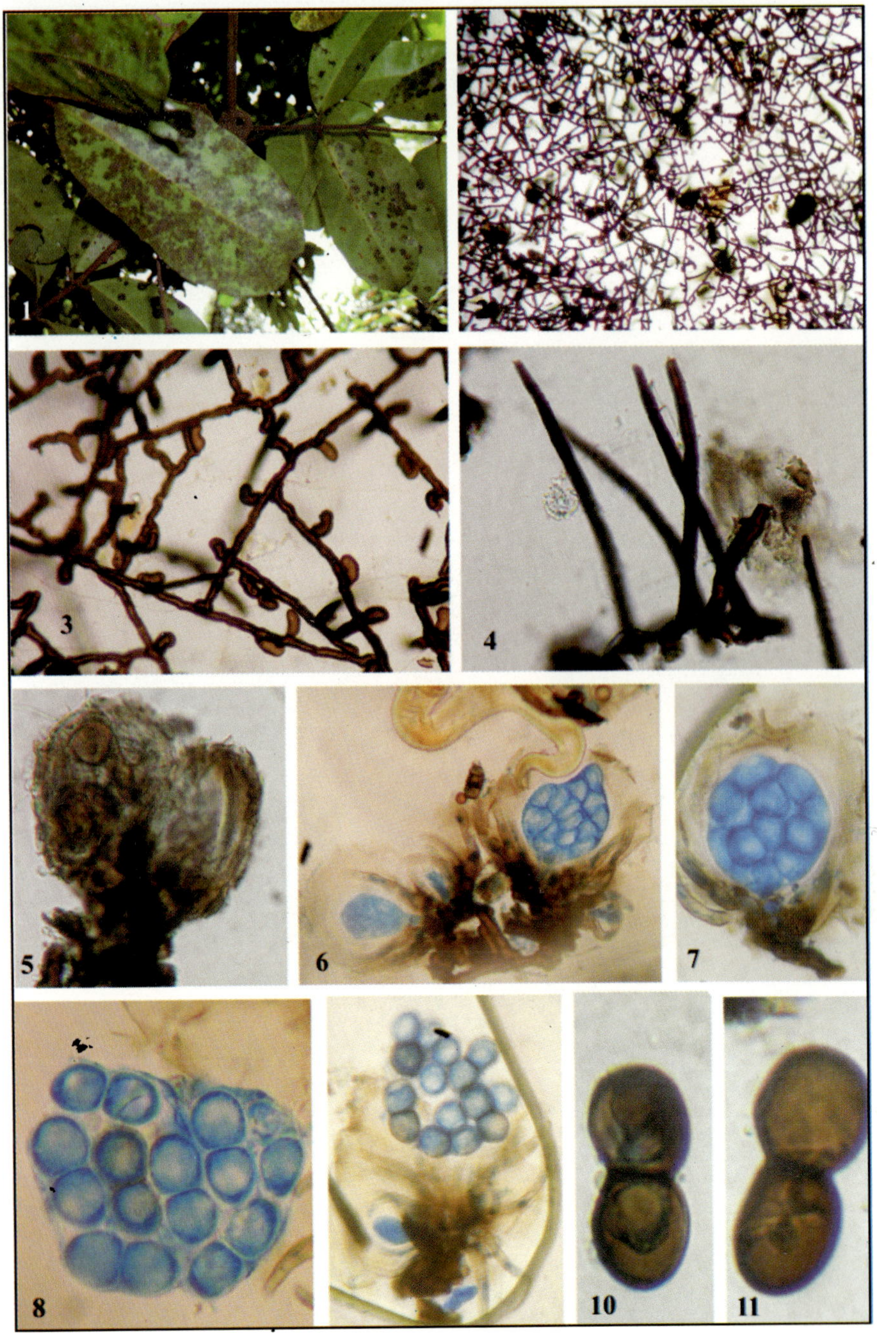

Plate 1.2: ***Balladyna*** **sp.**

1: Infected leaves; 2: Mycelial colony with perithecia; 3: Appressoriate branched mycelium; 4: Mycelial setae; 5: Stalked perithecium; 6–9: Asci; 10–11: Brown ascospores.

Plate 1.3: ***Dysrhynchis palmicola*** **(Sydow) Arx**

1: Habit of the plant; 2: Infected leaves; 3: Mycelium with perithecia; 5: Branched mycelium; 5–6: Mycelial seta; 7: Perithecia; 8: Broken perithecia showing asci; 9–10: Ascospores

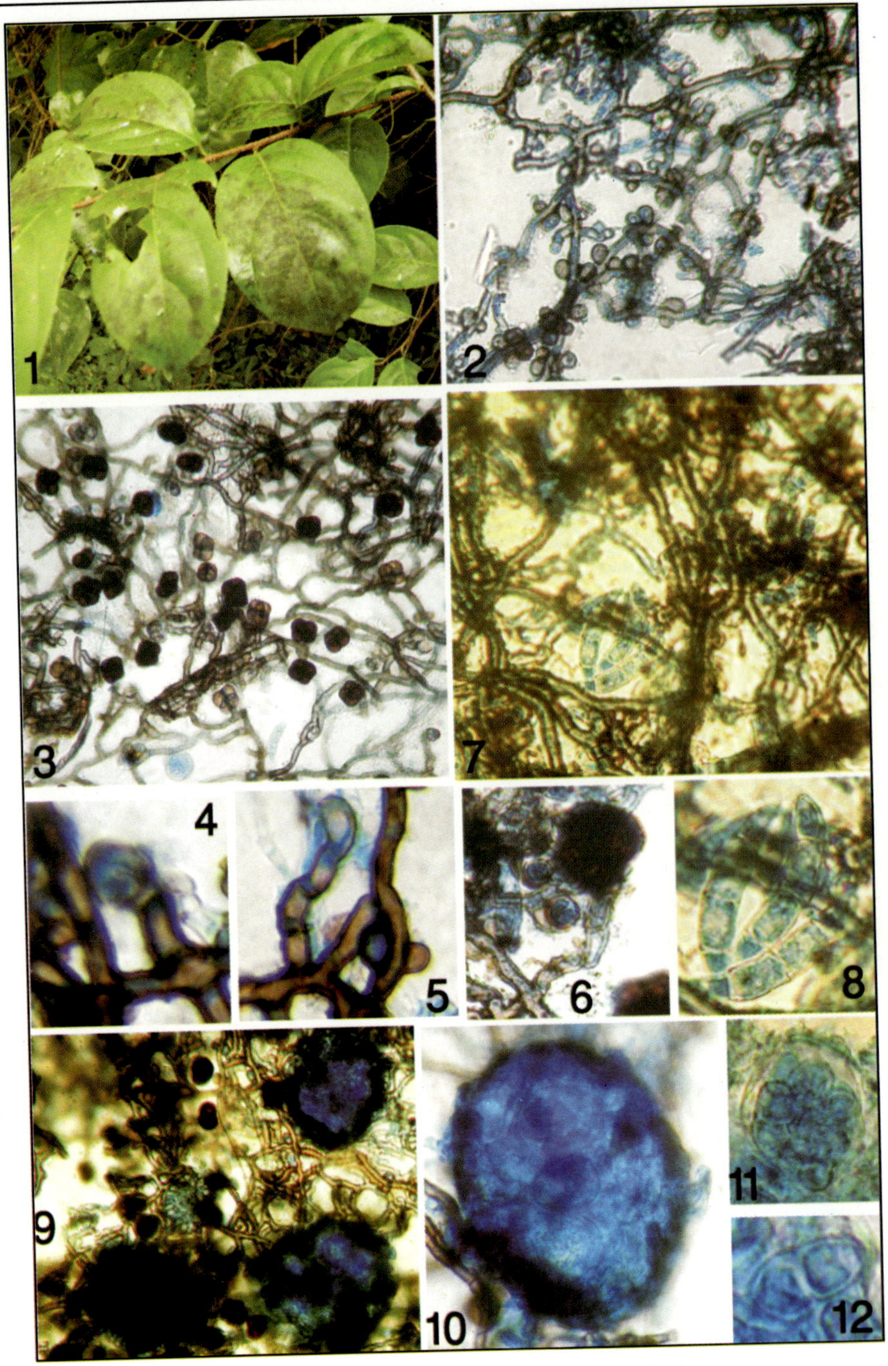

Plate 1.4: *Schiffnerula celastri* Hosag *et al.*

1: Infected leaves; 2: Appressoriate mycelium; 3: Sarciniform conidia; 4–6: Developing sarciniform conida; 7: Scattered conidia of *Questieriella*; 8: Spores of *Questieriella*; 9–10: Thyriothecia with exposed asci; 11: Ascus; 12: Ascospore

Asterolibertia

Single species ....*A. hydnocarpi*

Ishwaramyces

Single species ....*I. flacourtiae*

GENTIANACEAE

Asterina

Single species ....*A. enicostemmatis*

ICACINACEAE

Asterina

Single species ....*A. gomphandrae*

LAMIACEAE

Asterina

1. Ascospores less than 18µm long, wall smooth ....*A. plectranthi*
1. Ascospores more than 18µm long, wall punctate ....*A. pogostemonis*

LAURACEAE

Asterina

1. On *Cryptocarya* ....*A. cryptocariicola*
1. On the other hosts 2
2. On *Neolitsea* ....*A. neolitsiicola*
2. On *Cinnamomum* 3
3. Appressoria unicellular 4
3. Appressoria 2-celled ....*A. cinnamomi*
4. Appressoria globose, entire, sublobate ....*A. lauracearum*
4. Appressoria ampulliform, proliferate at the tip ....*A. cinnamomicola*

LOBELIACEAE

Asterina

Single species ....*A. lobeliacearum*

LORANTHACEAE

Asterina

1. Appressoria alternate only 2
1. Appressoria alternate and opposite ....*A. visci*
2. Appressoria ovate to globose ....*A. deightonii*
2. Appressoria oblong to clavate ....*A. loranthigena*

LYTHRACEAE

Asterina

Single species ....*A. lawsoniae*

MALVACEAE

Asterina

On *Sida* ....*A. diplocarpa*

On *Hibiscus* ....*A. hibisci*

MELASTOMATACEAE

Asterina

Single species ....*A. memecylonis*

MELIACEAE

Asterina

1. On *Cipadessa* ....*A. cipadessae*
1. On other hosts 2
2. Appressoria alternate, unilateral ....*A. chukrasiae*
2. Appressoria opposite to subopposite ....*A. aglaiae*

MENISPERMACEAE

Asterina

Single species ....*A. tinosporae*

Prillieuxina

Single species ....*P. anamirtae*

MYRISTICACEAE

Asterina

Single species ....*A. knemae- attenuatae*

MYRSINACEAE

Asterina

1. Appressoria entire ....*A. ardisiae*
1. Appressoria crenately lobate ....*A. ardisiicola*

MYRTACEAE

Asterina

1. On *Rhodomyrtus* ....*A. rhodomyrti*
1. On *Syzygium* 2
2. Appressoria unicellular ....*A. claviflori*
2. Appressoria 2- celled ....*A. jambolana*

OLACACEAE

Asterina

Single species ....*A. olacicola*

OLEACEAE

Asterina

1. Appressoria unicellular 2

1. Appressoria 2- celled 3
2. Ascospores more than 20 µm long ....*A. jasmini* var. *indica*
2. Ascospores less than 20 µm long ....*A. ligustricola*
3. Stalk cells cylindrical to cuneate ....*A. erysiphoides*
3. Stalk cells variously bulged, gibbous ....*A. pongalaparensis*

Prillieuxina

Single species ....*P. jasmini*

PASSIFLORACEAE

Asterina

Single species ....*A. adeniicola*

PIPERACEAE

Asterina

Single species ....*A. lepianthedis*

RANUNCULACEAE

Asterina

Single species ....*A. naraveliae*

RUBIACEAE

Prillieuxina

Single species ....*P. ixorigena*

RUTACEAE

Asterina

1. On *Atalantia* ....*A. atlantiae*
1. On other hosts 2
2. On *Murraya* ....*A. murrayae*
2. On other hosts 3
3. On *Toddalia* ....*A. toddaliae*
3. On other hosts 4
4. On *Glycosmis* ....*A. glycosmidis*
4. On *Melicope* 5
5. Appressoria alternate and opposite 6
5. Appressoria mostly opposite, rarely solitary ....*A. acronychiae*
6. Appressoria sublobate ....*A. melicopecola*
6. Appressoria uni to multilobate ....*A. clausenicola*

SABIACEAE

Asterina

1. Appressoria ovate, globose, mammiform ....*A. sabiacearum*
1. Appressoria cylindric, oblong ....*A. meliosma-simplicifoliae*

SANTALACEAE

Asterina

1. On *Santalum* .... *A. congesta*
1. On *Scleropyrum* .... *A. scleropyri*

SAPINDACEAE

Vishnumyces

Single species ....*V. otonepheli*

SAPOTACEAE

Asterina

Single species ....*A. mimusopsidicola*

Symphaster

Single species ....*S. mimusopsidis*

SIMAROUBACEAE

Asterina

Single species ....*A. samaderae*

STERCULIACEAE

Asterina

1. Thyriothecia more than 90 μm in diameter ....*A. leptalea*
1. Thyriothecia less than 90 μm in diameter ....*A. helicteridis*

Prillieuxina

Single species ....*P. pterigotae*

SOLANACEAE

Asterina

1. On *Solanum* 2
1. On *Lycianthes* .... *A. lycianthedis*
2. Appressoria unicellular ....*A. balakrishnanii*
2. Appressoria alternate, opposite ....*A. diplopoda*

SYMPLOCACEAE

Asterina

1. Asci globose, ascospores oblong, conglobate ....*A. indica*
1. Asci ovate, ascospores ellipsoidal, taper towards both ends ....*A. suttonii*

THEACEAE

Asterina

1. Appressoria alternate only 2
1. Appressoria alternate and opposite ....*A. theacearum*
2. Appressoria oblong, clavate .... *A. cannonii*
2. Appresoria not so 3

3. Appressoria ovate, globose, rounded towards apex ....*A. euryae*
3. Appressoria ovate, globose, angular, sublobate ... *A. theae*

TILIACEAE

Asterina

Single species ....*A. triumfetticola*

ULMACEAE

Asterina

Single species ....*A. dallasica*

URTICACEAE

Asterina

1. Appressoria unicellular 2
1. Appressoria 2- celled ....*A. girardiniae*
2. Appressoria alternate only 3
2. Appressoria alternate, opposite ....*A. oreocnidegena*
3. Ascospores more than 20 μm long ....*A. oreocnidecola*
3. Ascospores less than 20 μm long ....*A. elatostematis*

VERBENACEAE

Asterina

Single species ....*A. pusilla*

VITACEAE

Asterina

Single species ....*A. cissi*

XANTHOPHYLLACEAE

Asterina

Single species ....*A. xanthophylli*

The Family Lembosiaceae

Lembosiaceae Hosag., J. Mycopathol. Res. 39: 62, 2001.

Leaf parasites. Mycelium ectophytic, with or without appressoria, nutrient mycelium and leaf permating stroma present. Ascomata ectophytic, dimidiate, oval, ellipsoidal, "X" or "Y" shaped, elongated with radiating cells, astomatous, dehisce longitudinally at the center; asci globose, spherical, octosporous, bitunicate; ascospores two to many septate, conglobate, hyaline to brown.

Type genus: *Lembosia* Lev.

Key to the Genera of Lembosiaceae

1. Appressoria present 2
1. Appressoria absent 3
2. Appressoria intercalary ....*Cirsosia*
2. Appressoria lateral ....*Lembosia*

3. Conidia present 4
3. Conidia absent 5
4. Conidia 1-3 septate ....*Eupelte*
4. Conidia one to many septate ....*Maheshwaramyces*
5. Hypostroma present ....*Echidnodes*
5. Hypostroma absent ....*Echidnodella*

Key to the Species of Lembosiaceae

ANNONACEAE

Echidnodella

Single species ....*E. polyalthiae*

ARACEAE

Lembosia

Single species ....*L. malabarensis*

ARALIACEAE

Lembosia

Single species ....*L. araliacearum*

ARECACEAE

Cirsosia

Single species ....*C. globulifera*

CAESALPINIACEAE

Lembosia

1. Appressoria unicellular 2
1. Appressoria 1- 3 celled ....*L. humboldtiae*
2. Appressoria opposite, ovate, conoid ....*L. humboldtiicola*
2. Appressoria alternate, ovate, globose ....*L. humboldtiigena*

COMBRETACEAE

Lembosia

Single species ....*L. terminaliae-chebulae*

DIPTEROCARPACEAE

Echidnodella

Single species ....*E. vateriae*

FABACEAE

Lembosia

Single species ....*L. ormosiae*

HIPPOCRATACEAE

Lembosia

Single species ....*L. salaciae*

LAURACEAE

Lembosia

Single species ....*L. perseae*

LYTHRACEAE

Lembosia

Single species ....*L. lagerstroemiae*

MELASTOMATACEAE

Echidnodella

Single species ....*E. memecyli*

MENISPERMACEAE

Maheshwaramyces

Single species ....*M. pachygones*

MYRTACEAE

Lembosia

Single species ....*L. hosagoudarii*

OLEACEAE

Eupelte

Single species ....*E. amicta*

Lembosia

Single species ....*L. linocierae*

PANDANACEAE

Echidnodes

Single species ....*E. pandanicola*

SAPOTACEAE

Echidnodella

Single species ....*E. manilkarae*

SCHIFFNERULA

Schiffnerulaceous fungi flourish well in the tropics and have extended their distribution to subtropical to temperate regions of the world. The connection between teleomorphs and synanamorphs is well established. The genus *Schiffnerula* includes four synanamorphs: *Questieriella, Mitteriella, Digitosarcinella* and *Sarcinella.*

The Genus *Schiffnerula*

Schiffnerula Hohnel, Sber, Akad. Wiss. Wien, math. Nat.kl., I, 118: 867, 1909. Arx and Muller, Stud. Mycol. 9: 48, 1975; Hughes, Can. J. Bot. 61: 1763, 1983.

Clypeolella Hohnel, Sber.Akad.Wiss.Wien., math.- nat.kl. I, 119: 403, 1910.

Phaeoschiffnerula Thesis, Broteria 12: 21, 1917.

Questieria Arn. Les Asterinees 1: 186, 1918.

Diathrypton Sydow, Philippine J. Sci. 21: 137, 1922.

Coniosporiella Bat. Atas Inst. Univ. Recife 3: 113, 1966.

Hypahe superficial, colonies foliicolous, brown, appressoriate, appressoria unicellular. Ascomata arise from the short lateral branches, initially with radiating cells but the cells dissolve when the ascomata start resuming globose appearance. Asci few, bitunicate, broadly ellipsoid to globose, sessile, octosporous, exposed after deliqusing the ascomatal wall; ascospores brown, 1-septate, constricted at the septum.

Type: *S. mirabilis* Hohnel

Key to the Genera

1. Fruiting body present ....*Schiffnerula*
1. Fruiting body absent 2
2. Conidia cheiroid, with 4-5 closely appressed arms ....*Digitosarcinella*
2. Conidia not so, simple 3
3. Conidia globose, sarciniform, brown to black ....*Sarcinella*
3. Conidia not so 4
4. Conidia pale brown, falcate, 3-septate ....*Questieriella*
4. Conidia straight, ellipsoidal, 4-septate ....*Mitteriella*

Form Genus *Digitosarcinella*

Digitosarcinella Hughes, Can. J. Bot. 62: 2208, 1984.

Colonies foliicolous. Hyphae superficial, brown to dark brown, branched, appressoriate, appressoria sessile, lateral, unicellular. Conidiogenous cells lateral, sessile, monoblastic. Conidia cheiroid, with 4-5 closely appressed arms, up to 7-septate, constricted at the septa.

Type: *D. caseariae* Hughes

Form Genus *Mitteriella*

Mitteriella Sydow, Ann. Mycol. 31: 95, 1933; Ellis, Dematiaceous Hyphomycetes, p. 228, 1977.

Colonies black. Hyphae superficial, brown, branched, septate, appressoriate. Appressoria lateral, unicellualr. Conidiophores macronematous, mononematous, short, Simple. Conidiogenous cells polyblastic, integrated, terminal, sympodial, denticulate. Conidia solitary, simple, ellipsoidal to limoniform, black, 0-4-septate.

Type: *M. ziziphina* Sydow

Form Genus *Questieriella*

Questieriella Arn. ex Hughes, Can.J. Bot. 61: 1729, 1983.

Colonies black, hyphae superficial, brown, branched, septate, appressoriate. Appressoria lateral unicellular. Conidiophores micronematous, mononematous, lateral, 0-2-septate. Conidiogenous cells monoblastic to polyblastic, integrated, terminal, lateral or incorporated in the hyphae. Conidia blastic, terminal, solitary, narrowly ellipsoidal to obovoidal, curved, fulcate, sigmoid, truncate at the base, 3-septate.

Type: *Q. pulchra* Hughes.

Form Genus *Sarcinella*

Sarcinella Sacc, Michelia 2: 31, 1880; Ellis, Dematiaceous Hyphomycetes, p. 49, 1977.

Colonies black. Hyphae superficial, branched, septate, appressoriate. Appressoria lateral, unicellular. Conidiophores macronematous, semi-macronematous, simple to branched. Conidiogenous cells monoblastic, integrated, terminal, intercalary, determinate. Conidia solitary, acrogenous or acropleurogenous, subspherical, sarciniform, dark brown to reddish brown, smooth, constricted at the septa.

Type: *S. heterospora* Sacc.

Key to the Species

ANACARDIACEAE

Sarcinella ....*S. odinae*

APOCYNACEAE

Sarcinella

On *Catharanthus* ....*S. catharanthi*

On *Carissa* ....*S. oreophila*

On *Cryptostegia* ....*S. cryptostegiae*

On *Wrightia* ....*S. wrightiae*

ASTERACEAE

Sarcinella ....*S. vernoniae*

Schiffnerula

On *Eclipta* ....*S. ecliptae*

On *Spilanthus* ....*S. spilanthi*

On *Vernonia* ....*S. vernoniae*

On *Wedelia* ....*S. wedeliae*

BIGNONIACEAE

Sarcinella ....*S. bignoniacearum*

BUXACEAE

Questieriella ....*Q. sarcococcae*

CAESALPINIACEAE

Sarcinella

On *Cassia tora* ....*S. cassiae*

On *Cassia fistula* ....*S. cassiae-fistulae*

Schiffnerula ....*S. cassiae*

CELASTRACEAE

Schiffnerula ....*S. celastri*

Sarcinella

On *Celastrus* ....*S. palawanensis*

On *Elaeodendron* ....*S. shamboodarai*

On *Euonymous* ....*S. tandonii*

On *Gymnosporia* ....*S. gymnosporiae*

COMBRETACEAE

Questieriella ....*Q. terminaliae*

Sarcinella

On *Quesqualis* ....*S. quesqualidis*

On *Terminalia* ....*S. jarwaensis*

EBENACEAE

Sarcinella ....*S. gorakhpurensis*

ERICACEAE

Sarcinella ....*S. lyoniae*

EUPHORBIACEAE

Questieriella ....*Q. malloti*

Sarcinella

On *Glochidion* ....*S. glochidii*

On *Sapium* ....*S. indica*

Schiffnerula

On *Glochidion* ....*S. glochidii*

On *Ricinus* ....*S. ricini*

FABACEAE

Questieriella ....*Q. tephrosiae*

Sarcinella ....*S. dalbergiae*

FAGACEAE

Sarcinella

On *Quercus* ....*S. quercina*

On *Castenopsis* ....*S. castanopsidis*

FLACOURTIACEAE

Sarcinella ....*S. manilensis*

HIPPOCRATACEAE

Sarcinella ....*S. hippocrateae*

ICACINACEAE

Sarcinella ....*S. hughesii*

LAMIACEAE

Sarcinella ....*S. colebrookiana*

LAURACEAE

Schiffnerula ....*S. actinodaphnes*

LYTHRACCEAE

Schiffnerula ....*S. lagerstroemiae*

MELIACEAE

Sarcinella

On *Azadirachta* ....*S. azadirachtae*

On *Cipadessa* ....*S. cipadessae*

MORACEAE

Schiffnerula ....*S. fici*

MYRSINACEAE

Questieriella ....*Q. ardisiae*

MYRTACEAE

Sarcinella ....*S. kamalii*

SOLANACEAE

Sarcinella ....*S. raimundi*

OLEACEAE

Schiffnerula ....*S. pulchra*

PASSIFLORACEAE

Schiffnerula ....*S. mirabilis*

Questieriella ....*Q. passiflorae*

PERIPLOCACEAE

Schiffnerula ....*S. cryptolepidis*

RHAMNACEAE

Schiffnerula ....*S. ziziphi*

Mitteriella ....*M. ziziphina*

ROSACEAE

Sarcinella ....*S. prunicola*

RUBIACEAE

Questieriella ....*Q. braunii*

RUTACEAE

Sarcinella

On *Glycosmis* ....*S. glycosmidis*

On *Eagle* ....*S. fumosa*

Questeriella ....*Q. zanthoxyli*

SAPINDACEAE

Sarcinella ....*S. allophyli*

SOLANACEAE

Sarcinella ....*S. raimundi*

STRYCHNACEAE

Questieriella ....*Q. strychni*

THEACEAE

Sarcinella ....*S. theae*

Schiffnerula ....*S. camelliae*

TILACEAE

Questieriella ....*Q. grewiae*

TILACEAE

Questieriella ....*Q. grewiae*

ULMACEAE

Schiffnerula ....*S. hughesii*

URTICACEAE

Sarcinella

On *Oreocnide* ....*S. oreocnidecola*

On *Pouzolzia* ....*S. pouzolziae*

VERBENACEAE

Schiffnerula ....*S. tectonae*

Sarcinella

On *Gmelina* ....*S. gmelinae*

On *Vitex* ....*S. jabalpurensis*

S. vitecis

On *Tectona* ....*S. tectonae*

VITACEAE

Sarcinella ....*S. latifoliae*

Balladynaceous Fungi

Balladynaceous fungi produce ectophytic, septate, branched, appressoriate mycelium, with or without setae. Perithecia globose, vasiform, translucent, ostiolate; asci bitunicate, ovate to clave, mostly stipitate, octosporous; ascospores uniseptate, brown.

Key to the Genus *Balladyna* and Similar Genera

1. Mycelium non-appressoriate 4
1. Mycelium appressoriate 2
2. Only perithecial setae present ....*Balladynopsis*

2. Perithecial setae absent 3
3. Mycelial setae present ....*Balladyna*
3. Mycelial setae absent ....*Balladynocallia*
4. Conidia present ....*Alina*
4. Conidia absent ....*Dysrhynchis*

Genus *Balladyna*

Balladyna Racib., Parasit Algen und Pilze Javas 2: 3, 1900.

Mycelium superficial, brown, septate, branched, appressoriate, setose. Appressoria unicellular. Perithecia ovate, globose, stipitate, sessile, setose or mycelium like appendages present, ostiolate or widely opened at the apex; asci few, bitunicate, clavate, globose, ovoid, 4-8 spored; ascospores brown, uniseptate, cells may be equal or unequal in size.

Type: *B. velutina* (Berk. and Curt.) Hohnel

Key to the Species

1. On Monocots 2
1. On Dicots 4
2. Ascospores less than 25 μm long and up to 10 μm broad ....2. *B. butleri*
2. Ascospores more than 25 μm long and more than 10 μm broad 3
3. Mycelial setae less than 250 μm long ....5..*B. lelebe*
3. Mycelial setae more than 250 μm long ....9. *B. muroiana*
4. On Rubiaceae members 7
4. On others 5
5. On Verbanaceae members ....3. *B. callicarpicola*
5. On others 6
6. On Annonaceae members ....8. *B. melodori*
6. On *Strychnos* ....10. *B. strychni*
7. Mycelial setae less than 100 μm long ....12. *B. sydowii*
7. Mycelial setae more than 100 μm long 8
8. Ascospores more than 25 μm long 9
8. Ascospores less than 25 μm long 10
9. Appressoria sinuately lobate; mycelial setae straight and more or less curved; mycelial form of setae on perithecia present ....6. *B. leonensis*
9. Appressoria lobed; mycelial setae straight; mycelial form of setae on perithecia absent ....4 *B. deightonii*
10. Appressoria ampulliform ....1. *B. ajrekarii*
10. Appressoria otherwise 11
11. Perithecial setae or appendages present 12
11. Perithecial setae or appendages absent 13

12. Appressoria sinuately lobed, 4-7 µm long ....7 *B. magnifica*
12. Appressoria sinuous, angular to 1-4 times lobate ....11 *B. secedens*
13. Appressoria curved-cylindrical, entire to furcate ....13. *B. tenuis*
13. Appressoria globose, entire to lobate 14
14. Colonies inconspicuous, isolated, mycelial setae straight to slightly curved ....14 *B. terennae*
14. Colonies widely confluent; mycelial setae straight to slightly flexuous ....15 *B. velutinae*

Hosagoudar (2004) has dealt with this species. Sivanesan (1981) has provided an excellent account of the genera *Balladynopsis*, *Balladynocallia* and *Alina*. Muller and Arx (1962) have given a detailed account of the genus *Dysrhynchis*.

The Genus *Meliolina*

The genus *Meliolina* was proposed by Sydow and Sydow (1914) to accommodate the fungi similar to *Meliola* but have distantly septate hyphae, lack characteristic appressoria and having dichotomously branched phialophores. These fungi can be easily identified in the field by their velvet carbonaceous woolly colonies mostly on the lower surface of the leaves. Hughes (1993) revised the genus and gave an account of thirty-eight species and Hosagoudar (2002) provided key to species of this genus.

Key to the Species

1. On Melastomataceae ....*memecylonis*
1. On Myrtaceae 2
2. Pale bands in ascospores absent ....*novae- caledoniae*
2. Pale bands in ascospores present 3
3. Pale bands in ascospores are two 4
3. Pale bands in ascospores are more 6
4. Ascospores 45-65 µm long 5
4. Ascospores 60-80 (-90) µm long ....*shepherdii*
5. Ascospores 12-20 µm broad ....*lanceolata*
5. Ascospores 20-30 µm broad ....*demoulinii*
6. Pale bands in ascospores are four (including inconspicuous bands in the central cells) 7
6. Pale bands in ascospores are eight 36
7. Pale bands in ascospores are only four in the distal cells 8
7. Pale band in ascospores are four with one or two inconspicuous bands in the central cells 27
8. Ascospores 36-50 µm long 9
8. Ascospores more than 50 µm long 12
9. Terminal branches of phialophores are some strongly curved or coiled ....*octospora*
9. Terminal branches of Phialophores are straight only 10

10. On *Tristania* from Australia ....*baileyi*
10. On other hosts 11
11. On *Syzygium* from Philippines ....*luzonensis*
11. On *Leptospermum* from Australia ....*queenslandica*
12. Ascospores 45-65 µm long 13
12. Ascospores 60-80 (-90) µm long 21
13. Phialides unilaterally thick walled ....*africana*
13. Phialides uniformly thick walled 14
14. Ascospores (12-) 15-20 (-22.5) µm broad 15
14. Ascospores (17-) 20-29 µm broad 23
15. On *Melaleuca* from Papua New Guinea ....*melaleucae*
15. On other hosts 16
16. On *Metrosideros* from Samoa Islands ....*samoensis*
16. On other hosts 17
17. Ascospores cylindrical ....*neesiana*
17. Ascospores ellipsoidal 18
18. Stomatopodia rare or absent ....*australiensis*
18. Stomatopodia few to numerous 19
19. Colonies 3-6 mm in diameter ....*hainanensis*
19. Colonies more wider 20
20. Known from India ....*gorakhpurensis*
20. Known from Papua New Guinea ....*shawiae*
21. Hyphal cells 20-45 µm long ....*quinqueseptata*
21. Hyphal cells 45-110 µm long 22
22. Colonies rounded with more or less entire margin ....*arborescens*
22. Colonies with irregularly or rapidly lobed margin ....*radians*
23. End cells of ascospores subpapillate ....*subramanianii*
23. End cells other wise 24
24. Colonies 2-6 mm wide ....*khasiae*
24. Colonies more wider 25
25. Colonies 6-10 mm wide 26
25. Colonies 10-25 mm wide ....*yatesii*
26. Hyphal cells 20-45 µm long ....*ryukyuensis*
26. Hyphal cells 45-110 µm long ....*shepherdii*
27. On Metrosideros 28
27. On other hosts 31
28. Ascospores 45-65 µm long 29

28. Ascospores 60-80 (-90) μm long 30
29. Stomatopodia rare or absent ....*haplochaeta*
29. Stomatopodia rare to frequent ....*novae-zealandicae*
30. Ascospores ellipsoidal ....*hawaiiensis*
30. Ascospores cylindrical ....*metrosideri*
31. On *Leptospermum* 32
31. On other hosts 34
32. Ascospores 36-50 μm long 33
32. Ascospores 45-65 μm long ....*leptospermi*
33. Hyphal cells 20-45 μm long ....*sarawacensis*
33. Hyphal cells 45-110 μm long ....*queenslandica*
34. Colonies 2-6 mm in diameter ....*cladotricha*
34. Colonies 10-25 mm in diameter 35
35. On *Syzygium* ....*burmanica*
35. On *Lophostemon* ....*lophostemonis*
36. Ascospores 60-80 (-90) μm long ....*pulcherrima*
36. Ascospores smaller 37
37. Stomatopodia rare or absent 38
37. Stomatopodia scanty to abundant 39
38. End cells of ascospores rounded ....*degeneri*
38. End cells of ascospores subconical ....*stevensii*
39. Ascospores (12-) 15-20 (-22.5) μm broad ....*cookii*
40. Ascospores (17-) 20-29 μm broad ....*sydowiana*

When we examine the wooly colonies or overmatured colonies, actual parasites dominated by the hyperparasites and jeopardize the identity of the black mildews. Hansford (1946), Katumoto (1977, 1983), Ellis (1971, 1976), Deighton (1969) and Deighton and pirozynski (1972) have dealt these fungi in detail.

Acknowledgement

I thank Dr. A. Subramanian, Director, TBGRI, Palode for the necessary facilities.

References

Beeli, M. 1920. Note sur le Genre *Meliola*. Bull. Jard. Bot. Bruxelles 7: 89-160.

Cooke, M.C. 1878. Microscopic Fungi. Hardwicke, London.

Deighton, F.C. 1969. Microfungi. IV. Some hyperparasitic hyphomycetes, and a note on *Cercosporella uredinophila* Sacc. Mycol. Pap. 118: 1-41.

Deighton, F.C. and Pirozynski, K.A. 1972. Microfungi. V. More hyperparasitic hyphomycetes. Mycol. Pap. 128: 1-110.

Ellis, M.B. 1971. Dematiaceous Hyphomycetes. CMI, Kew, Surrey, England, pp. 608.

Forsyth, W. 1824. A Treatise on the Culture and Management of Fruit Trees. Longmans, London., 524 pp.

Hansford, C.G. 1946. The foliicolous Ascomycetes, their parasites and associated fungi. Mycol. Pap. 15: 1-240.

Hansford, C.G. 1961. The Meliolaceae. A Monograph. Sydowia. Beih 2: 1-806.

Hosagoudar, V. B. 1996. *Meliolales of India.* Botanical Survey of India, Calcutta, pp. 363.

Hosagoudar, V. B. 2008. *Meliolales of India. Vol. II.* Botanical Survey of India, Calcutta, pp. 390.

Hosagoudar, V. B. and Agarwal, D.K. 2008. *Taxonomic studies of Meliolales. Identification Manual.* International Book Distributors, Dehra Dun

Hosagoudar, V. B., Abraham, T. K. and Pushpangadan, P. 1997. *The Meliolineae–A Supplement.* Tropical Botanic Garden and Research Institute, Palode, Thiruvananthapuram, Kerala, India, pp. 201.

Hosagoudar, V.B. 2003. Digital formula for the identification of Meliolaceae. Sydowia 55:168-171.

Hosagoudar, V.B. 2003. The genus *Schiffnerula* and its synanamorphs. *Zoos´ Print J.* 18: 1071-1078.

Hosagoudar, V.B. and Kapoor, J. N. 1985. New technique of mounting meliolaceous fungi. Indian Phytopath. 38: 548-549.

Hosagoudar, V.B. 2002. Key to species of the genus *Meliolina. Zoos´* Print J. 17: 786-787.

Hosagoudar, V.B. 2004. Studies on foliicolous fungi- XI. The genus *Balladyna* Racib., based on literature. J. Econ. Taxon. Bot. 28: 202-208.

Hosagoudar, V.B., Abraham, T.K., Krishnan, P. N. and Vijayakumar, K. 1997. Biochemical changes in the leaves of Ebony tree affected with black mildew. Indian Phytopathol. 50: 439-440.

Hosagoudar, V.B., Krishnan, P.N. and Abraham, T.K. 1997. Biochemical changes in the Sandal tree infected with *Asterina congesta* Cooke. New Botanist 24: 27-32.

Hu, Y., Ouyang, Y. Song Bin and Jiang, G. 1996. *Flora Fungorum Sinicorum.* Vol. 4. Meliolales (1). Science Press, Beijing, pp. 270, plate IV.

Hu, Y., Song Bin, Ouyang, Y. and Jiang, G. 1999. *Flora Fungorum Sinicorum.* Vol. 11. Meliolales (2). Science Press, Beijing, pp. 252.

Hughes, S.J. 1976. Sooty moulds. Mycologia 68: 693-820.

Hughes, S.J. 1993. *Meliolina* and its excluded species. Mycol. Pap. 166: 1-255.

Hughes, S.J. 1995. *Pauhia*, a new genus of Meliolineae. Mycologia 87: 702-706.

Hughes, S.J. and Pirozynski, K.A. 1994. New Zealand fungi-34. *Endomeliola dingleyae*, a new genus and new species of Meliolaceae. New Zealand J. Bot. 32: 53-59.

Katumoto, K. 1977. Some hyperparasitic fungi on the black mildews from Japan. Trans. mycol. Soc. Japan 17: 280-285.

Katumoto, K. 1983. Hyperparasitic species of *Spiropes* in Japan. Trans. mycol. Soc. Japan 24: 249-258.

Knight, T.A. 1818. Trans. Hort. Soc. 2: 82-90.

Large, B.C. 1940. *The Advance of Fungi.* Holt, New York.

Mibey, R.K. and Hawksworth, D.L. 1997. Meliolaceae and Asterinaceae of the Shimba Hills, Kenya. Mycol. Pap. 174: 1-108.

Mibey, R.K. and Cannon, P.F. 1999. Biotrophic fungi from Kenya. Ten new species and some new records of Meliolaceae.–Cryptogamie, Mycol. 20: 249-282.

Muller, E. and Arx, J.A.von 1962. Die Gattungen der didymosporen Pyrenomyceten. Beitr. Kryptogamenfl. Schweiz 11:1-922.

Murray, J.A.A. (Ed.) 1908. *A New English Dictionary*. Clarendon, Oxford.

Rangaswami, G. 1975. *Diseases of crop Plants in India*. Prentice-Hall of India Pvt. Ltd., New Delhi

Sivanesan, A. 1981. *Balladynopsis, Balladynocallia* and *Alina*. Mycol. Pap. 146: 1-138.

Stevens, F.L. 1927. The Meliolineae- I. Ann. Mycol. 25: 405-469.

Stevens, F.L. 1928. The Meliolineae-II. Ann. Mycol. 26: 165-383.

Sydow, H. and Sydow, P. 1914. Diagnosen neuer Philippinischer Pilze. Ann. Mycol. 12: 545-576.

Wellman, F.L. 1972. Tropical American Plant Diseases.The Scarecrow Press, Metuchen,NJ. Chap.VII.

Yarwood, C.E. 1978. History and taxonomy of powdery mildews, pp. 37. In: The powdery mildews. (ed.)D. Spencer, Academic Press, New York.

Microbes: Diversity and Biotechnology (2012) *Pages* **37–42**
Editors: **Prof. S.C. Sati & Dr. M. Belwal**
Published by: **DAYA PUBLISHING HOUSE, NEW DELHI**

Chapter 2

Distribution of Endophytic Fungi in Different Parts of Rudraksh (*Elaeocarpus sphaericus*) Plants

A.K. Shukla[1], Y. Yongam[1] and P. Tripathi[2]
[1]Department of Botany, Rajiv Gandhi University, Rono Hills, Itanagar – 791 112
[2]Department of Botany, DAV-PG College, Kanpur – 208 001

ABSTRACT

Endophytic fungi were isolated from different parts of plant Rudraksh (*Elaeocarpus sphaericus*). A total number of 17 fungal taxa were isolated and maximum fungal species were found associated with leaves compare to stem and bark. The distribution of endophytic fungi in various parts of plant was not uniform. The number of fungal species isolated from leaves, stem and bark were 12, 10 and 10 respectively. Fungal species *Nigrospora sphaerica, Oidiodendron echinulatum, Oidiodendron griseum, Humicola fuscoatra and Arthroderma tuberculatum* were most dominating fungi. Whereas, *Mortierella minutissima* was found only with leaves, *Acremonium strictum, Humicola grisea* and *Mortierella hyaline* only with stem and *Acremonium butyric* was restricted to only bark of plants. Fungal species *Chaetomium globosum* and *Trichothesium roseum* were occurred with leaves and stem only.

Keywords: Endophytes, Fungi, Rudraksh.

Introduction

Endophytes are microbes that inhabit plant tissues at some stages in their life cycle without causing apparent harm to their host (Petrini, 1991). During 1940s these fungi were first noticed, but only at the turn of the 21st century the ubiquity of these fungi fully recognized. Endophytes can colonize virtually hundred percent of the host population, but many infect a far smaller fraction, while some are rare (Petrini, 1986). Endophytes colonized plants often grow faster than non-colonized ones (Cheplick *et al.*, 1989). Observed biodiversity of endophytes suggests that they can also be aggressive saprophytes or opportunistic pathogens at some stages (Strobel and Daisy, 2003). At present, endophytes are viewed as outstanding source of novel bioactive natural products because many of them occupying literally millions of unique biological niches (higher plants) and growing in so many unusual environments (Strobel *et al.*, 2004). A wide range of plants have now been examined for endophytes, and endophytes have been found in nearly all of them. Most of the fungi are uncommon and narrowly distributed, taxonomically and geographically. However a few fungi are widely distributed with the host, suggesting a long standing, close and mutually beneficial interaction. Some fungi are found in many different terrestrial hosts, especially endophytes of crop plants. While most information has been gathered from terrestrial ecosystems, fungi are also found in algae and sea grasses

Some species of endophytic fungi have been identified as sources of anticancer, antidiabetic, insecticidal and immunosuppressive compounds (Strobel *et al.*, 2004). Further, endophytic fungi may also produce metabolites with thermoprotective role. For example, plants in some volcanic areas in USA were found colonized by an endophytic fungus *Curvularia* sp. (Redman *et al.*, 2002). Whereas the plants grown from surface-sterilized seeds in sterile soil that had been inoculated with *Curvularia* sp. survived constant soil temperature of 50°C, the non-symbiotic plants died. Re-isolation of the fungus demonstrated that thermal protection was also provided to the fungus although the biochemical basis is presently not known. Because of their role in conferring plants the ability to adapt to stress conditions, and because they are proven or perceived sources of secondary metabolites with pharmaceutical importance, the study of fungal endophytes is expected to become an important component of fungal biology in the coming days.

Present study was carried out on *Elaeocarpus sphaericus* commonly known as Rudraksh. Plant grows in subtropical countries such as India, Nepal and Indonesia. *Elaeocarpus sphaericus* is known for many medicinal properties such as curing of epilepsy, Nervousness, insanity, lack of concentration, melancholia, mania, other mental disorders, insomnia, headache, jaundice, hypertension, fever, cholesterol lowering capabilities, diabetes, stress, and impotence. Defensive accomplishment of *E. sphaericus* (Rudraksha) extracts in experimental bronchial asthma, protection of guinea-pigs against bronchospasm, microbial activity against several microorganisms including against Salmonella and anticonvulsant activity etc. are the proven medicinal properties of rudraksh (Dasgupta *et al.*, 1984; Singh and Nath, 1999; Jones *et al.*, 2002). In ayurvedic treatments *Elaeocarpus sphaericus* fruits are used for mental diseases, epilepsy, asthma, hypertension, arthritis and liver diseases. Investigations and user testimonials through the centuries suggests beneficial properties of the rudraksha that requires more scientific investigations for the benefit of mankind.

A few studies are available on endophytic fungal association with plants from India. Some Indian workers have studied the endophytic fungi from medicinal plants such as *Azadirachta indica, Terminalia arjuna, Adhatoda zeylanica, Bauhinia phoenicea, Callicarpa tomentosa, Clerodendron serratum, Labelia nicotinifolia*, and *Crataeva magna* (Mahesh *et al.*, 2005; Nalini *et al.*, 2005; Raviraja, 2005; Tejesvi *et al.*, 2005). Present study was undertaken to find out the endophytic fungal species association with *Elaeocarpus sphaericus syn. E. ganitrus.*

Materials and Methods

Sampling Site

This study was conducted at the Rajiv Gandhi University campus (altitude 345masl, latitude 27° 09′00″N and longitude 93°46′15″E). The climate of area is subtropical. The minimum and maximum temperature rages between 9°C and 30°C respectively during the study period.

Plant Sample Collection

Mature healthy, asymptomatic plant materials (leaves, stem and bark) were collected by sampling different trees of *Elaeocarpus sphaericus* growing randomly in the campus premises of Rajiv Gandhi University campus, Itanagar. The sampling was performed on three trees of *Elaeocarpus sphaericus* and same repeated three times during this experiment. Bark samples were obtained by cutting tree bark at 150 cm above the ground level from a depth of 1–1.5 cm inwards with the help of sterile machete. Stem samples were obtained by cutting the portion of twigs. Small discs of leaves (0.5 cm diameter) were cut using sterile pinch cutter. Fifteen samples were taken from each tree, five each from inner bark, stem and leaves. From each sample 10 sub samples were prepared for further culturing to find out endophytic fungi. All samples collected in sterile polythene bags and were brought to the laboratory. Fungal isolation was carried out with in two hours of collection the samples.

Isolation and Culture of Endophytic Fungi

Samples were washed thoroughly in running tap water for 10 min to remove the debris adhered, and finally washed with double distilled water to minimize the microbial load from sample surface. The surface treatment was done adopting the methodology by Petrini (1991). Epiphytic mycelia were removed by immersing the tissues in 70 per cent ethanol for 1–3 min and in aqueous solution of sodium hypochlorite (4 per cent available chlorine) for 2–5 min followed by washing with 70 per cent ethanol for 5 s. The tissues were then rinsed in sterile distilled water and allowed to surface dry in sterile conditions. The outer bark was removed and the inner bark containing cortex was carefully dissected in to small pieces (1.0 · 0.5 cm). The pieces were placed on petri dishes containing rose Bengal agar medium and incubated for 15 days at 25 ± 2 ú C in BOD cum humidity incubator. Tissues were observed for fungal growth at 2 day intervals for 15 days. Actively growing fungal tips immerging from plant tissues were sub-cultured on PDA petri plates for identification and enumeration. The endophytic fungi were identified according to their macro and microscopic structures. Species level identification was done with the manual described by Barnett and Hunter (1998), Domsch *et al.* (1980).

The colonization frequency (per cent CF) of endophytic fungi was calculated using the formula given by Hata and Futai (1995). *per cent* CF = (N_{col}=N_t) x 100; where, N_{col} = Number of segments colonized by each fungus, N_t = Total number of segments studied.

Results and Discussion

A total number of 17 fungal taxa were isolated from different parts of plant (Table 2.1). Maximum fungal species were found associated with leaves compare to stem and bark. The distribution of endophytic fungi in various parts of plant was not uniform. The number of fungal species isolated from leaves, stem and bark were 12, 10 and 10 respectively. *Nigrospora sphaerica, Oidiodendron echinulatum, Oidiodendron griseum, Humicola fuscoatra and Arthroderma tuberculatum* fungal species were most dominating fungi. There were a number of fungal species which were restricting to the particular part of plant. Fungi *Mortierella minutissima* was found only with leaves, *Acremonium strictum, Humicola grisea* and *Mortierella hyaline* only with stem and *Acremonium butyric* was restricted to only bark of plants. Fungal species *Chaetomium globosum* and *Trichothesium roseum* were occurred with leaves and stem only.

Hyphomycetes of deuteromycotina are common fungal endophytes among plants inhabiting in temperate, tropical and rainforest vegetations (Bacon and White, 1994). Deuteromycetous fungal isolates, as endophytes in mangrove vegetations of costal Karnataka, Picchavaran and Pondicherry (India) were more prevalent than the ascomycotina (Maria and Sridhar, 2003; Suryanarayanan *et al.*, 1998).

The colonization frequency of endophytic fungi may vary according to seasons. The environmental conditions under which the host is growing also affect the endophytic population (Hata *et al.*, 1998). By definition, an endophytic fungus lives in mycelial form in biological association with the living plant, at least for some time. Therefore the minimal requirement before a fungus is termed an 'endophyte' should be the demonstration of its hyphae in living tissue.

Table 2.1: Colonization Frequency (per cent CF) of Endophytic Fungi Occurred with Plant Parts

Endophytic Fungi	*Plant Parts*		
	Bark	*Stem*	*Leaf*
Acremonium butyri	2	–	-
Acremonium strictum	–	3	-
Arthroderma tuberculatum	5	6	-
Chaetomium globosum	–	3	5
Cladosporium cladosporiodes	5	–	10
Cladosporium herbarum	–	-	5
Cochliobolus geniculatus	3	–	3
Humicola fuscoatra	12	14	-
Humicola grisea	–	3	-
Mortierella hyaline	–	8	-
Mortierella minutissima	–	-	10
Nigrospora sphaerica	10	–	25
Oidiodendron echinulatum	12	8	8
Oidiodendron griseum	5	8	3
Oidiodendron truncatum	–	3	3
Phoma leveillei	3	–	5
Trichothesium roseum	–	5	3
Sterile	3	–	3

Each plant is colonized by fungal propagules that arrive from the environment. The source of transmission has been determined in only a few cases. Propagules of some endophytes have been found in the body of insect pests of the host. Intriguingly, at least two entomopathogens have been documented as endophytic fungi. Thus insects may disperse some fungi from host to host. Aerial dispersal either in the wind or on vectors is probably the most common mechanism for fungal dispersal. Endophytic fungi colonize various parts of the plant. Many of the fungi sporulate in culture indicating the potential to release spores in the air. Indeed, sporulation is seen after senescence of plant tissues. However, few cases of dispersal have been documented in the wild and the various mechanisms remain unexplored.

Nutrients are cycled between the host and fungus. The endophytic fungus gains a predictable environment in which nutrients are readily available. Thus the benefits are clear for fungi establishing endophytic associations. The loss of plant resources to the fungus, and the potential of some fungi to grow rapidly, indicates that the host regulates development of colonizers. Each plant host has a range of physical, chemical, constitutive and induced controls over the spread of fungi within tissues. An enormous diversity of phenolic and other deterrent plant compounds are associated with the presence of endophytic fungi. In addition, presence of endophytes up regulates plant responses to pathogens. In the absence of plant controls, proliferation of endophytes through tissues would be expected. The reactions of the plant to endophytes suggest that the interaction is one of confinement by the plant. Colonisation by endophytes ranges from single cells (*Rhabdocline parkeri*) to patchy distribution through leaves and stems (*Chaetomium globosum*). Plants may benefit from the presence of endophytes in many ways. Potential plant benefits have been examined in only a few cases. *Rhabdocline parkeri* produces a compound that reduces needle attack by borers (Izumi, 2001). Metabolites produced by *Phomopsis* sp in cotton appear to deter larvae of *Helicoverpa* from feeding on leaves. The parallels with *Neotyphodium* are clear. In addition, aphids feeding on leaves of cotton may become colonised by *Lecanicillium lecanii*, when conditions permit. Thus the aphid may be killed or it may transfer the fungus to another leaf (Izumi, 2001).

Endophytes appear to have direct and induced effects on plant responses to biotic agents. The interaction with abiotic agents remains largely unexplored. The broader, ecological function of endophytic associations is still being debated. Many fungi that are associated with the initial stages of litter decomposition are found in healthy tissue of the same plants. It can be concluded that endophytes include fungi that have one or more of a variety of interactions with their host plant: some fungi are widespread and found on many different plant species; others are highly specific to single hosts in single environments. Further, diverse arrays of interactions between plant and fungus have been found (Clay, 1988). Given that a huge array of fungi may be isolated from any one host, it seems possible that endophytes will have one or more of a wide array of functions, most of which are unknown at present and it require further investigations concerning to production of bioactive compounds.

References

Bacon, C. W. and White, J. F. Jr. 1994. Biotechnology of endophytic fungi of grasses. CRC press, Boca Raton, Florida.

Barnett, H. L. and Hunter, B. B. 1998. Illustrated genera of imperfect fungi. Macmillan, Publ Co, NewYork.

Cheplick, G. P., Clay, K. and Masks, S. 1989. Interactions between infection by endophytic fungi and nutrient limitations in the grasses *Lolium perenne* and *Festuca arundinacea*. New Phytol. 111: 89–97.

Clay, K. 1988. Fungal endophytes of grasses: a defensive mutualism between plants and fungi. Ecology 69: 10–16.

Dasgupta, A., Agarwal, S. S. and Basu, D. K. 1984. Anticonvulsant activity of the mixed fatty acids of *Elaeocarpus ganitrus* roxb. (Rudraksh). Indian J Physiol Pharmacol. 28: 245-246.

Domsch, K. H., Games, W. and Anderson, T. 1980. Compendium of soil fungi. Academic Press, London.

Hata, K. and Futai, K. 1995. Endophytic fungi associated healthy Pine needle infested by Pine needle gall midge *Thecodiplosis japonensis*. Can J Bot. 73: 384–390.

Hata, K., Futai, K. and Tsuda, M. 1998. Seasonal and needle age-dependent changes of the endophytic mycobiota in *Pinus thunbergii* and *Pinus densiflora* needles. Can J Bot 76: 245–250.

Izumi, O. 2001. Taxonomy and ecology of endophytic fungi. Nipon Kin Gak. Kaiho 42: 149-161.

Jones, R. C., Mc Nally, J. and Rossetto, M. 2002. Isolation of microsatellite loci from a rainforest tree,

Elaeocarpus grandis (Elaeocarpaceae), and amplification across closely related taxa. Molecular Ecology Notes 2: 179-181.

Mahesh, B., Tejesvi, M. V., Nalini, M. S., Prakash, H. S., Kini, K. R., Subbiah, V. and Hunthrike, S. S. 2005. Endophytic mycoflora of inner bark of *Azadirachta indica* A. Juss. Curr Sci 88:218–219.

Maria, G. L. and Sridhar, K. R. 2003. Endophytic fungal assemblage of two halophytes from west coastal mangrove habitats, India. Czech Mycol 55:241–251.

Nalini, M. S., Mahesh, B., Tejesvi, M. V., Prakash, H. S., Subbaiah, V., Kini, K. R. and Shetty, H. S. 2005. Fungal endophytes from the three-leaved caper, *Crataeva magna* (Lour.) DC. (Capparidaceae). Mycopathologia 159:245–249.

Petrini, O. 1986. Taxonomy of endophytic fungi of aerial plant tissues. In Microbiology of the phyllosphere (Eds. Fokkema, N. J. and Den Heuvel, J. van) Pp. 175-187. Cambridge University Press, Cambridge.

Petrini, O. 1991. Fungal endophytes of tree leave. In Microbial ecology of leaves (Eds. Andrews, J. A. and Hirano, S. S.) Pp. 179-197. Springer-Verlag, New York.

Raviraja, N. S. 2005. Fungal endophytes in five medicinal plant species from Kudremukh Range, Western Ghats of India. J Basic Microbiol 45:230–235.

Redman, R. S., Sheehan, K. B., Stout, R. G., Rodriguez, R. J. and Henson, J. M. 2002. Science. 298: 1581–1582.

Singh, R. K. and Nath, G. 1999. Antimicrobial activity of *Elaeocarpus sphaericus*. Phytoth. Res. 13:448-450.

Strobel, G. and Daisy, B. 2003. Bioprospecting for microbial endophytes and their natural products. Microbiol Mol Biol Rev 67:491–502.

Strobel, G., Daisy, B., Castillo, U. and Harper, J. 2004. Natural products from endophytic microorganisms. J Nat Prod 67:257–268.

Suryanarayanan, T.S., Kumaresan, V., Johnson, J. A. 1998. Foliar endophytes from two species of the mangrove Rhizophora. Can J Microbiol 44:1003–1006.

Tejesvi, M. V., Mahesh, B., Nalini, M. S., Prakash, H. S., Kini, K.R., Subbiah, V. and Hunthrike S. S. 2005. Endophytic fungal assemblages from inner bark and twig of *Terminalia arjuna* W. and A. (Combretaceae). World J Microbiol Biotechno l 21:1535–1540.

Microbes: Diversity and Biotechnology (2012) *Pages* **43–62**
Editors: **Prof. S.C. Sati & Dr. M. Belwal**
Published by: **DAYA PUBLISHING HOUSE, NEW DELHI**

Chapter 3
Aspect and Prospect of Endophytic Fungi

K.R. Sridhar

Department of Biosciences, Mangalore University, Mangalagangotri, Mangalore – 574 199, Karnataka

ABSTRACT

Endophyte biology has become one of the hot topics of research in view of importance of endophytes in plant protection, natural products and existence of cryptic fungi. Mutualistic association between host plant species and endophytes has considerable importance in plant ecology, community structure, fitness and evolution. Exploration of rare and endangered habitats, habitat-dependent and host-dependent endophytic fungi needs special attention. Understanding the role of entomopathogenic endophytes are of immense value in biocontrol, plant protection and disease management. As endophytic fungal association with plants has evolutionary significance, reduction or elimination of endophytes in plant tissues by agricultural chemicals and human interference may be detrimental. Production, bioprospecting and utilization of bioactive metabolites from endophytic fungi need further investigation. There are several unsettled issues regarding endophytic fungi and their significance in host plant species.

Keywords: *Endophyte, Mutualistic association, Endophytic fungi, Bioactive metabolites.*

Introduction

Historical perspectives of endophyte study have been traced as early as 1898 (see Hyde and Soytong, 2008). De Bary (1866) first introduced the term 'endophytes' and defined as any organism occurring within plant tissues. Petrini (1991) defined endophytes as all organisms inhabiting plant organs that at some time in their life, can colonize internal plant tissues without causing apparent harm to the host. Schulz and Boyle (2005) denote that endophytes are broad range of inhabitants such as bacteria, fungi, algae and insects in healthy plant tissues. Definitions on endophytes have been viewed in different magnitude and revised time to time by several investigators (see Hyde and Soytong, 2008).

Explosion of research on fungal endophytes in the last three decades yielded tremendous results in understanding the diversity, systematics, ecology, evolution, metabolites, biological control and bioprospecting (*e.g.* Carroll, 1988; Fisher and Petrini, 1990; Petrini, 1991; Sieber, 1989, 2007; Schulz and Boyle, 2005; Arnold, 2007; Jones *et al.*, 2008; Suryanarayanan *et al.*, 2010; Zhang *et al.*, 2009). Besides publications of several books on endophytic fungi (*e.g.* Bacon and White, 2000; Schulz *et al.*, 2006; Cheplick and Feath, 2009), some issue of journals have been recently dedicated to project the magnitude of research on endophytic fungi (*Fungal Biology Reviews, 21* (2-3), 2007; *Fungal Diversity, 33* (6), 2008). The goal of the present chapter is to address briefly some recent developments in diversity, distribution and function of mainly non-clavicipitaceous endophytic fungi and their importance in selected habitats.

Diversity and Distribution

Endophytic fungi are polyphyletic group, primarily composed of ascomycetes. Details of diversity, life-history strategy, adaptation and ecological role of endophytic fungi have been well documented by Rodriguez *et al.* (2009). Based on phylogeny and life-history traits, endophytic fungi have been broadly classified into two categories: i) clavicipitaceous (grass-inhabiting); ii) non-clavicipitaceous (non-grass-inhabiting) (Schulz and Boyle, 2005; Rodriguez *et al.*, 2009). Clavicipitaceous endophytes have been studied extensively than non-clavicipitaceous endophytes (*e.g.* Clay, 1996, 1997). Based on host colonization, mechanism of transmission and ecological functions, the non-clavicipitateous endophytes have been differentiated into three functional groups: i) the first group establishes in both above- and below-ground tissues of host plants; ii) the second group confined to above-ground tissues; iii) the third group restricted to below-ground tissues. Although, these groups have a broad host range, the pattern of colonization of host tissues differs: i) the first and third groups extensively colonize the plant tissues, while the second group shows highly localized patchy colonization; ii) the second and third groups are transmitted to plant tissues horizontally, while the first group by both vertically and horizontally.

There are some evidences that saprobes are derived from endophytes and they are latent pathogens (*e.g.* Brown *et al.*, 1998; Sieber, 2007; Duong *et al.*, 2008). If so, Hyde and Soytong (2008) argue that endophytes colonize above-ground tissues are host- or tissue-specific and genus- or family-specific. Schulz and Boyle (2005) hypothesized that regardless of the life-history strategy of the fungal endophyte, the disease does not manifest until fungal virulence and the host defense reaction are balanced. Sieber (2007) suspects that the endophytic pathogens have been co-evolved with their host plant species. Molecular sequence data from 1403 endophytic fungal strains demonstrated increased diversity, incidence and host range from arctic to tropical regions (Arnold and Lutzoni, 2007). In higher latitudes, the endophytic fungi are characterized by a few species belonging to different classes of ascomycetes in contrast to very large number of endophytes by a small number of classes with a wide host ranges in tropics. Sampling endophytic fungi from selected angiosperms of different latitudinal gradient (northern boreal forest to tropical forest) revealed interesting facts: (*i*) higher diversity in tropical angiosperms; (*ii*) colonization of tissue segments decreased from tropics to temperate latitudes; (*iii*) tropics represented by fewer classes with dominance of Sordariomycetes, while boreal communities consists of several classes with dominance of Dothideomycetes; (*iv*) the host generalism was dominant in the tropics, whereas strong host-affinity was evident in boreal endophytes (Arnold, 2007; Arnold and Lutzoni 2007). The cultivated endophytes from tropical plant species showed dominance of a large number of rare species, whose host range is not clear (Arnold and Lutzoni, 2007).

World's average fungal resource was estimated by Hawksworth (1991, 2001) as 1.5 million species based on plant and fungus ratio. This estimation was based on plant and fungus ratio (1:6) out of 250,000 plant species. It was initially applied to temperate regions, while investigations on palm fungi in Queensland raised this ratio to 1:26 in the tropics, which has been further updated to 1:33 mainly based on palm fungi in Australia and Brueni Darussalam (Hyde, 1995; Fröhlich and Hyde, 1999). A variety of endophytic fungi have been reported from diverse palms (*e.g.* Fröhlich *et al.*, 2000; Lumyong *et al.*, 2009) and there is a strong belief that missing fungi are hidden as endophytes. However, the evolutionary origins and diversity of mutualistic endophytes are still not clearly understood.

Terrestrial Habitats

Leaves

A wide range of plant species surveyed composed of endophytic fungal symbionts in foliar tissues (Stone *et al.*, 2000). Endophytes inhabiting leaves (short-lived, photosynthetically versatile and easily subject to damage by herbivores) are under high selective pressure than those associated with persistent tissues (*e.g.* bark, xylem or other woody parts) (Arnold, 2007). Arnold *et al.* (2003) suspect that tropical plant species have the potential to develop differential endophytic symbiosis with diverse mycota by horizontal transmission. They have demonstrated that horizontally transmitted endophytes in woody angiosperms play an important role in host defense. On inoculation of endophyte-free leaves of cocoa (*Theobroma cacao*) with endophytic fungus (*Phytophthora* sp.) significantly decreased leaf necrosis and mortality (Arnold *et al.*, 2003). As endophyte in maize, *Fusarium verticillioides* reduced disease severity caused by *Ustilago maydis* (Lee *et al.*, 2009). Similarly, *Fusarium oxysporum* isolated from healthy tissues of tomato seedlings exhibited potent *in vivo* anti-oomycete activity against tomato late blight and *in vitro* anti-oomycete potential indicating its usefulness as biocontrol agent (Kim *et al.*, 2007).

Protection by endophytes was greater in mature than young leaves and *in vitro* experiments revealed that leaf chemistry mediate their host affinity. Thus, leaf chemistry assumes special importance in foliar endophytic assemblage, diversity and function. Due to biochemical diversity among the leaves of different plant species and high degree of competition among fungi to colonize foliage, Arnold (2007) suspects that endophytes of leaves are host-specific or facultative saprotrophs and possess different evolutionary origins than wood-inhabiting endophytes. It is likely the lifestyle of foliar endophytes dependent on nature of plant species such as deciduous or evergreen in contrast to those colonizing woody tissues although both substrates colonized by common endophytes (Arnold, 2007). Horizontal transmission of foliar endophyte (*Phialocephala scopiformis*) from white spruce trees (*Picea glauca*) to its seedlings has been studied recently by Miller *et al.* (2009). Dissemination of endophyte was up to 40 per cent in three years with mean rugulosin (anti-insect toxin) concentration up to 1 μg/g tissue. An interesting perspective has been given by Devarajan and Suryanarayanan (2006) in relation to the role of grasshoppers in the dispersal on non-grass endophytes. Grasshoppers preferred or avoided milkweed leaves covered by spore suspension of *Colletotrichum*. However, spores retained their viability through gut passage suggesting grasshoppers as potential vectors in dissemination of endophytic fungi.

Bark and Wood

Endophytic fungi in branches of many dicotyledonous trees possess several ascomycetes and basidiomycetes (Chapela and Boddy, 1988). Endophytic fungi in twigs of *Quercus ilex* and dominant fungi include: *Colletotrichum*, *Nodulisporium* and *Phyllosticta* (Fisher *et al.*, 1994). *Aureobasidium*, *Botryosphaeria* and *Cytospora* have been recovered from stems of *Eucalyptus* (Simeto *et al.*, 2005). Some studies have indicated that the diversity of endophytes in twigs is less diverse than other niches such as bark (Tejesvi *et al.*, 2005). Kumar and Hyde (2004) demonstrated tissue specificity of endophytes in woody angiosperms and conifers.

Endophytic fungal community from the twig xylem of Chinese medicinal plant (*Tripterygium wilfordii*) was most diverse than leaves, bark, root xylem and flowers (Kumar and Hyde, 2004). *Pestalotiopsis cruenta*, *Phomopsis* spp. were dominant in twig xylem and bark. In addition, 13 morphotypes were isolated from twig bark and twig xylem. Analysis of twigs and bark of *Terminalia arjuna*, a tropical woody angiosperm (used as an important ayurvedic medicinal plant) revealed greater diversity during monsoon season (Tejesvi *et al.*, 2005). Endophyte colonization was four-fold higher in bark than twigs. The most dominant endophyte genera were *Chaetomium*, *Myrothecium* and *Pestalotiopsis*. Recently, Gandadevi and Muthumary (2010) demonstrated production of anticancer drug, taxol by an endophytic fungus (*Chaetomella raphigera*) isolated from *Terminalia arjuna*. Freshwater hyphomycetes were also known as endophytes in aerial plant tissues (Sokolski *et al.*, 2006). Several typical freshwater hyphomycetes were colonizers of epiphytic fern (*Drynaria quersifolia*) in tree canopies of the west coast and Western Ghats of India (Sridhar *et al.*, 2006; Karamchand and Sridhar, 2009).

Roots

Leaves, berries, stems and roots of *Coffea arabica* from different geographic regions (Colombia, Hawaii, Mexico and Puerto Rico) yielded 843 isolates of fungal endophytes, which resulted in 257 unique genotypes (Vega *et al.*, 2009). Taxa belonging to *Colletotrichum*, *Fusarium*, *Penicillium* and Xylariaceae were most abundant. Species of dark-septate endophytic fungi are relatively few (Addy *et al.*, 2005), which colonize roots of many plant species and have been recognized as a functional group based on their melanized hyphae (Jumpponen and Trappe, 1998; Jumpponen, 2001, 2003). Although their diversity and ecological functions are not yet clear, it is known that they have a variety of lineages with Pezizales, Helotiales and Pleosporales (Jumpponen, 2001). The dark-septate endophytes are also found in foliar tissues (Jumpponen, 2001; Higgins *et al.*, 2007). These fungi have been examined in some plant species by Suryanarayanan and Vijaykrishna (2001) and Kumar and Hyde (2004). A brief account on colonization of roots by freshwater hyphomycetes is given in subsequent section.

Piriformospora indica is a cultivatable root-inhabiting basidiomycete, which has capability to establish in semi-defined media without a host (Varma *et al.*, 2001). It is known from the roots of a wide range of plant species including trees, agricultural/horticultural/medicinal plants, monocots, dicots and mosses (see Vadassery *et al.*, 2009). It colonizes the roots of *Arabidopsis thaliana* and promotes growth, development and seed production. The cell wall extract of *P. indica* promotes growth of *Arabidopsis* seedlings and also induces intracellular elevation of calcium roots. This fungus is also known to stimulate uptake of nutrients and helps to develop resistance to biotic and abiotic stresses (*e.g.* Verma *et al.*, 1998; Varma *et al.*, 1999, 2001; Pham *et al.*, 2004; Waller *et al.*, 2005; Shahollari *et al.*, 2007; Sherameti *et al.*, 2008a, 2008b).

Seeds

Relatively, seeds posses less number of endophytic fungi than other tissues probably due to vertical transmission. For instance, Ganley and Newcombe (2006) recovered over 2000 endophytic fungal isolates from foliage of *Pinus monticola*, while only 16 from 750 surface-sterilized seeds. Although foliage of *Theobroma cacao* associated with diverse endophytes, their seeds were devoid of endophytes (Arnold *et al.*, 2003). In contrast to vertical transmission of clavicipitaceous endophytes in grasses, multiple species of endophytes have been documented from seeds of plants other than grasses (Clay and Schardl, 2002; Arnold, 2007). The culture-independent methods (environmental PCR) revealed occurrence of diverse ascomycete endophytes in seeds of *Ceropia insignis* in tropics (Gallery *et al.*, 2007).

Coastal Sand Dunes

Leaves of *Suaeda fruticosa*, a beach halophyte consists of common endophytic fungi belonging to the genera *Acremonium, Alternaria, Cladosporium, Colletotrichum* and *Fusarium* (Fisher and Petrini, 1987). From 1512 root pieces of plant species belonging to coastal and inland soils, sandy soils and salt marshes of southwest Spain yielded 1830 isolates consisting of 142 identifiable endophytes (57 genera) and 177 morphospecies (Maciá-Vicente *et al.*, 2008). Leaves and rhizomes of grasses belonging to *Ammophila* and *Elymus* from 12 coastal sand dunes of the northern Spain yielded 103 isolates of endophytic fungi (Márquez *et al.*, 2008). Significant inverse relationship has been shown between similarities of endophytic assemblages and distance of host plant species.

In coastal sand dunes of southwest coast of India, two wild legumes (*Canavalia cathartica* and *C. maritima*) colonized by 46 taxa of endophytic fungi and morphospecies (Seena and Sridhar, 2004). Among different age (seed, seedling and mature plant) and tissues (root, stem, leaf, seed coat and cotyledon) classes, the highest number of endophytic fungi was recorded in the seedlings and mature plants with lowest in seeds. *Chaetomium globosum* exhibited single species dominance in root, stem and leaf segments of *C. maritima* and root segments of *C. cathartica*. *Chetomium* has been considered as a cosmopolitan genus and especially common as endophyte in desert plant species and suspected that it colonizes when exterior conditions are inhospitable (Hoffman and Arnold, 2007)

The genus *Chaetomium* is known to produce a variety of secondary metabolites (*e.g.* chaetomin, chaetoglobosins, chaetoquadrins, oxaspirodion, chaetospiron, orsellides, chaetocyclinones) (Sekita *et al.*, 1976; Loesgen *et al.*, 2007; Suryanarayanan *et al.*, 2010). Chaetoglobicins are cytotoxins, which are analogs of cytochalasin having inhibitory effect on actin polymerization (Yahara *et al.*, 1982; Suryanarayanan *et al.*, 2010). In addition, *C. globosum* also produces a nematicide called flavipin (Chitwood, 2002). Extensive colonization of root, stem and leaf of *Canavalia* by *C. globosum* in coastal sand dunes indicates its potential role against herbivores and nematodes.

Although beaches comprise of several marine fungi (*e.g.* saprophytes on woody litter, arenicolous fungi on sand grains), surprisingly only 3 per cent of marine fungi were endophytic in wild legumes (*Canavalia* spp.) (Seena and Sridhar, 2004). Similarly, marine fungi were also less dominant as root endophytes in other coastal sand dune halophytes (Beena *et al.*, 2000)

Aquatic Habitats

Freshwater

Currently, freshwater hyphomycetes are also well known endophytes in roots exposed to streams and aerial plant parts without producing disease symptoms (see Baerlocher, 2006; Sokolski *et al.*, 2006). Nemec (1969) reported *Tetracladium marchalianum* as root endophyte of *Fragaria* sp., while Watanabe (1975) found *Tetracladium setigerum* endophytic in roots *Fragaria* sp. and *Gentiana* sp. Many strains of *Gyoerffyella* have also been isolated from healthy roots of *Picea abies* (see Selosse *et al.*, 2008). Currently, about 50 species of freshwater hyphomycetes are known as endophytes in submerged roots belonging to approximately 25 plant species (angiosperms, gymnosperms and pteridophytes) growing in freshwater and brackish water habitats (*e.g.* Marvanová and Fisher, 1991; Fisher *et al.*, 1991; Sridhar and Baerlocher, 1992a, 1992b; Marvanová *et al.*, 1992, 1997; Raviraja *et al.*, 1996; Sati and Belwal, 2005; Sati *et al.*, 2008, 2009a, 2009b).

Fisher and Petrini (1989) first demonstrated the endophytic-phase of two typical aquatic hyphomycetes (*Campylospora parvula* and *Tricladium splendens*). Subsequently, Fisher *et al.* (1991) compared endophytic aquatic hyphomycete population in submerged and terrestrial roots and demonstrated higher colonization in submerged (30 per cent) than terrestrial (12 per cent) roots. Submerged macrophytes (*e.g. Potamogeton, Ranunculus, Apium*) were also constitute major substrate in the absence of submerged leaf litter in streams (Baerlocher, 1992). Studies on root endophytes may reveal additional anamorph-teleomorph connections (Webster, 1992; Sivichai and Jones, 2003). Interestingly, production of teleomorphic state by endophytic *Heliscus lugdunensis* upon subculturing was reported by Sridhar and Baerlocher (1992a). New species of aquatic hyphomycetes (*Filosporella fistucella, F. versimorpha, Fontanospora fusiramosa* and *Tetracladium nainitalense*) have also been described from endophytes of riparian roots (Marvanová and Fisher, 1991; Marvanová *et al.*, 1992, 1997; Sati *et al.*, 2009b).

Brackish and Marine

Although saprophytic fungi of mangrove plant species have attracted the attention of mycologists, studies on endophytic fungi have been initiated recently (Suryanarayanan *et al.*, 1998; Suryanarayanan and Kumaresan, 2000; Kumaresan and Suryanarayanan, 2001, 2002; Jones *et al.*, 2008; Sridhar, 2009). Endophytes belonging to the genera *Acremonium, Alternaria, Cladosporium, Colletotrichum* and *Fusarium* were common in mangrove plants (Suryanarayanan *et al.*, 1998, Suryanarayanan and Kumaresan, 2000; Kumaresan and Suryanarayanan, 2001, 2002) and seagrass, *Halophila ovalis* (Devarajan *et al.*, 2002). In addition, *Phomopsis* and *Phyllosticta* were also common foliar endophytes of mangrove plants (Suryanarayanan *et al.*, 1998; Suryanarayanan and Kumaresan, 2000; Kumaresan and Suryanarayanan, 2001, 2002; Ananda and Sridhar, 2002).

Sporormiella minima and *Cladosporium cladosporioides* were major foliar endophytes of mangrove plant species of southeast coast of India. Foliar endophytes were dominated by single species in many mangrove plant species: *Avicennia marina* (*Phoma* sp.), *Bruguiera cylindrica* (*Colletotrichum gloeosporioides*), *Rhizophora apiculata* (*Sporormiella minima*), *Rhizophora mucronata* (*Sporormiella minima*) and *Suaeda maritima* (*Camarosporium palliatum*) (Suryanarayanan *et al.*, 1998; Suryanarayanan and Kumaresan, 2000; Kumaresan and Suryanarayanan, 2001). However, *Lumnitzera racemosa* showed multiple species dominance by *Alternaria* sp., *Phomopsis* sp. and *Phyllosticta* sp. (Kumaresan and Suryanarayanan, 2001). Multiple species dominance of endophytic fungi was also seen in the roots of *Avicennia officinalis, Rhizophora mucronata* and *Sonneratia caseolaris* by Ananda and Sridhar (2002).

In mangrove associates (*Acanthus ilicifolius* and *Acrostichum aureum*), 25 endophytic fungi comprising three ascomycetes, 20 mitosporic taxa and two sterile morphotypes were recovered from the West Coast mangrove by Maria and Sridhar (2003). Overall colonization by endophytes was as high as 74.5-77.5 per cent. Out of four tissues screened (leaves, stem, rhizome, root), species richness and diversity were high in stems of *A. ilicifolius* and roots of *A. aureum*. In Mai Po Nature Reserve of Hong Kong, Pang *et al.* (2008) studied endophytic fungal association with bark, woody tissues and leaves of *Kandelia candel*. Endophytic assemblage was similar in bark and wood, but deferred in leaf samples. Dominant and cosmopolitan endophytes were *Guignardia* sp., *Pestalotiopsis* sp., *Phomopsis* sp. and *Xylaria* sp. In mangrove and mangrove associate plant species of the west coast of India, typical marine fungi as endophytes were low (up to 5 per cent) (Ananda and Sridhar, 2002; Maria and Sridhar, 2003; Anita and Sridhar, 2009; Anita *et al.*, 2009).

Seaweeds and Seagrasses

Live seaweeds and seagrasses constitute suitable substrates for colonization of fungi (Sugano *et al.*, 1994; Nielsen *et al.*, 1999; Devarajan *et al.*, 2002; Zhang *et al.*, 2009). Zuccaro *et al.* (2003, 2008) reported endophytic fungi from *Fucus serratus*. The endophytic fungus, *Mycophycias ascophylli* has been repeatedly isolated from thallus of *Ascophyllum nodosum* (Stanley, 1991). It grows mutually with *A. nodosum* and *Pelvetia canaliculata* (Kohlmeyer and Kohlmeyer, 1979; Kohlmeyer and Volkmann-Kohlmeyer, 1998; Ainsworth *et al.*, 2001) and remains associated with algal host throughout its life cycle (Stanley, 1991). Raghukumar (1996) has demonstrated association of several marine algae with endophytic fungi in Indian coast. For example, *Cladophora* sp. and *Rhizoclonium* sp. were frequently associated with the chytrid *Coenomyces* sp. Similarly, many filamentous fungi and thraustochytrid endophytes were isolated from *Centroceras clavulatum, Gelidium pusillum, Padina tetrastomatica, Sargassum cinereum, Valoniopsis pachynema* and *Ulva fasciata* (Raghukumar *et al.*, 1992; Raghukumar, 2008). Thraustochytrids and labyrinthulids (*e.g. Aplanochytrium minutum*) were often isolated from surface sterilized seaweeds (*P. tetrastomatica* and *S. cinereum*) (Sathe-Pathak *et al.*, 1993). Up to 26 species of endophytes have been reported in the seagrasses such as *Thalassia estudinum, Zostera japonica* and *Z. marina* (Alva *et al.*, 2002).

Endophytes from marine algae and seagrasses produce a variety of metabolites (see Raghukumar, 2008; Zhang *et al.*, 2009). Some novel metabolites were obtained from algal taxa belonging to different geographic origin (Schulz *et al.*, 2008). Marine fungus, *Dendryphiella salina* was associated with brown seaweeds is known to produce a range of bioactive compounds (Guerriero *et al.*, 1989). The obligate marine fungus *Aschochyta salicorniae* associated with *Ulva* sp. showed anti-plasmodial activity against *Plasmodium falciparum* (Osterhage *et al.*, 2000). However, many interesting algal-fungal associations have not been investigated in view of bioactive metabolite perspective.

Entomopathogens

Research on entomopathogenic endophytic fungi is fairly recent. Webber (1981) seems to be the first investigator to report elm tree protection by endophytic fungus (*Phomopsis oblonga*) against the beetle (*Physocnemum brevilineum*). Azevedo *et al.* (2000) extensively reviewed subsequent two decades of research on entomopathogenic endophytic fungi. Important and widely distributed endophytic fungi serve as entomopathogens include: *Beauveria* spp. (endophytic in maize, potato, cotton, tomato, cocoa, *Pinus*, banana, coffee) (Evans *et al.*, 2003; Ownley *et al.*, 2004; Arnold and Lewis, 2005; Ganley and Newcombe, 2006; Akello *et al.*, 2007; Posada *et al.*, 2007; Vega *et al.*, 2008), *Paecilomyces farinosus, P. varioti* and *Verticillium lecanii* (Petrini, 1981; Bills and Polishook, 1991; Ananda and Sridhar, 2002). Endophytic fungi are beneficial to host plant species to prevent herbivory and induce tolerance to stresses (*e.g.* heat, salt, disease, drought) and also increase below- and above-ground plant biomasses (Waller *et al.*, 2005; Tejesvi *et al.*, 2007). Survey of entomopathogenic fungal endophytes yielded several potential biocontrol agents against pests and possess potentialities to develop as future mycopesticides (see Kaewchai *et al.*, 2009). Vega *et al.* (2008) raised an interesting question: whether the metabolites of entomopathogenic fungal origin inoculated to agronomically important plants enter the food chain?

Natural Products

Medicinal Plants

Based on traditional morphological techniques, Huang *et al.* (2008) isolated 1160 endophytic fungi from 29 Chinese medicinal plants. Among these endophytes, *Alternaria, Colletotrichum, Phoma, Phomopsis*, Xylariales and sterile morphotypes were dominant. Medicinal plants of the Western Ghats of India endowed with a variety of endophytic fungi (*e.g.* Raviraja, 2005; Krishnamurthy *et al.*, 2008; Naik *et al.*, 2008). From 9000 leaf segments of 15 medicinal shrubs from seven locations of Western Ghats, 6125 fungal endophytes were isolated by Naik *et al.* (2008). Hyphomycetes and coelomycetes were dominant than mucorales and sterile morphotypes. *Alternaria, Chaetomium, Cladosporium, Colletotrichum, Fusarium, Penicillium, Phyllosticta* and *Xylaria* were most frequently isolated fungi. Number of isolates obtained during the winter season were significantly more than monsoon and summer seasons.

Active Metabolites

About 4,000 biologically active secondary metabolites have been described from fungi (mainly from *Acremonium, Aspergillus, Fusarium* and *Penicillium*) (Dreyfuss and Chapela, 1994; Hyde and Soytong, 2008). Endophytes are a good source of diverse novel metabolites and secondary metabolites including plant protectants (*e.g.* xanthones, phenols, isocoumarins, perylene derivatives, quinones, furandiones, terpenoids, depsipeptides, cytochalasines) (see Schulz *et al.*, 2002; Schulz and Boyle, 2005; Raj and Shetty, 2009). Endophytes have been screened for many novel metabolites such as antibiotics, anti-cancer drugs (*e.g.* paclitaxel, Hsp 90 inhibitors, sequoiatones, camptothecin), lactones, enalin, colletotrichic acid, myrocin and apiosporic acid, phomopsilactone, cyclopentanoids, (+)-ascochin; (+)-ascodiketone, chaetocyclinones, pestalotheols, isofusidienols and naphthoquinone spiroketals (see Suryanarayanan *et al.*, 2010). Schulz *et al.* (2002) studied 6,500 endophytic fungi for their natural products and found 51 per cent out of 135 metabolites as new bioactive compounds. Several endophytic fungal metabolites mimic the structure and function of host compounds (*e.g.* gibberellins: MacMillan, 2002; taxol: Stierle *et al.*, 1993; subglutinols: Lee *et al.*, 1995; Strobel, 2002). Such mimicking ability of associated endophytic fungi has been suspected to genetic recombination of the endophytes with the host during evolution of mutualism (Tan and Zou, 2001).

The US patents on active metabolites of endophytic fungal origin elevated steeply after discovery of taxol producing fungus, *Taxomyces andreanae* (Priti *et al.*, 2009). Certain metabolites produced by endophytic fungi are same or similar to that of host plant species (*e.g.* anticancer drug taxol by *Taxomyces andreanae* associated with yew plants *Taxus brevifolia*) (Stierle *et al.*, 1993). Similarly, endophytic fungus, *Pestalotiopsis microspora* associated with an endangered tree species, *Torreya taxifolia* produced torreyanic acid, which is a powerful apoptotic drug (Lee *et al.*, 1996). Endophytes have the ability to produce several compounds produced by the host plants (*e.g.* gibberellins, subglutinols) (Lee *et al.*, 1995; MacMillan, 2002; Strobel, 2002). Endophytic fungi derived from the oil palm (*Elaeis guineensis*) were also major source of several bioactive compounds and their association reduced the root decay (Rungjindamai *et al.*, 2008).

There seems to be multiple factors (*e.g.* host, tissue, age, season, environment, geographical location) influence the metabolite production by endophytic fungi (Moricca and Ragazzi, 2008; Shwab and Keller, 2008). Thus, there are failures to obtain commercially exploitable quantity of natural products from isolated endophytic fungi (Li *et al.*, 1998; Young *et al.*, 2006; Szewczyk *et al.*, 2008; Priti *et al.*, 2009). Priti *et al.* (2009) argue that the endophytic fungal metabolites are the product of interaction of host and fungus, which are regulated by several host and environmental factors. Lack of such host stimulus *in vitro* might attenuate the ability of endophytic fungi to generate desired metabolites. There is a need to develop appropriate methods to induce endophytic fungi to produce required metabolites *in vitro*.

Evaluation

Endopytic fungal evaluation in plant species needs special attention to avoid contaminants. Different methods of surface sterilization of plant tissues and isolation of endophytic fungi have been documented by Gallo *et al.* (2008). Schultz *et al.* (1998) advocated assessing sterilized tissue pieces by imprinting on a medium to ascertain efficiency of surface sterilization. Surface sterilization will be assumed effective only when no fungi grow on the medium on imprinting. It is also useful to evaluate fungi associated with plant species in both surface sterilized and unsterilized segments (Anita and Sridhar, 2009; Karamchand *et al.*, 2009). Similarly, Suryanarayanan *et al.* (2009) proposed that evaluation of endophytes should include different substrates to arrive at near-real species diversity.

Methodological constraints to study endophytic fungi and necessity to adapt traditional and molecular techniques have been addressed by Hyde and Soytong (2008). Endophytic fungi within leaves were ascertained by detecting β-D-glucans of fungal cell walls by Johnston *et al.* (2006). Fastidious endophytes may need different media than routine media for isolation (Guo *et al.*, 2001). Similarly, it is necessary to isolate slow-growing endophytes by preventing growth of fast-growing endophytes using differential media (Hyde and Soytong, 2008). Most endophyte isolation exercises usually result in recovery of several sterile morphotypes. A method to promote sporulation of palm-derived sterile morphotypes has been developed by Guo *et al.* (1998) and achieved sporulation of isolates up to 83.5 per cent against 47.8 per cent on conventional incubation on MEA agar medium. Some studies employed molecular techniques to place sterile morphotypes of endopytic fungi for appropriate taxonomic placement (Promputtha *et al.*, 2005; Wang *et al.*, 2005). Arnold (2007) advocated establishing repository of unknown cultures of endophytic fungi for future use. Interestingly, Hambleton and Sigler (2005) erected a new genus with three new taxa from sterile morphotypes based on gene sequence data. There may be several uncultivable endophytes in live plant tissues and appropriate to assess them by molecular techniques. Techniques like DNA cloning, DGGE, T-RFLP and stable isotope profiling (SIP) have been attempted by several investigators in order to detect total fungal communities in plant tissues (*e.g.* Nikolcheva and Baerlocher, 2005; Vandenkoornhuyse *et al.*, 2002; Seena *et al.*, 2008; Tao *et al.*, 2008). These methods are useful to evaluate endophytic fungal diversity more precisely.

Systemic fungicide treatment as foliar spray helps to obtain endophyte-free plants for experimental purpose (Cheplick, 1994; Hill and Brown, 2000). Recently, a systemic fungicide, hexaconazole [2-(2,4-dichlorophenyl)-1-(1H-1,2,5-trizol-1-yl) hexan-zol] treatment qualitatively and quantitatively decreased foliar endophytes of mango trees (*Mangifera indica*) (Mohandoss and Suryanarayanan, 2009). Elimination or differential/selective elimination of endophytes by fungicide treatment will be valuable to understand the role of endophytes in host susceptibility or resistance to diseases. Similarly, temperature treatment also reduced the endophytes of seeds in grass species, *Elymus virginicus* (Rudgers and Swafford, 2009) and proved to be an useful method to study the importance of vertically transmitted endophytes.

Outlook

According to Schulz and Boyle (2005) the endophytic continuum is a developmental and evolutionary process. Relatively, endophytes of foliar tissues have been studied extensively than other tissues. Are foliar endophytes transient flora? If so, what are the means by which such fungi will be attracted or accommodated by the host plant species for their benefits through horizontal transmission? Still there are several unsettled questions about the diversity, distribution, evolution and functions of endophytic fungi. Some fungi have multiple ecological roles such as endophyte, pathogen (opportunistic?) and or saprotroph (*e.g. Chaetomium globosum, Paecilomyces varioti*) (Ananda and Sridhar, 2002; Seena and Sridhar, 2004; Arnold *et al.*, 2007; Naik *et al.*, 2007; Vega *et al.*, 2008). Are such fungal features are overlapping? How these traits switchover? Is it depending on the substrate or environmental conditions? Possibly, fungicide and temperature treatments to reduce/eliminate endophytic fungi differentially/selectively in live tissues might facilitate to prove Koch's postulates to understand host range, pattern of colonization, mode of transmission (vertical or horizontal), anthropogenic/environmental influences, host vulnerability, impact of pathogens and ecological functions of endophytic fungi.

It is interesting to understand whether non-sporulating endophyte morphospecies also produce novel metabolites. For instance, rugulosin, a metabolite of a nonsporulating spruce endophyte is active against the spruce budworm (Miller *et al.*, 2009). Are sporulating and non-sporulating endophytes producing novel metabolites in duel culture *in vitro* due to mutual competition? It is worth understanding the impacts of host species and environmental factors in stimulating endophytic fungi to produce secondary metabolites for bioprospecting purpose.

Acknowledgements

I am grateful to Mangalore University for encouragement to carry out research on endophytic fungi and sanction of sabbatical leave during the academic year 2008-2009. I am indebted to Prof. Felix Baerlocher, Mount Allison University, Canada for his support and encouragement during course of investigation.

References

Addy, H. D., Piercey, M. M. and Currah, R. S. 2005. Microfungal endophytes in roots. Canadian Journal of Botany 83, 1–13.

Ainsworth, G. C., Bisby, G. R., Cannon, P. F., David, J. C., Staplers, J. A. and Kirk, P. M. 2001. Ainsworth and Bisby's Dictionary of the Fungi, 9th Edition. CAB International, Wallingford, UK.

Akello, J. T., Dubois, T., Gold, C. S., Coyne, D., Nakavuma, J. and Paparu, P. 2007. *Beauveria bassiana* (Balsamo) Vuillemin as an endophyte in tissue culture banana (*Musa* spp.). Journal of Invertebrate Pathology 96, 34–42.

Alva, P., McKenzie, E. H. C., Pointing, S. B., Pena-Muralla, R. and Hyde, K. D. 2002. Do seagrasses harbour endophytes? In: Fungi in Marine Environment (Ed. Hyde, K.D.), Fungal Diversity Research Series # 7, 167–178.

Ananda, K. and Sridhar, K. R. 2002. Diversity of endophytic fungi in the roots of mangrove species on the west coast of India. Canadian Journal of Microbiology 48, 871–878.

Anita, D. D. and Sridhar, K. R. 2009. Assemblage and diversity of fungi associated with mangrove wild legume *Canavalia cathartica*. Tropical and Subtropical Agroecosystems 10, 225–235.

Anita, D. D., Sridhar, K. R. and Bhat, R. 2009. Diversity of fungi associated with mangrove legume *Sesbania bispinosa* (Jacq.) W. Wight (Fabaceae). Livestock Research for Rural Development 21, Article # 67, http://www.lrrd.org/lrrd21/5/cont2105.htm

Arnold, A. E. 2007. Understanding the diversity of foliar fungal endophytes: progress, challenges, and frontiers. Fungal Biology Reviews 21, 51–66.

Arnold, A. E. and Lewis, L. C. 2005. Ecology and evolution of fungal endophytes, and their roles against insects. In: Insect-Fungal Associations: Ecology and Evolution (Ed. Vega, F.E. and Blackwell, M.), Oxford University Press, New York, 74–96.

Arnold, A. E. and Lutzoni, F. 2007. Diversity and host range of foliar fungal endophytes: are tropical leaves biodiversity hotspots? Ecology 88, 541–549.

Arnold, A. E., Mejía, L. C., Kyllo, D., Rojas, E., Maynard, Z., Robbins, N. and Herre, E. A. 2003. Fungal endophytes limit pathogen damage in a tropical tree. Proceedings of the National Academy of Sciences 100, 15649–15654.

Arnold, A. E., Henk, D. A., Eells, R. L., Lutzoni, F. and Vilgalys, R. 2007. Diversity and phylogenetic affinities of foliar fungal endophytes in loblolly pine inferred by culturing and environmental PCR. Mycologia 99, 185–206.

Azevedo, J. L., Maccheroni Jr. W., Pereira, J. O. and De Araújo, W. L. 2000. Endophytic microorganisms: a review on insect control and recent advances on tropical plants. Electronic Journal of Biotechnology 3, 40–65.

Bacon, C. W. and White, J. F. J. 2000. Microbial Endophytes. Marcel Dekker Inc., New York.

Baerlocher, F. 1992. Research on aquatic hyphomycetes: Historical background and overview. In: The Ecology of Aquatic Hyphomycetes (Ed. Baerlocher, F.). Springer-Verlag, Berlin, 1–15.

Baerlocher, F. 2006. Fungal endophytes in submerged roots. In: Microbial Root Endophytes (Ed. Schulz, B., Boyle, C. and Sieber, T.N.). Springer-Verlag, Berlin, 179–190.

Beena, K. R., Ananda, K. and Sridhar, K. R. 2000. Fungal endophytes of three sand dune plant species of west coast of India. Sydowia 52, 1–9.

Bills, G. F. and Polishook, J. D. 1991. Microfungi from *Carpinus caroliniana*. Canadian Journal of Botany 69, 1477–1482.

Brown, K. B., Hyde, K. D. and Guest, D. I. 1998. Preliminary studies on endophytic fungal communities of *Musa acuminata* species complex in Hong Kong and Australia. Fungal Diversity 1, 27–51.

Carroll, G. C. 1988. Fungal endophytes in stems and leaves: from latent pathogen to mutualistic symbiont. Ecology 69, 2–9.

Chapela, I. H. and Boddy, L. 1988. Fungal colonization of attached beech branches. I. Early stages of development of fungal communities. New Phytologist 110, 39–45.

Cheplick, G. P. 1994. Effect of endophytic fungi on the phenotypic plasticity of *Lolium pernne* (Poaceae). American Journal of Botany 84, 41–47.

Cheplick, G. P and Feath, S. 2009. *Ecology and Evolution of the Grass-Endophyte Symbiosis.* Oxford University Press, USA.

Chitwood, D. J. 2002. Phytopathological based strategies for nematode control. Annual Review of Phytopathology 40, 221–249.

Clay, K. 1996. Interactions among fungal endophytes, grasses and herbivores. Researches on Population Ecology 38, 191–201.

Clay, K. 1997. Fungal endophytes, herbivores and the structure of grassland communities. In: Multi-trophic Interactions in Terrestrial Systems (Ed. Gange, A.C. and Brown, V.K.). Blackwell, Oxford, 151–169.

Clay, K. and Schardl, C. 2002. Evolutionary origins and ecological consequences of endophyte symbiosis with grasses. The American Naturalist 160, S99–S127.

De Bary, A. 1866. Mrophologie und Physiologie der Pilze, Flechten, und Myxomyceten. In: Holfmeister's Handbook of Physiological Botany, Volume 2. Leipzig, Germany.

Devarajan, P. T. and Suryanarayanan, T. S. 2006. Evidence for the role of phytophagous insects in dispersal of non-grass fungal endophytes. Fungal Diversity 23, 111–119.

Devarajan, P. T., Suryanarayanan, T. S. and Geetha, V. 2002. Endophytic fungi associated with the tropical seagrass *Halophila ovalis* (Hydrocharitaceae). Indian Journal of Marine Sciences 31, 73–74.

Dreyfuss, M. M. and Chapela, I. H. 1994. Potential of fungi in the discovery of novel, low-molecular weight pharmaceuticals, In: The discovery of natural products with therapeutic potential (Ed. Gullo, V.P.). Butterworth-Heinemann, London, 44–80.

Duong, L. M., McKenzie, E. H. C., Lumyong, S. and Hyde, K. D. 2008. Fungal succession on senescent leaves of *Castanopsis diversifolia* in Doi Suthep-Pui National Park, Thailand. Fungal Diversity 30, 23–36.

Evans, H. C., Holmes, K. A. and Thomas, S. E. 2003. Endophytes and mycoparasites associated with an indigenous forest tree, *Theobroma gileri*, in Ecuador and a preliminary assessment of their potential as biocontrol agents of cocoa diseases. Mycological Progress 2, 149–160.

Fisher, P. J. and Petrini, O. 1987. Location of fungal endophytes in tissues of *Suaeda fruticosa*: a preliminary study. Transactions of the British Mycological Society 89, 246–249.

Fisher, P. J. and Petrini, O. 1989. Two aquatic hyphomycetes as endophytes in *Alnus glutinosa* roots. Mycological Research 92, 367–368.

Fisher, P. J. and Petrini, O. 1990. A comparative study of fungal endophytes in xylem and bark of *Alnus* species in England and Switzerland. Mycological Research 94, 313–319.

Fisher, P. J., Petrini, O. and Webster, J. 1991. Aquatic hyphomycetes and other fungi in living aquatic and terrestrial roots of *Alnus glutinosa*. Mycological Research 95, 543–547.

Fisher, P. J., Petrini, O., Petrini, L. E. and Sutton, B. C. 1994. Fungal endophytes from the leaves and twigs of *Quercus ilex* L. from England, Majorca, and Switzerland. New Phytologist 127, 133–137.

Fröhlich, J. and Hyde, K. D. 1999. Biodiversity of palm of fungi in the tropics: are global fungal diversity estimates realistic? Biodiversity and Conservation 8, 977–1004.

Fröhlich, J., Hyde, K. D. and Petrini, O. 2000. Endophytic fungi associated with palms. Mycological Research 104, 1202–1212.

Gallery, R., Dalling, J. W. and Arnold, A. E. 2007. Diversity, host affinity, and distribution of seed-infecting fungi: a case study with neotropical *Cecropia*. Ecology 88, 582–588.

Gallo, M. B. C., Chagas, F. O., Almeida, M. O., Macedo, C. C., Cavalcanti, B. C., Barros, F. W. A., De Moraes, M. O., Costa-Lotufo, L. V., Pessoa, C., Bastos, J. K. and Pupo, M. T. 2008. Endophytic fungi found in association with *Smallanthus sonchifolius* (Asteraceae) as resourceful producers of cytotoxic bioactive natural products. Journal of Basic Microbiology 48, 1–10.

Gangadevi, V. and Muthumary, J. 2010. A novel endophytic taxol-producing fungus *Chaetomella raphigera* oisolated from a medicinal plant, *Terminalia arjuna*. Applied Biochemistry and Biotechnology 160, DOI 10.1007/s12010-009-8532-0

Ganley, R. J. and Newcombe, G. 2006. Fungal endophytes in seeds and needles of *Pinus monticola*. Mycological Research 110, 318–327.

Guerriero, A., D'Ambrosio, M., Cumo, V., Vanzanella, F. and Pietra, F. 1989. Novel trinor-eremophilanes (dendryphiellin B, C and D), ermophilanes (dendryphiellin E, F and G), and branched C9-carboxylic acids (dendryphiellic acid A and B) from the marine deuteromycete *Dendryphiella salina* (Sutherland) Pugh et Nicot. Helvetica Chimica Acta 72, 438–446.

Guo, L. D., Hyde, K. D. and Liew, E. C. Y. 1998. A method to promote sporulation in palm endophytic fungi. Fungal Diversity 1, 109–113.

Guo, L. D., Hyde, K. D. and Liew, E. C. Y. 2001. Detection and taxonomic placement of endophytic fungi within frond tissues of *Livistona chinensis* based on *r*DNA sequences. Molecular Phylogenetics and Evolution 19, 1–13.

Hambleton, S. and Sigler, L. 2005. Meliniomyces, a new anamorph genus for root associated fungi with phylogenetic affinities to *Rhizoscyphus ericae* (*Hymenoscyphus ericae*), Leotiomycetes. Studies in Mycology 53, 1–27.

Hawksworth, D. L. 1991. The fungal dimension of biodiversity: magnitude, significance and conservation. Mycological Research 95, 641–655.

Hawksworth, D. L. 2001. The magnitude of fungal diversity: the 1.5 million species estimate revisited. Mycological Research 105, 1422–1432.

Higgins, K. L., Arnold, A. E., Miadlikowska, J., Sarvate, S. D. and Lutzoni, F. 2007. Phylogenetic relationships, host affinity, and geographic structure of boreal and arctic endophytes from three major plant lineages. Molecular Phylogenetics and Evolution 42, 543–555.

Hill, N. S. and Brown, E. 2000. Endophytic viability in seedling tall fiscue treated with fungicides. Crop Science 40, 1490–1491.

Hoffman, M. and Arnold, A. E. 2007. Geographic locality and host identity shape fungal endophyte communities in cupressaceous trees. Mycological Research 112, 331–344.

Huang, W. Y., Cai, Y. Z., Hyde, K. D., Corke, H. and Sun, M. 2008. Biodiversity of endophytic fungi associated with 29 traditional Chinese medicinal plants. Fungal Diversity 33, 61–75.

Hyde, K. D. 1995. Measuring biodiversity of microfungi in the wet tropics of North Queensland. In: Measuring and Monitoring Biodiversity of Tropical and Temperate Forests (Ed. Boyle, T.J.B. and Boontawee, B.). CIFOR, Bogor, Indonesia, 271–286.

Hyde, K. D. and Soytong, K. 2008. The fungal endophyte dilemma. Fungal Diversity 33, 163–173.

Johnston, P. R., Sutherland, P.W. and Joshee, S. 2006. Visualising endophyte fungi within leaves by detection of (1-3) β-D-glucans in fungal cell walls. Mycologist 20, 159–162.

Jones, E. B. G., Stanley, S. J. and Pinruan, U. 2008. Marine endophyte sources of new chemical natural products: a review. Botanica Marina 51, 163–170.

Jumpponen, A. 2001. Dark septate endophytes – are they mycorrhizal? Mycorrhiza 11, 207–211.

Jumpponen, A, 2003. Soil fungal community assembly in a primary successional glacier forefront ecosystem as inferred from rDNA sequence analyses. New Phytologist 158, 569–578.

Jumpponen, A, and Trappe, J. M. 1998. Dark septate endophytes: a review of facultative biotrophic root-colonizing fungi. New Phytologist 140, 295–310.

Kaewchai, S., Soytong, K. and Hyde, K. D. 2009. Mycofungicides and fungal biofertilizers. Fungal Diversity 38, 25–50.

Karamchand, K. S. and Sridhar, K. R. 2009. Association of water-borne conidial fungi with epiphytic tree fern (*Drynaria quercifolia*). Acta Mycologica 44, 19–27.

Karamchand, K. S., Sridhar, K. R. and Bhat, R. 2009. Diversity of fungi associated with estuarine sedge *Cyperus malaccensis* Lam. Journal of Agricultural Technology 5, 111–227.

Kim, H.Y., Choi, G. J., Lee, H. B., Lee, S. W., Lim, H. K., Jang, K. S., Son, S. W., Lee, S. O., Cho, K. Y., Sung, N. D. and Kim, J. C. 2007. Some fungal endophytes from vegetable crops and their anti-oomycete activities against tomato late blight. Letters in Applied Microbiology 44, 332–337.

Kohlmeyer, J. and Kohlmeyer, E. 1979. Marine Mycology–The Higher Fungi. Academic Press, New York.

Kohlmeyer, J. and Volkmann-Kohlmeyer, B. 1998. *Mycophycias*, a new genus for the mycobiont of *Apophlaea*, *Ascophyllum* and *Pelvetia*. Systema Ascomycetum 16, 1–7.

Krishnamurthy, Y. L., Naik, S. B. and Jayaram, S. 2008. Fungal communities in herbaceous medicinal plants from the Malnad Region, Southern India. Microbes and Environment 23, 24–28.

Kumar, D. S. S. and Hyde, K. D. 2004. Biodiversity and tissue-recurrence of endophytic fungi in *Tripterygium wilfordii*. Fungal Diversity 17, 69–90.

Kumaresan, V. and Suryanarayanan, T. S. 2001. Occurrence and distribution of endophytic fungi in a mangrove community. Mycological Research 105, 1388–1391.

Kumaresan, V. and Suryanarayanan, T. S. 2002. Endophyte assemblage in young, mature and senescent leaves of *Rhizophora apiculata*: Evidence for the role of endophytes in mangrove litter degradation. Fungal Diversity 9, 81–91.

Lee, J. C., Lobokovsky, N. B., Plam, N. B., Strobel, G. A. and Clardy, J. C. 1995. Subglutinol A and B: immunosuppressive compounds from the endophytic fungus *Fusarium subglutinans*. Journal of Organic Chemistry 60, 7076–7077.

Lee, J. C., Strobel, G. A., Lobkovsky, E. and Clardy, J. C. 1996. Torreyanic acid: a selectively cytotoxic quinone dimer from the endophytic fungus *Pestalotiopsis microspora*. Journal of Organic Chemistry 61, 3232–3233.

Lee, K., Pan, J. J. and May, G. 2009. Endophytic *Fusarium verticillioides* reduces disease severity caused by *Ustilagomaydis onmaize*. FEMS Microbiology Letters 299, 31–37.

Li, J. Y., Sidhu, R. S., Ford, E. J., Long, D. M., Hess, W. M. and Strobel, G. A. 1998. The induction of taxol production in the endophytic fungus–*Periconia* sp from *Torreya grandifolia*. Journal of Industrial Microbiology and Biotechnology 20, 259–264.

Loesgen, S., Schloerke, O., Meindl, K., Herbst-Irmer, R. and Zeeck, A. 2007. Structure and biosynthesis of chatocyclinones, new polyketides produced by and endosymbiotic fungus. European Journal of Organic Chemistry 2007, 2191–2196.

Lumyong, S., Techa, W., Lumyong, P., McKenzie, E. H. C. and Hyde, K. D. 2009. Endophytic Fungi from *Calamus kerrianus* and *Wallichia caryotoides* (Arecaceae) at Doi Suthep-Pui National Park, Thailand. Chiang Mai Journal of Science 36, 158–167.

Maciá-Vicente, J. G., Jansson, H. B., Abdullah, S. K., Descals, E., Salinas, J. and Lopez-Llorca, L. V. 2008. Fungal root endophytes from natural vegetation in Mediterranean environments with special reference to *Fusarium* spp. FEMS Microbiology Ecoligy 64, 90–105.

MacMillan, J. 2002. Occurrence of gibberellins in vascular plants, fungi and bacteria. Journal of Plant Growth Regulation 20, 387–442.

Maria, G. L. and Sridhar, K. R. 2003. Endophytic fungal assemblage of two halophytes from west coast mangrove habitats, India. Czech Mycology 55, 241–251.

Márquez, S. S., Bills, G. F. and Zabalgogeazcoa, I. 2008. Diversity and structure of the fungal endophytic assemblages from two sympatric coastal grasses. Fungal Diversity 33, 87–100.

Marvanová, L. and Fisher, F. 1991. A new endophytic hyphomycetes from alder roots. Nova Hedwigia 52, 33–37.

Marvanová, L., Fisher, P. J., Aimer, R. and Segedin, B. 1992. A new *Filosporella* from alder roots and from water. Nova Hedwigia 54, 151–158.

Marvanová, L. Fisher, P. J., Descals, E. and Baerlocher, F. 1997. *Fontanospora* sp. nov., a hyphomycete from live tree roots and from stream foam. Czech Mycology 50, 3-11.

Miller, J. D., Cherid, H., Sumarah, M. W. and Adams, G. W. 2009. Horizontal transmission of the *Picea glauca* foliar endophyte *Phialocephala scopiformis*. Fungal Ecology 2, 98 – 101.

Mohandoss, J. and Suryanarayanan, T. S. 2009. Effect of fungicide treatment on foliar fungal endophyte diversity in mango. Sydowia 61, 11–24.

Moricca, S. And Ragazzi, A, 2008. Fungal endophytes in Mediterranean oak forests: a lesson from *Discula quercina*. Phytopathology 98, 380–386.

Naik, B. S., Shashikala, J. and Krishnamurthy, Y. L. 2007. Study on the diversity of endophytic communities from rice (*Oryza sativa* L.) and their antagonistic activities *in vitro*. Microbiological Research 164, 90-296.

Naik, B. S., Shashikala, J. and Krishnamurthy, Y. L. 2008. Diversity of fungal endophytes in shrubby medicinal plants of Malnad region, Western Ghats, Southern India. Fungal Ecology 1, 89–93.

Nemec, S. 1969. Sproulation and identification of fungi isolated form root-rot in diseased strawberry plants. Phytopathology 59, 1552–1553.

Nielsen, J., Nielsen, P. H. and Frisvad, J. C. 1999. Fungal depside, guisinol, from a marine derived strain of *Emericella unguis*. Phytochemistry 50, 263–265.

Nikolcheva, L. G. and Baerlocher, F. 2005. Seasonal and substrate preferences of fungi colonizing leaves in streams: traditional versus molecular evidence. Environmental Microbiology 7, 270–280.

Osterhage, C., Kaminsky, R., Konig, G. M. and Wright, A. D. 2000. Ascosalipyrrolidonone A, an antimicrobial alkaloid from the obligate marine fungus *Ascochyta salicorniae*. Journal of Organic Chemistry 65, 6412–6417.

Ownley, B. H., Pereira, R. M., Klingeman, W. E., Quigley, N. B., Leckie, B. M. 2004. *Beauveria bassiana*, a dual purpose biocontrol organism, with activity against insect pests and plant pathogens. In: Emerging Concepts in Plant Health Management (Ed. Lartey, R.T. and Cesar, A.J.). Research Signpost, India, 255–269.

Pang, K. L., Vrijmoed, L. L. P., Goh, T. K., Plaingam, N. and Jones, E. B. G. 2008. Fungal endophytes associated with *Kandelia candel* (Rhizophoraceae) in Mai Po Nature Reserve, Hong Konga. Botanica Marina 51, 171–178.

Petrini, O, 1991. Fungal endophytes of tree leaves. In: Microbial Ecology of Leaves (Ed. Andrews, J.H. and Hirano, S.S.). Springer-Verlag, New York, 179–197.

Petrini, O., 1981. Endophytische Pilze in Epiphytischen Araceae, Bromeliaceae und Orchidiaceae. Sydowia 34, 135–148.

Pham, G. H., Kumari, R., Singh, A., Malla, R., Prasad, R., Sachdev, M., Kaldorf, M., Buscot, F., Oelmuller, R., Hampp, R., Saxena, A. K., Rexer, K. H., Kost, G. and Varma, A. 2004. Axenic cultures of *Piriformospora indica*. In: Plant Surface Microbiology (Ed. Varma, A., Abbott, L., Werner, D. and Hampp, R.). Springer-Verlag, Germany, 593–616.

Posada, F., Aime, M. C., Peterson, S. W., Rehner, S. A. and Vega, F. E. 2007. Inoculation of coffee plants with the fungal entomopathogen *Beauveria bassiana* (Ascomycota: Hypocreales). Mycological Research 111, 749–758.

Priti, V., Ramesha, B. T., Singh, S., Ravikanth, G., Ganeshaiah, K. N., Suryanarayanan, T. S. and Shaanker, R. U. 2009. How promising are endophytic fungi as alternative sources of plant secondary metabolites? Current Science 97, 477–478.

Promputtha, I., Jeewon, R., Lumyong, S., McKenzie, E. H. C. and Hyde, K. D. 2005. Ribosomal DNA fingerprinting in the identification of non-sporulating endophytes from *Magnolia liliifera* (Magnoliaceae). Fungal Diversity 20, 167–186.

Raghukumar, C. 1996. *Advances in Zoosporic Fungi* (Ed. R. Dayal), MD Publications Pvt. Ltd., New Delhi, 35–60.

Raghukumar, C. 2008. Marine fungal biotechnology: an ecological perspective. Fungal Diversity 31, 19–35.

Raghukumar, C., Nagarkar, S. and Raghukumar, S. 1992. Association of thraustochytrids and fungi with living marine algae. Mycological Research 96, 542–546.

Raj, N. and Shetty, H. S. 2009. Endophytic fungi – mutualism, bioactive metabolites and bioprospecting. In: Frontiers in Fungal Ecology, Diversity and Metabolites (Ed. Sridhar, K.R.). IK International Publishing House Pvt. Ltd., New Delhi, 169–183.

Raviraja, N. S. 2005. Fungal endophytes in five medicinal plant species from Kudremukh Range, Western Ghats of India. Journal of Basic Microbiology 45, 230–235.

Raviraja, N. S., Sridhar, K. R. and Baerlocher, F. 1996. Breakdown of introduced and native leaves in two Indian streams. Internationale Revue der Gesamten Hydrobiologie 81, 529–539.

Rodriguez, R. J., White Jr., J. F., Arnold, A. E., and Redman, R. S. 2009. Fungal endophytes: diversity and functional roles. New Phytologist 182, 314–330.

Rudgers, J. A. and Swafford, A. L. 2009. Benefits of a fungal endophyte in *Elymus virginicus* decline under drought stress. Basic and Applied Ecology 10, 43–51.

Rungjindamai, N., Pinruan, U., Choeyklin, R., Hattori, T. and Jones, E. B. G. 2008. Molecular characterization of basidiomycetous endophytes isolated from leaves, rachis petioles of the oil palm, *Elaeis guineensis* in Thailand. Fungal Diversity 33, 139–161.

Sathe-Pathak, V., Raghukumar, S., Raghukumar, C. and Sharma, S. 1993. Thraustochytrid and fungal component of marine detritus. I. Field studies on decomposition of the brown alga *Sargassum cinereum*. Indian Journal of Marine Sciences 22, 159–167.

Sati, S. C. and Belwal, M. 2005. Aquatic hyphomycetes as endophytes of riparian plant roots. Mycologia 97, 45–49.

Sati, S. C. and Belwal, M. and Pargaein, N. 2008. Diversity of water-borne conidial fungi as root endophytes in temperate forest plants of western Himalaya. Nature and Science 6, 59–65.

Sati, S. C., Arya, P. and Belwal, M. 2009a. *Tetracladium nainitalense* sp. nov., a root endophyte from Kumaun Himalaya, India. Mycologia 101, 692–695.

Sati, S.C., Pargaein, N. and Belwal, M. 2009b. Diversity of aquatic hyphomycetes as root endophytes on pteridophytic plants in Kumaun Himalaya. Journal of American Science 5, 179–182.

Schulz, B. and Boyle, C. 2005. The endophytic continuum. Mycological Research 109, 661–686.

Schulz, B., Guske, S., Dammann, U. and Boyle, C. 1998. Endoophyte-host interactions II. Defining symbiosis of the endophyte-host interaction. Symbiosis 25, 213–227.

Schulz, B., Boyle, C., Draeger, S., Roemmert, A. K. and Krohn, K. 2002. Endophytic fungi: a source of biologically active secondary metabolites. Mycological Research 106, 996–1004.

Schulz, B., Boyle, C. and Sieber, T. N. 2006. Microbial Root Endophytes. Springer-Verlag, Berlin.

Schulz, B., Draeger, S., Dela Cruz, T. E., Rheinheimer, J., Siems, K., Loesgen, S., Bitzer, J., Schloerke, O., Zeeck, A., Kock, I., Hussain, H., Dai, J. and Krohn, K. 2008. Screening strategies for obtaining novel, biologically active, fungal secondary metabolites from marine habitats. Botanica Marina 51, 219–234.

Seena, S. and Sridhar, K. R. 2004. Endophytic fungal diversity of 2 sand dune wild legumes from the southwest coast of India. Canadian Journal of Microbiology 50, 1015–1021.

Seena, S., Wynberg, N. and Baerlocher, F. 2008. Fungal diversity during leaf decomposition in a stream assessed through clone libraries. Fungal Diversity 30, 1–14.

Sekita, S., Yoshihira, K. and Natori, S. 1976. Structures of chaetoglobisins C, D, D, and F, cytotoxic indole-3-yl-(13) cytochalasans from *Chaetomium globosum*. Tetrahedron Letters 17, 1351–1354.

Selosse, M. A., Vohník, M. and Chauvet, E. 2008. Out of the rivers: are some aquatic hyphomycetes plant endophytes? New Phytologist 178, 3–7.

Shahollari, B., Vadassery, J., Varma, A. and Oelmueller, R. 2007. A leucine rich repeat protein is required for growth promotion and enhanced seed production mediated by the endophytic fungus *Piriformospora indica* in *Arabidopsis thaliana*. Plant Journal 50, 1–13.

Sherameti, I., Tripathi, S., Varma, A. and Oelmueller, R. 2008a. The root-colonizing endophyte *Piriformospora indica* confers drought tolerance in *Arabidopsis* by stimulating the expression of drought stress-related genes in leaves. Molecular Plant Microbe Interaction 21, 799–807.

Sherameti, I., Venus, Y., Drzewiecki, C., Tripathi, S., Dan, V. M., Nitz, I., Varma, A., Grundler, F. M. and Oelmueller, R. 2008b. PYK10, a beta-glucosidase located in the endoplasmatic reticulum, is crucial for the beneficial interaction between *Arabidopsis thaliana* and the endophytic fungus *Piriformospora indica*. Plant Journal 54, 428–439.

Shwab, E. K. and Keller, N. P. 2008. Regulation of secondary metabolite production in filamentous ascomycetes. Mycological Research 112, 225–230.

Sieber, T. N, 1989. Endophytic fungi in twigs of healthy and diseased Norway spruce and white fir. Mycological Research 92, 322–326.

Sieber, T. N. 2007. Endophytic fungi in forest trees: are they mutualists? Fungal Biology Reviews 21, 75–89.

Simeto, S., Alonso, R., Tiscornia, S. and Bettucci, L., 2005. Fungal community of *Eucalyptus globulus* and *Eucalyptus maidenii* stems in Uruguay. Sydowia 57, 246–258.

Sivichai, S. and Jones, E. B. G. 2003. Teleomorphic-anamorphic connections of freshwater fungi. In: Freshwater Mycology (Ed. Tsui, C.K.M. and Hyde, K.D.). Fungal Diversity Press, Hong Kong, 259–272.

Sokolski, S, Piché, Y., Chauvet, E. and Bérubé, J. 2006. A fungal endophyte of black spruce (*Picea mariana*) needles is also an aquatic hyphomycete. Molecular Ecology 15, 1955–1962.

Sridhar, K. R. 2009. Mangrove fungi of the Indian Peninsula. In: Frontiers in Fungal Ecology, Diversity and Metabolites (Ed. Sridhar, K.R.). IK International Publishing House Pvt. Ltd., New Delhi, India, 28–50.

Sridhar, K. R. and Baerlocher, F. 1992a. Endophytic aquatic hyphomycetes in spruce, birch and maple. Mycological Research 96, 305–308.

Sridhar, K. R. and Baerlocher, F. 1992b. Aquatic hyphomycetes in spruce roots. Mycologia 84, 580–584.

Sridhar, K. R., Karamchand, K. S. and Bhat, R. 2006. Arboreal water-borne hyphomycetes with oak-leaf basket fern *Drynaria quercifolia*. Sydowia 58, 309–320.

Stanley, S. J. 1991. The Autecology and Ultrastructural Interactions between *Mycosphaerella ascophylli* Cotton, *Lautita danica* (Berlese) Schatz, *Mycaureola dilseae* Maire et Chemin, and their Respective Marine Algal Hosts. Ph.D. thesis, University of Portsmouth, UK.

Stierle, A., Strobel, G. and Stierle, D. 1993. Taxol and taxane production by *Taxomyces andreanae*, an endophytic fungus of Pacific yew. Science 260, 214–216.

Stone, J. K., Bacon, C. W. and White Jr., J. F. 2000. An overview of endophytic microbes: endophytism defined. In: Microbial Endophytes (Ed. Bacon, C.W. and White Jr., J.F.), Marcel Dekker Inc., New York, 3–29.

Strobel, G. A. 2002. Microbial gifts from the rainforest. Canadian Journal of Phytopathology 24, 14–20.

Sugano, M., Sato, A., Iijima, Y., Furuya, K., Haruyama, H., Yoda K. and Hata, T. 1994. Phomactins, novel PAF antagonists from marine fungus *Phoma* sp. Journal of Organic Chemistry 59, 564–569.

Suryanarayanan, T. S. and Kumaresan, V. 2000. Endophytic fungi of some halophytes from an estuarine mangrove forest. Mycological Research 104, 1465–1467.

Suryanarayanan, T. S. and Vijaykrishna, D. 2001. Fungal endophytes of aerial roots of *Ficus benghalensis*. Fungal Diversity 8, 155–161.

Suryanarayanan, T. S., Kumaresan, V. and Johnson, J. A. 1998. Foliar fungal endophytes from two species of the mangrove *Rhizophora*. Canadian Journal of Microbiology 44, 1003–1006.

Suryanarayanan, T. S., Thirumalai, E., Prakash, C. P., Rajulu, M. B. G. and Thirunavukkarasu, N. 2009. Fungi from two forests of southern India: a comparative study of endophytes, phellophytes, and leaf litter fungi. Canadian Journal of Microbiology 55, 419–426.

Suryanarayanan, T. S., Thirunavukkarasu, N., Govindarajulu, M. B., Sasse, F., Jansen, R. and Murali, T. S. 2010. Fungal endophytes and bioprospecting. Fungal Biology Reviews 23, 9-18.

Szewczyk, E., Chiang, Y. M., Oakley, C. E., Davidson, A. D., Wang, C. C. C. and Oakley, B. R. 2008. Identification and characterization of the asperthecin gene cluster of *Aspergillus nidulans*. Applied and Environmental Microbiology 74, 7607–7612.

Tan, R. X. and Zou, W. X. 2001. Endophytes: a rich source of functional metabolites. Natural Products Report 18, 448–459.

Tao, G., Liu, Z. Y., Hyde, K. D. and Yu, Z. 2008. Whole rDNA analysis reveals novel and endophytic fungi in *Bletilla ochracea* (Orchidaceae). Fungal Diversity 33, 101–122.

Tejesvi, M. V., Mahesh, B., Nalini, M. S., Prakash, H. S., Kini, K. R., Subbiah, V. and Shetty, H. S. 2005. Endophytic fungal assemblages from inner bark and twig of *Terminalia arjuna* W. and A. (Combretaceae). World Journal of Microbiology and Biotechnology 21, 1535–1540.

Tejesvi, M. V., Kini, K. R., Prakash, H. S., Subbiah, V. and Shetty, H. S. 2007. Genetic diversity and antifungal activity of species of *Pestalotiopsis* isolated as endophytes from medicinal plants. Fungal Diversity 24, 37–54.

Vadassery, J., Ranf, S., Drzewiecki, C., Mithoefer, A., Mazars, C., Scheel, D., Lee, J. and Oelmueller, R. 2009. A cell wall extract from the endophytic fungus *Piriformospora indica* promotes growth of Arabidopsis seedlings and induces intracellular calcium elevation in roots. The Plant Journal 59, 193–206.

Varma, A., Verma, S., Sudha, Sahay, N. S., Butehorn, B. and Franken, P. 1999. *Piriformospora indica*, a cultivable plant growth promoting root endophyte. Applied and Environmental Microbiology 65, 2741–2744.

Varma, A., Singh, A., Sudha, Sahai, N. S., Sharma, J., Roy, A., Kumari, M., Rana, D., Thakran, S., Deka, D., Bharti, K., Turek, T., Blechert, O., Rexer, K. H., Kost, G., Hahn, A., Maier, W., Walter, M., Strack, D. and Kranner, I. 2001. *Piriformospora indica*: an axenically culturable mycorrhiza-like endosymbiotic fungus. In: Mycota IX. Springer Series, Germany, 123–150.

Vega, F. E., Posada, F., Catherine Aime, M., Pava-Ripoll, M., Infante, F. and Rehner, S. A. 2008. Entomopathogenic fungal endophytes. Biological Control 46, 72–82.

Vega, F. E., Simpkins, A., Aime, M. C., Posada, F., Peterson, S. W., Rehner, S. A., Nfante, R., Castillo, A., Arnold, A. E. 2009. Fungal endophyte diversity in coffee plants from Colombia, Hawaii, Mexico and Puerto Rico. Fungal Ecology 2, 149–159.

Vendenkoornhuyse, P., Bauldauf, S. L., Leyval, C., Straczek, J. and Young, J. P. W. 2002. Extensive fungal diversity in plant roots. Science *295*, 2051.

Verma, S. A., Varma, A., Rexer, K. H., Hassel, A., Kost, G., Sarbhoy, A., Bisen, P., Buetehorn, B. and Franken, P. 1998. *Piriformospora indica*, gen. et sp. nov., a new root-colonizing fungus. Mycologia, 90, 898–905.

Waller, F., Achatz, B., Baltruschat, H., Fodor, J., Becker, K., Fischer, M., Heler, T., Huckelhoven, R., Neumann, C., von Wettstein, D., Franken P. and Kogel, K. 2005. The endophytic fungus *Piriformospora indica* reprograms barley to salt-stress tolerance, disease resistance, and higher yield. Proceedings of the National Academy of Sciences 102, 13386-13391.

Wang, Y., Guo, L. D. and Hyde, K. D. 2005. Taxonomic placement of sterile morphotypes of endophytic fungi from *Pinus tabulaeformis* (Pinaceae) in northeast China based on rDNA sequences. Fungal Diversity 20, 235–260.

Watanabe, T. 1975. *Tetracladium setigerum*, an aquatic hyophomycete associated with gentian and strawberry roots. Transactions of the Mycological Society of Japan 16, 348–350.

Webber, J. 1981. A natural control of Dutch elm disease. Nature *292*, 449–451.

Webster, J. 1992. Anamorph-teleomorph relationships. In: The Ecology of Aquatic Hyphomycetes (Ed. Baerlocher, F.). Ecological Studies 94, Springer-Verlag, Berlin, 99–117.

Yahara, I., Harada, F., Sekita, S., Yoshihira, K. And Natori, S. 1982. Correlation between effects of 24 different cytochalasins on cellular structures and cellular events and those on actin *in vitro*. The Journal of Cell Biology 92, 69–78.

Young, C. A., Felitti, S., Shields, K., Spangenberg, G., Johnson, R. D., Bryan, G. T., Saikia, S., and Scott, B. 2006. A complex gene cluster for indole-diterpene biosynthesis in the grass endophyte *Neotyphodium lolii*. Fungal Genetics and Biology 43, 679–693.

Zhang, Y, Mu, J., Feng, Y., Kang, Y., Zhang, J., Gu, P. J., Wang, Y., Ma, L. F. and Zhu, Y. H. 2009. Broad-spectrum antimicrobial epiphytic and endophytic fungi from marine organisms: isolation, bioassay and taxonomy. Marine Drugs 7, 97–112.

Zuccaro, A., Schulz, B. and Mitchell, J. I. 2003. Molecular detection of ascomycetes associated with *Fucus serratus*. Mycological Research 107, 1451–1466.

Zuccaro, A., Schoch, C., Spatafora, J., Kohlmeyer, J., Draeger S. and Mitchell, J. 2008. Detection and identification of fungi intimately associated with the brown seaweed *Fucus serratus*. Applied and Environmental Microbiology 74, 931–941.

Microbes: Diversity and Biotechnology (2012) *Pages* **63–86**
Editors: **Prof. S.C. Sati & Dr. M. Belwal**
Published by: **DAYA PUBLISHING HOUSE, NEW DELHI**

Chapter 4

Phosphate Solubilizing Fungi and their Role in Improving Phosphorus Nutrition of Plants

Rachana Jain[1], Jyoti Saxena[2] and Vinay Sharma[1]
[1]Department of Bioscience and Biotechnology,
Banasthali University, Banasthali – 304 022, Rajasthan
[2]Biochemical Engineering Department, Bipin Tripathi Kumaon Institute of Technology,
Dwarahat, Ranikhet – 263 653, Uttarakhand

ABSTRACT

The use of phosphate solubilizing fungi as inoculants increases P uptake and crop yield in plants. Strains from the genera *Aspergillus* and *Penicillium* are the most powerful phosphate solubilizers. The principal mechanism for mineral phosphate solubilization is the production of organic acids. This paper reviews current knowledge of phosphate solubilizing fungi as bioinoculants. In general, phosphate solubilizing fungi show good activity in *in vitro* flask culture but they are not much successful as field inoculants. This paper also highlights the problems associated with the commercialization of phosphate solubilizing fungi.

Keywords: *Phosphate solubilizing fungi; Phosphate solubilization, Bioinoculant; Rhizosphere, Organic phosphate, Mineral phosphate.*

Introduction

Phosphorus (P) is the major nutrient after nitrogen (N) that limits plant growth (Gyaneshwar *et al.*, 2002; Anamika *et al.*, 2007; Fernandez *et al.*, 2007). It is one of the least available and the least mobile mineral nutrients for plants in the soil (Takahashi and Anwar, 2007). P content in soil is about 0.05 per cent of which only 0.1 per cent is available to plants (Scheffer and Schachtschabel, 1988). Deficiency of soil P is one of the most important chemical factors restricting plant growth in agricultural fields. As a result a large quantity of soluble forms of P-fertilizers is applied to achieve maximum plant productivity. However after application, a considerable amount of phosphorus is rapidly transformed into less available forms by forming a complex with Al or Fe in acid soils (Norrish and Rosser, 1983; Rengel and Marschner. 2005; Johnson and Loeppert, 2006)) or Ca in calcareous soils (Lindsay *et al.*, 1989) before the plant root has a chance to absorb it. Since the indiscriminate and excessive application of chemical fertilizers has led to health and environmental hazards (Shenoy *et al.*, 2005), agronomists are desperate to find alternative strategies that can ensure competitive yields while protecting the health of soil. In this context, use of microbial inoculants (biofertilizers) including phosphate-solubilizing microorganisms (PSMs) in agriculture represents an environmental friendly alternative to mineral fertilizers (Son *et al.*, 2006; Ayyadurai *et al.*, 2006; Chang *et al.*, 2009). Application of PSMs in the field has been reported invariably to increase crop yield (Nahas, 1996; Richardson, 2001; Niranjan Raj *et al.*, 2006; Saghir *et al.*, 2007).

Microorganisms are an important component of soil and directly or indirectly influence the soil health through their beneficial or detrimental activities. The microorganisms capable of solubilizing insoluble inorganic phosphates to the soluble form (Oberson *et al.*, 2001; Kang *et al.*, 2002; Pradhan and Shukla, 2005) that can be readily utilized by the plants are known as phosphate solubilizing microorganisms (PSMs). Phosphate-solubilizing microbes can transform the insoluble phosphorus to soluble forms $H_2PO_4^{1-}$ and HPO_4^{2-} by acidification, chelation, exchange reactions, and polymeric substance formation (Rodriguez *et al.*, 2004; Chung *et al.*, 2005; Delvasto *et al.*, 2006).

There arises a fairly large volume of information that has accumulated over the years on the characteristics of PSMs. There are several reports on the isolation of various PSMs from the rhizosphere, non-rhizospere, rhizoplane as well as from other environments such as rock phosphate deposits, marine environment etc. (Rakade and Patil, 1992; Thakker *et al.*, 1994). Many soil bacteria, actinomycetes and fungi have the ability to solubilize inorganic phosphatic compounds and make them available to growing plants (Subba-Rao 1982; Kucey *et al.*, 1989; Antoun *et al.*, 1998; Rodriguez and Frago 1999; Whitelaw, 2000; Oberson *et al.*, 2001; Gyaneshwar *et al.*, 2002; Egamberdiyeva *et al.*, 2003). In Indian alluvial soils, 14.3 per cent, 35 per cent and 30 per cent of the total bacterial, actinomycetes and fungal population, respectively possessed the ability of solubilizing tricalcium phosphate (TCP) (Banik and Dey, 1982). The PSMs isolated include fungi mostly from the genus *Aspergillus, Penicillium, Chaetomium, Trichoderma* and some yeast, bacteria from the genus *Bacillus, Micrococcus, Pseudomonas, Rhizobium, Burkholderia, Enterobacter, Achromobacter, Agrobacterium, Microccocus, Aerobacter, Erwinia* and *Arthrobacter* and *actinomycetes* mostly from the genus *Streptomyces* (Banik and Dey, 1982; 1983; Pandey and Palni, 1998; Rodriguez and Fraga, 1999; Igual *et al.*, 2001; Hamdali *et al.*, 2008).

PSMs are ubiquitous in nature and their number varies from soil to soil. They are present in both fertile as well as P deficient soil (Oehl *et al.*, 2001). In soil, phosphate solubilizing bacteria (PSB) generally constitute 1-50 per cent and phosphate solubilizing fungi (PSF) 0.5-0.1 per cent of the total respective population. Generally, PSB outnumber the PSF by 2-150 times in most of the soils (Kucey, 1983a). The high population of PSMs is concentrated in the rhizosphere and is known to be metabolically more active than those isolated from other sources (Vazquez *et al.*, 2000). Conversely, the salt, pH and temperature tolerant PSMs have been reported to be maximum in the rhizoplane followed by the rhizosphere and non rhizosphere in alkaline soils (Johri *et al.*, 1999).

Phosphate solubilizing fungi are superior to their bacterial counterpart for P solubilzation both on precipitated agar and in liquid (Kucey, 1983a). Fungal hyphae in liquid culture were found to be attached to P mineral particles as shown by scanning electron microscopy, whereas bacteria did not show such attachment (Chabot *et al.*, 1993). Furthermore, because of their hyphae, fungi are able to reach greater distances in soil more easily than bacteria. It has been observed that PSB upon repeated sub-culturing lose the phosphate solubilizing activity (Kucey, 1983b; Halder *et al.*, 1990 and Illmer and Schinner, 1992) but such losses have not been observed in PSF (Sperber, 1958; Kucey, 1983b). In general, PSF produce more acids and consequently exhibit greater phosphate solubilizing activity than bacteria in both liquid and solid media (Banik and Dey, 1982; Venkateswarlu *et al.*, 1984).

The aim of this review is to summarize all available studies that involve phosphate solubilizing fungi and plant nutrition and highlight the problems related to commercialization of phosphate solubilzing fungi.

Phosphorus in Soil System

Phosphorus is one of the major essential macronutrient for biological growth and development. It contributes to the biomass construction of micronutrients, the metabolic process of energy transfer, signal transduction, macromolecular biosynthesis, photosynthesis, and respiration chain reactions (Shenoy and Kalagudi, 2005). It is important for plant growth because it stimulates growth of young plants, promotes vigorous start and hastens maturity hence, when it is present in inadequate amount, the plant growth is diminished, maturity is delayed and yield reduced (Sawyer and Creswell, 2000).

Phosphorus is present at the level of 400-1200 mg kg^{-1} of soil (Fernandez and Novo, 1988). It exists in soil in organic and inorganic forms. Each form consists of many P compounds. Availability of P ranges from soluble phosphorus (plant available) to very stable (plant unavailable) compounds. There is a dynamic and complex relationship among different forms of P present in soil, plant and microorganisms.

Organic form of P constitutes 30-50 per cent of the total P in most soils, although it may range from as low as 5 per cent to as high as 95 per cent. Organic P compounds are found in humus and other organic materials including decayed plant, animal and microbial tissues. It is also a principal form of P in manure. Organic P in soil is largely in the form of inositol phosphate (soil phytate). It is synthesized by microorganisms and plants and is the most stable of the organic forms of P in soil, accounting for up to 50 per cent of the total organic P (Thompson and Troch, 1978). Other organic P compounds in soil are in the form of phosphomonoester, phosphodiester including phospholipids, nucleic acids and phosphotriester.

Phosphorus in labile organic compounds can be slowly mineralized (broken down and released) as available inorganic phosphate or it can be immobilized (incorporated into more stable organic materials) as part of the soil organic matter (Tate, 1984; Mckenzie and Roberts, 1990). The process of mineralization or immobilization is carried out by microorganisms and is highly influenced by soil moisture and temperature. Mineralization and immobilization are most rapid in warm and well drained soil (Busman *et al.*, 2002).

The second major form of soil P is inorganic. Most cultivated soils contain 70-80 per cent of inorganic P (Foth, 1990), a considerable part of which has accumulated as a consequence of regular application of phosphate fertilizers. In most soils, orthophosphate ions $H_2PO_4^{1-}$ and HPO_4^{2-} dominate at pH below 7 and above 7.2, respectively (Hinsinger, 2001). These negatively charged ions attach strongly to the surface of mineral containing positively charged ions such as iron (Fe^{3+}) and aluminum (Al^{3+}) in acidic soil via sorption/adsorption processes. Fe^{3+} and Al^{3+} act as absorption sites for the negatively charged P (Sato and Comerford, 2005). These ions are also precipitated with the Ca^{2+} in $CaCO_3$ mineral in calcareous soils forming relatively insoluble compounds. Both processes result in making P being fixed or bound, thus removed from the soil solution and become unavailable for plants (Banik and Dey, 1982; Foth, 1990; Schulte and Kelling, 1996). According to Lindsay superphosphate contains a sufficient amount of calcium to precipitate half of its own P, in the form of dicalcium phosphate or dicalcium phosphate dihydrated.

The conversion from stable P to labile P is a slow process and does not occur over the course of one growing season (Guo and Yost, 1998). In contrast, the conversion from labile P to plant available P is a rapid process (Tate and Salcedo, 1988). In neutral and calcareous soils, the pH ranges from 7.3 to 8.5 depending upon the amount of $CaCO_3$ present (Lindsay, 1979). With high levels of exchangeable Ca, available P ions react with solid phase $CaCO_3$ and precipitate on the surface of these particles to form Ca-P minerals: $Ca(H_2PO_4)_2$ (monocalcium P), $CaHPO_4.2H_2O$ (dicalcium phosphate dihydrate, DCPD, brushite), $CaHPO_4$ (dicalcium phosphate, DCP, monetite), $Ca_3(PO_4)_2$ (tricalcium phosphate, TCP), $Ca_4H(PO_4)_3.25H_2O$ (octacalcium P, OCP), $Ca_5(PO4)_3OH$ (hydroxyapatite) and least soluble apatites (Lindsay *et al.*, 1989). The finer the size of solid phase $CaCO_{3,}$ the higher is the fixation of P. The solubility of Ca-P minerals is generally accepted as DCPD > DCP > TCP > hydroxyapatite. In alkaline soils, the initial products of reaction of fertilizer triple superphosphate are mainly DCPD and DCP (Russell, 1980; Whitelaw *et al.*, 1999). Different phases of Ca-P compounds are transferable and, at a given pH, can be dissolved from unstable phase or to be precipitated as stable phase.

Phosphatic fertilizers can increase P availability initially, but after that will promote the formation of insoluble P minerals and consequently lead to P buildup. Therefore, P management is important both environmentally and economically. PSMs may be an answer for maintaining the supply of plant available P because they carry out the conversion from labile P to plant available form.

Phosphate Solubilizing Fungi

Phosphate solubilizing fungi have been found in most of the soils that have been investigated (Katznelson *et al.*, 1962; Kucey, 1983a; Chabot *et al.*, 1993; Wenzel *et al.*, 1994; Kim *et al.*, 1997). Among the fungal genera with the phosphate solubilization ability are *Aspergillus, Penicillium, Trichoderma, Mucor, Candida, Yeast, Discosia, Eupenicillium* and *Gliocladium* (Rahi *et al.*, 2009; Xiao *et al.*, 2008). The strains from the genera *Aspergillus* and *Penicillum* are among the most powerful phosphate solubilizers. The reported *Aspergillus* includes *A. niger, A. candidus, A. fumigatus, A. awamori, A. japonicus, A. foetidus, A. aculeatus, A. versicolor* (Singal *et al.*, 1994; Narisan and Patel, 2000; Rahi *et al.*, 2009). Whereas, *P. rugulosum, P. claviformis, P. aurantiogriseum, P. radicum, P. pinophillum, P. cyclopium, P. expansum, P. citrinum, P. italicum, P. variable, P. chrysogenum, P. decumbens, P. lanosum, P. purpurogenum, P. grisefulvum, P. simlicissinum, P. purpurogenum, P. pinophilum, P. pinetorum, P. oxalicum, P. aurantio- griseum, P. raistrickii* and *P. janthinellum* from the genus *Penicillium* solubilize phosphate efficiently (Gaur and Sacher, 1980; Banik and Dey, 1982; Vankateswarlu *et al.*, 1984; Roos and Luckener, 1994; Singal *et al.*, 1994; Basu, 1999; Reyes *et al.*, 1999b; Whitelaw *et al.*, 1999; Zaidi *et al.*, 2003; Babana *et al.*, 2006; El-azoouni, 2008; Mittal *et al.*, 2008; Pandey *et al.*, 2008; Xiao *et al.*, 2008 and Rahi *et al.*, 2009). *M. ramosissimus* from *Mucor*, *C. krissi* from genus *Candida*, *T. pseudokoningii* from *Trichoderma*, *G. roseum, G. virens, G. viride* from genus *Gliocladium* and *E. parvum and E. sherli*, from *Eupenicillium* were also reported as phosphate solubilizing fungi (Nahas, 1996; Xiao *et al.*, 2008 and Rahi *et al.*, 2009).

Phosphate solubilizing fungi could be detected in varying number in different soils (Salih *et al.*, 1989; Nahas, 1996). However, the influence of soil characteristics on these numbers has not been extensively assessed. Since fungi are heterotrophic microorganisms, it has been suggested that their frequency in different soils may be due to differences in organic matter content and soil properties (Vankateswarlu *et al.*, 1984; Narisan and Patel, 2009). It has also been suggested that legumes enhance the N_2 content in soil and also the number of solubilizer microorganisms (Paul and Sundara Rao, 1971). Kucey (1983a) found a correlation of the number of PSF with total P levels but not with the type of vegetation. On the contrary Barroso *et al.* (2005) did not observe any relation between total P or available P and the number of solubilizer fungi. Narisan *et al.* (1994) isolated 27 *Aspergilli*, 7 *Penicillia* and 1 *Rhizopus* sp. from rhizosphere of 24 crop plants, compost and garden soil of India. Nahas (1996) isolated and tested 31 bacteria and 11 fungi from soil of Brazil, out of which 8 bacteria *viz.*, 2 strains of *Bacillus subtilis*, 3 strains of *B. licheniformis* and one each of *B. macerans*, *B. megatherium*, *Psedomonas* spp. and seven fungi *viz.*, *A. niger*, *A. ochraceous*, *Eupenicillium sherli*, *Penicillium implicatum*, *P. minioluteum*, *P. viridicatum* and *P. purpurogenum* showed the activity, *Pseudomonas* and *Penicillium purpurogenum* being the most efficient phosphate solubilizers. Barroso *et al.* (2005) isolated 33 fungi that showed the ability to solubilize inorganic phosphate on agar plates. Of these 12 were isolated from pasture soil and remaining ones from the tropical rain forest, forests patch and corn field soil (seven from each). Of these, 14 were high or very high phosphate solubilizers based on the solubilization capacity > 1000 µg PO_4^{3-} ml^{-1}. Thirteen phosphate-solubilizing fungi were isolated by Rahi *et al.* (2009) and identified as *Aspergillus niger*, *A. versicolor*, *Discosia* sp., *Eupenicillium parvum*, *Gliocladium roseum*, *G. virens*, *G. viride*, *Penicillium auranteogrisium*, *P. decumbens*, *P. lanosum*, *P. purpurogenum*, *P. grisefulvum*, and *Trichoderma pseudokoningii* on the basis of cultural and microscopic features from tea rhizosphere. PSF were enumerated in 78 rhizosphere soil samples collected from various sites of Bhavnagar district with agroclimatic zone of hot, semi-arid region of Gujarat. 81 per cent samples were inhabited with indigenous phosphate solublizing fungi (Narsian and Patel, 2009).

Factors Determining the Phosphate Solubilization

Solubilization depends on the insoluble inorganic phosphate source as well as on culture conditions. For example, the ability of different organisms to solubilize Ca-phosphates is influenced by the source of carbon and nitrogen in the media, buffering capacity of the media and the stage at which cultures are sampled (Kucey 1983b; Illmer and Schinner 1995; Whitelaw *et al.*, 1999; Nahas 2007).

Effect of Different Type of Inorganic Phosphate

In most of the studies on P solubilizers, some type of calcium phosphate has been used, predominantly in the form of fluorapatite and hydroxyapatite (Bojinova *et al.*, 1997; Mba, 1997; Vassileva *et al.*, 1998a; Reyes *et al.*, 1999a and b), tricalcium phosphate (Sundara Rao and Sinha, 1963), calcium hydrogen phosphate (Cunningham and Kuiack, 1992) and freshly precipitated hydroxyapatite (Sperber, 1958; Nahas *et al.*, 1994). The reason behind higher number of Ca-P solubilizing fungi may be due to the initial screening procedure in which readily solubilized $CaPO_4$ or other Ca forms were used as insoluble phosphate sources in place of other mineral phosphate forms which are not so readily solubilized. However, some studies have demonstrated the ability of few fungi to solubilize Fe or Al phosphates (Banik and Dey, 1983; Antoun, 2002). Two fungi isolated from Scottish soils were effective in dissolving Fe phosphate but ineffective in solubilizing Ca phosphate on agar plates (Jones *et al.*, 1991). *A. niger* growing in vinasse medium solubilized 72 per cent of the total P from rock phosphate but only 8 per cent of the Al phosphate (variscite) (Nahas and Assis, 1992). P solubilization by *Penicillium rugulosum* was more efficient for hydroxyapatite than for $FePO_4$ or $AlPO_4$ (Reyes *et al.*, 1999b).

Solubilization depends on the microorganism and phosphate type (Nahas, 1996). The type of insoluble phosphate present in soil also affects the solubilization capacity of different microorganisms substantially. Xiao *et al.* (2008) studied the phosphate solubilization of different P sources by the different PSF and their results revealed that the maximum content of soluble P released by the isolates (*C. krissii*, *P. expansum* and *M. ramosissimus*) averaged 238.5 mg l^{-1} in presence of TCP, followed by aluminium phosphate (average 130.7mgl^{-1}), Yichang RP (average 96.0 mg l^{-1}), Baokang RP (average 81.2 mg l^{-1}), and Huangmei RP (average 77.7 mg l^{-1}). Interestingly, the total P content of RPs was also in the order of Yichang RP, Baokang RP, and Huangmei RP. The capability of RP solubilization was positively correlated with the grade of RP. It was also seen that *C. krissii* showed a higher soluble P releasing ability than other isolates with all the P sources. Bardiya and Gaur (1974) studied RP solubilization by some very efficient strains of *Aspergillus* and *Penicillium*. All of these strains showed lesser activity with other phosphate sources. This finding indicated that solubilization of phosphate varies greatly with the nature of phosphate and organisms. Three forms of most commonly found phosphates *viz.*, DCP, TCP and hydroxyapatite were chosen to show the effect of solubilization by 2 strains of *Aspergillus niger* and one strain each of *Penicillum claviformis* and *P. aurantiogriseum*. DCP was least efficiently solubilized by all the strains whereas TCP was solubilized maximally (Basu, 1999). According to Narsian *et al.* (1994), among the various phosphates tested for solubilization by *Aspergillus* and *Penicillium* spp., TCP supported the highest activity; DCP was next for all the fungi, followed by $AlPO_4$ and $FePO_4$. Rock phosphate solubilized to lesser extent than TCP but was superior to bone meal as a phosphate source. Arora and Gaur (1979) tested a range of bacteria and fungi for their ability to solubilize TCP, DCP, hydroxyapatite and RP. Species of *Pseudomonas, Bacillus, Escherichia, Aspergillus* and *Penicillium* effectively solubilized TCP and hydroxyapatite. Rock phosphate was more resistant to solubilization, relatively small amount being solubilized by most species.

Fungi were more effective than bacteria in general (Kucey, 1983a). Six strains of *A. awamori* were compared for their phosphate solubilizing activity using DCP, TCP, ferrous phosphate (Fe-P), aluminium phosphate (Al-P), and four types of RP. Four strains preferred TCP while 2 strains Al-P. Different RPs was utilized in the following decreasing order: Sonarai RP> Udaipur RP> Hirapur RP (Narsian *et al.*, 1994). *A. terreus* was also able to solubilize P from aluminum phosphate, phytate and lecithin (Oliveira *et al.*, 2008).

Phosphate solubilization by *P. radicum* was higher with Ca-P form than Fe-P or Al-P form of phosphate (Whitelaw *et al.*, 1999). *P. bilaiae* (*bilaiae* is a transliteration of the Russian scientist B. Bilai after whom it is named. Alternative spellings published in the literature are *bilaji* or *bilaii*) had superior ability of Ca-P solubilization (Kucey, 1988; Sanders, 2003). Lide and Frederikse (1998) found that the ability of the fungus to dissolve insoluble phosphate decreased in the following order Ca-P > Al-P > Fe-P. This might be explained by the difference in the solubilities of these phosphates in solution (Stumm and Morgan, 1995). Banik and Dey (1983) also observed the same tendency of several fungi from soil to solubilize Ca, Al and Fe phosphates. Leyval and Berthelin (1985) found that the content of solubilized P was higher with tricalcium phosphate than ferric phosphate. However, this relative order of phosphate solubilization may not be consistent when each isolate is analyzed individually. For example, isolate 55 (Barroso *et al.*, 2005) solubilized the phosphate in the Fe-P> Al-P> Ca-P order. The ability to solubilize $AlPO_4$ or $FePO_4$ by soil isolates was lower than $Ca_3(PO_4)_2$, probably due to the adaptive nature of the enzymes responsible for solubilizing earlier phosphates (Banik and Dey, 1982).

Besides the type of phosphates, their concentration also plays an important role. Phosphate solubilization activity with TCP decreased with the rise in its concentration (Narsian *et al.*, 1994). Gaur and Sache (1980) studied optimum amount of RP to be applied for maximum phosphate dissolution. The result showed that lower the quantity of phosphate supplied, greater was the conversion percentage. This may be because the treatment receiving lowest quantity of phosphate remained more acidic than treatment receiving higher amount of RP. The effect of soluble phosphate concentration on the solubilization of fluorapatite by *A. niger* was studied by Nahas and Assis (1992). Acid phosphatase production and fluorapatite solubilization decreased when the concentration of soluble phosphate was enhanced. It appears that the process of solubilzation by fungi is governed both by external phosphate levels as well as phosphatases. Xiao *et al.* (2008) isolated four fungal strains from phosphate mine in Hubei, China and found the maximum content of soluble phosphorus when 2.5g l^{-1} RP was added in the medium and inoculated with *Candida krissii* (108.9 mg l^{-1}), followed by *Penicillium expansum* (105.0 mg l^{-1}) and *Mucor ramosissimus* (100.1 mg l^{-1}).

The importance of soil microorganisms for increasing the availability of P from mineral phosphates to plants depends on the rate of P solubilization. However, few isolates possess a high potential to solubilize hardly soluble phosphates. The rate of tricalcium phosphate solubilization was in range of 0.21 to 0.90 mg ml^{-1} by the fungi isolated from forest soils (Surange, 1985). Whereas, from zero to 39 per cent by fungi associated with legume root nodule (Chhonkar and Subba-Rao, 1967). *A. niger* and *P. pinophillum* showed the maximum P solubilzing efficiency, solubilizing 66 and 44 per cent of the $Ca_3(PO_4)_2$ respectively (Venkateswarlu *et al.*, 1984). Pandey *et al.* (2008) reported the phosphate solubilization by 8 *Penicillum* species in the following order: *P. citrinum* (328 µg ml^{-1}) > *P. purpurogenum* (328 µg ml^{-1}) > *P. pinophilum* (287 µg $ml^{-1)}$ > *P. pinetorum* (285 µg ml^{-1}) > *P. oxalicum* (281 µg ml^{-1}) > *P. aurantio- griseum* (262 µg ml^{-1}) > *P. raistrickii* (226 µg ml^{-1}) > *P. janthinellum* (180 µg ml^{-1}). Four phosphate solubilizing fungi were isolated from Kanchanaburi, Thailand belonging to the genus *Aspergillus* strain numbers SA07P3332, SA22P3406, SA14P2418 and SA19P2120, solubilized TCP and showed 3.010, 2.993, 2.749 and 2.032 mg P_2O_5 ml^{-1} (Nopparat *et al.*, 2007). Bojinova *et al.* (2008) recorded that *A. niger* could be successfully used for solubilization of Morocco phosphorite. A maximum of 94.80 per cent total phosphate extraction was achieved which is almost complete extraction of phosphate in a form utilizable by plant.

Effect of Carbon Sources

Carbon source is an important factor for the active proliferation and organic acid production by microorganisms in phosphate dissolution. Microorganisms with the help of enzymes convert the sugar into intermediate metabolites including organic acids (Brock *et al.*, 1994). This enzyme system varies from microorganism to microorganism. The type and concentration of sugar also affects acid production (Gupta *et al.*, 1976; Cunningham and Kuiack, 1992; Mattey, 1992), and the type of carbon source indirectly regulates the rate of phosphate solubilization. *A. niger* solubilized 78 per cent of fluorapatite in culture medium containing fructose as opposed to 59 to 69 per cent solubilization in glucose, xylose, and sucrose based media (Cerezine *et al.*, 1988). *A. niger* solubilized more $CaHPO_4$ with maltose and mannitol than with sucrose in liquid medium (Barroso *et al.*, 2006). Mannitol was the best carbon source to solubilize phosphorus in case of *A. niger* strain SPV5, which was followed by glucose < sorbitol < fructose < sucrose < xylose, respectively (Seshadri *et al.*, 2004). Gaur and Gaind (1983) reported that *A.awamori* showed preference for sucrose followed by glucose, maltose and mannitol whereas, *A. aculeatus* preferred maltose followed by glucose, sucrose and mannitol. Most extensive growth of *A. niger* was found with maltose when compared to sucrose, lactose, glucose and fructose (Margaris *et al.*, 1974). Narsian and Patel (2000) reported maximum P solubilization by *A. aculeatus* with arabinose and glucose. Reyes *et al.* (1999a) reported that sucrose was the best carbon source for *P. rugulosum* for solubilization of hydroxyapatite and $FePO_4$. These studies suggest that different fungi use different carbon sources, and depending on the carbon sources, the fungi use alternative metabolic pathways to produce organic acids.

Nautiyal (1999) also found that not only glucose was necessary, but its concentration was also important for bacterial P solubilization in broth. Soluble P concentration increased with an increase in glucose. *P. radicum* favoured higher sucrose concentration (*e.g.* 30g l^{-1}) for phosphate solubilization

(Whitelaw *et al.*, 1999). The effect of varying dosage of glucose was investigated on rock phosphate solubilization by *A. awamori*. It increased solubilization due to addition of greater quantity of energy sources. Maximum solubilization was brought about by 3 per cent glucose (Gaur and Sachar, 1980).

In soil, high C concentration in the rhizosphere supported and enhanced microbial P solubilization activities (Lynch and Whipper, 1990), while decomposition of plant residues replenished the C source. The solubilization of two types of rock P increased significantly during decomposition of wheat straws and cattle urine (Singh and Amberger, 1991).

Effect of Nitrogen Sources

Microorganism for the synthesis of amino acids, protein and purine and pyrimidine nitrogenous bases need nitrogen. So the P-solubilizing ability of PSMs also depends on the nature of nitrogen sources used in the media. Phosphate solubilization was higher in the presence of NH_4^+ salts than when nitrate was used as nitrogen source. This has been attributed to the extrusion of proton to compensate for ammonium uptake, leading to a decreased extra cellular pH (Roos and Luckner, 1984). *P. radicum* released more P in the presence of NH_4^+-N compared to NO_3^--N (Whitelaw *et al.*, 1999). Similarly, *Aspergillus* sp. also preferred NH_4^+-N among NH_4^+-N, NO_3^--N, urea and casein as different N sources (Pradhan and Sukla, 2005). Nitrate was found to be an inferior N source than ammonium (Cunningham and Kuiack, 1992). *P. bilaiae* however, released more P from insoluble Ca-P in culture solution with NO_3^--N and sucrose as the C source (Cunningham and Kuiack, 1992). Ammonium nitrate has been reported to be a good N source for the growth of *A. niger* (Steinberg and Bowling, 1939). Sodium nitrate was one of the sources which stimulated Ca-P solubilization. However, Dixon-Hardy *et al.* (1998) reported that nitrate increased solubilization of several phosphates. In some cases however, NH_4^+ could lead to a decrease in phosphate solubilization (Reyes *et al.*, 1999a). Furthermore, the concentration of NH_4^+-N also affected the amount of P solubilization; higher concentration promoted P solubilization (Nautiyal, 1999).

From various studies it is evident that change in pH of the media is particularly important for solubilization of Ca phosphates, whereby cultures supplied with NH_4^+ are more effective than those with NO_3^- due to associated proton release and acidification of the media. Acidification is also commonly associated with the release of organic anions which have been widely reported for various microorganisms (*i.e.* citrate, oxalate, lactate, and gluconate being most common). Organic anions themselves may further increase the mobilization of particular forms of poorly soluble P (*e.g.* Al-P and Fe-P) through chelation reactions (Whitelaw, 2000).

Effect of pH and Buffered Condition

The type of organism and organic acid produced by the organisms may be affected differently by the buffering capacity of the substrate. *P. bilaiae* produced a ten fold increase in soluble P from rock phosphate in a buffered medium, compared to an unbuffered medium (Takeda and Knight, 2006). There were larger amounts of citric and oxalic acids produced, but the concentration of other organic acids such as malonic and succinic acids were higher in the non-buffered medium (Takeda and Knight, 2006). Organisms such as *P. bilaiae* might be less sensitive to pH changes and their P solubilization efficacy remained functional or even enhanced under highly buffered conditions. Singal *et al.* (1994) studied the rock phosphate solubilization by *A. japonicus* and *A. foetidus*. They found that RP solubilization was better at pH 6 than at pH 8 and 9. The content of soluble P was the highest when the initial pH for *C. krissii*, which was different from that of 7.5 in the medium inoculated with *M. ramosissimus* and 7.0 in the medium inoculated with *P. expansum* (Xiao *et al.*, 2008).

Mechanism of Phosphate Solubilization

Results of various experiments indicate that there must be several solubilization mechanisms regarding phosphate mobilization. Solubilization of mineral phosphate has been attributed to PSF by lowering of pH either by release of organic acids or protons. The organic acid secreted can either directly dissolve the mineral phosphate as a result of anion exchange of phosphate by acid anion or can chelate both Fe and Al ions associated with phosphate (Sperber, 1958; Katznelson and Bose, 1959; Bajpai and Sundara Rao, 1971; Bardiya and Gaur, 1972; Moghimi *et al.*, 1978). Chelation of cations can be important mechanism in the solubilization of P in the cases where organic acid structure favoured complexation (Swnson *et al.*, 1949; Johnston, 1956; Fox *et al.*, 1990).

The PSM produce organic acids such as acetate, lactate, oxalate, tartarate, succinate, citrate, gluconate, ketogluconate, glycolate, etc. (Loum and Webley, 1959; Sperber, 1958; Duff *et al.*, 1969; Taha *et al.*, 1969; Banik and Day, 1982; Goldstein, 1986; Cunningham and Kuiack, 1992; Mattey, 1992; Gyaneshwar *et al.*, 1998; Kim *et al.*, 1998, 1999). The nature and amount of organic acids excreted by fungi are mainly influenced by pH and the buffering capacity of the medium, the carbon, phosphorus, nitrogen sources and the presence or absence of certain metals and trace elements (Gadd, 1999; Gharieb and Gadd, 1999). The presence of insoluble metal phosphate in the growth medium also markedly influenced the production of gluconic acids by *A. niger*, increased the production of organic acids in the presence of $CO_3(PO_4)_2$ and citric acid in the presence of $Zn_3(PO_4)_2$ (Sayer and Gadd, 2001). Fomina *et al.* (2004) also reported secretion of succinic acid and acetic acid by the majority of the fungal strains during solubilization of zinc phosphate and pyromorphite. Reyes *et al.* (1999b) reported production of citric acid and gluconic acid by *P. rugulosum* and its mutant during solubilization of rock phosphate. The solubilization of $CaHPO_4$ by *P. bilaiae* was achieved at pH 4.5 in the presence of citrate, but no $CaHPO_4$ solubilzation occurred at the same pH in the presence of inorganic acid (Cunninghal and Kuiack, 1992), this indicates that the citric acid was chealating the Ca of $CaHPO_4$. The amount of collidal Al-P solubilized by gluconic acid and *P. radicum* was higher than that solubilized by HCl alone at the same pH could indicate that chelation of Al^{3+} by gluconate had occurred (Whitelaw *et al.*, 1999) but solubilization has also been reported due to the lowering of pH along with organic acid production (Banik and Dey, 1982; Kucey, 1988; Cunningham and Kuiack, 1992; Whitelaw, 2000; Pradhan and Sukla, 2005).

Abiotic P solubilization of two Ca phosphate minerals (hydroxyapatite and $CaHPO_4.2H_2O$) by gluconic acid was due to lowered pH alone and not because of chelation of Ca^{2+} by gluconate (Illmer *et al.*, 1995). The observation by Whitelaw *et al.* (1999) that solubilization of $Ca_3(PO_4)_2$ or $CaHPO_4$ with gluconate plus HCl was not higher than solubilization by HCl alone indicated that in *P. radicum*, Ca^{+2} was not complexed with gluconate and the main mechanism of phosphate solubilization was acidification.

Besides the production of organic acids, some other factors such as acidification of the medium as a result of H^+ ion efflux from hyphae during NH_4^+ uptake were also reported (Jakob *et al.*, 2002). H^+ excretion accompanying NH_4^+ assimilation is responsible for P solubilization. The NH_4^+-N had the lowest pH value among different N sources and was the most effective on P solubilization in liquid cultures by *P. bilaiae* (Cunningham and Kuiack, 1992). The reduction in pH in case of ammonium sulphate indicates the possibility of the operation of a NH_4^+/H^+ exchange mechanism acidifying the medium as reported by Roose and Luckener (1994) in *P. cyclopium*. Another mechanism of phosphate solubilization is the development of alkalinity in the liquid medium during the growth of phosphate solubilizing microorganisms. A few workers have reported alkali production in the medium by some TCP dissolving bacterial and fungal isolates (Chhonkar and Subba Rao, 1967; Ahmad and Jha, 1968).

PSF as Bioinoculants

Testing microorganisms for P solubilization in soil condition is an important step for confirmation of laboratory results and necessary for any meaningful application. The beneficial effect of PSF on various crops has been demonstrated with *Aspergillus* (Omar, 1998; Babana and Antoun, 2006), *Penicillium* (Kucey, 1987; Reyes *et al.*, 2002) and *Trichoderma* (Zayed and Motaal, 2005; Rudresh *et al.*, 2005). *P. bilaii* is considered most effective PSF based on the results of the field experiments (Kucey, 1988). Whitelaw *et al.* (1997) suggested that the inoculation of *P. radicum* increased the wheat yield and P uptake that could be partially due to phosphate solubilization in soil with lower available P by this fungus. Inoculation of maize that had been fertilized with Venezuelan Navay PR (phosphate rock) with *P. rugulosum* IR-94MF1 significantly increased shoot yield and P uptake by the plants as compared to the uninoculated control (Reyes *et al.*, 2002). Dry matter yield of wheat plant increased significantly after inoculation of rock phosphate solubilizing fungi *viz.*, *A. niger* and *P. citrinum* in pot soil and field soils (Omar, 1998). Asea *et al.* (1988) observed a 16 per cent increase in wheat plant matter yield and a 14 per cent increase in total plant uptake due to inoculation of *P. bilaji* in a greenhouse experiment. Dudeja *et al.* (1981) reported stimulatory effect of *A. awamori* strain on chickpea variety H208 in terms of nodule formation, nodule dry weight, nitrogenase activity, P/N uptake, and grain yield. Similar results were obtained by Mittal *et al.* (2008) who noticed the beneficial effect of *A. awamori* and *P. citrinum* on biomass, grain yield and P/N uptake of chickpea plant. *A. tubingenesis* and *A. niger* also improved the growth of maize plant in RP amended soil (Richa *et al.*, 2007). The results of nursery experiments showed that the growth of maize plant and shoot P level were significantly increased in soil amended by these fungi compared to control soil. The increased P uptake by plant was also reported due to inoculation of *A. niger* by Medina *et al.* (2007). Pot experiments showed that the dual inoculation of PSF (*A. niger* and *P. italicum*) significantly increased dry matter and yield of soyabean plant (*Glycine max L.*) compared to the control soil (El-azoouni, 2008). Rudresh *et al.* (2005) demonstrated positive effect of *Trichoderma* spp. inoculation on chickpea variety Annegeri-I on plant height, number of branches, biomass, and P/N-uptake by shoots/roots. Several studies have conclusively shown that PSF solubilzed the fixed soil phosphorus and applied phosphate, resulting in higher crop yield (Zaidi, 1999; Gull *et al.*, 2004). However, Zaidi *et al.* (2003) reported decline in nodule number, less increase in grain and straw yield of chickpea variety T3 using a PSF, *P. variabile*.

The alternative approach is to use these PSF along with other beneficial rhizosphere microflora to enhance crop productivity. In this context, simultaneous application of *Rhizobium* and PSF (Perveen *et al.*, 2002) and PSF and arbuscular mycorhhizal (AM) fungi (Zaidi *et al.*, 2003) has been shown to stimulate plant growth more than inoculation of each microorganism alone in certain situation when the soil is phosphorus deficient. AM fungi on the other hand, encourage the plant roots to rapidly absorb solubilized P. Accordingly, the increase in plant growth may be due to the release of certain plant growth promoting substances (Kucey *et al.*, 1989b) by the PSMs and/or AM development and mycorrhizal formation (Azcon-Aguilar and Barea, 1985). Many recent reports show synergistic interaction between PSMs and AM under different experimental conditions. In the field trails performed in southern Egypt, the highest significant effect on wheat (*Triticum aestivum L.*) yield and phosphorus content was observed when seeds were inoculated with the mixture of the AM fungus, *Glomus constrictum* with two Egyptian fungal isolates, *A. niger* and *P. citrinum* that solubilized rock phosphate (Omar, 1998). Inoculation of seeds with Tilemsi phosphate rock-solubilizing fungi, *A. awamori* (C1), *P. chrysogenum* (C13) and AM fungi (*G. intraradices*) under field conditions of Mali, increased the wheat (*Triticum aestivum* L.) grain yield which was comparable to that produced by using the expensive DAP (diammonium phosphate) fertilizer (Babana *et al.*, 2006).

In a study conducted by Dwivedi *et al.* (2004), it was found that the inoculation of *A. awamori* to wheat and rice seedlings increased the crop response to Musoorie rock phosphate over the non-inoculated treatments but was not statistically significant. It was suggested that possible reason for this outcome could be that soil organic carbon levels were very low and that temperature fluctuations were not ideal for enhanced PSM activity.

High carbon crop wastes have been recommended for use as fluctuation substrates in the microbial solubilization of RP before application to the field (Nahas *et al.*, 1990; Vassilev *et al.*, 1995; Vassilev *et al.*, 1996; Zayad and Motaal, 2005; Vassilev *et al.*, 2006). This method has advantage of creating more optimal conditions for microbial organic acid production as compared with conditions present in soil. For example, *A. niger* requires an excessive carbon source and sub-optimal concentrations of nitrogen and many trace metals to achieve maximal citric acid production (Grewal and Kalra 1995, Karaffa and Kubicek 2003). Greenhouse experiments were carried out in neutral calcareous soils aimed at evaluating the effectiveness of fermented products containing mineralized organic matter, soluble P and mycelial mass. Significant plant growth enhancement was observed in comparison to other treatments where the soil was amended with untreated combinations of sugar beet (SB) (*Beta vulgaris*) and RP (Vassilev *et al.*, 1996, 2002b; Vassileva *et al.*, 1998; Rodriguez *et al.*, 1999). Using an isotopic ^{32}P dilution technique, a lowering of the specific activity of the experimental plants was registered, thus indicating that plants benefit from P solubilized from RP by the microbial treatments (Vassilev *et al.*, 2002a). In a pot trial using Grenada (Spain) soil, Vassilev *et al.* (1996) studied the effect of adding SB (sugar beet) waste, pre-cultured with *A. niger* in the presence of Moroccan PR (phosphate rock) on the growth of white clover (*Trifolium repens*). The pretreatment of PR with the inoculated SB waste significantly improved crop dry weight and shoot P content compared to plants given the same amount of untreated PR and SB waste. In another pot trial, using the same soil and PR, Rodriguez *et al.* (1999) found similar results in an experiment measuring the yield of alfalfa (*Medicago sativa* L. cv Aragon). When PR was pre-incubated with *A. niger* inoculated SB waste, alfalfa dry matter yield increased significantly in comparison to the treatment that received non-inoculated SB residue and PR. Alguacil *et al.* (2003) tested the effect of the combined treatment of SB waste, pretreated with *A. niger* in the presence of Moroccan RP, when trying to revegetate degraded semi-arid land in the Mediterranean with *Cistus albidus* L. (rock rose shrub). It was found that the fermented SB and RP mixture significantly increased the available P, total N, and extractable K contents of the soil, with the greatest increase observed in available P content. Similarly Vassilev *et al.* (2006) used 4 agro-industrial wastes as a substrate for microbial solubilization of rock phosphate. Sugar beet wastes (SB), olive cake (OC) and olive mill wastewater (OMWW) were treated by *A. niger* and dry olive cake (DOC) was treated by *Phanerochaete chrysosporium.* A series of microcosm experiments were then performed in the greenhouse to evaluate effectiveness of the resulting fermented products. All amendments improved plant growth and P acquisition, which were further enhanced by mycorrhizal inoculation.

The fertilizing value of microbially treated OMWW, enriched with soluble phosphate, was tested under greenhouse and field conditions with white clover and wheat, respectively, as the experimental plants (Vassilev *et al.*, 1998a and b; Fenice *et al*,. 2002). The results of two crops under greenhouse conditions using sterile soil clearly demonstrated the phytotoxic nature of untreated OMWW, particularly for plants of the first crop. The plant-growth promoting effect of OMWW enriched with microbially solubilized P was evident in both greenhouse and field experiments. A high harvest index was registered for all plants grown in the field, and the plants also showed significant biomass increase (root, straw, seeds, and total biomass). Vassileva *et al.* (1998a) pretreated Moroccan PR with *A. niger* inoculated olive cake-based medium before application to greenhouse pots containing

P-deficient calcareous soil. It was found that white clover shoot dry weight and P content were significantly higher in the pots that received pretreated PR compared to the control that received untreated PR and olive cake waste. The authors concluded that the preincubation of the waste material was the key factor in the effectiveness of this system. Similar experiments have now been conducted in the field.

Kucey (1987) has carried out several experiments introducing fungal hyphae and spores of *P. bilaji*, previously grown on straw (which served as a carrier material), into an experimental calcareous soil. Under field conditions, a treatment containing RP at a rate of 20 kg P/ha soil, resulted in wheat yields and P uptake equivalent to increases due to the addition of mono-ammonium phosphate added at an equivalent rate of P. In another part of this study, greenhouse experiments were carried out in neutral calcareous soil aimed at evaluating the effectiveness of fermented products containing mineralized organic matter, soluble P and mycelial mass.

In general, the effect of treated wastes and RP on plant growth and P acquisition was more pronounced when arbuscular mycorrhizal fungi participated in the microcosm experiments. Mycorrhizal fungi are the dominant component of rhizosphere microbial communities that establish beneficial symbioses with the plants, thus providing a direct physical link between bulk soil and the plant root surface (Smith and Read 1997). In particular, AM fungi are known to significantly facilitate P uptake of the plants they colonize by increasing the absorbing area. It is now well established that cellulose stimulates mycorrhizal development (Gryndler *et al.*, 2002). Due to the presence of plant hormones additional plant growth stimulation could be expected when microbially treated liquid agro-industrial wastes such as vinase and OMWW are used as carbohydrate sources in biosolubilization processes. Therefore, the interactive effects of fermented products, rhizosphere microorganisms and soil-plant systems appear to play a key role in improvement of plant growth and soil quality, although a better understanding of all these interactions is needed to achieve the most appropriate management strategy.

In a few studies, the survival of inoculant strains was shown to be dependent upon the initial inoculum density (Jjemba and Alexander, 1999). The biotic factors play a very important role in the survival of the inoculated strains as the decline observed in non-sterile soils can often be abolished in sterile soils (Heijnen *et al.*, 1988; Heijnen and Van Veen, 1991). Additionally, an increase in the population of the introduced microbes can also be observed (Postma *et al.*, 1988).

Studies involving plants inoculated with PSF showed growth enhancement and increased P content but large variations were found in PSF effectiveness. Only less than 30 per cent experiments were successful and the yield was not comparable with superphosphate. The following reasons were proposed to account for the variations in the effectiveness of PSM inoculation on plant growth enhancement and yields (Kucey *et al.*, 1989).

1. Survival and colonization of inoculated PSF in the rhizosphere.
2. Competition with native microorganisms.
3. Nature and properties of soils and plant varieties (Bashan *et al.*, 1995; Heijnen *et al.*, 1988).
4. Insufficient nutrients in the rhizosphere to produce enough organic acids to solubilize soil phosphates.
5. Inability of PSMs to solubilize soil phosphate.
6. The laboratory conditions employed in the screening process for the isolation do not reflect the soil conditions (Gyaneshwar *et al.*, 2002).

First two reasons of such variation can be overcome by using the right inoculation preparation. Generally, it is observed that the strains that perform so well in *in vitro* conditions, does not show comparable results when they are tested *in vivo*. The reason for this reduction in performance is due to ineffectiveness of the inoculants due to abiotic and biotic stresses, as a result of which the number of fungal cells do not increase. The abiotic stresses normally include fluctuation in temperature, water content, pH, nutrient availability, along with potentially toxic pollutant levels in the case of contaminated soils. Biotic factors include mainly competition from indigenous organisms for nutrients, antagonistic interaction and predation by protozoa and bactriophages. Therefore, the success of microbial inoculant depends on high quality inoculants which can be prepared by increasing the initial inoculum density or by inoculating the microbe in the protective environment (immobilization). It is widely accepted that when applying the inoculant to a harsh environment such as soil, it may be desirable to use a carrier which provide protection and eventually, nutrition for the microbial cells (Gentry *et al.*, 2004).

The reasons 3rd, 4th and 6th can be partially overcome by isolating the PSF in conditions resembling those of soil. For example, addition of buffer in the screening medium proved that the PSF able to secrete large amount of acid were actually significantly few in number. The incorporation of a buffer allowed the isolation of PSMs capable of solubilizing P from rock phosphate under buffered medium conditions in contrast to those isolated under unbuffered screening medium.

Challenges during Commercialization of Phosphate Solubilizing Fungi

The commercialization of any bioformulation is a challenging process. It requires tremendous amount of work and efforts. Commercialization of any PSF is in itself more challenging process than other bioproducts. It requires extensive field trials to prove that the organism is effective over wide range of soil, environmental conditions and crop types. *Penicillium bilaiae* can be taken as case study to understand the challenges that are faced during its commercialization. It was isolated in 1982 and is successfully used as biofertilizer under the name of JumpsStart® in Canada. Its P solubilization capacity was proved in lab but for its commercialization it required further extensive research. The discovery and initial testing of *P. bilaiae* took about 6 person years[1]. The precommercialization stage required 32 person years, and since then, Philom Biose, a Canadian public company invested over 120 person years to improve the inoculant using new formulations and crops. Over 300 trials have been done in the last 10 years demonstrating an average yield increase of 7 per cent in crop yield due to inoculation. By this information it can be concluded that the commercialization of JumpsStart® was very challenging and required hard work. The success of this approach can be seen in market survey conducted every year. In 1990, most farmers had not heard of a phosphate inoculant. By 2001, JumpStart® was the most recognized inoculant product across the Canadian Pairies with 72 per cent of farmers aware of its use (Leggett *et al.*, 2002).

For preparation of microbial inoculant for environmental and agricultural purposes, the technique used should be user friendly, cheap, have high infectivity, able to survive during storage and desiccation after inoculation onto the seeds. Also, it should be able to face the natural competition in the rhizoshphere (Maurice *et al.*, 2001).

For the commercialization of any bioinoculant first there is need to develop a technique for the production of the large quantity of pure inoculum, free from pathogen, with high infectivity potential.

---

1 One person year is equal to 100 people working in a study for 1 year.

Kucey (1988) used a straw substrate to produce spores for greenhouse and field trials. This was effective for the small trials but impractical on a commercial scale. Also, this process was cumbersome and only a limited amount of material could be produced. So, for commercial scale production a liquid fermentation method was developed which produced sufficient spores to inoculate 25,000 hectares of wheat per batch. The spores were collected and than processed into dry powder, which had to be kept frozen to maintain viability for an effective inoculant. This frozen powder, PB50® was introduced in the market in 1990 but its viability was lost around 80 per cent during drying process. The development of a frozen liquid formulation, Provide® process solved the problem and raised the number of hectares treated to 126,000 per batch. The stability of formulation allowed the introduction of a room temperature stable powder of JumpStart® in fields. The move from the Provide® frozen liquid formulation to the dry room temperature stable JumpStart® increased the half life of the fungus on seeds from 10 to 35 days (Leggett *et al.*, 2002).

Once an inoculant is developed then the next step is quality assurance. Quality assurance is the assurity given to the costumer that the product will give them the value for money. Before commercialization, *P. bilaiae* spores had been produced and applied with the rate based on the amount of the dry material added per meter of furrow (Kucey, 1983b). This did not allow the development of a quality control procedure as the quality and the amount of fungus in a gram of substrate varied from batch to batch. Therefore for quality assurance reasons, seed plating method was recommended in place of furrow method. Further, it needed standardizing the minimum number of spores/seed required for efficacy.

Seed inoculation is the most commonly used method of inoculation. However, seeds applied with pesticides are also commonly used. Many seed treatment chemicals contain fungicides that though reduce the survival of fungal inoculant on seed but farmers need these materials to protect their crops from increased disease pressure and can not omit the fungicide in order to use the phosphate bioinoculant. Therefore, the ability of the inoculant organism to survive on fungicide treated seeds long enough to be effective in the field is of much concern. Hence, there is a need to develop a system to test the compatibility of *P. bilaiae* and for that matter other PSF with commonly used seed applied chemicals.

References

Ahmad, N. and Jha, K. K. 1968. Solubilization of rock phosphate by microorganisms isolated from Bihar soils. J. Can. Appl. Microbiol. 14: 89-95.

Alguacil, M. M., Caravaca, F., Azcon, R., Pera, J., Diaz, G. and Roldan, A. 2003. Improvements in soil quality and performance of mycorrhizal *Cistus albidus* L seedlings resulting from addition of microbially treated sugar beet residue to a degraded semiarid Mediterranean soil. Soil Use and Management. 19: 277– 283.

Anamika, Saxena, J. and Sharma, V. 2007. Isolation of tri-calcium phosphate solubilizing bacterial strains from semi-arid agricultural soil of Rajasthan. India. J. Pure and Appl. Micro. 1(2): 269-280

Antoun, H., Beauchamp, C. L., Goussard, N., Chabot, R. and Roger, L. 1998. Potential of *Rhizobium* and *Bradyrhizobium* species as plant growth promoting rhizobacteria on non-legumes: Effect on radish (*Raphanus sativus* L.). Plant and Soil. 204: 57-67.

Antoun, H. 2002. Field and greenhouse trials performed with phosphate-solubilizing bacteria and fungi. In: First International Meeting on Microbial Phosphate Solubilization, Salamanca, Spain, and 16-19 July. pp. 29-31.

Arora, D. and Gaur, A. C. 1979. Microbial solubilization of different inorganic phosphates. Ind. J. Exp. Biol. 17(11): 1258-1261.

Asea, P. E. A., Kucey, R. M. N. and Stewart, J. W. B. 1988. Inorganic phosphate solubilization by two *Penicillium* species in solution culture and soil. Soil Biol. Biochem. 20: 459-464.

Ayyadurai, N., Ravindra, N. P., Sreehari, R. M., Sunish, K. R., Samrat, S. K., Manohar, M., Sakthivel, N. 2006. Isolation and characterization of a novel banana rhizosphere bacterium as fungal antagonist and microbial adjuvant in micropropagation of banana. J. Appl. Microbiol. 100: 926–937.

Azcon-Aguilar, C. and Barea, J. M. 1985. Effect of soil microorganisms on formation of vesicular-arbuscular mycorrhizas. Trans. Brit. Mycol. Soc. 84: 536-537.

Babana, A. H. and Antoun, H. 2006. Effect of Tilemsi phosphate rock solubilizing microorganisms on phosphorus uptake and yield of field grown wheat (*Triticum aestivuml* L.) in Mali. Plant and Soil. 287: 51-58.

Bajpai, P. D. and Sundra Rao, W. V. B. 1971. Phosphate solubilizing Bacteria. Part 2. extracellular production of organic acids by selected bacteria solubilizing insoluble phosphate. Soil. Sci. Plant Nutr. 17: 44-45.

Banik, S. and Dey, B. K. 1982. Available phosphate content of an alluvial soil as influenced by inoculation of some isolated phosphate solubilizing microorganisms. Plant and soil. 69: 353-364.

Banik, S. and Dey, B. K. 1983. Phosphate solubilizing potentiality of microorganisms capable of utilizing aluminium phosphate as sole phosphate source. Zentrabl. Backteriol. Prasitenkd. Infektionskr. Hyg. II, 138:17-23.

Bardiya, S. and Gaur, A. C. 1974. Isolation and screening of microorganisms dissolving low grade rock phosphate. Folia Microbiol. 19: 386-389.

Bardiya, M. C. and Gaur, A. C. 1972. Rock phosphate dissolution by bacteria. Indian J. Microbiol. 12: 269-271

Brock, T. D., Madigan, M. T., Martinko, J. M. and Parker, J. 1994. Biology of microorganisms. pp 999. Prentice, New Jersey.

Barroso, C. B. and Nahas, E. 2005. The status of soil phosphate fraction and the ability of fungi to dissolve hardly soluble phosphates. Applied Soil Ecology. 29: 73-83.

Barroso, C. B., Pereira, G. T. and Nahas, E. 2006. Solubilization of $CaHPO_4$ and Al PO_4 by *Aspergillus niger* in culture media with different carbon and nitrogen sources. Brazilian J. Microbiol. 37: 434-438.

Bashan, Y., Puente, M. E., Rdriquea, M. N., Toledo, G., Holguin, G., Ferrera-Cerrato, R. and Pedrin, S. 1995. Survival of *Azospirillum brasilense* in the bulk soil and rhizosphere of 23 soil types. Appl. Environ. Microbiol. 61: 1938-1945.

Basu, P. 1999. Phosphate solubilization by different strains of *Aspergilli* and *Penicillia*, M. Sc. Dissertation Thesis, Banasthali Vidyapith, pp 64.

Bojinova, D., Velkova, R., Grancharov, I. and Zhelev, S. 1997. The bioconversion of Tunisian phosphorite using *Aspergillus niger*. Nutr. Cycl. Agroecosyst. 47: 227-232.

Bojinova, D., Velkova, R., and Ivanova, R. 2008. Solubilization of morocco phosphorite by *Aspergillus niger*. Bioresource Technology. 99(15): 7348-7353.

Busman, L., Lamb, J., Randall, G., Rehm, G. and Schmitt, M. 2002. The nature of phosphorus in soils. University of Minnesota Extension Service.

Cerezine, P. C., Nahas, E. and Banzatto, D. A. 1988. Soluble phosphate accumulation by *Aspergillus niger* from fluoroapatite. App. Microbiol. Biotech. 29: 501–505

Chabot, R., Antoun, H. and Cescas, M. P. 1993. Microbiological solubilization of inorganic P-fractions normally encountered in soils. *In* Phosphorus, Sulfur and Silicon p77-329.

Chang, Chen-Hsiung and Yang, Shang-Shyng 2009. Thermo-tolerant phosphate solubilizing microbes for multi-functional biofertilizer preparation. Bioresource Technology. 100: 1648-1658.

Chhaonkar, P. K. and Subba-Rao, N. S. 1967. Phosphate solubilization by fungi isolated with legume root nodules. Can. J. Microbiol. 33: 749-753.

Chung, H., Park, M., Madhaiyan, M., Seshadri, S., Song, J., Cho, H. and Sa, T. 2005. Isolation and characterization of phosphate solubilizing bacteria from the rhizosphere of crop plants of Korea. Soil Biol. Biochem. 37: 1970–1974.

Cunningham, J. and Kuiack, C. 1992. Production of citric and oxalic acids and solubilization of calcium phosphate by *Penicillium bilaii*. Appl. Environ. Microbiol. 58: 1451-1458.

Delvasto, P., Valverde, A., Ballester, A., Igual, J. M., Muñoz, J. A., González, F. Blázquez, M. L., García, C. 2006. Characterization of brushite as a re-crystallization product formed during bacterial solubilization of hydroxyapatite in batch cultures. Soil Biol. Biochem. 38: 2645–2654.

Dwivedi, B. S., Singh, V. K. and Dwivedi, V. 2004. Application of phosphate rock with or without *Aspergillus awamori* inoculation to meet phosphorus demands of rice-wheat systems in the Indo-Gangetic plains of India. Aust. J. Exp. Agr. 44: 1041– 1050.

Dixon-Hardy, J. E., Karamushka, V. I., Gruzina, T. G., Nikovska, G. N., Sayer, J. A. and Gadd, G. M. 1998. Influence of the carbon, nitrogen and phosphorus sources on the solubilization of insoluble metal compounds by *Aspergillus niger*. Mycol. Res. 102: 1050-1054.

Dudeja, S. S., Khurana, A. L. and Kundu, B. S. 1981. Effect of *Rhizobium* and phosphomicroorganisms on yield and nutrient uptake in chickpea. Current Science. 50: 503-505.

Duff, R. B., Webley, D. M. and Scott, R. O. 1969. Solubilization of minerals and related materials by 2-ketogluconic acid producing bacteria. Soil Sci. 95: 105-114.

Egamberdiyeva, D., Juraeva, D., Poberejskaya, S., Myachina, O., Teryuhova, P., Seydalieva, L. and Aliev, A. 2003. Improvement of wheat and cotton growth and nutrient uptake by phosphate solubilizing bacteria. 26th Southern Conservation Tillage Conference.

El-Azouni, Iman M. 2008. Effect of phosphate solubilizing fungi on growth and nutrient uptake of soyabean (*Glysine max L.*) plants. Journal of Applied Sciences Research. 4(6): 592-598.

Fenice, M., Federici, F., and Vassilev, N. 2002. Olive mill wastewaters enrichment with soluble phosphate by microbial treatment with *Aspergillus niger*: potential use in agriculture. In: Proceedings of International Workshop on Water in the Mediterranean Basin: resources and sustainable development, vol 1. 10–13 October 2002, Monastir, Tunisia, pp 206–209.

Fernandez, C. and Novo, R. 1988. Vida microbina en el Suelo, II. La Habana: Editorial Pueblo y Educacion.

Fernandez, L. A., Zalba, P., Gomez, M.A. and Sagardoy, M. A. 2007. Phosphate-solubilization activity of bacterial strains in soil and their effect on soybean growth under greenhouse conditions. Biol. Fertil. Soils. 43: 805-809.

Fomina, M., Alexander, I. J., Hillier, S. and Gadd, G. M. 2004. Zinc phosphate and pyromorphite solubilization by soil plant-symbiotic fungi. Geomicrobial. J. 21: 351-366.

Foth, H. D. 1990. Fundamentals of Soil Science. 8th John Wiley and Sons, New York, NY.

Fox, R., Comerford, N. B. and McFee, W. W. 1990. Phosphorus and aluminum released from a spodic horizon mediated by organic acids. Soil Sci. Soc. Amer. J. 54: 1763-1767.

Gadd, G. M. 1999. Fungal production of citric and oxalic acid: importance in metal speciation, physiology and biogeochemical processes. Advances in Microbial Physiology. 41: 47–92.

Gaur, A. C. and Gaind, S. 1983. Microbial solubilization of phosphate with particular reference to iron and aluminum phosphate. Sci. Cult. 49: 110.

Gaur, A. C. and Sachar, S. 1980. Effect of rock phosphate and glucose concentration on phosphate solubilization by *Aspergillus awamori*. Curr. Sci. 49: 553-554.

Gentry, T. J., Rensing, C. and Pepper, I. 2004. New approaches for bioaugmentation as a remediation technology. Crit. Rev. Environ. Sci. Technol. 34: 447-494.

Gharieb, M. M. and Gadd, G. M. 1999. Influence of nitrogen source on the solubilization of natural gypsum ($CaSO_4.2H_2O$) and the formation of calcium oxalate by different oxalic and citric acid producing fungi. Mycol. Res. 103: 473-481.

Goldstain, A. H. 1986. Bacterial phosphate solubilization: historical prospects. Am. J. Alt. Agric. 1: 57-65.

Grewal, H. S. and Kalra, K. L. 1995. Fungal production of citric acid. Biotechnol Adv. 13: 209– 234.

Gryndler,, M., Vosatka, M., Hrselova, H., Chvatalova, I. and Jansa, J. 2002 Interaction between arbuscular mycorrhizal fungi and cellulose in growth substrate. Appl. Soil Ecol. 19: 279–288

Gull, F. Y., Haheez, I., Saleem, M. and Malik, K. A. 2004. Phosphorus uptake and growth promotion of chickpea by co-inoculation of mineral phosphate solubilizing bacteria and a mixed rhizobial culture. Aust. J. Exp. Agric. 44: 623-628.

Guo, F. and Yost, R. S. 1998. Partitioning soil phosphorus into three discrete pools of differing availability. Soil Sci. 163:822-833.

Gupta, J. K., Heding, L. G. and Jorgensen, O. B. 1976. Effect of sugar, hydrogen ion concentration and ammonium nitrate on the formulation of citric acid by *Aspergillus niger*. Acta Microbiol. 23: 63-67.

Gyaneshwar, P., Naresh Kumar, G. and Parekh, I. J. 1998. Effect of buffering on the phosphate solubilizing ability of microorganisms. World J. Microbiol. Biotechnol. 14: 669-673.

Gyaneshwar, P., Kumar, G. N., Parekh, L. J. and Poole, P. S. 2002. Role of microorganisms in improving P nutrient of plants. Plant Soil. 245: 83-93.

Halder, A. K., Mishra, A. K., Bhattacharyya, P. and Chakrabartty, P. K. 1990. Solubilization of rock phosphate by *Rhizobium* and *Bradyrhizobium*. J. Gen. Appl. Microbiol. 36: 81-92.

Hamdali, H., Hafidi, M., Virolle, M. J. and Ouhdouch, Y. 2008. Rock phosphate solubilizing actinomycetes: screening for plant growth-promoting activities. World J. Microbiol. Biotechnol. 24: 2565-2575.

Heijnen, C. E., Van Elsas J. D., Kuikman, P. J. and Van Veen, J. A. 1988. Dynamics of *Rhizobium leguminosarum* bv. *trifolii* introduced into soil: the effect of betonite clay on predation by protozoa. Soil Biology and Biochemistry. 20: 483-488.

Heijnen, C. E. and Van Veen, J. A. 1991. A determination of protective microhabitats for bacteria introduced into soil. FEMS Microbiol. Ecol. 85: 73-80.

Hinsinger, P. 2001. Bioavailability of soil inorganic P in the rhizosphere as affected by root-induced chemical changes: a review. Plant and Soil. 237: 173-195.

Igual, J. M., Valverde, A., Cervantes, E. and Velazquez, E. 2001. Phosphate solubilizing bacteria as inoculants for agriculture: use of updated molecular techniques in their study. Agronomie. 21: 561-568.

Illmer, P. and Schinner, F. 1992. Solubilization of inorganic phosphate by microorganism isolated from forest soil. Soil Biol. Biochem. 24: 389-395.

Illmer, P., Barbato, A. and Schinner, F. 1995. Solubilization of hardly-soluble $AlPO_4$ with P-solubilizing microorganisms. Soil Biol. Biochem. 27: 265-270.

Illmer, P. and Schinner, S. 1995. Solubilization of inorganic calcium phosphates-Solubilization mechanisms. Soil Biol. Biochem. 27(3): 257-263.

Jjemba, P. K. and Alexander, M. 1999. Possible determinants of rhizosphere competence of bacteria. Soil Biol. Biochem. 31: 623-632.

Johri, J. K., Surange, S. and Nautiyal, C. S. 1999. Occurance of salt, pH and temperature tolerant phosphate solubilizing bacteria in alkaline soils. Curr. Microbiol. 39: 89-93.

Johnson, S.E. and Loeppert, R. H. 2006. Role of organic acids in phosphate mobilization from iron oxide. Soil Science Society of American Journal. 70: 222–234

Jones, D., Smith, B. F. L., Wilson, M. J. and Goodman, B. A. 1991. Phosphate- solubilizing fungi in a Scottish upland soil. Mycol. Res. 95: 1090-1093.

Johnston, H. W. 1956. Chelation between calcium and organic anion. New Zealand Journal of Science and Technology. 37: 522-537.

Jakob, H. Boswell, G. P. Ritz, K. *et al.*, 2002. Solubilization of calcium phosphate as a consequence of carbon translocation by *Rhizoctonia solani*. FEMS Microbiol Ecol. 40: 65-71.

Kang, S. C., Ha, C. G., Lee, T. G. and Maheswari, D. K. 2002. Solubilization of insoluble inorganic phosphate by a soil inhabiting fungus *Fomitopsis*.sp.102. Curr. Sci. 82: 439-442.

Karaffa, L. and Kubicek, C. P. 2003. *Aspergillus niger* citric acid accumulation: do we understand this well working black box? Appl. Microbiol. Biotechnol. 61: 189– 196.

Katznelson, H. and Bose, B. 1959. Metabolic acitivity and phosphate dissolving capability of bacterial isolates from wheat roots in the rhizoshere and non-rhizoshere. Can. J. Microbiol. 5: 79-85.

Katznelson, H., Peterson, E. A. and Rouatt, J. W. 1962. Phosphate dissolving microorganisms on seed and in the root zone of plants. Can. J. Bot. 40: 1181-1186.

Kim, K. Y., Jordan, D. and Krishnan, H. B. 1997. *Rahnella aquatills* a bacterium isolated from soyabean rhizosphere, can solubilize hydroxyapatite. FEMS Microbial Lett. 153: 273-277.

Kim, K. Y., Jordan, D. and Mcdonald, G. A. 1998. *Enterobacter agglomerans*, phosphate dissolving bacteria and microbial activity in soil; Effect of carbon sources. Soil Biol Biochem. 30: 995-1003.

Kim, K. Y., Mcdonald, G. A. and Jordan, D. 1999. Solubilization of hydroxyapatite by *Enterobacter agglomerans* and cloned *E.coli* in culture medium. Biol. Fert. Soil. 24: 347-352.

Kucey, R. M. N. 1983a. Phosphate solubilizing bacteria and fungi in various cultivated and virgin Alberta soils. Can. J. Soil Sci. 63: 671-678.

Kucey, R. M. N. 1983b. Effect of *Penicillium bilaji* on the solubility and uptake of phosphorus and micronutrients from soil by wheat. Can. J. Soil Sci.. 68: 261-270.

Kucey, R. M. N. 1987. Increased phosphorus uptake by wheat and field beans inoculated with a phosphor us solubilizing *Penicilliun bilaji* strain and vesicular arbscular mycorrhizal fungi. Appl. Enivion. Microbiol. 53: 2699-2703.

Kucey, R. M. N. 1988. Effect of *Penicilliun bilaji* on the solubility and uptake of P and micronutrients form soil by wheat. Can. J. Soil Sci. 68: 261-270.

Kucey, R. M. N., Janzen, H. H. and Leggett, M. E. 1989. Microbiologically mediated increases in plant-available-phosphorus. Adv. Agron. 42: 199-228.

Kucey, R. M. N. and Leggett, M. E. 1989. Increased yield and phosphorus uptake by westar canola (*Brassica napus L.*) inoculated with a phosphate solubilizing isolate of *Penicillium bilaji*. Can. J. Soil Sci. 69: 425-432.

Leggett, M., Cross, J., Hnatowich, G. and Holloway, G. 2002. Challenges in commercializing a phosphate solubilizing microorganism: *Penicillium bilaiae*, a case history. First International Meeting on Microbial Phosphate Solubilization. 215-222.

Leyval, C. and Berthelin, J. 1985. Comparison between the utilization of phosphorus from insoluble mineral phosphates by ectomycorrhizal fungi and rhizobacteria. In: 1st European Symposium on Mycorrhizae, Dijon, France, pp. 345–349.

Lide, D. R. and Frederikse, H. P. R. 1998. CRC Handbook of Chemistry and Physics. CRC, Boca Raton.

Lindsay, W. L. 1979. Chemical Equilibrium in Soils. Wiley-Interscience, New York, NY.

Lindsay, W. L., Vlek, P. L. G. and Chien, S. H. 1989. Phosphate minerals, In: Soil environment, 2nd ed., Dixon J. B., Weed, S. B., Soil Sci.Soc. America, Madison,1089-1130.

Louw, H. A. and D. M. Webley. 1959. A study of soil bacteria dissolving certain phosphate fertilizers and related compounds. J. Appl. Bacteriol. 22: 227-233.

Lynch, J. M. and Whipper, J. M. 1990. Substrate flow in the rhizosphere. Plant Soil. 128: 1-10.

Medina, A., Jakobsen, I., Vassilev, N., Azcon, R., and Larsen, J. 2007. Fermentation of sugar beet waste by *Aspergillus niger* facilitates growth and P uptake of external mycelium of mixed populations of arbuscular mycorrhizal fungi. Soil Biology and Biochemistry. 39(2): 485-492.

Margaris, N. S. Mitrakos, K. and Markou, S. 1974. Carbon source for *Aspergillus niger* growth under different shaking programmes. Folia Microbiol. 19: 394-396.

Mattey, M. 1992. The production of organic acids. Criti. Rev. Biotechnol. 12: 87-132.

Maurice, S. Beauclair P. Giraud, J-J, Sommer, G., Hartmann, A. and Catroux, G. 2001. Survival and change in physiological state of *Bradyrhizobium japonicum* in soyabean (*Glycine max L. Merril*) liquid inoculant after long-term storage. World J. Microbiol. Biotechnol. 17: 635-643.

Mba, C. C. 1997. Rock phosphate solubilizing streptosporangium isolates from casts of tropical earthworms. Soil Biol. Biochem. 29: 381-385.

McKenzie, R. H. and Roberts, T. L. 1990. Soil and fertilizers phosphorus update. *In* Proc. Alberta Soil Science Workshop Proceedings, Edmonton, Alberta. Feb. 20-22, 84-104.

Mittal, V., Singh, O., Nayyar, H., Kaur, J. and Tewari, R. 2008. Stimulatory effect of phosphate-solubilzing fungal strain (*Aspergillus awamori* and *Penicillium citrinum*) on the yield of chickpea (*Cicer arietinum* L. Cv. GPF2). Soil Biol. Biochem. 40: 718-727.

Moghimi, A., Tate, M. E. and Oades, J. M. 1978. Characterization of rhizospheric products especially 2-keto gluconic acid. Soil Biology and biochemistry. 10: 283-287.

Nahas, E. 2007. Phosphate solubilizing microorganisms: Effect of carbon, nitrogen, and phosphorus sources. In First International Meeting on Microbial Phosphate Solubilization, E. Velázquez (ed) 16-19 July 2002, Salamanca, Spain111-115..

Nahas, E. 1996. Factor determining rock phosphate solubilization by microorganism isolated from soil. World J. Microb. Biotech. 12: 18-23.

Nahas, E. and Assis, L. C. 1992. Solubilizac¸a˜ o de fosfatos de rocha por *Aspergillus niger* em diferentes tipos de vinhac¸a. Pesq. Agropec. Bras. 27: 325–331.

Nahas, E., Banzatto, D. A. and Assis, L. C. 1990. Fluorapatite solubilization by *Aspergillus niger* in vinasse medium. Soil Biol. Biochem. 22: 1097– 1101.

Nahas, E., Centurion, J. F. and Assis, L. C. 1994. Microorganisms solubilizadores de fosfato e produtores de fosfatases de va´rios solos. Rev. Bras. Ci. Solo. 18: 43–48.

Narisan, V., Thakkar, J. and Patel, H. H. 1994. Isolation and screening of phosphate solubilizing fungi. Indian J. of Microbiology. 34: 113-118.

Narisan, V. and Patel, H. H. 2000. *Aspergillus aculeatus* as rock phosphate solubilizers. Soil Boil. Biochem. 32: 559-569

Narisan, V. T. and Patel, H. H. 2009. Relationship of physico-chemical properties of rhizosphere soils with native population of mineral phosphate solubilizing fungi. Indian J. Microbiol. 49: 60-67.

Nautiyal, C. 1999. An efficient microbiological growth medium for screening phosphate solubilizing microorganisms. FEMS Microbiol. Lett. 170: 265-270.

Niranjan Raj, S., Shetty, H. S., Reddy, M. S. 2006. Plant growth promoting rhizobacteria: potential green alternative for plant productivity. In: Siddiqui, Z.A. (Ed.), PGPR: Biocontrol and Biofertilization. Springer, Netherlands, pp. 197–216.

Nopparat, C., Jatupornpipat, M. and Aree, R. 2007. Isolation of phosphate solubilizing fungi in soil from Kanchanaburi, Thailand. KMITL Sci. Tech. J. 7(2): 137-146.

Norrish, K. and Rosser, H. 1983. Mineral phosphate, In: Soils, an Australian viewpoint, Academic Press, Melbourne, CSIRO/London, UK, Australia, 335-361.

Oberson, A., Friesen, D. K., Rao, I. M., Buhler, S. and Frossard, E. 2001. Phosphorus transformations in an oxisol under contrasting land-use system: The role of the microbial biomass. Plant Soil. 237: 197-210.

Oehl, F., Oberson, M., Probst, A., Fliessbach, H., Roth, R. and Frossard, E. 2001. Kinetics of microbial phosphorus uptake in cultivated soils. Biol. Fertil. Soil. 34: 31-41.

Oliveira, C. A., Alves, V. M. C., Marriel, I. E., Gomes, E. A., Scotti, M. R., Carneiro, N. P., Guimara'es, C. T., Schaffert, R. E. and Sa', N. M. H. 2008. Phosphate solubilizing microorganisms isolated from rhizosphere of maize cultivated in an oxisol of the Brazilian cerradi biome. Soil Biology and Biochemistry. 1-6.

Omar, S. A. 1998. The role of rock phosphate solubilizing fungi and vesicular arbuscular mycorrhiza (VAM) in growth of wheat plants fertilized with rock phosphate. World Journal of Microbiology and Biotechnology. 14: 211-219.

Paul, N. B. and Sundara Rao, W. V. B. 1971. Phosphate-dissolving bacteria in the rhizosphere of some cultivated legumes. Plant Soil. 35: 127–132.

Pandey, A. and Palni, L. M. S. 1998. Isolation of *Pseudomonas corrugata* from Sikkim Himalaya. World J. Microbiol. Biotechnol. 14: 411-413.

Pandey, A., Das, N., Kumar, B., Rinu, K. and Trivedi, P. 2008. Phosphate solubilization by *Penicillium* spp. isolated from soil samples of Indian Himalayan region. World J. Microbiol. Biotechnol. 24: 97-102.

Perveen, S., Khan, M. S. and Zaidi, A. 2002. Effect of rhizospheric microorganisms on growth and yield of greengram (*Phaseolus radiate*). Ind. J. Agric. Sci. 72: 421-423.

Postma, J., Van Elsas, J. D., Govaert, J. M. and Van Veen, J. A. 1988. Dynamics of *Rhizobium laguminosarum* bv. *Trifolii* introduced into soil as determined by immunofluorescence and selective plating techniques. FEMS Microbiol. Ecol. 28: 281-290

Pradhan, N. and Sukla, L. B. 2005. Solubilization of inorganic phosphate by fungi isolated from agricultural soil. African J. Biotechnol. 5: 850-854.

Rahi, P., Vyas, P., Sharma, S., Gulati, A. Andb Gulati, A. 2009. Plant growth promoting potential of the fungus Discosia sp. FIHB571 from tea rhizosphere tested on chickpea, maize and pea. Indian J. Microbiol. 49: 128-133.

Rakade, S. M. and Patil, P. L. 1992. Phosphate solubilizing microorganisms: a review. J. Maharashtra Agric. Univ. 17 (3): 458-465.

Rengel, Z. and Marschner, P. 2005. Nutrient availability and management in the rhizosphere: exploiting genotypic differences. New Phytologist. 168: 305–312.

Reyes, I., Bernier, L., Simard, R., Tanguay, P. H. and Antoun, H. 1999a. Characteristics of phosphate solubilization by an isolate of a tropical *Penicillium rugulosum* and two UV-induced mutants. FEMS Microbiol. Ecol. 28: 291–295.

Reyes, I., Bernier, L., Simard, R. R. and Antoun, H. 1999b. Effect of nitrogen sourceS on the solubilization of different inorganic phosphates by isolate of *Penicillium rugulosum* and two UV- induced mutants. FEMS Microb. Ecol. 28: 281-290.

Reyes, I., Bernier, L. and Antoun, H. 2002. Rock phosphate solubilization and colonization of maize rhizosphere by wild and genetically modified strains of *Penicillum rugulosum*. Microbial. Ecol. 44: 39-48.

Richa, G., Khosla, B. and Reddy, M. S. 2007. Improvement of maize plant growth by phosphate solubilizing fungi in rock phosphate amended soils. World J. Agric. Sci. 3(4): 481-484.

Richardson, A. E. 2001. Prospects for using soil microorganisms to improve the acquisition of phosphorus by plants. Australian Journal of Plant Physiology. 28: 897–906.

Rodriguez, H. and Fraga, R. 1999. Phosphate solubilizing bacteria and their role in plant growth promotion. Biotechnol. Adv. 17: 319-339.

Rodriguez, R., Vassilev, N. And Azcon, R. 1999. Increases in growth and nutrient uptake of alfalfa grown in soil amended with microbially-treated sugar beet waste. Appl. Soil Ecol. 11: 9– 15.

Rodriguez, H., Gonzalez, T., Goire, I., Bashan, Y., 2004. Gluconic acid production and phosphate solubilization by the plant growth-promoting bacterium *Azospirillum* spp. Naturewissenschaften. 91: 552–555.

Roos, W. and Luckener, M. 1984. Relationships between proton extrusion and fluxes of ammonium ions and organic acids in *Penicillium cyclopium*. J. Gen. Microbiol. 130: 1007-1014.

Roos, W. and Luckener, M. 1994. Relationship between proton extrusion and fluxes of ammonium ions and organic acids on *Penicillum cyclopium*. J. Gen. Microbiol. 130: 1007

Rudresh, D. L., Shivprakash, M. K. and Prasad, R. D. 2005. Effect of combined application of *Rhizobium*, phosphate solubilizing bacterium and *Trichoderma* spp. on growth, nutrient uptake and yield of chickpea (*Cicer aritenium L.*). Appl. Soil Ecol. 28: 139-146.

Russell, E. W. 1980. Soil conditions and plant growth. 10th ed. Longman, London

Saghir Khan, M., Zaidi, A., Wani, P. A. 2007. Role of phosphatesolubilizing microorganisms in sustainable agriculture–a review. Agronomy for Sustainable Development 27: 29–43.

Sanders, E. M. 2003. Efficacy of *Penicillium bilaiae* for enhancing yield and phosphorus uptake of fall-seeded canola. M.Sc. Thesis. University of Saskatchewan, Saskatoon.

Salih, H. M., Yahya, A. Y., Abdul-Rahem, A. M. and Munam B. H. 1989. Availability of phosphorus in a calcareous soil treated with rock phosphate or super phosphate as affected by phosphate dissolving fungi. Plant and Soil. 120: 181-5.

Sato, S. and Comerford, N. B. 2005. Influence of soil pH on inorganic phosphorus sorption and desorption in a humid Brazilian Ultisol. Rev. Bras. Cienc. Solo. 29. at http://www.scielo.br/

Sawyer, J. and Creswell, J. 2000. Integrated crop management. *In* Phosphorus basics. Aug. 2000, Iowa State University, Ames, Iowa 182-183.

Sayer, J. A. and Gadd, G. M. 2001. Binding of cobalt and zinc by organic acids and culture filterates of *A. niger* grown in the absence and presence of insoluble cobalt or zinc phosphate. Mycol. Res. 105: 1261-1267.

Scheffer, F. and Schachtschabel, P. 1988. Lehrbuch der Bodenkunde. Enke, Stuttgart, (Original phosphate solubilization 2007, soil biology and biochemistry).

Schulte, E. E. and K. A. Kelling. 1996. Soil and applied phosphorus. In Undersatand plant nutrient research service, University of Wisconsin Extension, University of Wisconsin, Madison, Wisconsin.

Seshadri, S., Ignacimuthu, S. and Lakshminarasimhan, C. 2004. Effect of nitrogen and carbon sources on the inorganic phosphate solubilization by different *Aspergillus niger* strain. Chem. Eng. Comm. 191: 1043-1052.

Shenoy, V. V. and Kalagudi, G. M. 2005. Enhancing plant phosphorus use efficiency for sustainable cropping. Biotechnol. Adv. 23: 501–513.

Singal, R., Gupta, R. and Saxena, R. K. 1994. Rock phosphate solubilization under alkaline condition by *Aspergillus japonicus* and *Aspergillus foetidus*. Folia Microbiol. 39: 33-36

Singh, C. P. and Amberger, A. 1991. Solubilization and availability of phosphorus during decomposition of rock phosphate enriched straw and urine. Biol. Agric. Hort. 7: 261-269.

Smith, S. E. and Read, D. J. 1997. Mycorrhizal symbiosis, 2nd edn. Academic Press, San Diego

Son, H. J., Park, G. T., Cha, M. S., Heo, M. S. 2006. Solubilization of insoluble inorganic phosphates by a novel salt-and pH-tolerant Pantoea agglomerans R-42 isolated from soybean rhizosphere. Bioresour. Technol. 97: 204–210.

Sperber, J. I. 1958. The incidence of apatite solubilizing microorganisms in the rhizosphere and soil. Aust. J. Aric. Res. 9: 778-781.

Steinberg, R. A. and Bowling, J. D. 1939. Optimum solution as physiological reference standards in estimating nitrogen utilization by *A. niger.* J. Agr. Res. 58: 717-732.

Stumm, W. and Morgan, J. J. 1995. Aquatic Chemistry. Chemical eqilibria and Rates in Natural Waters, 3rd ed. John Wiley, New York.

Subha Rao, M. S. (1982). Advances in agricultural microbiology, In: Subha Rao M. S. (Ed.) oxford and IBH publ. Co. 229-305.

Sundara Rao, W. V. B. and Sinha, M. K. 1963. Phosphate dissolving micro-organisms in the soil and rhizosphere. Indian J. Agric. Sci. 33: 272–278.

Surange, S. 1985. Comparative phosphate-solubilizing capacity of some soil fungi. Curr. Sci. 54: 1134–1135.

Swenson, R. M. Cole, C. V. and Sieling, D. H. 1949. Fixation of phosphate by iron and aluminium and replacement by organic and inorganic anions. Soil Sci. 67: 3-22.

Taha, S. M., Mahmoud, S. A. Z., El Damtay, A. H. and El-Hafez. 1969. Activity of phosphate dissolving bacteria in soil. Plant Soil. 31: 149-160.

Takahashi, S. and Anwar, M. R. 2007. Wheat grain yield, phosphorus uptake and soil phosphorus fraction after 23 years of annual fertilizer application to an Andosol. Field Crops Res. 101: 160–171.

Takeda, M and Knight, J. D. 2006. Enhanced solubilization of rock phosphate by *Penicillium bilaiae* in pH-buffered solution culture. Can. J. Microbiol. 52: 1121-1129.

Tate, K. R. 1984. The biological transformation of P in the soil. Plant Soil. 76: 245-256.

Tate, K. R. and Salcedo, I. 1988. Phosphorus control of soil organic matter accumulation and cycling. Biogeochem. 5: 99-107.

Thakkar, J., Narsian, V. and Patel, H. H. 1994. Isolation and screening of phosphates solubilizing fungi. Ind. J. Microbiol. 34(2): 113-118.

Thompson, L. M. and Troch, F. R. 1978. Soils and soil fertility 4th ed. McGraw-Hill Inc., New York, NY.

Vankatewarlu, B., Rao, A. V., Rina, P. and Ahmad, N. 1984. Evaluation of phosphorus soubiization by microorganisms isolated from arid soils. J. Ind. Soc. Soil Sci. 32: 273-277.

Vassilev, N., Baca, M. T., Vassileva, M., Franco, I. and Azcon, R. 1995. Rock phosphate solubilization by *Aspergillus niger* grown on sugar-beet waste medium. Appl. Microbiol. Biotechnol. 44: 546–549.

Vassilev, N., Franco, I., Vassileva, M. and Azcon, R. 1996. Improved plant growth with rock phosphate solubilized by *Aspergillus niger* grown on sugar-beet waste. Bioresource Technol. 55: 237– 241.

Vassileva, M., Vassilev, N. and Azcon, R. 1998a. Rock phosphate solubilization by *Aspergillus niger* on olive cake-based medium and its further application in a soil-plant system. World J. Microbiol. Biotech. 14: 281–284.

Vassilev, N., Vassileva, M., Azcon, R., Fenice, M., Federici and Barea, J. M. 1998b. Fertilizing effect of microbially treated olive mill wastewater on *Trifolium* plants. Bioresour. Technol. 66: 133–137.

Vassilev, N., Vassileva, M., Azcon, R., Barea, J. M. 2002a. The use of ^{32}P dilution techniques to evaluate the effect of mycorrhizal inoculation on plant uptake of P from products of fermentation mixtures including agro-wastes, *Aspergillus niger* and rock phosphate. In: Assessment of soil phosphorus status and management of phosphoric fertilizers to optimize crop production. International Atomic Energy Agency, TECDOC-1272, Vienna, Austria, http://www-pub.iaea.org/MTCD/publications/ResultsPage.asp.

Vassilev, N., Vassileva, M., Medina, A. and Azcon, R. 2002b. Fungal solubilization of rock phosphate on media containing agroindustrial wastes. In: Proceedings of Microbial phosphate solubilization. Salamanca, Spain, p 37, http://webcd.usal.es/psm

Vassilev, N., Medina, A., Azcon, R. and Vassileva, M. 2006. Microbial solubiiization of rock phosphate on media containing agro-industrial wastes and effect of the resulting products on plant growth and P uptake. Plant and Soil. 287: 77-84.

Vazquez, P., Holguin, G., Puente, M., ELopez Cortes,A. and Bashan, Y. 2000. Phosphate solubilizing microorganisms associated with the rhizosphere of mangroves in a semi arid coastal lagoon. Biol. Fert. Soils. 30: 460-468.

Wenzel, C. L., Ashford, A. E. and Summerell, B. A. 1994. Phosphate solubilizing bacteria associated with protenoid root of seedlings of waratah (*Telopea speciosissima* (Sm.) R. Br.). New phytol. 128: 487-496

Whitelaw, M. A. 2000. Growth promotion of plants inoculated with phosphate solubilizing fungi. Adv. Agron. 69: 99-151.

Whitelaw, M. A., Harden, T. J. and Bender, G. L. 1997. Plant growth promotion of wheat inoculated with *Penicillium radicum* sp. nov. Aus. J. Soil Res. 35: 291-300.

Whitelaw, M. A., Harden, T. J. and Helyar, K. R. 1999. Phosphate solubilization in solution culture by the soil fungus *Penicillium radicum*. Soil Biol. Biochem. 32: 655–665.

Xiao, C., Chi, R., Huang, X., Zhang, W., Qiu, G. and Wang, D. 2008. Optimization for rock phosphate solubilizing fungi isolated from phosphate mines. Ecological Engineering. 33: 187-193.

Zaidi, A. 1999. Synergistic interactions of nitrogen fixing microorganisms with phosphate mobilizing microorganisms. Ph. D Thesis, Aligarh Muslim University, Aligarh.

Zaidi, A., Khan, S. and Amil, Md. 2003. Interactive effect of rhizotropic microorganisms on yield and nutrient uptake of chickpea (*Cicer arietinium L.*). Euro. J. of Agro. 19: 15-21.

Zayed, G. and Motaal, H. A. 2005. Bioactive compost from rice straw enriched with rock phosphate and their effect on the phosphorus nutrition and microbial community in rhizosphere of cowpea. Biores. Technol. 96: 929-935.

Microbes: Diversity and Biotechnology (2012) *Pages* **87–98**
Editors: **Prof. S.C. Sati & Dr. M. Belwal**
Published by: **DAYA PUBLISHING HOUSE, NEW DELHI**

Chapter 5

Botanicals in Control of Microbial Spoilage of Food Commodities

Priyanka Singh[1], R. Shukla[1], A. Kumar[1], B. Prakash[1], S. Singh[2] and N.K. Dubey[1]

[1]*Centre of Advanced Study in Botany, Banaras Hindu University, Varanasi – 221 005*

[2]*Allahabad Agricultural Institute Deemed University, Allahabad – 211 007*

ABSTRACT

Microbial contamination of harvested food produce is a serious problem in developing countries. Food spoilage due to mould includes off-flavours, mycotoxins contamination, discoloration, and rotting. The use of synthetic chemicals as antimicrobials in food protection comes with a cost for the environment, and the health of animals and humans. However, different plant products have been formulated for large scale application as botanical pesticides in eco-friendly management of plant pests and are being used as alternatives to synthetic pesticides in food protection. The present view deals with the exploitation of traditionally used plants, their components and essential oils, in protecting the stored food commodities as botanical preservatives to enhance their shelf-life.

Introduction

Spoilage of foods and feeds during storage is often the result of microbial activity from a variety of organisms. During storage, foods are severely destroyed by fungi, insects and other pests. In spite of the use of all available means of plant protection, about one-third of the yearly harvest of the world is destroyed by the pests. Losses at times are so severe so as to lead to famine in large areas of the world which are densely populated. Spoilage is a serious problem for the food industry because it renders the products unacceptable for consumption. The association of microbial flora depends on both intrinsic and extrinsic parameters, modes of processing and preservation and implicit parameters (Deak, 1991). Hence, there is urgent need to pay proper attention towards post-harvest storage of food commodities, particularly in humid tropical climates where nearly half of the food supply may be lost between harvest and consumption. A recent report from the ministry of Food and Civil Supplies, Govt. of India, New Delhi, shows that annual estimated post harvest loss of food grain is 20 mt. This would be enough to feed 380 million people for a month (Singh, 2000).

Amongst different pests, fungi are significant destroyers of foodstuffs during storage, rendering them unfit for human consumption by retarding their nutritive value and sometimes by producing mycotoxins. Production of mycotoxins by several fungi has added a new dimension to the gravity of the problem. Approximately 25–40 per cent of cereals world-wide are contaminated with mycotoxins produced by different storage fungi (Kumar *et al.*, 2007). Nearly 70 per cent of the total production of food grains in India is retained at farm level where the unscientific and faulty storage conditions enhance the chances of fungal attack and there by mycotoxin production. Fungi cause significant qualitative as well as quantitative losses of stored food stuffs rendering them unfit for human consumption by retarding their nutritive value and producing mycotoxins. Fungal toxins are low molecular weight chemical compounds which are not detected by body's antigens. Their effect is more often chronic rather than acute, hence they produce no obvious symptoms. Thus mycotoxins are insidious poisons (Pitt, 2002). Cereals and grains are major mycotoxin vectors because they are consumed by both humans and animals (Khattak, 1998). Generally, tropical conditions such as high temperatures and moisture, unseasonal rains during harvest, and flash floods lead to mycotoxin secretion. Poor harvesting practices, improper storage, and less than optimal conditions during transport and marketing can also contribute to proliferation of mycotoxins. Consequently, about five billion people in developing countries are at the risk of chronic exposure to mycotoxins through contaminated foods (Shephard, 2003; Williams *et al.*, 2004). Among mycotoxins, aflatoxins chiefly produced by strains of *Aspergillus flavus* are the most dangerous and about 4.5 billion people in underdeveloped countries are exposed to aflatoxicoses (Williams *et al.*, 2004; Srivastava *et al.*, 2008). Aflatoxins are potent toxic, carcinogenic, mutagenic, immunosuppressive agents, produced as secondary metabolites by the fungus *Aspergillus flavus* and *A. parasiticus* on variety of food products. Aflatoxin B1 is hepatotoxic in humans and animals and is nephrotoxic and immunosuppressive in animals.

The fungal growth may cause decrease in germinability (Sinha *et al.*, 1993), discolouration of grain, heating and mustiness, loss in weight, biochemical changes and production of toxins. Climatic conditions in India are most conducive for mould invasion, elaboration of mycotoxins. Unseasonal rains and flash floods are very common in India which enhance the moisture content of the grain making them more vulnerable for fungal attack (Srivastava, 1987). Fungi can grow on simple and complex food products and produce various metabolites (Khosravi *et al.*, 2007). Up to now, more than 100000 fungal species are considered as natural contaminants of agricultural and food products (Kacaniova, 2003). The quality and safety of food is of importance so that markets are not compromised by the sale of low quality or unsafe food.

Synthetic Antimicrobials in Control of Microbial Spoilage of Food Commodities

Synthetic antimicrobials are the most frequently employed storage technology to protect the stored commodities from fungal deterioration. Different synthetic chemicals have played great role in accomplishment of this goal but have also raised many environmental and toxicological problems. The repeated use of certain chemical fungicides in packaging has led to the appearance of fungicide-resistant populations of storage pathogens (Brent and Hollomon, 1998). In recent years there has been considerable pressure by consumers to reduce or eliminate chemical fungicides in foods. Further, the use of synthetic chemicals to control post-harvest biodeterioration has been restricted due to their carcinogenicity, teratogenecity, high and acute residual toxicity, hormonal imbalance, long degradation period, environmental pollution and their adverse effects on food and side effects on humans (Feng and Zheng, 2007; Unnikrishnan and Nath, 2002). There is increasing public concern over the level of pesticide residues in food. This concern has encouraged researchers to look for alternative solutions to synthetic pesticides. Their uninterrupted and indiscriminate use has not only led to the development of resistant strains but pressure of toxic residues on food grains used for human consumption has led to the health problems. (Sharma and Meshram, 2006). Different types of ecological problems have been reported from time to time by these xenobiotics as they lace the food with residue. The residual toxicity causes disturbances in food chain. Therefore, an effort is needed to find alternatives or formulations for improving currently used pesticides. At the same time, western society appears to be experiencing a trend of 'green consumerism' (Tuley de Silva, 1996; Smid and Gorris, 1999), desiring fewer synthetic food additives and products with a smaller impact on the environment.

Botanicals in Control of Microbial Spoilage of Food Commodities

In recent years, considerable attention has been directed towards the research and application of alternative sources of chemicals (botanical pesticides/phytopesticides) in place of synthetic pesticides. There has been a renewed interest in botanical antimicrobials because of several distinct advantages. Botanicals, being natural derivatives are biodegradable, so they do not leave toxic residues or by-products to contaminate the environment (Beye, 1978). Plant origin pesticides are much safer than conventionally used synthetic pesticides. Pesticidal plants have been in nature as its components for million of years without any ill or adverse effects on ecosystem. Some plants have more than one chemical as active principle responsible for their biological properties. The biological activity in such plants may be due to synergistic effects of different active principles. They may impart different mode of action during their pesticidal actions. These products may exhibit either one particular biological effect or may have diverse biological effects (Varma and Dubey, 1999). This in turn reduces the chances for multiple genomic mutations in insects and the subsequent development of resistance (Begon *et al.*, 1999). Higher plants contain a wide spectrum of secondary metabolites such as phenols, flavonoids, quinones, tannins, essential oils, alkaloids, saponins and sterols. Such plant-derived chemicals may be exploited for their different biological properties (Tripathi *et al.*, 2004). Terrestrial plants produce a bewildering array of natural products–terpenoids, phenolics, alkaloids–likely exceeding 100,000 novel chemical structures. Many of these are thought to serve an ecological function for the plants producing them, serving to defend the plants from herbivores and pathogens (Isman and Akhtar, 2006). Such defensive chemistry is thought to be extremely widespread among the plant kingdom. Aside from their protective aspect, such chemicals have no other apparent purpose in the physiology of the plant, and they are, therefore, called secondary plant metabolites. For many years plant secondary metabolites have been neglected in science. Gradually, recognition of the important role of these compounds has increased, particularly in terms of resistance to pests and diseases (Harborne, 1998). More than 30,000 of these secondary metabolites have been reported from plants so far (Harborne, 1997; Wink, 1993), but since only approximately 5-10 per cent of all higher plants have been analysed phytochemically in detail, the overall total number of secondary compounds will probably exceed 100,000 (Wink, 1993). Natural pest control using botanicals are safer to the user and the environment because they break down into harmless compounds within hours or days in the presence of sunlight. They are also very close chemically to those plants from which they are derived, so they are easily decomposed by a variety of microbes common in most soils. Hence, there is growing interest in the use of safer alternatives for pest control. Research and development cost of biologicals from discovery to marketing is reported to be less compared to chemical pesticides (Carlton, 1988; Woodhead *et al.*, 1990). Plants are an important source of useful chemicals, a fact has been recognised from ancient times.

It has been reported that natural plant products may successfully replace chemical fungicides and provide an alternative method to protect cereals, pulses and other agricultural commodities from aflatoxin B_1 production by *A. flavus* (Krishnamurthy and Shashikala, 2006; Mabrouk and El-Shayeb, 1992). Many tropical medicinal plants and spices have been used as pest control agents (Lale, 1992). Peasant farmers and researchers often claim successful use of plant materials in insect pest control including ash (Ajayi *et al.*, 1987), vegetable oils and powders of plant parts (Lajide *et al.*, 1998). Different crude extracts and other plant materials rich in polyphenols are becoming increasingly important in food industries because of their antifungal, antiaflatoxigenic and antioxidant activity (Kumar *et al.*, 2007). Hence, such plant chemicals can improve shelf-life, quality and nutritional value of stored food commodities (Tripathi and Dubey, 2004).

A perusal of literature shows that a number of higher plants belonging to different families and genera have been screened for their antifungal activity. The families reported to contain strong fungitoxic activity have been listed in Table 5.1 which shows that fungitoxicity is distributed randomly in various families of angiosperms irrespective of their taxonomic position. Variation of fungitoxicity from genus to genus in a family can also be observed through previous literature.

Thus, plants containing fungitoxicity are scattered throughout the flowering plants irrespective of their taxonomic positions. Therefore, the plants for screening should be selected randomly based on their availability.

The essential oils produced by different plant genera are many cases biologically active, endowed with antimicrobial, allelopathic, antioxidant and bio regulatory properties (Elakovich, 1988; Deans *et al.*, 1995; Vaughn and Spencer, 1991; Caccioni and Guizzardi, 1994; Holley and Patel, 2005). The volatility, ephemeral nature and biodegradability of flavour compounds of angiosperm will be specially advantageous if they are developed as pesticide (French, 1985). There may be least chance of residual toxicity by treatment of food commodities with volatile substances of higher plant origin. Concerns for residue of essential oil pesticides on food crops should be mitigated by the fact that different essential oil constituents acquired through the diet are actually beneficial to human health (Huang *et al.*, 1994).

Dubey *et al.* (1983) demonstrated the efficacy of essential oils of *Ocimum canum* and *Citrus medica* as volatile fungitoxicant in protection of some spices against their post harvest fungal deterioration. The essential oils of *Cymbopogon citratus, Caesulia axillaris* and *Mentha arvensis* have shown *in vivo* fumigant activity in the management of storage fungi of some cereals without exhibiting mammalian toxicity (Varma and Dubey, 2001). The essential oils are volatile substances generally composed of mono and sesqui terpinoids aldehydes, esters, acids, ketones, alcohols, coumarins and their composition varies within the same species as a result of genetic and environmental factors. Numerous studies have documented the antifungal (Suhr and Nielson, 2003; Mishra and Dubey, 1994; Elgayyar *et al.*, 2001) and antibacterial (Canillac and Mourey, 2001) effect of plant essential oils.

Examination of indigenous local herbs and plant materials have also been reported particularly from India (Ahmad and Beg, 2001), Australia (Cox *et al.*, 1998), Argentina (Penna *et al.*, 2001) and Finland (Rauha *et al.*, 2000). The practical applicability of essential oils and their combinations has also been tested as fumigants for the protection of store d wheat samples from biodeterioration caused by storage fungi. For complete protection of stored food commodities, efforts should be made to manage the infestation of every type of pest. A product having broad pesticidal efficacy would be preferred in control of post harvest biodeterioration of food commodities than the products having narrow or selective pest controlling efficacy.

Table 5.1: Angiospermic Families Reported to Contain Strongly Fungitoxic Plants

Families	*Worker(s)*
Acanthaceae	Mishra and Dixit, 1975
Amaryllidaceae	Masoko *et al.*, 2005
Anacardiaceae	Dixit *et al.*, 1978; Pandey *et al.*, 1982a
Annonaceae	Mishra and Dixit, 1975; Dikshit *et al.*, 1979; Marthanda *et al.*, 2005
Apiaceae	*Boyraz and Ozcan, 2005;* Meepagala *et al.*, 2005; Seung Won, 2005
Apocynaceae	Mishra and Dixit, 1975; Tripathi, 1976
Asteraceae	Dubey *et al.*1982; Kishore *et al.*, 1982a; Pandey *et al.*, 1982b
Brassicaceae	Gupta and Banerjee, 1970; Dixit *et al.*, 1978; Srivastava *et al.*, 1982
Chenopodiaceae	Dubey *et al.*, 1983; Tripathi *et al.*, 2004; Suhr and Neilsen 2003; Kumar *et al.*, 2007
Combretaceae	Masoko *et al.*, 2005
Convolvulaceae	Srivastava *et al.*, 1982
Euphorbiaceae	Mishra and Dixit, 1975; Onocha *et al.*, 2003; Masoko *et al.*, 2005
Fabaceae	Tripathi, 1977; Pandey *et al.*, 1982a
Geraniaceae	Kumar *et al.*, 2007
Lamiaceae	Mishra and Dixit, 1975; Pandey *et al.*, 1982a; Arras and Grella, 1992; Charai *et al.*, 1996; Dubey *et al.*, 2000; Lambert *et al.*, 2001; Suhr and Neilsen 2003; Boyraz and Ozcan, 2005; Silva *et al.*, 2005; Souza *et al.*, 2007
Lauraceae	Srivastava *et al.*, 2008
Liliaceae	Mishra and Dixit, 1975; Samuel *et al.*, 2000
Meliaceae	Abdelgaleil *et al.*, 2005; Carpinella *et al.*, 2005
Myrtaceae	Carson *et al.*, 1995; Nenoff *et al.*, 1996; Suhr and Neilsen 2003
Papilionaceae	Wang *et al.*, 2005
Poaceae	Pandey *et al.*, 1982a; Mishra, 1992; Suhr and Neilsen 2003
Polygonaceae	Fujita and Kubo, 2005
Portulaceae	Mishra and Dixit, 1975
Ranunculaceae	Bylka *et al.*, 2004
Rhamnaceae	Manojlovic *et al.*, 2005
Rosaceae	Tripathi *et al.*, 2004
Rubiaceae	Manojlovic *et al.*, 2005
Rutaceae	Dubey *et al.*, 1982; Mishra and Dubey, 1990; Suhr and Neilsen 2003; Tripathi *et al.*, 2004
Solanaceae	Mishra and Dixit, 1975; Srivastava *et al.*, 1982
Valeriaceae	Tripathi *et al.*, 2004
Verbenaceae	Tripathi *et al.*, 2004
Zingiberaceae	Mishra and Dubey, 1990; Tripathi *et al.*, 2004; Khattak *et al.*, 2005; Masoko *et al.*, 2005

Higher Plant Products as Aflatoxin Inhibitors

Aflatoxins are produced as secondary metabolites by *Aspergillus flavus* and *A. parasiticus* and are well known carcinogens, mutagens as well as immunosuppressive and teratogenic agents (Kumar *et al.*, 2008). Due to the hazards of aflatoxin exposure, the need for protection of food and feed against aflatoxin contamination is recognized worldwide and several approaches have been suggested. Amongst them, powders and extracts of many spices, herbs and higher plants have been reported to inhibit the production of aflatoxin (Paranagama *et al.*, 2003).

Natural plant products, such as cinnamon and clove oil (Bullerman *et al.*, 1977), phenols (Singh, 1983), some spices (Hasan and Mahmoud, 1993) and many essential oils (Mahmoud, 1994) have been reported as effective inhibitors of fungal growth and aflatoxin production. Essential oils isolated from *Cymbopogon citratus, Monodora myristica, Ocimum gratissimum, Thymus vulgaris* and *Zingiber officinale* were investigated for their inhibitory effect against food spoilage and mycotoxin producing fungi such as *Fusarium moniliforme, Aspergillus flavus* and *A. fumigatus*. Recently, the essential oils of *Cinnamomum camphora* (Singh *et al.*, 2008), *Thymus vulgaris* (Kumar *et al.*, 2008) and *Pelargonium graveolens* (Singh *et al.*, 2008) have been reported to inhibit aflatoxin B_1 secretion by different toxigenic strains of *A. flavus*. These effects against food spoilage and mycotoxin producing fungi indicated the potential of essential oils against food spoiling microorganisms.

Conclusion

Accordingly, use of green pesticides particularly for stored grain pests is being recommended globally and use of essential oils seems to be best answer. Studies have shown that essential oils are readily biodegradable and less detrimental to non-target organisms as compared to synthetic pesticides.

Investigations on the antimicrobial activities, mode of action and potential use of plant volatile oils have regained momentum. There appears to be a revival in the use of traditional approaches to protecting livestock and food from disease, pests and spoilage in industrial countries. This is specially true in regard to plant volatile oils and their antimicrobial evaluation, as can be seen from the comprehensive range of organisms against which volatile oils have been tested. Therefore there is urgent need to bioprospect the pesticidal property of different essential oils for the recommendation of their practical application as botanical antimicrobials for the control of post-harvest biodeterioration of food commodities and thereby enhancing the shelf life of the commodities.

References

Abdelgaleil, S. A., Hashinaga, F. and Nakatani, M. 2005. Antifungal activity of limonoids from *Khaya ivorensis*, Pest Management Science, 61, 186-190.

Ahmad, I. and Beg, A. Z. 2001. Antimicrobial and phytochemical studies on 45 Indian medicinal plants against multi-drug resistant human pathogens, Journal of Ethnopharmacology, 74, 113–123.

Ajayi, O. J. T., Arokoyo, J. T., Nesan, O. O., Olaniyan, M., Ndire Mbula, M. and Kannike, O. A. 1987. Laboratory assessment of the efficacy of some local plant materialsfor the control of storage insect pests. Samaru Journal of Agricultural Research, 5, 81-85.

Arras, G. and Grella, G. E. 1992. Wild thyme, *Thymus capitatus*, essential oil seasonal changes and antimycotic activity, Journal of Horticultural Sciences, 67, 197-202.

egon, M., Hazel, S. M., Baxby, D., Bown, K., Cavanagh, R.,Chantrey, J., Jones, T. and Bennett, M. 1999. Transmission dynamics of a zoonotic pathogen within and between wildlife host species. Proceedings of the Royal Society *B*: Biological Sciences, 266, 1939–1945.

Beye, F. 1978. Insecticides from the vegetable kingdom, Plant Research and Development, 7, 13-31.

Boyraz, N. and Ozćan, M. 2005. Antifungal effect of some spice hydrosols, Fitoterapia, 76, 661-665,

Brent, K.J. and Hollomon, D.W. 2007. Fungicide Resistance in Crop Pathogens: How can it be managed? 2nd Edition. Fungicide Resistance Action Committee, http://www.frac.info.

Bullerman, L. B., Lieu, F. Y. and Seier, S. A. 1977. Inhibition of growth and aflatoxin production by cinnamon and clove oil, cinnamic aldehyde and eugenol, Journal of Food Science, 42, 1107-1116.

Bylka, W., Szaufer, M., Matlawska, J. and Goslinska, O. 2004. Antimicrobial activity of isocytisoside and extracts of *Aquilegia vulgaris* L., Letters in Applied Microbiology, 39, 93-97.

Caccioni, D. R. L. and Guizzardi, M. 1994. Inhibition of germination and growth of fruit and vegetable post-harvest pathogenic fungi by essential oil components, Journal of Essential Oil Research, 6, 173-179.

Canillac, N., and Mourey, A. 2001. Antibacterial activity of the essential oil of *Picea excelsa* on *Listeria*, *Staphylococcus aureus* and coliform bacteria, Food Microbiology, 18, 261–268.

Carlton, B. C. 1988. Biotechnology for crop protection. American Chemical Society, 260"279.

Carpinella, M. C., Ferrovoli, C. G. and Palacious, S. M. 2005. Antifungal synergistic effect of scopoletin, a hydroxycoumarin isolated from *Melia azedarach* L. fruits, Journal of Agricultural and Food Chemistry, 53, 2922-2927.

Carson, C. F., Cookson, B. D., Farrelly, H. D. and Riley, T. V. 1995. Susceptibility of methicillinresistent *Staphylococcus aureus* to the essential oil of *Melaleuca alternifolia*, Journal of Antimicrobial Chemotherapy, 35, 421-424.

Charai, M., Mosaddak, M. and Faid, M. 1996. Chemical composition and antimicrobial activities of two aromatic plants: *Oreganum majorana L.* and *O. compactum Benth*, Journal of Essential Oil Research, 8, 657-664.

Cox, S. D., Gustafson, J. E., Mann, C. M., Markham, L., Liew, Y. C., Hartland, R. P., Bell, H. C., Warmington, J. R. and Wyllie, S. G. 1998.Tea tree oil causes K+ leakage and inhibits respiration in *Escherichia coli*, Letters in Applied Microbiology, 26, 355–358.

Deak, T. 1991. Food-borne yeasts, Advances in Applied Microbiology, 36, 179-278.

Deans, S. G., Noble, R. C., Hiltunen, R., Wuryani, W. and Penzes, L. G. 1995. Antimicrobial and antioxidant properties of *Syzygium aromaticum* (L.) Merr. and Perry: impact upon bacteria, fungi and fatty acid levels in ageing mice, Flavour and Fragrance Journal, 10, 323–328

Dikshit, A., Singh, A. K., Tripathi, R. D. and Dixit, S. N. 1979. Fungitoxic and phytotoxic studies of some essential oils, Biological Bulletin of India, 1, 45-51.

Dixit, S. N., Saxena, A. R. and Dikshit, A. 1978. Fungitoxic properties of some seedling extracts, National Academic Science Letters, 9, 219-221.

Dubey, N. K., Bhargava, K. S. and Dixit, S. N. 1983. Protection of some stored food commodities from fungi by essential oils of *Ocimum canum* and *Citrus medica*, International Journal of Tropical Plant Diseases, 1, 177-179.

Dubey, N. K., Dixit, S. N. and Bhargava, K. S. 1982. Evaluation of leaf extracts of some higher plants against some storage fungi, Indian Journal of Botany, 5, 20-22.

Dubey, N. K., Tiwari, T. N., Mandin, D., Andriamboavonjy, H. and Chaumont, J. P. 2000. Antifungal properties of *Ocimum gratissimum* essential oil (ethyl cinnamate chemotype), Fitoterapia, 71, 567-569.

Elakovich, S. D. 1988. Terpenoids as models for new agrochemicals In: Cutler HG. (Eds.) Biologically active natural products-potential use in agriculture, ACS Symposium. Series, 380, 250 -261.

Elgayyar, M., Draughon, F. A., Golden, D. A. and Mount, J. R. 2001. Antimicrobial activity of essential oils from plants against selected pathogenic and saprophytic microorganisms, Journal of Food Protection, 64, 1019–1024.

Feng, W. and Zheng, X. 2007. Essential oil to control *Alternaria alternata in vitro* and *in vivo,* Food Control, 18, 1126-1130.

French, R. C. 1985. The bioregulatory action of flavour compounds on fungal spores and other propagules, Annual Review of Phytopathology, 23, 173-199.

Fujita, K. and Kubo, I. 2005. Naturally occuring antifungal agents against *Zygosaccharomyces bailii* and their Synargism, Journal of Agricultural and Food Chemistry, 53, 5187-5191.

Gupta, S. and Banerjee, A. B. 1970. A rapid method of screening antifungal antibiotic producing plants, Indian Journal of Experimental Biology, 8, 148-149.

Harborne, J.B. 1997. Introduction to Ecological Biochemistry, Academic Press, London, UK.

Harborne, J.B. 1998. Phytochemical methods, a guide to modern techniques of plant analysis 3rd Eds, Academic Press, London, UK.

Hasan, H. A. and Mahmoud, A. L. E. 1993. Inhibitory effect of spice oils on lipase and mycotoxin production. Zentralblatt für Mikrobiologie, 148, 543-548.

Holley, A. H. and Patel, H. 2005. Improvement in shelf life and safety of perishable foods by plant essential oils and smoke antimicrobials, International Journal of Food Microbology, 22, 273-292.

Huang, M. T., Ferraro, T. and Ho, C.T., 1994. Cancer chemoprevention by phytochemicals in fruits and vegetables, *American Chemical* Society *Symp. Ser.* **546**, 2–15.

Ignacimuthu, S. 2002. Biological control of insect pests, Current Science, 82, 1196-1197.

Isman, M. B., and Akhtar, Y. 2007. Plant Natural Products as a Source for Developing Environmentally Acceptable Insecticides. In: I. Ishaaya, R. Nauen, and A. R. Horowitz (Eds.), *Insecticides design using advanced technologies* Berlin, Heidelberg: Springer-Verlag, 235-248.

Kacaniova, M. 2003. Feeding soybean colonization by microscopic fungi, Trakya University Journal of Science, 4, 165-168.

Khattak, S. 1988. Mycotoxins in food (1) cereal grains and their products, *Pakistan Journal of* Pharmaceutical Sciences, 67-74.

Khattak, S., Rehman, S., Shah, H. U., Ahmed, W. and Ahmed, M. 2005. Biological effects of indigenous medicinal plants *Curcuma longc* and *Alpenia galangal*, Fitoterapia, 76, 254-257.

Khosravi, A. R., Shokri, H. and Ziglari, T. 2007. Evaluation of fungal flora in some important nut products (Pistachio, Peanut, Hazelnut and Almond) in Tehran, Iran, Pakistan Journal of Nutrition, 6, 460-462.

Kishore, N., Dubey, N. K., Tripathi, R.D. and Singh, S.K. 1982 a. Fungitoxic activity of leaves of some higher plants, National Academic Science letters, 5, 9-10.

Krishnamurthy, Y. L. and Shashikala, J., 2006. Inhibition of Aflatoxin B_1 production of *Aspergillus flavus*, isolated fronm soyabean seeds by certain natural plant products. Letters in Applied Microbiology, 43, 469-474.

Kumar, A., Shukla, R., Singh, P., Prasad, C. S. and Dubey, N. K., 2008. Assessment of *Thymus vulgaris* L. essential oil as a safe botanical preservative against post-harvest fungal infestation of food commodities, Innovative Food Science and Emerging Technologies, 9, 575-580.

Kumar, R., Mishra, A. K., Dubey, N. K., and Tripathi, Y. B. 2007. Evaluation of *Chenopodium ambrosioides* oil as a potential source of antifungal, antiaflatoxigenic and antioxidant activity, International Journal of Food Microbiology, 115, 159-164.

Lajide, L. C. O., Adedire, C. O., Muse, W. A. and Agele, S. O., 1998. Insecticidal activity of powders of some Nigerian plants against the maize weevil, *Sitophilus zeamais* Motsch. In: *Entomology and the Nigerian Economy: Research Focus in the 21st century; ESN Occasional Publication 31*, (Lale, N. E. S., Molta, N. B., Donli, P. O., Dike, M. C., and Aminu-Kano, M. Maiduguri eds), 227-235.

Lale, N. E. S., 1992. A laboratory study of the comparative toxicity of products from three spices to the maize weevil, Post-harvest Biology and Technology, 2, 612-664.

Lambert, R. J. W., Skandamis, P. N., Coote, P. and Nychas, G. J. E. 2001. A study of the minimum inhibitory concentration and mode of action of oregano essential oil, thymol and carvacrol, Journal of Chemical Ecology, 25, 1319-1330.

Mabrouk, S. S. and El-Shayeb, N. M. A. 1992. Inhibition of aflatoxin production in *Aspergillus flavus* by natural coumerins and chromones, World Journal of Microbiology and Biotechnology, *8*, 60-62.

Mahmoud, A. L. E., 1994. Antifungal action and antiaflatoxigenic properties of some essential oil constituents, Letters in Applied Microbiology, **19**, 110-113.

Manojlovic, N. T., Solujic, S., Sukdolak, S., and Milosev, M, 2005. Antifungal activity of *Rubia tinctorium*, *Rhamnus frangula* and *Caloplaca cerina*, Fitoterapia, 76, 244.

Marthanda, M., Subramanyan, M., Hima, M. and Annapurna, J. 2005. Antimicrobial activity of clerodane diterpenoids from *Polyalthia longifolia* seeds, Fitoterapia, 76, 336-339.

Masoko, P., Picard, J. and Eloff, J. N. 2005. Antifungal activities of six South African *Terminalia* species (Combretaceae), Journal of Ethnopharmacology, 99, 301-308.

Meepagala, K. M., Schrader, K. K., Wedge, D. E., and Duke, S. O. 2005. Algicidal and antifungal compounds from the roots of *Ruta graveoiens* and synthesis of their analogs, Phytochemistry, 66, 2689-2695.

Mishra, A. K. and Dubey, N. K. 1994. Evaluation of some essential oils for their toxicity against fungi causing deterioration of stored food commodities, Applied and Environmental Microbiology, 60, 1101-1105.

Mishra, A. K. and Dubey, N. K. 1990. Fungitoxic properties of *Prunus persica* oil, Hindustan Antibiotics Bulletin, 32, 91-93.

Mishra, A. K. 1992. Evaluation of some higher plant products as natural preservatives against biodeterioration of stored food commodities. Ph.D. Thesis, Banaras Hindu University, Varanasi, India.

Mishra, S. B., and Dixit, S. N. 1975. Antifungal activity of plant extracts against orange rust, Symposium on Physiology of Microorganisms, 221-224.

Nenoff, P., Haustein, U. F. and Brandt, W. 1996. Antifungal activity of the essential oil of *Melaleuca alternifolia* (tea tree oil) against pathogenic fungi *in vitro*, Skin, Pharmacology, 9, 388-394.

Onocha, P. A., Opegbemi, A. O., Kadri, A. O., Ajayi, K. M. and Okorie, D. A., 2003. Antimicrobial evaluation of Nigerian Euphorbiaceae plants: *Phyllanthus amarus* and *Phyllanthus muelleranus* leaf extracts, Nigerian Journal of Natural Products and Medicine, 7, 9-12.

Pandey, D. K., Chandra, H. and Tripathi, N. N. 1982a. Volatile fungitoxic activity in higher plants with special reference to that of *Callistemon lanceolatus* D. C., Phytopathologische-Zeitschrift, 105, 175-182.

Pandey, D. K., Tripathi, N. N., Tripathi, R. D. and Dixit, S. N. 1982b. Fungitoxic and phytotoxic properties of the essential oils of *Hyptis suaveolens* (L.) Poir. Z., Pflkrarkh Pflschutz, 89, 344-349.

Paranagama, P. A., Abeysekera, K. H., Abeywickrama, K. and Nugaliyadde, L. 2003. Fungicidal and anti-aflatoxigenic effects of the essential oil of *Cymbopogon citratus* (DC.) Stapf.(lemongrass) against *Aspergillus flavus* Link. isolated from stored rice, Letters in Applied Microbiology, 37, 86-90.

Penna, C., Marino, S., Vivot, E., Cruanes, M. C., Munoz, J. de D., Cruanes, J., Ferraro, G., Gutkind, G. and Martino, V. 2001. Antimicrobial activity of Argentine plants used in the treatment of infectious diseases. Isolation of active compounds from *Sebastiania brasiliensis*, Journal of Ethnopharmacology, 77, 37-40.

Pitt, J. I. 2000. Toxigenic fungi: which are important?, Medical Mycology, 38, 17-22.

Rauha, J. P., Remes, S., Heinonen, M., Hopia, A., Kahkonen, M., Kujala, T., Pihlaja, K., Vuorela, H. and Vuorela, P. 2000. Antimicrobial effects of Finnish plant extracts containing flavonoids and other phenolic compounds, International Journal of Food Microbiology, 56, 3-12.

Samuel, J. K., Andrews, B. and Shylajebashree, H. 2000. *In vitro* evaluation of the antifungal activity of *Allium satium* bulb extract against *Trichophyton rubram*, a human skin pathogen, World Journal of Microbiology and Biotechnology, 16, 617-620.

Seung Won, S. 2005. Antifungal activities of essential oils from *Glehnia littoralis* alone and in combination with ketoconazole, Natural Product Sciences, 11, 92-96.

Sharma, K. and Meshram, N. M. 2006. Bioactivity of essential oils from *Acorus calamus* Linn. and *Syzygium aromaticum* Linn. against *Sitophilus oryzae* Linn. in stored wheat, Biopesticide International, 2, 144-152.

Shephard, G. S. 2003. Aflatoxin and Food Safety: Recent African Perspectives, Journal of Toxicology 22, 267-286.

Silva, M. R., Oliveira, J. G. Jr., Fernandez, O. F., Passos, X. S., Costa, C. R., Souza, L. K., Lemos, J. A. and Paula, J. R. 2005. A study of the minimum inhibitory concentration and mode of action of oregano essential oil, thymol and carvacrol, Mycoses, 48, 172.

Singh, D., 2000. Bioinsecticides from plants, Current Science, 78, 7-8.

Singh, P. 1983. Control of aflatoxin through natural plant extracts. In: K. S. Bilgranii, T. Prasad, and K. K. Sinha (Eds.), *The Proceeding of Symposium on Mycotoxin in Food and Feed*. Bhagalpur: Allied Press, 307–315.

Singh, P., Srivastava, B., Kumar, A. and Dubey, N. K. 2008. Fungal contamination of raw materials of some herbal drugs and recommendation of *Cinnamomum camphora* oil as herbal fungitoxicant, Microbial Ecology, 56, 555-560.

Singh, P., Srivastava, B., Kumar, A., Kumar, R., Dubey, N. K., and Gupta, R., 2008. Assessment of *Pelargonium graveolens* oil as plant based antimicrobial and aflatoxin suppressor in food preservation, Journal of the Science of Food and Agriculture, 88, 2421-2425.

Sinha, K. K., Sinha, A. K., Ggajendra, P. and Prasad, G. 1993. The effect of clove and cinnamon oils on growth of and aflatoxin production by *A. flavus*, Letters in Applied Microbiology, **16**, 114–117.

Smid, E. J. and Gorris, L. G. M. 1999. Natural antimicrobials for food preservation. In: M. S. Rehman (Ed.), *Handbook of food preservation*. New York: Marcel Dekker, 285-308.

Souza, E. L., Stamford, T. L. M, Lima, E. O. and Trajano, V.N. 2007. Effectiveness of *Origanum vulgare* L. essential oil to inhibit the growth of food spoiling yeasts, Food Control, 18, 409–413.

Srivastava, B., Singh, P., Shukla, R. and Dubey, N. K. 2008. A novel combination of the essential oils of *Cinnamomum camphora* and *Alpinia galanga in* checking aflatoxin B_1 production by a toxigenic strain of *Aspergillus flavus*, World Journal of Microbiology and Biotechnology, 24, 693–697.

Srivastava, J. L. 1987. Mycotoxin problems in food in India. Paper presented at the Joint FAO/ WHOIUNDP Second International Conference on Mycotoxins held at Bangkok, Thailand from September 28-October 3.

Srivastava, O. P., Pandey, D. K., Tripathi, R. N. and Tripathi, N. N. 1982. Fungitoxic evaluation of some higher plants, Environmental Indica, 5, 51-54.

Suhr, K. I. and Nelson, P. V. 2003. Antifungal activity of essential oils evaluated by two different application techniques against rye bread spoilage fungi, Journal of Applied Microbiology, 94, 665-674.

Tripathi, P. and Dubey, N. K. 2004. Exploitation of natural products as an alternative strategy to control post harvest fungal rotting of fruits and vegetables, Post-harvest Biology Technology, 32, 235-245.

Tripathi, P., Dubey, N. K., Banerji, R. and Chansouria, J. P. N. 2004. Evaluation of some essential oils as botanical fungitoxicants in management of post-harvest rotting of Citrus fruits. World J. Microbiology and Biotech., 20, 317-321.

Tripathi, R. D. 1977. Assay of higher plants for antifungal antibiotics and some aspects of mode of action of three active principle. Ph.D. Thesis, University of Gorakhpur, Gorakhpur, India.

Tripathi, S. C. 1976. Antifungal activity in flowers of some higher plants. Ph.D. Thesis, University of Gorakhpur, India.

Tuley de Silva, K., A. 1996. manual of the essential oil industry. United nation industrial development organisation, Vienna.

Unnikrishnan, V. and Nath, B. S. 2002. Hazardous chemicals in foods, *Indian Journal of Dairy Science*, **11**, 155–158.

Varma, J. and Dubey, N. K., 1999. Prospectives of botanical and microbial products as pesticides of tomorrow, Current Science, 76, 172-179.

Varma, J. and Dubey, N. K., 2001. Efficacy of essential oils of *Caesulia axillaris* and *Mentha arvensis* against some storage pests causing biodeterioration of food commodities, International Journal of Food Microbiology, 68, 207-210.

Vaughn, S. F. and Spencer, G. F. 1991. Volatile monoterpenes inhibit potato tuber sprouting, Potato Journal, 68, 821-831.

Wang, S.Y., Chen, P. F. and Chang, S.T. 2005. Antifungal activities of essential oils and their constituents from indigenous cinnamon (*Cinnamomum osmophloeum*) leaves against wood decay fungi, Bioresource Technology, 96, 813-818.

Williams, H. J., Phillips, T. D., Jolly, E. P., Stiles, K. J., Jolly, M. C. and Aggarwal, D. 2004. Human aflatoxicosis in developing countries: a review of toxicology, exposure, potential health consequences and interventions, American J.Clinical *Nutrition*, 80, 1106–1122.

Wink, M., 1993. Production and application of phytochemicals from an agricultural perspective.In: Van Beek, T. A. and Breteler, H. (Eds.) Phytochemistry and Agriculture. Proceedings of the Phytochemical Society of Europe. Clarendon Press, Oxford, UK, 171-213.

Woodhead, S. H., O'Leary, A. L. and Rabatin, S. C. 1990. Discovery, development, and registration of a biocontrol agent from an industrial perspective, Canadian Journal of Plant Pathology, 12, 328-331.

Microbes: Diversity and Biotechnology (2012) *Pages* **99–117**
Editors: **Prof. S.C. Sati & Dr. M. Belwal**
Published by: **DAYA PUBLISHING HOUSE, NEW DELHI**

Chapter 6

Biological Control of Fungal Phytopathogens by *Trichoderma* sp.: Mechanisms of Action

**Lakshmi Tewari[1], Bipin Chandra[2] and Jitendra Saini[1]*
[1]Department of Microbiology, CB SH, GB. Pant University of Agriculture and Technology, Pantnagar – 263 145, US Nagar
[2]Department of Plant Pathology (MRTC) College of Agriculture, G.B. Pant University of Agriculture and Technology, Pantnagar – 263 145, US Nagar

ABSTRACT

Biological control of phytopathogens by antagonistic organisms is a potential, non-chemical and more eco-friendly tool for crop protection. *Trichoderma* species are considered as one of the most important biocontrol fungi for improving plant growth and protecting crops from several fungal phytopathogens. It is most extensively used for the management of plant pathogens affecting seed, root and aerial plant parts. These are avirulent, rhizosphere competent filamentous soil fungi that readily colonize roots of several plants and are highly interacting with other rhizosphere micro flora in root, soil and foliar environment. They are known to produce a wide range of antibiotic substances and parasitize other fungal phyto-pathogens. *Trichoderma* sp. uses a variety of mechanisms to provide protection against several plant pathogens and/or plant diseases such as mycoparasitism and/or antibiosis, adverse effects on the growth and development of the pathogen by competing for the nutrients, oxygen or space, alteration in fitness of the pathogen, induce systemic plant resistance, enhance plant growth and its tolerance to stress, metabolize plant exudates supporting pathogen, and/or inactivate the enzymes produced by the pathogens and synthesize cell wall degrading enzymes (lytic enzymes) that degrade the cell wall of pathogen. Some strains establish long lasting colonization of root surfaces and penetrate into the epidermis. Root colonization by *Trichoderma* sp. also enhances root growth and development, crop productivity and nutrients uptake.

Some strains are known to synthesize siderophore like metal chelating agents that bound metal ions, such as iron, and make them available for plant growth; some can solubilize bound nutrients and enhance their bioavailability in soil.

Keywords: *Biological control, Phyto-pathogens, Trichoderma, Mycoparasitism.*

Introduction

During recent years, there has been a growing concern for environmental hazards caused by chemical methods used for the control of plant diseases and their pathogens. Chemical control methods cause considerable environmental pollution, imbalance in nature and ecological disruption, thus posing a threat to the health of living organisms. These chemical pesticides are not always economical and are not effective for many diseases. Moreover, they have other limitations too, such as low cost: benefit ratio, poor availability, selectivity, temporary effects, efficacy affected by physico-chemical and biological properties, development of resistance in pathogens and resurgence of pathogen. Considering these limitations, there has been a growing concern to develop such management practices which alone or in combination with other practices could bring about a reasonably good degree of reduction of pathogenicity and virulence of the pathogen and at the same time ensure sustainability of production, cost, effectiveness and healthy ecosystem. Biological control is an important approach in this direction. Over the recent times, Biological control gained an increased scientific interest and now it has already attained a status of commercial venture. Commercialized system for biological control of plant disease has been reported to be as effective as chemical pesticide control without any residual toxic effect (Harman, 1991).

Rhizosphere, the immediate environment of plant roots and the most dynamic region of soil, harbours a variety of microorganisms. These Rhizosphere microorganisms have diverse metabolic potentials and may be classified as harmful including pathogens, beneficial and neutral to plants. These soil borne plant pathogens cause several severe diseases in plants and heavy yield loss. Therefore, protection of crops from harmful pathogens requires chemical control methods, which are neither always economical nor effective. Moreover, most of the chemicals used to control disease causing pathogens are highly toxic and harmful for the human beings and are also not eco-friendly causing environmental hazards. An effective and attractive alternative or supplement to these chemical pesticides is "*Biological Control*". Biological Control can be defined as "*Reduction of the amount of inoculum or disease producing activity of a pathogen accomplished by or through one or more organisms other than man*" (Cook and Baker, 1983). Biological control strategy is highly compatible with sustainable agriculture and has a major role to play as a component of integrated pest management (IPM) programme. It involves the use of beneficial rhizosphere competent microorganisms, such as specialized fungi and bacteria to attack and control plant pathogens and disease they cause. Numerous rhizosphere microorganisms, having antagonistic potential against several disease causing fungal pathogens have been identified; *Trichoderma* sp. is one of them which has been identified as a potential antagonist and also having other plant growth promotory properties. Commercialized system for biological control of plant disease has been reported to be as effective as chemical pesticide control without any residual toxic effect (Harman, 2000).

Rhizosphere Microorganisms having Biocontrol Potential

Several soil microorganisms found in the rhizosphere of various plants have been identified to have antagonistic potential against several plant pathogens. Several biocontrol agents are currently being used for plant disease control and crop protection. The first attempt at control of a root disease with the introduction of microorganisms into soil was by C. Hartley in 1921.He introduced twelve isolates of saprophytic fungi and one of bacteria which significantly reduced severity of damping off of pine seedling caused by *Pythium debaryanum* (Baker *et al.*, 1984). Thus a better initial emergence and seedling survival was obtained by use of biocontrol agents. After that G.B.Stanford in 1926 reported biocontrol potential of some bacteria and R.Weindling in 1932 reported *Trichoderma* sp. as a biocontrol agent. The potential biocontrol agents are rhizosphere competent antagonists, which are capable of inducing plant growth responses either by controlling minor pathogens or by producing growth stimulating factors. Their mechanism of action includes various attributes such as competition, antibiosis, antagonism, predation etc.

Though biocontrol agents include all classes or groups of organisms existing in an ecosystem, maximum emphasis for developing biocontrol programmes was given to fungal and bacterial systems primarily because of ease of their mass multiplication and formulation. The important genera of fungi studied as biocontrol agents are species of *Trichoderma, Gliocladium, Penicillium, Neurospora, Aspergillus Chaetomium Glomus etc.* A number of bacterial species/strains studied for their plant growth promoting activity and biocontrol potential, include, *Pseudomonas, Bacillus, Agrobacterium, Enterobacter, Erwinia, Streptomyces* etc. Among several groups of plant diseases, major amount of work has been done on biological control of soil borne fungal pathogens by using fungal antagonists like *Trichoderma sp.* and/or *Gliocladium* sp. Among bacteria, plant growth promoting rhizobacteria – the fluorescent Pseudomonads are under intensive research because of their wide natural occurrence, biocontrol potential against fungal and nematode diseases as well as inducing host defense and plant growth promoting activities. The prominent antagonists have been developed as bio-pesticides and commercialized today.

Trichoderma: The Beneficial Soil Fungi

Trichoderma is considered as one of the most important biocontrol fungus for improving plant growth. The genus *Trichoderma* consists of asexually reproducing soil fungi that are common in nearly all types of soils, root ecosystems and other natural habitats especially those containing high organic matter throughout the world. These are free-living fungi that are highly interacting with other rhizosphere micro flora in root, soil and foliar environment. Mycoparasitic activity and antibiotic production potential were first demonstrated in *Trichoderma lignorum* by Weindling (1932). One of the most interesting aspects of studies on *Trichoderma* is its potential to employ varied mechanisms for disease control. In general the fungus exhibits a preference for wet soil. However, there are reports that individual species of *Trichoderma* exhibit different preferences for soil temperature and soil moisture. Apart from these factors, the iron content of the soil, HCO_3^- salt and organic matter content, presence and absence of other microbes in soil are also important determinants of micro site preference by *Trichoderma* sp. They show a high level of genetic diversity, and can be used to produce a wide range of products of commercial and ecological interest. The genus *Trichoderma* was introduced by Persoon almost 200 years ago and was isolated primarily from soil and decomposing organic matter. *Trichoderma,* for the most part, classified as *imperfect fungi,* in that they produce only asexual spores. The sexual stage, when found, is within the *Ascomycetes* in the genus *Hypocrea* (Harman, 2002).These fungi also colonize woody and herbaceous plant materials, in which the sexual teleomorph (genus *Hypocrea*) has most often been found. Rifai (1969) outlined the speciation concept within the genus *Trichoderma* and described nine species aggregates: *T. piluliferum* Webster and Rifai, *T. polysporum* (Link) Rifai, *T. virens* Gidden and Foster, *and T. hamatum* (Bon.) Bain, *T. koningii* Oudem. Apud Oudem. Et Koning, *T. aureoviride* Rifai, *T. harzianum* Rifai, *T. longibrachiatum* Rifai, *T. pseudokoningii* Rifai and *T. viride*

Pers: Fr. However, with the use of molecular approaches particularly sequence polymorphism with internal transcribed spacer (ITS) regions of nuclear ribosomal DNA (rDNA), the taxa recently have gone from nine to at least 35 species (Hayes *et al.*, 1994). The molecular technique, randomly amplified polymorphic DNA (RAPD) analysis has also been used to differentiate numerous fungi including *Trichoderma* sp. The mycelium is hyaline with septate, profusely branched and smooth walled hyphae (Figures 6.1,a and b). Chlamydospores are present in most species. The conidiophores are highly ramified and phialides are flask shaped or ovoidal (Hermosa *et al.*, 2000).

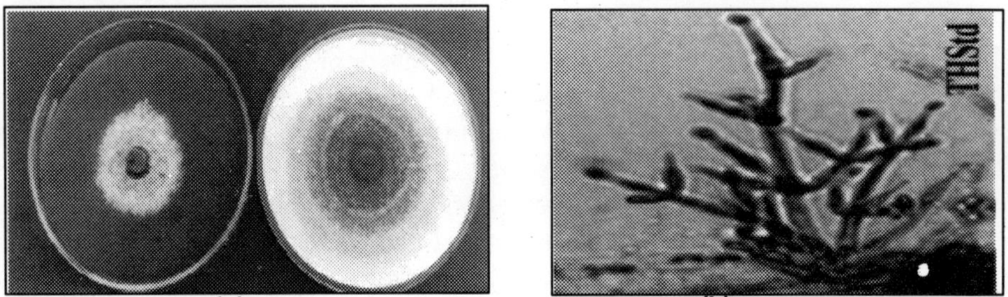

(a) (b)

Figure 6.1: Morphological Features of *Trichoderma harzianum*
(a) Colony morphology on culture medium,
(b) Microscopic view of mycelia with conidiophores and conidiospores.

Biocontrol Potential of *Trichoderma* sp.

Since the pioneer work of Weindling, several reports on successful biocontrol of fungal phytopathogens by *Trichoderma* sp. have accumulated. Among *Trichoderma* sp., most widely reported and commonly used biocontrol species are *T. harzianum, T.viride and T. virens.* They have been reported to inhibit many soil borne pathogenic fungi such as *Fusarium, Pythium, Sclerotium, Rhizoctonia,Sclerotinia, Macrophomia* sp. *etc.*, which are the major wilt causing fungal pathogens of various crops (Table 6.1).

T. harzianum has potential for biological control of sheath blight of rice by antagonizing the pathogen *Rhizoctonia solani* (Tewari and Singh, 2005). Today *Trichoderma* strains are used for biological control, either alone or in combination with other microbes or chemical adjuvants. They are known to produce a wide range of antibiotic substances and parasitize other fungal phyto-pathogens. They also compete with other soil microorganisms for key exudates from seed and roots that stimulate germination of propagules of plant pathogenic fungi in soil, nutrients and space. They are also known to produce certain lytic enzymes that degrade the cell wall of the pathogen. Besides these direct inhibitory effects on pathogens, *Trichoderma* sp. exerts beneficial effects on plant growth and development. These versatile fungi are highly efficient producers of many extra cellular enzymes like cellulases, chitinases, glucanases, proteases etc. *Trichoderma* sp. is most extensively used fungal biocontrol agent for the management of plant pathogens affecting seed, root and aerial plant parts. A large number of its formulations are available in market for use by the farmers. Treatment with *Trichoderma* generally increases root and shoot growth, reduces the activity of deleterious microorganisms in the rhizosphere of plants and improves the nutrient status of the plant. Growth enhancement by *Trichoderma* sp. has been observed even in the absence of any detectable disease and in sterile soil and is not considered to be a side effect of suppression of disease or minor plant pathogens.

Table 6.1: Management of Plant Pathogens/Diseases by *Trichoderma* sp.

Crop	Disease	Pathogen	Trichoderma sp.	Mode of Application
Chickpea	Gray mould	*Botrytis cinerea*	*Trichoderma* spp.	Spray
Chickpea	Collar rot	*S. rolfsii, R. solani*	*T. harzianum, T. viride*	Seed Treatment
Cotton	Root rot	*M.phaseolina*	*T. harzianum, T. virens*	Seed Treatment
Egg Plant	Collar rot	*Sclerotinia sclerotium*	*T. virens, T. viride*	Soil Treatment
French bean	Root rot	*R. solani*	*T. harzianum,T. virens, T. viride, T. hamatum*	Seed Treatment
Ginger	Rhizome rot	*F. oxysporum, F. Zingiberi*	*T. harzianum, G. virens*	Rhizome treatment
Mulberry	Cutting rot	*F. solani*	*T. harzianum, T. virens, T. pseudokoningii*	Cutting and soil treatment
Pea	White rot	*Sclerotinia sclerotium*	*T. harzianum, T. virens, T. viride*	Soil Treatment
Pigeonpea	Wilt	*F. udum*	*T. viride, T. hamatum, T. koningii*	Seed and seedling treatment
Potato	Black scurf	*R. solani*	*T. virens, T. viride*	Tuber treatment
Rice	Sheath Blight	*R. solani*	*T. harzianum, T. viride, T. atroviride*	Soil treatment, Seed treatment Foliar spray
Sugarcane	Root rot	*Pythium graminicola*	*T. viride*	Soil treatment
Sugarcane	Grey mould	*B. cinerea*	*T. harzianum*	Foliar spray
Tomato	Root knot	*Meloidogyne incognita*	*T. harzianum*	Soil Treatment
Wheat	Karnal bunt	*Neovossia indica*	*T. viride, T. hamatum*	Seed Treatment

Source: Singh *et al.*, 2004.

Mechanisms Involved in Biocontrol of Fungal Phytopathogens by *Trichoderma* sp.

Trichoderma sp. uses a variety of mechanisms to provide protection against several plant pathogens and/or plant diseases and enhance plant growth, such as it may (*i*) directly kill the pathogen by mycoparasitism and/or antibiosis, (*ii*) adversely affect the growth and development of the pathogen by competing for the nutrients, oxygen or space (*iii*) alter fitness of the pathogen, (*iv*) induce systemic plant resistance, (*v*) enhance plant growth and its tolerance to stress, (*vi*) metabolize plant exudates supporting pathogen, and/or (*vii*) inactivate enzymes produced by the pathogens and (*viii*) synthesize cell wall degrading enzymes (lytic enzymes) that degrade the cell wall of pathogen.

1. Mycoparasitism/Hyperparasitism

As an antagonist, *Trichoderma* may directly kill the pathogen either by antibiosis or by mycoparasitism. The simplest technique to test the antagonism between a biocontrol agent and fungal plant pathogen is dual culture method. *Mycoparasitism* is a complex process that involves tropic growth of biocontrol agent towards the target (pathogen) organism and finally attack and dissolution of the pathogen's cell wall by the activity of various enzymes, which may be associated with physical penetration of cell wall. Thus, in this process, the antagonist exists in intimate association with the other target fungi from which it derives some or all its nutrients. *Trichoderma* sp. parasitizes a range of other fungi and the events leading to mycoparasitism of fungal plant pathogens are complex, and take place as follows: first, *Trichoderma* strains detect other fungi and grow tropically towards them; remote sensing is at least partially due to the sequential expression of cell-wall-degrading enzymes (Chet, 1987). Different strains can follow different patterns of induction, but the fungi apparently always produce low levels of an extra-cellular exo-chitinase. Diffusion of this enzyme catalyses the release of cell-wall oligomers from target fungi, and this in turn induces the expression of fungitoxic endochitinases (Brunner *et al.*, 2003), which also diffuse and begin the attack on the target fungus before contact is actually made (Zeilinger *et al.*, 1999).Once the fungi come into contact, *Trichoderma* sp. attaches to the host and can coil around it. Attachment of *Trichoderma to* target fungus may involve *haustoria* formation, which help in penetration of target fungus and nutrient absorption or lectin mediated coiling for attachment of *Trichoderma* hyphae or both. Attachment is followed by a series of degenerative events which promote osmotic imbalances triggering cell disruption; host cell disruption may involve one or more of the following fungal cell wall degrading (hydrolytic) enzymes (CWDEs): cellulases, glucanases, chitinases, proteases and/or xylanases and probably also peptaibol antibiotics (Schirmböck *et al.*, 1994)). The combined activities of these compounds result in parasitism of the target fungus and dissolution of the cell walls (Figure 6.2).

There are at least 20–30 known genes, proteins and other metabolites that are directly involved in this interaction, which is typical of the complex systems used by these fungi in their interactions with other organisms. From very beginning mycoparasitism have been ascribed as major mechanism involved in disease suppression by *Trichoderma* (Elad, 1995; Howell, 2003). *Trichoderma harzianum* hyphae may form loops, coil around, penetrate and colonize second stage larvae of the root knot nematode, *Meloidogyne*. Some of the strains may also colonize egg masses (Sharon *et al.*, 2001). *T. harzianum* exhibits excellent mycoparasitic activity against *Rhizoctonia solani* hyphae (Figure 6.3) whereas *T. virens* relies more on antibiosis against hyphae of this pathogen. Nature of antagonism not only depends on antagonist but also on target pathogen. Numerous separate genes and gene products have been proposed to be involved in mycoparasitism by *Trichoderma* species.

2. Antibiosis

Antibiosis is the second major mechanism for biocontrol strategy employed by *Trichoderma* in the biocontrol of several fungal phyto-pathogens. Antibiosis occurs during interactions involving low-molecular- weight diffusible compounds or antibiotics produced by *Trichoderma* strains that inhibit the growth of other microorganisms. It releases more than 43 substances that have antibiotic activity. Most *Trichoderma* strains produce volatile and nonvolatile toxic metabolites that impede colonization by antagonized microorganisms. Weindling (1934) characterized the 'lethal principal' excreted by a strain of *T. lignorum* into the medium as "gliotoxin" and demonstrated that it was toxic to both *Rhizoctonia solani* and *Sclerotinia americana.* Mutation studies with *Trichoderma* strains have revealed that mutants deficient for antibiotic production often lack the ability to control *Pythium* damping off disease. At present *Trichoderma* species are reported to produce a number of antibiotics, of these, following have frequently been associated with biocontrol activity:

1. Gliotoxin, gliovirin and glioviridin from *T. virens,*
2. Viridin, alkyl pyrones, isonitriles, polyketides, peptaibols, diketopiperazines, sesquiterpenes and some steroids from other *Trichoderma* species.

The mutants deficient in gliovirin production lost the capacity to control damping-off of cotton while mutants with enhanced production of gliovirin were no more effective than wild type in

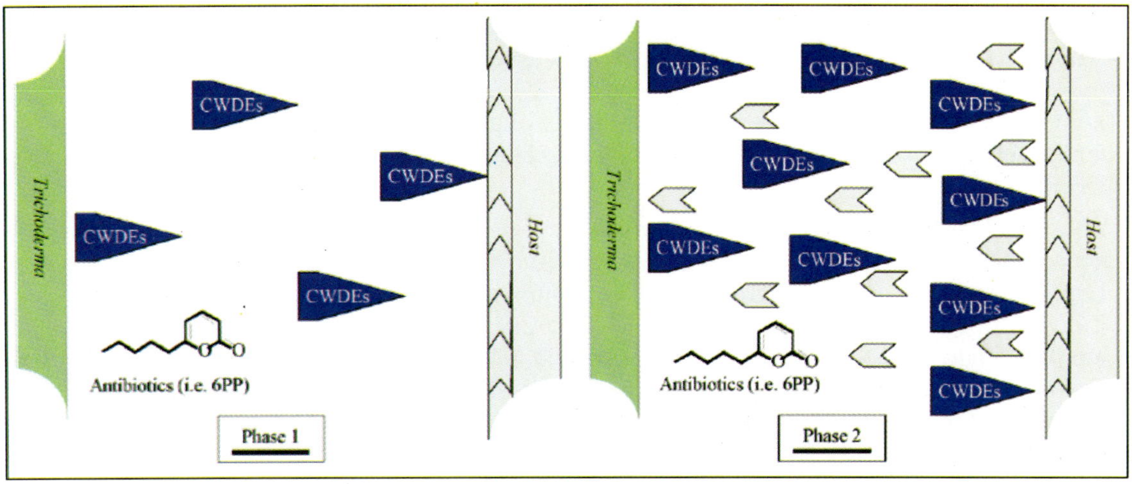

Figure 6.2: Phase 1: the mycoparasite produces high molecular weight compounds that reach the host. Phase 2: low molecular weight-degradation products that are released from the host cell walls reach the mycoparasite and activate the mycoparasitic gene expression cascade (Schirmbock *et al.,* 1994)

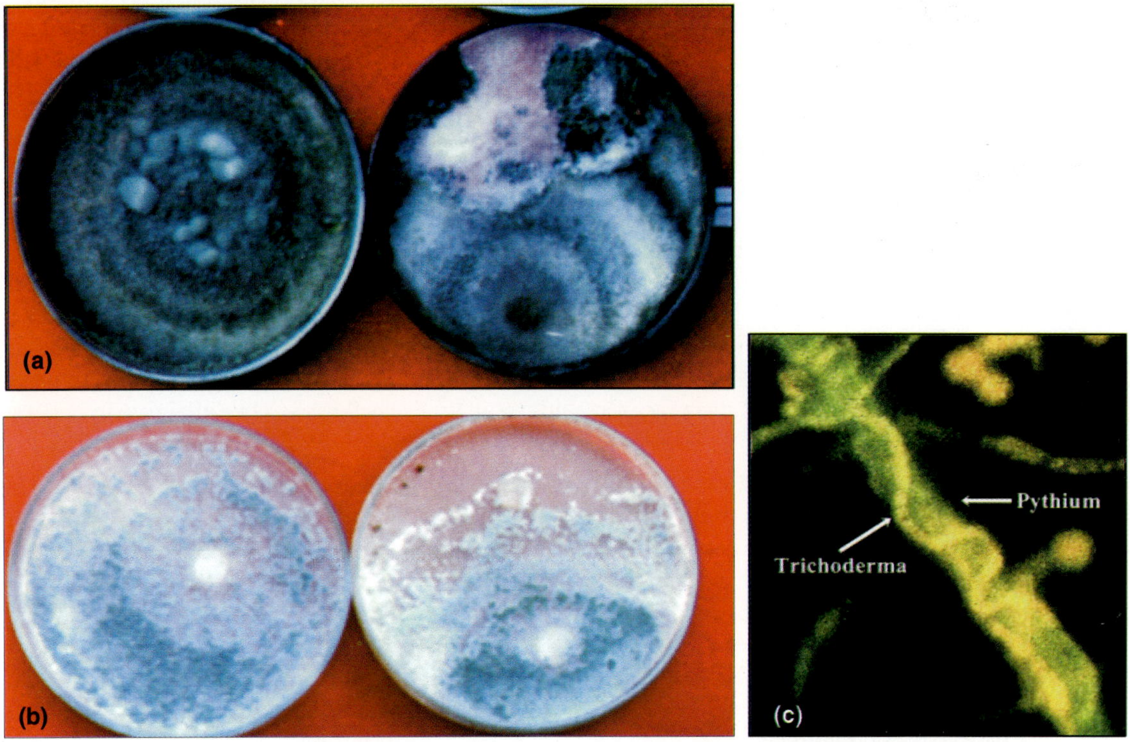

Figure 6.3: Mycoparasitism and Inhibition of Growth of Pathogen by *Trichoderma* sp. (a) Colonization of *Fusarium oxysporum* on dual culture plate, (b) Colonization of *Rhizoctonia solani* on dual culture plate and (c) Coiling of *Trichoderma* hyphae around the pathogen (*Pythium* sp.)

controlling the disease (Howell, 2003). The production of secondary metabolites by *Trichoderma* sp. is strain dependent and includes antifungal substances belonging to a variety of chemical compounds (Figure 6.4). They have been classified into three categories: (*i*) volatile antibiotics, *i.e.* 6-pentyl-α-pyrone (6PP) and most of the isocyanide derivates; (*ii*) water-soluble compounds, *i.e.* heptelidic acid or koningic acid; (*iii*) peptaibols, which are linear oligopeptides of 12–22 amino acids rich in α-aminoisobutyric acid, N-acetylated at the N-terminus and containing an amino alcohol at the C-terminus.

The chemical structures of *Trichoderma* antibiotics may suggest two different mechanisms of action. The production of low molecular weight, non-polar, volatile compounds (*i.e.* 6PP) results in a high concentration of antibiotics in the soil environment that have a relatively long distance range of influence on the microbial community. On the contrary, a short distance effect may be due to the polar antibiotics and peptaibols acting in close proximity to the producing hyphae. Although the role and the effects of peptaibols are clear, the mode of action of other *Trichoderma* secondary metabolites (*i.e.* pyrones), and their possible synergisms with other compounds have not yet been elucidated (Howell, 1998).

A unique class of linear hydrophobic polypeptides called *peptaibols* is produced by most species of *Trichoderma*. Peptaibols form a class of antibiotics known for their high α-aminoisobutyric acid form and their synthesis as a mixture of isoforms ranging from 7 to 18 amino acids in length. Many peptaibols, such as trichorzianins, trichokindins, trichorzins, trichorozins, harzianins, hypomuricins, trikoningins, trichokonins, trichogins etc. exhibit a broad range of bioactivities related to cell membrane perturbations (Wiest *et al.*, 2002). They are thought to act on the membrane of the target fungus to inhibit membrane-associated enzymes involved in cell wall synthesis. Peptaibols have been shown to act synergistically with cell wall degrading enzymes to inhibit growth of fungal pathogens.

3. Competition for Nutrients or Space

Competition is considered as a *'classical' mechanism* of biological control that involves competition between antagonist and plant pathogen for space and nutrients (Chet, 1987). It seems to be an important mechanism of biocontrol, but it is difficult to assess its actual contribution in biological control. The omnipresence of *Trichoderma* in agricultural and natural soils throughout the world proves that it is an excellent competitor for space and nutritional resources. However, the competitive ability and therefore its biocontrol potential are influenced by soil characteristics. *Trichoderma* when applied as seed treatment for the management of seedling and root diseases in chickpea, french-bean and tomato, is more effective than *Pseudomonas fluorescens* in acidic soil, whereas later is more effective in neutral and alkaline soil.

Neither antibiosis nor mycoparasitism is mainly involved in biocontrol of seedling disease in cotton, but competition is the main mechanism in this case. Starvation is the most common cause of death for microorganisms, so that competition for limiting nutrients results in biological control of fungal phytopathogens. For instance, in most filamentous fungi, iron uptake is essential for viability, and under iron starvation, most fungi excrete low-molecular-weight ferric-iron specific chelators, termed siderophores, to mobilize environmental iron. A biocontrol agent cannot compete for space and nutrients, if it is unable to colonize rhizosphere, a characteristic termed as *rhizosphere competence*. *Trichoderma* sp, either added to soil or applied as seed treatment, grows readily along with the developing root system of the treated plant (Harman, 2000). However, it may not be among the principal mechanisms of biocontrol. The strains of *T. koningii* are excellent root colonizers but still exhibit little or no biocontrol activity against *Rhizoctonia solani* on cotton seedlings. Nevertheless, rhizosphere competence is valuable for the effective biocontrol strains.

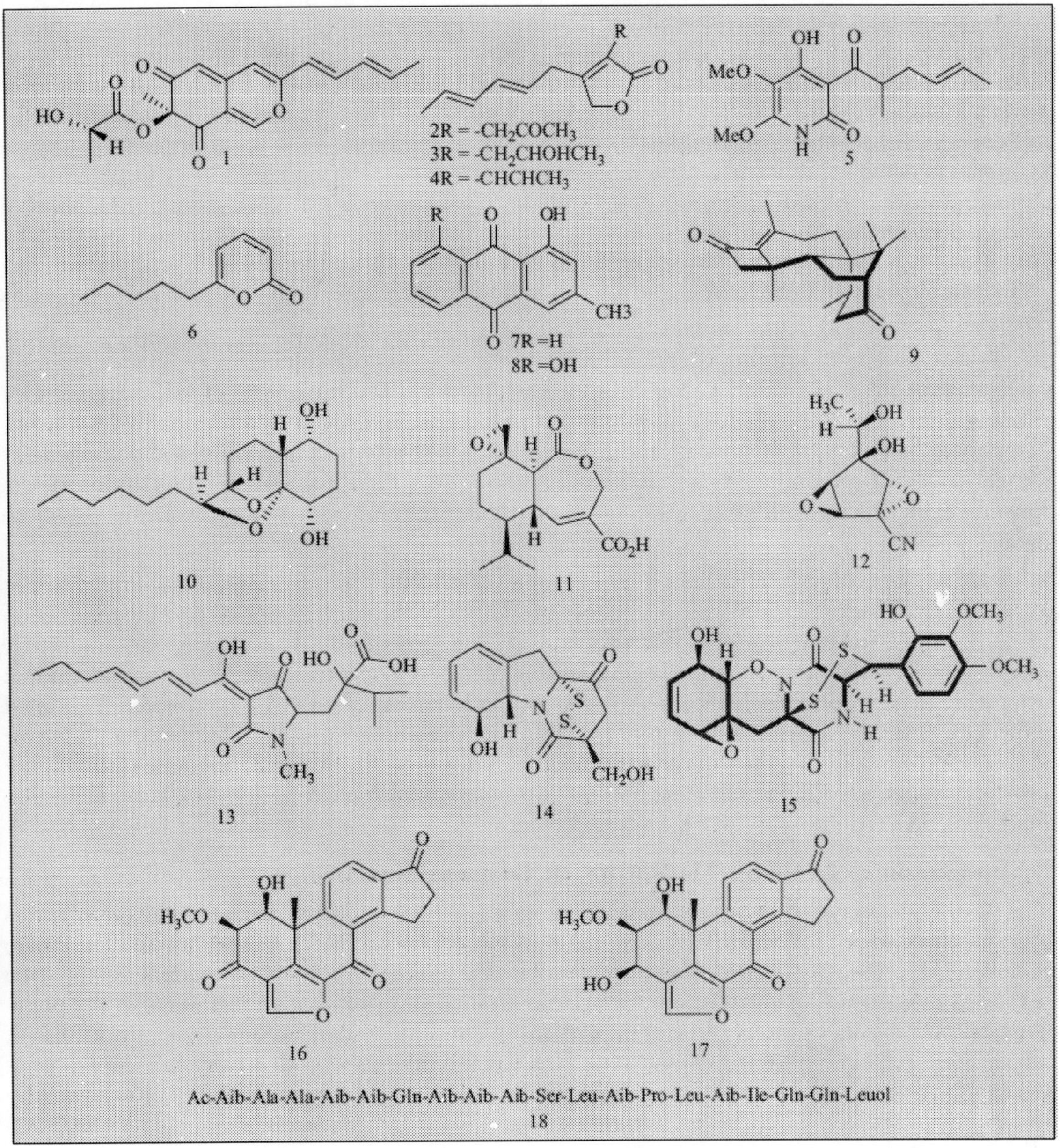

Figure 6.4: Chemical Structures of Secondary Metabolites of *Trichoderma* sp.

1: T22 azaphilone; 2: butenolide; 3: harzianolide; 4: dehydro harzianolide; 5: harzianopyridone; 6: 6-pentyl-a-pyrone; 7: 1-hydroxy-3- methyl-anthraquinone; 8: 1,8-dihydroxy-3-methyl-anthraquinone; 9: harziandione; 10: koninginin A; 11: heptelidic acid; 12: trichoviridin; 13: harzianic acid; 14: gliotoxin; 15: gliovirin; 16: viridin; 17: viridiol; 18: trichorzianines (Source: Wiest *et al.,* 2002).

4. Induction of Systemic Resistance (ISR) in Plants

Trichoderma sp. also have the ability to induce systemic and localized resistance in many plants and this acts as an indirect biocontrol mechanism. The induction of plant defence responses mediated by the antagonistic fungus has been well documented (Hanson and Howell, 2004). Most examples in Table 6.2 show systemic resistance because disease control occurs at a site that is distant from the location of *Trichoderma*. Various plants, mono- and dicotyledonous species, showed increased resistance to pathogen attack when pre-treated with *Trichoderma* (Harman *et al.*, 2004). There are three general pathways of induced resistance in plants. Two pathways involve direct production of pathogenesis related (PR) proteins; in one pathway PR protein production occurs due to attack by pathogenic microorganism while in other pathway it is the result of wounding or necrosis inducing plant pathogens and these pathways involve signaling molecules, salicylic and jasmonic acids respectively.

The PR proteins include antifungal chitinase, glucanases, and thaumatins; and oxidative enzymes such as peroxidase, polyphenol oxidase and lipoxygenases. The third type of ISR is induced by rhizobacteria (RISR) that is phenotypically similar but functionally different from above two pathways. Plant colonization by *Trichoderma* sp. reduces disease caused by one or more different pathogens, at the site of inoculation (induced localized acquired resistance, ILAR), as well as when the biocontrol fungus was inoculated at different times or sites than that of the pathogen (induced systemic resistance or ISR).

The induction of plant resistance by colonization with some *Trichoderma* species is similar to that elicited by rhizobacteria (RISR), which enhance the defense system but do not involve the production of pathogenesis-related proteins (PR proteins) (Harman *et al.*, 2004). Recently, the molecular mechanisms of resistance induction at gene level in plants by *Trichoderma hamatum* 382 was studied using a high-density oligonucleotide microarray approach. *Trichoderma*-induced genes were associated with biotic or abiotic stresses, as well as RNA, DNA, and protein metabolism. The genes codifying for extensin and extensin-like proteins were found to be induced by the BCA, but not those codifying for proteins belonging to PR-5 family (thaumatin-like proteins), which are considered the main molecular markers of SAR (Alfano *et al.*, 2007).

5. Biochemical Elicitors/Mediators of Disease Resistance

During interaction of *Trichoderma* with the plant, different classes of metabolites may act as elicitors or resistance inducers (Woo *et al.*, 2006; Woo and Lorito, 2007). In addition to their innate antimicrobial effect, their action may also stimulate the biological activity of resident antagonistic microbial populations or introduced *Trichoderma* strains, and promote an ISR effect in the plant. Trichoderma strains produce three main classes of compounds that induce resistance in plants, which are: (*i*) proteins with enzymatic activity (*ii*) avirulence-like gene products able to induce defence reactions in plants (*iii*) low-molecular-weight compounds released from fungal or plant cell walls by the activity of *Trichoderma* enzymes. Recently, Djonovic´ *et al.* (2006) identified a small protein (Sm1) elicitor secreted by *T. virens*, and demonstrated its involvement in the activation of plant defense mechanisms and the induction of systemic resistance.

1. *Proteins with enzymatic activity*: Xylanase from *Trichoderma* sp. is reported to be responsible for ISR in cotton, tobacco, grapevine, etc. A 22-KDa xylanase secreated by several Trichoderma sp. has been shown to induce ethylene production and plant defence responses. Recently, terpenoid phytoalexin biosynthesis and peroxidase activity in cotton was shown to be induced by series of proteins and peptides produced by strains of *Trichoderma virens*.

Table 6.2: Evidence for, and Effectiveness of, Induced Resistance in Plants by *Trichoderma* Species for Biocontrol of Pathogens and Protection of Plant from Diseases

Sp. and Strain	*Plant*	*Pathogens*	*Effects*	*Time after Application*	*Efficacy*
T. virens G-6, G-88 and G-11	Cotton	*Rhizoctonia solani*	Protection of plant by induction of fungitoxic terpenoid phytoalexin	4 days	78 per cent reduction in disease, ability to induce phyto-alexin required for maximum bio-control activity
T. harzianum T-39	Bean	*Colletotrichum* sp. *Botrytis cinerea*	Protection of leaves when T-39 was present on roots	10 days	42 per cent reduction in lesion area number of spreading lesions reduced
T. harzianum T-39	Tomato, Bean, Tobacco, Lettuce	*B. cinerea*	Protection of leaveswhen T-39 was present on roots	7 days	25-100 per cent reduction in grey-mould symptoms
T. aspergillum T-203	Cucumber	*Pseudomonas syringae*	Protection of leaves when T-39 was present on roots, production of anti-fungal compoundsin leaves	5 days	Upto 80 per cent reduction in disease on leaves, 100 per cent reduction in pathogennic bacterial cell density in leaves
T. harzianum T-22 *T. atroviridei*P1	Bean	*B. cinerea, Xanthomonas campestris*	Protection of leaves when T-22 or P1 was present on roots,production of anti-fungal compoundsin leaves	7-10 days	69 per cent reduction in grey mold (*B. cinerea*) symptoms withT-22; 54 per cent reduction inBacterial disease symptoms; Lower level of control with P1
T. harzianum T-22	Tomato	*Alternaria solani*	Protection of leaves when T-22 or P1 was present on roots,	3 months	Upto 80 per cent reduction in early blight symptoms from natural field infection
T. harzianum T-22,T-1 and *T. virens* T3	Cucumber	Green-mottle mosaic virus	Protection of leaves when *Trichoderma*-strains were presention roots	7 days	Disease-induced reduction in growth eliminated
T. harzianum	Pepper	*Phytophthora capsici*	Protection of stems when *Trichoderma*-strains were present only on roots, enhanced production of the phytoalexin capsidiol	9 days	40 per cent reduction In lesion length
T. harzianum NF-9	Rice	*Xanthomonas oryzae;* pv oryzae	Protection of leaves when *Trichoderma*-strain NF-9 was present only on roots	14 days	34-50 per cent reduction in disease

Source: Harman *et al*., 2004.

2. *Avr Homologues*: Avirulence (avr) -like gene products able to induce defence reactions in plants are synthesized by certain biocontrol agents. These function as race- or pathovar-specific elicitors that can induce hypersensitive responses and other defence related reactions in plant cultivars containing corresponding genes.
3. *Oligosaccharides and Low-Molecular Weight Compounds*: Some low molecular weight compounds and oligosaccharides produced during *Trichoderma*- fungal pathogen -host plant root interaction, function as inducers of antagonistic gene expression cascade in *Trichoderma* and some as elicitors of plant defence mechanism. These are generally released from fungal or plant cell walls by the activity of *Trichoderma* enzymes (Harman *et al.*, 2004). Some of the low-molecular-weight degradation products released from fungal cell walls were purified and characterized, and found to consist of short oligosaccharides comprised of two types of monomers, with and without an amino acid residue (Woo *et al.*, 2006). These compounds elicited a reaction in the plant when applied to leaves or when injected into root or leaf tissues. Further, they also stimulated the biocontrol ability of *Trichoderma* by activating the mycoparasitic gene expression cascade.

6. Inactivation of Pathogens' Enzymes

Some pathogens depend upon production of plant cell wall degrading enzymes to infect living plants. *Botrytis cinerea* depends upon production of pectinolytic, cutinolytic, and cellulolytic enzymes to infect living plants. Secondary metabolites or proteolytic enzymes produced by biocontrol agents may inactivate these pathogens' enzymes resulting in reduced ability of the pathogen to infect the host plant. Proteolytic activity of *T. viride* was claimed to be involved in biocontrol of *Sclerotium rolfsii* in autoclaved soil. Conidia of two strains of *T. harzianum* (T39 and NCIM 1185), when applied to the leaves, produce a serine protease that is capable of degrading the plant cell wall degrading enzymes produced by the pathogen, thereby reducing the ability of the pathogen to infect the plant. Proteases from *T. harzianum* may not only inactivate pathogen's primary modes of ingress into plants, but they might be directly toxic to germination of the pathogen propagules and may inactivate its enzymes (Elad and Kapat, 1999).

7. Synthesis of Fungal Cell Wall Degrading (Lytic) Enzymes

Attachment of *Trichoderma* to the host (pathogen) is followed by a series of degenerating events and degradation of cell wall of the pathogen by synthesizing various cell wall degrading enzymes (CWDEs), among them, chitinases, glucanases and proteinases are the major ones involved in antagonism.

Glucanases

Trichoderma strains generally produce β-1, 3 and β-1, 6-glucanases that hydrolyze the glucan polymer of the cell wall of the pathogen. Out of these, β-1,3-glucanases are the main glucanolytic enzymes that can act via two possible mechanisms identified by the products of hydrolysis: (i) Exo- or endwise- splitting, β-1,3 glucan glucanohydrolase (EC3.2.1.58) and (ii) Endo- or random splitting, β-1,3- glucan glucanohydrolase (EC 3.2.1.39). *T. harzianum* produces four β-1, 3-glucanases iso-enzymes under different *in vitro* culture conditions. Some *Trichoderma* strains also secrete β-1, 6-glucanases which are also involved in cell lysis, alongwith chitinase and proteinase activity.

Chitinases

These enzymes act on chitin, a major component of fungal cell wall. Chitin is an unbranched homopolymer of β-1, 4-linked N-acetyl-D-glucosamine (Glc NAc) and are also found in several plants, bacteria and invertebrates.Chitinolytic enzymes have been divided into three main categories: (*a*). β- 1,4 -N-acetyl glucosaminidase (*b*). Endo-chitinase and (*c*). Exo-chitinase.

Proteinases

Fungal cell wall contains chitin and/or β-glucan fibrils that are embedded into protein matrix. Thus, extra-cellular proteases, synthesized by *Trichoderma*, hydrolyze these proteins present in pathogens' cell walls and play a significant role in mycoparasitism. Proteases play an important role in biocontrol of *B. cinerea* by *T. harzianum*.

Synergism of Hydrolytic Enzymes: Extra-cellular hydrolytic enzymes of *Trichoderma, e.g. T. harzianum*, act synergistically as shown by *in vitro* studies. Chitinolytic enzymes and glucan β1,3-glucosidases from this organism not only act synergistically among themselves but also with chitinolytic enzymes from *Gliocladium virens* and with biocontrol bacterium *Enterobacter cloacae* (Lorito *et al.*, 1993). Zeilinger *et al.* (1999) reported that the regulation of individual genes presumably involved in mycoparasitism differs. They showed that 42-kDa endo-chitinase is induced before *T. harzianum* comes into contact with *Botrytis cinerea*, while 72-kDa N-acetyl-hexosaminidase is induced only after the two fungi are in direct contact. A single step in the mycoparasitic process of *T. harzianum* may involve more than 20 separate genes and gene products under complex regulatory control. Most of these gene products are synergistic with one another. Given this entire arsenal of synergistic gene products that are part of only one mechanism by which *Trichoderma* species attack and gain nutrition from other fungi, it is not surprising that this genus has been reported to be pathogenic against, and provide control of, very diverse groups of fungi (Harman, 2000).

Interaction of *Trichoderma* sp. with Plants

Trichoderma sp.is very well known for its ability to colonize roots and enhancing plant growth and productivity. Many plant diseases can also be controlled by the application of their conidial formulations to fruit, flower and foliage. They are highly interactive in root, soil and foliar environments. They are favoured by the presence of high levels of plant organic matter and plant roots, which they colonize readily. Some *Trichoderma* strains can colonize only local sites on roots, but rhizosphere competent strains colonize entire root surfaces for several weeks or months, In few cases, like mycoparasitism, their hyphae invade the root epidermis of host, but penetration of the root tissue unusually, limited to the first or second layers of cells. However, some strains, such as *Trichoderma stromaticum* are endophytic in the vascular system in cocoa. This outer root cell-invasion by *Trichoderma* strains can result in ISR and a systemic vascular colonizing strain might be even more effective. One of the most interesting factors is that although *Trichoderma* sp. and other root colonizing fungi infect roots, they are usually not plant pathogens. However, in rare cases, their particular strains are pathogenic to plants, but these strains limit their infection to superficial cells in plant roots. Moreover, they are also producers of certain enzymes including those that are necessary to degrade plant cell wall. Thus, *Trichoderma* sp. infect roots and have the intrinsic ability to be plant pathogen but limit their infection to superficial cells in plant roots, is a remarkable phenomenon. Strong interaction between *T. harzianum* (T-22) and the nitrogen fixing bacterium *Bradyrhizobium japonicum* has also been reported.

Effect of Root Colonization on Plant Metabolism

Several studies have shown that root colonization by *Trichoderma* strains results in increased levels of defense related plant enzymes such as peroxidases, chitinases, β-1,3- glucanases and lipoxygenase pathway hydroperoxide lyase enzymes. Change in plant metabolism can accumulate antimicrobial compounds, for example, cucumber root colonization by *Trichoderma aspergillum* strain 203 causes an increase in phenolic glucoside levels in leaves and their glycans (formed by removal of carbohydrate moieties) are strongly inhibitory to a range of bacteria and fungi. The ability of *T. virens* to induce phytoalexin production and localized resistance in cotton has already been discussed. *Trichoderma* strains not only produce antibiotic substances directly, they also strongly stimulate plants to produce their own antimicrobial compounds. Root colonization by these fungi, therefore induces significant changes in the plant metabolic machinery. These fungi strongly modify plant metabolism which in most cases benefits the plant.

Many resistance inducing fungi and bacteria increase both root and shoot growth. At least some non pathogenic root colonizing fungi such as some strains of *Trichoderma* and arbuscular *mycorrhiza* (*Glomus fasciculatum, G. mosseae,G. caledonicum, Gigaspora margarita etc.*) have also similar plant growth promotory activities and are considered as plant growth promoting rhizo–fungi. Although many *Trichoderma* strains are likely to have this ability, the greatest long term effects probably occur with rhizosphere- competent strains.

Direct Plant Growth Promotory Activities of *Trichoderma* sp.

Trichoderma sp. like other beneficial root colonizing microorganisms, also enhance plant growth and productivity. Responses to application of *Trichoderma* sp. are characterized by increased germination percentage, plant height and dry weight and a shorter germination time in vegetables. These responses may be due to one or more of the following mechanisms:

1. Vitamin production or conversion of materials to a form useful to the plants
2. Solubilization of nutrients and their release from the soil or organic matter
3. Production of growth stimulating factors (hormones or growth factors)
4. Suppression of deleterious root micro flora including those not causing obvious disease
5. Increased uptake and translocation of minerals
6. Synthesis of metal chelating agents – siderophores

Root colonization by rhizosphere competent strains of *Trichoderma* results in increased development of root and/or aerial systems and crop yields.

1. Solubilization and Sequestration of Plant Nutrients

In soil, both micro- and macro-nutrients undergo a complex dynamic equilibrium of soluble and insoluble forms which is greatly influenced by soil pH and microflora. Numerous soil microorganisms have the ability to mineralize and mobilize nutrients from soil, such as *Trichoderma* strongly influence the complex transitions of various plant nutrients from insoluble forms to soluble forms, thereby enhancing their accessibility and absorption by plant roots. During *in vitro* solubilization, the culture growth and pH of the broth are indirectly correlated (Saravanan *et al.*, 2007). Several species of *Trichoderma e.g. T. harzianum, T. virens, T. viride* and *T. atroviride* have been reported to solubilize various forms of inorganic plant nutrients and thus play an important role in nutrient management in soil. Three possible mechanisms for *in vitro* solubilization of some insoluble or sparingly soluble minerals by plant growth promoting and biocontrol fungus *T. harzianum* Rifai have been proposed, these are: (*a*) Acidification of the medium (*b*) Production of chelating metabolites, and (*c*) Redox activity (Altomare *et. al.*, 1999).

Phosphorus is one of the key macro-nutrients limiting plant growth and metabolism as approximately 95 to 99 per cent is present in the form of insoluble phosphates in soil and can not be utilized by plants. Moreover, a major portion (more than 80 per cent) of the phosphatic fertilizers added to soil becomes immobile and unavailable for plant uptake because of adsorption, precipitation or conversion to insoluble fixed inorganic form. It is generally fixed as tricalcium phosphate (TCP) in alkaline soil (at pH above 7.0); as ferric phosphate ($FePO_4$) and $AlPO_4$ in acidic soil (at pH d" 5.0), which therefore needs to be solubilized where phosphate solubilizing microorganisms play an important role. Phosphate solubilizing microorganisms have been shown to produce monocarboxylic acid (acetic, formic); monocarboxylic hydroxy acids (lactic, gluconic, glycolic); monocarboxylic keto acids (2-keto gluconic); dicarboxylic acid (oxalic, succinic); dicarboxylic hydroxy acids (malic, maleic) and tricarboxylic hydroxyl acids (citric) acids in liquid media from simple carbohydrates. Therefore, release of organic acids that sequester cations and acidify the micro-environment near root zone is thought to be a major mechanism of solubilization of nutrients such as phosphorus, manganese, iron and zinc by several phosphate solubilizing microorganisms (PSM).

Trichoderma sp. can solubilize and store phosphate in its biomass that is released in readily available form in close proximity of roots after lysis of the mycelium with age. *Trichoderma harzianum* is a good solubilizer of phosphorus but different strains show wide variation in their ability to solubilize phosphorus. *T. harzianum* T-22 has been shown to solubilize MnO_2, metallic zinc, and rock phosphate (mostly calcium phosphate) in a liquid sucrose-yeast extract medium. Iron and manganese in particular have been investigated with regard to both microbial solubilization of oxidized forms of these elements and their influence on plant disease (Altomare *et al.*, 1999).

2. Synthesis of Metal Chelating Agents-Siderophores

Siderophores (Gr. "iron- bearers") are defined as "low molecular weight, virtually ferric specific ligands, the biosynthesis of which is carefully regulated by iron and the function of which is to supply iron to the cell" (Neilands and Leong, 1986). The structural diversity among different siderophores is quite considerable and depends on the producing microorganism. However, a common feature of all siderophores is that they form six co-ordinate octahedral complexes with ferric ion. Several bacteria and fungi are known to produce these compounds that can specifically chelate a metal ion such as ferrous, zinc, aluminium etc. The role of siderophores in plant growth promotion and biological control is well established. Siderophores help in improving antagonistic activities, rhizosphere competence and plant growth. The amount of naturally utilizable iron, the fourth most abundant element of the earth crust, is very limited, as most of this metal exists in extremely insoluble complexes of ferric hydroxide in nature, and thus unavailable to organisms in soil solution. Siderophores chelate Fe^{3+} which is virtually insoluble in presence of O_2, and therefore remains unavailable for plant and microbial growth. The microbial membrane receptor proteins specifically recognize and take up the siderophore-Fe complex. Thus, iron is made unavailable to other rhizosphere microorganisms including plant pathogens, which produce less siderophores or different siderophores with lower binding coefficients. This also results in decreasing pathogen infection.

Trichoderma virens is reported to produce three types of hydroxymate siderophores:

1. Monohydroxymate (cis- and trans-fusarinines),
2. Dipeptide of trans fusarinine (dimerium acid), and
3. Trimer dis-depsi peptide (copragen).

Siderophore mechanisms will only be relevant under conditions of low iron availability. As soil pH decreases below 6, iron availability increases and siderophores become less effective. Manganese is a microelement essential for diverse physiological functions in plants, including both plant growth and disease resistance. Only Mn^{2+} form of this element is soluble, the more highly oxidized forms are insoluble. Some strains of *Trichoderma* could solubilize rock phosphate, Zn-metal, Mn^{4+}, Fe^{3+} and Cu^{2+} by producing a large number of chemicals. An increase in the micro-element content (*viz.* copper, phosphorus, iron, zinc, manganese, and sodium) of *Trichoderma* treated plants has been reported (Harman, 2000). Enhanced root development is also helpful in tolerating the biotic and abiotic stresses by plants. There was a significant increase in germination, seedling length and weight when rice seeds were treated with powdered formulations of *T. harzianum, T. virens and Aspergillus niger*. The enhanced rooting by *T. harzianum* T-22 induces tolerance to drought and enhanced nitrogen utilization. It also induces tolerance to *Phytophthora*, a pathogen it does not directly control (Harman, 2000).

3. Other Mechanisms for Plant Growth Promotion

Other mechanisms by which *Trichoderma* sp. directly promote plant growth, are not fully understood, but are thought to involve- the ability to produce or change concentration of the plant hormones like Indole acetic acid, gibberellic acid, cytokinins and ethylene- the mechanism shown by several other plant growth promotory microorganisms. The first report on discovery of gibberellin was made from the fungus *Gibberella fujikmoi.* Plants are influenced heavily by growth hormones. Auxins, gibberllins, abscissic acid and ethylene are some of the plant growth hormones which are produced in rhizosphere by certain plant growth promotory organisms. Indole acetic acid (IAA) is produced by *Pseudomonas fluoroscens*. Also, an increase of IAA, gibberellin and cytokinin level was observed in *G. fasciculatum* inoculated *Prosopis juliflora.*However, these mechanisms are not well studied in case of *Trichoderma*.

Metabolism of Germination Stimulants

One unique mechanism employed by *Trichoderma* species to effect biological control that does not fit nearly into any of the categories previously mentioned was recently discovered. Howell (2003) recently implicated this mechanism in the biological control of pre-emergence damping off of cotton seedlings incited by *Pythium ultimum* and/or *Rhizopus oryzae*. He reported that control by *T. virens* strains or protoplast fusants of *T. virens*/*T. longibrachiatum* was due to metabolism of germination stimulants released by the cotton seed. Germination stimulants are the compounds which normally induce pathogen propagules to germinate. Disease control could be affected by wild-type strains/ mutant strains deficient for mycoparasitism/antibiotic production/induction of terpenoid synthesis. However, he observed that if pathogen propagules were induced to germinate by artificial means none of the above treatments gave effective control.

Trichoderma as a Major Component of Integrated Disease Management System

An integrated pest management (IPM) system entails simultaneous or sequential use of several methods for pest and disease control. Biological control is of particular interest as a component of IPM and can best be exploited within the framework of this management system. Biocontrol agents have distinct advantages in being compatible with most of the agricultural practices and hence, can be successfully utilized as a part of total crop management practices or broadly, as a component of the agro-ecosystem management. Different biocontrol agents have been integrated with cultural practices, soil solarization, fungicides and disease resistant varieties for managing different crop diseases.

Combination of the seed/root application of *T. harzianum* or *P. fluorescens* with soil solarization was very effective in management of seed and seedling diseases of tomato, brinjal and capsicum in nursery at farmers' field. Wilt and root-rot complex of chickpea, lentil and pigeon pea were successfully managed by integration of *T. harzianum* or *T. virens* with carboxin. Integration of fertilizers or herbicides with biocontrol agents to control plant diseases has also been attempted. Integrating biocontrol agents with reduced doses of fungicides seems to be a very promising way of controlling pathogens with minimal interference with biological equilibrium. This would not only reduce the use of fungicides but also improve the efficacy of a biocontrol system with reduced cost and lessen the chances of development of fungicide resistant strains of the pathogen. Integration of biocontrol agents and fungicides for seed treatment is very effective even against high population of fast growing pathogens like *Rhizoctonia solani* in soil. Under such conditions biocontrol agent alone may not be very effective as by the time it gets activated in soil, pathogen is able to penetrate the host. Integration of fungicide with biocontrol agents helps in early protection by fungicide and later protection by biocontrol agent. Fungicide also provide congenial environment to insensitive biocontrol agent to multiply and colonize spermosphere and rhizosphere by suppressing other microbes sensitive to the fungicide.

Conclusion and Future Prospects

Although, it is fact that use of biocontrol for disease management may not totally replace chemicals in near future, but, judicious use of biocontrol agents, such as *Trichoderma* sp., can significantly reduce our dependence on chemical pesticides and thereby contribute to sustainable development of agriculture. Not only this, the biocontrol agents by replacing noxious carcinogenic chemical pesticides and also being more eco-friendly in nature, help in clean up environment. Use of biocontrol agent as a component of IPM, both under normal and organic cultivation will have additive effects. In addition to biocontrol and growth promoting effects, *Trichoderma* sp. plays an important role in decomposition of compost. It is also reported to suppress *Phalaris minor*, an important weed of wheat crop, in farmers' fields if applied through compost. Therefore, there is need to promote its application as much as possible.

In India, biocontrol research has got well established, yet, it has to become an integral part of protection technology. There are some areas which require some further research and breakthroughs:

- ☆ Improvement of efficacy of biocontrol strains by using powerful molecular tools.
- ☆ Integration of biocontrol with other management practices under field conditions
- ☆ Improvement in mass multiplication techniques and formulation development
- ☆ Enhancement of shelf life of formulations and use of efficient delivery system(s) for their application in fieids as bio-pesticides
- ☆ Selection of indigenous strains having high antagonistic as well as plant growth promoting and plant defence inducing properties.

References

Alfano, G., Lewis Ivey, M. L., Cakir, C., Bos, J. I. B., Miller, S. A., Madden, L. V., Kamoun, S. and Hoitink, H. A. J. 2007. Systemic modulation of gene expression in tomato by *Trichoderma hamatum* 382, Phytopathology, 97: 429–437.

Altomare, C., Norvell, W. A., Bjorkman, T. and Harman, G. E. 1999. Solubilization of phosphates and micronutrients by the plant growth promoting biocontrol fungus *Trichoderma harzianum* Rifai 1295-22. Appl. Environ. Microbiol. 65: 2926-2933.

Baker, R., Elad, Y. and Chet, I. 1984. The controlled experiment in the scientific method with special emphasis on biological control. Phytopathology. 74 1019-1021.

Brunner, K., Peterbauer, C. K., Mach, R. L., Lorito, M., Zeilinger, S. and Kubicek, C. P. 2003.The Nag1 N-acetylglucosaminidase of *Trichoderma atroviride* is essential for chitinase induction by chitin and of major relevance to biocontrol. *Current Genetics* **43:** 289-295

Chet, I. 1987. *Trichoderma*: Applilcation, mode of action and potential as a bioconrol agent of soil borne plant pathogenic fungi. *In:* Innovative Approaches to Plant Disease Control. (ed, I.Chet,). John Wiley and Sons, New York, pp137-160.

Cook, R. J. and Baker, K. F. 1983. The nature and practices of biological control of plant pathogens. APS Books, St. Paul. MN, U.S.A., pp 599.

Djonovic, S., Pozo, M. J., Dangott, L. J., Howell, C. R. and Kenerley, C. M. 2006. Sm1, a proteinaceous elicitor secreted by the biocontrol fungus *Trichoderma virens* induces plant defense responses and systemic resistance. Mol. Plant–Microbe Interact., 19: 838–853.

Elad, Y. 1995. Mycoparasitism. In: *Pathogenesis and Host Specificity in plant diseases: Histopathological, Biochemical, Genetic and Molecular Basis*, Eukaryotes, Vol. II. Pergamon, Elsevier, Oxford, UK, pp. 289-307.

Elad, Y. and Kapat, A. 1999. Role of *Trichoderma harzianum* protease in the biocontrol of *Botrytis cineria*. Eur. J. Plant Pathol. 105: 177-189.

Hanson, L. E. and Howell, C. R. 2004. Elicitors of Plant Defense Responses from Biocontrol Strains of *Trichoderma virens*. Phytopathology, 94 (2): 171-176.

Harman, G. E.1991. Seed treatments for biological control of plant disease, *Crop Pro.* 10: 166- 171

Harman, G. E. 2000. Myths and dogmas of Biocontrol: Changes in the perceptions derived from research on *Trichoderma harzianum* T-22. Plant Dis. 84: 377-393.

Harman, G. E. 2002. *Trichoderma* sp. including, *T. harzianum, T.viride, T.koningii, T. hamatum* and other spp. Deurteromycetes, Moniliales (asexual classification system). Biological Control. Cornell University, Geneva, NY. 144-56.

Harman, G. E., Howell, C. R., Viterbo, A., Cget, I. and Lorito, M. 2004. *Trichoderma* sp. opportunistic, avirulent plant symbionts. Nature Reviews Microbiology. 2 (1): 43-56.

Hayes, C. K., Klemsdal, S., Lorito, M., Di Pietro, A., Peterbauer, C., Nakasa, J. P., Tronsmo, A. and Harman, G. E. 1994. Isolation and sequence of an endochitinase –encoding gene from a cDNA library of *Trichoderma harzianum*. Gene 138: 143-148.

Hermosa, M. R., Grondona, I., Iturriaga, E. A., Diaz-Minguez, J. M., Castro, C., Monte, E. and Garcia-Acha, I. 2000. Applied and Environmental Microbiology. 66: 1890-1898.

Howell, C. R. 1998. The role of antibiosis in biocontrol. *In: Trichoderma* and *Gliocladium*, Vol. 2, (eds, G. E. Harman and C.P. Kubicek,). Taylor and Francis, London. pp 173-184

Howell, C. R. 2003. Mechanism employed by *Trichoderma* species in the biological control of plant diseases: The history and evolution of cureent concepts. Plant Disease 87:4-1

Lorito, M., Harman, G. E., Hayes, C. K., Broadway, R. M., Tronsmo, A., Woo SL and Di Pietro, A. 1993. Chitinolytic enzymes produced by *Trichoderma harzianum*: Antifungal activity of purified endochitinase and chitobiosidase. Phytopathology 83:302-307.

Neilands, J. B. and Leong, J. 1986. Sideropohores in relation to plant growth and diseases. *Annu. Rev.* Plant Physiol. 37: 187-208.

Rifai, M. A. 1969. A revision of genus *Trichoderma.* Mycological Papers 116:1-56

Saravanan, R., Pavani, V., Devi, Shanmugam, A., and Sathish, Kumar, D. 2007. *Isolation and Partial Purification of Extracellular Enzyme (1, 3)- D Glucanase from Trichoderma reesei (3929).* Biotechnology, 6: 86-98.

Schirmbock, M., Lorito, M., Wang, Y.L., Hayes, C.K., Artisan-Atac, I., Scala, F., Harman, G.E., and Kubicek, C.P. 1994. Parallel formation and synergism of hydrolic enzymes and peptaibol antibiotics, molecular mechanisms involved in the antagonistic action of *Trichoderma harzianum* against phytopathogenic fungi. Appl. Environ. Microbiol. 60:4364-4370.

Sharon, E., Bar-Eyal, M., Chet, I., Herrera-Estrella, A., Kleifeld, O., and Spiegel, Y. 2001. Biological control of the root knot nematode *Meloidogyne javanica* by *Trichoderma harzianum*. Phytopathology 91: 687-693.

Singh, U.S., Zaidi, N. W., Joshi, D., Khan, D., John, D. and Bajpai, A. 2004. *Trichoderma*: A microbe with multifaceted activity. Annu. Rev. Plant Pathol. 3: 33-75.

Tewari, L. and Singh, R. 2005. Biological control of sheath blight of rice by *Trichoderma harzianum* using different delivery systems. Indian Phytopath. 58 (1): 35-40.

Weindling, R. 1932. *Trichoderma lignorum* as a parasite of other soil fungi. Phytopathology 22: 834-845.

Weindling, R. 1934. Studies on a lethal principle effective in the parasitic action of *Trichoderma lignorum* on *Rhizoctonia solani* and other soil fungi. Phytopathology 24: 1153-1169.

Wiest, A., Grezagorski, D., Xu, B., Goulard, C., Rebuffat, S., Ebbole, D. J., Bodo, B., and Kenerly, C. 2002. Identification of peptibols from *Gliocladium virens* and cloning of a peptibol synthetase. J. Biol. Chem. 277 (23): 2086-2088.

Woo, S. L. and Lorito, M. 2007. Exploiting the interactions between fungal antagonists, pathogens and the plant for biocontrol. In: Vurro, M., Gressel, J. (Eds.), Novel Biotechnologies for Biocontrol Agent Enhancement and Management. IOS, Springer Press, Amsterdam, theNetherlands, pp. 107–130.

Woo, S.L., Scala, F., Ruocco, M. and Lorito, M. 2006. The molecular biology *of* the interactions between *Trichoderma sp., phytopathogenic* fungi, and plants, Phytopathology **96**: 181–185.

Zeilinger, S., Galhaup, C., Payer, K. L., Woo, S., Mach, R., Fekete-Csaba, L., Lorito, M. and Kubicek, C. P. 1999. Chitinase gene expression during mycoparasitic interaction of *Trichoderma harzianum* with its host. Fungal Genet. Biol. 26: 131-140.

Microbes: Diversity and Biotechnology (2012) *Pages* 119–135
Editors: Prof. S.C. Sati & Dr. M. Belwal
Published by: DAYA PUBLISHING HOUSE, NEW DELHI

Chapter 7

Arbuscular Mycorrhizal Fungi as Symbiotic Bioengineers

Meenakshi Mishra[1], Pradeep Kumar Singh[2] and Deepak Vyas[1]
[1]*Lab of Microbial Technology and Plant Pathology, Department of Botany, Dr. H.S. Gour University, Sagar – 470 003, M.P.*
[2]*Goverment Naveen College, Janakpur, Korea, C.G.*

ABSTRACT

Soils is not a pile of dirt but are treasure house of earthy material and home of many micro-organisms as it was rightly said that "we think that we are standing on the earth but reality is that we are standing on the roof of another world". Rhizospheric region of the soil is one of the important rooms where plant, soil and microbes interact. However, these interactions are of many types but among these, mutulisitic interactions are of great importance. An Arbuscular mycorrhizal symbiosis is one of the mutualistic relationships among the plant, soil and other microbes. In this review we have tried to summerize some of the multifacet role performed by the AMF fungi in the soil for the benefit and better plant growth, soil remediation, dialogue among the co-organisms and their importance in the modern biology. The use of microbial inoculates must take into account the importance of retaining microbial diversity in rhizosphere ecology, and in achieving realistic and effective biotechnological applications ('rhizosphere technology'). The improvement of molecular biology – based approaches will be fundamental for analyzing microbial diversity and community structure, add to predict responses to microbial inoculation/processes in the environment (ecological engineering).

Keywords: Arbuscular mycorrhizal, Rhizosphere ecology, Bioengineers.

Introduction

AM symbioses are the most widespread mycorrhizal association, with a very long evolutionary history. Fossil survey and molecular data have demonstrated that the evolutionary history of AM fungi goes back at least to the Ordovician, about 460 million years ago (Redecker, 2002). The earliest land plants show an association with fungi that formed vesicles and arbuscules that was very similar to today's AM fungi (Nicolson, 1975; Remy *et al.*, 1994). Recently Vyas *et al.* (2007-2008) reported occurrence of arbuscules and vesicles in bryophytes. About 83 per cent of dicotyledonous and 79 per cent of monocotyledonous plants is associated with mycorrhizal fungi (Wilcox, 1991). Mycorrhizal associations are found in a very wide range of habitats, including aquatic ecosystems, deserts, lowland tropical rain forests, high altitudes, high latitudes and in canopy epiphytes (Allen, 1991) Bi-directional movement of nutrients characterizes the fungus-plant symbioses, where: carbon flows to the fungus and inorganic nutrients move to the plant, thereby providing a critical linkage between the plant root and soil. This association is a survival mechanism for both the fungi and plants, allowing each to survive in different environments (Gupta *et al.*, 2000). Mycorrhizal plants, in comparison with non mycorrhizal plants, have greater nutrient uptake because they possess a network of external hyphae (Sanders and Sheikh, 1983). The hyphae are the interface between soil and plant and have a large surface area that acts as an extension of the root absorbing area (Rhodes and Gerdemann, 1975; Owusu-Bennoah and Wild, 1979; Li *et al.*, 1991). This not only increases the volume of soil from which nutrients are absorbed, but also overcomes problem of depletion of nutrients (Nurlaeny *et al.*, 1996; Smith and Read, 1997) and water (Marulanda *et al.*, 2003) depletion close to actively absorbing roots, and plays a significant role in stabilizing soil structure. Mycorrhizal fungal associations have several advantages for their hosts, including increased growth and yield and reproductive success due to enhanced nutrient acquisition (Diederichs, 1990; Lewis and Koide, 1990; Stanley *et al.*, 1993). They may also increase disease and pest resistance; improve water relations (Allen and Allen, 1986; Davies *et al.*, 1993; Subramanian *et al.*, 1997), soil structure (Tisdall and Oades, 1979; Thomas *et al.*, 1986; Degens *et al.*, 1994; Beaden and Petersen, 2000) and tolerance of extreme pH (Sidhu and Behl, 1997; Douds *et al.*, 2000). This literature review will concentrate on structure and function of AM and their potential role in revegetating saline soils.

AMF in Carbon Sequestration

Arbuscular mycorrhizal fungi (AMF) comprising fungi in order Glomales (Zygomycota), ubiquito symbionts of the majority of higher plants. AMF have been shown to have numerous effect on plant physiology and plant communities (Allen 1991; Smith and Read, 1997), which can lead to indirect effect on soil carbon storage. AMF are also very important in the process of soil aggregate stabilization. Relatively labile carbon can be protected inside soil aggregates us which means AMF have yet another indirect influence on soil carbon storage (Miller and Jastrow, 1992).

Recently, a glycoprotein named glomalin was discovered using specialized protocols for soil that revealed amount up to several mg proteins per gram of soil (Wright and Upadhyaya, 1996). Glomalin is produced by hyphae of all members of AM genera, but not by other groups of soil fungi so for tested. Although concentration of glomalin in soil seem to be responsive to global change factors such as elevated atmospheric CO_2 (Rillig *et al.*, 1999).

(a) ERH (Extra Radical Hyphae) and Soil aggregation

Indeed the influences of ERH on soil aggregation might be even more important to the carbon stock than the influence of the hyphal standing crop along (Miller *et al.*, 2000). Through their role in soil microaggregate stabilization, the ERH appear to contribute to the certain of an aggregate hierarchy. In doing so they help to create mechanism for increasing the residence time of organic debris within soil macroaggregates.

Mechanistically, the ERH contribution to soil aggregation can be viewed as 'Stioky string bag' mechanism, in which the hyphae help to etangle and enmesh soil particles to form macroaggregate structures (Miller *et al.*, 2000). The physical dimensions of ERH allow them to grow and to remify through soil pores the size of those between macroaggregates. The contribution of AMF hyphae to carbon cycling lies not only with ERH but also, with exudates from hyphae. Arbuscular mycorrhizal (AM) fungi take up photosynthetically fixed carbon from plant roots and translocate it to their external myceliumAMF are also responsible for directing the movement of huge quantities of photosynthate to the soil (for review, Douds *et al.*, 2000; Graham, 2000). Carbon in the root flows from plant to fungus in the form of sugars (Shachar-Hill *et al.*, 1995; Solaiman and Saito, 1997), and together with the transfer of mineral nutrients from fungus to root (Koide and Schreiner, 1992; George *et al.*, 1995; Jakobsen, 1995), this is the nutritional mainstay of what is arguably the world's most important mutualistic symbiosis. AM fungi obtain most or all of their carbon within the host root. Here, they acquire hexose and trans-form it into trehalose and glycogen, typical fungal carbohydrates (Shachar-Hill *et al.*, 1995). Triacylglyceride (TAG) is the main form of stored carbon in AM fungi (Beilbyand Kidby, *1980;* Jabaji-Hare, 1998), and this is mostly or exclusively made in the intraradical mycelium (IRM; Pfeffer *et al.*, 1999). Some of this storage lipid flows from the IRM to the extraradical mycelium (ERM; Pfeffer *et al.*, 1999), and in vivo microscopic observations indicate that the rate of export is sufficient to account for the high levels of stored lipid in the ERM (Bago *et al.*, 2002). The glyoxylate cycle is active in the ERM (Lammers *et al.*, 2001), and this pathway appears to be important in using exported TAG to make carbohydrate in the ERM.

The structure of glomalin has not been 'completely defined. The molecule appears to be a complex of a repeated monomeric structures bound together by hydrophobic interactions (Nichols, 2003) that, attaches to soil to help stabilize aggregates. The mole- cule contains tightly bound iron (0.04-8.8 per cent) (Rillig *et al.*, 2001; Nichols, 2003; Wright and Upadhyaya, 1998), and does not contain phenolic compounds such as tannins (Rillig *et al.*, 2001). Preliminary evidence (Wright and Upadhyaya, 1998; Nichols, 2003) suggests that cations are bound to glomalin in amounts that vary *for* different soils.

(*b*) Rhizodeposition

The release of carbon compounds from living plant roots rhizodeposition into the surrounding soil is a ubiquitous phenomenon (Curl and Trueglove, 1986). The loss of C from root epidermal and cortical cells leads to a proliferation of microorganisms within (endorhizosphere) and on the surface (rhizoplane) and outside the root ectorhizosphere. C release also result in the rhizosphere having different chemical physical and biological characteristics to the bulk soil. Theoretically, almost any soluble component present inside the root can be lost to the rhizosphere; however current evidence suggests that exudation is dominated by low molecular weight solutes such as sugars, amino acids and organic acids that are present in the cytoplasm at high concentrations (Farrar *et al.*, 2003). Influence of AM formation upon roots exudates following. AM colonization root exudation patterns may be expected *to* alter because the AM fungus is a considerable C sink (Douds *et al.*, 2000; Graham, 2000). AM colonization alters the carbohydrate metabolism of the roots (Shachar-Hill *et al.*, 1995; Douds *et al.*, 2000) and increases root respiration (Douds *et al.*, 2000; Shachar-Hill *et al.*, 1995). Hyphal exudation from the AMF will also occur.

AMF in Nutrient Squestration

Contact of Mycorrhizas with Substrate

Most mycorrhizas posses a considerable hyphal connexion with the soil as simple branching hyphae or as hyphal strands which ramify in the substrate.(Those which do not have an extensive hyphal system, may be smooth or posses short setae or spines,and will be shortly considered later). The extensive hyphae were at first viewed as simple extensions of the absorbing surface. It is clear that the concentration of any poorly mobile nutrient in the soil will fall in the environs of an absorbing surface, if its rate of absorption exceeds the rate at which it diffuses or moves towards the surface. It will continue to fall till the rate of uptake is equalled by its rate of replacement at the absorbing surface. In extreme the 'a zero sink' or deficiency zone will devlop about the absorbing surface, and then no physiological property of the living system can increase the rate of uptake, which is entirely dependent on the rate of movement of the substrate through the soil. The hyphae of mycorrhizal fungi serve to exploit a wider volume of soi outside the deficiency zone and translocate absorbed nutrients to root.

Subsequent work, especially in P.B. Tinker's laboratory has served to show that the rates of translocation especially of phosphate in the hyphae of vesicular arbuscular mycorrhizal fungi were great enough to accommodate the rate of inflow into the root which had been observed. Indeed the subject of translocation of inorganic substances and carbohydrates in mycorrhizal hyphae has been widely investigated, and the kinds of substances and their rates of translocation have confirmed that the hyphae are the essential vehicles for transport of nutrients into and from the mycorrhizal roots.

Although the hyphae of mycorrhizas serve to cross the 'zero sink' and exploit the soil further and wider, they will also in time devlop similar deficiency zones about themselves. However the provision of a new absorbing surface by the growth of hyphae to contact a new soil zone is much less expensive in material terms than the growth of a root.

As an example let us assume that the dry weight per unit volume of roots and hyphae are similar and that the radius of a hyphae is 2μm and of a root 200μm. Then the surface areas per unit volume (or weight) are;

$$2\pi rl/\pi R^2 L = 2/R$$

For a given length (L). Hence the surface area of the root per unot volume (or weight) is 0.01μm^2 and for hypha 1 μm^2 that is they differ by the factor of 100.On these assumptions and they are conservatives, it takes 100 times as much material to produce an equal surface area of exploitation by root growth as by hyphal growth.

Enhanced uptake of P is generally regarded as the most important benefit that AMF provide to their host plant and plant P is often the main in plant fungal relationship (Thompson 1987; Smith and Read 1997; Graham 2000). AMF can play a significant role in crop P nutrition, increasing total uptake and in some cases P use efficiency (Koide *et al.*, 2000). This may be associated with increased growth and yield (Ibibijen *et al.*, 1996; Koide *et al.*, 2000). Where colonization by AMF is disrupted, uptake of P, growth and in some cases yield can be significantly reduced (Thompson 1987, 1991, 1994).

In many cases, this is due to a high concentration of (phyto) available soil P (Bethlenfalvay and Barea, 1994; Hetrick *et al.*, 1996; Thingstrup *et al.*, 1998; Soren Sen *et al.*, 2005). Under such conditions, the colonization of roots by AMF is often suppressed (Jensen and Jakobsen, 1980; Ali-Karaki and Clark, 1999; Kahiluoto *et al.*, 2001). Where strong AMF colonization still occurs under condition of high soil P concentration it may reduce crop growth (Gavito and Varela 1995; Kahiluoto *et al.*, 2001).

AMF play significant role in the uptake of other nutrients by the host plant, Zinc (Zn) nutrition is most commonly reported as being influenced by the AM association, though uptake of copper (Cu), iron, N, K, calcium (Ca) and Mg have been reported being as being enhanced (Smith and Read, 1997; Clark and Zeto, 2000). AMF may also enhance plant uptake of N from organic sources (Hodge *et al.*, 2001). In many cases AMF cause a change in the absorption of soil nutrients by host simultaneously, though the effect on different nutrients is rarely the some (Lambert *et al.*, 1979, Kothari *et al.*, 1990; Wellings *et al.*, 1991; Thompson, 1987, 1991, 1994; Azaizeh *et al.*, 1995; Smith and Read, 1997; Srivastava *et al.*, 2002; Mohammad *et al.*, 2003).

The apparently contradictory evidence regarding the effect of AMF on plant nutrient absorption may be connected to the increasing realization that there is a degree of selectivity between host and the fungi and that different AMF have varying effects on different plant species, from strongly positive increases in nutrient uptake and or growth to strongly negative. (Monzon and Azcon, 1996; Bever *et al.*, 2001; Vander Heijden, 2002;.Munkvold *et al.*, 2004). There is a good evidence for a substantial capacity of external hyphae to absorb NH_4-N and deliver it to the host plant. In the hyphosphere as in the rhizosphere, uptake of NH_4-N is associated with substrate acidification.)

(*a*) Increase of K/Na Ratio

Some previous studies on the effects of salinity on mycorrhizal plants have shown that AM roots had higher Na concentrations but also higher K concentrations and thus maintained a high K/Na compared to non-mycorrhizal plants. Scientists have found that mycorrhizal roots of salt grass plant (*Distichlis spicata*) had higher sodium, potassium and phosphorus concentrations than non-mycorrhizal roots. In contrast, the results of other studies have shown that Na uptake decreased in AM plants compared to controls. Sodium content of shoots of mycorrhizal halophytic *Aster tripolium* plants was lower than non-mycorrhizal plants under salinity stress (Rozema *et al.*, 1986). K/Na ratio increased in mycorrhizal barley (moderate salinity tolerant plant) at high levels of salinity by decreasing Na concentration, rather than by increasing K concentration (Mohammad *et al.*, 2003). Dwevedi 2003 and Soni 2006) reported potassium may an important role in mycorrhization in wheat, legumes and rice plants.

(*b*) Increased Uptake of Other Mineral Nutrients

Concentrations of Ca, Mg and Zn in onion plants inoculated with *Glomus fasciculatum* increased in saline conditions and improved the nutritional status, which was at least partially responsible for increased plant growth (Ojala *et al.*, 1983). The improved growth and nutrient acquisition (P, K, Zn, Cu and Fe) in tomato demonstrate the potential of AM fungi for protecting plants against salt stress in arid and semiarid areas (AI-Karaki, 2000; AI-Karaki and Hammad, 2001). Although effects of AM fungi in increasing some toxic elements (*e.g.*, Na, Cl and Mn) have been reported (Pfeiffer and Bloss, 1988; Cantrell and Linderman, 2001), a direct effect of AM fungi in rednenon of others has also been reported. Mycorrhizal fungi decreased sodium concentration in barley (*Hordeum vulgar*) when grow in saline conditions (Mohammad *et al.*, 2003), and Mn can be reduced in mycorrhizal plants when it occurs at toxic levels compared to non-mycorrhizal plants (Sanders and Fitter, 1992; Cardoso *et al.*, 2003).

AMF in Metal Squestration

The use of living organism for the remediation of soils contaminated with heavy metals, radionuclide or polycyclic hydrocarbon is known as bioremediation Kumar *et al.*, 1995; Brooks and Robinson, 1998; Salt *et al.*, 1998, Baker *et al.*, 2000). AM fungi are involved in bioremediation through phytoremediation, the technique based on the use of plants for soil remediation (Leyval *et al.*, 1997; Turnau *et al.*, 2005). Depending on the type of pollutant different strategy for phytoremediation, such as phytostabilization, phytodegradation and phytoextraction has been used. For phytoremediation of soil polluted with heavy metals, the phytostabilization strategy involves the immobilization of heavy metals in the soil by establishing plants.

Table 7.1: Extension and Interconnectedness of Extraradical Mycelial Networks Produced by AM Fungi Living

Plant Species/Fungal Species	*Hyphal Density (mm mm⁻²)*	*No. of Anastomoses per Hyphal Length*	*Anastomosis Frequence (per cent)*	*Reference*
Allium porum/Glomus mosseae	2.7	4.6	75.0	Giovannetti M *et al.*, 1999
Allium porum/Glomus mosseae	3.5	3.8	59.3	Giovannetti M *et al.*, 2004
*Daucus carota/Gigaspora margarita**	-	0.0075	9.8	De la Providencia IE *et al.*, 2005
*Daucus carota/Gigaspora rosea**	-	0.012	4.2	De la Providencia IE *et al.*, 2005
*Daucus carota/Glomus hoi**	-	0.057	100	De la Providencia IE *et al.*, 2005
*Daucus carota/intraradices**	-	0.076	100	Giovannetti M *et al.*, 2004
Daucus carota/Glomus mosseae	3.9	2.5	45.5	Giovannetti M *et al.*, 2004
*Daucus carota/Glomus proliferum**	-	0.066	100	De la Providencia IE *et al.*, 2005
*Daucus carota/Scutellospora reticulate**	-	0.0079	5.2	De la Providencia IE *et al.*, 2005
Gossypium hirsutum/Glomus mosseae	6.8	6.2	53.1	Giovannetti M *et al.*, 2004
Lactuca sativa/Glomus mosseae	2.9	3.0	63.8	Giovannetti M *et al.*, 2004
Petrosslinum cripsum/Glomus caledonium	3.8	1.5	18.6	$
Petroselinum cripsum/Glomus intradices	2.3	5.5	56.9	$
Petroselinum cripsum/Glomus mosseae	3.5	4.7	62.3	$
Prunus cerasifera/Glomus mosseae	2.4	5.1	64.0	Giovannetti M *et al.*, 2001
Solanum melongena/Glomus mosseae	4.1	2.1	47.0	Giovannetti M *et al.*, 2004
Thymus vulgaris/Glomus mosseae	2.1	5.1	78.0	Giovannetti M *et al.*, 2001

(*) Ri-TDNA transformed carrot root.

($) Unpublished Data.

(M Giovanetti, L Avio, P Fortuna, EP Sbrana, P Sterani, 2006).

AMF can help phytoremediation activities, particularly in phytostabilization (Goncalves *et al.*, 1997; Leyval *et al.*, 1997, 2002; Orlowska *et al.*, 2002; Regvar *et al.*, 2003; Turnau *et al.*, 2005). Among possible mechanism by which AM fungi improve the resistance of plants to heavy metal is the ability of AM fungi to sequester heavy metals through the production of chelates or by absorption AM plants typically less translocate heavy metal to their shoots than the corresponding non AM control. Among the diverse type of mycorrhizosphere interactions known to benefit plant growth and health, those related to phytoremediation process as rhizabacteria and AM fungi interact synergistically to the benefit of phytoremediatio.

A key point in phytoremediation is the use of heavy metal adapted microbes, soil microbial diversity and activity both negative affected by excessive concentratic of heavy metals. Indigenous bacterial population (Giller *et al.*, 1998) and AM fungi (Del Val *et al.*, 1999) must be adapted to metal toxicity and have evolved abilities to enable them to survive in polluted soil.

Various authors have reported isolating spores of arbuscular mycorrhizal fungal texa such as *Glomus* and' *Gigaspora* associated with most of the plants growing inheavy metal polluted habitats (Chaudhry *et al.*, 1999). Identified *Glomus and Gigaspora spp.* in the mycorrhizospheres of fourteen plant species colonising a magnesite mine spoil in India. Whereas *Glomus mosseae* was isolated only and Duek *et al.* (1986) isolated *Glomus fasciculatum* alone from the heavy metal polluted soils. Pawlowska *et al.* (1996) surveyed a calamine spoil mound rich in Cd, Pb and Zn in Poland and recovered spores of *Glomus aggregatum, G. fasciculatum* and *Entrophospora spp.* from the mycorrhizospheres of the plants growing on spoil. Joner and Leyval (1997) reported that extra-radical hyphae of AM fungus G. *mosseae* can transport Cd from soil to subterranean clover plants growing in compartmented pots, but that transfer from fungus to plant is restricted due to fungal immobilization. It was also reported also reported no restriction of fungal hyphal growth into soil with high extractable Cd levels. It have been also showed very little, if any, translocation of Zn absorbed by mycorrhizal maize seedlings grown in contaminated soil, to the shoots. It was studied that localization of heavy metals within the fungal mycelium and mycorrhizal roots of *Euphorbia cyparis-sias* from Zn contaminated wastes and found higher concentrations of Zn as crystaloids deposited within the fungal mycelium and cortical cells of mycorrhizal roots. Studies by various researchers (Galli *et al.*, 1994; Hetrick *et al.*, 1994) have shown that mycorrhizal fungal ecotypes from heavy metal contaminated sites seem to be more tolerant to heavy metals (and have developed resistance) than reference strains from uncontaminated soils Galli *et al.* (1995) reported that although there was an increase in the contents of cystein, gamma EC and GSH in the mycorrhizal maize roots grown in quarto sand with added Cu, no differences in Cu uptake were detected between non-mycorrhizal and mycorrhizal, plants. These results do not support the idea that AM fungi protect maize from Cu-toxicity. Mycorrhizae are also known to produce growth stimulating substances for plants, thus encouraging mineral nutrition and in-creased growth and biomass necessary for phytoremediation to become commercially viable strategy for decontamination of polluted soils. In addition to the damaging effects on plants, the effect of heavy metals on the soil microorganisms and soil microbial activity also need to be considered. Various soil factors such as the clay contents and mobility of heavy metals effect plants and soil biota. As metal uptake by plant roots depends on soil and their associated symbionts, it is important to monitor metal mobility and availability to plant and its symbionts when assessing the effect of soil contamination on plant uptake and related phytotoxic effects. The prospect of symbionts existing in heavy metal contaminated soils has important implications for phytoremediation. The potentials of phytoremediation of contaminated soil can be enhanced by inoculating hyper-accumulator plants with mycorrhizal fungi most appropriate for contaminated site. It is further suggested that the potential of phytoremediation of contaminated soil can be enhanced by inoculating hyper-accmnulator plants with mycorrhiza.

AMF as Communicators

The most important AM fungal structure for plant nutrition is represented by the extra radical mycelium spreading from mycorrhizal roots into the surrounding soil which is able to uptake nutrients N, P, S, Ca, K, Fe, Cu, Zn and to transfer them to root cells (Smith SE and Read, 1997; Cox *et al.*, 1980). Mycorrhizal mycelium has been investigated in different experimental studies, based on either destructive extraction from soil or root observation chambers or in vitro system which yielded only qualitative data on its structure and growth (Jakobson *et al.*, 1990; Jones *et al.*, 1998).

The first visualization of intact AM mycelium extending from mycorrhizal roots into the extraradical environment was obtained by means of a bidimensional model system which utilized two cellulose esters members "sandwiched" around the roots of individual plantlets. After only seven days growth, a fine network of extramatrical hyphae growing on the membranes was visible to the naked eye, and its length extended from 5169 to 7471 mm (hyphal length in *Thymus vulgaris* and *Allium porum*, respectively. The experimental system divided to visualize the mycorrhizal mycelium also evidenced that the mechanism allowing the formation of the network was self recognition and hyphal anastomosis.

It is important to stress that the viability of mycorrhizal networks was 100 per cent and that all the anastomoses showed protoplasmic continuity and nuclear occurrence in hyphal bridges, confirming the occurrence of nuclear exchange also during fusion between extraradical symbiotic hyphae. AM fungi are known to infect a wide range of host species. They have a large geographical distribution being found even in the Arctic tundras and the Antarctic region (DeMars and Boerner 1995b). Unlike most ectomycorrhizal species, AM are not host specific. This enables them to form associations with a large number of plant species.

AM fungi regulate plant communities by affecting competition, composition and succession (Allen and Allen 1984; Kumar *et al.*, 1999). Limited resources and the struggle of the plants for a share of these is the primary selection pressure operating on plant species. In competition between plants, mycorrhizae in the soil favour the growth of one species and are detrimental to other competing species. It was demonstrated that in competition between two grasses *Lolium perenne* and Holcus *lanatus*. Inoculation with mycorrhizae favoured the growth of *H.lanatus*. This was an indirect effect as infection with mycorrhizae reduced the root length of *L.perenne* by 40 per cent. AM may regulate competition between plants by making available to mycorrhizal plants, resources that are not available to non-mycorrhizal neighbours (Allen and Allen 1984). AM symbiosis increases intraspecific competition (Facelli *et al.*, 1999). As a result, density of individuals of a single species would be reduced thereby allowing the co-existence of individuals of different species. This would lead to an increase in species diversity.

Mycorrhizae govern species composition in communities by influencing plant fitness at the establishment phase. AM preventing non-mycorrhizal plants from growing in soils colonized by them. This has a selective advantage for the fungus. Maintaining a high proportion of compatible host species at the expense of non-compatible species provides the fungus with an undisturbed carbon supply (Francis and Read 1994).

Succession is a chain of predictable processes whose course is influenced by nutrient availability. Mycorrhizae, owing to their role in nutrient uptake, may play an important part in determining the rate and direction of the process (Smith and Read 1997a). They influence the outcome of succession by amending the composition of species or by affecting species diversity.

AM fungi have been reported to be active in mediating nutrient transfer among plants (Chiariello *et al.*, 1982; Francis and Read 1984; Grime *et al.*, 1987; Watkins *et al.*, 1996; Graves *et al.*, 1997; Lerat S *et al.*, 2002) mainly through the extensive mycelial networks, which due to the lack host specificity may link the roots of contiguous plant species (Graves, *et al.*, 1997; Van der Heijden 1998). Recent studies showed a novel mechanism by which plants may become interconnected, that is hyphal fusion between extraradical hyphae originating from different individual plant root systems of different species, genera and families (Giovannetti, 2004).

AMF in Biotechnology

Introduction with AMF on micropropagate plantlets improves its establishment and growth in the field. Unfortunately not much work has been conduced in his line. But the few convincing reports decisively prove that successful hardening and *ex vitro* establishments of plantlets could be achieved by inoculation with AM fungi at the time of plants out.

One of the earliest report one the effect of effect of growth and leaf mineral content of two apple clones propagated *in vitro* were increased substantially. Similarly enhanced effect on the growth of in vitro cultured strawberry planlets could be achieved due to the association of AMF. The rooting of planlets of garlic regenerated from called was significantly enhanced due to inoculation with *Glomous mosseae.* The transplant success and growth of *Robus idaeus* and *Pistacia integerrima* were achieved with mycorrhiza inoculation.

Effect of AMF on these culture flowers is also encouraging in a different study at Vellayani, evaluated the effect of AMF in enhancing of survival, growth characteristic and uptake of nutrients by micropropagated Anthurium. AMF was incorporated into the rooting medium at the time of planting out through infected root bits. The study was conducted with *Glomus constrictum* and *G.etunicatum.* Micropropagated avocado plants generally exhibition low survival and very slow rate of growth during acclimatization. Inoculation of *G.fasciculatum* on micropropagated avocado planlets improved the formation of well devolved root system, shot growth, shoot root ratio and NPK content in plant tissue which helped plants to tolerate environmental stress transplanting. It was concluded that tolerate inococulation of AMF seems to be the key factor for subsequent growth and development of micropropagated plants of avocado.

Conclusion

1. Soils are the major sink of carbon and AM fungi have shown that, they have also potential to sequester the carbon from the soil.
2. Deposition of heavey metal in the soil is troublesome affair which resulted into the toxicity, which is not only harmful for the plants but also unsafe for organism dependent on their food from plants. Here also AMF fungi provide release from the heavy metal toxicity by sequestring heavy metals from the soil,
3. As it has been established that microbes and plants have very strong communication in the from of signal transduction,here also AMF provided evidences that they are best communicators in microbial world and have developed wood wide-web for their communication.
4. Biotechnology is the hour and therefore role of AMF fungi have also been explored and recognised in the field of morden biology. And have been tried to elucidate their relevance in biotechnology, molecular biology, genomics, proteomics and fluxomics.

References

Al-Karaki, G. N. and Clark, R.B. 1999. Varied rates of mycorrhizal inoculum on growth and nutrient acquisition by barley grown with drought stress. J. Plant Nutrion, 22: 1775-1784.

Al-Karaki, G. N. and Hammad, R. 2001. Mycorrhizal influence on fruit yield and mineral content of tomato grown under salt stress. J. Plant Nutrion, 24: 1311-1323.

Al-Karaki, G. N. 2000. Gwowth of mycorrhizal tomato and mineral acquisition under salt stress, Mycorrhiza, 10: 51-54.

Allen, E. B. and Allen, M. F. 1984. The competition between plants of different successional stages: mycorrhizal as regulators. Can. J. Bot., 62: 2625-2692.

Allen, E. B and Allen, M. F. 1986. Water relations of xeric grasses in the field: interaction of mycorrhizas and competition. New Phytol, 104: 559-571.

Allen, R. and Read, D. J. 1994. The contribution of mycorrhizal fungi to the determination of plant community structure. Plant and soil 159:11-25.

Allen, M. F. 1991. TheEcology of Mycorrhizae.Cambridge: Cambridge University Press.

Azaizeh, H. A., Marschner, H., Romheld, V. and Wittenmayer, L. 1995. Effects of vesicular arbuscular mycorrhizal fungus and other soil microorganisms on growth, mineral nutrient acquisition and root exudation of soil grown maize plants. Mycorrhiza, 5: 321-327.

Bago. B, Pfeffer, P. E. and Shachar-Hill, Y. 2000. Carbon metabolism and transport in arbuscular mycorrhizas. Plant Physiol, 128: 949-957.

Bago, B., Zipfel, W., Williams, R., Jun, J., Arreola, R., Lammer, P., Pfeffer, P. E. and Shachar-Hill, Y. 2002. Translocation and utilization of fungal lipid in the arbuscular mycorrhizal symbiosis. Plant Physiol, 128: 108-124.

Baker, A. J. M., MC Grath, S. P., Reeves, R. D. and Smith, J. A. C. 2000. Metal hyperaccumulator plants: a review of the ecology and physiology of a biological resource for phytoremediation of metal polluted soils. In: Phytoremediation of contaminated soil and water. (eds., Terry N, Banuelos G, Vangronsveld J,) Boca Ratton, Fl, USA: CRC Press, 85-107.

Barber, S. 1975. Soil nutrient bioavailability: a mechanistic approach.Newyork USA: John Wiley and Sons.

Beaden, B. N. and Peterson, L. 2000. Influence of arbuscular mycorrhizal fungi on soils structure and aggregate stability of vertisol. Plant and Soil. 218: 173-188.

Beilby, J. P. and Kidby, D. K. 1980. Biochemistry of ungerminated and germinated spores of the vesicular arbuscular mycorrhizal fungus, *Glomus caledonium*: changes in neutral and polar lipids. J. Lipid Res, 21: 739-750.

Bethlenfalavy, G. J. and Barea, J. M. 1994. Mycorrhizae in sustainable agricultural. J. Effects on seed yield and soil aggregation. Am. J. Agric., 9:157-161.

Bever, J. D., Schulatz, P. A., Pringle, A. and Morton, J. B. 2001. Arbuscular mycorrhizal fungi: more diverse than meets the eye, and the ecological tale of why. Bioscience, 51: 923-931.

Brooks, R. R. and Robinson, B. H. 1998. The potential use of hyperaccumulators and other plants for phytomining. In: Brooks RR, ed., Plants that hyperaccumulate heavy metals their role in phytoremediation, microbiology, archeology, mineral exploration, and phytomining. Cambridge: CAB International, 327-356.

Cantrell, I. C. and Linderman, R. G. 2001. Preinoculation of lettuce and onion with VA mycorrhizal fungi reduce deleterious effects of soil salinity. *Plant and Soil* 223: 269-281.

Cardoso, E., Navarro, R. B. and Nogueira, M. A. 2003. Charges in manganese uptake and translocation by mycorrhizal soyabean under increasing Mn doses. Revista Brasileira De Ciencia Do Solo, 27: 415-423.

Chaudhary, T. M., Hill, Khan A. G. and Duek, C. 1999. Colonization of iron and zinc-contaminated dumped filter-cake waste by microbes, plants and association mycorrhizae. In: Remediation and management of Degraded Land. (eds Wong MH, Wong JWCBaker, AJM) CRC Press, Boca, Chap, 27:275-283.

Chiariello, N., Hickman, J. C. and Mooney, H. A. 1982. Endomycorrhizal role for interspecific transfer of phosphorous in a community of annual plants, Science 217: 941-943.

Clark, R. B. and Zeto, S. K. 2000. Mineral acquisition by arbuscular mycorrhizal plants. J. Plant Nutron. 23: 867-902.

Cox, G., Moran, K. J., Sanders, F., Nockolads, C. and Tinker, P. B. 1980. Translocation and transfer of nutrients in vesicular-arbuscular nutrients in vesicular arbuscular mycorrhizas III Polyphosphate granules and phosphorus translocation. New Phytol. 84: 649-659.

Curl, E. A. and Trueglove, B. 1986. The rhizosphere, advanced series in agriculture science 15. Berlin. Germany: Springe –Verlag.

Davies, F. T., Jr. Potter, J. R. and Linderman, R. G. 1993. Drought resistance of mycorrhizal pepper plants independent of leaf P concentration –response in gas exchange and water relations. Physiologia- Plantarum 87:45-53.

De la Providenică, I. E., DeSouza, F.A., Fernandez, F., Sejalon Delmas, N., Declerck, F., Sejalon Delmas, N and Declerck, S. 2005. Arbuscular mycorrhizal fungi reveal distnict patterns of anastomosis formation and hyphal healing mechanism between different pathogenic groups. New phytol. 165:2 61-271.

De mars, B. G. and Boerner, R. E. J. 1995. Arbuscular mycorrhizal development in three crucifers. Mycorrhiza. 5:4 05-408.

Degans, B.P., Sparling, G. P. and Abott, L. K. 1994. The contribution from hyphae,roots and organic carbon constituients to the aggregation of a sandy loam under long term clover based and grass pastures., European J. Soil Science 45: 459-490.

Del Val, C., Barea, J. M. and Azcon-Aguilar, C. 1999. Diversity of arbuscular mycorohizal fungus population in heavy metal contaminated soil. Appl. Environ. Microbiol. 65: 718-723.

Diedrich, C. 1990. Improved growth of *Cajanus cajan* (L.) Millsp. In an unsterile tropical soil by three mycorrhizal fungi. Plant and soil 123: 261-266.

Douds, D. D., Pfeffer, P. E. and Shachar-Hill, Y. 2000. Carbon partitioning, cost eds. Arbuscular mycorrhizae. In: Douds DD, Kapulnik Y, eds. Arbuscular mycorrhizas physiology and function. Dordrecht, The Netherlands: Kluwer Academic Publishers, 197-130.

Douds, D. D., Gadkar, V. and Adholeya, A. 2000. Production of VAM fungus biofertilizer. In: Mukerji KG, Chamola BP, Singh J, Mycorrhizal Biology. New York Academic Publisher.

Duek, T. A., Viser, P., Ernst, W. H. O. and Schat, H. 1986. Vesicular arbuscular- mycorrhizae decrease zinc toxicity to grasses growing in zinc-polluted soil. Soil Biol. Biochem. 18: 331-333.

Dwivedi, O. P. 2003. Studies in soil microorganisms with special reference to vesicular-arbuscular mycorrhizal (VAM) fungal association with wheat crop of Sagar region. Ph.D. Thesis. Dr. H.S. Gour University, Sagar (M.P.).

Facelli, E., Facelli, J. M., Smith, S. E. and Mclaughlin, M. J. 1999. Interaotive effects of arbuscular mycorrhizal symbiosis intraspecific competition and resource availability on Trifolium subterranean cv. Mt. Barker. New Phytol. 141: 535-547.

Farrar, S. C., Hawes, M., Jones. D. and Lindo, S. 2003. How roots control the flux of carbon to the rhizosphere. Ecology 84: 827-833.

Francis, R. and Read, D. J. 1994. The contributions of mycorrhizal fungi to the determination of plant community structure. Plant and soil 159:11-25.

Francis, R and Read, D. J. 1984. Direct transfer of carbon between plants connected by vesicular-arbuscular mycorrhizal mycelium Nature; 307:53-56

Galli, U., Schuepp, H. and Brunold, C. 1994. Heavy metal binding by mycorrhizal fungi. Physiol. Plantarum 92: 364-368.

Gavito, M. E. and Varela, L. 1995. Response of criolla maize to single and mixed species inoculation of arbuscular mycorrhizal fungi. Plant soil. 176: 101-105.

George, E., Marschner, H. and Jakobsen, I. 1995. Role of arbuscular mycorrhizae fungi in uptake of phosphorus and nitrogen from soil. Crit Rev. Biotechnol 15: 257-270.

Giller, K. 1998. Toxicity of heavy metals to micro-organisms and microbial processes in agricultural soils: A review: Soil Biol. Biochem. 30: 1389-1414.

Giovannetti, M., Sbrana, C., Avio, L., Citernesi, A. S. and Logi, C. 1993. Differential hyphal morphogenesis in arbuscular mycorrhizal fungi during pre–infection stages. New Phytol. 125: 587-593.

Giovannetti, M., Sbrana, C. and Avio, L. 2004. Patterns of below ground plant interconnection established by means of arbuscular mycrrhizal networks. New Phytol. 164: 175-181.

Giovannetti, M., Azolini, D. and Citernesi, A. S. 1999. Anastomosis formation and nuclear and protoplasmic exchange in arbuscular mycorrhizal fungi. Appl. Eviron. Microbiol. 69: 616-624.

Giovannetti, M., Fortuna, P., Citernesi, A. S., Morini, S. and Nuti, M. P. 2001.The ocurrence of anastomosis formation and nuclear exchange in arbuscular mycorrhizal network. New phytol. 151: 171-124.

Goncalves, S. C., Goncalves, M. T., Freitas, H. Martin, S. and Loucao, M. A. 1997. Mycorrhizae in a Portuguese serpentine community. In: The ecology of ultarmafic and metalliferous areas, (eds. Jaffre T, Reeves RD, Becquer T), Proceedings of the second International Conference on Serpentine Ecology in Noumea, 87-89.

Graham, J. H. 2000. Assessing costs of arbuscular mycorrhizal symbiosis in agroecosystems. In: Podila GK, Douds DD, eds. Current advances in mycorrhizae research. St. Paul, MN, USA: American Phytopathological Society Press, pp 127-140.

Graves, J. D., Walkins, N. K., Fitter, A. H., Robinson, D. and Scrimgeour, C. 1997. Interspecific transfer of carbon between plants linked by a common mycorrhizal network. Plant Soil: 192: 153-159.

Grime, J. P., Mannkey, J. M. L., Hillersh, Read, D. J. 1987. Floristic diversity in a model system using experimental microcosms, Nature, 328: 420-422.

Gupta, V., Satyanarayana, T. and Sandeep, G. 2000. General aspects of mycorrhiza.In: Mukerji KG, Chamola BP, Singh J, Mycorrhizal Biology. New York: Kluwer

Hetrick, B. A. D., Wilson, G. W. T. and Todd, T. C. 1996. Mycorrihiza response in wheat cultivars, relationship to phosphorus, Can. J. Bot. 74: 19-25.

Hetrick, B. A. D., Wilson, G. W. T. and Figge, D. H. 1994. The influence of mycorrhizal symbiosis and fertilizer amendment on establishment of vegetation in heavy metal mine spoil. Environ. Pollut. 86: 171-179.

Hodge, A., Campbell, C. D. and Fitter, A. H. 2000. An arbuscular mycorrhizal fungus accelerates decomposition and acquires nitrogen directly from organic material. Nature. 413: 297-299.

Ibijbijen, J., Urguiaga, S., Ismaili, M., Alves, B. J. R. and Boddey, M. R. 1996. Effect of arbuscular mycorrhizas an uptake of nitrogen by *Brachiaria arrecta* and *Sorghum vulgaris* from soils labeled for several years with ^{15}N. New phytol.134: 132-139.

Jabaji- Hare, S. 1998. Lipid and profiles of some vesicular arbuscular mycorrhizal fungi: Contribution to taxonomy. Mycologia 80: 622-629.

Jackson, I. and Rosendahl, L. 1990 Carbon flow into soil and external hyphae from roots of mycorrhizal cucumber plants. New phytol.115: 77-83.

Jakobsen, I. 1995. Transport of phosphorous and carbon in VA mycorrhizae: structure, function, molecular Biology and Biotechnology. Springer- Verlag, Berlin. pp, 297-323.

Jensen, A. and Jakobsen, I. 1980. The occurance of vesicular-arbuscular mycorrhiza in barley and wheat grow in some Danish soils with different fertilizer treatments, Plant soil. 55: 403-414.

Joner, E. L. and Levyal, C. 1997. Uptake of ^{109}Cd by roots and hyphae of a Glomus mosseae/Trifolium subterranean mycorrhiza frm soil amended with high and low concentrations of cadamium. New Phytol. 135: 353-360.

Jones, M. D., Durall, D. M. and Tinker, P. B. 1998. Comparison of arbuscular and ectomycorrhizal *Eucalyptus coccifera*: Growth response, phosphorus uptake efficiency and external hyphal production. New Phytol 140: 125-134.

Kahiluoto, H., Ketoja, E., Vestberg, M. and Saarela, I. 2001. Promotion of AM utilization through reduced P fertilization Z. Field studies. Plant Soil 231:65-79.

Koide, R. T., Goff, M. D. and Dickie, I. A. 2000. Component growth efficiencies of mycorrhizal and non-mycorrhizal plant. New Phytol. 148: 163-168.

Koide, R. T. and Schreiner, R. P. 1992. Regulation of the vesicular-arbuscular mycorrhizal symbiosis. Annu Rev. Plant Physiol Plant Mol Biol 43: 557-581.

Kothari, S. K., Marschner, H. and Romheld, V. 1990. Direct and Indirect effects of VAMF and rhizospheric-microorganism on acquisition of mineral nutrients by maize (*Zea mays L.*) in calcarecus soil. New Phytol. 116: 637-645.

Kumar, P., Duschenkov, V., Motto, H. and Raskin I. 1995. Phytoextraction: the use of plants to remove heavy metals from soils. Environatl. Sci. Technol. 29: 1232-1238.

Lambert, D. H., Baker, D. J. and Code, H. 1979. The role of mycorrhize in the interactions of phosphorus with Zn, Cu and Other elements. Soil Sci. Am. J. 43: 976-980.

Lammers, P. J., Jun, J., Abubaker, J., Arreola, R., Gopalan, A., Bago, B., Hernandez-Sebastia, C., Allen, J. W., Douds, D. D. and Pfeffer, P. E. 2001. The glyoxylate cycle in an arbuscular mycorrhizal fungus: gene expression and carbon flow. Plant Physiol 127: 1287-1298.

Lerat, S., Gauci, R., Catford, J. G., Vierheilig, H., Pichey, Y. and Lapointe, L. 2002. C-14 transfer between the spring epehemeral *Erythronium americanum* and sugar maple saplings via arbuscular mycorrhizal fungi in natural strands. Oceologia. 132: 181-187.

Lewis, J. D. and Koide, R. T. 1990. Phosphorus supply, mycorrhizal infection and plant offspring vigour. Functl. Ecol. 4: 695-702.

Leyval, C., Turnau, K. and Haselwandter, K. 1997. Effect of heavy metal pollution on mycorrhizal colonization and function: physiological, ecological and applied aspects, Mycorrhiza 7: 139-153.

Li X-L, George, E. and Marschner, H. 1991. Extension of the phosphorous, depletion zone in VA-mycorrhizal white clover in calcareous soil. Plant and Soil 136: 41-48.

Marulanda, A., Azcon, R. and Ruiz-Lozano, J. M. 2003. Contribution of six arbuscular mycorrhizal fungal isolates to water uptake by *Lactuca sativa* plants under drought stress. Physiologia-Plantarum 19:1-8.

Miller, R. M. and Jastrouw, J. D. 2000. Mycorrhizal fungi influence soil structure. In Arbuscular Mycorrhizae: Physiology and Function, 4-8, Kluwer Academic Publications.

Mishra, M., Dubey, A., Singh, P.K. and Vyas, D. 2006. VAM association in *Catharanthus rosea*. J. Basic. Appl. Mycol. 5(I and II): 57-59.

Mohammad, M. J., Malkwai, H.I. and Shibli. 2003. Effects of mycorrhizal fungi and phosphorus fertilization on growth and nutrient uptake of barley grown on soils with different levels of salts. J. Plant Nuter. 26: 125-137.

Monzon, A. and Azcon, R. 1996. Relevance of mycorrhizal fungaql origin and host plant genotype to inducing growth and nutrient uptake in *Medicago Species*. Agric. Ecosyst. Envion. 60, 9-15.

Mosse, B. 1973. Advances in the study of vesicular arbuscular mycorrhizal fungi; Annu. Rev. Phytopathol. 11: 171-196.

Munkvold, L., Kjoller, R., Vestberg, M., Rosendahl, S. and Jakobsen, I. 2004. High functional diversity with in species of arbuscular-mycorrhizal fungi, New phyto. 164: 357-364.

Nichols, K. 2003. Characterization of Glomalin-A Glycoprotein Produced by Arbuscular- mycorrhizal fungi. Ph.D. Dissertation, University of Maryland, College Park, Maryland.

Nicoloson, T. H. 1975. Evolution of vesicular-arbuscular mycorrhizas In: Sanders E, MosseB, Tinker PB, Endomycorrhizas. Landon: Academic press. 25-34.

Nurlaeny, N., Marschner, H. and George, E. 1996. Effects of liming and mycorrhizal colonization on soil phosphate depletion and phosphate uptake by maize (*Zea mays* L.) and soyabean (*Glycine max* L.) grown in two tropical soils. Plant and Soil 181: 275-285.

Oades, J. M. and Waters, A. G. 1991. Aggregate hierarchy in soils. *Aust. J. Soil. Res.*, 29: 815-828.

Ojala, J. C. and Linderman, R. G. 2001. Preinoculation of Lettuce and onion with VA mycorrhizal fungi reduces deleterious effect of soil salinity. Plant and Soil. 223: 269-281.

Orlowska, E., Zubek, S. Z., Jurkiewic, Z. A. and Szarek- Lukaszewska, G. 2002. Influence of restoration on arbuscular mycorrhiza of *Biscutella laevigata L.* (Brassicaceae) and *plantago lanceolata L.* (Plantaginaceae) from calamine spoil mounds. Mycorrhiza 12:153-160.

Owusu-Bennoah, E. and Wild, A. 1979. Autoradiography of the depletion zone of phosphate around onion roots in the presence of vesicular-arbuscular mycorrhiza. New Phytol. 82:133-140.

Pawlowska, T. B., Blazkowski, J. and Rhiling, A. 1996. The myorrhiza status of plants colonizing a colamine spoil mound in southern Poland. Mycorrhiza. 6: 499-505.

Pfeffer, P. E., Douds, D. D., Becard, G. and Shachar-Hill, Y. 1999. Carbon uptake and the metabolism and transport of lipids in and arbuscular mycorrhiza. Plant Physiol 120: 587-598.

Pfeiffer, C. M. and Bloss, H. E. 1988. Growth and nutrition of guayule (*Parthenium argenatum*) in a saline soil as influenced by esicular- arbuscular mycorrhiza and phosphorus fertilizer. New phytol. 108: 315-321.

Redecker, D., Kodner, R. and Graham, L. E. 2000. Glomalean fungi from Ordovician. Science. 289: 1920-1921.

Regvar, M., Vogel, K., Irgel, N., Wraber, T., Hildebrandt, U., Wilde, P. and Bothe, H. 2003. Colonization of pennycresses (*Thlaspi* spp.) of the Brassicaceae by arbuscular mycorrhizal fungi. J. Plant Physiol. 160: 615-626.

Remy, W., Taylor, T. N., Hass, H. and Kerp, H. 1994. Four hundred –million year –old vesicular arbuscular mycorrhizae. Proc. Natl. Acad. Sciences, USA 91: 841-843.

Rhodes, H. and Gerdemann, J. W. 1975. Phosphate uptake zones of mycorrhizal and non-mycorrhizal onions. New Phytol. 75: 555-561.

Rilling, M. C., Field, C. B. and Allen, M. F. 1999. Soil biota responses to long term atmospheric CO_2 enrichment in two califorina annual grasslands. Oceologia 119: 572-577.

Rilling, M. C., Wright, S. F., Nichols, K. A., Schmidt, W. F and Torns, M. S. 2001. Large contribution of arbuscular mycorrhizal fungi to soil carbon pools in tropical forest soils. Plant soil 233: 167-177.

Rozema, J., Arp, W., Vandiggelen, J., Vanesbroek, M., Broekman, R. and Punte, H. 1986. Occurrence and ecological significance of vesicular arbuscular mycorrhizae in the salt Marsh environment. Acta Botanica Neerlandica 35: 457-467.

Salt, D. E., Smith, R. D. and Raskin, I. 1998. Phytoremediation. Ann. Rev. Plant Physiol. Plant Molec. Biol. 49: 643-668.

Sanders, F. E. and Sheikh, N. A. 1983. The development of vesicular-arbuscular mycorrhizal in plant root systems. Plant and Soil 71: 223-246.

Sanders, I. R. and Fitter, A. H. 1992. The ecology and functioning of vesicular arbuscular mycorrhizas in coexisting grassland species I. Seasonal patterns of mycorrhizal occurrence and morphology. New Phytol. 120: 517-524.

Shachar-Hill, Y., Pfeffer, P. E., Douds, D., Osman, S. F, Doner, L. W. and Ratcliffe, R. G. 1995. Partitioning of intermediate carbon metabolism in VAM colonized leek. Plant Physiol 108: 7-15.

Sidhu, O. P. and Behl, H. M. 1997. Response of three *Glomus* species on the growth of *Prosopis Juliflora* Qwartz at high pH levels. Symbiosis 23: 23-24.

Smith, S. E and Read, D. J. 1997. Mycorrhizal symbiosis. Academic Press, London.

Solaimain, M. D.and Satio, M. 1997. Use of sugars by intraradical hyphae of arbuscular mycorrhizal fungi revealed by radiorespirometry. New phytol. 136: 533-538.

Soni, P. 2008. Studies on some medicinal herbs with special reference to nutrient uptake by VAM fungi. Ph.D. Thesis. Dr. H.S.G. University, Sagar.

Sorensen, N., Larsen, J. and Jakobsen, I. 2005. Mycorrhiza formation and nutrient concentration in looks (*Allium porrum*) in relation to previous crop and cover crop management on high P soils. Plant soil. 273: 101-114.

Soni, P. and Vyas, D. 2006. Effect of native VA -mycorrhizal fungi on growth, biomass and essential oil content of *Mentha arvensis*. J. Botanic. Soc. University Sagar, 41: 134-141.

Srivastava, A. K., Singh, S. and Marathe, R. A. 2002. Organic citrus, soil fertility and plant nutrition. Sustain agric, 19: 5-29.

Stanley, M. R., Koide, R. T. and Shumway, D. L. 1993. Mycorrhizal symbiosis increases growth, reproduction and recruitment of Abutilon theophrasti Medicin the field. Oecologia 94: 30-35

St-Arnaud, M., Hamel, C., Vimard, B., Caron, M. and Fortin, A. 1996. Enhanced hyphal growth and spore production of the arbuscular mycorrhizal fungus *Glomus intraradices* an *in vitro* system in the absence of host roots. Mycol. Res. 100: 328-332.

Subramanian, K. S., Charest, C., Dwyer, L. M. and Hamilton, R. I. 1997. Effects of arbuscular mycorrhizae on leaf water potential, sugar content and phosphorus content during drought and recovery of maize. Can. J. Bot. 75: 1582-1591.

Thingstrup, I., Rubaek, G., Sibbesen, E. and Jakobsen, I. 1998. Flax (*Linum uritatiusim* L.) depends on arbuscular mycorrhizal fungi for growth and P uptake at intermediate but not high soil P levels in the field. Plant soil. 203: 37-46.

Thomas, R. S., Dakessian, S., Ames, R. N., Brown, M. S. and Bethlenfalvay, G. J. 1986. Aggregation of a silty clay loam soil by mycorrhizal onion roots. Soc. America J. 50: 1494-1499.

Thompson, J. P. 1987. Decline of vesicular arbuscular mycorrhizae in long fallow disorder of field crops and its expression in phosphorus deficiency of sunflower, Aust. J. Agric. Res., 38: 847-86.

Thompson, J. P. 1994. Inoculation with vesicular-arbuscular mycorrhizal fungi from cropped soil overcomes long-fallow disorder of linseed (*Linum usitatissimum L.*) by improving P and Zn uptake, Soil Biol. Biochem. 26: 1133-1143.

Thompson, J. P. 1991. Improving the mycorrhizal condition on the soil through cultural practices and effects on growth and P uptake in plants. In: Phosphorous Nutrition of Grain Legumes in the Semi Arid Tropics, (eds Johansen C, Lee KK and Sahrawat KL). International Crops Research Institute for Semi Arid Tropics, Andhra Patancheru, India, pp. 117-137.

Tisdall, J. M. and Behl, H. M. 1997. Organic matter and water stable aggregates in soils. J. Soil Science. 33: 141-163.

Turnau, K., Jurkiewicz, A., Lingua, G., Barea, J. M. and Gianinazzi- Pearson, V. 2005. Role of arbuscular mycorrhiza and associated microorganisms in phytoremédiation of heavy metal polluted sites. Trace Element in the Environment Biochemistry, Biotechnology and Bioremediation. CRC Press/ Lawis Publishers.

Van der Heijden. 2002. Arbuscular mycorrhizal fungi as determinant of plant diversity, in search for underlying mechanisms and general principles. In: *Mycorrhizal Ecology*. (eds, Van der Heijden MGA and Sanders IR), Ecological studies. 157. Springer Verlag, Berlin, Germany, 243-265.

Vyas, D., Mishra, M. K., Singh, P. K. and Soni, P. 2006. Studies on mycorrhizal association in wheat. Indian Phytopath. 59:174-179.

Vyas, D. and Dwiedi, O. P. 2002. Effect of potassium on the occurrence of VAM fungi. *J. Basic.Appl.Mycol.*1: 233-235.

Vyas, D. and Soni, P. 2007. Arbuscular mycorrhizal fungi associated with important medicinal plants of Sagar. Indian Phytopathol. 60: 52-57.

Vyas, D., Dubey, A. and Singh, P. K. 2006. VA Mycorrhizal Fungi in tropical monsoonic grassland. J.Basic.Mycol. 5: 78-81.

Vyas, D., Dubey, A., Soni, A., Mishra, M. and Singh, P. K. 2007. Arbuscular mycorrhizal fungi in early land plants. Mycorrhiza News.19: 21-23.

Vyas, D. Soni, A. and Singh, P. K. 2007. Differential effects of pesticides on occurrence of Vam fungi associated with some leguminous plants. J. Basic Appl. Mycol. 6: 143-150.

Watkins, N. K., Fitter, A. H., Graves, J. D. and Robinson, D. 1996. Carbon transfer between C_3 and C_4 plants linked by a common mycorrhizal network quantified using stable carbon isotopes. Soil Biol Biochem. 28: 471-477.

Wellings, N. P., Wearing, A. H. and Thompson, J. P. 1991. Vesicular-arbusculr mycorrhizal (VAM) improve phosphorus and zinc nutrition and growth of pigeonpea in vertisol. Austt J. Agric Res. 42: 835-845.

Wilcox, H. E. 1991. Mycorrhizae. In: Waisel Y, Eshel A, Kafkafi U. The Plant Root, the Hidden Half. New York: Marcel Dekker. 731-765.

Wright, S. F. and Upadhayaya, A. 1996. Extraction of an abundant and unusual protection from soil and comparison with hyphal protection of arbuscular mycorrhizal fungi. Soil Sci., 161: 575-586.

Wright, S. F. and Upadhyaya, A. 1998. A survey of soil for aggregate stability and glomlin, a glycoprotein produced by hyphae of arbuscular- mycorrhizal fungi. Plant and Soil, 86: 97-107.

Microbes: Diversity and Biotechnology (2012) *Pages* **137–149**
Editors: **Prof. S.C. Sati & Dr. M. Belwal**
Published by: **DAYA PUBLISHING HOUSE, NEW DELHI**

Chapter 8

Studies on Micro Fungal Diversity under Variable Habitats in Arunachal Pradesh

M. Majumder and A.K. Shukla

Department of Botany, Rajiv Gandhi University,
Rono Hills, Itanagar – 791 112

ABSTRACT

Micro fungal diversity was studied in various habitats covering six districts of Arunachal Pradesh. A total number of 124 micro fungal species were recorded from different habitats. Higher fungal diversity was found with decomposing litter of plants (75 species) than soil (64 species). Out of recorded 124 micro fungal species 57 belong to ascomycetes, 53 deuteromycetes and 14 species to phycomycetes class. Maximum diversity was recorded in case of fungi *Penicillium* and for which 14 species were isolated from different habitats. In terms of diversity *Penicillium* was followed by *Aspergillus* (8 species), *Oidiodendron* (7 species), *Verticillium* (6 species) and *Mortierella* (6 species). Some economically important fungal species were also recorded during the study. As far as occurrence of fungal species in different land use system is concerned 62 species were isolated from paddy field soil, 48 from natural forest, 42 from other sedentary cultivation, 33 from plantation forest, 18 from tea garden and 16 from jhum cultivation system.

Keywords: *Fungi, Fungal diversity, Habitat, Plant litter, Soil.*

Introduction

Soil biodiversity reflect the variability among living organisms ranging from the myriad of invisible microorganisms like bacteria and fungi to the macro-fauna which are essential for the maintenance of system. Soil is an essential component of terrestrial ecosystem, support plant growth and provides a habitat for diverse and interacting population of soil organism. In soil after bacteria, fungi represent the second largest group of soil micro-organism. The saprophytic fungi represent the largest proportion of fungal species and they performed crucial role in the decomposition of plant structural polymers such as cellulose, hemicellulose and lignin. Thus they contribute a major role in maintenance of the global carbon cycle. Bacteria have diverse metabolic capabilities that allow exploitation of wide range of energy sources in soil. Bacteria along with fungi are the primary agent of biogeochemical transformation and decomposition of dead plants and animals remains, which recycle nutrient of plant growth. Microorganisms in soil are influence by number of factors such as temperature, moisture, pH and minerals nutrient concentration present in the soil (Shukla *et al.*, 1989). The catabolic activities of bacteria and fungi enable them to grow on diverse substrate. There have been a number of studies on the distribution of soil micro fungi. Some of these have dealt with the influence of plant community type on the soil mycoflora (Christensen, 1969), other have examine the effect of soil depth (Widden, 1979; Bissett and Parkinson, 1979; Shukla and Mishra, 1992). The present work deals with the study of micro fungal diversity under variable habitats in Arunachal Pradesh.

Materials and Methods

The study was carried out in six districts (Papumpare, East Kameng, West Kameng, Upper Siang, East Siang and Changlang) of Arunachal Pradesh by collecting soil samples from different habitats like jhum cultivation system, sedentary cultivation systems, various forest types, tea garden as well as decomposing leaf litter of crops and weed species.

Litter Decomposition

The crop and weed residues were collected immediately after harvest of crops. Plant samples were air-dried for three weeks, sorted out into leaves and stems litter for the *C. frutescens*, *E. coracana* and *P. typhoides*, *A. conyzoides*, *S. paniculata* and *E. odoratum*, however, for *O. sativa* and *B. campestris*, the whole aboveground parts were considered as foliage materials and sliced to 5 cm long pieces. Litter bag technique was used to study the decomposition rate (Bocock *et al.*, 1963). For each type of residues, 100 litter bags (1 mm mesh; 15 cm x 15 cm) were prepared. Ten g of air-dried sample was placed in each bag; subsamples were oven dried for moisture correction. The bags were placed randomly in the field. At monthly intervals, the litter bags were retrieved (7 replicate bags of each litter type) and brought to the laboratory in sterilized polythene bags.

Isolation of Fungal Diversity

The litter was then cut into small pieces of 10 mm and placed into a 250 ml conical flask, containing 100 ml of sterilized distilled water and then shaken on a horizontal shaker (120 throws min^{-1} and 1.5 cm displacement) to form a homogenous suspension. Fungal population was determined by dilution plate method (Johnson and Curl, 1972) using Martin's Rose Bengal agar medium. To isolate micro fungi present in soil the samples were collected from different places at the depth of 0-15 cm and brought to laboratory. In case of soil also serial dilution plate method was followed. Five replicate was maintained for each sample. The Petri dishes were inoculated at 25° C for a period of five days. To calculate the population of fungi, colonies developed on Petri dishes were counted with the help of digital colony counter. The final counts have been expressed on the basis of g^{-1} dry weight of soil as well as litter. Representative isolates of fungi were identified under microscope with the help of manuals of Domsch *et al.* (1980), Barnett and Hunter (1972) and Subramanian (1971). Relative frequency, relative density and relative abundance of individual species were calculated and the sum of these represented the importance value index (IVI) for various species.

Soil Chemical Properties

Soil pH was determined electrometrically by a digital pH meter in 1:2.5 W/v H_2O (Allen *et al.*, 1974). Soil organic carbon was determined by rapid titration method. Total Kjeldhal nitrogen was determined following semi-micro Kjeldahl procedure by acid digestion, distillation and titration (Allen *et al.*, 1974). Available phosphorus was determined spectrophotometrically using molybdenum blue method as outlined in Allen *et al.* (1974).

Result and Discussion

During the study period 124 species of micro fungi were isolated from different habitats of Arunachal Pradesh, which belongs to 60 genera. Out of these 57 belongs to Ascomycetes, 53 Deuteromycetes and 14 to Phycomycetes class. Fungal diversity was recorded more in decomposing litter than soil. Total of 124 micro fungi 64 were isolated from soil and 75 species from decomposing leaf litter of different crops and weed plants. Fungal diversity in the present systems were higher than those (33 species) reported from subtropical humid forest soils in northeast India (Arunachalam *et al.*, 1997), 26, 21 and 27 species respectively from soil of valley, terrace and slope land of Meghalaya (Shukla and Mishra, 1992), 80 species from agricultural land soil (Majumder and Shukla, 2008) and 41 species isolated from South Dakota grassland soil (Dennis and Christensen, 1981). Conversely, the present species number was much lower as compared to 287 species of fungi reported from tropical and temperate habitats (Taylor *et al.*, 2000). Maximum species diversity was recorded incase of genus *Penicillium* for which fourteen species were isolated. The micro fungal species *P. chrysogenum* known for production of antibiotic penicillin was isolated from the decomposing leaf litter of *Oryza sativa*. Eight species of *Aspergillus* were reported and one of that includes economically important *A. fumigatus*. This fungal species synthesized alkaloids of clavine group (used for abortion) which is also produced by *Claviceps sclerotia*. A maroon coloured pigment, fumigatin was also produced by this fungi. High species diversity was also observed for *Oidiodendron* (seven species), *Verticillium* (six species) and *Mortierella* (six species). Polyunsaturated fatty acids (PUFA) are produced by *Mortierella alpine* with solid substrate fermentation. Rice bran is the most effective substrate for PUFA production, followed by peanut meal residue, wheat bran and sweet potato residue (Jang *et al.*, 2000). *Trichoderma* harzianum, *T. koningii* and *T. viride* were isolated from soil of mixed forest, teak plantation, degraded forest, different agricultural systems; decomposing leaf litter of *Brassica campestris* and *Ageratum conyzoides* (Table 8.1). The antagonistic property of this fungus has led to their use as biological control agent of some plant pathogenic fungi. *Metarhizium anisopliae* isolated from jhum field of Papumpare district is an imperfect, entomopathogenic fungus. This fungus has been recognized as a biocontrol agent against four groups of insect pests (termites, locusts, spittlebugs and beetles) and is currently being targeted for control by *M. anisopliae* (Zimmermann, 1993). *M. anisopliae* is applied as spores or mycelia in various formulations. Similarly, *Beauveria* and *Fusarium* also kill insects under appropriate environmental conditions and are being studied for use as biological insecticides. During the study *Beauveria* and *Fusarium* fungi were found growing with decomposing leaf litters of plants.

In soil highest population was estimated for *Aspergillus* followed by *Penicillium*, *Oidiodendron*, *Cladosporium* etc (Figure 8.1), whereas, in decomposing leaf litter *Penicillium* represent the highest population followed by *Cladosporium*, *Geotrichum*, *Mortierella*, *Aspergillus*, *Verticillium*, *Oidiodendron*, sterile forms etc (Figure 8.2). The fungal population found to be influenced by the physico-chemical properties of soil (Shukla *et al.*, 1989). Physico-chemical properties of soil are given in Table 8.2.

The diversity of fungal population was greater in *P. typhoides* foliage (26), followed by stem of *C. frutescens* (25) and foliage of *A. conyzoides* (24) while lowest was observed in foliage materials of *C. frutescens* (9). From the undecomposed leaf materials, *Cladosporium macrocarpum*, *Chrysosporium merdarium*, *Metarrhizium* sp., *Mortierella elongata*, *Cladosporium cladosporioides*, *Penicillium canescens*, *Aspergillus niger*, *Geotrichum candidum*, *Aspergillus flavus*, *Geotrichum candidum*, *Penicillium fellutanum*,

Table 8.1: List of Fungal Species Isolated from Different Habitats of Arunachal Pradesh

Fungal Species	*Area*	*Habitat*
Absidia corymbifera	Harmoti	Teak plantation
Absidia glauca	Namdapha National Park	Decomposing leaf litter of *Eleusine coracana*
Acremonium butyri	Namdapha National Park	Decomposing leaf litter of *Ageratum conyzoides*
Acremonium kiliense	Namdapha National Park	Decomposing leaf litter of *Capsicum frutescens*
Acremonium murorcum	Harmoti	Soil of tea garden; decomposing leaf litter of *Pennisetum typhoides*
Acremonium sp.	Harmoti	Soil of maize field
Apiospora montagnei	Namdapha National Park	Decomposing leaf litter of *Eupatorium odoratum*
Arthoderma insingulare	Doimukh	Soil of forest
Arthoderma quadifidum	Namdapha National Park	Decomposing leaf litter of *Pennisetum typhoides*
Arthoderma tuberculatum	Namdapha National Park	Decomposing leaf litter of *Ageratum conyzoides*
Aspergillus candidus	Namdapha National Park	Decomposing leaf litter of *Pennisetum typhoides*
Aspergillus clavatus	Harmoti, Kameng, Kamki, Simong, Yinkiong and Midpu	Soil of jhum field after burning, maize field, tea garden, homgarden, paddy field, bamboo forest, millet field, jhum
Aspergillus flavus	Namdapha National Park, Doimukh, Midpu, Harmoty, Karsingsha, Pasighat, Kameng, Kamki, Simong, Yinkiong, Zero	Soil of mixed forest, paddy field, wheat field, mixed cropping system, before burning, maize field, tea garden, teak plantation, homegarden, degraded forest, jhum fallow, agroforestry system, bamboo forest, paddy cum fish cultivation system, pine forest; decomposing leaf litter of *Ageratum conyzoides*, *Oryza sativa*, *Eleusine coracana*, *Pennisetum typhoides*
Aspergillus fumigatus	Namdapha National Park, Kamki	Soil of degraded forest; decomposing leaf litter of *Ageratum conyzoides*
Aspergillus japonicus	Midpu	Paddy field
Aspergillus niger	Namdapha National Park, Midpu, Harmoty,Pasighat, Yinkiong, Simong, Zero,	Soil of paddy field, maize field, jhum before burning, mixed cropping system, agroforestry system, millet field, bamboo forest, mixed forest, pine forest; homegarden, teak plantation decomposing leaf litter of *Spilanthes paniculata*, *Eupatorium odoratum*, *Oryza sativa*, *Brassica campestris*, *Eleusine coracana*
Aspergillus oryzae	Harmoty	Soil of tea garden
Aspergillus parasiticus	Harmoty	Soil of tea garden
Aspergillus sp.	Pasighat, Zero	Soil of maize field, paddy field, bamboo forest, paddy cum fish cultivation system, pine forest
Aurobasidium pullulans	Harmoty	Soil of tea garden, jhum field, maize field.

Contd...

Table 8.1–Contd...

Fungal Species	*Area*	*Habitat*
Aurobasidium sp.	Namdapha National Park, Harmoty	Soil of teak plantation; decomposing leaf litter of *Eleusine coracana*, *Pennisetum typhoides*
Beauveria bassiana	Namdapha National Park	Decomposing leaf litter of *Eleusine coracana*, *Pennisetum typhoides*
Blastomyces sp.	Namdapha National Park	Decomposing leaf litter of *Oryza sativa*
Botrichum sp.	Harmoty	Soil of mixed cropping system
Botrytis cinera	Namdapha National Park	Decomposing leaf litter of *Spilanthes paniculata*, *Brassica campestris*
Ceratocystis moniliformis	Doimukh	Soil of mixed forest
Cercospora sp.	Namdapha National Park	Decomposing leaf litter of *Ageratum conyzoides*
Chrysosporim merdarium	Namdapha National Park	Decomposing leaf litter of *Ageratum conyzoides*, *Eupatorium doratum*, *Pennisetum typhoides*
Cladorrhinum foecudissimum	Doimukh, Midpu, Namdapha National Park	Soil of mixed forest, maize field; decomposing leaf litter of *Ageratum conyzoides*, *Brassica campestris*
Cladorrhinum sp.	Harmoti	Soil of tea garden
Cladosporium cladospoidies	Doimukh, Harmoty, Simong, Midpu, Namdapha National Park	Soil of maize field, mixed forest, tea garden, teak plantation, jhum field decomposing leaf litter of *Ageratum conyzoides*, *Spilanthes paniculata*, *Eupatorium odoratum*, *Oryza sativa*, *Brassica campestris*, *Pennisetum typhoides*
Cladosporium herbarum	Harmoty, Namdapha National Park	Soil of mixed cropping system; decomposing leaf litter of *Ageratum conyzoides*
Cladosporium macrocarpum	Doimukh, Harmoty, Namdapha National Park	Soil of mixed forest, mixed cropping system; decomposing leaf litter of *Ageratum conyzoides*, *Brassica campestris*, *Capsicum frutescens*
Cladosporium sp.	Namdapha National Park	Decomposing leaf litter of *Eleusine coracana*
Cochelibolus spiifera	Doimukh, Midpu, Harmoti	Soil of paddy field, homegarden, degraded forest
Connigmella elegans	Doimukh, Midpu, Harmoti	Soil of paddy field, degraded forest, teak plantation, tea garden
Cylendrocarpon magnusianum	Botanical garden.	Decomposing root litter of *Lantana camera*
Doretomyces sp.	Botanical garden	Decomposing root litter of *Bidens pilosa*
Didymostible sp.	Harmoty	Soil of tea garden
Doratomyces microsporus	Harmoty	Soil of maize field
Exophiala jeanselmei	Namdapha National Park	Decomposing leaf litter of *Brassica campestris*
Fusarium flocciferum	Yinkiong, Namdapha National Park	Soil of millet field; decomposing leaf litter of *Spilanthes paniculata*
Fusarium sporotrichioides	Namdapha National Park	Decomposing leaf litter of *Eleusine coracana*

Contd...

Table 8.1–Contd...

Fungal Species	*Area*	*Habitat*
Geotrichum candidum	Midpu, Harmoty, Karsingsha, Doimukh, Namdapha National Park	Soil of mixed forest, tea garden, maize field, paddy field, homegarden; decomposing leaf litter of *Ageratum conyzoides*, *Spilanthes paniculata*, *Eupatorium odoratum*, *Oryza sativa*, *Brassica campestris*, *Capsicum frutescens*, *Eleusine coracana*, *Pennisetum typhoides*
Glomerella cingulata	Namdapha National Park	Decomposing crop stem
Gonytrichum macrocladum	Harmoty	Soil of maize field
Humicola fuscoatra	Doimukh, Namdapha National Park	Soil of mixed forest, maize field; decomposing leaf litter of *Ageratum conyzoides*, *Spilanthes paniculata*, *Eupatorium odoratum*, *Brassica campestris*, *Eleusine coracana*
Humicola gresium	Harmoty	Soil of mixed cropping system
Humicola sp.	Harmoty	Soil of mixed cropping system
Hyalopycnis sp.	Midpu	Soil of degraded forest
Hypomyces sp.	Harmoti	Soil of jhum field before burning
Isariopsis sp.	Botanical garden	Decomposing root litter of *Lantana camera*
Mammaria echinobotritis	Midpu	Soil of maize field
Marianaea elegans	Harmoty	Soil of tea garden
Metarrhizium anisopliae	Simong	Soil of jhum field
Metarrhizium sp.	Namdapha National Park	Decomposing leaf litter of *Ageratum conyzoides*
Monilia sp.	Namdapha National Park	Decomposing leaf litter of *Ageratum conyzoides*, *Spilanthes paniculata*, *Brassica campestris*, *Capsicum frutescens*, *Pennisetum typhoides*
Mortierella alpina	Botanical garden	Decomposing leaf litter of *Bidens pilosa*
Mortierella bisporales	Namdapha National Park	Decomposing leaf litter of *Oryza sativa*, *Capsicum frutescens*
Mortierella elongata	Namdapha National Park	Decomposing leaf litter of *Ageratum conyzoides*, *Spilanthes paniculata*, *Oryza sativa*, *Pennisetum typhoides*
Mortierella gamsii	Namdapha National Park	Decomposing leaf litter of *Ageratum conyzoides*, *Oryza sativa*, *Pennisetum typhoides*
Mortierella minutisma	Harmoty	Soil of tea garden
Mortierella sp.	Harmoty	Soil of teak plantation
Mucor hiemalis	Harmoty, Namdapha National Park	Soil of maize field; decomposing leaf litter of *Ageratum conyzoides*

Contd...

Table 8.1–Contd...

Fungal Species	*Area*	*Habitat*
Mucor racemosus	Adjoining area of Namdapha National Park	Decomposing leaf litter of *Eupatorium odoratum*
Necteria radicicola	Namdapha National Park	Decomposing leaf litter of *Pennisetum typhoides*
Necteria ventricosa	Harmoty, Simong, Namdapha National Park	Soil of teak plantation, jhum field; decomposing leaf litter of *Eupatorium odoratum*, *Brassica campestris*, *Capsicum frutescens*
Necteria sp.	Derang	Soil of mixed forest
Nigrospora sphaerica	Namdapha National Park	Decomposing leaf litter of *Pennisetum typhoides*
Oidiodendron echinulatum	Harmoty, Doimukh, Namdapha National Park	Soil of tea garden, paddy field; decomposing leaf litter of *Pennisetum typhoides*
Oidiodendron griseum	Doimukh, Harmoti, Midpu, Namdapha National Park	Soil mixed forest, homegarden, maize field; decomposing leaf litter of *Spilanthes paniculata*, *Eleusine coracana*, *Pennisetum typhoides*
Oidiodendron moniliades	Namdapha National Park	Decomposing leaf litter of *Pennisetum typhoides*
Oidiodendron rhodogenum	Harmoty	Soil of mixed cropping system
Oidiodendron sp.	Namdapha National Park	Decomposing leaf litter of *Pennisetum typhoides*
Oidiodendron tenuissimum	Harmoty	Soil of maize field
Oidiodendron trunkatum	Harmoty	Soil of mixed cropping system
Pecelomyces sp.	Harmoty	Soil of teak plantation
Penicillium bravicompactum	Doimukh, Midpu, Harmoty, Pasighat, Derang, Kameng, Kamki, Simong, Yinkiong, Zero, Karsingsha, Namdapha National Park	Soil of mixed forest, homegarden, jhum field, maize field, tea garden, teak plantation, paddy field, degraded forest, jhum fallow, mixed cropping system, agroforestry system, bamboo forest, paddy cum fish cultivation system, pine forest; decomposing leaf litter of *Ageratum conyzoides*, *Spilanthes paniculata*, *Eupatorium odoratum*, *Oryza sativa*, *Brassica campestris*, *Eleusine coracana*, *Pennisetum typhoides*
Penicillium canascens	Harmoty, Midpu, Namdapha National Park	Soil of tea garden, paddy field, teak plantation; decomposing leaf litter of *Ageratum conyzoides*, *Spilanthes paniculata*, *Eupatorium odoratum*, *Capsicum frutescens*, *Pennisetum typhoides*
Penicillium crysogenum	Namdapha National Park	Decomposing leaf litter of *Oryza sativa*
Penicillium citrinum	Namdapha National Park	Decomposing crop stem
Penicillium coccosporum	Namdapha National Park	Decomposing leaf litter of *Spilanthes paniculata*
Penicillium daleae	Karsingsha, Simong	Soil of mixed forest, bamboo forest

Contd...

Table 8.1–Contd...

Fungal Species	*Area*	*Habitat*
Penicillium fellutanum	Namdapha National Park	Decomposing leaf litter of *Spilanthes paniculata*, *Brassica campestris*, *Capsicum frutescens*
Penicillium frequentans	Namdapha National Park	Decomposing leaf litter of *Ageratum conyzoides*, *Eupatorium odoratum*
Penicillium granulatum	Yinkiong	Soil of millet field
Penicillium jensenii	Botanical garden	Decomposing leaf litter of *Bidens pilosa*
Penicillium janthilinum	Yinkiong	Soil of millet field
Penicillium lividum	Namdapha National Park	Decomposing crop stem
Penicillium nigricans	Botanical garden	Decomposing crop stem/root of *Lantana camera*
Peniillium rubrum	Namdapha National Park	Decomposing leaf litter of *Pennisetum typhoides*
Penicillium sp.	Harmoti, Pasighat, Simong, Zero	Soil of jhum fallow, paddy field, maize field, paddy cum fish cultivation system
Phialophora americana	Namdapha National Park	Decomposing crop stem
Phoma eupyrena	Botanical garden	Decomposing stem litter of *Lantana camera*
Ramichloridium schulzers	Namdapha National Park	Decomposing leaf litter of *Pennisetum typhoides*
Rhizopus oryzae	Doimukh, Harmoty, Pasighat, Kameng, Simong, Namdapha National Park	Soil of mixed forest, paddy field, mixed cropping system, tea garden, teak plantation, homegarden, bamboo forest, maize field; decomposing leaf litter of *Oryza sativa*
Rhizopus stolonifer	Harmoty	Soil of jhum field
Rhizopus sp.	Harmoti	Soil of maize field
Scopulariopsis brumptii	Doimukh, Karsingsha, Midpu	Soil of paddy field, mixed forest, homegarden, degraded forest, maize field
Sesquicillium candelabrum	Botanical garden	Decomposing root litter of *Lantana camera*
Spegazzinia sp.	Midpu	Soil of mixed forest
Sporonima phacidiodes	Midpu	Soil of degraded forest
Staphylotrichum coccosporiodes	Namdapha National Park	Decomposing leaf litter of *Spilanthes paniculata*
Stephanoma tetracoccum	Namdapha National Park	Decomposing leaf litter of *Pennisetum typhoides*
Syncephalastrum sp.	Namdapha National Park	Decomposing crop stem
Torula herbarum	Doimukh, Harmoty	Soil of tea garden, mixed forest, jhum field, mixed cropping system
Trichoderma hargianum	Harmoty, Doimukh, MIdpu, Simong, Namdapha Yinkiong	Soil of jhum field, teak plantation, paddy field, degraded forest, maize field, millet field; decomposing leaf litter of *Brassica campestris*
Trichoderma koningii	Namdapha National Park	Decomposing leaf litter of *Ageratum conyzoides*

Contd...

Table 8.1–Contd...

Fungal Species	*Area*	*Habitat*
Trichoderma viride	Midpu, Yinkiong	Soil of mixed forest, homegarden, millet field
Trichoderma sp.	Zero	Soil of paddy cum fish cultivation system
Trichocladium asperum	Botanical garden	Decomposing leaf litter of *Lantana camera*
Trichocladium opacum	Botanical garden	Decomposing leaf litter of *Lantana camera*
Trichosporella cerebriformis	Botanical garden	Decomposing leaf litter of *Bidens pilosa*
Verticillium albo-atrum	Midpu, Doimukh, Namdapha National Park	Soil of mixed forest; decomposing leaf litter of *Ageratum conyzoides*, *Eupatorium odoratum*, *Oryza sativa*, *Brassica campestris*, *Eleusine coracana*
Verticillium clamidosporium	Midpu	Soil of homegarden, maize field
Verticillium dahliae	Harmoty	Soil of tea garden
Verticillium lecanii	Harmoty, Namdapha National Park	Soil of teak plantation; decomposing leaf litter of *Pennisetum typhoides*
Verticillium nigrescens	Namdapha National Park	Decomposing leaf litter of *Ageratum conyzoides*
Verticillium sp.	Simong, Midpu, Namdapha National Park	Soil of jhum field, homegarden, mixed forest; decomposing leaf litter of *Spilanthes paniculata*, *Eupatorium odoratum*, *Capsicum frutescens*, *Pennisetum typhoides*
Wardomyces columbinus	Harmoty	Soil of mixed cropping system
Wardomyces simplex	Doimukh	Soil of mixed forest
Zygorrhinchus heteroganum	Harmoty	Soil of jhum field
Zygosporium sp.	Namdapha National Park	Decomposing leaf litter of *Eleusine coracana*.

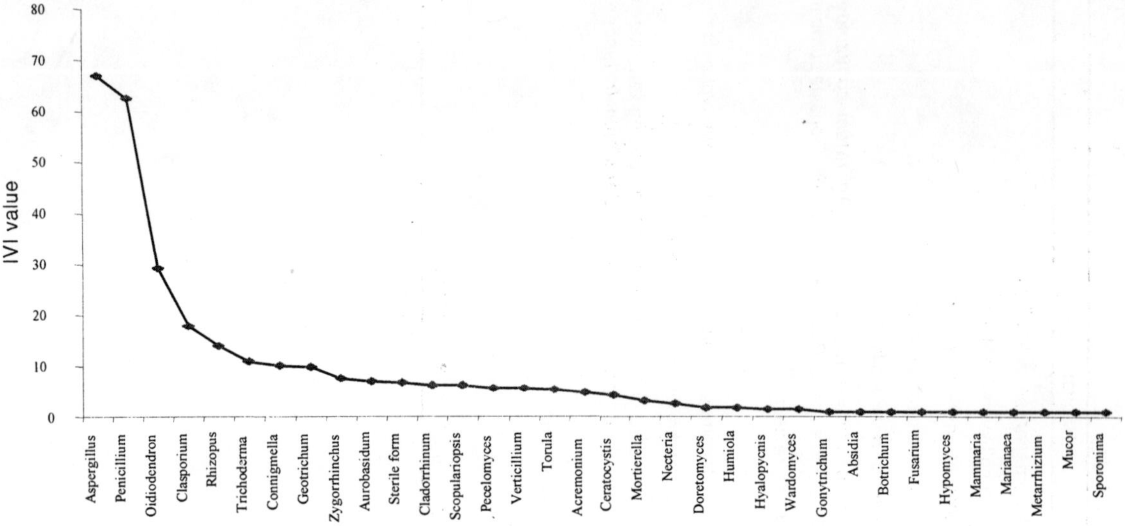

Figure 8.1: IVI of some Fungi Isolated from Soil

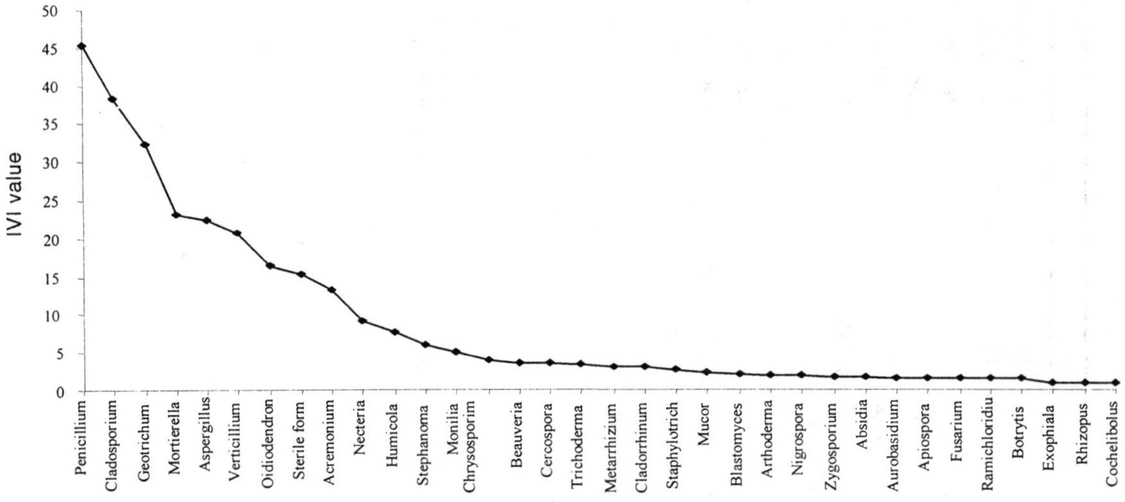

Figure 8.2: IVI of some Fungi Isolated from Decomposing Leaf Litter

Trichoderma herzianum, Nectria ventricosa, Verticillium sp. were isolated. Population and diversity of fungal population was comparatively lower in undecomposed plant materials. Initially, along with the existing fungal population of the plant surface, some weak parasitic fungi colonized on the litter *e.g. Geotrichum candidum, Mortierella* sp., which invade living tissues and pave the way for saprophytes. The saprophytes colonize litter in successive and overlapping layers (Sadaka and Ponge, 2003). Each fungus that colonizes the litter alters the microhabitat and paves the way for the next. In this context, Shukla *et al.* (1990) reported *Fusarium* sp., *Mucor* sp., *Penicillium* sp. and *Rhizopus* sp. as the early successional micro fungi in decomposing crop residues form subtropical environment. In this study, the early successional fungal community comprised *Mortierella* sp., *Geotrichum candidum, Aspergillus*

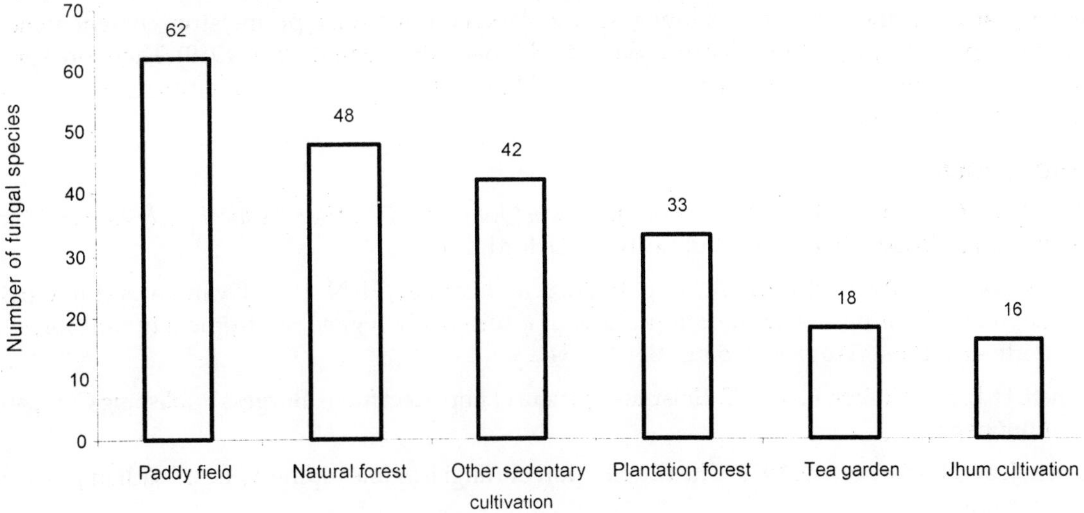

Figure 8.3: Fungal Species in Different Land Use System

sp., *Cladosporium* sp., *Penicillium* sp. and *Verticillium* sp. The large number of other fungal species colonizes sporadically to the litter during later stage. Among the late colonizers were *Mucor racemosus, Verticillium albo-atrum, V. nigrescens, A. niger, P. chrysogenum, Botrytis cinera, Fusarium sporotrichoides, Oidiodendron griseum, M. bisporalis, Aurobasidium pullulans, Cladorrhinum foecudissimum, Conninghmella elengans, Fusarium oxysporim, Glomerela cingulata, Humicola fuscoatra, Penicillium citrinum, P. claviformae, P. granulatum, P. libidum, P. nigrens,* Sterile form and *Nectria ventricosa* appeared in later part of study. *Penicillium brevicompactum, Geotrichum candidum, Verticillium* sp., *Cladosporium cladosporioides,* etc. were common and were isolated frequently from all types of decomposing residues.

Table 8.2: Physico-chemical Properties of Soil Under Different Land Use Systems of Arunachal Pradesh

Sites	*pH*	*Moisture Content (per cent)*	*Available-N* $\mu g\ g^{-1}$	*Available-P* $\mu g\ g^{-1}$	*Organic C (per cent)*
Natural Forest	4.09–6.72	15.8–35.9	48.9–105.2	15.4–62.0	0.79–4.0
Plantation forest	4.74–6.13	13.0–23.4	67.2–94.7	15.7–18.6	1.03–1.57
Paddy field	4.60–6.92	15.2–56.3	50.7–98.9	7.1–54.9	0.56–1.98
Jhum	5.41–6.33	17.0–35.0	66.0–127.9	15.8–62.7	1.06–3.41
Tea garden	4.39–5.44	17.1–21.6	57.8–158.5	18.4–41.3	0.81–1.15
Other agricultural fields	4.62–6.86	13.6–31.4	48.4–113.5	12.8–54.4	0.72–3.59

For a given community one or few species are numerically predominant and may strongly affect environmental conditions for other species (Wardle and Parkinson, 1991). The species like *Penicillium brevicompactum, Aspergillus flavus, Cladosporium cladosporioides, Verticillium alboatrum, Geotrichum candidum, Mortierella bisporales* etc. that seem to be tolerant to a wide range of environmental conditions were common in all the systems. These species were isolated frequently at relatively high density.

Many of the important chemical and physical properties of soil vary over a short period. It has been demonstrated that the occurrence of fungal species depends upon soil type, moisture content, mineral nutrients, and soil temperature (VanVuurda and Schippers, 1980; Shukla *et al.*, 1989). Therefore spatial variation in soil properties and micro climate would play an important role in influencing the micro fungal diversity in soil.

References

Allen, S. E., Grimshaw, H. M., Parkinson, J. A. and Quarmby, C. 1974. *Chemical analysis of ecological materials*. Blackwell Scientific Publications, Oxford. U.K.

Arunachalam, K., Arunachalam, A., Tripathi, R. S. and Pandey, H. N. 1997. Dyanamics of microbial population during the aggradation phase of a selectively logged subtropical humid forest in north-east India. Tropical Ecology 38: 333-341.

Barnett, H. L. and Hunter, B. B. 1972. Illustrated genera of imperfect fungi. Burgess Publishing Company, Minneapolis.

Bissett, J. and Parkinson, D. 1979. The distribution of fungi in some alpine soils. Canadian J. Bot. 57: 1609-1629.

Bocock, K. L. 1963. Changes in the amount of nitrogen in decomposing leaf litter sessile oak (*Quercus petraca*). J. Ecol. 51: 555-566.

Christensen, M. 1969. Soil microfungi of dry to mesic conifer-hardwood forests in northern Wisconsin. Ecology 50: 9-27.

Dennis, C. and Christensen, M. 1981. The soil microfungal community of a South Dakota grassland. Can. J. Bot. 59:1950-1960.

Domsch, K. H., Gams, W. and Anderson, T. H. 1980. *Compendium of soil fungi*. Academic Press, London.

Jang, H. D., Lin, Y. Y. and Yang, S. S. 2000. Polyunsaturated fatty acid production with *Mortierella alpine* by solid substrate fermentation. Bot. Bull. Acad. Sin. 41: 41-48.

Johnson, L. E. and Curl, E. A. 1972. *Methods for research on ecology of soil borne plant pathogens*. Burgess publishing Company, Minneapolis.

Majumder, M. and Shukla, A. K. 2008. Micro fungal diversity in soil of three different agricultural systems. In *Advances in Applied Microbiology* (Eds. Parihar, P. and Parihar, L.) Pp. 279-286. Agrobios (India), Jodhpur.

Sadaka, N. and Ponge, J. F. 2003. Fungal colonization of phyllosphere and litter of *Quercus rotundifolia* Lam. in a holm oak forest (High Atlas, Morocco). Biol. Fert. Soils 39: 30-36.

Shukla, A. K., Tiwari, B. K. and Mishra, R. R. 1989. Temporal and depthwise distribution of microorganisms, enzymes activities and soil respiration in potato field soil under different agricultural systems in north-eastern hill region of India. Revue a Ecologie et de Biologie du Sol. 26: 249-265.

Shukla, A. K., Tiwari, B. K. and Mishra, R. R.. 1990. Decomposition of potato litter in relation to microbial population and plant nutrients under field conditions. Pedobiologia 34: 287-298.

Shukla, A. K. and Mishra, R. R. 1992. Influence of soil management systems on the microfungal communities of potato field. Cryptog. Mycol. 13: 135-144.

Subramanian, C. V. 1971. Hyphomycetes an account of Indian species except *Cercospora*. Indian Council of Agricultural Research Pub., New Delhi.

Taylor, J. E., Hyde, K. D. and Jones, E. B. G. 2000. The biogeographical distribution of microfungi associated with three palm species from tropical and temperate habitats. Journal of Biogeography 27: 297-310.

VanVuurde, J. W. L. and Schippers, B. 1980. Bacterial colonization of seminal wheat roots. Soil Biol Biochem 12:559–565

Wardle, D. A. and Parkinson, D. 1991. Analysis of co-occurrence in a fungal community. Mycol. Res. 95: 504-507.

Widden, P. 1978. Fungal populations from forest soil in southern Quebec. Canadian Journal of Botany 57: 1324-1331.

Zimmermann, G. 1993. The entomopathogenic fungus *Metarhizium anisopliae* and its potential as a biocontrol agent. Pesticide Science 37: 375-379.

Microbes: Diversity and Biotechnology (2012) *Pages* 151–158
Editors: Prof. S.C. Sati & Dr. M. Belwal
Published by: DAYA PUBLISHING HOUSE, NEW DELHI

Chapter 9

Culture of *Cordyceps sinensis* (Berk) Sacc.: A High Value Rare Medicinal Fungus of High Altitude Himalayan Region

R.C. Gupta[1] *and P.S. Negi*[2]

[1]*Department of Botany, Kumaun University, S.S.J. Campus, Almora - 263 601, Uttarakhand*

[2]*Defence Agricultural Research Laboratory, Pithoragarh – 262 501, Uttarakhand*

ABSTRACT

Cordyceps sinensis (Caterpillar mushroom) is a high value medicinal, nontoxic entomogenous fungus, which is found growing in the higher hills of Himalayas including Nepal, China, Tibet and India. This fungus is having a high medicinal value and is used in traditional remedies for various physiological disorders. *Cordyceps* is an important anti-aging medicine. It also inhibits the formation of active monoamine oxidase, an enzyme responsible for aging in man. It is found beneficial in case of climatic age illness, impotence, emission, neurasthemia, rheumatoid arthritis, cirrhosis, flabby waist and knee. *Cordyceps* has been in traditional use for the treatment of various diseases like chronic bronchitis, insomnia, hypertension, pneumonia, tuberculosis, pulmonary emphysema, anaemia, night sweats and cough.

Pure culture of *Cordyceps sinensis* has been raised in the laboratory after culturing tissues from stroma region of the living specimen. The culture of this fungus is possible on different types of the culture media like potato dextrose agar (PDA), beef extract dextrose agar (BEDA), casein hydrolysate dextrose agar (CHDA), soybean extract dextrose agar (SEDA) and rice extract dextrose agar (REDA). Optimum rate of mycelial growth was observed on casein hydrolysate dextrose agar medium followed by beef extract dextrose agar, potato dextrose agar, soybean extract dextrose agar and rice extract dextrose agar. The fungus prefers acidic pH (5 to 5.5) for its optimum mycelial growth and low temperature condition between 10–15°C for its optimum growth under *in vitro* condition.

Keywords: *Cordyceps sinensis, Caterpillar mushroom, 'Yarsa Gamboo', Medicinal mushroom.*

Cordyceps sinensis or Caterpillar Mushroom is a high value medicinal, nontoxic, entomogenous fungus, which is found growing in the higher hills (10,000 ft to 14,000 ft altitude) of Himalayas including Nepal, China, Tibet and India. Locally this fungus is also known as Yarsha Gumba, Yarsha Gamboo or *"Kira Ghas"* since it parasitizes on the Lepidopteran insect larvae of the caterpillar *Hepialus armoricanus* Oberthuer, family Hepialidae under the soil (Huang, 1999). The first ever report on this fungus dates back to the eighteenth century when Torrubia, a Franciscian friar in Cuba described it as the trees growing out of the bellies of wasps. That is why the genus *Cordyceps* is sometimes known as *Torrubia* in the honour of its inventor (Christensen, 1975). The combination of an insect larvae or rather the mummy of a larvae converted into fungus mycelium along with the fungus stalk was known as "Chinese plant worm" in the ancient time and it has since been named as *Cordyceps sinensis.* In China this caterpillar mushroom is known as *"Dong Chong Xia Cao"* which means *"summer grass, winter worm"*. This fungus is known as *Tochukas* in Japan.

The fungus is having a high medicinal value and is used in traditional remedies for various physiological disorders. Owing to its high efficacy and potency in curing various diseases, it was recommended by the ancient medicinal practitioners as the *"Panacea of all ills"*. However, it attracted the attention of the people since 1993 when a group of nine Chinese women athletes who have been taking *Cordyceps,* shattered nine world records in world outdoor track field championships in Germany. In Tibet the potent medicinal qualities of this fungus dates back to 500 years ago when it was used by Emperor's physicians in the Ming Dynasty for the treatment of royal ailments. The initial record of *Cordyceps* as medicine is as old as the Qing Dynasty during 1757 which appeared in *New Compilation of Materia Medica* (Zhu, Halpern and Jones, 1998).

Having wide medicinal properties of *Cordyceps* it is highly prized in the markets world over. Traditionally the fungus is traded in China for its weight in silver and gold and the market prize is astronomically high as of December 1999, it was sold at a whole sale price of Rupees one lakh per kilogram. The trade and market channel in India is not a transparent one. It is being smuggled out to China and Tibet across the boarder lines. In these boarder lines the cost of per kg mushroom varies from Rupees 80,000 to 1, 30,000. It is believed that in international market this fungus may fetch a price between one to two million rupees per kg (US $ 20,000 – 40,000). However, the cost of this mushroom varies among the trade channels starting from field gatherers to broker and their agents who collect the material from different locations and sell it at a higher rate. The current prevailing price per kg of *Cordyceps sinensis* in the markets of Tibet ranges between Rs 68, 000 to Rs 80, 000, in Nepal between Rs 80, 000 to Rs 90, 000 and in India between Rs 1,25000 to Rs 1,30,000 (Sharma,2004).

In India, this fungus has got popularity almost half decade earlier when it was collected from the high altitude hills of Dharchula (Uttaranchal) in the Central Himalayas by some local people called Khambas (a Tibetan race). The specimens of this fungus are very rare, occurring in the far reaches of high, cold and arid hills at an altitude between 10,000 ft to 14,000 ft above the sea level. A very few locations namely Brahamkot, Ultapara, Ghwardhap, Chhipalakot and Najari which are spread in about 8 km area in Dharchula region of Uttarakhand have been identified as a place of occurrence of *Cordyceps sinensis* till date. But there is possibility of its occurrence in several other places having the similar type of climatic conditions in the Central Himalayas.

The upright stalk, visible above the ground is technically called a stroma or fruiting body which grows from insect larva buried under the ground. Normally a single stroma emerges from the backside of the head but very rarely two stromae have also been found emerging from the head of the larva. The structure of this stalk is like a blade of grass which is measuring 8 to 11 cm in length, sometimes measuring 17cm. The length of the caterpillar is 5-5.5cm and thus the total length of the mushroom is

13-15.5cm. The diameter of the stalk is 3-3.5mm and that of stroma is 4-4.5mm (Figure 9.1). This iridescent blue coloured stroma of *Cordyceps* is cylindrical in shape with a tapering apex. The apex of the stoma bears numerous round shaped whorls of perithecia. Each perithecium contains two asci inside it and each ascus has 8 filiform ascospores. The underground portion of *Cordyceps* is represented by a compact mass of mycelium filled inside the mummy of caterpillar. However, Sharma (2004) reported the length of the stroma to be 2.5cm and the length of caterpillar 1.5cm only.

Cordyceps sinensis belongs to the fungal subdivision Ascomycotina which is mainly characterized by the presence of sexually produced spores called ascospores formed within an organ called ascus. Having the characteristic features like ascomata surrounded by a peridial wall, unitunicate asci, flasked shaped perithecia and long narrow asci having thread like ascospores, this fungus is kept under the family Clavicipitaceae of class Pyrenomycetes. The genus *Cordyceps* has about 200 species

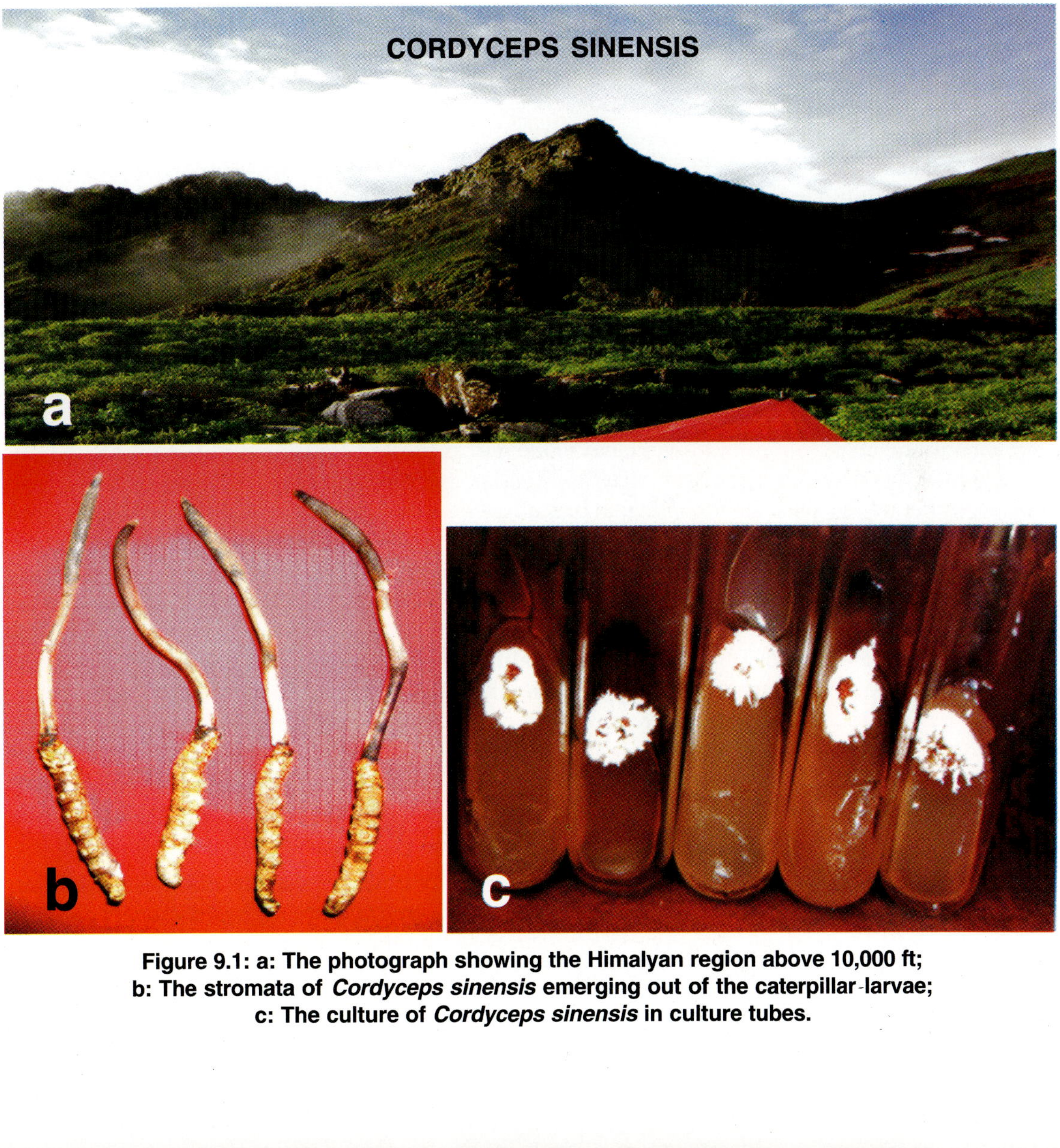

Figure 9.1: a: The photograph showing the Himalyan region above 10,000 ft; b: The stromata of *Cordyceps sinensis* emerging out of the caterpillar-larvae; c: The culture of *Cordyceps sinensis* in culture tubes.

(Dubey, 1983). mostly parasitic on insects of one sort or another like flies, wasps, ants, scale insects, spiders, moths, and assorted bugs and beetles.The species of *Cordyceps* found in abundance include *Cordyceps sinensis, C. gansuensis, C. militaris, C.capitata, C. memorabilis, C. ophioglossoides, C. nepalensis, C. multiaxialis* and *C. cicadae*. However, Christensen (1975) has reported only 100 species of *Cordyceps*. A very few species of this fungus are also found growing on the underground fruiting bodies of another fungus (Christensen 1975). Out of the seven such mycophilous species of *Cordyceps*, two species are found growing on *Claviceps* and rest five grow on *Elaphomyces* under the ground (Sharma,2004). A new classification of *Cordyceps* species has also been suggested on the bases of chemotaxonomy of spatial nucleotide sequences of 18S rDNA obtained from four different species (Ito, and Hirano, 1997).

The infection in the body of insect larvae is initiated by the ascospores or their fragments. This inoculum gives out a germination tube which penetrates the body integuments and develops into the hyphae or mycelium within the body tissue of caterpillar. This fungus produces an enzyme called cutinase which helps in the digestion of larval body tissue (Huber,1958). The hyphae breakup into fragments and get distributed in the haemocoel. Further these fragments bud off to produce more propagules which give rise to complete hyphae. Thus, the body tissue of the insect larvae get stuffed with the compact mass of mycelium. The mycelium modifies into the sclerotium which remains covered by the integument of dead insect. The corpse of the insect enlarges in size and does not decay due to the cordycepin, an antibiotic produced by *Cordyceps*. The mummy of the insect having endosclerotium lies buried underground until the favourable climatic conditions for the germination of sclerotium is available. As soon as the snow melts the slerotium starts germinating in the form of a structure resembling to the blade of grass which comes out of the insect body from the backside of the head. The club shaped mature stromata bears a stalk and a head above the ground. The head region of this mushroom contains numerous peripheral perithetia. The sexual reproductory organs (ascus and ascospores) are present inside the perithetia. These filiform ascospores normally divide into the small fragments which spread over the earth surface by the wind.

This fungus is having a high medicinal value and is used in traditional remedies for various physiological disorders. Owing to its high efficacy and potency in curing various diseases, this fungus was recommended by the ancient medicinal practitioners as the *"Panacea of all ills"*. Being a hormone stimulator, Cordyceps is an important anti-aging medicine. It also inhibits the formation of active monoamine oxidaze, an enzyme responsible for aging in man. The frequent use of this fungus may prevent the senile disorder. Important information about the medicinal effects of *Cordyceps sinensis* have been published by the various workers (Mizuno,1999; Hobbs, 1995; Halpern, 1999). *Cordyceps sinensis* contains crude protein cordycepin, an antibiotic (3'- deoxyadenosine) (C10 $H_{13}0_{3}N_{6}$) and *d*-mannitol Today this fungus is in wide popularity throughout the world as a nutriceutical having potential therapeutic applications as well as an important nourishing tonic and adaptogen, which increases the physical stamina. The research findings have indicated that this fungus enhances the oxygen supply to the brain and heart and thus improves the resistance to hypoxia. *Cordyceps* is found beneficial in case of climatic age illness, impotence, emission, neurasthemia, rheumatoid arthritis, cirrhosis, flabby waist and knee.

In China *Cordyceps* has been in traditional use for the treatment of various diseases like chronic bronchitis, insomnia, hypertension, pneumonia, tuberculosis, pulmonary emphysema, anaemia, night sweats and cough (Zhu *et al.*, 1998; Xie *et al.*, 1988). Based on the recent advances in literature, the major pharmacological actions by *Cordyceps sinensis* on hepatic, renal, cardiovascular and endocrine systems, anticancer activity, immunomodulation and hypoglycemic activity, and effects on erythropoiesis and haematopoiesis have been reviewed (Bao *et al.*, 1988: Wang-Sheng *et al.*, 2000). *Cordyceps sinensis* works as an immunomodulatory agent and thus confirms the presence of immunosuppressive ingredients in it (Zhu, 1990; Kuo *et al.*, 1996C). Restoration of cellular immunological function is observed during the trials on patients suffering from advanced stage of cancer (Degliantoni, *et al.*, 1985).

The various products of *Cordyceps* regulate the normal functioning of different bodyparts, strengthen the immune system and promote the overall vitality and longevity. This fungus increases IL-1, interferon and TNF production under *in vitro* Kupffer cells of rats and elevate serum level of these substances (Zhu *et al.*, 1996)). *Cordyceps* also exhibits antitumor and antiviral activities (Bok *et al.* (1999)). This fungus has a diuretic effect and thus is good for the prevention of nephralgia (Chiou *et al.*, 2000). *Cordyceps* is also useful in preventing arteriosclerosis, coronary heart diseases and other diseases related to the blood vessels of brain (Chiou. *et al.*, 2000). The water soluble solution of *Cordyceps sinensis* is found to have the stimulatory effect in the production of corticosteroids (Wang *et al.*, 1998). *Cordyceps sinensis* has shown the outstanding activity against the human leukemia (Chen *et al.*, 1997). It is also beneficial to the patients suffering from hyperglycemia and hypertension since it has the potential to reduce the blood sugar as well as the blood cholesterol (Hsue and Lo, 2002; Kiho *et al.*, 1996).

In India, this fungus has got popularity almost half decade earlier when it was collected from the high altitude hills of Dharchula (Uttarakhand) in the Central Himalayas by some local people called Khambas (a Tibetan race). The specimens of this fungus are very rare, occurring in the far reaches of high, cold and arid hills at an altitude between 12000 fts. to 16000 fts. above the sea level. A very few locations namely Dharchula bioshere in Kumaun and Nandadevi biosphere in Garhwal division of Uttaranchal state have been identified as a places of occurrence of *Cordyceps sinensis* till date. But there is possibility of its occurrence in several other places having the similar type of climatic conditions in the Central Himalayas.

In vitro micropropagation of *Cordyceps sinensis* in the laboratory is quite important since there are frequent complaints regarding contamination to the natural specimens with grass sticks or some heavy metals. This contamination is mainly done by the poor village folks who sometimes slip a small stick up the tail end of caterpillar to increase the weight of their product. Sometimes the contamination is also done by inserting the small lead bits into the insect body. Two cases of lead poisoning have already been reported in Taiwan (Wu *et al.*, 1996).

The live specimens of *Cordyceps sinensis* were collected carefully from their natural habitat in the high altitude region above 10,000ft of Central Himalayan hills. Collection was done during May-June, 2002. The specimens were wrapped inside the moss plants and then packed in the ice cube. The specimen was washed with tap to remove the adhering dust particles on it. The stroma of the fresh specimen of *Cordyceps* was washed 2 –3 times in double distilled water and dipped in 0.1 per cent $HgCl_2$ solution for 5 minutes. Again the stroma was washed with sterile distilled water and surface dried by pressing between sterilized filter paper. In order to propagate the mycelium *in vitro*, tissues were taken from different parts of the fungus body like spores, stalk tissue and tissue from stroma region. These tissues were excised from the fungus body with the help of a sterilized blade inside a laminar flow and cultured on to various culture media. Eight different types of culture media were also prepared to get the pure culture of the fungus.

The culture media utilized for the pure mycelium culture are potato dextrose agar (PDA), casein hydrolysate dextrose agar (CHDA), beef extract dextrose agar (BEDA), soybean seed extract dextrose agar (SEDA), rice extract dextrose agar (REDA, mushroom extract dextrose agar (MEDA), soyabean extract dextrose agar (SBEDA) and black soya seed extract dextrose agar (BSEDA). The chemical composition of these 8 culture media is given in Table 9.1. pH of these media varied from 4.5 to 6.5. The cultures were incubated at various range of temperatures (10°C to 25°C) inside an incubator. Observations were taken for the mycelium spread on the different media. It is observed that the tissue taken from the stromata region of the fungus is the most suitable inoculum to get the mycelial run in the culture media. However, spores and stalk tissue did not response at all. Out of these eight culture media the mycelial growth was possible on five culture media namely potato dextrose agar (PDA), beef extract dextrose agar (BEDA), casein hydrolysate dextrose agar (CHDA), soybean extract dextrose agar (SEDA) and rice extract dextrose agar (REDA) (Table 9.1).

Table 9.1: Chemical Composition of Various Culture Media

Constituents	*Media (gm/litre)*							
	PDA	*CHDA*	*BEDA*	*SEDA*	*REDA*	*MEDA*	*SBEDA*	*BSEDA*
Peptone	10		10	–	–	–	–	–
Dextrose	40	40	40	40	40	40	40	40
Casein hydrolysate	–	10	–	–	–	–	–	–
Beef extract	–	–	3	–	–	–	–	–
Sodium chloride	–	–	5	–	–	–	–	–
Mushroom powder	–	–	–	–	–	50	–	–
Soybean powder	–	–	–	80	–	–	–	–
Rice powder	–	–	–	–	100	–	–	–
Black soya powder	–	–	–	–	–	–	–	80
Soya bari Powder	–	–	–	–	–	–	40	–
Agar powder	15	15	15	15	15	15	15	15

During the experiment it was also observed that the optimum growth of the fungus occurs under low temperature condition between 10 to 15°C and more acidic pH (5 to 5.5). However, sclerosis is observed in the mycelium obtained on all the types of the culture media. With the result mycelia having numerous spores are observed under the microscope. The cultures are being maintained in the laboratory after sub-culturing these from time to time.

The availability of the specimens in nature is meager and it involves a high labour cost to collect the fungus from the natural habitat. Under such circumstances, laboratory culture of this fungus is the only solution to fulfill the demand of such a high value medicinal and highly prized fungus. Hence, standardization of laboratory culture technique of this fungus needs to be given the prime importance. The laboratory production of the mycelium of *Cordyceps* will definitely prove a great success in preparation of various products from the dried mycelium, which has numerous potential therapeutic applications. Much effort by several groups in China has been made in recent years to culture the fungus and other species of *Cordyceps,* and to determine their medicinal properties The 'Catterpillar Fungus' has become a popular subject in China. More successes in pure culture or growing the fungus through the suitable larva can be expected in the near future (Pegler *et al.*, 1994).

Acknowledgements

We thank the Director, Defence Agricultural Research Laboratory, Pithoragarh and the Head, Department of Botany, Kumaun University, SSJ Campus, Almora for providing necessary research facilities and the Director, Vivekanand Parvatiya Krishi Anusandhan Sansthan, Almora for library facilities. Thanks are also due to UGC, New Delhi for providing financial assistance.

References

Bao, T. T., Wang, G. F. and Yang, J. L. 1988. Pharmacological actions of *Cordyceps sinensis*. Chang His I Chieh Ho Tsa Chih, 8(6): 352-354.

Bok, J. W., Lermer, L., Chilton, J., Klingeman, H.G. and Towerrs, G. H. 1999. Antitumor sterols from the mycelia of *Cordyceps sinensis*. Phytochem., 51(7): 891-898.

Chen, Y. J., Shiao, M. S., Lee, S. S. and Wang, S. Y. 1997. Effect of *Cordyceps sinensis* on the proliferation and differentiation of human leukemic U937 cells. Life Sci., 60 (25): 2349-2359.

Chiou. W. F., Chang, P. C., Chou, C. J. and Chen, C. F. 2000. Protein constituent contributes to the hypotensive and vasorelaxant activities of *Cordyceps sinensis*. Life Sci., 66(14): 1369-1376.

Christensen, C. M. 1975. Fungus Predators and Parasites. In: Moulds, Mushrooms and Mycotoxins, Christensen, C. M. (Ed.), pp 164. University of Minnesota Press, Minneapolis.

Degliantoni, G., Murphy, M., Kobayashi, M., Francis, M. K., Erussia, B.P. and Trinchieri, B. 1985. Natural killer (NK) cell derived hematopoietic colony inhibiting activity and NK cytotoxic factor. Relationship with tumor necrosis factor and synergism with immune interferon. J. Exp. Med, 162: 1512-1530

Dubey, H. C. 1983. An introduction to fungi, pp 231-232. Vikas Publishing House, New Delhi.

Halpern, G. M. 1999. *Cordyceps*:China's Healing Mushroom. New York Avery Publishing Group, New York, USA, 116.

Hobbs, C. 1995. Medicinal Mushrooms: An exploration of tradition, healing and culture. Botanica Press, Santa Cruz, CA, pp 81-86

Hsue, T. H. and Lo, H. C. 2002. Biological activity of *Cordyceps* (Fr) Link species (Ascomycetes) derived from a natural source and from fermented mycelia on diabetes in STZ induced rates. Intl. J. Med. Mush.4: 11-125.

Huang, K. C. 1999.Tonics and supporting herbs. In: The Pharmacology of Chines Herbs. Huang, K. C. (Ed.), pp 263-264. CRC Press, Boca Raton, London, New York,Washington D.C Huber,J. (1958). (In German). Arch. Microbiol. 29: 257-276

Ito, Y. and Hirano, T. 1997. The determination of the partial 18S ribosomal DNA sequences of *Cordyceps* species. Lett. Appl. Microbiol.25: 239-242.

Kiho, T., Yamane, A., Hui, J., Usui, S. and Ukai, S. 1996. Polysaccharides in fungi. XXXVI. Hypoglycemic activity of a polysaccharide (CS-F30) from the cultural mycelium of *Cordyceps sinensis* and its effect on glucose metabolism in mouse liver. Biol. Pharm. Bull. 19 (2): 294-296.

Kuo, Y. C., Tsai, W. J., Shiao, M. S., Chen C. F. and Lin, 1996C. *Cordyceps sinensis* as an immunomodulatory agent. Am. J. Chin. Med, 24(2): 111-125.

Liu, P., Zhu, J., Huang, Y and Liu, C., Chung Kuo Chung Yao Tsa Chih, 1996. 21(6): 367-369.

Mizuno, T. 1999. Medicinal effects and utilization of *Cordyceps* (Fr.) Link (Ascomycetes) and *Isaria* Fr. (Mitosporis fungi) Chinese caterpillar fungi, "Tockukaso" (Review). Intl. J. Med. Mush., 1: 251-262.

Pegler, D. N., Yao, J. and Li, Y. 1994. The Chinese 'Catterpillar Fungus'. Mycologist 8: 3-5.

Sharma, S. 2004. Trade of *Cordyceps sinensis* from high altitude of the India Himalayas: Conservation and Biotechnological priorities. Current Science, 86: 1614-1619.

Wang –Sheng, Y, Shiao-Ming, Shi and Wang, S.Y. 2000. Pharmacological functions of Chinese medicinal fungus *Cordyceps sinensis* and related species. J. Food and Drug Analysis, 8 (4): 248-257.

Wang S. M., Lee L. J., Lin, W.W. and Chung C.M. J. 1998. Cell Biochem., 15: 69(4): 483-489

Wu, T. N., Yang. K. C., Wang, C. M., Lai, J.S., Ko, K. N., Chang, P. Y. and Liou, S. 1996. H.. Lead poisoning caused by contaminated *Cordyceps,* a Chinese herbal medicine: Two case reports. Sci. Total Environ., 182: 193-195.

Xie, Z., Huang, X., Lou, Z., Li, S., Zhou, L., Yang, Z., and Tang, Z. 1988. Dictionary of Traditional Chinese Medicine. The Commercial Press Limited, Hong Kong,.

Zhu, J. S., Halpern, G. M. and Jones, K. 1998. The scientific rediscovery of an ancient Chinese herbal medicine: *Cordyceps chinensis* J. Altern. Complement Med., 4(3): 289-303.

Zhu, X., 1990. Immunosuppressive effect of cultured *Cordyceps sinensis* on cellular immune response. Chin. J. Modern Devs. Trad. Med., 10: 485-487.

Microbes: Diversity and Biotechnology (2012) *Pages* **159–176**
Editors: **Prof. S.C. Sati & Dr. M. Belwal**
Published by: **DAYA PUBLISHING HOUSE, NEW DELHI**

Chapter 10

Mass multiplication and Development of Formulations of the Biocontrol Agent: *Trichoderma* sp.

***Lakshmi Tewari*[1], *Rekha*[1], *Ruby*[1] *and Bipin Chandra*[2]**

[1]*Department of Microbiology, College of Basic Sciences and Humanities,*
[2]*Department of Plant Pathology, College of Agriculture,*
G.B. Pant University of Agriculture and Technology,
Pantnagar – 263 145, US Nagar, Uttarakhand

ABSTRACT

Biological control with beneficial microorganisms is in its infancy and has a promising future. The renewed interest in biocontrol is due to its environmental friendliness, long lasting effects and safety features. Some of the bacterial and fungal antagonists have, however, also been found to show distinct plant growth promoting effects. Fungicidal control of pathogens under field conditions is beset with practical problems on account of cost to user and the environmental pollutions and persistence of chemicals. Besides increased use of potentially hazardous pesticides has become a matter of growing concern among both environmentalists and public health authorities. Considering the cost of chemical pesticides and hazards involved biological control of plant disease is now increasingly being practiced by microbiologists. Further, biological control strategy is highly compatible with sustainable agriculture and has a major role to play as a component of integrated pest management (IPM) programme. General unavailability of suitable commercial methods for growth, formulation and deliveries of antagonists are the major constraints in the implementation of biocontrol. Therefore, efforts are being made by scientists for selection of a suitable cheaper substrate for mass production, product formulation and application methods of the selected antagonist. Development of acceptable easily prepared and cost effective formulations for delivery should be a major goal. Considering these facts, in the present chapter we have discussed various methods used for mass production, formulation development and application of the potential fungal antagonists in fields for biological control of fungal phytopathogens.

Keywords: *Trichoderma sp., Biological control, Fungal antagonists.*

Introduction

Biological control by potential antagonistic organisms, such as *Trichoderma* sp., is a non-chemical and eco-friendly approach for crop protection against plant pathogenic fungi. But the lack of cost effective commercial methods for inoculum development are the major constraints in the implementation of biocontrol agents. Research on biocontrol is in its infancy and work on commercialization of potential antagonists is being carried out throughout the world. Several fungal antagonists, such as *Trichoderma* sp., have been developed into several commercial biological control products (bio-pesticides) and are being used in field crops and greenhouse systems (Harman, 2000). These products are known to control numerous soil-borne diseases. There is some degree of host-specificity in biocontrol agents even at sub-species level that may partially account for the reported inconsistent performance of biocontrol agent preparations. Single biocontrol agent is not likely to be active in all soil environments or against all pathogens that attack the host plant. Control of a wide spectrum of pathogens under a wide range of environmental conditions by applied antagonists largely remains an unfulfilled goal for biological control. There are four main approaches to achieve this goal- (1) select strain of biocontrol agent with wide host range, (2) modify the genetics of the biocontrol agent to add mechanisms of disease suppression that are operable against more than one pathogen, (3) alter the environment to favour the biocontrol agent and to disfavour competitive micro-flora, and (4) develop strain mixtures with superior biocontrol activity (Janisiewicz, 1988). Several strategies for developing mixtures of biocontrol agents or microbial consortium could be envisioned including mixtures of organisms with different plant colonization patterns, mixtures of antagonists that control different pathogens or mixtures of antagonists with different mechanisms of disease suppression or consortium of antagonists with different optimum temperature, pH, or moisture conditions for rhizosphere/phyllosphere colonization. Thus, microbial consortia comprising of two or more effective antagonists are considered to account for protection in disease suppressive soils (Chaube and Singh, 1991). However, there also are reports of combinations of biocontrol agents that do not result in improved suppression of disease compared with the separate antagonists. Incompatibility of the co-inoculants can arise because biocontrol agents may also inhibit each other as well as the target pathogen(s). Thus an important prerequisite for successful development of strain mixtures appears to be the compatibility of the co-inoculated microorganisms (Baker, 1990). Over the past one hundred years, research has repeatedly demonstrated that phylogenetically diverse microorganisms can act as natural antagonists of various plant pathogens.

Developing appropriate formulation and delivery systems/mode of application are the prerequisite for implementing biological control using microbial antagonists. Formulation of biological control agents depends upon rapid and cost effective method of biomass production and maintaining viability at the end of the process. Several commercial formulations of *Trichoderma* sp. mainly based on inert carriers are available for controlling plant diseases (Lewis and Papavizas, 1985). Multiplying *Trichoderma* sp. on easily biodegradable substrates with long shelf-life would be beneficial for field application. Growing concerns about environmental health and safety have led to substantial regulatory changes in the past several years, including the Food Quality Protection Act of 1996.

Major Types of Microbial Antagonists of Phytopathogens

Although, biocontrol agents, antagonistic to fungal plant pathogens, include all classes or groups of organisms existing in an ecosystem, maximum emphasis for developing bio control programme has been given to fungal and bacterial antagonists primarily because of ease of their multiplication and formulation. The most important genera of fungi studied as biocontrol agents are *Trichoderma*,

Gliocladium, Penicillium, Neurospora, Aspergillus, Glomus, etc. Among fungi, species of *Trichoderma* and *Gliocladium* are widely used as potential antagonists. A number of bacterial species/strains have been studied for their plant growth promoting activity and biocontrol potential. The potential bacterial antagonists include *Pseudomonas, Bacillus, Erwinia, Agrobacterium, Streptomycese, Actinoplanes, Serratia,* etc.

Plant growth promoting rhizobacteria- especially the fluorescent Pseudomonads and the fungal antagonists *Trichoderma* sp.are under intensive research because of their wide natural occurrence and biocontrol potential against fungal and nematode diseases. However, development of a new soil microbial antagonist begins with discovery of a useful naturally occurring rhizosphere competent organism. It requires an extensive screening of natural soil isolates for selection of strains having desired traits. The isolates are then screened for their activity against the target pathogens in laboratory (*in vitro*) and green house (*in vivo*) conditions as well as for their optimum growth conditions (Lewis *et al.*, 1990). Once an isolate with desired trait is obtained, it is evaluated under field conditions. The selected potential indigenous strain may either be used as a commercial strain or it is improved further for certain desired traits using various techniques before formulation and developing it as a biopesticide.

Isolation and Screening of *Trichoderma* Strains for Antagonistic Potential

Selection of an effective biocontrol strain plays a prime role in disease management. Isolation is generally done from the pathogen suppressive soils either by dilution plate technique or by baiting the soil with fungal structures like sclerotia of pathogen. For isolation of *Trichoderma* sp., soil samples are collected from the rhizosphere of healthy plants. Serially diluted soil sample is mixed with 20 ml of melted agar medium in Petri plates by giving a gentle swirling motion to the plate and allowed them to incubate at room (28±2°C) temperature (Islam *et al.*, 2008). The isolated fungal strains are further purified by single spore method and identified on the basis of their morphological, biochemical and molecular characters. The purified fungal isolates are stored on PDA slants at 4°C until being used. All the strains isolated from different cropping systems have to be ascertained for their virulence and broad spectrum of action against different plant pathogens causing serious economic threat to cultivation. Selection of an effective strain decides the viability of the technology. Hence a proper yardstick should be developed to screen the antagonistic potentiality of the biocontrol agents.

Thus, for developing a successful bio-pesticide formulation, the selected antagonistic organism should possess following characteristics (Jeyarajan and Nakkeeran, 2000):

1. High rhizosphere competence
2. High competitive saprophytic ability
3. Plant growth promotory ability
4. Ease for mass multiplication
5. Broad spectrum action
6. Excellent and reliable control
7. Safe to environment
8. Compatible with other rhizobacteria
10. High tolerance to desiccation, heat, oxidizing agents and UV radiations.

Dual Culture Plate Technique is used for testing *in vitro* antagonistic potential of the isolates. Dual culture plates inoculated with the test isolate and the test pathogen are evaluated periodically for suppression of pathogen's growth as compared with the growth of the pathogen on control plate; inhibition percent of growth of pathogen is calculated using the following formula (Sallam *et al.*, 2008):

Growth reduction (per cent) = (Growth in control – Growth in treatment/ Growth in control) × 100

However, *in vitro* screening of the antagonists through dual culture plate technique alone could not be an effective method for strain selection. To be an effective antagonist, it should possess a high level of competitive saprophytic ability, antibiosis, should have the ability to secrete increased level of cell wall lytic enzymes (chitinases, glucanases and proteinases) and plant growth promotory potential. Hence the yardstick should be developed, comprising of above-mentioned components. Each component should be given weightage depending upon their role in disease management. This type of rigorous and meticulous screening will lead to identification of an effective biocontrol strain suited for commercialization (Berg *et al.*, 2001). The biocontrol efficacy should then be tested under glass house conditions and finally field trials should also be conducted before selection of the potential biocontrol strain. The plant, pathogen and antagonists are co-exposed to controlled environmental conditions. Exposure of the host to the heavy inoculum-pressure of the pathogen along with the antagonist will provide ecological data on the performance of the antagonist under controlled conditions. Promising antagonists from controlled environment are tested for their efficacy under field conditions along with the standard recommended fungicides. Since the variation in the environment under field condition influence the performance of biocontrol agent, trials on the field efficacy should be conducted for at least 15 – 20 locations under different environmental conditions to promote the best candidate for mass multiplication and formulation development (Jeyarajan and Nakkeeran, 2000). Inoculum of selected bioagent is raised on seeds/grains such as wheat, sorghum or jhingora. Grains inoculated with spore suspension of *Trichoderma* are incubated for 10-12 days and colonized seeds are used as inoculum for mass multiplication of the bioagent.

Improvement of Strains for Bio-efficacy

The selected strains of the antagonist may be improved further for certain desirable characteristics like better antagonistic ability, wider host range, tolerance to pesticides, survival ability, rhizosphere competence, tolerance to stress environmental conditions, faster growth rate and longer shelf life. Different methods are used today for strain improvement such as – mutagenesis, irradiation using gamma or UV radiations, protoplast fusion or some other genetic engineering technique as discussed below:

(*i*) Mutation

Several genetic variants or mutants of fungal biocontrol agents have been developed by using radiations such as UV rays or γ-rays or through chemical mutagenesis. Benomyl tolerant mutants of *Trichoderma viride* have been developed by chemical mutagenesis (Mukherjee *et al.*, 1997). The stable mutants produced more antifungal substances and were equally effective as the wild type strain in disease control potential. Several stable mutants of *T.virens* have also been developed using 125 k rad of gamma radiation. Selvakumar and co workers (2000) developed carboxin tolerant mutants of *Trichoderma viride* by exposing the cultures to UV light and ethyl methan sulphonate.

(*ii*) Protoplast Fusion

Protoplast fusion is one of the most important methods that can be used for development of improved strains of bioagents well suited for the modern integrated diseases management system.

This method is of significant importance for developing improved strains because *Trichoderma* is well known for producing various cell wall degrading enzymes such as chitinase, cellulase, glucanases, proteinases and various other mycolytic enzymes. Many genetic variants have been developed by protoplast fusion of two different species of *Trichoderma- T. harzianum* and *T. virens;* some of the fusants have been reported to exhibit improved biocontrol potential over parental strains.

(iii) Recombinant DNA and Genetic Engineering Techniques

Several genetically modified microorganisms (GMMs) or genetically engineered microorganisms (GEMs) have been developed through gene cloning and genetic engineering techniques. These transgenic microorganisms constructed to express foreign genes from another organism whose products are inhibitory to the plant pathogens, are likely to have high utility as bio control agents. Biocontrol efficiency of *Trichoderma* has been improved by transformation with genes *prb1*(basic protease,), *egl 1* (β-1,4-glucanase) and *chit* 33 (chitinase) (Migheli *et al.*, 1998). The best documented success story of biocontrol has been management of crown gall caused by *Agrobacterium tumefaciens* using genetically modified strain of *Agrobacterium radiobacter* strain K84. Strain K84 contains the plasmid pAgK84, that harbours genes encoding agrocin 84 biosynthesis and agrocin 84 resistance (Ryder *et al.*, 1987). This plasmid is readily mobilizable into virulent strains of *A. tumefaciens* by conjugation so that virulent strain would become resistant to agrocin 84 resulting in failure of biocontrol. Therefore, conjugal transfer deficient deletion mutant, designated as Strain K1026, were constructed by deleting a part of the plasmid. The resulting strain K 1026 possessed all the traits of the parental K84 strain with the exception of its inability to transfer the plasmid pAgK84 (Jones and Kerr, 1989). This strain K1026 was registered in Australia in 1988 under the trade name *'Nogali'* for the control of crown gall on stone fruits, nuts and roses. Today, *A radiobacter* strain K1026 is the only genetically engineered biocontrol agent registered for the management of plant diseases.

Mass Scale Production Strategies

Commercial success of *Trichoderma* sp. based biocontrol formulations would also require economically feasible mass scale production processes. One of the major constraints in the development of biological control is lack or scarcity of suitable method and substrate for mass culturing (as 35–40 per cent costs of production depends on raw material) and mode of application of the bio-agent (Papavizas and Lewis, 1982; Mukhopadhyay, 1994). The ultimate objective of any BCA lies in its feasibility of economical mass production which also holds true for *Trichoderma* sp based formulations. The most significant problem in developing bio-pesticides is that these formulations contain living propagules, such as vegetative cells, spores and/or mycelial fragments which must be able to withstand the process of formulation and should have longer shelf life retaining viability for certain period of time until it reaches the farmers (Vidhyasekaran and Mutha milan, 1995).Almost all available *Trichoderma* sp. based BCA products contain spores as active ingredients (Batta, 2004). This could be attributed to the physiological aspects of their three microbial propagules, namely, mycelia, conidia, and chlamydospores. The three propagules possess distinct physiological characteristics in terms of production, stability and BCA activity. Therefore, it is imperative to select the best suitable form of *Trichoderma* sp. propagules in order to efficiently execute their biocontrol potential. For commercial production of these antagonists different technologies have been adopted on industrial scale (Sen, 2000). Fermentation technology is widely employed for mass multiplication of these bioagents using bioreactors.

Of the two broad categories of fermentative processes, namely, solid state fermentation (SSF) and liquid State fermentation (LSF), LSF has been widely adopted by several researchers despite lower sporulation rates. Labour, scale-up, process control, productivity, material handling (pumping, pressurized lines), compatibility with pre-existing large scale facilities are some positive features of LSF that encourage most researchers to pursue LSF in lieu of SSF. Unfortunately, accurate information regarding factors directing sporulation process in SLF of fungi, especially, for *Trichoderma* sp. is scarce and incomplete. Therefore, production of spores in LSF still remains a challenging task and warrants considerable research inputs. Basically three types of fermentation technologies are used for mass production of biocontrol agents: (*i*) Liquid state fermentation or surface culture method, (*ii*) Solid state fermentation and (*iii*) Submerged fermentation. In most cases adequate and effective biomass formation has been accomplished by modification of liquid (deep tank) or solid (semi-solid) fermentation.

(*i*) Solid-State Fermentation (SSF)

Wide range of organic substrates could be used for the solid-state fermentation and mass multiplication of the bioagents. Media used for mass culturing of the biocontrol agent through solid state fermentation generally consist of inert carriers with food bases (Lewis, 1991). Solid substrates include straws, wheat bran, sawdust, moistened bagasse, sorghum grains, paddy chaff, and decomposed coir pith, farmyard manure and other substrates rich in cellulose individually or in combination for inoculum production. Solid or semi solid fermentation is being used almost exclusively for experimental production of fungal biomass. In this technique, the culture medium is impregnated in a career such as bagasse, wheat straw, wheat bran, potato pulp, etc. and the organism is allowed to grow on it. This method allows greater surface area for fungal growth. Solid state fermentations are also suitable for production of fungi which either do not sporulate in liquid cultures or do not survive in the liquid fermentation process. This process is appropriate for countries like ours where agricultural wastes are available in plenty amount, elaborate technical facilities are limited and labour is abundant. Various grain seeds (*e.g.* sorghum, wheat, soya, barley and other minor millets), bagasse, straws, wheat bran, saw dust, spent tea leaf waste, coffee husk, diatomaceous earth granules impregnated with molasses, cow dung, Farm Yard manure, individually or in combination have been used as substrate in solid fermentation. However, SSF is often preferred to LSF, when production scale is of moderate range and labour force is cheap (Molla *et al.*, 2004). Additionally, recent advancements in industrial automatization have encouraged major BCA producers to consider SSF as a viable mass scale option.

Trichoderma sp. have been successfully grown on different seeds and grains such as wheat and sorghum, but it is not desirable to use food grains or seed for this purpose because of low availability, moreover the process has to be cost effective. Therefore, various cheaper cellulosic agro wastes are being used for mass multiplication of the bioagent. *Trichoderma* is well known for its cellulose degrading potential and grows well on these substrates (Figure 10.1). Conidial counts in spore powders of *T. harzianum* raised on various cellulosic substrates ranged from 0.51 to 4.95 x 10^8/g powder in 5 to 20 d old cultures as shown in Table 10.1 (Tewari and Bhanu, 2004).

Table 10.1: Conidiospore Yield (cfu/g) of *Trichoderma harzianum* on Different Cellulosic Agrowastes

Substrates	*($x10^8$) cfu/g in 20 d Old Cultures*
Wheat Straw	4.86
Paddy Straw	4.95
Maize cob	3.67
Paper waste	1.16
Saw dust	2.29
Sugarcane bagasse	3.73
Farm yard manure	0.65
Spent compost of mushroom	0.55
Spent straw	3.03

Figure 10.1: Radial Growth and Production of Conidiospores of *Trichoderma harzianum* on Different Cellulosic Substrates

For development of conidia based powdery formulations, a suitable cheaper substrate favouring enormous quantities of conidiospores production is needed. Higher mycelial growth but lower conidial yield on organic manures suggests that a mixed substrate combination of various categories of substrates might be used to achieve higher conidial yield in lesser time. Faster mycelial growth on wheat and paddy straw might be attributed to easily available nutrients and lower C: N ratio than saw dust, paper waste and shelled maize cob and enhanced cellulase enzyme activity (Lewis and Papavizas, 1985). Supplementation of cellulosic substrates with nitrogen and phosphorus source further enhance conidial yield of *Trichoderma harzianum* (Papavizas *et al.,* 1984). Successful multiplication of the bioagents on ground mesocarp fibre of oil palm in plastic bags has also been reported (Sawangsri *et al.,* 2007).

(*ii*) Liquid State Fermentation (LSF)–Surface Culture Method

In this method, the antagonist is multiplied in liquid media. The broth cultures are grown without agitation that is no additional aeration is provided for the growth of aerobic cultures. Surface culture methods therefore, result in only surface growth of the organism. After appropriate incubation period, biomass of the antagonist is separated from culture filtrate and is used for formulation development. Liquid fermentation for biomass production of bacterial antagonists (*e.g. Pseudomonas, Bacillus* sp.) is the preferred approach. This method is time consuming and needs large area or space. This method is not suitable for cultivating aerobic microorganisms. For successful fermentation, not only appropriate substrates are used but also sufficient biomass containing adequate amount of effective propagules must be obtained.

(*iii*) Submerged Fermentation

The conventional method of large scale production typically employs the use of a submerged culture. The use of this method comprises preparing an inoculant of a desired strain of *Trichoderma* by submerged culture in a suitable liquid culture medium under continuous aeration and placing the inoculant in a sufficient volume of a suitable liquid medium to permit the production of *Trichoderma*. In this process, the organism is grown in a liquid medium which is vigorously aerated and agitated in large tanks called fermentors/bioreactors. The fermentor could be either an open tank or a closed tank and may be a batch type or a continuous type and are generally made of non- corrosive type of metal or glass lined or of wood. In batch fermentation, the organism is grown in a known amount of culture medium for a defined period of time and then the cell biomass is separated from the liquid and used for formulation development. In continuous culture, the culture medium is withdrawn depending on the rate of product formation and the inflow of fresh medium. Most fermentation industries today use the submerged process for the production of biomass. However, optimization of fermentation conditions using suitable substrate, such as temperature, pH, and aeration, amount of inoculum and supplementation of substrate for higher biomass yield is highly required. There are several additional factors to be considered like the rate at which an effective biomass is produced, cost of production as well as contamination and viability. For mass multiplication the selected medium should be inexpensive and readily available with appropriate nutrient balance (Manjula and Podile, 2001). Labour, scale-up, process control, productivity, material handling (pumping, pressurized lines), compatibility with pre-existing large scale facilities are some positive features of LSF that encourage most researchers to pursue this method.

For economical (low cost raw material) and efficient production (higher sporulation) of *Trichoderma* sp. based BCAs, a novel array of substrates have to be explored. Akin to prior mentioned wastes, experimenting with wastewater and wastewater sludge (source of carbon, nitrogen, phosphorus, and other essential nutrients for many microbial processes) could provide probable viable solution to combat the raw material cost and enhance sporulation. In this context, Verma *et al.* (2007) have reported mass multiplication of *T. viride* on municipal wastewater from BCA point of view. The studies advocated potential utilization of municipal wastewater sludge and industrial wastewaters as shown in the mass balance flow chart (Figure 10.2). Fermented broths of conventional media/wastewaters/wastewater sludge will be mixed with either dry Talc/silica powder, or dry dewatered sludge powder as carrier material in the ratio of 1:99 (w/w), respectively. Thus, *Trichoderma* sp. production on sludge would not just serve as potent BCA, but also as a novel technique for sustainable sludge management. *Trichoderma* can be grown on simple and less expensive locally available substrates both by liquid and solid fermentation

(*iv*) Combined Process

To overcome limitations of SSF and LSF, many researchers also suggested hybrid strategies involving both SSF and LSF. Normally, LSF is followed by SSF in many industrial production processes. In a typical *Trichoderma* sp. production process, 2–3 days old broth of LSF is used as inoculum for solid substrates, *e.g.*, rice bran, grain-husk and others. The solid substrates thus inoculated are incubated for further 2–12 days, followed by addition of formulation agents, *e.g.*, carboxy methyl cellulose, silica, talc and moderate temperature (about 20–40ºC) air-drying below 8–10 per cent moisture content (Verma *et al.*, 2007).

Formulation Development

Technologies become viable only when the research findings are transferred from the lab to field, and therefore, even the best technology is of no value unless it is transferred in the consumer's hand for use. Ideally, formulation ensures protection of active ingredients (spores, conidia, mycelial germlings of antagonistic fungi) from extreme pH, lower humidity, chemicals and UV radiation. Major research on biocontrol is centered with the use of cell suspensions of the bioagent directly to seed. Though these

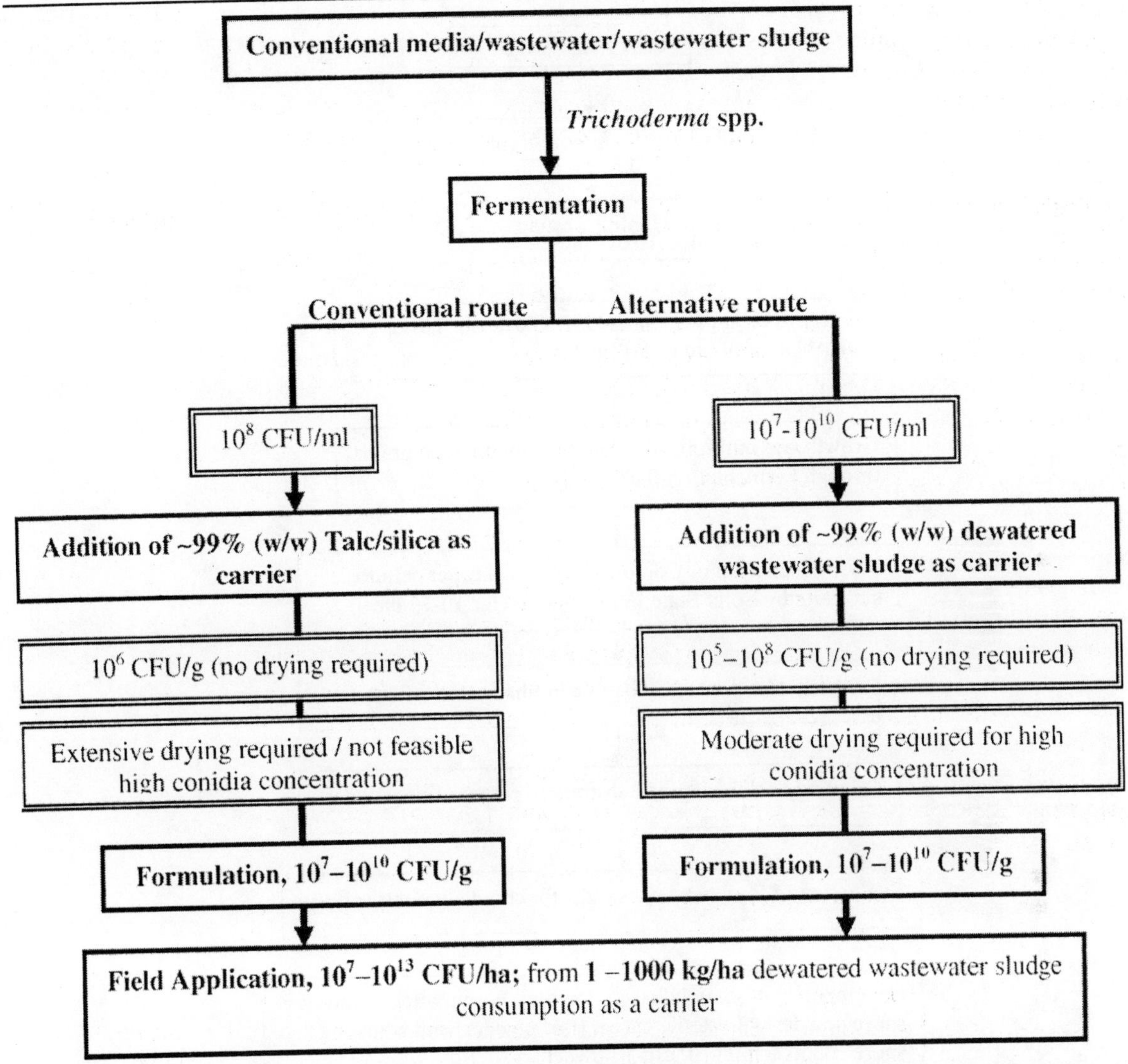

Figure 10.2: Comparative Flow Chart Showing Mass Production of the Bio-agent (*T. viride*) through Conventional and an Alternative Route Using Municipal Wastewater and Wastewater Sludge (*Source*: Verma *et al.*, 2007)

biocontrol agents have a very good potential in the management of pests and diseases, it could not be used as cell suspension under field conditions. Hence, the cell suspensions of bioagents should be immobilized in certain carriers and should be prepared as formulations for easy application, storage, commercialization and field use. For obvious reasons, formulation development of BCAs is one of the most important steps in the overall production process (Figure 10.3).

Formulated BCAs exhibit antagonistic action without being affected by the adverse environmental factors. Hence, developing a safe, easy to use, cost-effective formulation that will keep *Trichoderma*

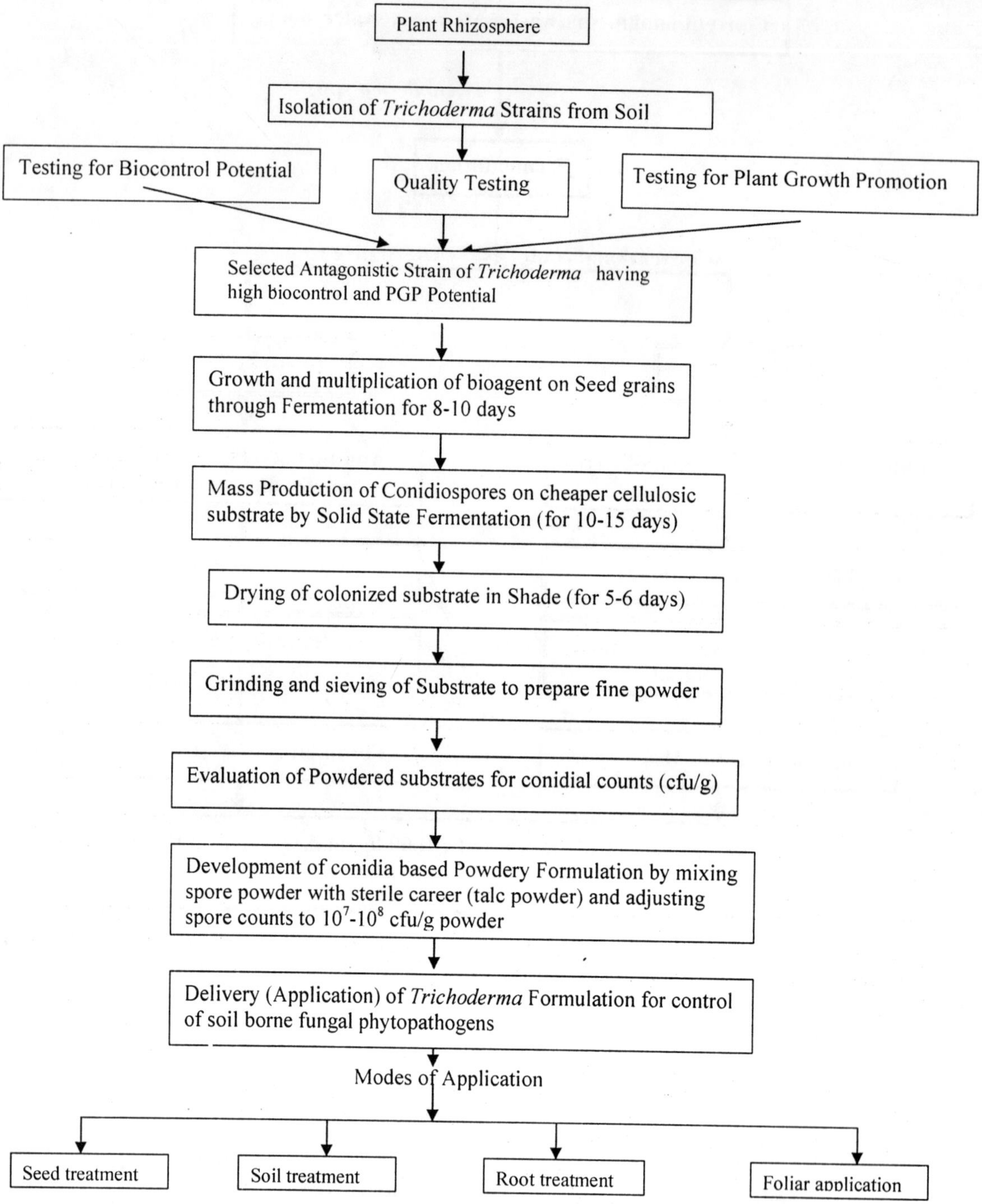

Figure 10.3: Mass Multiplication on Cellulosic Substrates, Development of Formulation and Modes of Application of the Bioagents–(*Trichoderma* sp.) for Management of Plant Diseases and Fungal Phytopathogens

alive for longer time is one of the most important steps in developing final product. Moreover, for any crop protection agent, an efficient formulation is essential to translate laboratory activity into adequate field performance. Formulation is the blending of active ingredients (*Trichoderma* propagules and hyphae) with inert carriers such as talc, peat and peat plus additives, coal and coal plus additives, inorganic soils, plant materials (compost made from bagasse, saw dust, rice husk, plant compost, corncob, cellulose, soybean and coir dust), FYM powder, cow dung powder or inert materials (*e.g.* vermiculite, vermiculite plus additives, perlite, ground rock phosphate, calcium phosphate, polyacrylamide gel, entrapped alginate beads etc.). Carboxy methyl cellulose (CMC) @ 1 per cent is used with talc powder. Little progress has been made in this area despite the fact that formulations will improve field efficacy of fungal BCAs and expand their market opportunities beyond high value, niche markets. New, more effective formulation components must be found (*e.g.* U V protectants, carriers, virulence-enhancing factors and "stickers"). In all cases, it is essential that these formulants be compatible with other BCAs (viruses, bacteria, and entomophilic nematodes) which may be sprayed at the same time as the fungal BCA (Bateman and Luke, 2000).

Many researchers have reported different types of formulations of *Trichoderma, e.g.*,invert emulsion; cane molasses amendment; seed coating; pregelatinized starch-flour granules; wettable powder; alginate pellets; gluten matrix. However, in general, conidia without any amendments were ineffective. Furthermore, Bae and Knudsen and McLean *et al.* (2005) have demonstrated that when introduced in soil, the biocontrol activity of *Trichoderma sp.* was formulation dependent. Characteristics of an ideal formulation are given in Box-1.

Box 1: Characteristics of an Ideal BCA–formulation

An ideal formulation should have following characteristics (Jeyarajan and Nakkeeran, 2000):

a. Should have increased shelf life
b. Should not be phyto-toxic to the crop plants and animals
c. Should dissolve well in water
d. Should tolerate adverse environmental conditions
e. Should be cost effective and should give reliable control of plant diseases
f. Should be compatible with other agrochemicals
g. Carriers must be cheap and readily available for formulation development

Modern formulations of bioagents are liquid based. They are made usually in mineral oil along with emulsifiers. Type of formulation may influence field efficacy. Wettable powder (WP) formulations are most popular but less efficient and environmentally hazardous particularly when used as foliar spray. Liquid formulations are better for foliar application. Requirement for humid conditions for the fungus to work can be avoided by spraying fungal spores in oil.

Type of carriers used in formulations may influence shelf-life. Addition of FYM powder to talc + CMC based formulation of *T. harzianum* enhanced temperature and UV radiation tolerance of the biocontrol agent. One of the critical factors that affect quality of formulated product is process of drying and moisture content of the formulation. Drying should be quick and safe without compromising on the bioagents' viability and efficacy. Adverse drying conditions lead to the death of bioagent's propagules, which may serve as a source of nutrition to pathogens and will lead to inconsistent results during tests. Ideally moisture content of formulated product should be 6 to 8 per cent. Development of some of *Trichoderma* based formulations is described below:

Preparation of Pellets

Conidia of *Trichoderma* sp. are collected from the surface of colonies on Potato Dextrose Agar (PDA) plates by washing. The concentration of conidia is determined with a haemocytometer prior to pellet preparation (Lewis and Papavizas, 1985). Sodium alginate is dissolved in distilled water using a magnetic hotplate stirrer. The zeolite is mixed in a blender with distilled water, autoclaved for 30 min and cooled. Then *Trichoderma* isolate and sodium alginate solution are added to it and mixed for 30 min at high speed. The concentration of conidia added should be of the order of 10^7- 10^8 propagules per litre. This final mixture containing *Trichoderma propagules*, alginate and zeolite are added dropwise into gellant solution (0.25 M $CaCl_2$, pH 5.4). After some time, beads are formed in the gellant solution which are separated by gentle filtration, washed and dried for 24 h at 28°C. For determining conidial counts, the pallets were disintegrated in water and serially diluted. Serial dilutions of the homogenate were spread on TSM agar plates, colonies were counted and populations were reported as colony-forming-units (cfu) per g of pellet (Kucuk and Kivanç., 2005).

Preparation of Powdery Formulations

For preparing dry powdery formulations, the substrates colonized with *Trichoderma* spores are air dried, ground and sieved to prepare fine powder. The powder is mixed with some inert carrier like charcoal, talc, peat etc. and subjected to conidial counts (cfu/g substrate). These powdery formulations are stored in dry air tight bottles in refridgerator till use and are evaluated periodically for shelf life.

Shelf Life of Formulations

Since bio-pesticidal preparations contain living propagules which loose their viability during storage period, therefore, lower shelf life *i.e.* loss of viability of the biocontrol agents over time is one of the critical obstacles to commercialization of a bio-pesticidal preparation.The final formulation should have a minimum shelf-life of 2 years at room temperature, be easy to handle, insensitive to abuse and must be stable over a range of temperature (–5 to 35°C). This is the ideal situation and should be kept as an ultimate aim for developing a formulation. Failure to meet these rigid standards, however, should not stop the commercialization of biologicals.

Several attempts have been made to determine the viability of biocontrol agents in their preparations when stored at room temperature and in refrigerator. Most of the results are variable and it appears that shelf life is dependent on several factors like type of propagules, species/isolate/strain, culture media, formulating media, moisture content of the formulation etc. Effect of storage period on biological activity of the formulated antagonists has been tested by several workers. The powder formulations of *Trichoderma* sp. ($3x10^7$cfu/g) stored for four months in sealed polyethylene bags at room temperature (25-30°C) were evaluated for cfu/g of powder formulation (Sallam *et al.*, 2008). Mixed formulations exhibit longer shelf life, compared to individual ones when stored at room temperature, while no significant difference among them was observed at low temperature. The higher shelf life of mixed formulation at room temperature on coated seeds might be due to the reason that the biocontrol agents are reported to stimulate exudation from the seeds that may serve as nutrients for biocontrol agents.

Delivery System (Mode of Application)

Efficacy of a developed *Trichoderma* formulation also depends on its mode of application to control the target pathogen; therefore, delivery system (mode of application) of the formulation has been one of the major areas of biocontrol research. Like chemical pesticides *Trichoderma* formulations are applied through seed, soil, root and/or foliage. It is necessary to have an efficient, cost effective and ecologically viable mode of application of the bio-agent in soil ecosystem.

(*a*) Seed treatment

Seed treatment is one of the most commonly used methods of *Trichoderma* application and has emerged as a feasible way of delivering the antagonist for the management of plant diseases. A large number of seed, seedling, root and foliar diseases have been suppressed by seed treatment with antagonists. Even internally seed-borne disease like loose smut of wheat is suppressed by seed treatment with *Trichoderma* (Selvakumar et al., 2000). This could be due to action of metabolites produced by *Trichoderma*, which may penetrate seeds to affect growth of internally seed-borne pathogen-inoculum. When applied through seeds, antagonists not only colonize spermoplane and spermosphere, but also rhizoplane and rhizosphere. Spore/cell suspension as well as dry powder has been used to coat the seeds with potential antagonists (Harman, 2000). For commercial purpose dry powder of antagonists are used @ 3 to 10 g powder per kg seed, based on seed size and formulations of antagonist (Mukhopadhyay, 1994). *Trichoderma hamatum, T. harzianum, T. virens* and *T. viride* are effective seed protectants against *Pythium* sp. and *Rhizoctonia solani*. One approach for this might be supplying the coated seeds to the farmers directly by the seed companies/agencies.

(*b*) Seedling (Root) Treatment

This method is generally used for the vegetable crops, rice etc. where transplanting is practiced. Seedling roots can be treated with spore or cell suspension of antagonists either by drenching bioagent to nursery beds or by dipping roots in bioagent spore suspension before transplanting. There are also reports on the reduction of sheath blight disease severity in rice by treatment of seedlings in nursery before transplanting (Tewari and Singh, 2005). Root dipping in antagonist's suspension not only reduces disease severity but also enhances seedling growth in rice, tomato, brinjal and chili.

(c) Seed Biopriming

Seed biopriming refers to 'treatment of seeds with biocontrol agents followed by its incubation under warm and moist conditions until just prior to radicle emergence. This method has potential advantages over simple coating of seeds as it results in rapid and uniform seedling emergence. *Trichoderma* conidia germinate on the seed surface and form a layer around bioprimed seeds. Such seeds are better tolerant to adverse soil conditions.

(d) Soil Treatment

The bioagent-formulation can be added to the soil during sowing/plantation or just before plantation for disease management. There are several reports on the application of biocontrol agents to the soil and other growing media either before or at the time of planting for control of a wide range of soil-borne fungal pathogens. Such applications are ideally suited for green house and nursery but because of the bulk requirement, cost and problem of uniform distribution, feasibility of field application is less. Granular or pellet preparations have been used directly for soil application and they have provided effective control of diseases both under green house and field conditions. Prior to soil application, *Trichoderma* formulation is added to moist FYM, incubated for few days to allow colonization and colonized FYM is distributed in field followed by irrigation. It is most effective method of application of *Trichoderma* particularly for the management of soil borne diseases.

(e) Foliar Application

Application of antagonistic microorganisms through foliar spray is an effective mode for control of fungal phyto-pathogens. Foliar spray of *Trichoderma harzianum* either alone or in combination with other delivery systems has been reported as an effective mode in controlling the disease caused by *Rhizoctoni solani* (sheath blight) in rice plants (Tewari and Singh, 2005). There are several reports showing effectiveness of biocontrol agents applied as foliar spray against different plant pathogens. However, success of any antagonist on leaf/sheath surface depends largely on its ability to colonize these surfaces. Even environmental factors like humidity, temperature and sun light affecting this colonization also affect bio-efficacy of antagonists. Because of these reasons foliar application is preferred during evening hours and it is more successful when applied as two sprays at an interval of around fifteen days.

Factors Affecting Field Efficacy

Biocontrol potential of an antagonist is highly influenced by several factors. One of major complaints against bioagents including *Trichoderma* has been inconsistent field performance. Strain of the bioagents and soil factors are the main factors influencing field performance of BCA. Although *Trichoderma* species are relatively broad spectrum in their antagonism but most effective strains against different pathogens and crops may be different. It is not only the degree of antagonism against pathogen but its temperature and pH tolerance and rhizosphere/phyllosphere colonization ability which may affect field performance of *Trichoderma*. Also there is wide spread belief that resident strains are more effective. Soil moisture, pH and microbial composition influences field efficacy. *Trichoderma harzianum* was more effective than *Pseudomomas fluorescens* in acidic soil against pre- and post-emergence damping off and root rots in a number vegetable crops.

For more consistent performance under field condition:

- ✰ *Trichoderma* should be used as a component of IPM, not as a stand alone treatment.
- ✰ More than one method of application (*e.g.* through seed and colonized FYM) should be invariably adopted.
- ✰ If possible it should be used along with nutrient base

Trichoderma Based Commercialized Products (Biopesticides)

Today several *Trichoderma* based commercial products are available in market that can be used as biopesticide and biofertlizers. The formulation not only control most of the soil borne diseases but also increases the yield of several test crops like sunflower, mustard, soybean, chrysanthemum etc. These biopesticides are eco-friendly, safe and biodegradable. The ultimate objective of any BCA lies in its feasibility of economical mass production which also holds true for *Trichoderma* sp. based BCAs. Further, from Figure 10.4, it is obvious that *Trichoderma* sp. based BCAs are commercially viable as numerous commercial products exist in market. A vast majority are rather being promoted as soil enhancer and/or growth promoter. Almost all available *Trichoderma* sp,. based BCA products contain spores as active ingredients (Batt *et al.*, 2004), this could be attributed to the physiological aspects of their three microbial propagules, namely, mycelia, conidia, and chlamydospores. The three propagules possess distinct physiological characteristics in terms of production, stability and BCA activity. Therefore, it is imperative to select the best suitable form of *Trichoderma* sp. propagules in order to efficiently execute their BCA action.

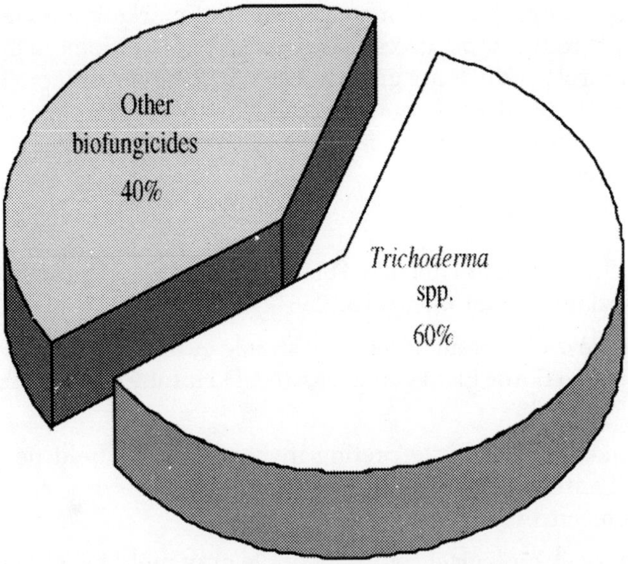

Figure 10.4: *Trichoderma* spp. Based Biofungicide Market. Other biofungicides include bacteria, nematodes and virus.

Table 10.2: Some of the Commercial *Trichoderma* sp. Based Biocontrol Products

Trade Name	*Active Bio-agent*	*Formulation*	*Target Pathogen/Effects*	*Manufacturer/Distributor*
SUPRESIVIT	*Trichoderma harzianum* PV 5736-89	Wettable Powder (WP)	Damping off of ornamentals, fungal diseases of peas	FYTovita, Czech Republic
TRI 002	*T. harzianum* KRL AG-2	Granules (G)	Plant growth stimulation, Strengthening of plants against pathogens	Bioworks Inc., Geneva, USA;PLANT SUPPORT, B.V., The Netherlands
ECOFIT	*T.viride*	WP	Species of *Fusarium Pythium,* and *Rhizoctonia*	Hoechst Schering AgrEvo, Mumbai,India
BINAB T	*T. harzianum* ATCC 20476 and *T.polysporum*-ATCC 20475	WP (biorational fungicide)	Fungal pathogens that infect tree wounds of ornamental shade andforest trees	BINAB Bio-innovation EFTRAB, Florettgatan 5,SE-254 67 Helsingborg, Sweden
SoilGard	*Gliocladium virens* GL-21	Granules	Damping off and root rot pathogens especiaiiy *R. solani* and *Pythium* sp. of ornamentals and food crops	Certis USA LLC, 9145 Guilford Road, Columbia
T-22, Root shield, Plant Shield	*T. harzianum* Rifai Strain KRL-AG2 (T-22)	Granules and wettable powder	*Pythium, Rhizoctonia* and *Fusarium* sp. for vegetables, shrubs etc.	Bioworks, Inc., 345 Woodcliff Dr., I Floor NY, USA

There is a growing demand for sound, biologically-based plant disease management practices. Currently a number of biologically based products are being sold all over the world including India. A growing number of companies are also developing new products that are in the process of being registered. Scientific, regulatory, business management and marketing issues must all be handled effectively for a biocontrol product to be successful in the marketplace. Some of the common *Trichoderma* based commercial biocontrol products are given in Table 10.2; however more information regarding these bio-products can be obtained from a number of sources (from web pages). Increased demand for organic produce by pesticide-worry urban populations has indicating a very bright market potential and future of the biocontrol products.

References

Bae Y. S., and Knudsen G. R.. 2005. Soil microbial biomass influence on growth and biocontrol efficacy of Trichoderma harzianum, Biol. Control 32 236–242.

Baker, R. 1990. An overview of current and future strategies and models for biological control. *In:* Biological Control of Soil Borne Plant Pathogens. (ed, D.Hornby,) CAB International Wallingford, United Kingdom. Pp 375-388

Bateman, R. P. and Luke, B., 2000. Interpreting mycoinsecticide field performance: an uneasy relationship with chemical pesticides,In Predicting Field Performance in Crop Protection, 89-100, BCPC, Farnham, Surrey

Batta Y. A., 2004. Postharvest biological control of apple gray mold by *Trichoderma harzianum* Rifai formulated in invert emulsion, Crop Prot. 2319–26.

Berg, G., Fritze, A., Roskot, N. and Smalla, K. 2001. Evaluation of potential biocontrol rhizobacteria from different host plants of *Verticillium dahliae* Kleb. J. Appl. Microbiol. 91: 963-971.

Chaube, H. S. and Singh, U. S. 1991. Plant Disease Management: Principles and Practices. CRC Press, Boca Raton, U.S.A.

Harman, G. E. 2000. Myths and dogmas of Biocontrol: Changes in the perceptions derived from research on *Trichoderma harzianum* T-22. Plant Dis. 84: 377-393.

Islam M. S., Saha A. K., Mosaddeque H. Q. M., Amin M. R. and Islam M. M. 2008. In Vitro Studies on the Reaction of Fungi Trichoderma to Different Herbicides Used in Tea Plantation. Int. J. Sustain. Crop Prod. 3(5): 27-30

Janisiewicz, W. J. 1988. Biocontrol of post-harvest diseases of apples with antagonists mixtures. Phytopathology 78: 194-198.

Jeyarajan, R. and Nakkeeran, S. 2000, Exploitation of microorganisms and viruses as biocontrol agents for crop disease mangement. In: Biocontrol Potential and their Exploitation in Sustainable agriculture, (Ed. Upadhyay *et al.*) Kluwer Academic/Plenum Publishers, USA;pp95-116.

Jones,D. A. and Kerr, A. 1989. *Agrobacterium radiobacter* K1026, a genetically engineered derivative of strain K84, for biological control of crown gall. Plant Disease 73: 15-18.

Küçük Ç and Kivanç M. 2005. Effect of formulation on the viabilýty of biocontrol agent, *Trýchoderma harzýanum* conidia, African Journal of Biotechnology Vol. 4 (5), pp. 483-486

Lewis, J. A., 1991, Formulation and delivery system of biocontrol agents with emphasis on fungi Beltsville symposia in Agricultural Research. In: The rhizosphere and plant growth (Keister, D. L. and Cregan, P. B. eds.,) 14 pp. 279-287.

Lewis J. A., Barksdale T. H. and Papavizas G. C. 1990. Greenhouse and field studies on the biological control of tomato fruit rot caused by *Rhizoctonia solani*. Crop Prot. (9): 8–14.

Lewis J. A. and Papavizas G. C. 1985. Characterization of alginate pellets formulated with *Trichoderma* and *Gliocladium* and their effect on the proliferation of the fungi in soil. Plant Pathol. 34: 571-577.

Manjula, K. and Podile, A. R., 2001, Chitin supplemented formulations improve biocontrol and plant growth promoting efficiency of Bacillus subtilis AF1. Can. J. Microbiol. 47: 618-625.

McLean K. L., Swaminathan J., Frampton C. M., Hunt J. S., Ridgway H. J. and Stewart A. 2005. Effect of formulation on the rhizosphere competence and biocontrol ability of *Trichoderma atroviride* C52 212–218.

Migheli, Q., Gonzalez-Candelas, L. Dealessi, Camponogara, A. and Raman-Vidal, D. 1998. Transformants of *Trichoderma longibrachiatum* overexpressing the β-1,4-glucanase gene egl1 show enhanced biocontrol of *Pythium ultimum* on cucumber. Phytopathology 88: 673-677.

Molla A. H., Fakhru'l-Razia, A., Hanafi, M. M. and Alam, M.Z. 2004. Optimization of process factors for solid-state bioconversion of domestic wastewater sludge, Int. Biodeterior. Biodegrad. 53: 49–55.

Mukherjee, P. K., Haware, M.P. and Raghu, P. 1997. Induction and evaluation of benomyl-tolerent mutants of *Trichoderma viride* for biological control of botrytis gray mould of chickpea. Indian Phytopathol. 50: 485-489.

Mukhopadhyay, A. N. 1994. Biological control of Soil borne fungal plant Pathogens- current status, future prospects and potential limitations. Indian Phytopathol. 47: 119-126.

Papavizas, G. C. 1985. *Trichoderma* and *Gliocladium*: Biology and potential for biological control. Annu. Rev. Phytopathol. 23: 23-54.

Papavizas, G. C., Dunn, M. T., Lewis, J. A. and Beagle-Ristaino, J. 1984. Liquid fermentation technology for experimental production of biocontrol fungi. Phytopathology 74:1171-1175.

Papavizas G. C. and Lewis J. A. 1982. Introduction and augmentation of microbial antagonists for the control of soil borne plant pathogens. In: Biological Control in Crop Production, (ed., Papavizas,G.C., Allenhald), Osmum, Granada, USA., pp. 305-322.

Ryder, M. H., Slota, J. E., Scarim, A. and Farrand, S. K. 1987. Genetic analysis of agrocin 84 production and immunity in *Agrobacterium* spp. J.Bacteriol. 169: 4184-4189.

Sallam N. M. A., Abo-Elyousr K. A. M. and Hassan M. A. E. 2008. Evaluation of *Trichoderma* Species as Biocontrol Agents for Damping-Off and Wilt Diseases of *Phaseolus vulgaris* L. and Efficacy of Suggested Formula, Egypt. J. Phytopathol., Vol. 36, No.1-2, pp. 81-93

Sawangsri, P., Pengnoo, A., Suwanprasert, J. and Kanjanamaneesathian, M. 2007 Effect of *Trichoderma harzianum* biomass and *Bradyrhizobium* sp. strain NC 92 to control leaf blight disease of bambara groundnut (*Vigna subterranea*) caused by *Rhizoctonia solani* in the field Songklanakarin J. Sci. Technol., 29(1): 15-24.

Selvakumar, R., Srivastava, K. D., Aggarwal, Rashmi, Singh, D.V. and Dureja, Prem. 2000. Studies on development of *Trichoderma viride* mutants and their effect on *Ustilago segetum tritici*. Indian Phytopathol. 53: 185-189.

Sen, B. 2000. Biological control: A success story. Indian Phytopathol. 53: 243-249.

Tewari, L. and Bhanu, C. 2004. Evaluation of agro-industrial wastes for conidia based inoculum production of biocontrol agent: *Trichoderma harzianum*. J. Sci. Indus Res. 63: 807-812.

Tewari,L. and Singh,R. 2005. Biological control of sheath blight of rice by *Trichoderma harzianum* using different delivery systems. Indian Phytopath. 58 (1): 35-40.

Verma M., Brar S. K., Tyagi R. D., Surampallib R.Y. and Val´ero J.R. 2007 Antagonistic fungi, *Trichoderma* spp.: Panoply of biological control. Biochemical Engineering Journal 37: 1–20

Vidhyasekaran, P. and Muthamilan, M. 1995. Development of Formulations of *Pseudomonas fluorescens* for control of chickpea wilt. Plant Disease. 79: 782-786.

Microbes: Diversity and Biotechnology (2012) *Pages* **177–196**
Editors: **Prof. S.C. Sati & Dr. M. Belwal**
Published by: **DAYA PUBLISHING HOUSE, NEW DELHI**

Chapter 11

Diversity of Arbuscular Mycorrhizal Fungi in Wheat Agro-climatic Regions of India

Reena Singh and Alok Adholeya*

Centre for Mycorrhizal Research, Biotechnology and Management of Bioresources Division, The Energy and Resources Institute (TERI), Indian Habitat Centre, Lodhi Road, New Delhi – 110 003

ABSTRACT

AM (Arbuscular mycorrhizal) fungi are important yet unknown components of biodiversity in the agricultural fields in India. To study their diversity and habitat relationships, we characterized and enumerated spores in 55 fields of wheat from 11 agro-climatic regions of India, varying in climatic and edaphic characteristics. The AM fungal spore count, species richness, most frequent species, and intra-radical colonization were studied in various samples drawn from these regions. A total of 165 samples were collected at the time of the wheat harvest. These samples were used as trap cultures and multiplied in a green-house for a period of one year, which yielded 34 species scattered over 6 genera. The genera *Glomus* Tulasne and Tulasne occurred most frequently, constituting 89.1 per cent of the total species. The number of species in a given region ranged from 1–9. *Glomus albidum* Walker and Rhodes and *G. macrocarpum* Tulasne and Tulasne were found to be the most commonly occurring species.

Keywords: AM fungi, Diversity, Wheat, Agro-climatic regions.

---

* Corresponding Author.

Introduction

AM (Arbuscular mycorrhizal) fungi are ubiquitous in soils around the globe and have been associated with improved plant growth for over 100 years (Trappe 1987). These symbiotic fungi are the main components of the soil micro biota in most of the agro-ecosystems and account for 25 per cent of the biomass of the soil micro flora and micro fauna combined (Hamel 1996). They are generally known to increase the absorption and translocation of mineral nutrients from the soil to the host plant (George *et al.*, 1995), to improve the tolerance of the host plant towards biotic (Singh *et al.*, 2001) and abiotic stresses (Gaur and Adholeya 2004), and to build up the macro-porous structure of the soil that allows penetration of water as well as air and prevents erosion (Miller and Jastrow 1992).

The beneficial effects of AM fungi on plants and soil health prompted great interest in their research (Sharma and Adholeya 2004). The diversity of AM fungi has been extensively studied in natural ecosystems (review by Allen *et al.*, 1995) and its importance for plant diversity, productivity and ecosystem processes has also been recognized (van der Heijden *et al.*, 1998). However, little is known about AM fungal diversity in agricultural lands. Agricultural lands are artificial ecosystems that are subjected to constant human intervention. Whereas, in natural ecosystems the internal regulation of functions is a product of plant biodiversity through the flow of energy and nutrients, and this form of control is progressively lost under agricultural intensification (Altieri 1999). Modern agriculture implies simplifying the structure of the environment over vast areas, and replacing nature's diversity with a small number of cultivated plants and domesticated animals. Several workers noted that the diversity of AM fungal communities tends to decrease when natural ecosystems are converted into agro-ecosystems and their diversity decreases as the intensity of agricultural inputs increases (Schenck *et al.*, 1989; Oehl *et al.*, 2003).

India presents a range and diversity of climate, soils, flora and fauna, with few parallels in the world and hence affords scope for much diversity in agriculture. However, only 14 AM fungal species were recorded from the agricultural fields of India (Dalal and Hippalgaonkar 1995; Singh and Pandya 1995) with only 4–5 species being from the wheat fields (Singh and Pandya 1995). These studies were either conducted at single or relatively few study sites. Although such findings are invaluable while invaluable in increasing knowledge about the ecology of AM fungi, they are limited in application as they are largely based on site-specific research. Studies on a larger scale are needed to allow robust prediction of the relative importance of different agro-climatic regions for these fungi.

In India, wheat, the second important food crop, is grown on 27 million hectares out of the total 114 million hectares of land under cultivation (Ladha *et al.*, 2000). The wheat region occupies most of the northern, western and central India and fields from these regions were selected for the study. To the best of our knowledge, there is no detailed study on the diversity of AM fungi in these regions. Our study primarily aims at improving the understanding of the broad-scale distribution of AM fungi in different wheat-growing agro-climatic regions of India. To achieve this, we sampled the spores of AM fungi in 55 study fields, and analyzed the distribution of some of the more commonly recorded species and the overall number of species in relation to different agro-climatic regions. This study will help in exploiting the potential of these fungi in sustainable agriculture, particularly in wheat.

Materials and Method

General Study Area

The study was established within the northern, western, and central regions of India (Figure 11.1), which form the major wheat-growing belt and contribute two-thirds of the 45 million tonnes of

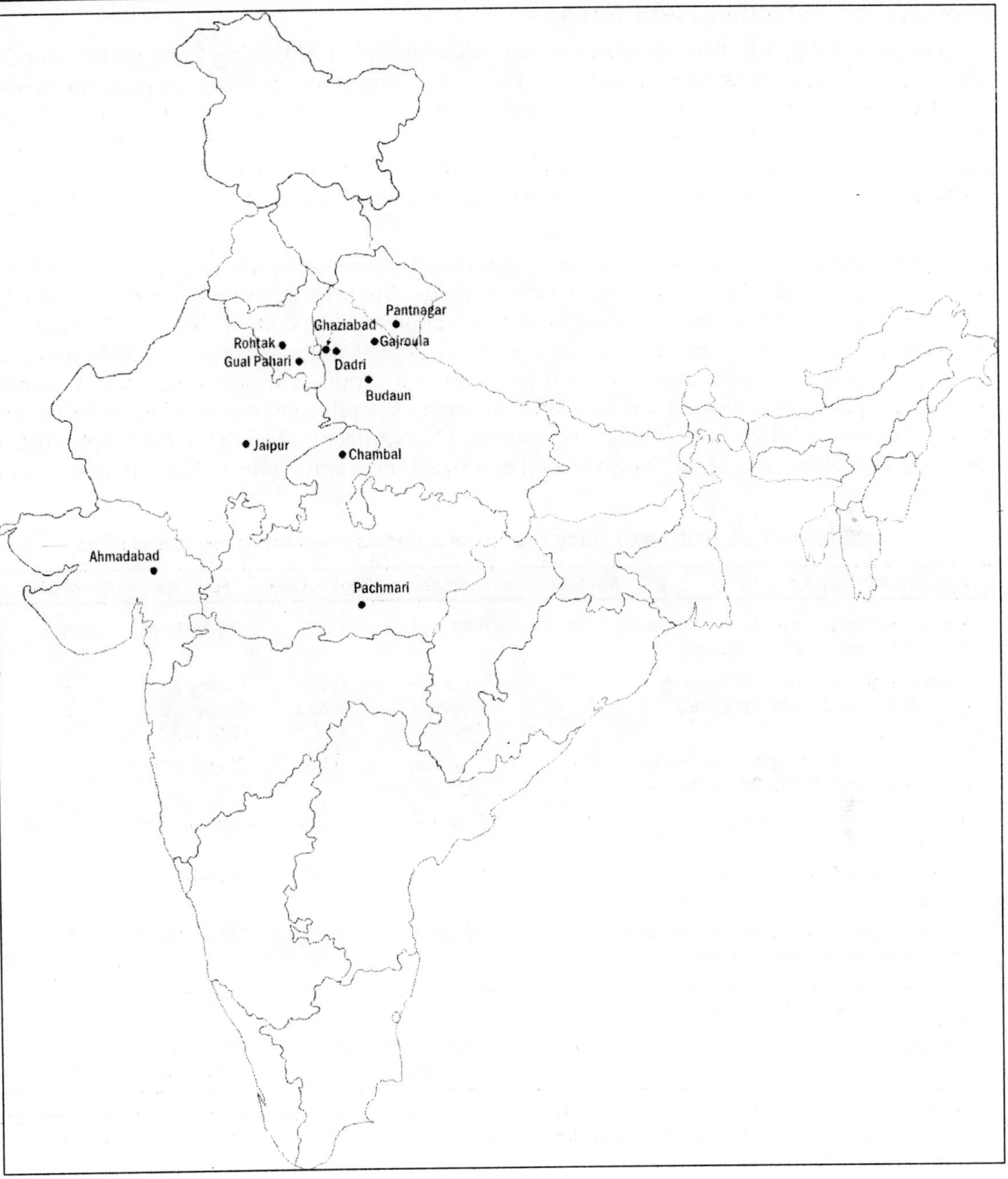

Figure 11.1: Sample Collection Sites in Northern, Western, and Central India

the total production of wheat (Ministry of Finance 2000). The area included a range in soil types and climate. More specifically, the study area formed a rectangle bound by the longitudes 72°35′E and 82°09′E, and the latitudes 21°14′N and 28°54′N, encompassing agro-climatic regions of the States in India of Uttar Pradesh, Madhya Pradesh, Haryana, Rajasthan, and Gujarat.

Strategy for Selecting Field Sites

Out of 18 states, which are divided into agro-climatic regions (Planning Commission, working group report on agricultural research and education for the formulation of 8th five year plan, government of India 1989), we identified 5 states, which are major wheat- growing zones, and a total of 11 agro-climatic regions were selected. The basis of the division of agro-climatic regions is essentially based on climate, soils, and existing cropping patterns of each state as a unit. These parameters have been shown to influence patterns in the occurrence of plants and animals (Margules *et al.*, 1987; Lindenmayer *et al.*, 1996).

From each region, one or two replicates were selected and for each replicate, five collection fields were chosen (Table 11.1), thus totalling 55 collection fields. Wherever possible, it was possible that the fields were at least 500m apart to allow for the collection of independent data. The location and elevation of each site was determined from fine scale (1:25 000) topographic maps. To minimize heterogeneity with respect to agricultural practices and inputs, only the fields having similar agricultural practices (wheat under cultivation for more than five years and tillage with tractor), fertilizer inputs (120–150 kg/ha [kilogram per hectare] N [Nitrogen], 40-60 kg/ha P [Phosphorus], 40-60 kg/ha K [Potassium], 25 kg/ha Zn [Zinc]) and irrigation inputs (four to five irrigations) were selected.

Table 1 Distribution of the 55 study sites in relation to agro-climatic regions of India

Agro-climatic Regions	*Replicates**	*Place*	*Region Code*	*Soil Type*	*No. of Sites*
Bhabar and Tarai Zone of Uttar Pradesh (28°22'N latitude 79°25'E longitude)	1	Pantnagar	Bht	Clayey loam	5
Western Plain Zone of Uttar Pradesh (28°40'N latitude 77°43'E longitude)	3	Dadri	Wep 1	Loamy sand	5
		Gajroula	Wep 2	Sandy loamy	5
		Ghaziabad	Wep 3	Silty loamy	5
Mid-Western Plain Zone of Uttar Pradesh (28°02'N latitude 79°07'E longitude)	1	Budaun	Miw	Clayey loam	5
Gird Zone of Madhya Pradesh (21°14'N latitude 81°38'E longitude)	1	Chambal	Giz	Light alluvium	5
Malwa Plateau Zone of Madhya Pradesh (22°05'N latitude 82°09'E longitude)	1	Pachmari	Mal	Medium alluvium	5
Semi-Arid Eastern Main Zone of Rajasthan (26°55'N latitude 75°48'E longitude)	1	Jaipur	Samz	Loamy sand	5
South Gujarat Zone of Gujarat (23°02'N latitude 72°35'E longitude)	1	Ahmedabad	Sgz	Clayey	5
Western Zone of Haryana (28°54'N latitude 76°35'E longitude)	2	Rohtak	Whz 1	Fine loam	5
		Gual-Pahiri	Whz 2	Alluvium	5

* Within each combination of the agro-climatic region we selected five sites. Hence, for each of these regions sampled only once, there were five sites. For two of the agro-climatic regions, we replicated twice or thrice in each of the agro-climatic categories.

Sampling AM Fungi

For sampling purposes, each collection field was divided into four blocks. From each block, five undisturbed core samples (500 g each) were collected (soil and roots) from the rhizosphere of wheat plants from a depth of 0–30 cm using a core sampler at the time of wheat harvest (April 2001). Thus, a total of 20 soil cores (5´4) were collected from each collection field. The samples were air-dried in the shade to the point where there was no free moisture and were placed into zip bags, and stored at 4°C in a cold room until processed. The samples were used for three different purposes: (1) propagation of AM fungal isolate of each collection field for their characterization, (2) analysis of AM fungal parameters, and (3) analysis of soil chemical parameters.

Propagation of AM Fungi in Trap Cultures

Previous studies of studying the distribution and abundance of AM fungi have largely been based on recording AM fungal species at the time of sample collection. Several issues must be considered when traditional taxonomic identification of spores is used to describe the AM fungal community diversity. First, the number of spores in the soil may not reflect the relative amount of colonization of roots by this fungus or the amount and distribution of hyphae in the soil. Second, non-sporulating species may be present (Clapp *et al.*, 1995). A fungus may be a significant member of the 'vegetative' community, but because of the date of sampling, local environment, or host plant regulation of carbon expenditure, be unable to produce spores yet able to persist to the following year as infective hyphae in the roots or the soil. Third, spores collected from the field may be difficult to identify as a result of degradation of the spore walls (Morton *et al.*, 1993). The non-sporulating species can often be coaxed to sporulate in 'trap cultures'. As these fungi are obligatory mycotrophs, propagation of AM fungal cultures require their growth in association with a living plant.

The trap cultures were established using modified methodology (pers. comm. by Prof. Andres Wiemken) and Oehl *et al.* (2003). Plastic trays (460´290´240 mm^3) were used to establish AM fungal cultures in a greenhouse using soil samples from all collection fields. For each collection field, four plastic trays (comprising four blocks) having 10mm hole at the bottom were prepared. A 20 mm thick drainage mat (Enkadrain ST, Schoellkopf AG, CH-8057 Zurich, Switzerland) was placed at the bottom of each tray and the tray was filled with 25 kg of substrate (50 per cent Terragreen [American aluminium oxide, oil dry US special, Type IIIR] and 50 per cent soil sediment having Olsen P = 1.56 ppm; Organic C = 0.28 per cent; Total N = 0.052 per cent and K = 52.66 ppm). The substrate was autoclaved at 120°C for one hour at 15 psi (pounds per square inch) before filling. Substrate cores (500 g) were taken out from five different places (four corners and centre) in each tray and were replaced by five undisturbed soil cores (500 g, containing collection field's AM fungi) to inoculate the hosts. Seeds of *Allium cepa* (onion), *Tagetus* spp. (marigold), *Daucus carotus* (carrot), *Medicago sativa* (alfalfa), and *Trifolium alexandrianum* (berseem) were pre-germinated. Five pre-germinated seeds of each species were placed on top of the five soil cores. The plants were watered to a moisture level of approximately 60 per cent of the water-holding capacity and were grown in a greenhouse at 20 ± 5 °C with 60 per cent relative humidity. The pots were arranged on a greenhouse bench in a completely randomized design with three replications. Half-strength Hoagland's nutrient solution (Hoagland and Arnon 1938) was provided to the plants at fortnightly intervals. After four months of the growth cycle, the pots were left to dry undisturbed at a fairly stable temperature so that the drying period would not be too rapid. After completion of the growth cycle the dried shoots were cut at the ground level without disturbing the substrate and pre-germinated seeds of different hosts; *Gossypium* spp. (Cotton), *Vetiveria zizanioides* (Vetiver), *Vigna radiata* (mungbean), *Sorghum* spp., and *Tagetus* spp. (marigold) were sown again. After completion of each growth cycle, rhizosphere soil cores were taken from the vicinity of plants growing in trap cultures at a depth of 0–15cm and species characterization was done.

AM Fungal Spore Isolation, Quantification, and Characterization

The spores were quantified and characterized at three different time intervals at the time of sample collection (collection field soil), after the first and second trap culture cycle. The spores were isolated following the modified technique of wet sieving and decanting (Gerdemann and Nicolson 1963) and were quantified as given by Smith and Skipper (1979). The diagnostic slides of different species of AM fungal spores were prepared as described by Schenck and Perez (1990). The initial observation of spores (colour, shape, etc. in water) was recorded under a stereomicroscope (Leica). The main morphological variables used for their characterization were (Giovannetti and Gianinazzi-Pearson 1994): (1) sporocarp occurrence, shape, colour and size; (2) peridium occurrence, and characteristics; (3) spore colour, size and shape; (4) spore wall number, colour, thickness, and ornamentation; and (5) hyphal attachment, shape and type of occlusions.

The measurement of AM fungal spores (spore diameter, wall thickness, hyphal thickness and thickness at the attachment point) was done using an image analyzer system (Image Pro-Plus, 4.0 version) attached to a compound microscope (Olympus BH 2). Once the data on the spores was generated they were characterized up to the level of species using the manual for the identification of VA mycorrhizal fungi (Schenck and Perez 1990). Some species were also characterized by the monograph released by Hall (1984), the INVAM (http://invam.caf.wvu.edu/fungi/taxonomy/speciesID) and BEG (http://www.kent.ac.uk/bio/beg/) websites, and the existing published taxonomic literature. Collections, which could not be thus determined, were characterized up to the genus level and labelled sp. 1, sp. 2, and so on. The collections, which were characterized up to the species level, were deposited at CMCC (Centre for Mycorrhizal Culture Collection), TERI, India, which has been designated as the mycorrhizal germplasm culture collection by the Department of Biotechnology, Government of India.

Intraradical Colonization by AM Fungi

The roots were cleared and stained by the technique by Phillips and Hayman (1970) and percentage colonization was calculated according to Biermann and Lindermann (1981) at three intervals of time: at the time of sample collection (collection field soil) and after the first and second trap-culture cycle.

Soil Chemical Parameters

The collection field soils were analyzed at the time of sample collection for their chemical parameters. A soil suspension of 1:2.5 (soil-to-water mixture) was made. The pH of the soil suspension was measured by a digital pH meter (Expandable Ion Analyser EA 940, Orion Research) and the electrical conductivity was measured by a digital electrical conductivity meter (controlled dynamics). A protocol by Datta *et al.* (1962) was followed for measuring the percent organic carbon. The percent total nitrogen was calculated using Kjeldahl's method by Bremner (1960). The available phosphorus was determined using Olsen's method (Olsen *et al.*, 1954) and the estimation of available potassium was done using a flame photometer with filters (Wood and Deturk 1940).

Data Analysis

We analyzed AM fungi community structure using the following ecological parameters: population abundance at each site, mean population abundance, total population abundance at each site, species richness at each site, species diversity at each site, species evenness at each site, and species dominance at each site. Population abundance at each site was defined as the sum of individuals of a particular species, counted at each site during all the observations. Mean population abundance was obtained by averaging the population abundance of sites only where a specific species was found. Total population abundance was calculated as the population abundance of all component species at each site. Species richness was expressed by the number of species found in each study site during the observation period. Species diversity at each site was expressed by Shannon-Wiener function. $H' = -S^{S}_{i=1}pi$, Where S is the number of species in each site. Species evenness was expressed by the Shannon equitability index, $J' = H'/\text{In } S$, where H' is the Shannon-Wiener function and S is the number of species in each site. Species dominance was expressed by McNaughton's dominance index, $D1 = (n_1+n_2)/N$, where n_1 and n_2 is the population abundance of the first and second dominant species, respectively.

Inter-regional differences in richness, diversity, dominance and the effects of soil chemical properties on their scores, were analysed by means of ANOVA (Analysis of Variance) with the software costat and means were separated using DMRT (Duncan's Multiple Range Test) at 5 per cent level and significance (LSD, P=0.05). It should be noted that data are spatially autocorrected {that is, two points close to each other will be less independent of each other than two points located at a larger distance from each other (Legendere 1990; Borcard *et al.*, 1992)}. This would lead to a pseudoreplication problem (Hurlbert 1984) if each point were to consider as an independent unit. Multiple regression procedures allow for the generation of models for conditioning on the contribution of macronutrients of sampling units to the variation in the studied variables (Borcard *et al.*, 1992). Coefficient of correlation (r^2) was calculated between different soil chemical features and mycorrhizal parameters.

Results

Differences in the AM Fungal Community Structure Between Different Agro-climatic Regions of India

Overall, 34 species and 165 isolates of AM fungi were recorded in the 55 study fields scattered in 11 different wheat-growing agro-climatic regions of India. These regions exhibited variable trends. 15 species of AM fungi were recorded in the Arid Zone of Madhya Pradesh, which is characterized by light alluvial soil and only one species of AM fungi was recorded from the Western-Plain Zone (Wep 2, Gajroula) and the Mid-Western Plain Zone of Uttar Pradesh, associated with sandy loam soil (Figure 11.2). Table 11.2 shows the results of soil chemical analysis.

Occurrence of AM Fungi

The rhizosphere samples of all the plant species were found to contain AM fungal spores. Spore population ranged from 0.1-19 g^{-1} with an overall average of 3.1 g^{-1} (LSD [0.05] = 167.22). The spores were found to be maximum in the Giz (Chambal) region (up to 24.14 g^{-1}), followed by Whz 1 (Rohtak) region (6.47 g^{-1}). (data not given). The spore count was poor in Miw (Budaun) (0.94 g^{-1}) and Wep2 (Gual Pahiri) (0.35 g^{-1}) regions in fields. The average rate of sporulation increased in the trap cultures. However, most species did not sporulate untill the trap cultures passed through a minimum of two successive cycles. The trap cultures of Wep 2 (Gajroula), Miw, Samz (Jaipur) and Shne (Ahmedabad) showed no sporulation in first culture cycle although the roots were colonized with the mycorrhiza (data not given). In other trap cultures also, the number of species sporulated were less during the first cycle. In contrast, the total number of AM fungi increased to maximum of 6 to 7 species in the second cycle of multiplication. Although total sporulation increased dramatically from the first to the second cycle, the organisms that were sporulating varied with site and culture cycle.

The genus *Glomus* Tulasne and Tulasne was recorded as the most frequently occurring (89.1 per cent), followed by *Gigaspora* Gerdemann and Trappe emend. Walker and Sanders (10.9 per cent),

Table 11.2: Comparison of Soil Chemical Properties in Different Agro-climatic Regions

Region Code	*pH*	*Electrical Conductivity (dSm⁻¹)*	*% Organic Carbon*	*% Total Nitrogen*	*Olsen's Phosphorus (ppm)*	*Available Potassium (ppm)*
Bht	7.45 b	0.30 b	0.68 cd	0.07 a	5.73 cd	57.80 d
Wep1	8.62 a	0.75 a	0.30 d	0.02 c	11.11 b	75.80 cd
Wep 2	7.39 b	0.54 b	1.78 a	0.02 c	19.23 a	216.60 a
Wep 3	7.55 b	0.37 cd	0.66 cd	0.02 c	7.56c	52.40 d
Miw	6.65 c	0.22 d	0.39 d	0.02 c	6.16 cd	92.40 cd
Mal	7.14 b	0.19 d	1.14 a	0.02 c	4.85 d	61.40 d
Giz	7.41 b	0.16 d	1.72 a	0.02 c	2.66 d	133.20 b
Samz	8.22 a	0.20 d	0.43 d	0.03 c	5.91 cd	104.20 bc
Sgz	7.26 b	0.82 a	0.87 bc	0.04 b	5.36 cd	71.20 cd
Whz 1	7.41 b	0.21 d	0.55 cd	0.03 b	4.89cd	72.80 cd
Whz 2	7.45 b	0.42 bc	0.58 cd	0.02 c	5.49 cd	70.60 cd
LSD (P=0.05)	0.43	0.18	0.36	0.01	3.33	35.41
F value	11.97	14.21	16.33	17.34	14.89	14.62
Level of significance	***	***	***	***	***	***

Bht: Pantnagar; Wep 1: Dadri; Wep 2: Gajroula; Wep 3: Ghaziabad; Miw: Budaun; Mal: Pachmari; Giz: Chambal; Samz: Jaipur; Sgz: Ahmedabad; Whz 1: Rohtak; Whz 2: Gual Pahiri.

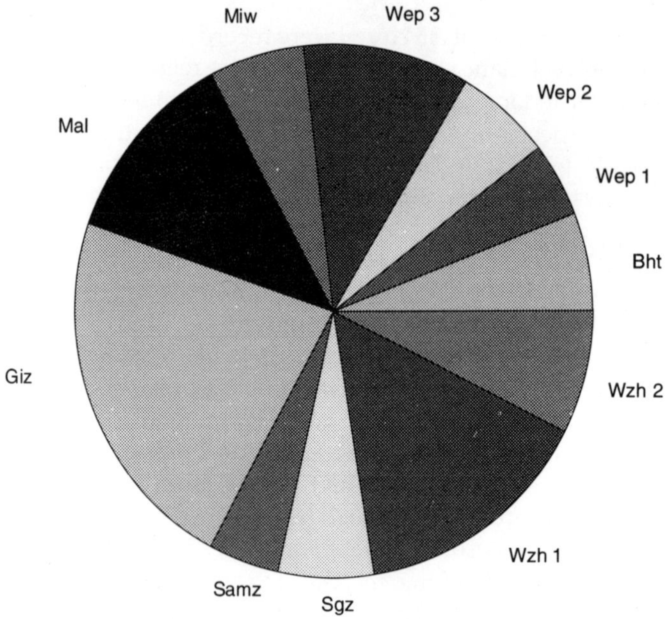

Figure 11.2: AM Fungal Species Richness Observed in Different Agro-climatic Regions of India

Bht: Pantnagar; Wep 1: Dadri; Wep 2: Gajroula; Wep 3: Ghaziabad; Miw: Budaun; Mal: Pachmari; Giz: Chambal; Samz: Jaipur; Sgz: Ahmedabad; Whz 1: Rohtak; Whz 2: Gual Pahiri.

Scutellospora Walker and Sanders (9.09 per cent), and *Entrophospora* Ames and Schneider (3.6 per cent). Spores of *Glomus albidum* Walker and Rhodes (30 per cent), *G. macrocarpum* Tulasne and Tulasne (25 per cent) and *G. mosseae* (Nicolson and Gerdemann) Gerdemann and Trappe (20 per cent) were amongst the frequently occurring species (Figure 11.3). Table 11.3 depicts the AM fungal species recorded in different agro-climatic regions of India. Each region contained some characteristic species and displayed a characteristic pattern of population.

Table 10.3: AM Fungal Species Diversity in Samples Collected from Different Regions of India at the Time of Sample Collection (*), After the First Trap Culture Cycle (♦) and After the Second Trap Culture Cycle (J)

AM Fungal Species Diversity	*CMCC Accession Number*	*Collection Fields where AM Fungal Species were Reported*
Entrophospora infrequens (Hall) Ames and Schneider	AM WE01	Wep 3a*, Wep 3e*
Gigaspora gigantea (Nicolson and Gerdemann) Gerdemann and Trappe	AM WGI02	Sgz b*J, Sgz c*♦J, Sgz d*, Sgz e*♦J
Gigaspora margarita Becker and Hall	AM WGI03	Wep 2d*, Samz aJ, Samz b*, Samz c*J, Samz d*, Samz eJ
Glomus aggregatum Schenck and Smith emend. Koske	AM WG04	Mal a*J, Mal b*J, Mal d*♦J, Mal e*J, Giz b♦J, Giz c♦J, Giz d*♦J, Giz e♦J, Wzh 2b♦, Wzh 2cJ
G. albidum Walker and Rhodes	AM WG05	Bht dJ, Bht e*, Wep1a♦, Wep1b*, Wep1dJ, Wep 1e*, Wep 2a*♦, Wep 2e*♦, Miw b*, Miw dJ, Wzh 1a*J, Wzh 1c*, Wzh d*♦J, Wzh 2a*, Wzh 2c*, Wzh 2d*J, Wzh 2e*J
G. ambisporum Smith and Schenck	AM WG06	Giz a*J, Giz b *J, Giz cJ, Giz d♦J, Giz e*J
G. botryoides Rothwell and Victor	AM WG07	Giz a*♦J, Giz b*J, Giz c J, Giz d*♦, Giz e*J
G. caledonium (Nicolson and Gerdemann) Trappe and Gerdemann	AM WG08	Wep 3a♦J, Wep 3b*♦J, Wep 3c♦J, Wep 3d♦, Wep 3e*♦J, Mal a*♦J, Mal b*♦J, Mal d♦J, Mal e♦J, Giz a*♦, Giz b♦J, Giz c*♦J, Giz dJ, Wzh 1a*J, Wzh 1e*J
G. claroideum Schenck and Smith	AM WG09	Wep 1a*, Wep 1b*, Wep 1c*, Wep 1d*, Wep 1e*
G. clarum Nicolson and Schenck	AM WG10	Mal a*
G. clavisporum (Trappe) Almeida and Schenck	AM WG11	Giz a*J, Giz d*, Wzh 1a*♦, Wzh 1b*, Wzh 1c*J
G. constrictum Trappe	AM WG12	Wzh 1c*
G. dimorphicum Boyetchko and Tewari	AM WG13	Giz aJ, Giz b*
G. etunicatum Becker and Gerdemann	AM WG14	Sgz a♦J, Sgz b♦J, Sgz c♦J, Sgz d*♦J, Sgz eJ, Wzh 1a*♦, Wzh 1c*♦J, Wzh 1d*J
G. fasciculatum (Thaxter) Gerdemann and Trappe emend. Walker and Koske	AM WG15	Bht a*J, Bht b*J, Bht c*J, Bht d*J, Bht e*J, Wep 2b*J, Wep 2cJ, Wep 2eJ, Wep 3c*J, Wep 3d♦J, Wep 3e♦J, Miw a*, Miw bJ, Miw cJ, Miw dJ, Wzh 1e*♦J
G. fulvum (Berk. and Broome) Trappe and Gerdemann	AM WG16	Mal aJ, Mal d*, Giz aJ, Giz c*, Giz d*
G. geosporum (Nicolson and Gerdemann) Walker	AM WG17	Mal bJ, Mal dJ, Mal eJ, Giz a*♦J, Giz c*♦J, Giz dJ, Giz e*♦J, Samz a♦J, Samz b♦J, Samz c♦, Samz d♦J, Samz e♦
G. heterosporum Smith and Schenck	AM WG18	Giz a*, Giz c*J

Contd...

Table 11.3–Contd...

AM Fungal Species Diversity	*CMCC Accession Number*	*Collection Fields where AM Fungal Species were Reported*
G. intraradices Schenck and Smith	AM WG19	Bht a[•J], Bht b[•J], Bht c[•J], Bht d[•J],Bht e[•J] Wep 1a[•J], Wep 1b[•J], Wep 1c[•J], Wep 1d[•J], Wep 1e[•J], Wep 3a[•J], Wep 3b[•J], Wep 3c[•J], Wep 3d[•], Wep 3e[•J], Miw a[J], Miw b[J], Miw c[J], Miw d[J], Miw e[J], Mal a[•J], Mal b[J], Mal c[J], Mal d[•J], Mal e[•J], Giz a[•], Giz b[•J], Giz c[•J], Giz d[•J], Giz e[•J], Samz a[•J], Samz b[*J], Sam c[*J], Samz d[*J], Samz e[*], Wzh 1a[J], Wzh 1b[•J], Wzh 1c[•], Wzh 1d[•], Wzh 1e[*•J], Wzh 2a[•J], Wzh 2b[•J], Wzh 2c[•], Wzh 2d[•], Wzh 2e[*•J]
G. macrocarpum Tulasne and Tulasne	AM WG20	Giz a[J], Giz b[J], Giz c[*J], Giz c[*J], Giz d, Wzh 1a[*J], Wzh 1b[*•J], Wzh 1c[*•J]
G. microcarpum Tulasne and Tulasne	AM WG21	Giz a[*]
G. monosporum Gerdemann and Trappe	AM WG22	Mal a[*], Mal e[*]
G. mosseae (Nicolson and Gerdemann) Gerdemann and Trappe	AM WG23	Bhtc[J], Bhtd[J], Bhte[*J], Wep 2c[*J], Wep 2d[J], Wep 2e[J], Wep 3a[J], Wep 3b[J], Wep 3c[J], Wep 3e[J], Wzh 1a[J], Wzh 1b[*•J], Wzh 1c[*•J], Wzh 1d[•J]
Glomus pallidum Hall	AM WG24	Wep 3d[*]
G. pubescens (Sacc. and Ellis) Trappe and Gerdemann	AM WG25	Giz c[*]
Scutellospora calospora (Nicol. and Gerdemann) Walker and Sanders	AM WS26	Wep 3c[*], Wep 3d[*], Wep 3e[•], Miw c[*], Miw d[*], Miw e[*], Mal e[J], Giz c[*•J], Giz d[J], Giz e[•J], Wzh 1b[*J], Wzh 1d[*]
S. coralloidea (Trappe, Gerd. and Ho) Walker and Sanders	AM WS27	Giz d[*], Wzh 2a[*•J], Wzh 2b[*•J], Wzh 2c[*J], Wzh 2d[•], Wzh 2e[*•J]

The spores of *G. fulvum* (Berk. and Broome) Trappe and Gerdemann, *G. clarum* Nicolson and Schenck and *G. dimorphicum* Boyetchko and Tewari were very rare which were observed in only one sample out of the 55 samples studied. *G. ambisporum* Smith and Schenck and *G. botryoides* Rothwell and Victor were observed only in Giz region. Likewise, each region contained its characteristic most frequent species and varied with respect to the agro-climatic region. *Glomus intraradices* Schenck and Smith was found to sporulate in most of the trap cultures even if it is absent at the time of sample collection. The photographs of the recorded AM fungal species are documented in Figures 11.4–11.6.

The spores of *G. fulvum* (Berk. and Broome) Trappe and Gerdemann, *G. clarum* Nicolson and Schenck and *G. dimorphicum* Boyetchko and Tewari were very rare which were observed in only one sample out of the 55 samples studied. *G. ambisporum* Smith and Schenck and *G. botryoides* Rothwell and Victor were observed only in Giz region. Likewise, each region contained its characteristic most frequent species and varied with respect to the agro-climatic region. *Glomus intraradices* Schenck and Smith was found to sporulate in most of the trap cultures even if it is absent at the time of sample collection. The photographs of the recorded AM fungal species are documented in Figures 11.4–11.6.

Discussion

This is the first study that has attempted to describe the diversity and distribution of AM fungi in the wheat agro-climatic regions of India. Because the sampling effort was standardized, conducted within a relatively short time frame during the wheat harvesting period and the trap cultures established

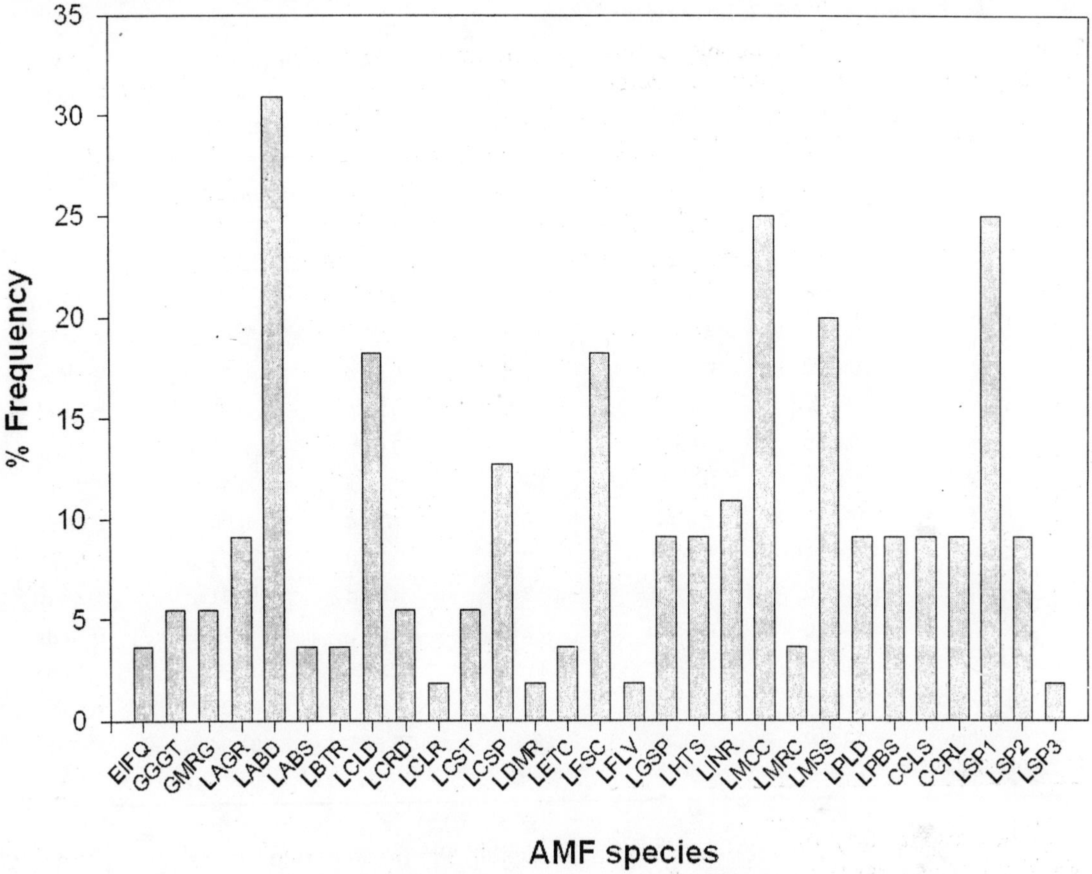

Figure 11.3: Percentage Frequency of AM Fungal Species Observed in Different Agro-climatic Regions of India

EIFQ: *Entrophospora infrequens;* GGGT: *Gigaspora gigantea;* GMRG: *Gigaspora margarita;* LAGR: *G. aggregatum;* LABD: *G. albidum;* LABS: *G. ambisporum;* LBTR: *G. botryoides;* LCLD: *G. caledonium;* LCRD: *G. claroideum;* LCLR: *G. clarum;* LCSP: *G. clavisporum;* LCST: *G. constrictum;* LDMR: *G. dimorphicum;* LETC: *G. etunicatum;* LFSC: *G. fasciculatum;* LFLV: *G. fulvum;* LGSP: *G. geosporum;* LHTS: *G. heterosporum;* LINR: *G. intraradices;* LMCC: *Glomus macrocarpum;* LMRC: *G. microcarpum;* LMNS: *G. monosporum;* LMSS: *G. mosseae;* LPLD: *G. pallidum;* LPBS: *G. pubescens;* CCLS: *Scutellospora calospora;* CCRL: *Scutellospora coralloidea;* LSP1: *Glomus* species 1; LSP2: *Glomus* species 2; LSP3: *Glomus* species 3.

were studied for successive trap culture cycles, the occurrence and the number of fungal taxa across the agro-climatic regions can be compared meaningfully. Although isolation from the spores has the advantage that these may belong to an identified fungus and result in a single-species isolate. However, not all AM fungi produce sufficient quantities of spores in field soils to allow for isolation or identification, so trap cultures can reveal species not observed to sporulate in soils (Stutz and Morton 1996). Trap culturing methods often produce more healthy spores than the soils from which they originated, but they usually result in a mixture of species which changes with subsequent cultures generations (Morton *et al.*, 1993; Bever *et al.*, 1996).

Table 11.4: Comparison of Population Abundance, Intraradical Colonization, Species Richness, and Diversity Indices (Shannon Wiener's diversity index, simpson's equitability index, dominance index) of Arbuscular Mycorrhizal (AM) Fungi in Different Agro-climatic Regions

Regions	*Total Population Abundance*	*% AMF Intraradical Colonization*	*Species Richness*	*Shannon Wiener's Diversity Index*	*Simpson's Equitability Index*	*Dominance Index*
Bht	5.48 bc	47.8 ab	2.8 c	0.82 a	0.82 a	0.89 ab
Wep1	0.88 d	40.0 ab	2.6 c	0.83 d	0.91 a	0.90 ab
Wep 2	0.35 d	37. 0 abc	1.0 d	0.00 e	0.00 b	1.00 a
Wep 3	3.72 bcd	54.8 a	2.2 c	0.68 d	0.76 a	0.91 ab
Miw	0.94 d	26.6 c	1.0 d	0.00 e	0.00 b	1.00 a
Mal	5.41 bc	30.4 bc	4.4 b	1.25 bc	0.84 a	0.70 cd
Giz	14.24 a	56.8 a	6.8 a	1.73 d	0.90 a	0.46 ab
Samz	2.37 cd	26.2 c	2.2 c	0.73 d	0.93 a	0.95 ab
Sgz	1.35 d	56.8 a	2.8 c	0.93 d	0.94 a	0.84 b
Whz 1	6.47 b	47.6 ab	4.2 b	1.36 b	0.96 a	0.59 de
Whz 2	4.8 b	49.2 ab	3.0 c	0.97 cd	0.88 a	0.79 bc
LSD (P=0.05)	3.1623	18.2624	0.8847	0.2931	0.1827	0.1345
F value	12.907	3.3069	28.39	25.5781	31.9405	13.5179
Level of significance	***	**	***	***	***	***

In the present study, direct field sampling indicated a richness of one to six species per sample. Comparable numbers of species were recovered from one cycle of trap cultures. However, some of the species observed after two propagation cycles were not detected in the first trap culture cycle. This is in agreement with the observation by Stutz and Morton (1996) that additional cycles of pot cultures are required for the fungi colonizing root to sporulate in numbers sufficient for detection.

The host plants are known to affect the diversity of the AM fungi in the fields (Bever 2001). Many host plants were chosen as trap plants. Grasses and legumes are known hosts for the successful propagation of AM fungi and hence were included as trap plants. *Sorghum* was used mainly because of its compatibility with many fungal species in all genera (except *Sclerocystis*) originating from a wide range of habitats (Morton *et al.*, 1993). Marigold was chosen as a trap plant as it gives a good root biomass and is a good host for the propagation of AM fungi. These host plants have been used historically in the experimentation of the mycorrhizal phenomena. Onion and subterranean clover (*Trifolium subterraneum*) are also used widely as a culture host (Pearson and Schweiger 1994). Sometimes, the number of species isolated into trap cultures exceeded those identified from field-collected spores, suggesting the inaccuracy of fungal surveys based solely on spore observations. However, some cultures were unsuccessful and produced single species either because of the failure of spores to germinate, or germination but failure to colonize the roots, or colonization but no sporulation under the growing conditions used. Stürmer and Bellei (1994) also observed that only two of the twelve species of AM fungi from the sand dune soils in Brazil were successfully cultured. Sporulation occurs

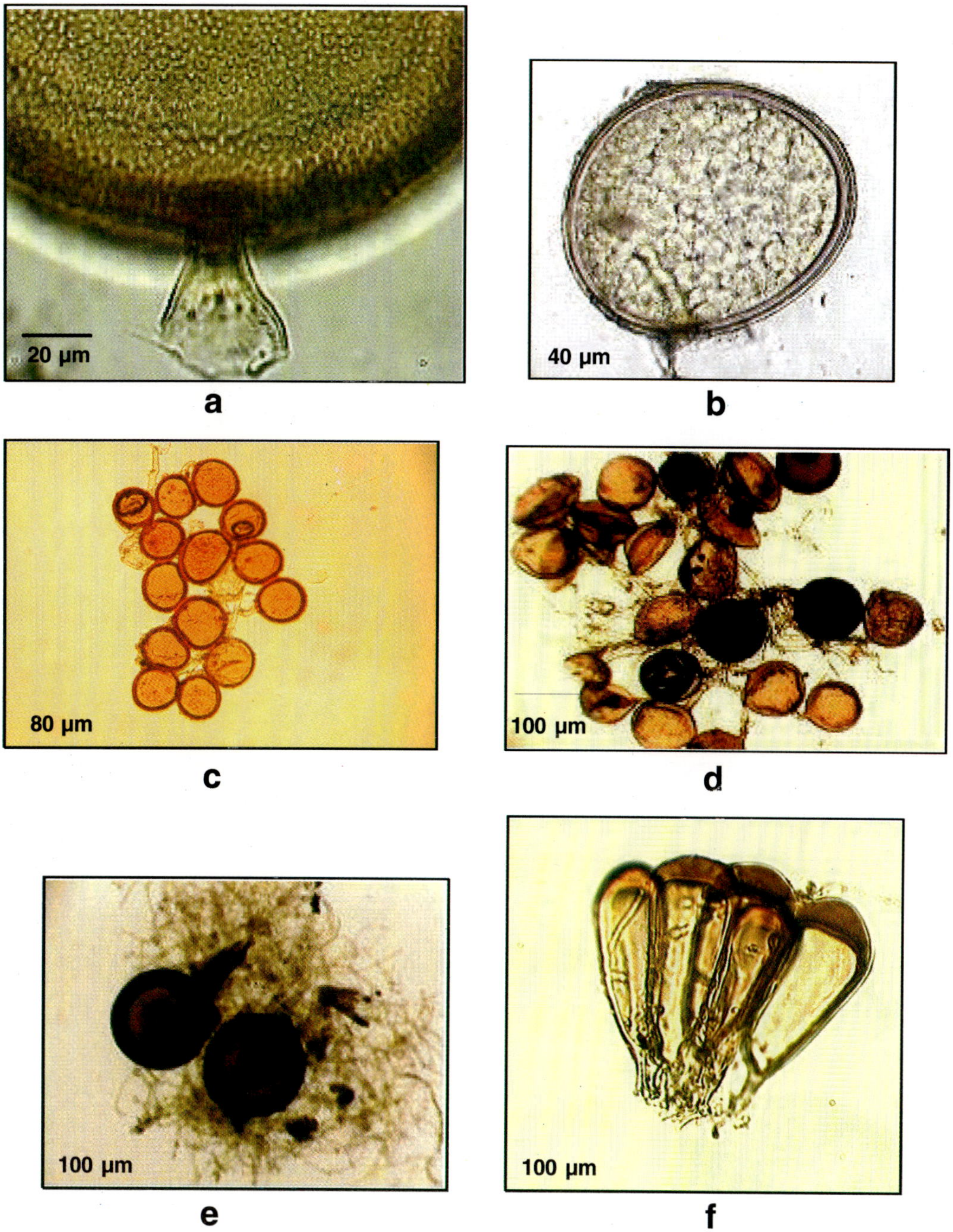

Figure 11.4: Representative AM Fungal Species Observed in Different Agro-climatic Regions of India (*a*) Azygospore of *Entrophospora infrequens* (40×); (*b*) Chlamydospore of *Glomus albidum* (20×); (*c*) Loose sporocarp of *G. aggregatum* (4×); (*d*) Sporocarp of *G. ambisporum* (10×); (*e*) Chlamydospores of *G. botryoides* (10×); (*f*) Sporocarp of *G. clavisporum* (10×)

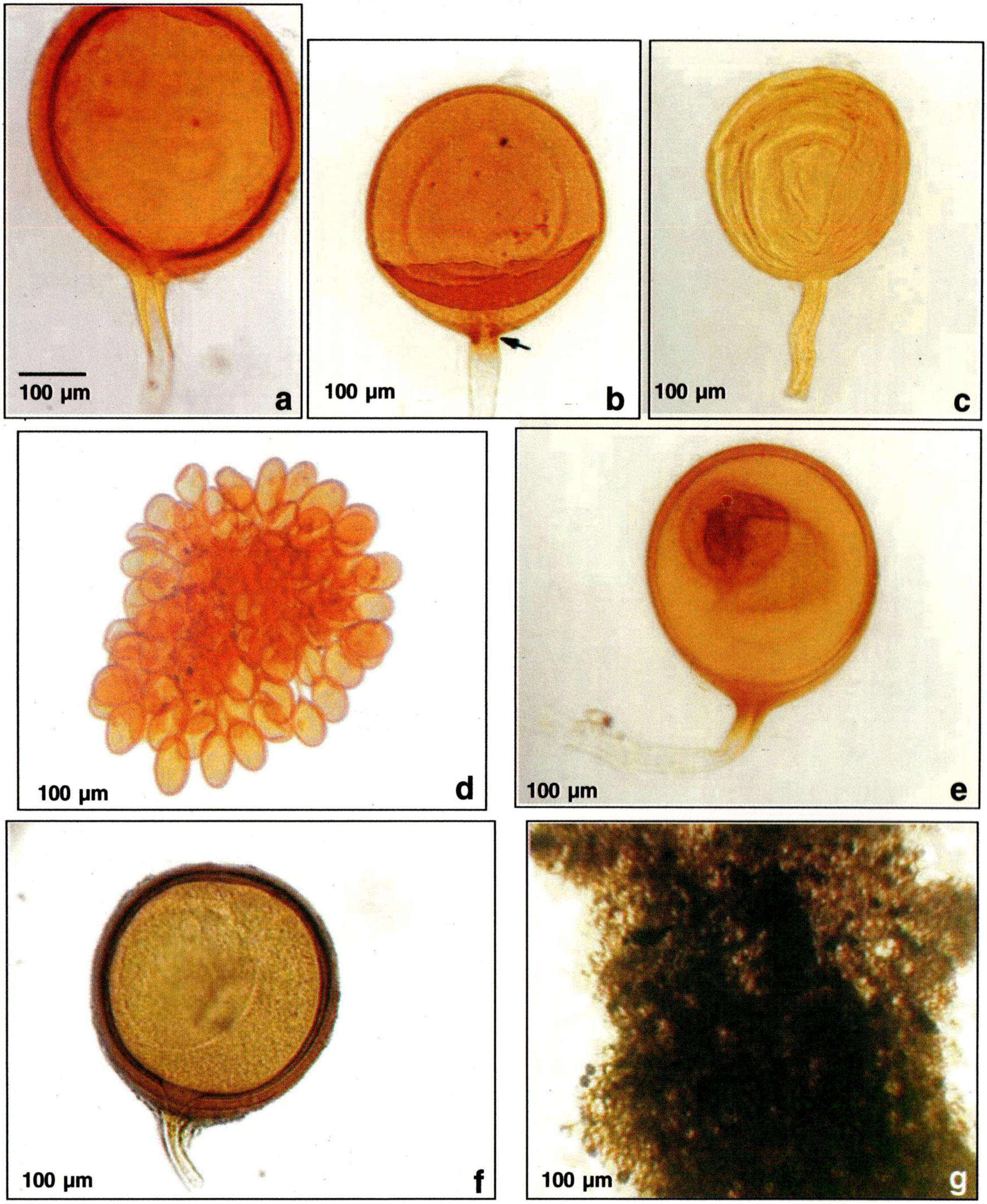

Figure 11.5: Representative AM Fungal Species Observed in Different Agro-climatic Regions of India, (*a*) Chlamydospore of *G. caledonium* (10×); (*b*) Chlamydospore of *G. constrictum* (10×); (*c*) Chlamydospore of *G. fasciculatum* (10×); (*d*) Sporocarp of *G. fulvum* (4×); (*e*) Chlamydospore of *G. macrocarpum* (10×); (*f*) Chlamydospore of *G. mosseae* (10×); (*g*) Sporocarp of *G. pubescens* (10×)

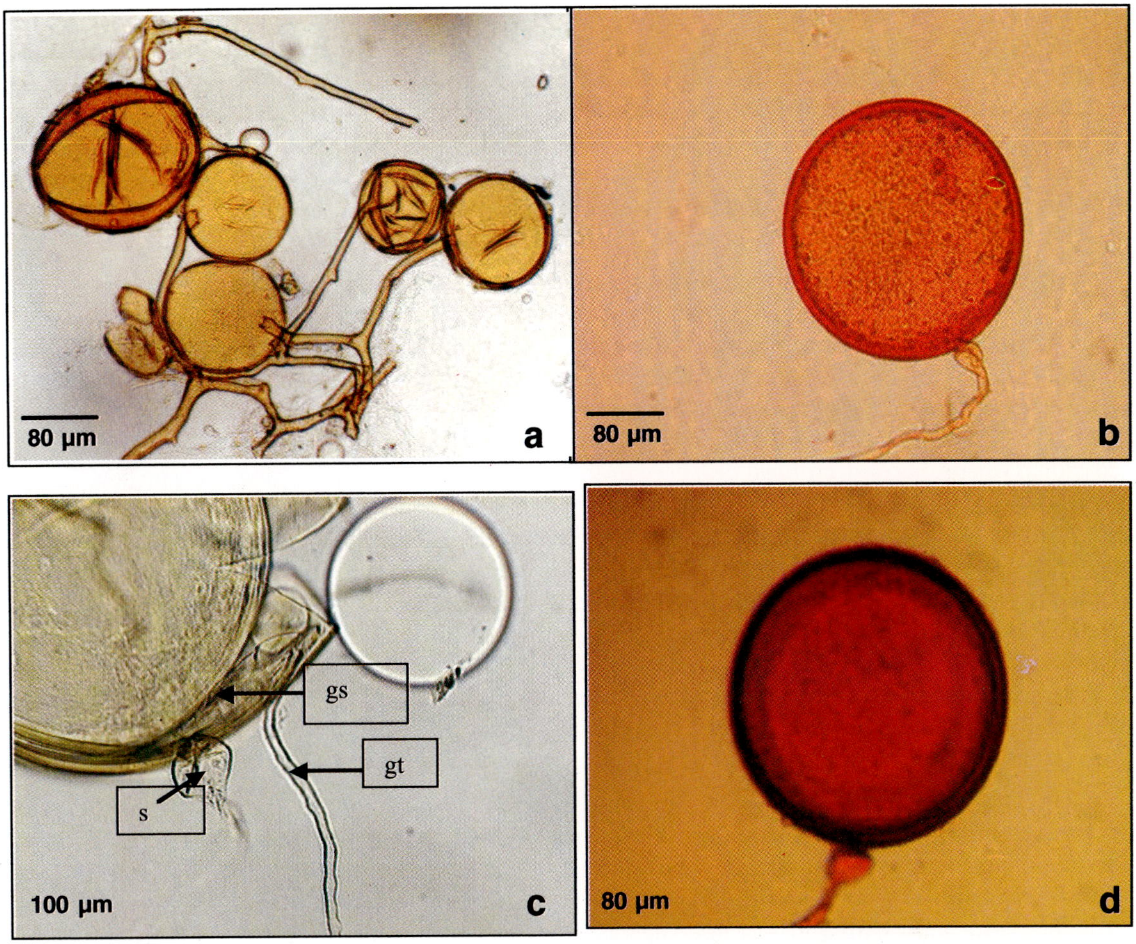

Figure 11.6: Representative AM Fungal Species Observed in Different Agro-climatic Regions of India (*a*) Broken sporocarp of *G. intraradices* (4×); (*b*) Azygospore of *Gigaspora margarita* (4×); (*c*) Azygospore of *Scutellospora calospora* (10×); (*d*) Azygospore of *Scutellospora coralloidea* (4×)

at varying rates depending on the fungus, host, and environment. When there is a mixture of fungal species trying to occupy the same root space, all are likely to find room to become established (unless inoculum is layered and distribution of all available propagules is severely constrained) and then obtains enough carbon to survive and flourish to varying degrees. The fungi which establish most quickly and then grow aggressively will be the first to have the excess carbon to sporulate. Even after plants have ceased root growth, only the dominant colonizers are likely to sequester the carbon to sporulate profusely. Poor colonizers may inhabit roots for numerous propagation cycles. Sometimes two or three cycles of trap cultures are needed before the spores of some species appear (Stutz and Morton 1996) and this would be a function of their growth relative to other fungal occupants.

A major highlight of our study is the low diversity of AM fungi in the agricultural fields. Although a total of 34 AM fungal species were recorded from 11 agro-climatic regions of India, an average of only 3–4 species per field was found. Low diversity in the arable fields has earlier been reported by

Helgason *et al.* (1998; 2002) and Daniell *et al.* (2001). This observation is in contrast to the high species richness (upto 20 species from a single site) in the natural ecosystems in the world (Eom *et al.*, 2001; Bever *et al.*, 2001; Mangan *et al.*, 2004) as well as in India (Mukerji 1995). It has been observed that the diversity of AM fungi in temperate grassland soil is limited to ten species and in cultivated soils to even a fewer species (Johnson 1993). A higher diversity of AM fungal communities in the woodland ecosystem than on arable land was suggested from the analysis of the AM fungal sequences amplified from the root samples from those two ecosystems (Helgason *et al.*, 1998). Another striking observation was the ubiquity and adaptability of a few genera and species in the field.

These findings indicate that the species present in the disturbed habitat (*e.g.*, agricultural fields which are constantly subjected to human intervention) are a restricted subset of those which would occur naturally in the region. Two possible explanations could account for this pattern. First, the diversity of AM fungal communities has been related to the diversity of plant communities (Rabatin and Stinner 1989). Although most natural ecosystems contain a great variety of plant species, intensively managed agro-ecosystems contain very few plant species, generally a single crop per field with occasional weeds. Second, cultural practices probably exert strong selective pressures on AM fungal communities (Jansa *et al.*, 2002). All the fields chosen for study are conventionally managed fields, which integrate the use of commercial seed-bed preparation, mechanized planting and inorganic inputs i.e, chemical fertilizers, pesticides, etc. All these cause changes in the habitat and substrate availability that may discourage the growth of selected microorganisms, so fungal species and strains most tolerant to these stresses proliferate. During the present study, the genera *Glomus* was found to be ubiquitous. Abbott and Robson (1977), Blaszkowski (1993) and Talukdar and Germida (1993) also reported the prevalence of *Glomus* sp. in agriculturally used soils, in contrast to the rich AM fungal communities containing *Gigaspora* spp., *Scutellospora* spp., and *Acaulospora* sp. in uncultivated soils (Blaszkowski 1993).

The comparatively higher frequency of some species, *G. albidum* (33 per cent) and *G. macrocarpum* (25 per cent) among the spore communities indicates their adaptability to varied soil conditions, whereas other species showed a narrow range of their host/environment adaptation. Johnson (1993) also earlier observed a decrease in AM fungal community diversity due to application of mineral fertilizers, associated with a loss of non-*Glomus* fungi and an increase in the population of *Glomus intraradices*. Some AM fungal species and isolates are more efficient in colonizing the same host plants species than the others (Helgason *et al.*, 2002) and those fungi (*e.g. Scutellospora calospora*) which utilize a larger proportion of host carbon substrates than other fungi show lower root colonization.

The role of soil nutrient concentration on the diversity and colonizing ability of AM fungi was also investigated during the course of the study. Amongst different soil conditions that can affect mycorrhizal colonization, differences in soil fertility, particularly in the phosphorus content, is well known to stimulate differential sporulation and colonization by AM fungal species in the field (Hayman 1975; Brundrett *et al.*, 1999). In the present study, the spore density was found to be related to the phosphorus concentrations (r^2=0.51). Other possible reasons for a low infection level or for the lack of infection might be the inability of the fungi to compete with other soil microorganisms (Bhatia *et al.*, 1996).

There was no correlation between the AM fungal colonization percentage and the spore numbers. This may suggest that propagules other than spores are causing the colonization or it may also represent differences in spore viability or maturity at the time the samples were taken. There was a positive correlation between spore density and the number of AM fungal species. This positive relationship has also been observed by Kulkarni *et al.* (1997) in tropical coastal sand dunes. However, whether the increased spore population or increased diversity confers more benefits to plants is a matter of further investigation.

Acknowledgements

We wish to thank financial contributions for the project (SA 6) from Swiss Agency for Development and Cooperation, Government of Switzerland and the DBT, Govt of India under the Indo-Swiss collaboration in Biotechnology. The publication does not constitute an endorsement by the Governments of Switzerland and India. We also thank to the Ministry of Environment and Forests, Govt of India. We are grateful to Prof. Andres Wiemken and Dr Fritz Oehl, University of Basel, Switzerland for their guidance. We thank Dr R K Pachauri, Director General, TERI, for providing the infrastructure for carrying out the research work. We also thank Mr U Gangi Reddy for soil analysis and Mr Dinesh Vajapayee for maintenance of trap cultures.

References

Abbott L. K. and Robson A. D. 1977. Growth stimulation of subterranean clover with vesicular arbuscular mycorrhizas. Australian Journal of Agricultural Research 28: 639–49

Allen E. B., Allen, M. F., Helm, D. J., Trappe, J. M., Molina, R. and Rincón, E. 1995. Patterns and regulation of mycorrhizal plant and fungal diversity. Plant and Soil 170: 47–62

Altieri, M. A. 1999. The ecological role of biodiversity in agroecosystems. Agriculture Ecosystems and Environment 74: 19–31.

Bever, J. D., Morton, J. B., Antonovics, J. and Schultz, P. A. 1996. Host-dependent sporulation and species diversity of arbuscular mycorrhizal fungi in a mown grassland. Journal of Ecology 84: 71-82.

Bever, J. D., Schultz, P.A., Pringle, A. and Morton J. B. 2001. Arbuscular mycorrhizal fungi: More diverse than meets the eye, and the ecological tale of why. Bioscience 51: 923–931.

Bhatia, N. P., Sundari, K. and Adholeya, A. 1996. Diversity and selective dominance of vesicular-arbuscular mycorrhizal fungi. In: Mukerji KG (ed) Concepts in Mycorrhizal Research. Kluwar Academic Publishers, Netherlands, pp 133–178.

Biermann, B. and Linderman, R. G. 1981. Quantifying vesicular-arbuscular mycorrhizae: Proposed method towards standardization. New Phytologist 87: 63–67.

Blaszkowski, J. 1993. Comparative studies on the occurrence of arbuscular fungi and mycorrhiza (Glomales) in cultivated and uncultivated soils of Poland. Acta Mycologia 28: 93–140.

Borcard, D., Legendre, P. and Drapeau, B. 1992. Partialling out the spatial component of ecological variation. Ecology 73: 1045–1055.

Bremner, J, M. 1960. Determination of nitrogèn in soil by Kjeldahl method. Journal of Agricultural Science 55: 11–13.

Brundrett, M. C., Jasper, D. A. and Ashwath, N. 1999. Glomalean mycorrhizal fungi from tropical Australia II. The effect of nutrient levels and host species on the isolation of fungi. Mycorrhiza 8: 315–321.

Clapp, J. P., Young, J. P. W., Merryweather, J. W. and Fitter, A. H. 1995. Diversity of fungal symbionts in arbuscular mycorrhizas from a natural community. New Phytologist 130: 259–265.

Dalal, S. and Hippalgaonkar, K. V. 1995. The occurrence of vesicular-arbuscular mycorrhizal fungi in arable soils of Konkan and Solapur. Proceedings of the third National Conference on Mycorrhizae. In: Adholeya A, Singh S (eds) Mycorrhizae: Biofertilizers for the future, pp.3-7.

Daniell, T. J., Husband, R., Fitter, A. H. and Young, J. P. W. 2001. Molecular diversity of arbuscular mycorrhizal fungi colonising arable crops. FEMS Microbiology Ecology 36: 203–209.

Datta, N. P., Khera, M. S. and Saini, T. R. 1962. A rapid calorimetric procedure for the determination of the organic carbon in soils. Journal of Indian Society of Soil Science 10: 67–74.

Eom, A. H., Hartnett, D. C. and Wilson, G. W. T. 2000. Host plant species effects on arbuscular mycorrhizal fungal communities in tallgrass prairie. Oecologia 122: 435–444.

Gaur, A. and Adholeya, A. 2004. Prospects of AM fungi in Phytoremediation of heavy metal contaminated soils- Mini-review. Current science 86: 528–534.

George, E., Marschner, H. and Jakobsen, I. 1995. Role of arbuscular mycorrhizal fungi in uptake of phosphorus and nitrogen from soil. Critical Reviews of Biotechnology 15: 257–270.

Gerdemann, J. W. and Nicolson, T. H. 1963. Spores of mycorrhizal *Endogone* species extracted from soil by wet sieving and decanting. Transactions of British Mycological Society 46: 235–244.

Giovannetti, M. and Gianinazzi-Pearson, V. 1994. Biodiversity in arbuscular mycorrhizal fungi. Mycological Research 98: 705–715.

Hall, I. R. 1984. Taxonomy of VA mycorrhizal fungi. In: Powell CL, Bagyaraj DG (eds) VA Mycorrhiza. CRC Press, Boca Raton Fl, pp 234.

Hamel, C. 1996. Prospects and problems pertaining to the management of arbuscular mycorrhizae in agriculture. Agriculture Ecosystems and Environment 60: 197–210.

Hayman, D. S. 1975. The occurrence of mycorrhizas in field crops as affected by soil fertility. In: Sanders FE, Mosse B, Tinker PB (eds) Endomycorrhizas. Academic Press, NewYork and London, pp 495–509.

Helgason, T., Merryweather, J. W., Denison, J., Wilson, P., Young, J. P. W., Fitter, A. H. 2002. Selectivity and functional diversity in arbuscular mycorrhizas of co-occurring fungi and plants from a temperate deciduous woodland. Journal of Ecology 90: 371–384.

Helgason, T. Daniell, T. J., Husband, R., Fitter, A. H. and Young, A. P. W. 1998. Ploughing-up the wood-wide web? Nature 394: 431.

Hoagland, D. R. Arnon, D. I. 1938. The water culture method of growing plants without soil. California Agricultural Experiment Stations, circular 347, Berkeley California.

Hurlbert, S. H. 1984. Pseudoreplication and the design of ecological field experiments. Ecological Monographs 54: 187–211.

Jansa, J., Mozafar, A., Anken, T., Ruh, R., Sanders, I. R., Frossard, E. 2002. Diversity and structure of AMF communities as affected by tillage in a temperate soil. *Mycorrhiza 12: 225–234.*

Johson, N. C. 1993. Can fertilization of soil select less mutualistic mycorrhizae? Ecological Applications 3: 749–757.

Kulkarni, S. S., Raviraja, N. S. and Sridhar, K. R. 1997. Arbuscular mycorrhizal fungi of tropical sand dunes of west coast of India. Indian Journal of Coastal Research 13: 931–936.

Ladha, J. K., Fischer, K. S., Hossain, M., Hobbs, P. R. and Hardy, B. 2000. Improving the productivity and sustainability of rice-wheat systems of the Indo-Gangetic Plains: A synthesis of NARS-IRRI partnership research. IRRI Discussion Paper Series No. 40. IRRI, Los Banos Philippines.

Legendre, P. 1990. Quantitative methods and biogeographic analysis. In: Garbary DJ, South GR (eds) Evolutionary biogeography of the marine algae of the North Atlantic. NATO ASE Series. G 22, Springer-Verlag, Berlin, pp 9–34.

Lindenmayer, D. B., Mackey, B. G. and Nix, H. A. 1996. The bioclimatic domains of four species of commercially important eucalypts from south-eastern Australia. Australian Forestry 59: 74–89.

Mangan, S. A., Eom, A. H., Adler, G. H., Yavitt, J. B. And Herre, E. A. 2004. Diversity of arbuscular mycorrhizal fungi across a fragmented forest in Panama: insular spore communities differ from mainland communities. Oecologia: 141: 687–700.

Margules, C. R., Nicholls, A. O. and Austin, M. P. 1987. Diversity of Eucalyptus species predicted along a multivariable environmental gradient. Oecologia 71: 229–232.

Miller, R. M. and Jastrow, J. D. 1992. The application of VA mycorrhizae to ecosystem restoration and reclamation. In: Allen MF (ed) Mycorrhizal functioning. Chapman and Hall Ltd., London England, pp 438–467.

Ministry of Finance 2000, Economic Survey, 1999–2000, Economic Division, Ministry of Finance, New Delhi.

Morton, J. B., Bentivenga, S. P. and Wheeler, W. W. 1993. Germplasm in the International Collection of Arbuscular and Vesicular-arbuscular Mycorrhizal Fungi (INVAM) and Procedures for Culture Development, Documentation and Storage. Mycotaxon 48: 491–528.

Mukerji, K. G. 1996. Taxonomy of endomycorrhizal fungi. In: Mukerji KG, Mathur B, Chamola BP, Chitralekha P (eds) Ashish Publishing House, Delhi, pp 211–219.

Oehl, F., Sieverding, E., Ineichen, K., Mader, P., Boller, T. and Wiemken, A. 2003. Impact of land use intensity on the species diversity of arbuscular mycorrhizal fungi in agro-ecosystems of Central Europe. Applied and Environmental Microbiology 69: 2816–2824.

Olsen, S. R., Cole, C. V., Watanabe, F. S. and Dean, L. A. 1954. Estimation of available phosphorus in soils by extraction with Sodium bicarbonate. Circular 939 US Department of Agriculture, Washington, DC.

Pearson, N. J. and Schweiger, P. 1994. *Scutellospora calospora* (Nicol. and Gerd.) Walker and Sanders associated with subterranean clover produces non-infective hyphae during sporulation. New Phytologist: 697-701.

Phillips, S. J. M. and Hayman, D. S. 1970. Improved procedures for clearing roots and staining parasitic and vesicular arbuscular mycorrhizal fungi for rapid assessment of infection. Transactions of British Mycological Society 55: 158–160.

Rabatin, S. C. and Stinner, B. R. 1989. The significance of vesicular-arbuscular mycorrhizal fungal-soil macro-invertebrate interactions in agroecosystems. Agriculture Ecosystems and Environment 27: 195–204.

Schenck, N. C., Perez, Y. 1990. Manual for the identification of VA mycorrhizal fungi. Third edition. INVAM, Synergistic Publications, Gainesville, USA.

Schenck, N. C., Siqueira, J. O. and Oliveira, E. 1989. Changes in the incidence of VA mycorrhizal fungi with changes in ecosystems. In: Vancura V, Kunc F (eds) Interrelationships between microorganisms and plants in soil. Elsevier, New York, pp 125–129.

Sharma, M. P., Adholeya, A. 2004. Effect of arbuscular mycorrhizal fungi and phosphorus fertilization on the post *vitro* growth and yield of micropropagated strawberry grown in a sandy loam soil. Canadian Journal of Botany 82: 322–328.

Singh, R., Pandya, R. K. 1995. The occurence of vesicular-arbuscular mycorrhiza in pearl millet and other hosts. Proceedings of the Third National Conference on Mycorrhizae. In: Adholeya A, and Singh S (eds) Mycorrhizae: Biofertilizers for the future. pp 56–58.

Singh, R., Adholeya, A. and Mukerji, K. G. 2001. Mycorrhiza in Control of Soil Borne Pathogens. In: Mukerji KG, Chamola BP, Singh J (eds) Mycorrhizal Biology. Kluwer Academic Publishers, New York, pp 173–196.

Smith, G. W. and Skipper, H. D. 1979. Comparison of methods to extract spores of VAM fungi. Soil Science Society of American Journal 43: 722–725.

Sturmer, S. L. and Bellei, M. M. 1994. Composition and seasonal variation of spore populations of arbuscular mycorrhizal fungi in dune soil on the island of Santa Catarina, Brazil. Canadian Journal of Botany 72: 359-363.

Stutz, J. C. and Morton, J. B. 1996. Successive pot cultures reveal high species richness of arbuscular endomycorrhizal fungi in arid ecosystems. Canadian Journal of Botany 74: 1883-1889.

Talukdar, N. C., Germida, J. J. 1993. Occurence and isolation of vesicular arbuscular mycorrhizae in cropped field soils of Saskatchewan, Canada. Canadian Journal of Microbiology 39 567–575.

Trappe, J. M. 1987. Phylogenetic and ecological aspects of mycotrophy in the angiosperms from an evolutionary standpoint. In: Safir GR (ed) Ecophysiology of VA Mycorrhizal Plants. CRC Press, Boca Raton Fl, pp 5–25.

van der Heijden, M. G., Boller, T., Wiemken, A. and Sanders, I. A. 1998. Different arbuscular mycorrhizal fungal species are potential determinants of plant community structure. Ecology 79: 2082–2091.

Wood, L. K. and Deturk, E. E. 1940. The adsorption of potassium in soils in replaceable form. Soil Science American Proceedings 5:152–161.

Microbes: Diversity and Biotechnology (2012) *Pages* **197–203**
Editors: **Prof. S.C. Sati & Dr. M. Belwal**
Published by: **DAYA PUBLISHING HOUSE, NEW DELHI**

Chapter 12

Saprolegniasis in Fishes of Kolleru Lake in Andhra Pradesh

S.A. Mastan

Post Graduate Department of Biotechnology, D.N.R. College, P.G. Courses and Research Centre, Bhimavaram – 534 202, Distt. W.G., Andhra Pradesh

ABSTRACT

In the present study a total of 16 isolates of fungi were obtained from infected fishes which belongs to five species namely *Saprolegnia parasitica, Saprolegnia ferax, Saprolegnia hypogyana, Saprolegnia diclina.* and *Achlya americana*. All these fungi were isolated from five different species of fishes *viz. Chaanna gachua, Channa punctatus, Chaanna stratius, Clarias batrachus* and *Mystus cavasius*. The parasitic ability of all the 16 isolates obtained from infected fishes was confirmed by conducting pathogenecity tests under laboratory conditions using healthy fishes of the same species from which fungi were isolate. All the species of fungi were found to be pathogenic to fish. But *Saprolegnia parasitica* is more vigorous showing infection with in eight hours.

Keywords: *Saprolegniasis, Fungus, Fishes.*

Introduction

Fungal infection of fish by *Oomycetes* commonly known as watermolds are wide spread in fresh water and represents the most important fungal group affecting wild cultured fish and their eggs. The *saprolegniaceae* are responsible for significant infections, involving both living and dead fish eggs, particularly in aquaculture. *Oomycetes* are saprophytic opportunists multiplying on fish that are physically injured, stressed or infected (Pickering and Willoughby,1982a). Members of this group are generally considered agents of secondary infections arising from conditions such as bacterial infection, poor husbandry practices, and infestations by parasites and social interactions. However, there are several reports of *Oomycetes* as primary infection agents of fish (Willoughby 1978, Pickering and

Christie 1980) and their eggs (Walser and Phelps, 1993, khulbe *et al.*, 1995), Vikassalgotra *et al.* (2005) Mastan, (2008). The present paper reports the incidences of Saprolegniasis in fishes of Kolleru lake, AP.

Materials and Methods

Incidences of Saprolegniasis were recorded during the winter months of 2007 – 2008 from kolleru lake, A.P. A total of 1,250 fishes were screened. The fungal infected fishes were brought to the laboratory in the living condition and kept in glass aquarium, of the size of 90x45x45 cm filled with clean fresh water. The dead as well as living fishes were examined grossly for lesions and ulcerations.

Isolation of fungi from infected fishes were carried out by taking small pieces of muscles about 2 mm in diameter from infected portions of the body. They were then washed thoroughly with sterilized distilled water to remove the unwanted micro organisms adhered on the surface. These tissues were then inoculated over the plates containing different agar media. Alternatively small pieces of mycelia taken out from infected parts of fish body, were washed thoroughly with distilled water. They were placed in a Petri dish containing 20-30 ml distilled water and baited on different baits *viz.* Hemp seeds, and Mustered seeds. These Petri dishes were incubated at 15° to 22° C temperatures for a week. Pure and bacteria free cultures were prepared by using the methods of Coker (1923) Johnson (1956) and Scott (1961). Identification of fungi was done on the basis of their vegetative and reproductive characters using the monographs of Coker (1923).

In order to demonistrate the pathogenecity of the isolates obtained from the naturally infected fishes, experimental infection trails were conducted in the laboratory. Isolated species of fungi *viz. Saprolegnia parasitica Saprolegnia ferax, Saprolegnia hypogyana, Saprolegnia diclina and Achyla americana* were tested separately on the fingerlings of different species of fishes, having average size and weight 8.16 ± 0.13 cm and 12.5 ± 0.28 g, respectively.

The pathogenecity tests were carried out by employing the methodology of Scott and O' warren (1964). Covered glass troughs (12" x 9"), wrapped in aluminium foils were sterilized in hot air oven at 120° C temperatures for 24 hours. Filtered sterile lake water was filled aseptically in to each trough. An aerator was used to aerate the water throughout the experiment. Six fungal inoculated blocks (1.0 cm^2) of SPS agar/Potato Dextrose Agar (PDA) medium were placed at different sites in the trough. Six uninoculated blocks of the same agar medium were placed in another trough which was used as control. After 48 hours, when spores developed, experimentally injured fishes were placed in these troughs. Four fishes of each species were kept in each trough. All the experiments were conducted at 20.0°C to 25.0° C temperature in triplicate sets.

Water samples were collected from lake for analysis of various Physico-chemical parameters as per the methods given in APHA (1995).

Results

In the present study, a total of 16 isolates were obtained from the fishes investigated. These isolates represents five species and belonged to two genus namely *Saprolegnia* and *Achlya* (Table 12.1).

Saprolegnia declina

Saprolegnia diclina was isolated twice from infected fishes. One isolate was collected from *Channa gachua* and one from *Channa straitus.*

Table 12.1: Fungi Isolated from Infected Fishes of Kolleru Lake, (A.P.)

Sl.No.	Date of Collection	Isolate No.	Name of Fungi	Host Fish
1.	10-11-07	KL/S.d/1	*Saprolegnia diclina*	*Channa striatus*
2.	11-11-07	KL/S.d/2	*Saprolegnia diclina*	*Channa gachua*
3.	02-12-07	KL/S.f/3	*Saprolegnia ferax*	*Clarias batrachus*
4.	10-12-07	KL/S.f/4	*Saprolegnia ferax*	*Channa striatus*
5.	15-12-07	KL/S.f/5	*Saprolegnia ferax*	*Channa gachua*
6.	18-12-07	KL/S.h/6	*Saprolegnia hypogyana*	*Channa gachua*
7.	20-12-07	KL/S.h/7	*Saprolegnia hypogyana*	*Channa gachua*
8.	28-12-07	KL/S.p/8	*Saprolegnia parasitica*	*Channa gachua*
9.	30-12-07	KL/S.p/9	*Saprolegnia parasitica*	*Channa gachua*
10.	02-01-08	KL/S.p/10	*Saprolegnia parasitica*	*Heteropneustis fossilis*
11.	05-01-08	KL/S.p/11	*Saprolegnia parasitica*	*Heteropneustis fossilis*
12.	10-01-08	KL/S.p/12	*Saprolegnia parasitica*	*Heteropneustis fossilis*
13.	20-01-08	KL/S.p/13	*Saprolegnia parasitica*	*Mystus cavasius*
14.	25-01-08	KL/S.p/14	*Saprolegnia parasitica*	*Mystus cavasius*
15.	05-02-08	KL/A.m/15	*Achlya americana*	*Heteropneustis fossilis*
16.	10-02-08	KL/A.m/16	*Achlya americana*	*Mystus cavasius*

Saprolegnia ferax

A total of three isolates of *Saprolegnia ferax* were obtained, one was collected from *Clarias batrachus*, one from *Channa straitus* and one from *Channa gachua.*

Saprolegnia hypogyana

One isolate of *Saprolegnia hypogyana* was collected from infected *Channa gachua.*

Saprolegnia parasitica

Saprolegnia parasitica is the most frequently occurring parasite of fish. A total of six isolates of this species were obtained from infected fishes. Three isolated from *Heteropneustis fossilis*, Two from *Mystus cavasius* and one from *Channa punctatus.*

Achlya americana

A total of four isolates of this species were obtained from infected fishes. Two each from *H. fossilis* and *M.cavasius.*

The maximum percentage of infection was recorded to be 3.2 in the month of December, 2007.While the minimum percentage of infection was recorded to be 0.6 in the month of February, 2008.

In case of fish species the highest percentage of infection was reported in *Channa gachua* while lowest infection was reported in *Clarias batrachus.*

The experimental infection trails were conducted with fungi isolated from naturally infected fish to test their pathogenecity under laboratory conditions.Each isolate was tested on that particular species of fish from which it was originally isolated. The results are summarized in Table 12.2.

Table 12.2: Experimental Infection Trails with Various Species of Fungi Isolated from Diseased Fishes

Sl.No.	Isolate No.	Fungi Inoculated	Experimental Fish	No. of Fish Used	Mycosis Evident (hrs)	Death (hrs)
1.	KL/S.d/2	*Saprolegnia diclina*	*Channa gachua*	6	24	48
2.	KL/S.h/6	*Saprolegnia hypogyana*	*Channa gachua*	6	48	72
3.	KL/S.p/8	*Saprolegnia parasitica*	*Channa gachua*	6	8	36
4.	KL/S.p/9	*Saprolegnia ferax*	*Channa gachua*	6	24	48
5.	KL/S.d/1	*Saprolegnia diclina*	*Channa striatus*	6	16	24
6.	KL/S.h/13	*Saprolegnia parasitica*	*Mystus cavasius*	6	40	36
7.	KL/S.f/3	*Saprolegnia ferax*	*Clarias batrachus*	6	48	96
8.	KL/S.f/4	*Saprolegnia ferax*	*Channa striatus*	6	36	72
9.	KL/S.f/5	*Saprolegnia ferax*	*Channa striatus*	6	48	72
10.	KL/S.p/10	*Saprolegnia parasitica*	*Heteropneustis fossilis*	6	8	42
11.	KL/S.p/11	*Saprolegnia parasitica*	*Heteropneustis fossilis*	6	8	42
12.	KL/A.m/16	*Achlya americana*	*Heteropneustis fossilis*	6	38	72
13.	KL/S.p/14	*Saprolegnia parasitica*	*Mystus cavasius*	6	18	48
14.	KL/A.m/15	*Achlya americana*	*Heteropneustis fossilis*	6	38	72
15.	KL/S.p/12	*Saprolegnia parasitica*	*Heteropneustis fossilis*	6	9	40
16.	Kl/S.h/7	*Saprolegnia hypogyana*	*Channa gachua*	6	46	72

All the isolates of genus *Saprolegnia* are found to be pathogenic to fish. Hyphal growth of fungi was clearly visible at the site of injured areas of experimental fishes with in 8 – 48 hours after inoculation. All the test fishes died within 24 – 96 hours after catching infection (Table 12.2). It is observed that although all the isolates has potentiality to parasitize the fish but *Saprolegnia parasitica* is more vigorous showing infection within 8 hours (Table 12.2).

A wide range of fluctuations were noticed in various water quality parameters of affected lake (Table 12.3)

Discussion

Fungal infection in fish was first reported during mid – eighteen century (Arderon, 1748). Later on some other workers reported several pathogenic fungi from different species of fish and fish eggs. [Sati (1982), Hatai *et al.* (1994). Fraser *et al.* (1992). Hatai and Hoshiai (1992, 1993), Roberts *et al.* (1993), Chinnabut *et al.* (1995) and Willoughby *et al.* (1995).Khulbe *et al.* (1995) and Mastan (2008)].

In India the mycological studies were initiated by Chidambaram (1942) who observed

Table 12.3: Showing Water Quality Parameters of Lake during Study Period

Sl.No.	Parameters	Values in Range
1.	Water temperature (°C)	17-26
2.	Conductivity (μscm)	280-300
3.	pH	6.8-8.5
4.	FCO_2 (mg/l)	1.0-3.6
5.	Dissolved Oxygen (mg/l)	5.8-8.6
6.	Total Alkalinity (mg/l)	78-196
7.	Total hardness (mg/l)	69-175
8.	Chloride (mg/l)	10-36

red patches on the body of *Osphronemus gouramy* due to *Saprolegnia* species. Tiffney (1939b) was the first to demonstrate the ability of *Saprolegnia parasitica* (Coker) to parasitize a wide range of fishes and amphibians and emphasized the fact that the injury greatly lowers the resistance of hosts to fungal infections. Vishniac and Nigrelli (1957) conducted laboratory experiments and demonstrated the parasitic ability of sixteen species of aquatic fungi belonging to seven genera of *Saprolegniaceae.* Scott (1964) demonstrated that *Saprolegnia parasitica, Saprolegnia ferax, Saprolegnia diclina,, Saprolegnia monoica, Achyla bisexualis* and some non fruiting isolates of *Saprolegnia* could parasitize wounded platy *fish* under controlled conditions. Sati and Khulbe (1983) carried out host range studies with Saprolegnia *diclina* on nine species of cold water fish's *viz. Barilius bendelisis, Carassius auratus, Cyprinus carpio, Nemachelius rupicola, Puntius conchonius, Puntius ticto, Schizothorax palgiostomus, Saprolegnia richardsoni and Tor tor.* The experimental infection of Saprolegnia on different species of fishes has also been reported by Qureshi *et al.* (2002). Roberts *et al.* (1993) and Hatai *et al.* (1994) reported the pathogenecity of *Aphanomyces* species on *Channa* species and Dwarf gourami.

In the present study mycological examination of infected fishes revealed the presence of sixteen isolates of five species *viz. Saprolegnia parasitica, Saprolegnia ferax, Saprolegnia hypogyana, Saprolegnia diclina, and Achlya americana.* All the species of *Saprolegnia* are found to be virulent for fishes. This observation is in agreement with the finding of Scott and O.Bier (1962) who have reported that the species of fungi, *Saprolegnia parasitica* is found to be the most destructive. This finding confirms with the reports of Hatai and Hoshai (1992) who have reported that the infection caused by *Saprolegnia parasitica* in salmon resulted mass mortality. Both the scaly and non – scaly fishes were found to be equally susceptible to the species of fungi tested. *Saprolegnia hypogyana* was isolated from and tried on *Channa straitus,* also showed its wide range on fishes. The same is also reported by Chauhan and Qureshi (1994). Qureshi *et al.* (2000) have conducted pathogenicity studies with various species of *Saprolegnia* on different species of fishes of Central India.

References

Arderon, W. 1748. The Substance of a letter from Mr. William Arderon, F.R.S.,Phil. Trans. Res.Soc., 45 (487): 321 – 323.

APHA. 1995. Standard methods for the Examination of Water and Waste Water published jointly by American Public Health Association and American Water works Association and Pollution Control Federation, New York (10th Ed.) pp -1- 1268.

Chauhan, R. and Qureshi, T. A. 1994. Host range studies of *Saprolegnia ferax* and *Saprolegnia hypogyana.* J. Inland Fish. Soc. India 26(2):99-106.

Chidambaram, K. 1942. Fungus disease of gourami (*Osphromenus goramy, Lacepede*) in a pond at Madras. Cur. Sci., 11: 288-289.

Chinnabut, S., Roberts, R. J., Willoughby G. R. and Pearson, M. D. 1995. Histopathology of snake head, *Channa striatus* (Bloch) experimentally infected with the specific *Aphanomyces* fungus associated with Epizootic Ulcerative Syndrome (EUS) at different temperatures. J. Fish Dis., 18: 41-47.

Coker, W. C. 1923. The Saprolegniaceae, with notes on other wate molds. University of North California Press, Chapel Hill, North Carolina, 201pp.

Fraser, G. C., Callinan, R. B. and Calder, L. M. 1992. *Aphanomyces* species associated with red spot disease; an ulcerative disease of estuarine fish from eastern Australia. J.Fish disease., 15, 173-181.

Hatai, K. and Hoshai, G. 1992. Characteristics of two *Saprolegnia* species from coho Salmon with Saprolegniosis. J. Aqu. Anim. Health, 5: 115-118.

Hatai, K. and Hoshai, G. I. 1993. Characteristics of two *Saprolegnia* species from coho Salmon with Saprolegniosis. J. Aqu. Anim. Health., 5, 115-118.

Hatai, K., Nakamura, K. Rha, S. A., Yuasa, K. and Wada, S. 1994. *Aphanomyces* infection In the *dwarf gourami* (*Colisa Lalia*). J. Fish Patho., 29, 95-99.

Johnson, T. W. Jr. 1956. The Genus *Achlya*, Morphology and Taxonomy. University of Michigan Press, Ann Arbor, 180pp.

Khulbe, R. D., Joshi, C. and Bisht, G. S. 1995. Fungal diseases of fish in Nanak Sagar, Nainital, India. Mycopathol., 130; 71-74.

Mastan, S. A 2008. Incidences of Dermatomycosis in Fishes of Larpur reservior, Bhopal, (M.P). J. Herbal Med. and Toxicol., 2(1): 37- 40.

Pickering, A. D. and Willoughby, L. G. 1982a. *Saprolegnia*, infections of salmonid fish. In: 50th Annual Report, Institutes of freshwater Ecology, Windermere Laboratory, England, pp. 38-48.

Pickering, A. D. and Christie, P. 1980. Sexual differences in the incidence and severity of ectoparasitic infestation of the Brown trout, *Salmo trutta* L. J. fish Biol., 16: 669-683.

Qureshi, T. A., Chaunan, R. and Mastan, S. A. 2002. Experimental infection of *Saprolegnia* species on different species of fishes. J. Nat Cons., 14 (2): 385-388.

Qureshi, T. A., Chauhan, R., Prasad, Y. and Mastan, S. A. 1995. Fungi isolated from EUS affected Fishes of Hataikheda reservoir, Bhopal. Indian J. Appl. and Pure Biol., 10 (2): 153–157.

Qureshi, T. A., Chauhan, R., Prasad, Y. and Mastan, S. A. 1999. Association of Fungi with Epizootic Ulcerative Syndrome of Fishes. Indian J. Appl. and Pure Biol., 14 (1): 45 – 49.

Qureshi, T. A., Prasad, Y., Mastan, S. A. and Chauhan, R. 2000. Involvment of Fungal and Bacterial pathogens in EUS of Fishes. Biotech Consorium India Limited, pp- 125-139.

Roberts, R. J., Willoughby, L.G. and Chinnabut, S. 1993. Mycotic aspects of Epizootic Ulcerative Syndrome (EUS) of Asian fishes. J. Fish Dis., 16: 169-183.

Sati, S. C. and Khulbe, R. D. 1983. A host range of *Saprolegnia diclina* Humphery on certain cold water fishes of India. Proc. Nat. Acad. Sci. India, 53 (8): IV 309-312.

Sati S. C 1982. Aquatic fungi of Kumaun in relation to fish infection. Ph.D. Thesis, Kumaun University, Nainital. India, pp 1-.189.

Scott, W. W. 1961. A monograph of the genus *Aphanomyces* Var. Agar. Expt. Station. Tech. Bull., 151: 1-95.

Scott, W. W. and O' Bier, A. H. 1962. Aquatic fungi associated with diseased fish and fish eggs. Progressive Fish Culturist, 24: 3-15.

Scott, W. 1964. Fungi associated with fish disease. Dev. Ind, Microbiol., 5: 109-123.

Scott, W. W. and O' Warren, C. O. 1964. Studies of the Host Range and Chemical Control of Fungi Associated with Diseased Tropical Fish. Technical Bulletin, 171, Virginia Agriculture Experimental Station, Blacks burg, 24 pp.

Tiffney, W. N. 1939b: The identity of certain species of the *Saprolegnia parasitica* to fish. J Elisha Mitchell Scient. Soc.55: 143-151.

Vishniac, H. S. and Nigrelli, R. F. 1957. The ability of the *Saprolegniaceae* to parasitize platy fish. Zoologica, 42: 131-134.

Vikas Salgotra, Mastan, S. A. and Qureshi, T. A. 2005. Incidences of Saprolegniasis in fishes of Hataikhada reserviour,Bhopal. Indian J. fish., 52 (3):367-37

Walser, C. A. and Phelps, R. P. 1993. The use of Formalin and iodine to control. *Saprolegnia* infections on *Channel catfish, Talurus punctatus* eggs. J. Appl. Aqu. 3: 269-278.

Willoughby, L. G., Roberts, R. J. and Chinnabut, S. 1995. *Aphanomyces invaderis* sp. Nov. The fungal pathogen of freshwater tropical fish affected by Epizootic ulcerative syndrome. J. Fish Dis., 18: 273-275.

Willoughby, L. G. 1978. *Saprolegniasis* of Salmonid fish in Windermere, a critical analysis. J. Fish Dis., 1: 51-67.

Microbes: Diversity and Biotechnology (2012) *Pages* **205–218**
Editors: **Prof. S.C. Sati & Dr. M. Belwal**
Published by: **DAYA PUBLISHING HOUSE, NEW DELHI**

Chapter 13

Arbuscular Mycorrhizal (AM) Fungi Associated with Wild Medicinal Plants Exhibit Variation in Phosphorus Concentration during Growth Developmental Stages

Nair Radhika and B.F. Rodrigues*
Department of Botany, Goa University, Goa – 403 206

ABSTRACT

Three herbaceous medicinal plants (*Rauwolfia serpentina*, *Catharanthus roseus* and *Andrographis paniculata*) growing in the wild were studied for AM fungal association and phosphorus (P) concentration at different growth stages. All the plant species were found to be colonized by AM fungi with variation in the extent of colonization and sporulation among the plant species during developmental stages. Twenty-one AM fungal species were recovered from the rhizosphere soil at different growth stages. *Glomus fasciculatum* was found to be the most commonly occurring AM fungal species in all the growth stages of the three plant species. The results of the present study revealed that an increase in the P concentration during flowering stages of *A. paniculata* and *C. roseus* is directly related to the presence of arbuscules. In addition, the study confirm that higher P levels in plants decrease AM fungal spore production in the rhizosphere soil.

Keywords: *AM fungi, Frequency of occurrence, Medicinal plants, Phosphorus concentration, Relative abundance.*

---

* Corresponding Author: E-mail: felinov@gmail.com

Introduction

Arbuscular mycorrhizal fungi are ubiquitous and form a symbiotic association with most terrestrial plant species. The characteristics and dynamics of occurrence of AM fungi under natural conditions are important for the evaluation of the inoculum potential and root colonization in the process of understanding their behaviour in the soil and determining their symbiotic efficiency (Bethlenfalvay and Linderman, 1992).

Phenology is the study of the timing of vegetative activities, flowering and fruiting and their relationship to environmental factors. A growing plant may experience different stages in mineral nutrition, based on the balance among internal and external nutrient supplies and crop demand for nutrients. Plants require adequate P from the very early stages of growth for optimum biomass production (Grant *et al.*, 2001). The prevalent form of plant available P in the environment is the oxidized anion phosphate and it is estimated that on average, P diffuses only 0.5mm, so that only phosphate within close proximity of a plant root is positionally available for absorption (Grant *et al.*, 2001). Soluble minerals such as nitrogen (N) move through the soil via bulk flow and diffusion, whereas inorganic P (Pi) moves by a slow rate of diffusion in the soil solution creating a Pi-depletion zone around the root (Jungk, 2001). Therefore, the low availability of Pi in the bulk soil affects its uptake into roots (Rausch and Bucher, 2002).

Although mycorrhizal fungi have been shown to enhance growth and P nutrition of plants in pots (Cooper and Tinker, 1978), few studies have shown a functional relationship of AM fungi to plants in the wild (Miller, 1987). Examination of the functional significance of AM fungi through studies of P uptake is also difficult because the need for P is not constant during the life cycle of most plants (Fitter, 1985). Mycorrhizae may benefit plants only during times of P demand, *i.e.* during flowering or seed development. No previous studies have reported the P concentration in wild medicinal plants in relation to phenology. Thus, the present study was undertaken to study the variation in P concentration in different developmental stages of three selected medicinal plant species growing in the wild.

Materials and Methods

Study Area and Selection of Plant Species

The site undertaken for the study is located in Sanguem taluka situated in South Goa. Sanguem taluka has a geographical position marked at 15°48′ 00" N to 14°53′ 54" N latitude and 73°E to 75°E longitude. The climate is tropical with three main seasons *viz.*, monsoon, winter and summer. The soil is moderately drained, gravelly with a silty clay loam texture, pH ranging from 5.6 to 6.2 and is low in nutrients especially P (16kg ha^{-1}) and total N (0.24 per cent). Three medicinally important herbaceous plant species *viz.*, *Rauwolfia serpentina* (L.) Benth. (Apocynaceae), *Catharanthus roseus* L. (Apocynaceae) and *Andrographis paniculata* Nees. (Acanthaceae) growing wild in the forest community were selected for the study from the locality. Of these, two plant species, *R. serpentina* and *C. roseus* are listed as endangered species by IUCN red data list.

Rauwolfia serpentina, an erect perennial shrub commonly known as Indian snakeroot or Sarpagandha contains a number of bioactive chemicals including ajmalicine, deserpidine, rescinnamine, serpentine and yohimbine. Reserpine is an alkaloid first isolated from *R. serpentina*, widely used as an antihypertensive drug (Lewis and Lewis, 2003).

Catharanthus roseus (Madagascar periwinkle), a perennial herb, has been cultivated for herbal medicine and as an ornamental plant. The substances vinblastine and vincristine extracted from the plant are used in the treatment of leukemia (Leveque and Jehl, 2007).

Andrographis paniculata an erect annual herb commonly known as "King of Bitters" has been used for centuries in Asia to treat upper respiratory infections, fever, Herpes, sore throat and other chronic and infectious diseases. Some of the beneficial properties include analgesic, anti-inflammatory, antibacterial, antipyretic, cancerlytic, antiviral and vermicidal. The primary medicinal component of *A. paniculata* is Andrographolide, which is a diterpene lactone. The other active components include 14 deoxy 11, 12- di dehydroandrographolide, homoandrographolide, andrographan, andrographosterin and stigmasterol (Siripong *et al.*, 1992).

Sample Collection

Collections were made in different growth the stages of the plant *viz.*, vegetative stage (June-August), flowering stage (September-December), and fruiting stage (January- April). Three plants of each species were collected from the same locality (radius of 100m). For each plant species, three rhizosphere soil samples were mixed to form a composite sample, packed in polyethylene bags, labeled and brought to the laboratory. Root samples were freshly processed, whereas soil samples were stored at 4°C until analyzed.

Estimation of Root Colonization and Spore Density

For processing of roots, trypan blue staining technique (Koske and Gemma, 1989) was used and percent colonization (proportion of root length colonized) was determined according to slide method (Giovannetti and Mosse, 1980). Isolation of AM fungal spores was carried out by a wet sieving and decanting method (Gerdemann and Nicolson, 1963). Spore density was calculated by number of spores present per 100g of rhizosphere soil.

Taxonomic Identification of Spores

Taxonomic identification of spores was carried out by using the Manual for Identification of VAM Fungi by Schenck and Perez (1990) and various taxonomic papers *viz.*, Walker and Vestberg (1998), Redecker *et al.* (2000), Morton and Redecker (2001). Taxonomic identification of spores was also carried out by matching the descriptions provided by the International Culture Collection of Vesicular Arbuscular Mycorrhizal Fungi (http://invam.caf. wvu.edu.).

Phosphorus Estimation by Colorimetric Method

Root and shoot tissues of three medicinal plant species collected during different growth stages were previously analyzed by dry ash digestion procedure and assessed for the estimation of total P concentration (ppm) using Vanadomolybdate phosphoric yellow colour method (Chapman and Prat, 1961).

Relative Abundance and Frequency of Occurrence

Relative abundance and Frequency of occurrence of AM fungi was calculated in each plant species at different growth stages using the following formulae (Beena *et al.*, 2000).

$$\text{Relative abundance (per cent)} = \frac{\text{Number of AM fungal spores of particular species}}{\text{Total number of AM fungal spores of all species}} \times 100$$

Frequency of occurrence (per cent) =

$$\frac{\text{Number of soil samples containing spores of particular species}}{\text{Total number of soil samples screened}} \times 100$$

Statistical Analysis

Pearson correlation analysis was carried out to assess the relationship between colonization and spore density. Results of P concentration in roots and shoots of each plant species at different growth stages was analyzed by One Factorial Analysis of Variance (ANOVA) using WASP 1.0 (Web based Agricultural Statistical Package). For all the analyses, differences were considered significant when $P<0.05$.

Results

Mycorrhizal colonization was observed in all the plant species samples studied. Hyphal and vesicular colonization was observed in all the three medicinal plant species at different growth stages whereas arbuscular colonization was observed only during the flowering stage in *C. roseus* and *A. paniculata*. Percent root colonization differed in all the plant species at different growth stages.

In *R. serpentina* and *C. roseus* maximum percent AM colonization was observed during the vegetative stage (58.33 per cent and 70 per cent) followed by flowering stage (50 per cent and 44.44 per cent) and least in fruiting stage (35.71 per cent and 30.7 per cent) respectively whereas in *A. paniculata* flowering stage showed maximum colonization (55.50 per cent) followed by fruiting stage (42.50 per cent) and least in vegetative stage (38.50 per cent).

Spore density showed variation in different growth stages of all the plant species studied. In *R. serpentina* and *C. roseus* maximum spore density was recorded in the fruiting stage (251 and 56 spores $100g^{-1}$ soil) whereas in *A. paniculata* highest spore density was observed during the flowering stage (147 spores $100g^{-1}$ soil). Among the plant species, least spore density was recorded in *R. serpentina* (8 spores $100g^{-1}$ soil) in the flowering stage. A non-significant negative correlation was observed between percentage colonization and spore density (r= -0.1, $P<0.05$).

Twenty AM fungal species belonging to five genera *viz., Glomus, Acaulospora, Scutellospora, Gigaspora* and *Ambispora* were identified from the rhizosphere soil samples. *Glomus* was the dominant genus and *G. fasciculatum*, the dominant species, was recorded in all stages of development in the three plant species.

In *A. paniculata*, five AM fungal species belonging to three genera *viz., Acaulospora, Glomus* and *Ambispora* were recorded in different growth stages. *G. fasciculatum* was the most commonly occurring AM fungal species in all the growth stages and was more abundant during flowering and fruiting stages of the plant *A. scrobiculata* was more abundant during the vegetative stage. Out of five AM fungal species, *Am. leptoticha* recorded the least in relative abundance (2.04 per cent) and frequency of occurrence (25 per cent), both observations made during flowering stage of the plant (Figures 13.1–13.3).

Thirteen AM fungal species belonging to five genera *viz., Acaulospora, Ambispora, Glomus, Gigaspora* and *Scutellospora* were recorded in all the growth stages of *R. serpentina*. *Glomus fasciculatum* was recovered in all the growth stages and found to be the dominant species in terms of relative abundance (66.53 per cent) and frequency of occurrence (87.5 per cent) recorded during the fruiting stage.

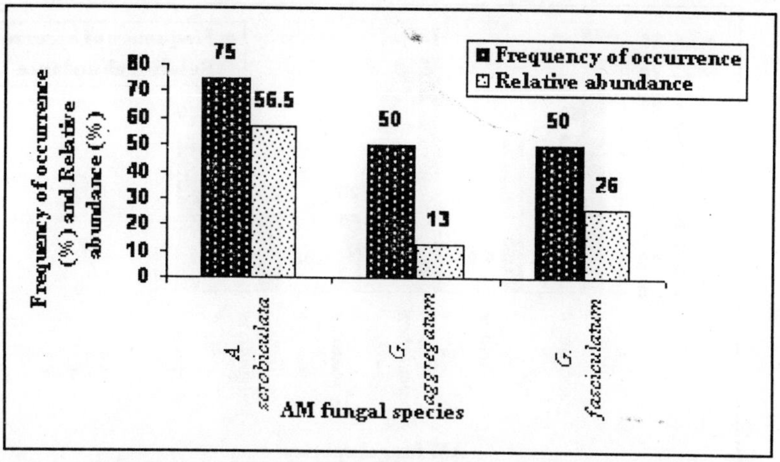

Figure 13.1: Frequency of Occurrence (%) and Relative Abundance (%) of AM Fungal Species in Vegetative Stage of *A. paniculata*

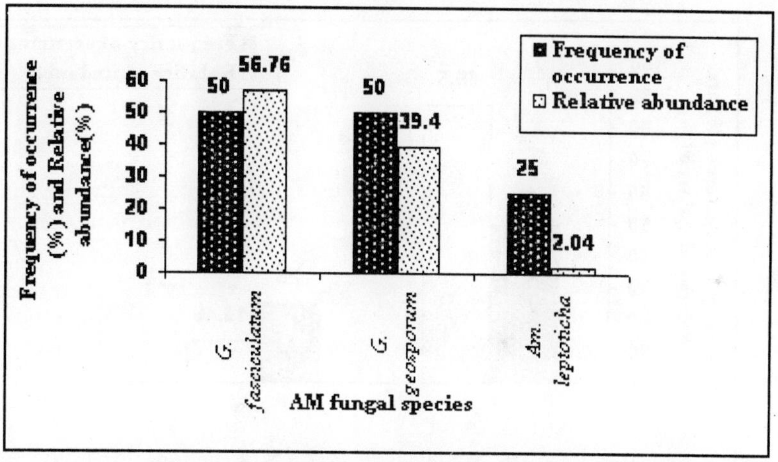

Figure 13.2: Frequency of Occurrence (%) and Relative Abundance (%) of AM Fungal Species in Flowering Stage of *A. paniculata*

G. geosporum was the most abundant and frequently occurring AM fungal species during the vegetative and flowering stages (Figures 13.4–13.6).

In *C. roseus*, 13 AM fungal species belonging to four genera *viz., Acaulospora, Glomus, Gigaspora* and *Scutellospora* were recorded in the rhizosphere soil samples in all the three growth stages. *Glomus fasciculatum* and *G. maculosum* occurred in all the growth stages. *G. fasciculatum* was the most abundant (79.5 per cent) and frequently occurring (80 per cent) species and was recorded during the flowering stage whereas *G. maculosum* was the most abundant and frequently occurring species during the vegetative and fruiting stages (Figures 13.7–13.9)

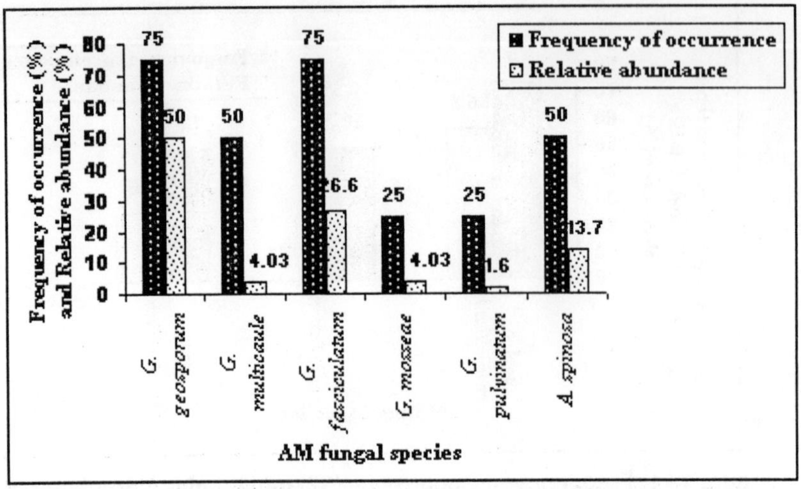

Figure 13.3: Frequency of Occurrence (%) and Relative Abundance (%) of AM Fungal Species in Vegetative Stage of *R. serpentina*

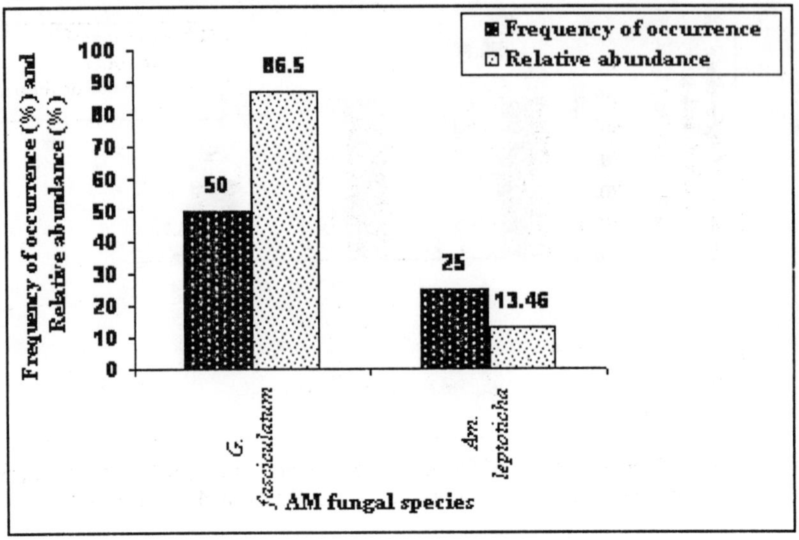

Figure 13.4: Frequency of Occurrence (%) and Relative Abundance (%) of AM Fungal Species in Fruiting Stage of *R. serpentina*

Maximum species richness (13) was recorded in the rhizosphere soil samples of *R. serpentina* and *C. roseus* and least in *A. paniculata* (4).

The P concentration studies revealed variation in both root and shoot tissues of the three medicinal plant species at different growth stages. In *R. serpentina,* maximum P concentration in root (39.3ppm) and shoot tissues (53.8ppm) was recorded during the vegetative stage of the plant (Figure 13.12) whereas in *A. paniculata* and *C. roseus* (Figures 13.10 and 13.11) maximum P concentration was recorded during the flowering stage which suggests the presence of arbuscules. An increase in shoot P concentration was also observed after the formation of arbuscules.

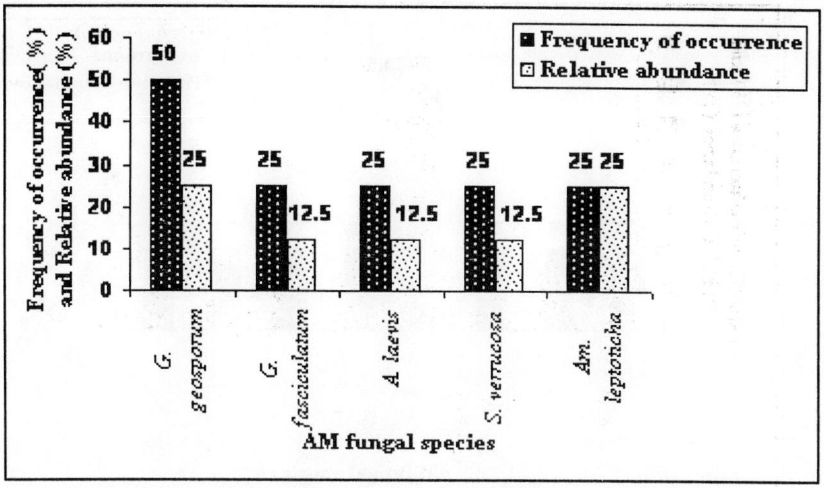

Figure 13.5: Frequency of Occurrence (%) and Relative Abundance (%) of AM Fungal Species in Flowering Stage of *R. serpentina*

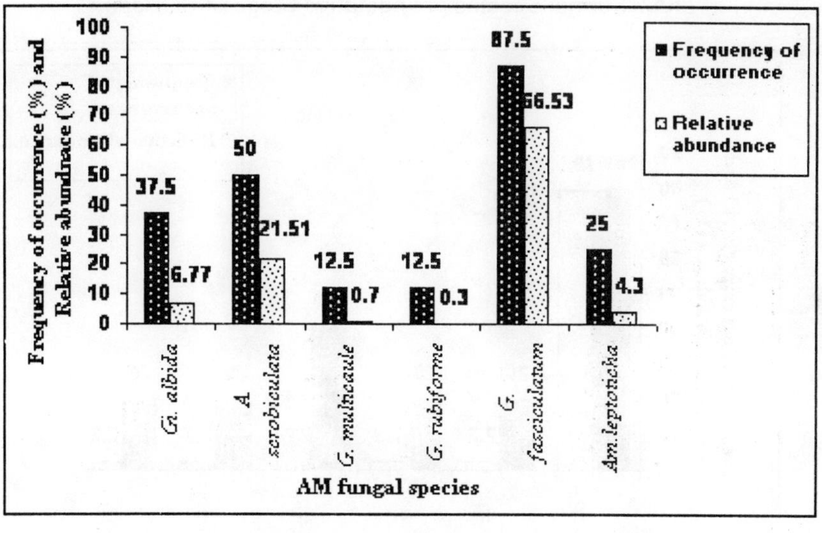

Figure 13.6: Frequency of Occurrence (%) and Relative Abundance (%) of AM Fungal Species in Fruiting Stage of *R. serpentina*

Results of P concentration in roots and shoots were analyzed using ANOVA. Root and shoot tissues of *R. serpentina* showed a significant difference in P concentration between the growth stages (F=0.000, df=5, P<0.05). A significant decrease in P concentration from vegetative to flowering stage followed by an increase in fruiting stage was observed. Roots and shoots of *A. paniculata* showed a significant increase in P concentration from vegetative to flowering stage of the plant, and then showed a gradual decrease in the fruiting stage (F=0.004, df=5, P<0.05). In *C. roseus*, a significant increase from vegetative to flowering stage was observed in the shoots followed by a decrease in the fruiting stage,

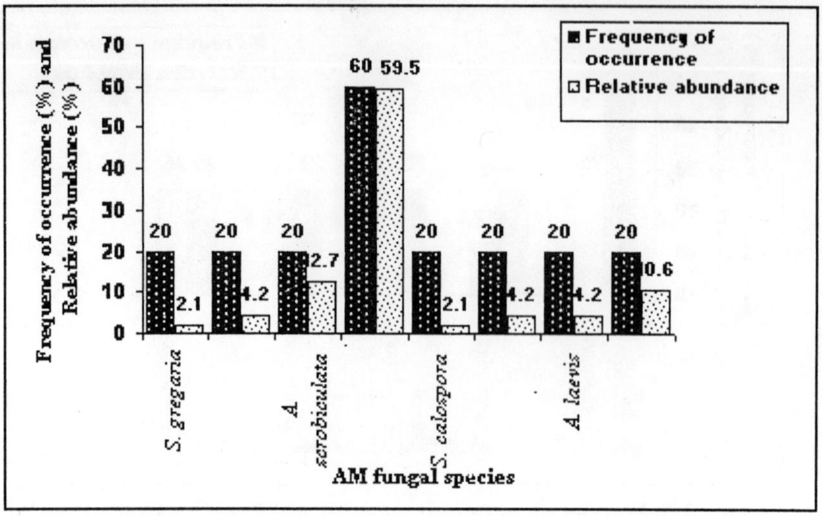

Figure 13.7: Frequency of Occurrence (%) and Relative Abundance (%) of AM Fungal Species in Vegetative Stage of *C. roseus*

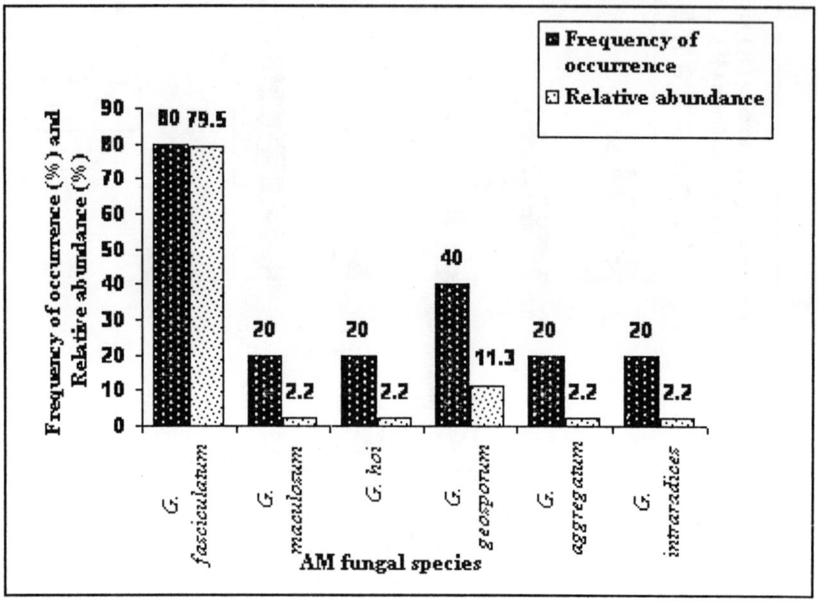

Figure 13.8: Frequency of Occurrence (%) and Relative Abundance (%) of AM Fungal Species in Flowering Stage of *C. roseus*

whereas in roots, P concentration increased significantly from flowering stage onwards (F=0.003, df=5, P<0.05). Negative correlation was observed between percentage colonization and P concentration in roots and shoots (r = -0.1, -0.2, P<0.05).

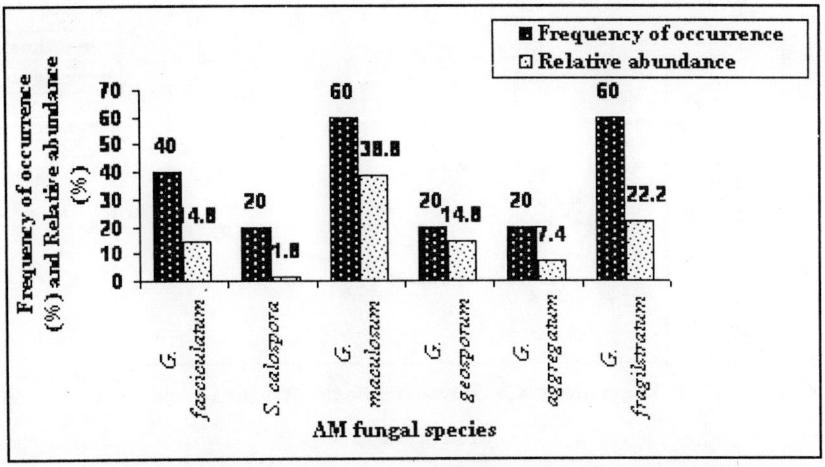

Figure 13.9: Frequency of Occurrence (%) and Relative Abundance (%) of AM Fungal Species in Fruiting Stage of *C. roseus*

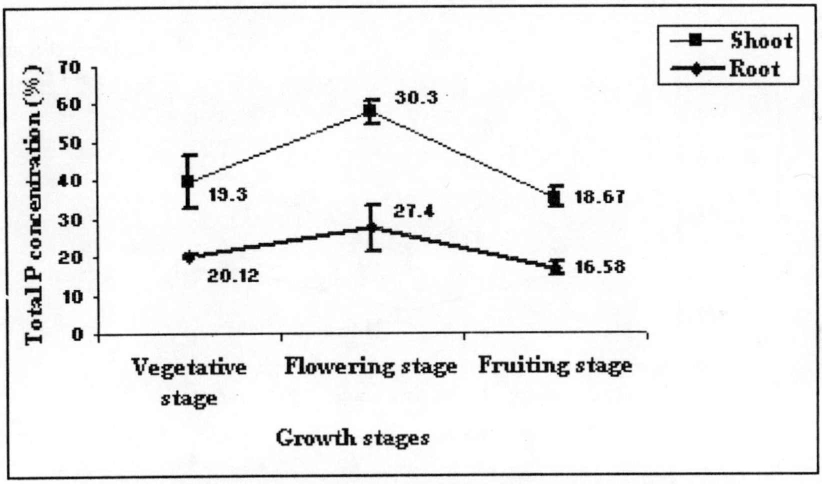

Figure 13.10: Phosphorus Concentration in Roots and Shoots of *A. paniculata* at Different Growth Stages

Discussion

The study indicates AM fungi could play an important role in growth of medicinal plants. Muthukumar and Udaiyan (2000) reported the presence of AM fungal colonization in *A. paniculata* and *R. serpentina* but absence of colonization in *C. roseus*. However, in the present study AM fungal colonization was observed in all the growth stages of *C. roseus*. The study also revealed that *Glomus* was the most dominant genus in the rhizosphere soil of three medicinal plants. The predominance of *Glomus* in tropical soils has been reported by other workers (Thapar and Khan, 1985; Ragupathy and Mahadevan, 1993). They also reported the presence of 20 AM fungal species from the rhizosphere soil of three medicinal plant species in different growth stages. Johnson *et al.* (1991) reported similar findings in their study on plant and soil controls of mycorrhizal communities. They recorded the

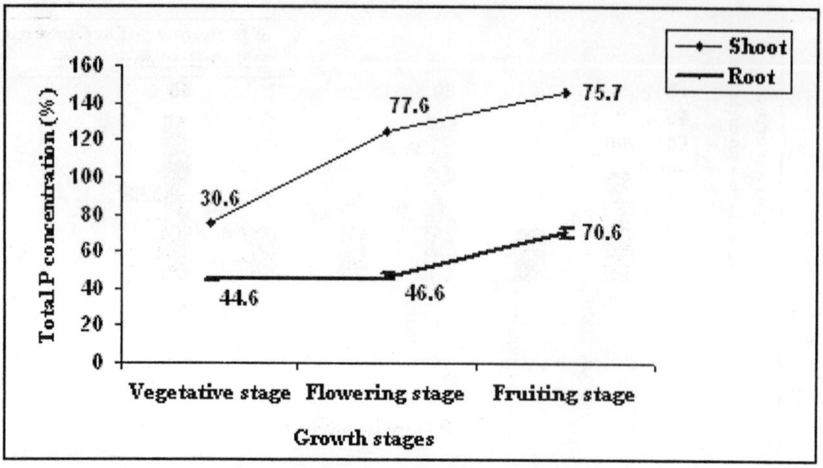

Figure 13.11: Phosphorus Concentration in Roots and Shoots of *C. roseus* at Different Growth Stages

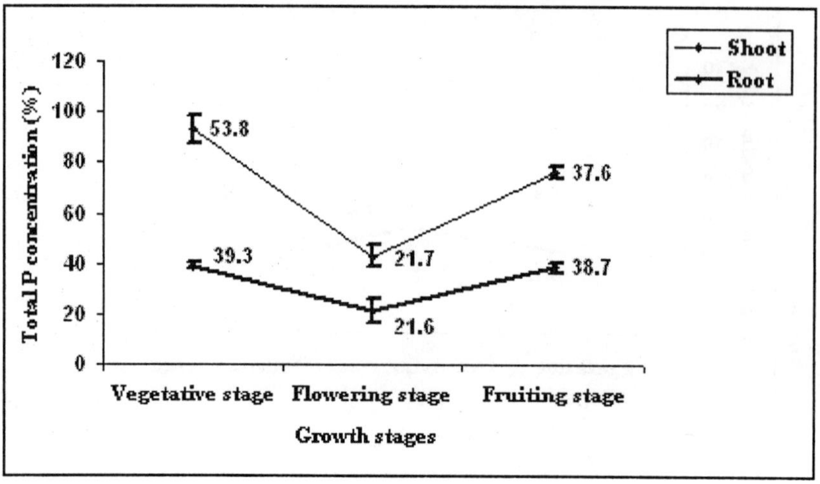

Figure 13.12: Phosphorus Concentration in Roots and Shoots of *R. serpentina* at Different Growth Stages

presence of 12 to 22 different species of AM fungi per study site. Muthukumar *et al.* (2001) however reported only 35 AM fungal species from the rhizosphere soil of about 329 plant species from the Western Ghats of Southern India. Recovery of a relatively high number of species in the present study is in agreement with Francis and Read (1994) who suggested that high species diversity, characteristic of phosphorus-deficient grassland ecosystem dominated by plant species with AM fungi, may be attributed to a low level of host specificity.

Large variations in the spore numbers (8 to 251 spores $100g^{-1}$ rhizosphere soil) recorded in plant species at different growth stages in the present study can be attributed mainly to the interspecific competition (Brundrett and Kendrick, 1990) and by the subsequent variation in the timing of spore production associated with host plants, suggesting that the competition between fungi and environmental factors probably influences spore production in natural communities (Gemma and Koske, 1988).

In the present study, increase in the P concentration during flowering stage of *A. paniculata* and *C. roseus* is directly related to the presence of arbuscules as flower initiation demands extra uptake of P. An increase in the shoot P concentration was observed after the formation of arbuscules. Similar observations have been reported in *Ranunculus adoneus* (Mullen and Schmidt, 1993).

Sanders (1990) studied AM fungal development in natural grassland species but did not record nutrient accumulation and found no relationship from one year to the next in developmental patterns. Other studies of mycorrhizal development in natural populations (Brundrett and Kendrick, 1988; Sanders, 1990) have reported only mycorrhizal colonization. Reinhardt and Miller (1990), working on temperate grasslands, found that AM colonization reached a peak in the growing season (April) and they assessed the levels of arbuscules from root cores containing roots of the entire plant community, not relating colonization to phenology. However, the presence of arbuscules recorded during the flowering stages in two plant species *viz.*, *A. paniculata* and *C. roseus* indicated that arbuscules are essential for P uptake. In *R. serpentina* no arbuscules were observed in any of the growth stages, maximum P concentration being observed during the vegetative stage indicating that this species requires P during early growth stage for optimum yield. A similar observation was made earlier (Grant *et al.*, 2001) and in this study supports the fact that the need for P is not constant during the life cycle of the plant (Fitter, 1985). In *A. paniculata* a significant decrease in P concentration was observed during the fruiting stage as P concentration in tissue of annual plants declines with advancing plant age or stage of growth. As the plant matures an increasing proportion of its dry weight is composed of low P in structural and storage tissues. Similar observations were recorded earlier (Belanger and Richards, 1999).

In the present study, vesicles were observed in all the growth stages of the plants. The high number of vesicles in the roots indicated that conditions were favourable for their formation. Dunne and Fitter (1989) who examined the P budget of strawberry plants in the field, found that flower initiation required extra uptake of P. However, Dodd and Jeffries (1986) in their study of mycorrhizal development in winter wheat quantified arbuscule production and found that levels were highest immediately before seed formation, presumably a time of high P demand. Fitter (1991) suggested that wild plants may have different patterns of P uptake than crop plants.

In the present study, *C. roseus* showed high P concentration in plant tissues, but recorded fewer number of spores in the rhizosphere soil. This observation supports the earlier study by De Miranda and Harris (1994) that higher tissue P in the plant reduces the production of spores , and secondary external hyphae (Bruce *et al.*, 1994). Root exudates from host plants secrete signal molecules that are known to enhance hyphal branching when there is P limitation in host roots (Nagahashi and Douds, 2000). Therefore, increasing P status in the root may reduce the secretion of these signal molecules, thus reducing hyphal branching and mycorrhizal colonization. Phosphorus status of the roots may affect membrane phospholipids, thus influencing membrane permeability and the release of carbohydrates from the roots which nourish the fungi (Schwab *et al.*, 1991). Thus, when P concentration in the plant is low, carbohydrate exudation will encourage mycorrhizal association, which will then enhance the uptake of P from the soil (Grant *et al.*, 2001).

Increase in P concentration during flowering stage indicates that presence of arbuscules corresponds to active P accumulation in wild medicinal plant species. An increase in shoot P concentration was observed after arbuscule formation which could be due to Pi being readily available.

More P transported to shoots leads to P luxury consumption (Chapin, 1980) or Pi storage that could be used in the future to support long term growth (Aerts and Chapin, 2000). Further studies pertaining to the beneficial effects on growth in medicinal plant species and understanding the importance of the relationships to each of the symbionts needs to be undertaken.

Acknowledgements

The authors thank the financial support provided by the Planning Commission, Government of India, New Delhi for carrying out this study.

References

Aerts, R. and Chapin, F. S. 2000. The mineral nutrition of wild plants revisited: A re-evaluation of processes and patterns. Advances in Ecological Research, 30: 1-67.

Beena, K. R, Raviraja, N. S and Arun, A. D and Sridhar, K. R. 2000. Diversity of arbuscular mycorrhizal fungi on coastal sand dunes of the west coast of India. Current Science, 79: 1459-1465.

Belanger, G and Richards, J. E. 1999. Relationship between P and N concentrations in timothy. Canadian Journal *Plant* Science, 79: 65-70.

Bethlenfalvay, G. J. and Linderman, R. G. 1992. Mycorrhizae in sustainable agriculture. ASA Spec Publ, Madison.

Bruce, A., Smith, W. E. and Tester, M. 1994. The development of mycorrhizal infection in cucumber: effects of P supply on root growth, formation of entry points and growth of infection units. New Phytologist, 1276: 507-514.

Brundrett, M. C. and Kendrick, B. 1988. The mycorrhizal status, root anatomy and phenology of plants in a sugar maple forest. Canadian Journal of Botany, 66: 1153-1173.

Brundrett, M. C. and Kendrick, B. 1990. The roots and mycorrhizas of herbaceous woodland II. Structural aspects of morphology. New Phytologist, 114: 469-479.

Chapman, H. D. and Prat, P. F. 1961. Methods of analysis for soils, plants and waters. University of California, Divison of Agriculatural Sciences, Berkeley.

Cooper, K. M. and Tinker, P. B. 1978. Translocation and transfer of nutrients in vesicular arbuscular mycorrhizas II. Uptake and translocation of phosphorus, zinc and sulphur. New Phytologist, 81: 43-52.

Chapin, F. S. 1980. The mineral nutrition of wild plants. Annual Review of Ecology and Systematics 11: 233-260.

De Miranda, J. C. C. and Harris, P. J. 1994. Effect of soil phosphorus on spore germination and hyphal growth of arbuscular mycorrhizal fungi. New Phytologist, 128: 103-108.

Dodd, J. C. and Jeffries, P. 1986. Early development of vesicular arbuscular mycorrhizal in autumn-shown cereals. Soil Biology Biochemistry, 18: 149-154.

Dunne, M. J. and Fitter, A. H. 1989. The phosphorus budget of a field grown strawberry (*Fragaria* x *ananassa* cv.Hapil) crop: evidence for a mycorrhizal contribution. Annals of Applied Biology, 114: 185-193.

Fitter, A. H. 1985. Functioning of vesicular arbuscular mycorrhizas under field conditions. New Phytologist, 99: 257-265.

Fitter, A. H. 1991. Cost and benefits of mycorrhizas: Implications for functioning under natural conditions. Experientia, 47: 350-355.

Francis, R. and Read, D. J. 1994. The contribution of mycorrhizal fungi in determination of plant community structure. Plant and Soil, 159: 11-25.

Gemma, J. N. and Koske, R.E. (1988). Seasonal variation in spore abundance and dormancy of *Gigaspora gigantea* and in mycorrhizal inoculum potential of a dune soil. *Mycologia*. 80: 211-216.

Gerdemann, J. W. and Nicolson, T. H. 1963. Spores of mycorrhizal *Endogone* species extracted from soil by wet sieving and decanting. Transactions of British Mycological Society, 46: 235-244.

Grant, C. A., Flaten D. N., Tomasiewicz, D. J. and Sheppard, S. C. 2001. The importance of early season P nutrition. Canadian of Journal of Plant Sciences, 81: 211-224.

Giovannetti, M. and Mosse, B. 1980. An evaluation of techniques for measuring vesicular arbuscular mycorrhizal infection in roots. New Phytologist, 84: 489-500.

Johnson, N. C., Tilman, D. and Wedin, D. 1991. Plant and soil controls on mycorrhizal fungal communities. Ecology, 61: 151-162.

Jungk, A. 2001. Root hairs and the acquisition of plant nutrients from soil. Journal of Plant Nutrition and Soil Science, 164: 121-129.

Koske, R. E. and Gemma, J. N. 1989. A modified procedure for staining roots to detect VA mycorrhizas. Mycological Research, 92 (4): 486-505.

Leveque, D. and Jehl, F. 2007. Molecular pharmacokinetics of *Catharanthus* (*Vinca*) alkaloids. Journal of Clinical Pharmacology, 47 (5): 579-588.

Lewis, W. H. and Elvin-Lewis, M. P. F. 2003. Medical Botany. Hoboken, Wiley. 286.

Mullen, R. B. and Schmidt, S. K. 1993. Mycorrhizal infection, phosphorus uptake, and phenology in *Ranunculus adoneus*: Implications for the functioning of mycorrhizae in alpine systems. Oceologia, 94: 229-234.

Miller, R. M. 1987. The ecology of vesicular-arbuscular mycorrhizae in grass and shrubland. In: Ecophysiology of VA mycorrhizal plants. (Ed. Safir, G.). pp 135-170, CRC Boca Raton Florida.

Morton, J. B. and Redecker, D. 2001. Concordant morphological and molecular characters reclassify five arbuscular mycorrhizal fungal species into new genera *Archaeospora* and *Paraglomus* of new families Archaeosporaceae and Paraglomaceae respectively. Mycologia, 93: 181-195.

Muthukumar, T. and Udaiyan, K. 2000. Arbuscular mycorrhizas of plants growing in Western Ghats region, Southern India. Mycorrhiza, 9: 297-313.

Muthukumar, T., Udaiyan, K. and Manian, S. 2001. Vesicular arbuscular mycorrhizal association in the medicinal plants of Maruthumalai Hills, Western Ghats, Southern India. Journal of Mycology and Plant Pathology, 31: 180-184.

Nagahashi, G. and Douds, D. D. Jr. 2000. Partial separation of root exudates and their effects upon the growth of germinated spores of AM fungi. Mycological Research, 104: 1453-1464.

Ragupathy, S. and Mahadevan, A. 1993. Distribution of vesicular arbuscular mycorrhizae in the plants and rhizosphere soil of the tropical plains Tamil Nadu, India. Mycorrhiza, 3: 123-136.

Rausch, C. and Bucher, M. 2002. Molecular mechanisms of phosphate transport in plants. Planta, 216: 23-37.

Redecker, D., Morton, J. B. and Bruns, T. D. 2000. Ancestral lineages of arbuscular mycorrhizal fungi (Glomales). Molecular Phylogenetics and Evolution, 14: 276-284.

Reinhardt, D. R. and Miller, R. M. 1990. Size classes of root diameter and mycorrhizal fungal colonization in two temperate grassland communities. New Phytologist, 116: 129-136.

Sanders, F. E., Tinker, B. P., Black, R. L. B. and Palmerely S. M. 1977. The development of endomycorrhizal root systems. I. Speed of infection and growth-promoting effects with four species of vesicular-arbuscular endophyte. New Phytologist, 78: 257-268.

Sanders, I. R. 1990. Seasonal patterns of vesicular arbuscular mycorrhizal occurrence in grasslands. Symbiosis, 9: 315-320.

Sanders, F. E., Tinker, B. P., Black, R. L. B. and Palmerely, S. M. 1977. The development of endomycorrhizal root systems. I. Speed of infection and growth-promoting effects with four species of vesicular arbuscular endophyte. New Phytologist, 78: 257-268.

Schenck, N. C. and Perez, Y. 1990. Manual for identification of VAM mycorrhizal fungi. INVAM, University of Florida, Gainesville pp. 1-283.

Schwab, S. M., Menge, J. and Tinker, P. B. 1991. Regulation of nutrient transfer between host and fungus in vesicular and arbuscular mycorrhizas. New Phytologist, 117: 387- 398.

Siripong, P., Kongkathip, B., Preechanukool, K., Picha, P., Tunsuwan, K., Taylor, W. C. 1992. Cytotoxic diterpenoid constituents from *Andrographis paniculata* Nees leaves. Journal of Scientific Society of Thailand, 18: 187-194.

Thapar, H. S. and Khan, S. N. 1985. Distribution of VA mycorrhizal fungi in forest soils of India. Indian Journal of Forest, 8: 5-7.

Walker, C. and Vestberg, M. 1998. Synonymy amongst the arbuscular mycorrhizal fungi *Glomus claroideum*, *Glomus maculosum*, *Glomus multisubstensum* and *Glomus fistulosum*. Annals of Botany, 82: 601-624.

Microbes: Diversity and Biotechnology (2012) *Pages* 219–229
Editors: Prof. S.C. Sati & Dr. M. Belwal
Published by: DAYA PUBLISHING HOUSE, NEW DELHI

Chapter 14

Isolation and Mass Multiplication of Vesicular Arbuscular Mycorrhizal Inoculum Using Monocot Hosts

Vipin Parkash[1] *and Ashok Aggarwal*[2]
[1]*Rain Forest Research Institute (ICFRE),*
Autonomous Council of Ministry of Environment and Forests,
Govt. of India, Deovan, Jorhat – 785 001, Assam
[2]*Botany Department, Kurukshetra University,*
Kurukshetra – 136 119, Haryana

ABSTRACT

Inoculum production or starter culture of mixed VAM, *G. mosseae* and *G. fasciculatum* on host sorghum proved best because the percentage mycorrhizal root colonization, total VAM spore number and seed germination were more in treated sterilized soil in comparison to control after 60 days. All the three hosts *i.e.* maize (*Zea mays* L.), sorghum (*Sorghum vulgare* Pers.) and wheat (*Tritichum aestivum* L.) for mass multiplication of VAM spores proved good but maize (*Zea mays* L.) proved best host for mass production of mixed VAM, *Glomus mosseae. Glomus fasciculatum* mycorrhizae as this monocot host had maximum percentage mycorrhizal root colonization and mycorrhizal spore number after 90 days in comparison to control and other host plants.

Keywords: Mass multiplication, Monocot hosts, Glomous mosseae, Glomus fasciculatum, Starter culture, VAM.

Introduction

There are many cultures, techniques and systems for mass production of vesicular arbuscular mycorrhizal inoculum such as root organ culture, pot culture, nutrient film techniques, circulation hydroponic culture system and aeroponic culture system (Mungier and Mosse, 1987; Menge and

Timmer, 1982; Singh and Tilk, 2002). Soilless cultures and adaptations of hydroponics produce high quality inoculum with propagule numbers many times greater than inoculum used in the past. The lower cost of inoculum and ease of production are making these techniques applicable in less developed agricultural areas as well as in highly industrialized agricultural systems that currently use phosphate fertilizers. A greater understanding of the biology of VAM fungi, coupled with empirical trials, have shown several innovative application techniques to be compatible with current production protocols. VA mycorrhizal fungi have great potential for use as biofertilizers in agriculture, floriculture, horticulture and forestry as evident from laboratory, greenhouse and limited field studies by a large number of mycorrhizal scientists (Rani *et al.*, 1997; Singh, 1999; Parkash, 2004). The potential of using VA mycorrhizal fungi on a large scale depends upon 1., the technique by which axenic VAM fungi are grown; 2., the economic production of a large volume and high quality of inoculum; 3., formulation of VAM inoculants preparation with an extended shelf-life and easy handling characteristics and 4., development of growth promoting strains superior to indigenous soil VAM fungi (Hua, 1990). For bulk production, the natural choice is to develop a dual culture of host and VA-mycorrhizal fungi. These are three prerequisites for obtaining such dual cultures *i.e.* an efficient VA-mycorrhizal fungus, a suitable host and a suitable substrate (Singh, 2002).

Pot culture of VAM fungi on host in sterilized soil using selected spores, roots and infested soil as inoculum has been the most used technique for increasing propagule numbers (Menge, 1983, 1984; Wood, 1984). Pot cultures of vesicular asbuscular mycorrhizal inoculum are usually harvested after two months period. This inoculum consisting of chopped roots, infested soil and selected spores are air dried and packed in sterilized polythene bags for storage (Ferguson, 1981). Selection of a suitable substrate and host plant for mass inoculum multiplication is also an important factor for mass production of VAM fungi (Sreenivasa and Bagyaraj, 1988; Liyange, 1989; Thompson, 1986; Mehrotra and Mehrotra, 1999; Harikumar and Potty, 2002).

Mass inoculum production of VAM fungi with efficient strains in the field for cultivated crops has usually met with little success because of the inability to culture the fungus axenically (Harikumar and Potty, 2002) and also VAM fungi are difficult to mass production because these are obligate symbionts (Jeffries and Dodd, 1991; Sharma *et al.*, 2000). Large-scale production of VAM inoculum for field inoculation is technically feasible through pot culture using an appropriate host (Wood, 1984; Harikumar and Potty, 2002). Soils of low fertility have proved particularly effective in supporting inoculum production. Production of high quality mycorrhizal inoculum necessitates the exclusion of unwanted organisms like pathogenic fungi, nematodes and insects. This can be easily achieved by the use of agrochemicals, which do not have deleterious effects on AM fungi. A wide range of agrochemicals have been used earlier to study this effect on VAM fungi. Some of the agrochemicals proved deleterious and some quite compatible with VAM fungi (Kumar and Bagyaraj, 1999).

In view of the above, the objective was undertaken in which an effort was made for inoculum production and mass multiplication of dominant VAM spores *i.e.* mixed VAM (*Glomus, Acaulospora, Gigaspora* and *Sclerocystis*), *G. mosseae, G. fasciculatum* using different monocot hosts to see their efficacy in these hosts for further inoculation experiments.

Materials and Methods

Soil Sample Collection

Rhizospheric soil samples from the roots were collected by digging out a small amount of soil close to plant roots up to the depth of 15-30 cm and these samples were kept in sterilized polythene bags at 5-10°C for further processing in the laboratory for mycorrhizal quantification and root colonization.

Isolation of Spores

Isolation of VAM spores was done by wet sieving and decanting (Gerdemann and Nicolson, 1963) technique. Approximately 100gm of soil were suspended in 1 liter or more of water. Heavier particles were allowed to settle for a few seconds and the liquid was decanted through sieves of different sizes in the order 150μm, 120μm, 90μm, 63μm, 45μm which removes large particles of organic matter, but course enough to allow the desired spores to pass through. The seivings retained on different sieves were collected on different Petri dishes then the trapped spores were transferred to Whatman filter paper no. 1 by repeating washing with water. The spores were picked by hydrodermic needles under stereo- binocular microscope.

Mycorrhizal Quantification

For quantitative estimation of VAM spores, Gaur and Adholeya modified method (1994) was used. The filter paper was divided into many small sectors by marking with a ball pen. The total number of spores was counted by adding the number of spores present in each sector under stereo-binocular microscope.

Identification of VAM Fungi

For identification of VAM spores the following criteria were used like conventional morphological character *i.e.* colour, size, shape wall structure, surface, ornamentation of spores, nature and size of subtending hyphae, bulbous suspensor, the number and arrangement of the spores in the sporocarp. These VAM spores were identified by using the keys of Schenck and Perez (1990), Morton and Benny (1990), and Mukerji (1996).

Percent Mycorrhizal Root Colonization

It was studied by rapid clearing and staining method by Phillips and Hayman (1970). The root segments were washed with water to remove soil particles. It was then cut into 1cm small pieces. Root segments were washed with water and placed in 10 per cent KOH solution at 90°C for half an hour or for 24 hours at room temperature. KOH was decanted and it was washed with water till the brown color was cleared. They were acidified with 1 per cent HCl for 3-5 minutes. The acid was poured out and root segments were submerged in 0.5 per cent, Trypan blue for 24 hours. After 24 hours the segments were destained with Lactophenol. The roots were mounted in lactic acid or lactic acid: glycerol (1:1) solution. The percentage mycorrhizal root colonization was studied by following formula.

$$\text{Percent mycorrhizal root colonization} = \frac{\text{Total no. of infected root segments}}{\text{Total no. of root segments examined}} \times 100$$

Inoculum Production of VAM Spores (Starter Culture)

In the present investigation for mass production of VAM inoculum efficient strains of VAM fungi (*G. mosseae* and *G. fasciculatum*) were isolated by 'wet sieving and decanting technique' (Gerdemann and Nicholson, 1963) as discussed above. For inoculum production, single spore was taken and multiplied on sterilized substrate *i.e.* sand: soil (10: 30 gm) mixture by funnel technique. In this technique, glass funnels/earthen funnels are taken and germination of seeds was made in such a way that roots of the seedlings must touch the inoculum of VAM fungi. The seedlings were raised up to 30 days in the

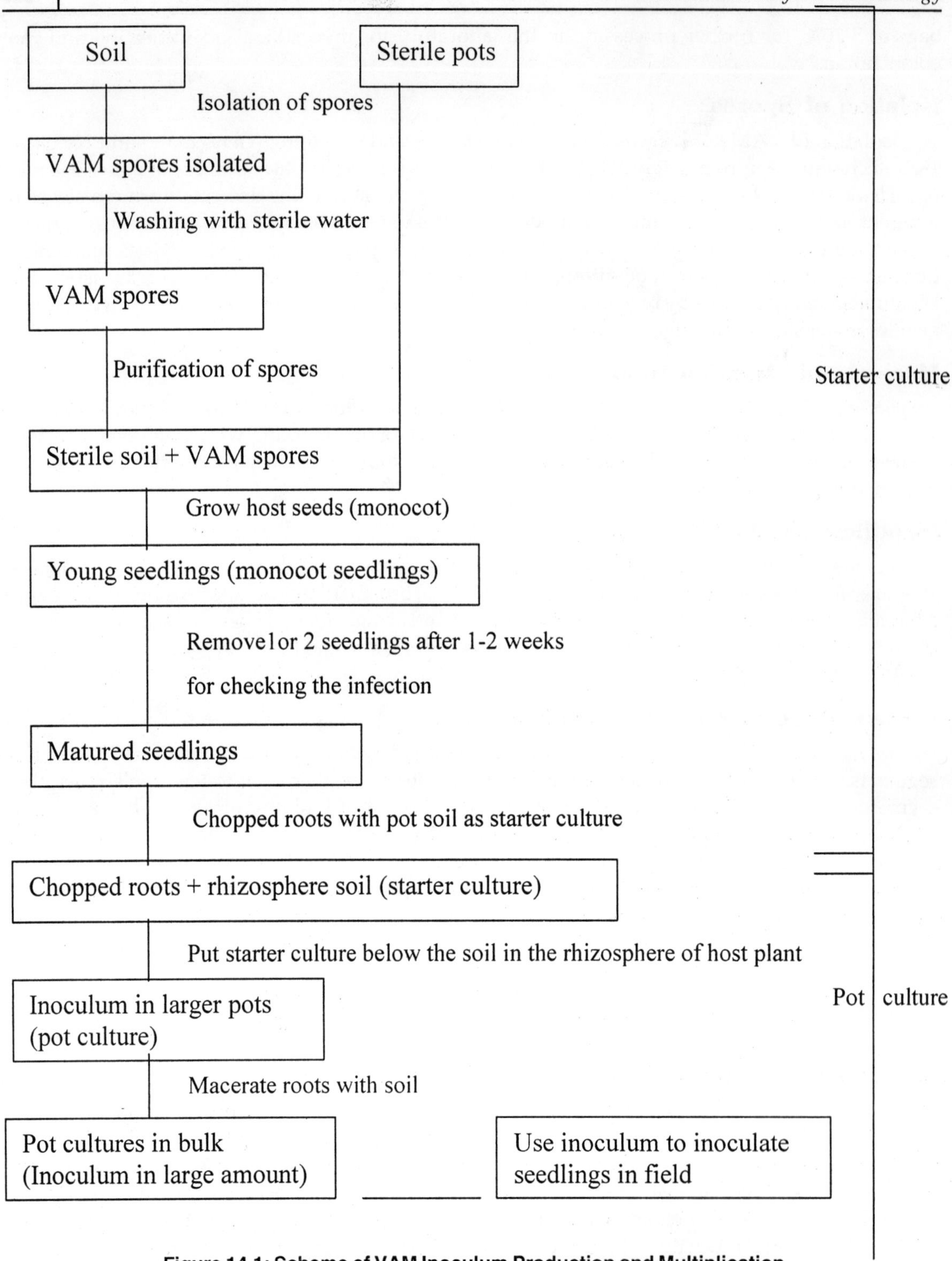

Figure 14.1: Scheme of VAM Inoculum Production and Multiplication

glass/earthen funnels. Three hosts were selected for starter culture of inoculum production *i.e.* sorghum, wheat and onion. The seedlings were raised up to 60 days in the earthen funnels containing sterilized sand: soil (40: 120 gm.). In each treatment two gm. of inoculum (mixed VAM, *Glomus mosseae* and *Glomus fasciculatum*) per earthen funnel containing 25 viable mycorrhizal spores was added. After 60 days, these VAM spores were collected again by wet sieving and decanting technique. These spores were finally used for mass production by using different hosts in bigger earthen pots for further study.

Mass Multiplication of VAM Spores (Pot Culture)

For pot cultures of mycorrhizal spore production, three hosts were selected *i.e.* maize, (*Zea mays* L.), wheat (*Tritichum aestivum* L.) and Jower (*Sorghum vulgare* Pers.) for different seasons. The mass production of mixed VAM fungi (*Glomus, Acaulospora, Gigaspora* and *Sclerocystis*), *G. mosseae* and *G. fasciculatum* was done in earthen pots of size 25 x 25cm. The substrate used for pot cultures was sand: soil (1: 3) by gm. ratio in sterilized conditions. Three replicates of each treatment were taken and analyzed after 90 days of experiments. In control set (three replicates) no inoculum was added.

The pot cultures were maintained for several days on different hosts. The pots were given Hoagland's solution except KH_2PO_4 solution once in a fortnight. The soil containing mycorrhizal spores, mycelium and colonized roots in pot cultures can be used to inoculate seedlings for further growth experiments and to prepare another pot cultures for further use.

Results and Discussion

For starter cultures of mixed VAM, *G. mosseae* and *G. fasciculatum*, three hosts *i.e.* sorghum, wheat and onion were selected. Out of the three hosts, sorghum was selected as a host for further multiplication of mixed VAM, *G. mosseae* and *G. fasciculatum* spores as it proved best among rest of hosts. In the next experiment all the three VAM cultures were grown in earthen funnels with sorghum as host. Table 14.1 showed the inoculum production (starter culture) of mixed VAM, *G. mosseae* and *G. fasciculatum* on host sorghum by gm. ratio of sterilized soil and sand (3: 1) in earthen funnels after 60 days in comparison to control in which no inoculum was added. In Table 14.1, it was observed that percentage mycorrhizal root colonization (75.5±2.02) and VAM spore number (80.5 ± 5.56) were more in comparison to control after 60 days when mixed VAM inoculum was used. Similarly, maximum percentage mycorrhizal root colonization (80.0± 0.0) and VAM spore number (85.5 ± 6.35) were reported after 60 days when *G. mosseae* was used. Percentage mycorrhizal root colonization 80 ± 0.0) and VAM spore number (82.5 ± 2.89) were more in comparison to control when *G. fasciculatum* was used for starter culture on sorghum as a host.

Table 14.1: Inoculum Production of Mixed VAM Spores, *Glomus mosseae, Glomus fasciculatum* on Host Sorghum in Earthen Funnels After 60 Days (Starter cultures)

Sl.No.	Treatment	Inoculum 2gm/Funnel (25 Viable Spores)	Sand : Soil gm: gm	*Percentage Seed Germination	*Mycorrhizal Spore Number/ 10 gm Soil	*Percentage Mycorrhizal Root Colonization
1.	T_1	25	40 : 120	80 ± 0.0	80.5±5.56	75.5±2.02
2.	T_2	25	40 : 120	90 ± 0.0	85.5±6.35	80.0 ±0.0
3.	T_3	25	40 : 120	70 ± 0.0	82.5±7.89	80.0 ±0.0
4.	C	25	40 : 120	70 ± 0.0	14.0±0.0	25.5 ±2.14

C: Control; T_1: Mixed VAM; T_2: *Glomus mosseae;* T_3: *Glomus fasciculatum.*

*: Average of three replicates in each column.

For mass multiplication of mixed VAM, *G. mosseae* and *G. fasciculatum* in pot cultures again three hosts were selected *i.e.* maize, sorghum and wheat. After 90 days, VAM spore number and percentage mycorrhizal root colonization were noticed for all cultures (Tables 14.2–14.4). Table 14.2 showed that maize was best host for mass production of mixed VAM than as it showed maximum percentage mycorrhizal root colonization (92.5 ± 4.78) and mycorrhizal spore number (149.0 ± 4.78) after 90 days in comparison to control and other host plants (Table 14.2).

Table 14.2: Mass Production of Mixed VAM Spores on Monocot Hosts in Pot Culture After 90 Days

Sl.No.	*Name of Host*	*Presenceof*		*Per cent Mycorrhizal Root Colonization*	*Mycorrhizal Spore Number/10gm Soil*
		Arbuscules	*Vesicles*		
1.	Wheat	+	+	*82.5 ± 5.95	*137.66 ± 2.41
2.	Sorghum	+	+	90.0 ±4.08	125.00 ± 17.60
3.	Maize	+	+	92.5 ±4.78	149.00 ± 4.78
4.	Control	+	+	30.0±8.76	19.00 ± 0.87

*: Mean of five replicates each.

±: Standard error; +: Present, –: Absent.

Mass production of *Glomus mosseae* spores on host maize was also high than other hosts (Table 14.3). VAM spore production and percentage root colonization were also high on host maize after 90 days *i.e.* (121.6 ± 12.7) and (90.0 ± 7.07) than control in which percentage mycorrhizal root colonization (30.0 ± 7.07) and VAM spore number (42.0 ± 1.41) were also low (Table 14.3).

Table 14.3: Mass Production of *Glomus mosseae* Spores on Monocot Hosts in Pot Culture After 90 Days

Sl.No.	*Name of Host*	*Presenceof*		*Per cent Mycorrhizal Root Colonization*	*Mycorrhizal Spore Number/10gm Soil*
		Arbuscules	*Vesicles*		
1.	Wheat	+	+	*90.00 ± 7.07	*121.66 ± 12.73
2.	Sorghum	+	+	90.00 ± 7.07	141.33 ± 2.05
3.	Maize	+	+	93.33 ± 8.20	161.0 ± 3.74
4.	Control	+	+	30.00 ± 7.07	42.0 ± 1.41

*: Mean of five replicates each.

±: Standard error; +: Present, –: Absent.

Table 14.4 also showed that mass production of *Glomus fasciculatum* spores was more on host maize than other host plants. In this case also spore production (123.6 ± 10.06) and percentage root infection (96.6 ± 2.72) were higher on host maize than control in which these were (42.0 ± 1.41) and (30.0 ± 7.07) respectively (Table 14.4). Keeping in view these observations, maize was used as a best host for mass production in further studies.

In this study, maize (*Zea mays* L.) was selected as a host for mass production for all of the VAM fungal species under the used growth conditions. In general, monocots are the best hosts for VAM spore production (Sundra Babu *et al.*, 2001). Sieverding and Leihner (1984) have reported that graminaceous and leguminous crops generally tend to increase VAM population whereas mycotrophic plants decrease the population of VAM fungi.

Table 14.4: Mass Production of *Glomus fasciculatum* Spores on Monocot Hosts in Pot Culture After 90 Days

Sl.No.	*Name of Host*	*Presence of*		*Per cent Mycorrhizal Root Colonization*	*Mycorrhizal Spore Number/10gm Soil*
		Arbuscules	*Vesicles*		
1.	Wheat	+	+	*96.66 ± 2.72	*123.66 ± 10.60
2.	Sorghum	+	+	96.66 ± 9.81	150.66 ± 4.25
3.	Maize	+	+	98.33 ± 2.72	151.00 ± 2.88
4.	Control	+	+	30.0 ±7.07	42.00 ± 1.41

*: Mean of five replicates each.

±: Standard error; +: Present, –: Absent.

Al. Raddad, (1995) used five crops for inoculations with *G. mosseae* and crops were grown for 10 weeks to assess mycorrhizal infection and sporulation. For all hosts, the percentage of the root length infected by the VAM fungus increased rapidly up to 10 weeks after sowing. The highest spore number was achieved in the rhizosphere of barley plants, followed by chickpea and beans. The type of the crop as well as the harvest date greatly influenced the spore popuiation and the extent of root colonization by *G. mosseae* (Al. Raddad, 1995).

Sreenivasa and Bagyaraj (1987) screened seven hosts *i.e. Panicum maximum, Chrysopogon fulvus, Themeda triandra, Chlorius gayana, Brachiaria brizantha, Passoalum serobiculatum* and *Eleusine coracana* to find a suitable host for mass production of the VAM fungus. *C. gayana* (Rhode grass) was found to be the best host with the highest percentage of mycorrhizal colonization (94 per cent), sporulation (547 spores per 50 ml substrate) and inoculum potential (1.65 x 10^7 per g).

When both soil and spore inoculum at a similar range of spore densities were used to inoculate Sudan grass roots they were infected more rapidly and subsequent spore production was greater with soil inoculum than with spore inoculum (Ferguson, 1981). Many hosts, which show great susceptibility to VAM infection, are Bahia grass, Rhode grass, Guinea grass, Coleus and Clover (Mehrotra, 1992).

Studies conducted at Forest Research Institute, Dehradun (Division of Forest Pathology) showed that *Paulowinia* was an ideal plant species for mass multiplication of VAM fungi as its seedlings in poly pots develop massive and dense root systems with extensive branching of lateral roots (Mehrotra, 1996). In the present investigation also maize showed massive and dense root system in comparison to wheat and sorghum. It can be said if the root system is well developed then the percentage mycorrhizal root colonization is maximum and so is the VAM spore number.

The VAM fungus-plant symbiosis is a complex system and the extent of endophyte- host interactions is not fully understood. The host involvement in sporulation and colonization is still unknown. There are evidences that suggest the type of root system and supply of carbohydrates may be important. Reduction in carbohydrate supply to the roots by defoliation or lower light intensities caused reduction in sporulation (Daft and El. Giahmi, 1978; Menge, 1984). But in this investigation, maize showed more dense and complex rootings and has more percentage mycorrhizal root colonization and VAM spore number. It can be due to the environmental conditions and other biotic factors.

Water has been shown to affect VAM fungal sporulation. Non-saturated and non- stressed water conditions are best for spore production in high and low – P conditions (Nelson and Safir, 1982). A correlation between water content and spore number existed across a natural soil moisture gradient in the field (Andreson *et al.*, 1984). Evidences from aeroponic and hydroponic cultures have demonstrated the benefits of well-watered and aerated rhizosphere for the large number of spore production in several *Glomus* spp. (Sylvia and Hubbel, 1986).

Sporulation is also positively correlated with temperature from 15°C to 30°C for many VAM fungi. Increased temperature decreases the lag phase of colonization and growth response. At higher temperature, the host is stressed and sporulation may decrease (Scenck and Smith 1982).

In the ensuing years research on VAM spore production has been limited to physiological manipulation such as light intensity manipulations, drought, superphosphate drenches and defoliations (Ferguson and Woodhead, 1982). Studies of plant species for spore production by VAM fungi have concentrated on one or two VAM species with limited efforts made to compare different VAM species and plant species. Ferguson (1981) found that there was no difference in spore production by *Glomus fasciculatum* with seven different plant species. A similar investigation showed that spore production by *Glomus fasciculatum* was greatest with Sudan grass when compared to four other plant species with no difference in spore production by *Glomus macrocarpum* or *Glomus mosseae* on the five plant species used in the experiment (Daniels and Bloom, 1986). Bagyaraj and Manjunath (1980) found that spore production by *Glomus fasciculatum* was greatest with Guinea grass when compared to seven other grass species.

Several areas of inoculum production of VAM fungi, other beneficial microorganisms and application technology merit further investigation. Further research on the storage properties of inocula would spur commercial production and development of new formulations. New and innovative inoculation technologies are needed to provide the most efficient application of available inocula in wide array of crop production systems. Co- inoculation of VAM fungi with other beneficial microorganisms has great merit and should allow more holistic approaches to rhizosphere health in future.

References

Al-Raddad, A. M. 1995. Mass production of *Glomus mosseae* spores. Mycorrhiza, 5 (3): 229-231.

Anderson, R. C., Liberta, A. E. and Dickman, L. A. 1984. Interaction of vascular plants and vesicular-arbuscular mycorrhizal fungi across a soil moisture-nutrient gradient. Oecologya, 64: 111-117.

Bagyaraj, D. J. and Manjunath, A. 1980. Selection of a suitable host for mass production of VA mycorrhizal inoculum. Plant and Soil, 55: 495-498.

Daft, M. J. and El. Giahmi, A. A. 1978. Effect of arbuscular mycorrhizae on plant growth. VIII. Effects of defoliation and light on selected hosts. New Phytol., 80: 365-372.

Daniels Hetrick, B. A. and Bloom, J. 1986. The influence of the plant on production and colonization ability of vesicular arbuscular mycorrhizal spores. Mycologia, 78: 32-36.

Ferguson, J. J. and Woodhead, J. J. 1982. Production of endomycorrhizal inoculum. A. Increase and maintenance of vesicular arbuscular mycorrhizal fungi. In: Method and principals of mycorrhizal research. (Ed. Schenck, N. C.), The American Phytopathological Society, St. Paul MN, pp. 47-54.

Ferguson, J. J. 1981. Inoculum production and field application of vesicular arbuscular mycorrhizal fungi. Ph.D. Thesis, Uni. of California, Riverside. Pp. 117.

Gaur, A. and Adholeya, A. 1994. Estimation of VAMF spore in soil. Mycorrhiza News 6(1): 10-11.

Gerdemann J. W. and Nicolson, T. H. 1963. Spores of mycorrhizal Endogone species extracted from soil by wet sieving and decanting. Trans. Br. Mycol Soc.(46):235-244

Hari Kumar, V. S. and Potty, V. P. 2002. Technology for mass multiplication of arbuscular mycorrhizal (AM) fungi for field inoculation to sweet potato. Mycorrhiza News, 14 (1): 11-13.

Harni Kumar, K. M. and Bagyaraj, D. J. 1988. Effect of crop rotation on native vesicular arbuscular mycorrhizal propagules in soil. Plant and Soil, 110: 77-80.

Hua, S.S.T. 1990. Prospects for axenic growth and feasibility of genetic modification of vesicular arbuscular mycorrhizal (VAM) fungi. In: Innovation and hierarchical integration, p. 145, (Ed. Allen M. F.) Wyoming: University of Wyoming, 324 pp.

Jarstfer, A. Y. and Sylvia, D. M. 1993. Inoculum production and inoculation strategies for vesicular-asbuscular mycorrhizal fungi. In: Soil Microbial Ecology, Application in Agriculture and Environmental Management. (Eds. Metting, F.B. Jr., Marcel Dekker), Inc./New York, Basal, Hong Kong, pp. 349-377.

Jeffries, P. and Dodd, J. C. 1991. The use of mycorrhizal inoculants in agriculture. In: Handbook of Applied Mycology Vol. – 1. (Eds. Arora, D. K., Rai, B., Mukerji, K. G. and Knudson, G.R.), Mercel Dekker, Inc. Madison, Ave., New York, pp. 156-186.

Kumar, K. A. P. and Bagyaraj, D. J. 1998. Effect of plant growth promoting microorganisms on mass production of arbuscular mycorrhizal fungus, *Glomus mosseae*. Indian J. Microbiol., 38: 33-35.

Kumar, P. K. A. and Bagyaraj, D. J. 1999. Mass production of arbuscular mycorrhiza as influenced by some agrochemicals. Proc. Nat. Acad. Sci. India, 69 (B), I: 61-66.Liyanage, H. D. 1989. Effects of phosphorus nutrition and host species on root colonization and sporulation by vesicular-arbuscular (VA) mycorrhizal fungi in sand-vermiculite medium. M.Sc. Thesis, University of Florida, Gainesville.

Mehotra, M. D. 1996. Multiplication of VAMF on *Paulownia*- a veritable possibility. Indian Forester, 122 (9): 858-860.

Mehrotra, M. D. 1992. Endomycorrhizae. In: Mycorrhizae of Indian forest tress. Pub. F.R.I. Dehradun, pp. 151-232.

Mehrotra, M. D. and Mehrotra, A. 1999. Suitability of potting mixture for VAM infection and spore population in root trainer raised seedlings. Indian J. of Forestry, 22 (1): 49-52.

Menge, J. A. 1983. Utilization of VAM fungi in agriculture. Can. J. Bot., 61: 1015-1024.

Menge, J. A. and Timmer, L. W. 1982. Procedure for inoculation of plants with VAM in the laboratory greenhouse and field. In: Methods and Principle of Mycorrhiza Research. (Ed. Schenck, N. C.), American Phytopathology Society, St. Pauls, pp. 59.

Menge, J. A. 1984. Inoculum production. In: VA Mycorrhiza. (Eds. Powell, C. L. and Bagyaraj, D. J.), CRC Press Boca Raton, pp. 187-203.

Morton, J. B. and Benny, G. L. 1990. Reviesd classification of arbuscular mycorrhizal fungi (zycomycetes): A new order, Glomates, two new suborders Glominae and Gigasporineae and two new families Acaulosporaceae and Gigasporaceae, with an emendation of Glomaceae. Mycotaxon 37: 471-491.

Mukerji, K. G. 1996. Taxonomy of endomycorrhizal fungi. In: Advance in Botany (Eds. Mukerji K.G.,Mathur B., Chamola B.P. and Chitralekha P.). pp. 213-221. APH corporation, New Delhi, India.

Mungier, J. and Mosse, B. 1987. Vesicular arbuscular mycorrhizal infection in transformed root including T-DNA root grown axenically. Phytopathol., 77: 1045-1050.

Nelsen, C. E. and Safir, G. R. 1982. Increased drought tolerance of mycorrhizal onion plants caused by improved phosphorus nutrition. Planta, 154: 407-413.

Papavizas, G. C. 1985. *Trichoderma* and *Gliocladium*: biology, ecology and potential for biocontrol. Ann. Rev. Phytopathol., 23: 23-54

Parkash, V., Aggarwal, A., Sharma, S. and Sharma, D. 2004. Effect of endophytic mycorrhizae and fungal agent on the growth and development of *Eucalyptus saligna* Sm. Seedlings. Bull. Nat. Inst. Ecol. 15, 127-131.

Phillips, J. M. and Hayman, D. S. 1970. Improved produces for clearing roots and staining parasitic and VAM fungi for rapid assessment of infection. Trans. Brit. Mycol. Soc. 55,158-161.

Rani, P., Aggarwal, A. and Mehrotra, R. S. 1997. Role of mycorrhizal fungi in forestry. 84th Session of Indian Science Cong. Delhi, pp: 37.

Safir, G. R., Coley, S. C., Siqueria, J. O., and Carlson, P. S. 1990. Improvement and synchronization of VA mycorrhizal fungal spore germination by short-term storage. Soil. Biol. Biochem., 22: 109-111.

Schen N. C. and Perez, Y. 1987. Manual for identification of VAM fungi. University of Florida, Gainseville, pp.245.

Scenck, N. C. and Smith, G. C. 1982. Responses of six species of vesicular arbuscular mycorrhizal fungi and their effects in soybean at four soil temperatures. New Phytol., 92: 193-201.

Severding, E. and Leihner, D. E. 1984. Influence of crop rotation and intercropping of cassava with legumes on VA mycorrhizal symbiosis of cassava. Plant Soil, 80: 143-146.

Sharma, A. K., Singh, and Akhauri, P. 2000. Mass culture of arbuscular mycorrhizal fungi and their role in biotechnology. Proceedings of the Indian National Science Academy, Part B, Reviews and tracts – Biological Sciences, 66(4/5): 223-237.

Singh, G. and Tilk, K. V. B. R. 2002. Techniques in AM fungi inoculum production. In: Techniques in mycorrhizal studies, (Eds. Mukerji, K. G., Manoharachary, C. and Chamola, B. P.), Kluwer Acad. Press. Nederland, pp. 273-285.

Singh, S. 1999. Role of mycorrhiza in tree nurseries Part-II. Inoculation of nursery soil/plants with mycorrhizal fungi. Mycorrhiza News, 10 (4): 2-11.

Singh, S. 2002. Mass production of AM fungi. Part-I. Mycorrhiza News, 14(3): 2-10.

Sreenivasa, M. N. and Bagyaraj D. J. 1987. Selection of a suitable substrate for mass multiplication of *Glomus fasciculatum.* In: Mycorrhiza Round Table: Proceedings of workshop, (Eds. Verma, A. K., Oka, A. K., Mukerji, K. G., Talk, K. V. B. R. and Janak Raj), New Delhi, India, pp. 592-599.

Sreenivassa, M. N. and Bagyaraj, D. J. 1988. Selection of a suitable substrate for mass multiplication of *Glomus fasciculatum*. Plant and Soil, 109: 125-127.

Sundra Babu, R., Poornima, K. and Suguna, N. 2001. Mass production of vesicular arbuscular mycorrhizae using different hosts. Mycorrhiza News, 13 (1): 20-21.

Sylvia, D. M. and Hubble, D. H. 1986. Growth and sporulation of vesicular arbuscular mycorrhizal fungi in aeroponic and membrane systems. Symbiosis, 1: 259-267.

Thompson, J. P. 1986. Soilless cultures of VA mycorrhizae of cereals: Effects of nutrient concentration and nitrogen source. Can. J. Bot., 64: 2282-2294.

Wood, T. 1984. In: Proceedings of 6th American Conference on Mycorrhizae. (Ed. Molina, R.), Forest Research Laboratory, Oregon State University, Corvallis. OR; pp. 84.

Microbes: Diversity and Biotechnology (2012) *Pages* **231–267**
Editors: **Prof. S.C. Sati & Dr. M. Belwal**
Published by: **DAYA PUBLISHING HOUSE, NEW DELHI**

Chapter 15

Seed Mycoflora of Some Oil Yielding Plants from Dharwad

Ch. Ramesh and Avitha K. Marihal
Department of Botany, Karnatak University,
Dharwad – 580 003, Karnataka

ABSTRACT

Seeds are basic input for crop production. Fungi form a major group of pathogens that can be seed-borne or transmitted through seeds. Seed are great economic interest and also constitute a major part of diet thus play a vital role in associating microorganisms which prove hazardous for the seeds from their ecological point of view they may be either field fungi or storage fungi. It is well know fact that several fungi are know to cause considerable damage to seeds in storage and also produce various activities. Keeping in view the importance of stored oil/seeds the mycological investigations were attempted from Dharwad region. In and around Dharwad the major oilseeds grown are Soybean, Sunflower, Groundnut, Safflower, Linseed and Niger. In the present investigation isolation of seed mycoflora of six different oilseeds, comparison of seed mycoflora by Standard Blotter Method and Agar Plate Method, Effect of culture filtrate of dominant fungi on seeds germination and seedling vigour and lastly the effect of plant extract on some of dominant seed Mycoflora were studied.

Keywords: *Seed Mycoflora, Oilseeds, Seed-borne externally, Seed-borne internally.*

Introduction

Seeds are basic input for crop production. Seed is a fertilized mature ovule consisting of an embryonic plant, stored food material and protective seed coat. About 90 per cent of world food crops are sown and propagated by seeds. Hence, the importance of seed has been recognized since time human practices crop husbandry.

Fungi form a major group of pathogens that can be seed-borne or transmitted through seeds. Fungi are multicellular plants without roots, leaves or chlorophyll. Therefore, they must live off other materials including grains. The vegetative parts of fungi produce enzymes that interact with grain tissue to extract the nutrients needed for growth. Fungi reproduce primarily by means of small, light airborne spores that are easily distributed by the wind.

The significance of sustainable agricultural production is hidden in the use of quality seed. It is the most crucial and vital input for enhancing productivity. Since, seed is the custodian of the genetic potential of the cultivar thus the quality of the seed determines the limits of productivity to be realized in a given cropping system. One of the outstanding achievements of Indian agriculture during green revolution era was the production of sufficient food to meet the needs of growing population. But the ratio of growth of food grain production to population growth has entered a period of decline over past decades. The shortage of quality seed, absence of techno-infrastructure facilities needed for processing, storage and distribution of seed are amongst the major considerable attributes for such outbreaks.

Seeds are of great economic interest and also constitute a major part of diet, they play a vital role in associating microorganisms, which prove hazardous for the seed or the new plant created from it. So any infectious agent (bacteria, fungi, nematode etc), which is associated with seeds having potential agent of causing a disease of a seedling or plant, is termed Seed-borne pathogen (Agarwal and Sinclair 1987). The associating microorganisms may be pathogenic, weak parasites or saprophytes. They may be associated internally or externally with the seed or as concomitant contamination as sclerotia, galls, fungal bodies, bacterial ooze, infected plant parts, soil particles etc are mixed with the seed. Where as seed-borne mycoflora is a generalized term indicating the association of fungi with seeds that includes pathogenic and saprophytic mycoflora. A pathogen may be both externally and internally seed-borne. The terms "externally" and "internally" seed borne refer to the location of the pathogen in relation to the seed. If a pathogen is located on the outside of the functional part of the seed it is externally seed-borne but if it occurs inside the seed it is internally seed-borne.

From the standpoint of their ecology the conditions that permit fungi to invade seeds, the fungi are divided into two groups.

1. Field fungi
2. Storage fungi

Field Fungi

Field fungi invade seeds developing on the plants in the field or after the seeds have matured and the plants are either still standing or are cut and swathed, awaiting threshing. The main field fungi require an equilibrium relative humidity of 95 to 100 per cent. In some regions a combination of wet weather during harvest, lack of drying facilities or lack of sufficient drying capacity and lack of transport may result in much grain of high moisture content being piled on the ground. If such high moisture content grain is of a temperature that allows microflora to grow; it will deteriorate rapidly.

Storage Fungi

Seeds are seldom infected with storage fungi in the field. Instead the infection takes place at the elevator where the grain is dried and stored. Unlike the field fungi, storage fungi grow at equilibrium relative humidity as temperatures down to 23°F.

It is well known fact that several fungi are known to cause considerable damage to seeds in storage and also produce various activities.

1. Seed-borne infection generally reduces germination capability of seeds, which may result in seed decay and/or pre or post emergence damping off.
2. Field fungi may cause weakening or death of embryos while storage fungi slowly kills the embryos of the seeds they invade.
3. Infected seeds loose their market value due to discoloration.
4. Invasion of seeds by pathogens may result in biochemical deterioration and change in the quantity of seed nutrients.
5. Seed-borne fungi reduce the processing quality of seeds.
6. Production of toxins which are injurious to man and domestic animals.

All the above information made the researcher to switch on to the mycofloral investigation of the stored Oilseeds of Dharwad, which is one of the districts in Karnataka, India. The location of Dharwad unit is at 15°30′ N latitude and 75°30′ E longitude which is in the semi malnad tracts, has been built on hillocks which provide excellent scenic features.

The climate of Dharwad is characterized by tropical monsoon, which indicates the seasonal rhythm of weather. All the weather elements like temperature, pressure, wind, relative humidity exhibit well marked seasonal variation.

Oilseeds one of the great importance of our diet, are used for various domestic purposes. India is one of the leading countries in the production and marketing of some of the major oilseeds.

In and around Dharwad the major oilseeds grown are Soybean, Sunflower, Groundnut, and Safflower even Linseed and Niger are grown in lower percentage. These seeds are marketed by farmers in Agriculture Produce Market Committee (APMC), Dharwad where these seeds are stored in gunny bags in warehouses and storage houses of Dalal shops in this market area.

The objectives of the present work incorporates the results of following investigations,

1. Isolation of seed mycoflora of six different stored oilseeds namely Soybean (*Glycine max* L Merr.), Sunflower (*Helianthus annuus* L.) Safflower (*Carthamus tinctorius* L.), Groundnut (*Arachis hypogea* L.), Linseed (*Linum usitatissimum* L.) and Niger (*Guizotia abyssinica* L.)
2. Comparison of seed mycoflora by Standard Blotter Method and Agar Plate Method with respect to season.
3. Effect of culture filtrate of dominant fungi on seed germination and seedling vigour.
4. Effect of plant extract on some of dominant seed mycoflora.

Materials and Methods

Collection of Samples

Seeds of unknown variety were collected in polyethylene bags from Agriculture Produce Market Committee (APMC), Dharwad. Farmers from in and around Dharwad sell their seeds. These seeds are usually stored in gunny bags. The seeds are kept in warehouses till the end of the marketing period or they are transported. The samples were collected once in two months from 2001-2003.

Isolation of Mycoflora from Different Oilseeds

The mycoflora associated with different oilseeds in the study were isolated by following methods (ISTA 1999) *viz.,*

1. Standard Blotter Method (SBM).
2. Agar Plate Method (APM).

Standard Blotter Method (SBM)

Sterile glass petriplates (9 cm) were used in the study; their discs of blotter were moistened with sterilized water and were placed at the bottom of each petriplates. Excess water was drained from the blotters.

Total 200 seeds were taken randomly from the lots for the study. 100 seeds were used to isolate the external mycoflora of the oilseeds. For isolation of internal mycoflora of seeds, another 100 seeds were surface sterilized with 2 per cent sodium hypochlorite solution for one to two minutes, and then ten seeds were placed at equidistance on blotter paper in each petriplates. For better growth and sporulation of fungal flora 12h artificial light was provided by placing the plates below the Philips 40w tubes and alternated 12h darkness. After incubation of the plates for seven days fungal growth was seen under stereo Binocular Microscope and Compound Microscope (Plate 15.1).

Agar Plate Method (APM)

To detect the mycoflora of the seed, two hundred seeds were taken at random from the seed lots; to isolate external mycoflora hundred seeds were used where as to isolate internal mycoflora another hundred seeds were surface sterilized with 2 per cent sodium hypochlorite. Sterilized media was poured in pre-sterilized petriplates under aseptic condition. After solidification the plates were kept inverted for 12 hrs, contaminated plates were rejected. Ten seeds were inoculated in each petridish containing nutrient media. The plates were incubated for seven days by providing 12h light 12h darkness alternatively. The plates were examined for the mycoflora under stereo-binocular microscope and compound microscope (Plate 15.2).

Media Used

The selection of satisfactory media for a stimulating growth and sporulation of seed fungi was difficult. The different media were used to get maximum number of fungi as suggested by Booth (1971) in his book "Methods in Microbiology". The media used were Potato dextrose agar (PDA), Malt extract agar (MEA) and Richard's broth media (RBM).

Counting of Infected Seeds

The incubated plates were observed under stereo binocular microscope. In case of Agar Plate Method seeds were examined with naked eye. Number of seeds infected with fungi was entered in the data sheet.

Transferring of Fungi

There is keen competition for development of various fungi in the initial medium. Colonies appearing on media probably don't grow and develop to the full extent. Hence, single colonies of different fungi were transferred from the petriplates to the slants. This transfer was done by chromatin wire or loop dipped in rectified sprit. At the time of transfer slant containing the organisms was held

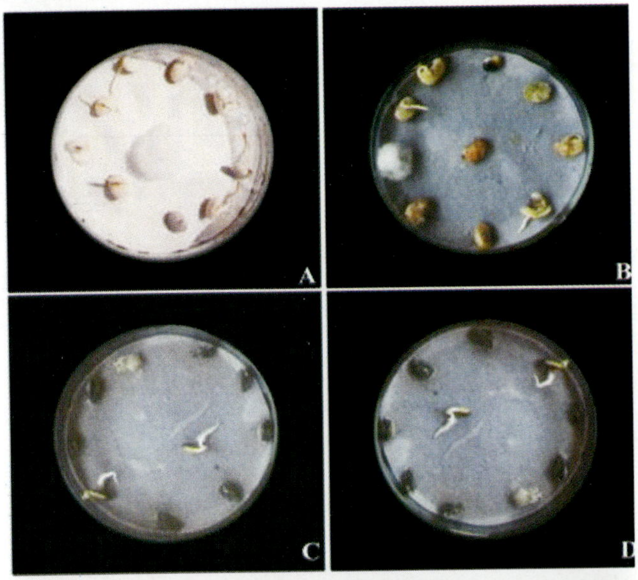

Plate 15.1: A: Growth of seed mycoflora on sterilized seeds of soybean by standard botter method; B: Growth on seed mycoflora on unsterilized seeds of soybean by standard botter method; C: Growth of seed mycoflora on sterilized seeds of sunflower by standard botter method; D: Growth of seed mycoflora on unsterilized seeds of sunflower by standard botter method

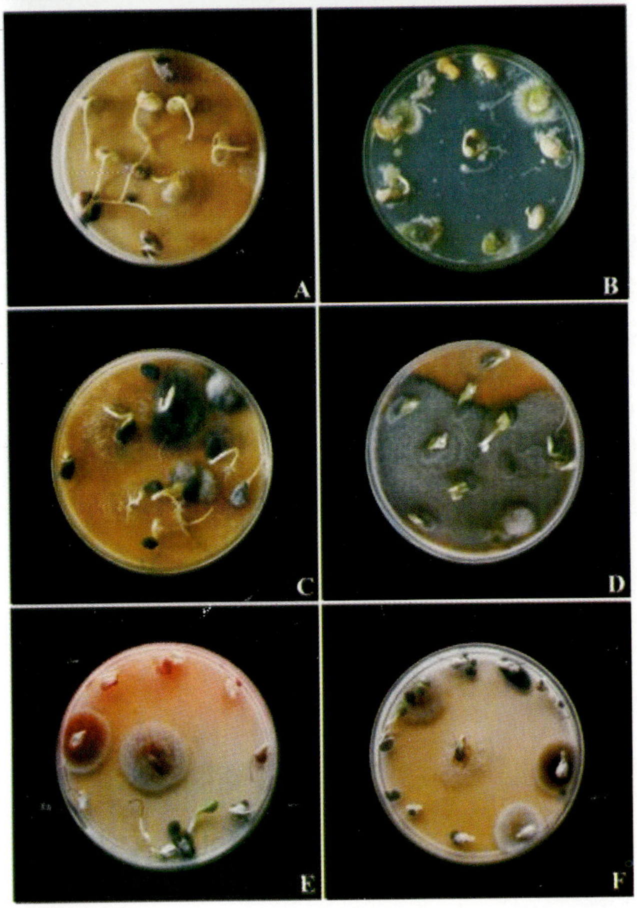

Plate 15.2: A: Growth of seed mycoflora on sterilized seeds of soybean by agar plate method; B: Growth of seed mycoflora on unsterilized seeds of soybean by agar plate method; C: Growth of seed mycoflora on sterilized seeds of sunflower by agar plate method; D: Growth of seed mycoflora on unsterilized seeds of sunflower by agar plate method; E: Growth of seed mycoflora on sterilized seeds of safflower by agar plate method; F: Growth of seed mycoflora on unsterilized seeds of safflower by agar plate method

near the flame. The entire process was done under the Laminar air flow chamber. This process is to be done quickly to minimize the contamination and to have the pure culture under aseptic condition.

Pure Culture

The cultures obtained were further purified by single spore isolation. A dilute spore suspension 100 spores ml^{-1} prepared in sterile water was poured on water agar (0.2 per cent) in culture plate and was allowed to settle down on the agar just to form a very thin layer over the surface of the agar. This enables the spores to settle quite apart from one another. A single spore was selected and lifted with the help of inoculating needle along with the agar and transformed to the culture plates containing PDA. After obtaining good growth, subcultures of the fungus were done.

Identifications

Identification of the described fungi up to generic and species level and their reports from Karnataka and India were confirmed with the help of available literatures, such as Barnett (1955), Ainsworth and Bisby (1971), Ainsworth and Sussman (1965), Ainsworth and Sussman (1966), Ainsworth and Sussman (1968), Ainsworth, Sparrow and Sussman (1973), Bessey (1950), Bilgrami *et al.* (1979, 1981, 1991), Booth (1971), Burnett (1976), Dickinson and Preece (1976), Ellis (1971, 1976), Ellis and Ellis (1985), Funder (1961), Garraway and Evans (1984), Gilman (1967), Kendrick (1971) Preece and Dickinson (1971), Smith and Berry (1975), Stevens (1974), Subramanian (1971), Tandon (1968), Von Arx (1981), Wicklow and Carroll, (1981) and other relevant literature.

All the specimens and the preparations are deposited, under the code number SMF (Seed Mycoflora), in the Mycology Laboratory, Department of Botany, Karnatak University, Dharwad, Karnataka, India.

Effect of Culture Filtrate of Different Fungi of Seed Mycoflora

Seven different fungi isolated from the oilseeds were used to study its effect of culture filtrate on seed germination and seedling vigour. The following fungi were selected for this purpose, *Alternaria alternata, Aspergillus flavus, Aspergillus niger, Cladosporium herbarum, Drechslera hawaiiensis, Fusarium moniliforme* and *Rhizopus oryzae.*The isolated fungi were grown on Richards broth medium. For this study, method followed by Singh *et al.* (2004) was employed.

In Erylen mayer flaks (250ml) 50ml of Richard broth medium was taken and sterilized. The discs from seven days old culture were inoculated to the flask and incubated at 21d at temperature 25ºC. With the help of Whatman filter paper No. 44 the content of the flask were filtered. To study the effect of these filtrate on seed germination and seedling vigour. One hundred healthy seeds were surface sterilized with 2 per cent sodium hypochlorite and then all the different oilseeds were soaked separately in these culture filtrates for 12 hours. Seeds soaked in sterilized distilled water served as control.

Seeds soaked in culture filtrate were inoculated in sterilized plates containing moist sterile blotter paper. Ten seeds were inoculated in each plate. Petriplates were incubated at 25±2 for seven days. After incubation period, per cent germination and seedling vigour were recorded.

Screening of Plant Extracts (Botanicals)

The following plants have been used to test their antifungal properties against seed mycoflora of oilseeds. The dominant mycoflora *Alternaria alternata, Aspergillus flavus* and *Aspergillus niger* were used as test fungi.

Name of the Plant	Family	Parts Used
Calotropis gigantia R.Br	Asclepiadceae	Leaves
Carica papaya Linn.	Passsifloraceae	Leaves
Nerium odorum Soland	Apocynaceae	Leaves
Vinca rosea Linn.	Apocynaceae	Leaves
Vitex negundo Linn	Verbanaceae	Leaves

Preparation of Extracts

For preparation of the plant extracts the methods of Thiribhuvanamala and Narasimhan (1998) were followed and to check the antifungal principles of the extracts filter paper disc method was employed (Dhingra and Sinclair 1986).

Preparation of Crude Extract

For crude extraction 10gm of plant material was washed with tap water and sterile water. It was then crushed with pestle and mortar by adding 10ml of sterilized distilled water. The crushed material was then filtered through double layered muslin cloth and filter paper (Whatman No. 40) and the filtrate obtained was standard plant extract (100 per cent). This standard extract was heated to 50°C for 10min. and it is used for further studies.

Inhibitory Activity of Plant Extracts against Seed Mycoflora

In a culture plate about 15ml of a culture medium (PDA medium) was poured in 9cm uniformly flat-bottomed culture plates. After solidification uniformly distribute 4ml of 1.5 per cent water agar seeded with the test pathogen. Sterile Filter paper discs of 1 to 2 diameter are soaked in the plant extract and distilled water was used as control. They are dried separately and then placed on the seeded agar medium at least 1 to 1.5cm from periphery of the plate. Three discs were placed on single plates and incubated for 7 days. After incubation the zones of inhibition around the filter paper discs are measured.

Results

Quantitative Studies on Externally and Internally Borne Mycoflora

The present investigation deals with the isolation of fungal flora from six different types of oilseeds collected from agriculture produce marketing committee, Dharwad during 2001-2003. The nature and distribution of the seed mycoflora incidence (per cent) of different fungi on these seed samples were determined by two different methods. Studies on effects of cultural filtrate of seven different fungi on seed germination and seedling vigour were made. Experiments were conducted on the control of seed mycoflora using botanicals separately and the results are presented here.

Standard Blotter Method

Internal Seed Mycoflora

Among the surface sterilized seeds plated over Blotter the percent incidence of seed found carrying the fungus internally varied from seed to seed.

In Soybean overall varied twenty-eight fungal organisms were isolated. Nine fungal species were isolated from Zygomycotina six species from Ascomycotina and rest thirteen species were isolated from the Mitosporic fungi. In Zygomycotina, *Rhizopus oryzae* (6.72 per cent) showed maximum appearance while *Mucor rouxinous* (0.53 per cent) showed minimum appearance. In Ascomycotina *Chaetomium aurangabadense* (4.03 per cent) and *Chaetomium indicum* (4.03 per cent) both showed maximum incidence, where as *Chaetomium apiculatum* (1.25 per cent) occurred with minimum incidence (Figure 15.1a). In Mitosporic fungi, *Cercspora kikuchii* (23.42 per cent) showed maximum incidence followed by *Aspergillus ustus* (17.47 per cent) where as *Fusarium lactis* (1.08 per cent) occurred with minimum incidence. (Figure 15.1b).

Sunflower seeds exhibited the association of thirty-eight fungal species on sterilized seeds by Blotter method. Six fungal species were recorded in Zygomycotina, seven species from Ascomycotina and rest twenty-five species from Mitosporic fungi. In Zygomycotina, *Rhizopus nigricans* (3.51 per cent) occurred in maximum incidence while *Mucor javanicus* (2.08 per cent) with minimum incidence, *Chaetomium indicum* (3 per cent) appeared with maximum incidence in the Ascomycotina group, where as *Chaetomium globosum* (1.04 per cent) occurred in minimum incidence (Figure 15.2a). In Mitosporic fungal group, *Alternaria alternata* (15.32 per cent) occurred with highest incidence followed by *Aspergillus niger* (12.5 per cent) and *Aspergillus flavus* (9.71 per cent) where as *Cladosporium cucumerinum* (0.58 per cent) occurred with lowest incidence (Figure 15.2b).

In case of Safflower seeds thirty-six fungal species were isolated. Eight species belongs to Zygomycotina, four species belongs to Ascomycotina and twenty-four belonged to Mitosporic fungi.

Among the members of Zygomycotina, *Rhizopus oryzae* (3.93 per cent) occurred with highest percentage of incidence where as *Mucor rouxinous* (0.82 per cent) showed its least incidence on sterilized seeds. In Ascomycotina group, *Chaetomium indicum* (1.43 per cent) showed maximum incidence where as *Emericella nidulans* (0.6 per cent) with least incidence (Figure 15.3a). In case of Mitosporic fungi, *Aspergillus niger* (153.83 per cent) and *Alternaria alternata* (9.53 per cent) occurred with maximum incidence while *Fusarium sporotrichoids* (0.90 per cent) and *Paecilomyces fusisporas* (0.75 per cent) occurred with minimum incidence (Figure 15.3b).

Twenty-three fungal species associated with Groundnut seeds. Out of twenty-three species seven belongs to Zygomycotina, three belongs to Ascomycotina and the rest belonged to Mitosporic fungi. In Zygomycotina group, *Rhizopus arrhizus* (1.82 per cent) showed maximum incidence while *Mucor praini* (0.67 per cent) and *Mucor racemosus* (0.67 per cent) showed minimum incidence. In the members of Ascomycotina group, *Chaetomium aurangabadense* (2.99 per cent) showed maximum incidence while *Chaetomium globosum* (1.86 per cent) occurred with minimum incidence (Figure 15.4a). In case of Mitosporic members, *Aspergillus niger* (28.93 per cent) occurred with maximum incidence followed by *Aspergillus flavus* (19.63 per cent) while *Fusarium moniliforme* (0.97 per cent) showed minimum incidence (Figure 15.4b).

Surface sterilized Linseeds were tested; twenty-two different species were isolated. Three belongs to Zygomycotina, four species belongs to Ascomycotina and the remaining fifteen species belonged to Mitosporic fungi. *Rhizopus arrhizus* (1.90 per cent) showed maximum incidence, *Absidia spinosa* (0.42 per cent)occurred in minimum incidence in Zygomycotina group. In Ascomycotina, *Chaetomium funicola* (1.38 per cent) has shown maximum incidence while *Chaetomium aurangabadense* (0.99 per cent) appeared minimum incidence (Figure 15.5a). In the members of Mitosporic fungi, *Aspergillus niger* (7.01 per cent) showed highest incidence while *Drechslera hawaiiensis* (0.47 per cent) showed least incidence compared to other organisms (Figure 15.5b).

In case of Niger seeds, the association of fungi varied from group to group. Totally twenty-seven species were isolated. From Zygomycotina group. *Rhizopus arrhizus* (1.74 per cent) showed maximum percentage of incidence and the species to occur with minimum incidence was *Syncephalastrum*

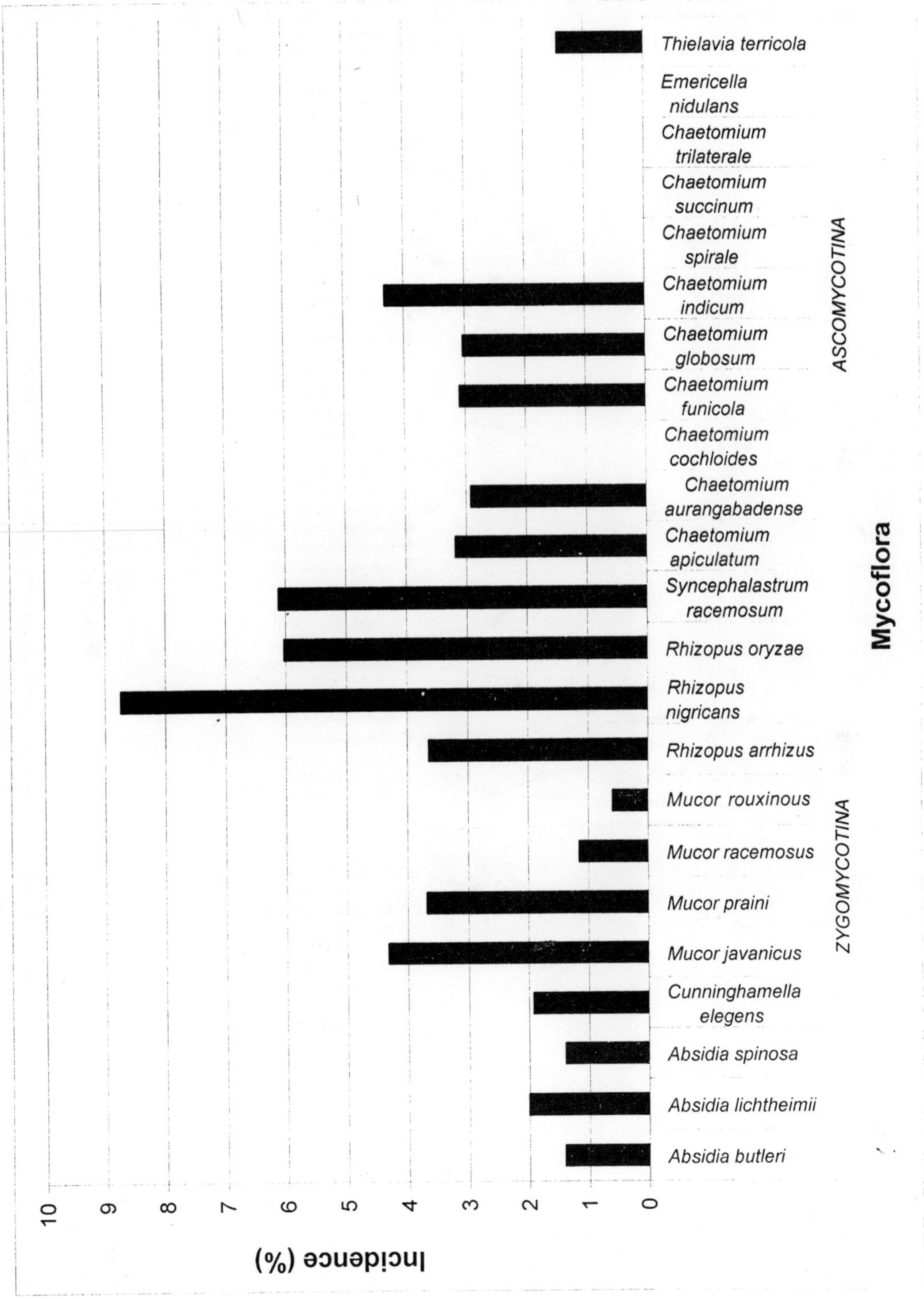

Figure 15.1a: Incidence (per cent) of Seed Mycoflora of Zygomycotina and Ascomycotina on Soybean

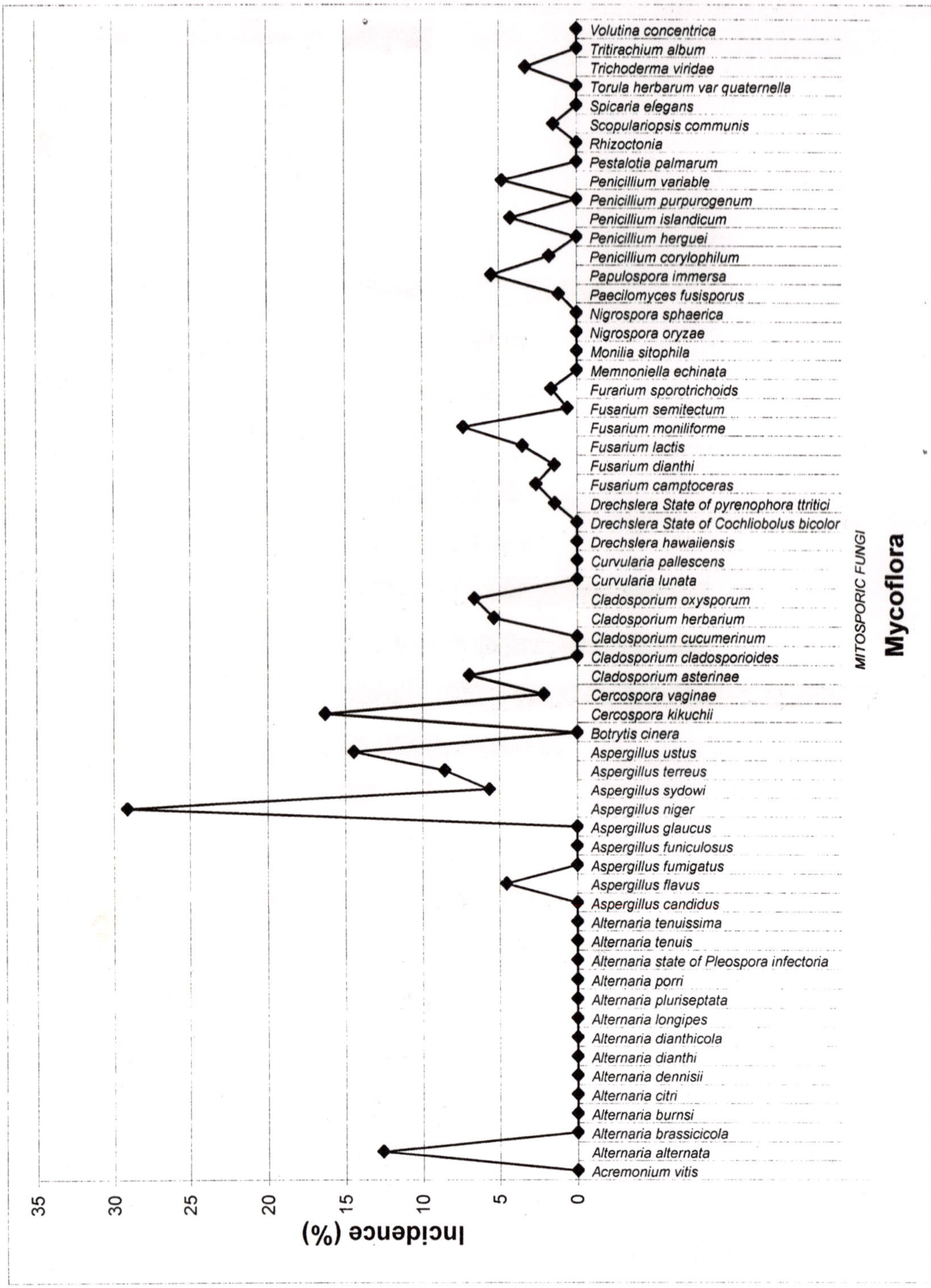

Figure 15.1b: Incidence (per cent) of Seed Mycoflora of Mitosporic Fungi on Soybean

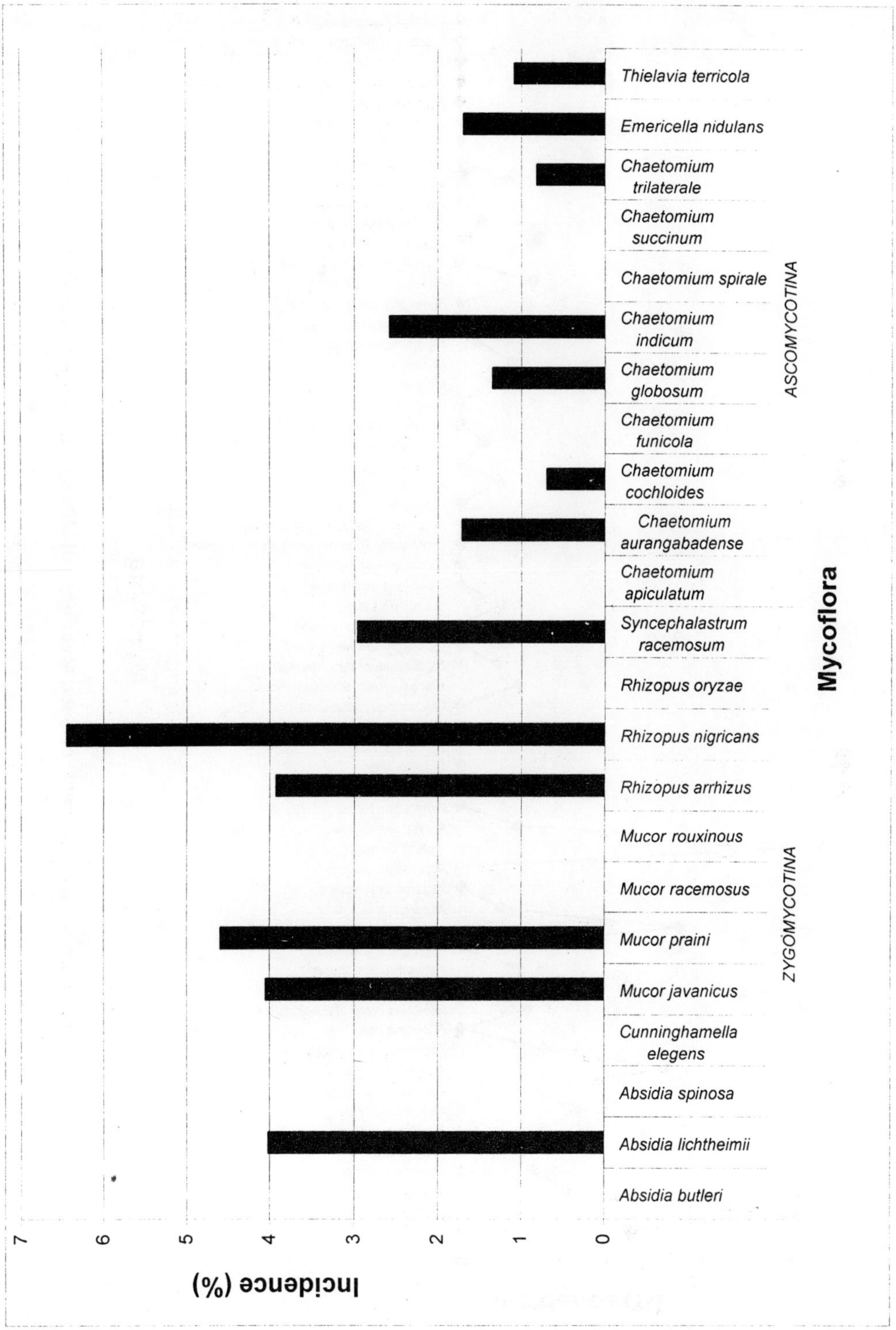

Figure 15.2a: Incidence (per cent) of Seed Mycoflora of Zygomycotina and Ascomycotina on Sunflower

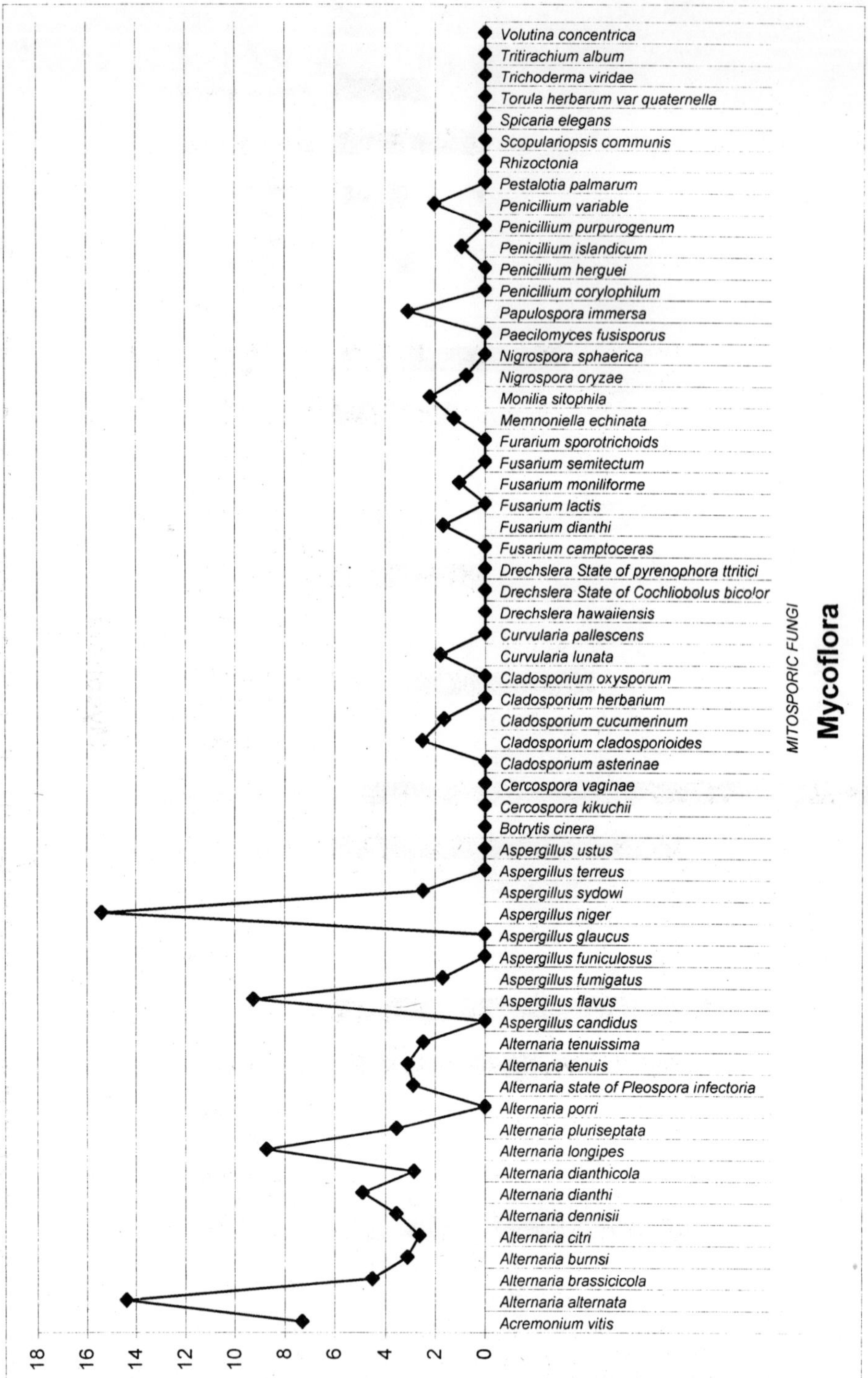

Figure 15.2b: Incidence (per cent) of Seed Mycoflora of Mitosporic Fungi on Sunflower

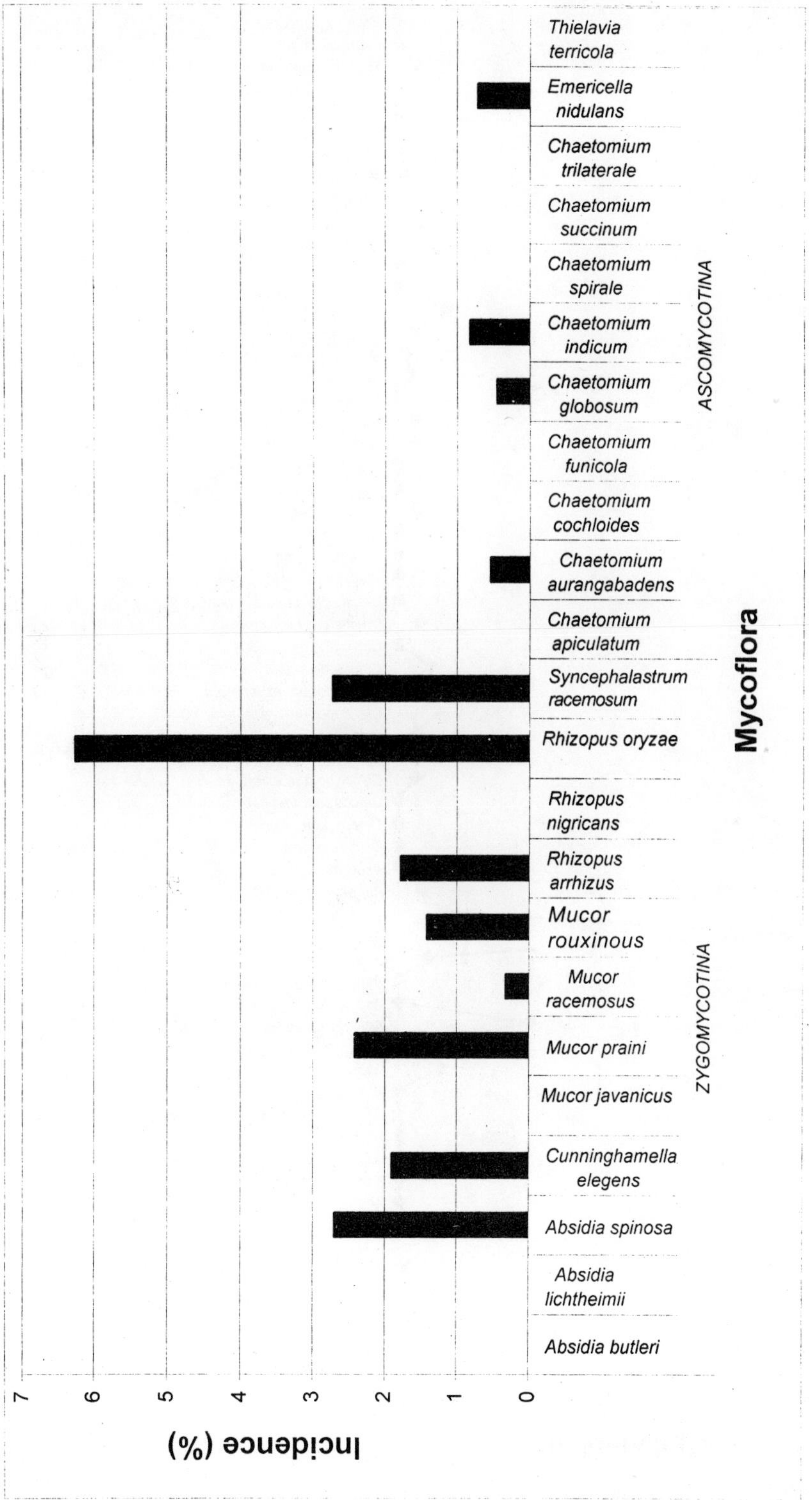

Figure 15.3a: Incidence (per cent) of Seed Mycoflora of Zygomycotina and Ascomycotina on Safflower

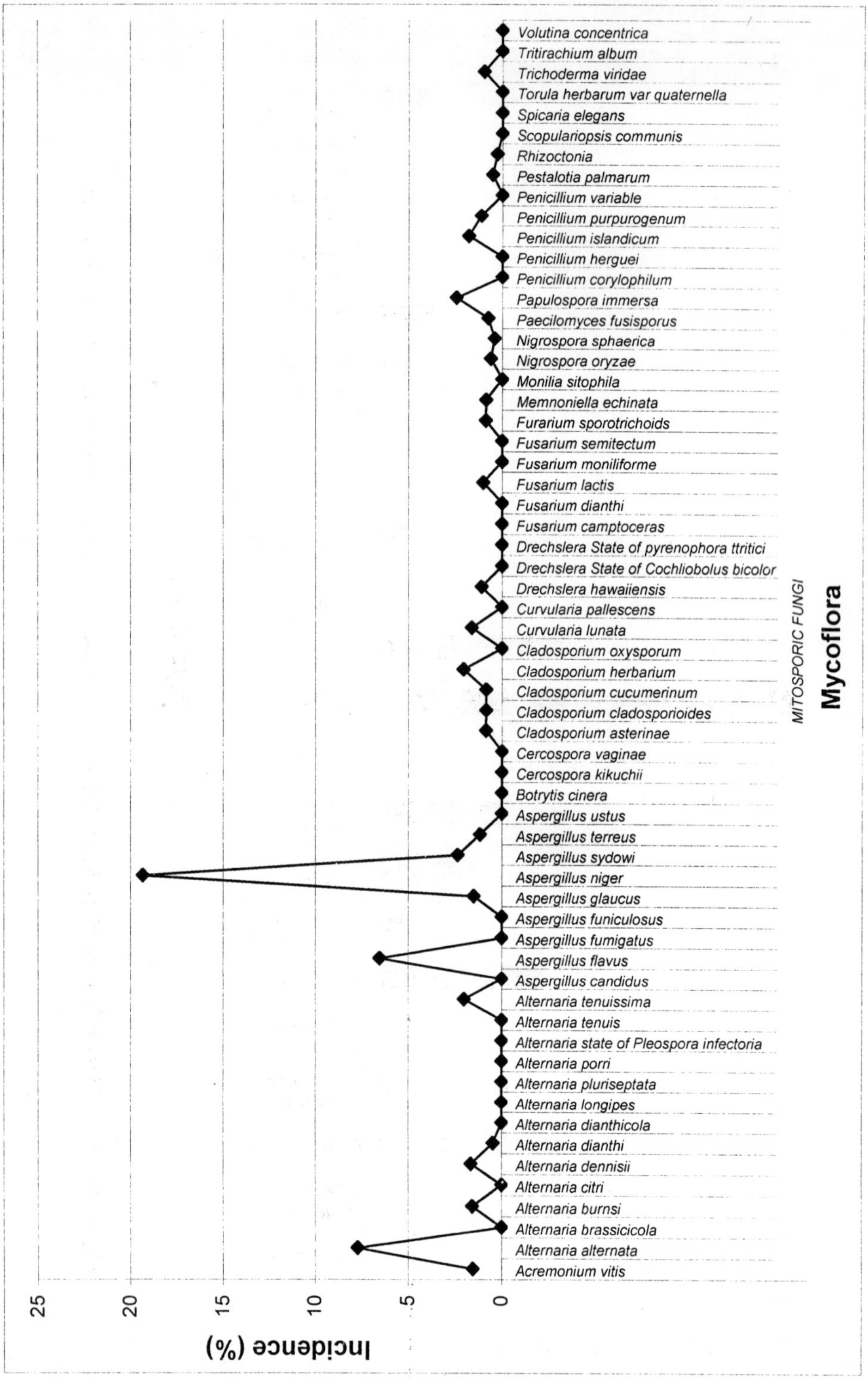

Figure 15.3b: Incidence (per cent) of Seed Mycoflora of Mitosporic Fungi on Safflower

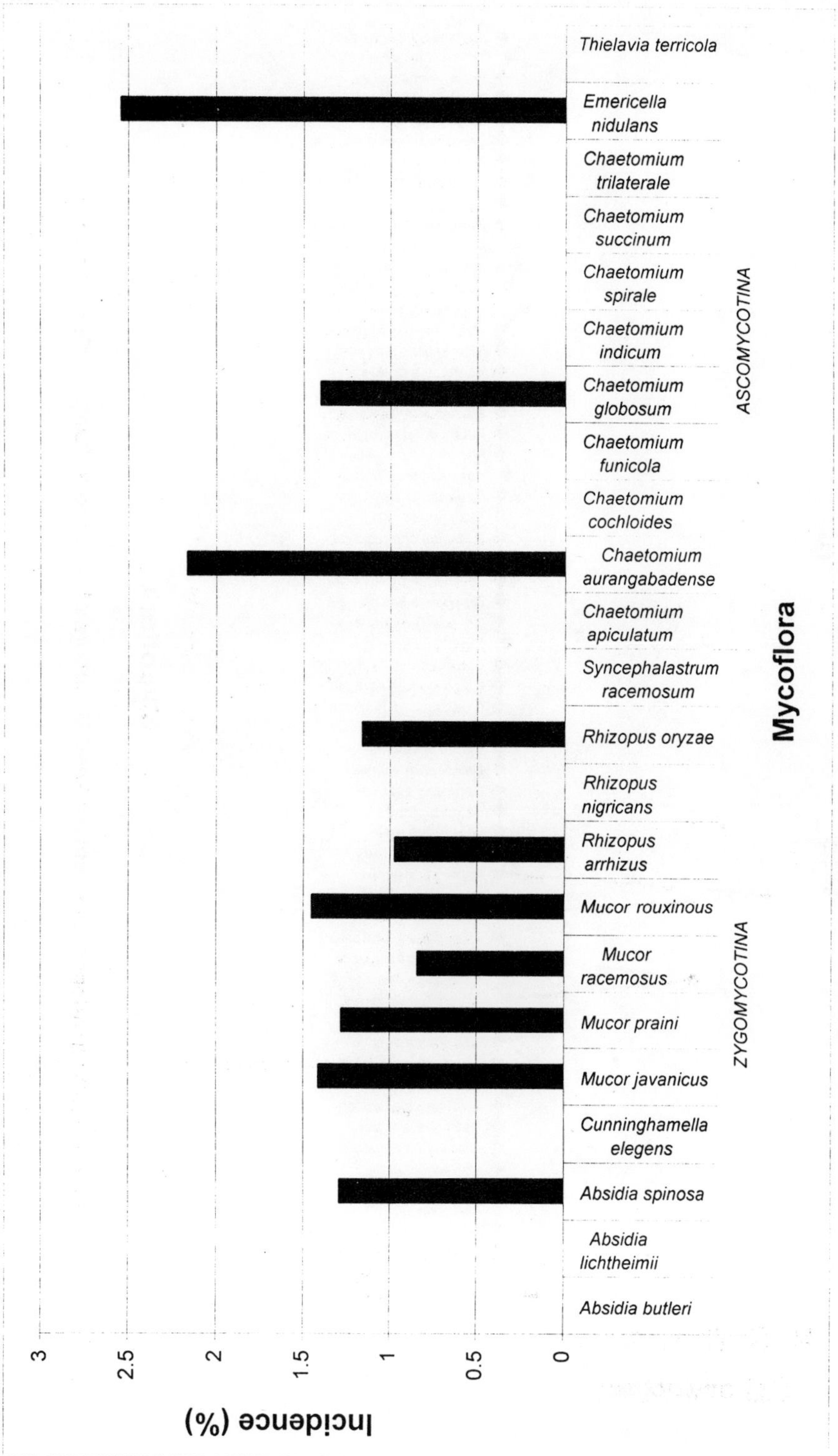

Figure 15.4a: Incidence (per cent) of Seed Mycoflora of Zygomycotina and Ascomycotina on Groundnut

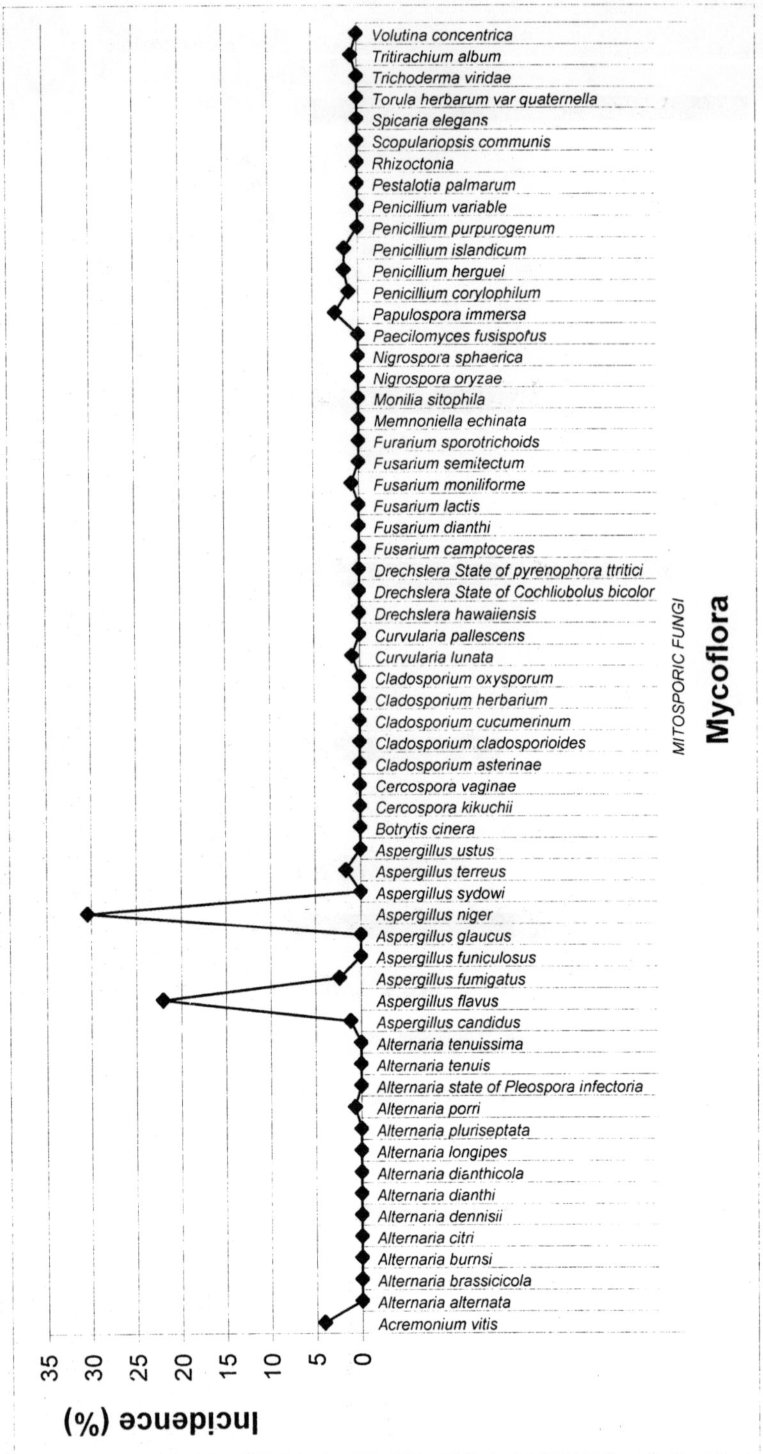

Figure 15.4b: Incidence (per cent) of Seed Mycoflora of Mitosporic Fungi on Groundnut

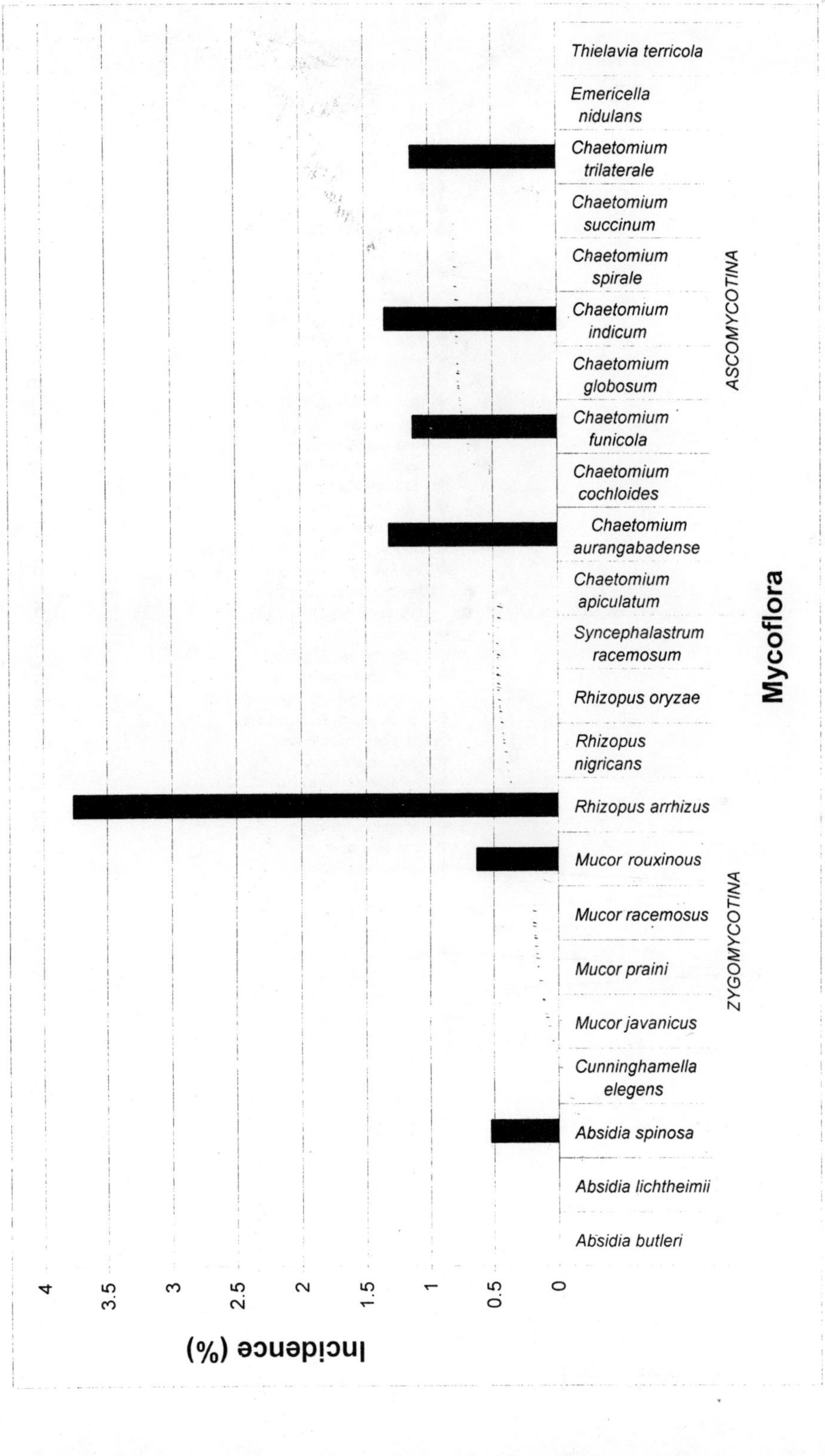

Figure 15.5a: Incidence (per cent) of Seed Mycoflora of Zygomycotina and Ascomycotina on Linseed

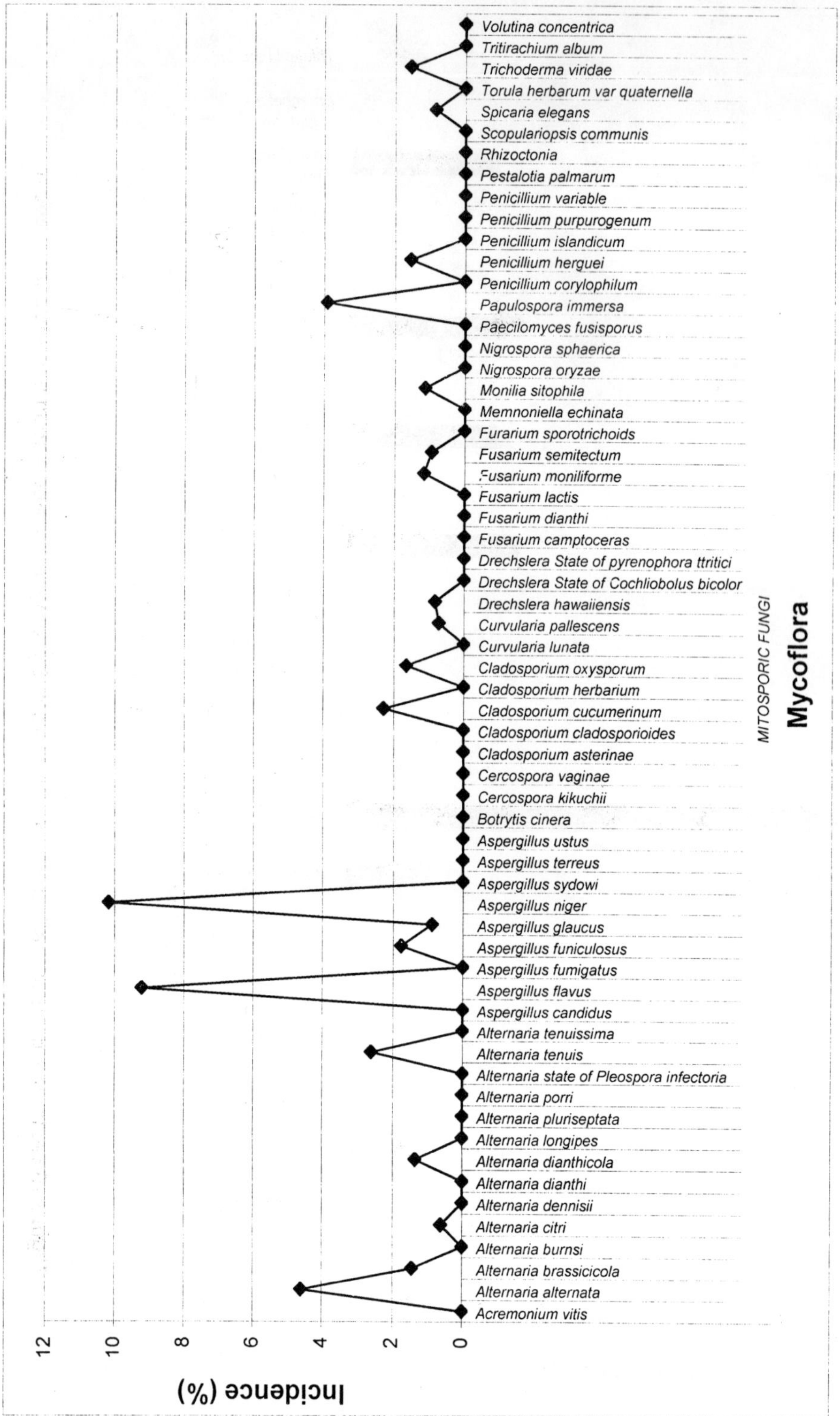

Figure 15.5b: Incidence (per cent) of Seed Mycoflora of Mitosporic Fungi on Linseed

racemosum (0.88 per cent) similarly in Ascomycotina *Chaetomium globosum* (1.21 per cent) showed maximum incidence while *Chaetomium succinum* (0.88 per cent) showed minimum percentage of incidence (Figure 15.6a). *Aspergillus niger* (5.96 per cent) and *Aspergillus flavus* (4.29 per cent) occurred with maximum percentage of incidence while *Alternaria porri* (0.64 per cent) and *Volutina concentrica* (0.76 per cent) occurred with minimum percentage of incidence compared to other members of the Mitosporic fungi group (Figure 15.6b).

Standard Blotter Method

External Seed Mycoflora

From Soybean in all seeds tested total of 39 different fungal species were isolated from unsterilized seeds by Blotter method. Of these 10 species belongs to Zygomycotina, six species belonging to Ascomycotina and twenty-three belonged to Mitosporic fungi. The percentage of incidence of all the species isolated on Soybean is shown in *Rhizopus nigricans* (9.60 per cent) showed highest incidence followed by *Syncephalastrum racemosum* (9.39 per cent) and *Mucor rouxinous* showed minimum percent of incidence (1.47 per cent) among the Zygomycotina group, while among the members of Ascomycotina *Chaetomium indicum* shows maximum incidence (5.97 per cent) followed by *Chaetomium aurganbadaense* (4.69 per cent) where as *Thielavia terricola* (1.51 per cent) shows minimum incidence. The members of Mitosporic fungi shows varying incidence. *Aspergillus niger* is the fungi in this group with maximum incidence of 38.49 per cent followed by *Cercospora kikuchii* 29.36 per cent of incidence where as *Fusarium dianthi* (2.36 per cent) and *Paceliomyces fusispora* (2.75 per cent) shows ever lowest in their appearance.

In Sunflower seeds, total of 36 fungal species were isolated from unsterilized seeds by blotter method. Six species are members of Zygomycotina, seven species belonged to Ascomycotina and twenty-three species belonged to Mitosporic fungi. Among the members of Zygomycotina *Rhizopus nigricans* (14.69 per cent) showed the maximum incidence while *Absidia lichtheimii* showed the minimum incidence 4.64 per cent, remaining species showed moderate in their appearance in Ascomycotina species of *haetomium* were highest among which *Chaetomium indicum* (4.35 per cent) showed maximum incidence while rest all species showed moderate in their appearance. While *Chaetomium trilaterale* (1.40 per cent) showed minimum incidence Among the members of Mitosporic fungi *Aspergillus niger* (17.99 per cent) showed the highest incidence followed by *Alternaria alternata* (17.65 per cent) while *Cladosporium cucumerinum* (0.94 per cent) and *Curvularia lunata* (1.15 per cent) showed least appearance.

From Safflower seeds tested for external mycoflora from unsterilized seeds 25 species were isolated by Blotter method. Six species were isolated from Zygomycotina, four species from Ascomycotina and fifteen from Mitosporic fungi. In case of Zygomycotina group *Rhizopus oryzae* (9.10 per cent) appeared with maximum incidence while *Cunninghamella elegans* (1.14 per cent) showed minimum incidence.

Among the members of Ascomycotina, *Emericella nidulans* (1.04 per cent) showed maximum number incidence as external fungi while *Chaetomium globosum* (0.49 per cent) showed minimum incidence. In case of Mitosporic fungi among fifteen species isolated *Aspergillus niger* (27.76 per cent) showed maximum incidence as external fungi followed *Aspergillus flavus* (8.10 per cent), while *Alternaria burnsii* (0.81 per cent) and *Penicillium islandicum* (0.88 per cent) showed minimum incidence as external mycoflora.

From Groundnut seeds, twenty fungi were isolated of these six belonged to Zygomycotina, three belongs to Ascomycotina and eleven belongs to Mitosporic fungi. Among the members of Zygomycotina *Mucor rouxinous* (2.92 per cent) showed maximum incidence followed by *Rhizopus oryzae* (1.88 per cent) while *Rhizopus arrhizus* (0.99 per cent) showed minimum incidence. In Ascomycotina *Chaetomium*

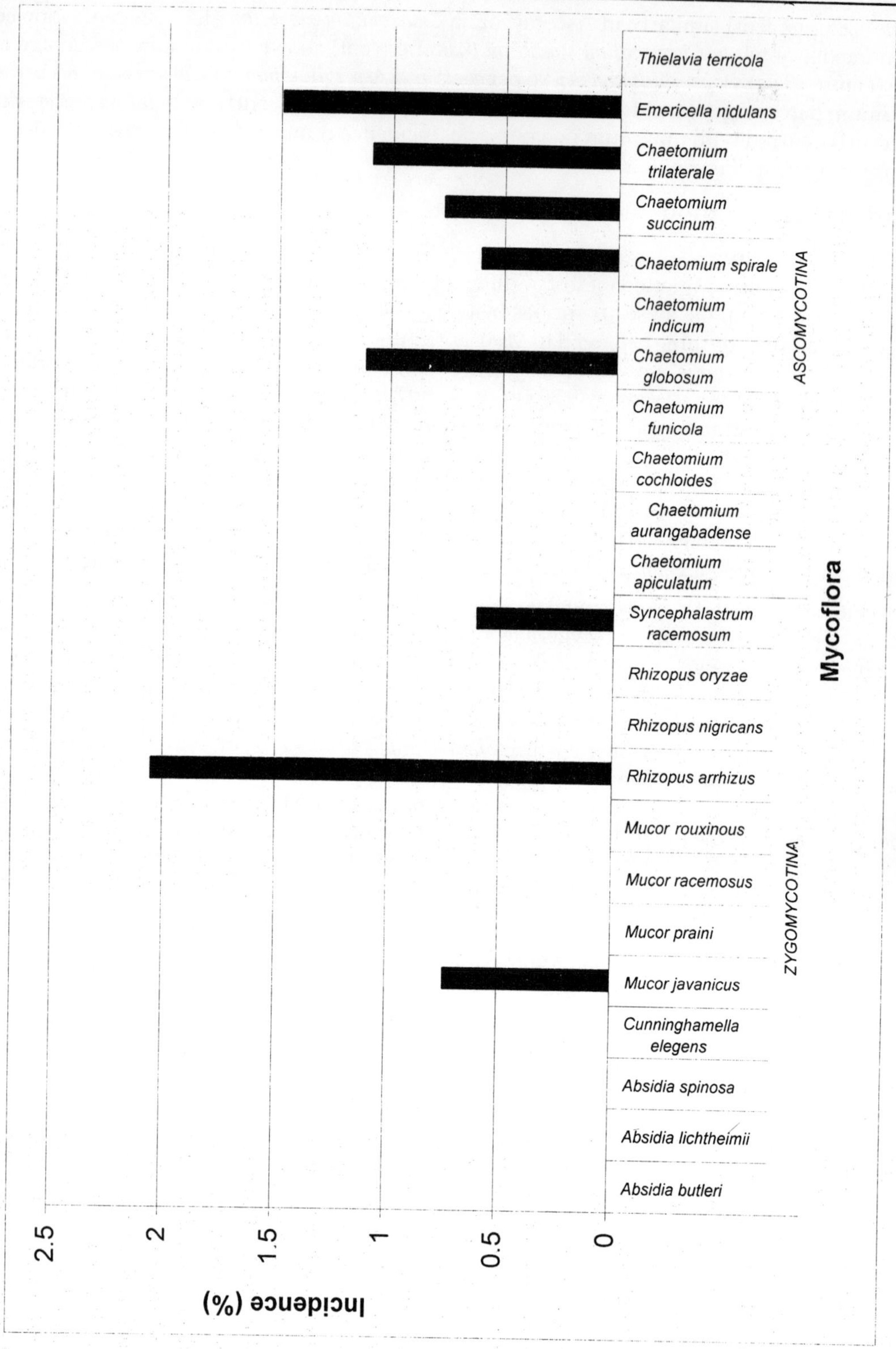

Figure 15.6a: Incidence (per cent) of Seed Mycoflora of Zygomycotina and Ascomycotina on Niger

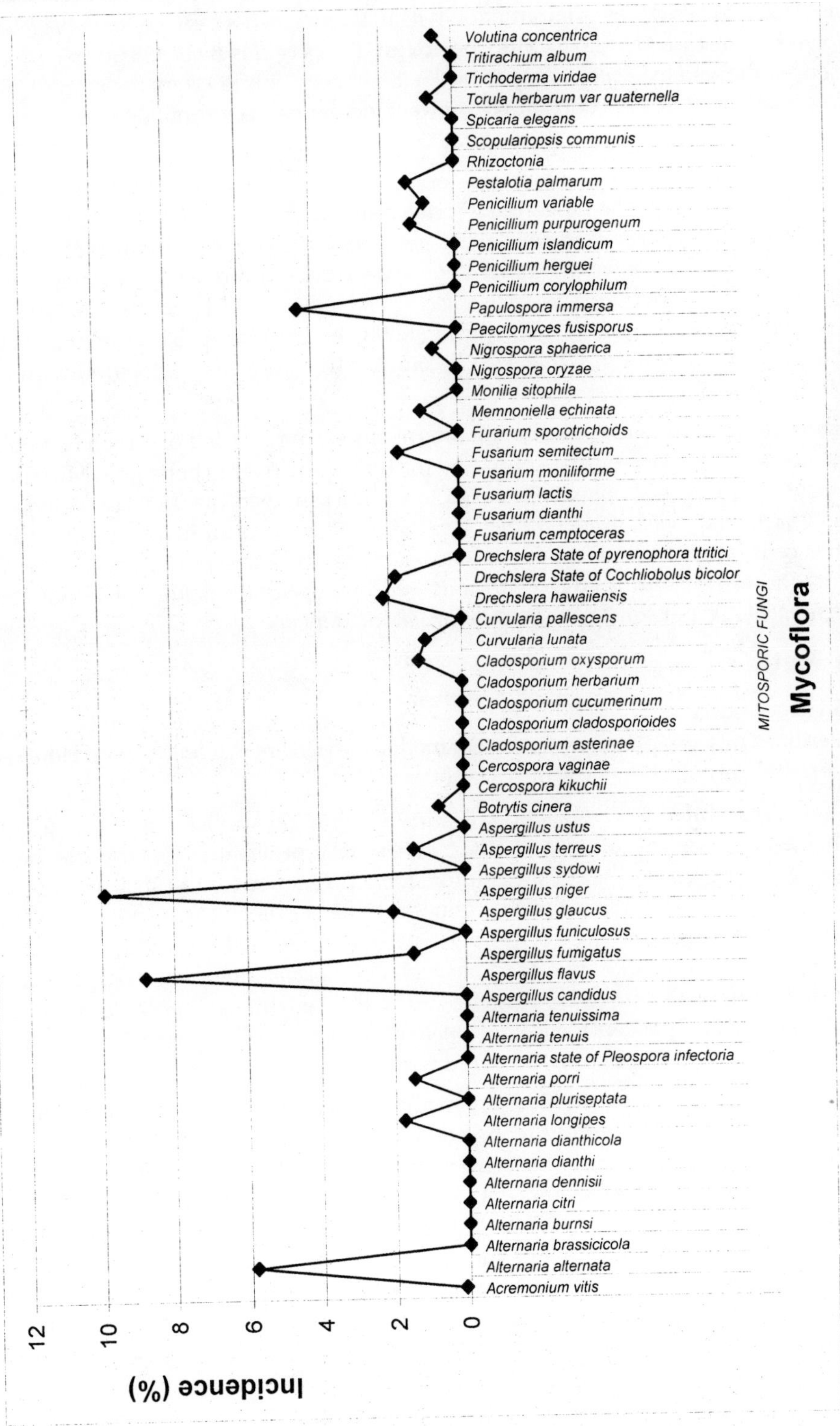

Figure 15.6b: Incidence (per cent) of Seed Mycoflora of Mitosporic Fungi on Niger

aurangabadense (3.21 per cent) was predominant in its incidence, whereas *Chaetomium globosum* (1.54 per cent) showed lowest incidence. In Mitosporic fungi *Aspergillus niger* (38.90 per cent) showed highest incidence followed by *Aspergillus flavus* (28.38 per cent) while *Alternaria porri* (0.78 per cent) and *Aspergillus candidus* (0.81 per cent) showed lowest incidence as external flora.

Linseed exhibited the association of nineteen fungal species from unsterilized seeds by Blotter method. Three species belongs to Zygomycotina, four species belonging to Ascomycotina while twelve species are members of Mitosporic fungi. In Zygomycotina, *Rhizopus arrhizus* (6.18 per cent) showed maximum incidence while *Absidia spinosa* (0.92 per cent) showed minimum incidence. Among Ascomycotina members, *Chaetomium auragabadense* (3.60 per cent) showed maximum incidence followed by *Chaetomium funicola* (3.10 per cent) whereas *Chaetomium indicum* (2.54 per cent) showed minimum incidence. From Mitosporic fungi *Aspergillus flavus* (12.93 per cent) was predominant in its incidence compared to other dominant fungi while *Monilia sitophila* (0.92 per cent) appeared with minimum incidence from unsterilized seeds of Linseed.

From Niger seeds, in all twenty-six fungal organisms were recorded of which two belonged to Zygomycotina, four belonged to Ascomycotina and the remaining twenty belongs to Mitosporic fungi. From Zygomycotina group *Rhizopus arrhizus* (3.23 per cent) showed maximum incidence while in Ascomycotina *Chaetomium globosum* (1.99 per cent) showed maximum incidence and *Chaetomium spirale* (1.36 per cent) showed minimum incidence. Among the Mitosporic fungi *Aspergillus niger* (14.60 per cent) showed maximum incidence followed by *Aspergillus flavus* (14.49 per cent) while *Memnoniella echinata* (1.08 per cent) appeared in lowest incidence.

Agar Plate Method

Internal Seed Mycoflora

Surface sterilized oilseeds plated over nutrient media, the percent incidence of seed found carrying the fungus internally varied from one group to another.

Soybean seeds, exhibited the association of thirty-two fungal species internally. Among these species, nine belongs to Zygomycotina, six belongs to Ascomycotina and fifteen species belonged to Mitosporic fungi. It is observed that of the total incidence 2.51 per cent of incidence occurred by Zygomycotina species, 1.51 per cent by Ascomycotina and 1.53 per cent by Mitosporic fungi *Aspergillus niger* (23.10 per cent), *Alternaria alternata* (11.74 per cent), *Aspergillus sydowi* (10.11 per cent), *Fusarium moniliforme* (6.71 per cent) *Aspergillus ustus* (6.64 per cent) and *Rhizopus nigricans* (6.72 per cent) showed maximum incidence on sterilized seeds by Agar plate method where as *Cladosporium herbarum* (2.03 per cent) *Fusarium lactis* (2.17 per cent) *Mucor rouxinous* (0.36 per cent) *Cunninghamella elegans* (1.26 per cent) and *Thielavia terricola* (1.81 per cent) occurred with minimum incidence.

All seeds of Sunflower were tested, thirty-nine fungal species were isolated of which six species belongs to Zygomycotina, five belongs to Ascomycotina and twenty-eight belongs to Mitosporic fungi. The over all incidence of isolated species in each group is Zygomycotina 0.92 per cent, Ascomycotina 0.57 per cent and Mitosporic fungi 1.63 per cent. *Aspergillus niger* (16.15 per cent), *Alternaria alternata* (10.46 per cent), *Aspergillus flavus* (10.60 per cent), *Alternaria longipes* (6.65 per cent), followed by *Alternaria pluriseptata* (5.29 per cent), were showed maximum incidence, where as *Cladosporium cladosporiodes* (0.82 per cent), *Curvularia lunata* (1.43 per cent), *Fusarium dianthi* (1.57 per cent), *Memnoniella echinata* (0.72 per cent) *Penicillium islandicum* (1.28 per cent), *Rhizopus nigricans* (1.11 per cent) and *Thielavia terricola* (0.58 per cent) showed minimum percentage of incidence.

There are thirty seven fungal species which were isolated from Safflower seeds, out of these seven species were from Zygomycotina, three belongs to Ascomycotina and rest twenty eight species were belongs to the Mitosporic group. The over all incidences of these species varied in each group, in Zygomycotina 1.57 per cent, Ascomycotina 0.08 per cent and Mitosporic fungi 1.04 per cent. The percent incidence of seed carrying individual species also varied. It was seen that *Aspergillus niger* (12.71 per cent), *Aspergillus flavus* (4.99 per cent) and *Alternaria alternata* (9.99 per cent) showed maximum incidence followed by *Cunninghamella elegans* (3.16 per cent), *Rhizopus oryzae* (4.15 per cent) where as *Acremonium vitis* (1.25 per cent), *Aspergillus terreus* (0.71 per cent) *Mucor rouxinous* (0.33 per cent) and *Pestalotia palmarum* (0.44 per cent) had exposed minimum incidence.

Groundnut seeds were associated with twenty-two species of fungi on sterilized seeds. Among these, seven species belonged to Zygomycotina, three species belonged to Ascomycotina and the remaining twelve belonged to Mitosporic fungi. The overall incidence of each group in sterilized seeds of groundnut comprised Zygomycotina (0.58 per cent), Ascomycotina (0.51 per cent) and Mitosporic fungi (1.08 per cent). The species of *Aspergillus niger* (23.40 per cent) and *Aspergillus flavus* (24.04 per cent), have shown maximum incidence whereas *Absidia spinosa* (1.17 per cent), *Chaetomium aurangabadense* (1.49 per cent), *Aspergillus terreus* (0.96 per cent), *Curvularia lunata* (0.69 per cent), *Fusarium monoliforme* (0.92 per cent), *Penicillium herquei* (0.92 per cent) have shown minimum incidence.

In case of Linseed, totally twenty-three species were isolated, two species belong to Zygomycotina, two species belong to Ascomycotina and nineteen species belong to Mitosporic fungi. Percent incidence of species isolated in each group is, Zygomycotina (0.19 per cent), Ascomycotina (0.11 per cent) and Mitosporic fungi (0.54 per cent). When compared with the incidence of species, *Aspergillus niger* (7.13 per cent), *Aspergillus flavus* (3.82 per cent), *Aspergillus glaucus* (1.79 per cent), *Papulospora immersa* (1.99 per cent), *Penicillium herquei* (1.10 per cent) and *Torula herbarum quaternella* (2.28 per cent) showed maximum occurrence whereas *Absidia spinosa* (0.75 per cent), *Chaetomium aurangabadense* (1.24 per cent), *Chaetomium succinum* (0.65 per cent), *Alternaria citri* (0.64 per cent), *Alternaria dianthi* (0.64 per cent), *Aspergillus funiculosus* (0.58 per cent) occurred with minimum incidence.

Totally twenty-three species were isolated that were associated with sterilized seeds of Niger. Three species were isolated from Ascomycotina and twenty from Mitosporic fungi. About 0.57 per cent of fungal incidence was in Mitosporic fungi, and 0.26 per cent in Ascomycotina. Among these species isolated *Aspergillus flavus* (5.82 per cent), *Aspergillus niger* (4.13 per cent) and *Alternaria alternata* (3.43 per cent) and *Papulospora immersa* (2.26 per cent) occurred with maximum percentage of incidence. While *Acremonium vitis* (1.08 per cent), *Alternaria porri* (1.25 per cent), *Memnoniella echinata* (0.69 per cent), *Pestalotia palmarum* (0.72 per cent) and *Penicillium variable* (0.86 per cent) occurred with lowest incidence.

Agar Plate Method

External Seed Mycoflora

To study the external seed mycoflora unsterilized seeds were also used for Agar Plate method.

In Soybean, totally thirty-two species were isolated by agar plate method. From total isolated species ten belonged to Zygomycotina six belongs to Ascomycotina and sixteen belonged to Mitosporic fungi. The percent incidence of each group is Zygomycotina (3.15 per cent), Ascomycotina (1.54 per cent) and Mitosporic fungi (2.50 per cent) of all the species recorded were *Alternaria alternata* (11.86 per cent), *Aspergillus niger* (39.13 per cent), *Aspergillus sydowi* (12.58 per cent), *Aspergillus terreus* (10.36 per cent), *Aspergillus ustus* (8.08 per cent), *Cercospora kikuchii* (8.75 per cent), *Cladosporium oxysporum* (9.25

per cent), *Fusarium moniliforme* (7.99 per cent), *Papulospora immersa* (8.75 per cent), and *Rhizopus nigricans* (12.97 per cent).

In Sunflower total of thirty-seven species were isolated, among which six species belong to Zygomycotina, four belongs to Ascomycotina and twenty-seven belongs to Mitosporic fungi. From the overall incidence in each group the frequency of species varied in Zygomycotina 2.91 per cent, in Ascomycotina 0.49 per cent and Mitosporic fungi 2.16 per cent was seen. The maximum incidence of species occurring on sunflower seeds were *Aspergillus niger* (14.16 per cent), *Alternaria alternata* (14.07 per cent) followed by *Alternaria longipes* (13.43 per cent), *Acremonium vitis* (10.18 per cent), *Alternaria pluriseptata* (6.71 per cent), *Mucor praini* (8 per cent), *Rhizopus arrhizus* (6.31 per cent) and *Rhizopus nigricans* (6.43 per cent) where as *Alternaria tenuis* (2.04 per cent) *Aspergillus fumigatus* (1.15 per cent), *Curvularia lunata* (1.60 per cent), *Nigrospora oryzae* (0.72 per cent) and *Thielavia terricola* (1.14 per cent) occurred in minimum incidence.

Safflower exhibited the association of thirty-one species, out of which six belongs to Zygomycotina; two from Ascomycotina and twenty-three belongs to Mitosporic fungi. The incidence of each group bearing the species is Zygomycotina 1.85 per cent, Ascomycotina 0.12 per cent and Mitosporic fungi 1.08 per cent. Among the species isolated, the maximum incidence of seed is by *Aspergillus niger* (23.93 per cent), *Aspergillus flavus* (8.43 per cent), *Alternaria alternata* (5.26 per cent), *Mucor praini* (3.21 per cent), *Rhizopus oryzae* (7.85 per cent), *Absidia spinosa* (4.25 per cent) and *Mucor rouxinous* (3.32 per cent) while minimum incidence is by *Chaetomium indicum* (0.59 per cent), *Aspergillus sydowi* (0.85 per cent), *Aspergillus terreus* (1.24 per cent), *Cladosporium herbarum* (1.19 per cent), *Fusarium sporotrichoids* (0.94 per cent), *Nigrospora spherica* (0.65 per cent), *Pestalotia palmarum* (0.24 per cent) and *Trichoderma viride* (0.51 per cent).

Seeds of Groundnut were tested for the external mycoflora, twenty different species were found. Six species belongs to Zygomycotina taking into account of 0.82 per cent of the total incidence, three species belong to Ascomycotina of 0.37 per cent of the total incidence and eleven species belongs to Mitosporic fungi comprised of 1.09 per cent of the total incidence. *Aspergillus niger* (30.40 per cent), *Aspergillus flavus* (15.21 per cent), *Alternaria alternata* (6.36 per cent) and *Emericella nidulans* (2.46 per cent) occurred with maximum incidence, while *Aspergillus candidus* (0.60 per cent), and *Curvularia lunata* (0.86 per cent) occurred by means of minimum percent of incidence.

Total seventeen species were isolated from Linseed. Only one species each was isolated from Zygomycotina and Ascomycotina group respectively, and the rest fifteen species belongs to Mitosporic fungi. The total incidence of each group is Zygomycotina consist of 0.45 per cent, Ascomycotina 0.16 per cent and Mitosporic fungi 1.12 per cent. *Alternaria alternata* (7.04 per cent), *Aspergillus flavus* (14.63 per cent), *Aspergillus niger* (15.26 per cent), *Cladosporium oxysporium* (5.26 per cent), *Papulospora immersa.* (5.49 per cent), *Penicillium herquei* (4.10 per cent) and *Rhizopus arrhizus* (5.38 per cent) occurred in highest percent of incidence, while *Aspergillus funiculosus* (0.92 per cent), *Aspergillus glaucus* (0.69 per cent), *Fusarium semitectum* (1.99 per cent), *Monilia sitophila* (1.69 per cent) and *Spicaria elegans* (2.21 per cent) occurred with lowest percent of incidence.

Niger seed exhibited the association of totally twenty-three species. Out of these two belongs to Zygomycotina, three belongs to Ascomycotina and eighteen belongs to Mitosporic fungi. The total incidence exhibited by each group is Zygomycotina 0.42 per cent, Ascomycotina 0.52 per cent and Mitosporic fungi 1.17 per cent. The species showing maximum incidence are *Alternaria alternata* (8.43 per cent), *Aspergillus flavus* (10.60 per cent), *Aspergillus glaucus* (4.21 per cent), *Aspergillus niger* (14.93), *Cladosporium oxysporum* (2.35 per cent), *Drechslera hawaiiensis* (2.69 per cent), *Drechslera* State of

Cochliobolus bicolor (2.75 per cent), *Papulospora immersa* (7.03 per cent), *Emericella nidulans* (3.21 per cent) and *Rhizopus arrhizus* (3.49 per cent); while the species of *Aspergillus terreus* (2.21 per cent), *Penicillium variable* (1.47 per cent) and *Syncephalastrum racemosum* (1.54 per cent) occurred in minimum incidence.

It is evident that quite a good number of mycoflora were isolated frequently with different seeds. The mycoflora varied in each group compared with to their frequency of incidence in different seasons and the methods (ABM and SBM) complied. Maximum numbers of species were recorded in rainy season followed by winter and summer (Figures 15.7–15.12). Among Zygomycotina, species like *Rhizopus oryzae* and *Mucor javanicus* occurred rather consistently throughout the year. However in Ascomycotina *Chaetomium aurangabadense, Chaetomium indicum, Thielavia terricola* throughout the year in few seeds only.

In Mitosporic fungi many species of *Alternaria, Aspergillus, Cladosporium, Fusarium* and *Penicillium* were observed to be adaptable to various temperatures fluctuations and hence were found in all the seasons. Forms like *Alternaria burnsii, Alternaria longipes, Alternaria tenuis, Fusarium dianthi, Penicillium islandicum, Penicillium variable, Pestalotia palmarum* etc. were adaptable to winter and rainy but on some seeds they were isolated in summer also.

Discussion

In the present study, six oilseeds that are economically important and are commonly used in Dharwad have been studied for nearly two years regarding taxonomical and certain physiological aspects of the isolated seed mycoflora. The seeds that are stored in the storehouses of Agriculture Produce Market Committee (APMC), Dharwad are usually used for consumption, extraction of oil or transported.

The examination of six different oilseeds resulted in the isolation of 84 species belonging to 31 genera. An analysis of the composition of the mycoflora reveal that although a great majority of fungi isolated belonged to Mitosporic fungi and other groups, are represented in composition of mycoflora is as follows:

Group	Genera	Species
Zygomycotina	5	12
Ascomycotina	3	11
Mitosporic fungi	23	61

Apart from these several, interesting species previously either unreported or of rare and very limited distribution on other sources were also recorded. An analysis of the results reveals that of the total 84 species, some of the species may be common to all seeds and intensively. These may be common to 1 or 2 or 3 or even upon 6 seeds. Taking all six different oilseeds interestingly it was found that most of the species were common to all of them. There are some, which were specific seed to seed.

Among the different groups, 12 species of Zygomycotina, 11 species of Ascomycotina and 61 species of Mitosporic fungi were isolated.

From the methods employed, it is observed that in standard blotter method the overall mean percent incidence of mycoflora of all the groups varied in all the seeds tested. The highest overall mean percent incidence of Zygomycotina was observed in Soybean seeds (4 per cent) while lowest in Niger

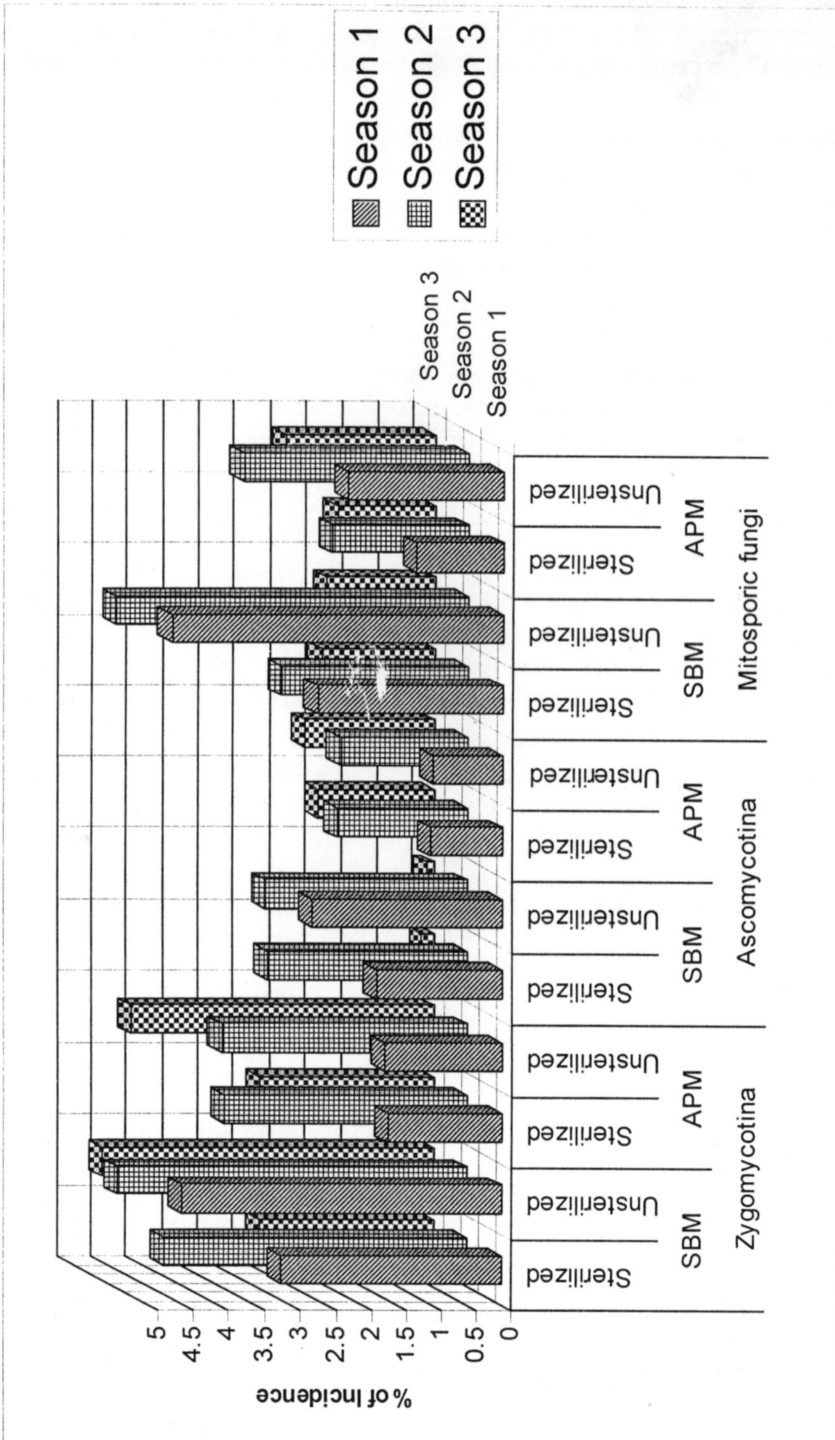

Figure 15.7: Distribution of Different Groups of Seed Mycoflora of Soybean in Relation to Seasons and Method

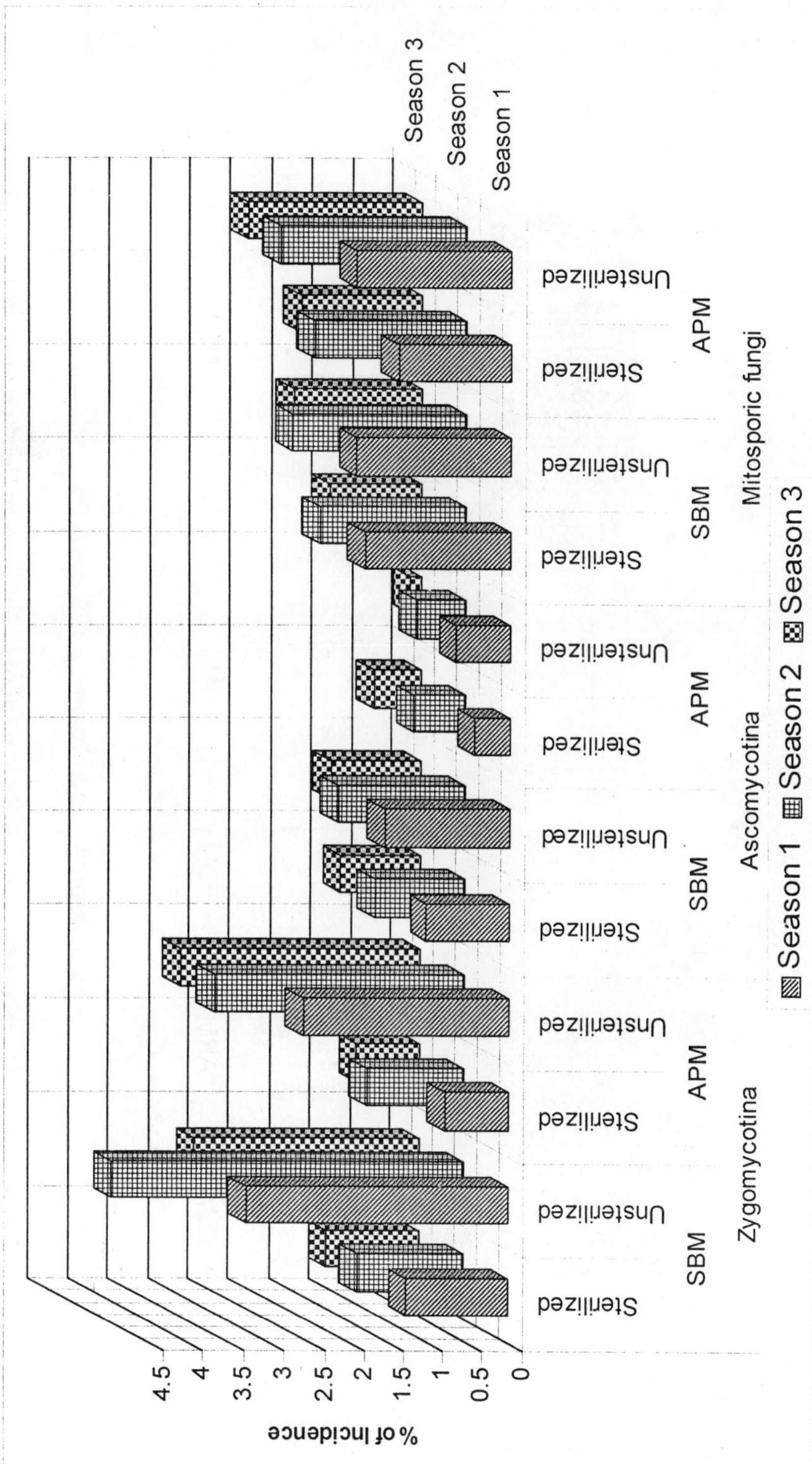

Figure 15.8: Distribution of Different Groups of Seed Mycoflora of Sunflower in Relation to Seasons and Methods

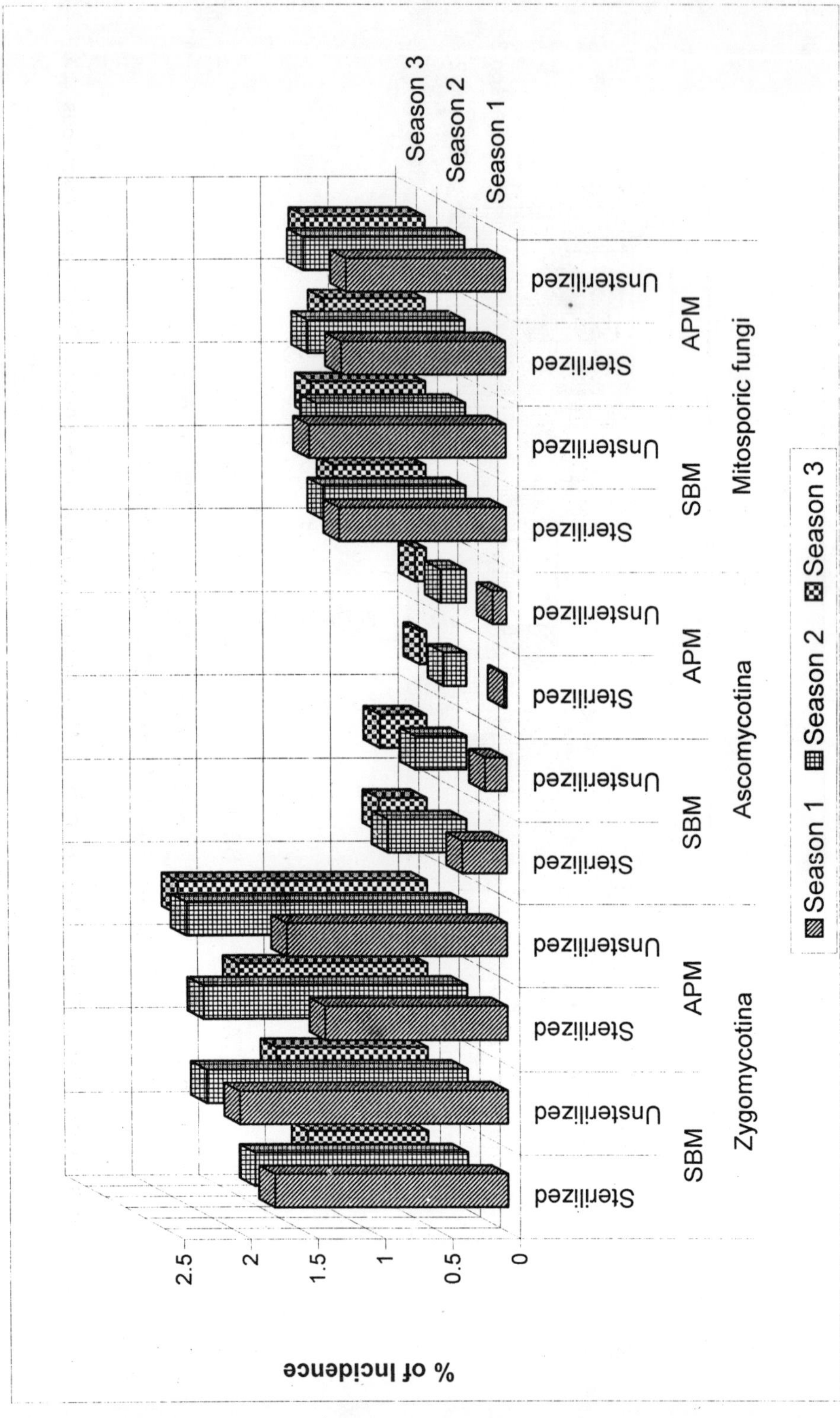

Figure 15.9: Distribution of Different Groups of Seed Mycoflora of Safflower in Relation to Seasons and Methods

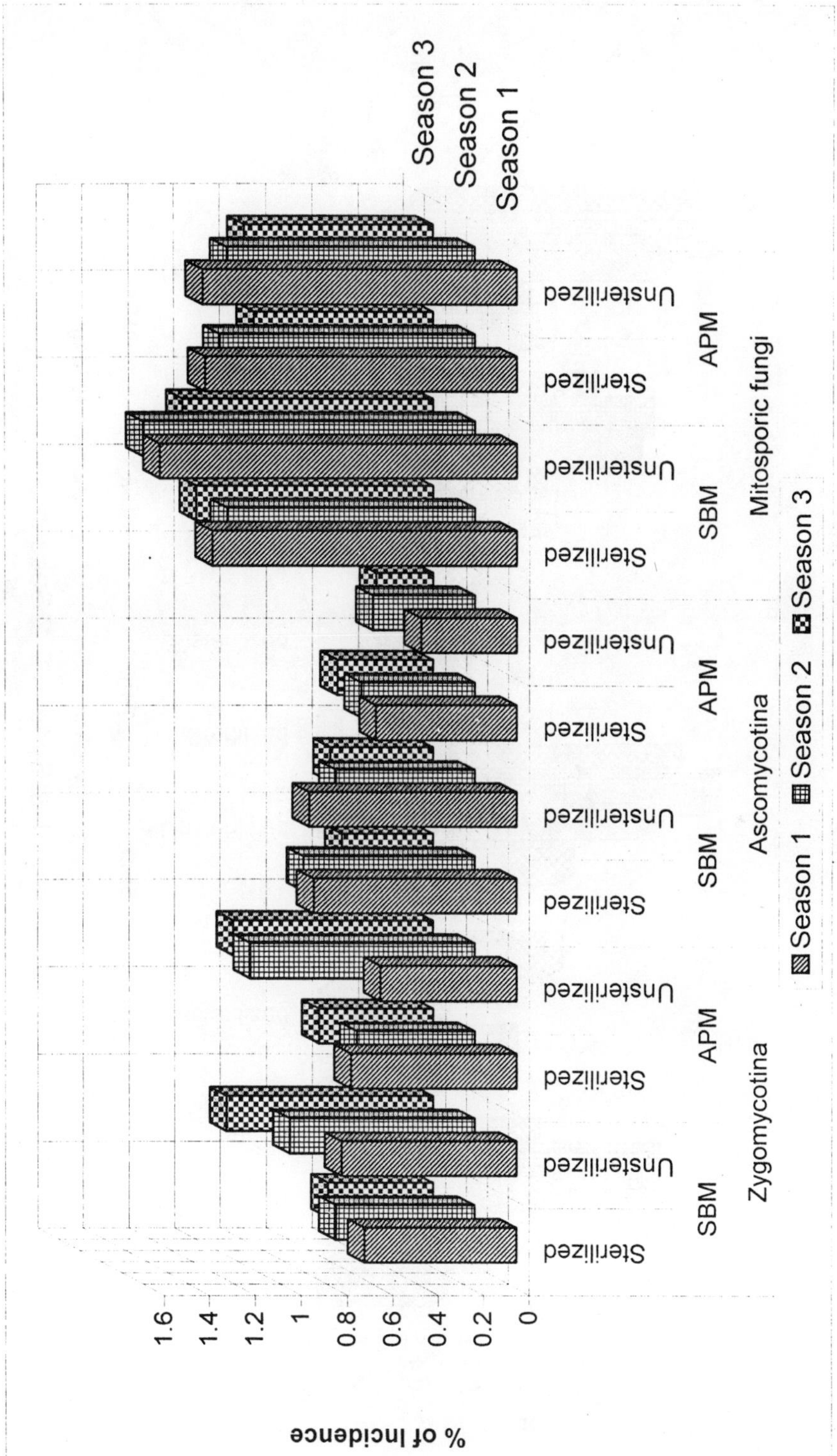

Figure 15.10: Distribution of Different Groups of Seed Mycoflora of Groundnut in Relation to Seasons and Methods

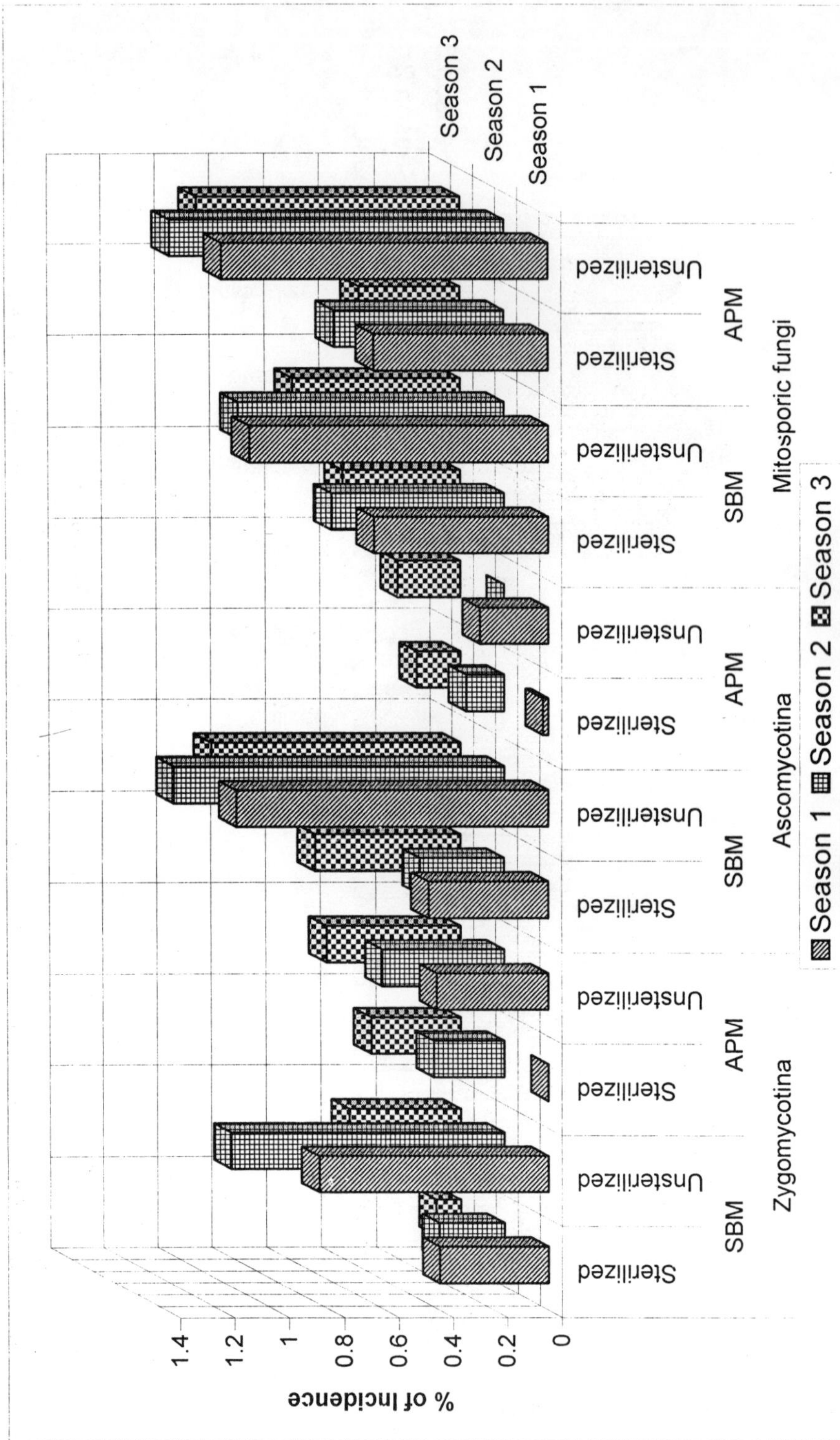

Figure 15.11: Distribution of Different Groups of Seed Mycoflora of Linseed in Relation to Seasons and Methods

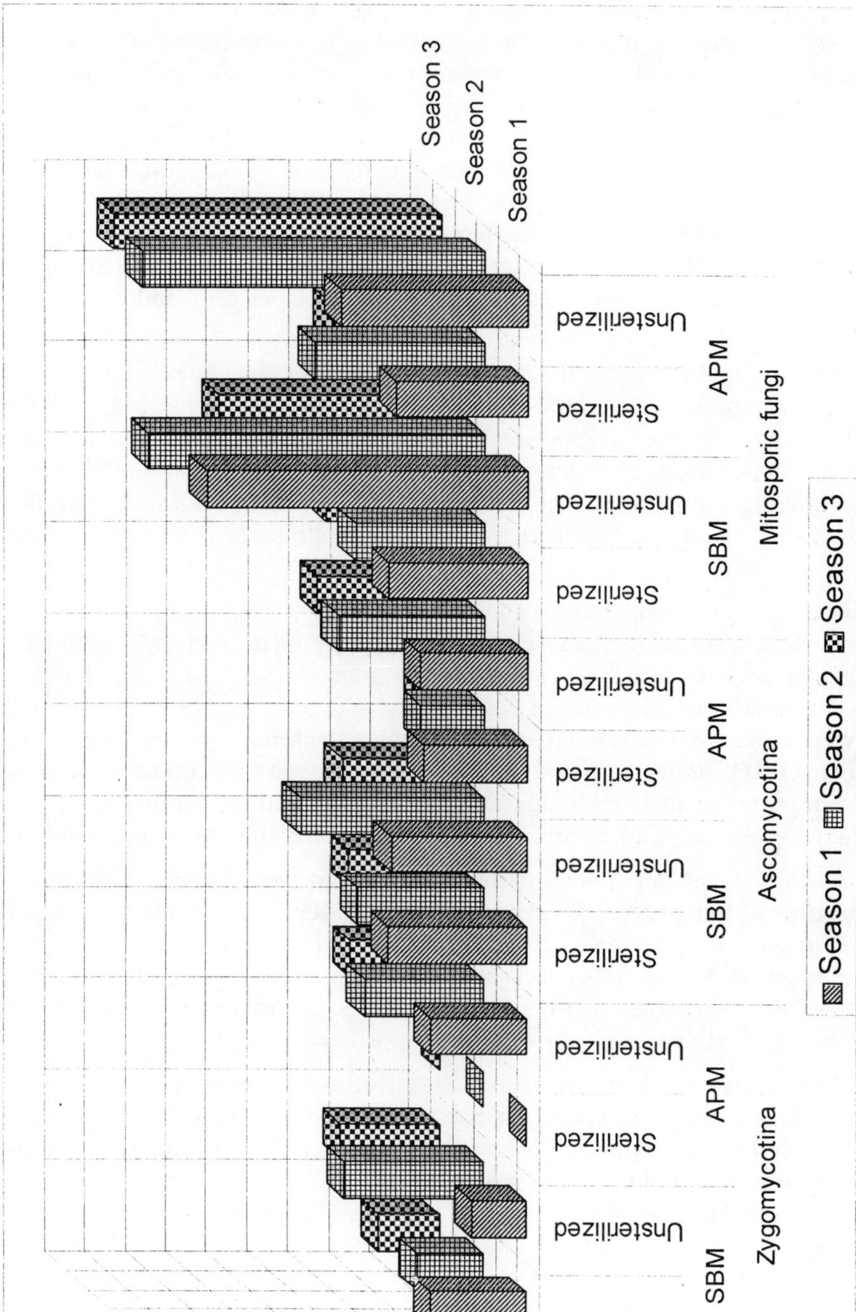

Figure 15.12: Distribution of Different Groups of Seed Mycoflora of Niger in Relation to Seasons and Methods

seeds (0.36 per cent). In Ascomycotina overall mean percent incidence was observed maximum in Soybean (1.74 per cent) where as minimum in Safflower (0.29 per cent) similarly the overall percent incidence of Mitosporic fungi was observed maximum in Soybean (1.91 per cent) whereas minimum incidence in Linseed and Niger (0.55 per cent). In agar plate method also there was variation in overall mean incidence in each group tested for all the seeds. In Zygomycotina, the overall mean percent incidence was maximum in Soybean seeds (2.83 per cent), followed by sunflower and Safflower while minimum incidence of this group was observed in Niger seeds (0.21 per cent). The Ascomycotina group was found maximum in overall mean incidence in Soybean seeds (1.52 per cent) while minimum in Safflower (0.10 per cent). The Mitosporic fungi group showed overall mean incidence maximum in Soybean seeds (3.11 per cent) and minimum in Safflower (1.11 per cent) followed by Linseed and Niger seeds.

From the above findings it is noted that there was variation of mycoflora in the methods used and also varied from seed to seed. Some of the species were isolated by one method while some species were isolated by both the methods. Therefore these two methods proved standard for isolation of number of species. As per the rules of International Seeds Testing Association (ISTA, 1999) after an extensive comparative seed health test schemes organized by the Plant Diseases Committee of ISTA, it is noted that the two tests are inevitable for getting a complete picture of the fungal information/ association with the seeds.

Seed health testing primarily aimed at locating the infesting/infecting organisms during the period of storage. Results showing that some of the fungal organisms were encountered throughout the year. These organisms obviously show that they are more versatile in the food requirement and also they have the capacity of tolerance for varying environment (Bilgrami *et al.*, 1979). While some species which were restricted to particular season reflect greater sensitive to the storage temperatures. Mathur and Neergaard (1970) has reported the relationship between the amount of seed infection and the severity of symptoms at different temperatures. The fungi with minimum per cent of incidence show that there is more demanding nature for them for nutrients and environmental conditions.

The occurrence of particular species varied from seed to seed. The variation might be due to difference in nutritional constituents of each seed type. Besides these, dust from vegetative parts of plants, soil thrown up during harvesting, residue in harvesting. Further increase in the mycoflora occurs during transportation and in storage (Kanchanlata, 1991), other possibility for such divergence might be attributed to a comparatively rapid growth of the saprophytes adhering to the seed surface and marking of slow growth pathogenic forms (Neergaard and Saad, 1962).

Maximum appearance of saprophytes in unsterilized seeds reduced the appearance of deep seated fungi. In surface sterilized seeds the incidence of genera like *Aspergillus, Mucor, Penicillium* and *Rhizopus* were reduced when compared to unsterilized seeds in both the methods. This shows that these saprophytes were mostly located on seed surface. It is claimed by Holfman-Meiri (1992) that *Rhizopus arrhizus* is mostly located on seed surface, where as *Alternaria* sp. is deep seated in seed tissue perhaps this is the reason why the saprophytes reduced greatly by surface sterilization and there by made easy for the isolation of some internal fungi.

The results obtained reveal that *Aspergillus* (storage fungus) invaded all types of seeds. Field fungi *e.g. Alternaria, Fusarium* and *Curvularia* designated, as "Miscellaneous fungi" were also isolated *Aspergillus flavus* and *Aspergillus niger* were well adapted to all the types of seed since they were predominant during storage, the information of these reduced in the population of field fungi (Kashinath and Subarata, 2002).

The works on seed-borne microorganisms of Soybean emanating from Soybean growing countries of the world situated in diversified climatic conditions is quite adequate (Karmakar *et al.*, 1982; Zad 1982, Nik 1983, Papoola and Kueshi, 1982; Hussain *et. al.*, 1989; Singh 1991; Jordan *et. al.*, 1986, Garcia *et al.*, 1993). It is also seen from the observation that totally 9 species of *Chaetomium* were isolated. But the highest incidence of these species was recorded by blotter method that proves valuable for the isolation of *Chaetomium. Chaetomium succinum* is the only species that is isolated on Niger seed. *Chaetomium* species that were isolated occur on many seeds but the isolation of so many species has shown their diversity on storage seeds. One can consider these fungi as stored fungi. Bale and Khare (1982) and Saxena (1985) reported that *Chaetomium* species reduced seeds germination heavily infected seeds.

The percent incidence of *Cercospora kikuchii* in Soybean seeds was maximum and was isolated by both the methods. The presence of this species in Soybean causes purple seed stain in Soybean (Ilyas *et. al.*, 1975) and was observed in almost all the seeds tested.

The presence of *Alternaria* species is also of much concerned. Totally 13 species were isolated but the maximum number of species were isolated on Sunflower seeds. Earlier works have also proven that the species of *Alternaria* constitute an important group of plant pathogens of oil producing crops as they cause characteristic leaf spots on them (Mehrotra *et al.*, 1992). Narain and Swarup (1989) have isolated seven species of *Alternaria* that were associated with oilseed crops in U.P.

The combined effects of the abiotic and biotic environments are many and complex. The physical environment will always exert on influence directly on the organisms in a community. An influence of season would appear to be reflected in isolation. Seasonal fluctuation in the fungal population showed most of the members of Zygomycotina and Mitosporic fungi were isolated during all the seasons on all the seeds tested.

Maximum numbers of species were recorded in rainy season followed by winter and summer (Figures 15.7–15.12). Among Zygomycotina, species like *Rhizopus oryzae* and *Mucor javanicus* occurred rather consistently throughout the year. However in Ascomycotina *Chaetomium aurangabadense, Chaetomium indicum, Thielavia terricola* through out the year in few seeds only.

In Mitosporic fungi many species of *Alternaria, Aspergillus, Cladosporium, Fusarium* and *Penicillium* were observed to be adaptable to various temperatures fluctuations and hence were found in all the seasons. Forms like *Alternaria burnsii, Alternaria longipes, Alternaria tenuis, Fusarium dianthi, Penicillium islandicum, Penicillium variable, Pestalotia palmarum* etc. were adaptable to winter and rainy but on some seeds they were isolated in summer also. It is observed that environmental factors act upon some of the species for their appearance in particular season and disappearance in other seasons. Jordan *et al.* (1986) found environmental factors on the incidence of seed-borne fungi of Soybean in the North and South of Illinois as over riding factors in predisposition of seed to fungal infection regardless of soil types.

Effect of Culture Filtrate of Seed Mycoflora

The plant pathogenic fungi and microorganisms produce toxic substances when cultivated on synthetic culture media. Hence the fungi associated with seeds are responsible for causing damage to the seeds by producing phytotoxic substances. In the present study however culture filtrate of most of the test fungi caused reduction in percent seed germination and seedling vigour seeds of sunflower and soybean were most susceptible to the culture filtrate of all the test fungi and affected seedling vigour while those of Linseed and Niger were resistant. The culture filtrates of *Aspergillus flavus* and

Aspergillus niger were most toxic and inhibited the germination and seedling vigour. The result presented here is in confirmative with Chandra *et al.* (1985). They have reported the culture filtrate effect grown in liquid medium and Hawker's medium but it was reported that Richards's medium was invariably found to be more suitable for production of toxic metabolite (Mridula and Srivastava, 1991). In Groundnut seeds *Aspergillus flavus* reduced seed germination and seedling vigour. Singh and Swami (2004) have also reported that *Aspergillus flavus* culture filtrate reduced seed germination as well as plumule and radicle length of pearl millet seeds. From overall results it was observed the culture filtrate *Aspergillus niger* inhibited germination and also seedling vigour of almost all the oilseeds. Singh and Swami (2004) reported similar results in case of pearl millet seeds. Many investigators like Subbaraja (1973) have reported that the culture filtrate of *Curvularia lunata, Fusarium moniliforme, Aspergillus* species and *Cladosporium* inhibited the seed germination, root and shoot elongation. *Rhizopus oryzae,* which shows very much effective on Soybean, is also known to affect adversely the germination of a number of oilseeds (Chandra *et. al.,* 1985). The observation on seed germination and seedling vigour clearly indicates that the culture filtrate of different seed mycoflora under investigation invariably caused inhibition in germination to varying degree and also seedling vigour was also poor as compared to those of control. The fungus affected the seed germination directly either by lowering the viability of the seed, by making it nutritionally poor, or by screening the mycotoxic substances unfavourable to the seed. Even when the seeds germinate, the toxin substance produced by the fungus may affect the seedlings resulting in stunted growth and thus the seedling vigour is reduced.

Plant Extracts

Efficacy of plant extracts on the production of inhibition zone against *Alternaria alternata, Aspergillus flavus* and *Aspergillus niger* revealed that *Calotrophis gigantia* and Vinca rosea were more effective. It has been observed that *Aspergillus flavus* was inhibited maximum by *C. gigantia* and minimum by *Nerium odorum* While Alternaria alternata showed maximum inhibition against *Vinca rosea* and minimum against *Vitex negundo*. For *Aspergillus niger, Vinca rosea* showed maximum inhibition followed by *Calotropis gigantia* and *Nerium odorum*. Least inhibition zone was obtained by *Carica papaya* against *Aspergillus niger.*

The suppression of pathogen growth by plant extract treatment may be attributed to the activity of antimicrobial compounds present in plants. The synergistic action of antifungal compounds may also lead to effective inhibition in the growth of the pathogen. The action of different antifungal compounds present in the extract might destroy the germ tube formation or disintegrate the hyphal elongation (Raju 2003).

Avdhesh and Satapathy (1977) found the leaf extract of *Vinca rosea* were more antifungal against *Helminthosporium nodulosum, Sclerotium rolfsii, Pestalotia* sp., *Fusarium oxysporum, Colletotrium* sp., and *Aspergillus niger*. The extracts of *A. marmelos,P. pinnata* and *Vitex negundo* were effective against *F. moniliforme* (Menna and Mariappan, 1993). Thiribuvanamala and Narasimhan (1998) found that the extracts of *A. marmelos, P.glabra, O.sanctum* and *V. negundo* effectively inhibited growth of *A,helianthi, M. phaseolina* and *F. solani*.

Conclusions

The present investigation showed that the mycoflora is very rich and represented by various types of fungi on seeds. This is because of unscientific storage conditions adversely affect the preservation of oilseeds under storage conditions. In the present investigation various saprophytic as well pathogenic fungi were isolated, but there are many more fungi, which are hidden underneath the ecological niches to be surveyed. Our dedicated approach is essential in studying them, classifying them, testing their pathogenicity and controlling them

So far the survey on seed mycoflora has been more or less neglected from the present study area. A very few reports of these fungi have been reported from this part of the country. In the present work extensive collections of the species of three groups *viz;* Zygomycotina, Ascomycotina and Mitosporic fungi were made. As such this is more or less pioneering work of this part of the state.

References

Agarwal, V. K., Sinclair, J. B. 1987. Principles of seed pathology Vol. I. CRC Press. New Delhi. pp. 1-2.

Ainsworth, G. C. and Bisby's. 1971. Dictionary of the fungi – 6th ed CMI, Kew, Surrey.

Ainsworth, G. C. and Sussman, A. S. 1965. The fungi: An advanced treatise; Vol. 1, Academic Press, New York and London.

Ainsworth, G. C. and Sussman, A. S. 1966. The fungi, Vol-II, Academic Press, New York.

Ainsworth, G. C. and Sussman, A. S. 1968. *The fungi,* Vol-III, Academic Press, New York.

Ainsworth, G. C., Sparrow, F. K. and Sussman, A. S. 1973. The fungi an advanced treatise, Vol. IV A. A taxonomic reviews with keys; Ascomycetes and fungi imperfectii, Academic Press. 621pp.

Bale, M. S. and Khare, M. N. 1982. Seed-borne fungi of sorghum in Madhya Pradesh and their significance. Indian Phytopath., 35: 676-678.

Barnett, H. L. 1955. Illustrated genera of Imperfect fungi, Burges Publishing Company, 2nd ed., p. 221.

Bessey, E. A. 1950. Morphology and Taxonomy of fungi. The Blakiston Company, Toronto, p. 757.

Bilgrami, K. S., Jamaluddin, S. and Rizwi, M. A. 1979. Fungi of India, Part I, Today and Tomorrow's Printer and Publishers, New Delhi, p. 467.

Bilgrami, K. S., Jamaluddin, S. and Rizwi, M. A. 1981. Fungi of India, Part II, Today and Tomorrow's Printer and Publishers, New Delhi, p. 268.

Bilgrami, K. S., Jamaluddin, S. and Rizwi, M. A. 1991. Fungi of India, Part III, Today and Tomorrow's Printer and Publishers, New Delhi, p. 798.

Bilgrami, K. S., Prasad, T. and Sinha, R. K. 1979. Changes in nutritional components of stored seeds due to fungal associations. Today and Tomorrow's Printers and Publishers, New Delhi. p. 59.

Booth, C. 1971. Methods in Microbiology Vol. 4, Academic Press, New York, pp. 795.

Burnett, J. H. 1976. Fundamentals of Mycology. 2nd ed, Edward Arnold, p. 665.

Chandra, S. Narang, M. and Srivastava, R. K. 1985. Studies on mycoflora of oilseeds in India mycoflora in relation to pre and post emergence mortality. Seed Sci. and Technol., 13: 537- 541.

Dhingra, O. D. and Sinclair, J. B. 1986. Basic plant pathology methods. CRC Press Inc. Bola Raton, Florida. pp. 250.

Dickinson, C. H. and Peece, T. F. 1976. Microbiology of Aerial plant surfaces. Academic Press, New york and London, p. 659.

Ellis, M. B. 1971. Dematiaceous Hyphomycetes. CMI. Publn., pp. 608.

Ellis, M. B. 1976. More Dematiaceous Hyphomycetes. CMI. Publn., pp. 507.

Ellis, M. B. and Ellis, J. P. 1985. Microfungi on land plants. Croomttelm Ltd., Beckenham, p. 818.

Funder, S. 1961. Practical Mycology: Manual for identification of fungi. A. W. Broggers Boktrykkeri A/ S, Oslo-Norway, p. 139.

Garcia, J. L., Diaz, C. H. and Gonzales, L. A. 1993. Major pathogens observed on soybean seeds. Rev. Plant Pathol., 72: 872.

Garraway, M. O. and Evans, R. C. 1984. *Fungal Nutrition and Physiology*, Wiley, New York, pp. 385.

Gilman, G. C. 1967. A manual of soil fungi. Oxford and IBH Publishing Co., New Delhi, p. 441.

Hussain, S., Hassan, S. and Khan, B. A. 1989. Seed-borne mycoflora of soybean in NWFP Pakistan Sarhad. J. Agri., 5: 421-424.

Ilyas, M. B., Dhingra, L. D., Ellis, M. A. and Sinclair, J. B. 1975. Location of mycelium of *Diaporthe phaseplorum* var.sojae and *Cercospora kikuchii* in infected soybean seed. Plant Dis. Rep., 59: 17.

ISTA. 1999. International Seed Testing Rules, International Seed Testing Association. Seed Sci. and Technol., 13: 299-355.

Jordan, E. G., Manandhar, J. B., Thapliyal, P. N. and Sinclair, J. B. 1986. Factors affecting in soybean seed quality in Illinois. Plant Dis., 70: 246-248.

Karmakar. S and Subuddhiyopaphya, S. B. B. 1982. Studies on seed-borne mycoflora. Indian J. Microbial., 20: 236-238.

Kashinath Bhatttacharya and Subrata Raha 2002. Deteriorative changes of maize, groundnut and soybean seeds by fungi in storage. Mycopathologia, 155: 135-141.

Kendrick, B. 1971. Taxonomy of fungi imperfectii. Univ. of Toronto Press, Toronto, p. 307.

Mathur, S. B. and Paul Neergaard, 1970. Seed health testing of rice II. Seed-borne of fungi of rice in Philippines, India, Portugal and Egypt. Investigations of *Trichonis padwickii*. Proc. I. Int. Symp. Plant. Path., 1967: 69-81.

Mehrotra, R. S., Neera Rashmi and Navneet. 1992. Seed mycoflora of oilseed crops with special reference to mycotoxin production. Current concepts in Seed Biology (Prof. S.N. Dixit Festschrift Volume) Edited by K.G. Mukerji Naya Prakash. Publisher. 206, Bidhan Sarani, Calcutta. 700 006. India pp 171-199.

Mridula Roy, and Srivastava, H. P. 1991. Toxic effect of the culture filtrate of *Penicillium citrinum* on moth bean seeds. Proc. Nat. Acad. Sci. India, 61(B) III: 357-361

Narain, U. and Swarup, J. 1989. Studies on *Alternaria* spp. associated with oilseed crops.in U. P. Indian Phytopath., 42: 317.

Neergaard, P. and Saad, A. 1962. Seed health testing of Rice. A contribution to development of laboratory routine testing methods. Indian Phytopathol., 15: 85-111.

Nik, W. Z. 1983. Seed-borne fungi of soybean and their control. Pertanica, 3: 125-132.

Preece, T. F. and Dickinson, C. H. 1971. Ecology of leaf surface Microorganisms. International Symposium, Academic Press, New York, p. 631.

Raju, M. 2003. Studies on Biological control of *Alternaria* Blight of sunflower Ph.D. Thesis. Osmania University, Hyderabad.

Saxena, R. M. 1985. Seedling mortality of *Eucalyptus spp.* caused by seed mycoflora. Indian Phytopath., 38(1): 151-152.

Siddaramaiah, A. L., Bhat, R. P. and Desai, S. A. 1980. Mycoflora of a cultivar of *Dolichos lablab.* Curr.Res., 9: 24.

Singh S. D and Swami S. D. 2004. Pathogenic potential of seed mycoflora of Pear Millet (*Pennisetum glaucum* (L.) R. Br). J. Mycol. Pl. Pathol. Vol. 34. No.1.

Singh, D. P. 1991. Seed-borne diseases of soybean and their control. Seeds and Farms, 17: 12-14.

Singh, S. D. 2004. Effect of fungal metabolites on seed germination and seedling vigour of Radish. J. Mycol. Pl. Pthol., 34, (1): 127

Smith, J. E. and Berry, D. R. 1975. The filamentous fungi. -l. I, Industrial Mycology. pp. 340.

Steven, R. B. 1974. Mycology guide book. University of Washington Press, p. 705.

Subbaraja, K. J. 1973. Sorghum News letter 16: 37.

Subramanian, C. V. 1971. Hyphomycetes. ICAR, New Delhi, pp. 903.

Tandon, R. N. 1968. Mucorales of India. ICAR Publication, New Delhi, pp.120.

Thiribhuvanamala G. and Narasimhan. 1998. Efficacy of plant extract on seed-borne pathogens of sunflower. Madras Agricultural Journal, 85(5,6): 227-236.

Wicklow, D. T. and Carroll, G. C. 1981. Fungal Community: its organization and role in the ecosystem. Marcel Dekker: New York and Basel.

Zad, 1982. Mycoflora of Soybean seeds. Review of Plant Pathol., 61: 40.

Microbes: Diversity and Biotechnology (2012) *Pages* **269–276**
Editors: **Prof. S.C. Sati & Dr. M. Belwal**
Published by: **DAYA PUBLISHING HOUSE, NEW DELHI**

Chapter 16

Studies on the Fungal Flora in Rhizosphere of Pea Plant in Subtropical Region

***B.S. Bhandari*[1], *S. Saxena*[2] *and S. Guleri*[1]**

[1]*Department of Botany, Birla College Campus, H N B Garhwal University, Srinagar – 246 174, Uttarakhand*

[2]*Department of Botany, Shri Guru Ram Rai Post Graduate College, Dehradun – 248 001, Uttarakhand*

ABSTRACT

Quantitative and qualitative studies on the rhizosphere mycoflora of pea plant (*Pisum sativum* L.) were carried out in agricultural crop of garden pea in Doon valley of Uttarakhand. Almost all samples showed a sandy to sandy-loam texture. pH of the soil samples ranged from 7 to 10 and water content ranged from 8 to 9 per cent. 18 species of fungi belonging to 10 genera were isolated from the rhizosphere soil of Pea by direct plating method, whereas, 14 fungal species belonging to 8 genera were isolated by serial dilution. Results showed that direct plating isolated the greatest number of fungi. Deuteromycota dominated the rhizosphere mycoflora followed by Zygomycota. Among these *Alternaria fasciculata, Sclerotium rolfsii, Fusarium oxysporum* were common in all the sampling sites.

Keywords: Rhizosphere soil, Mycoflora, Pea, Doon valley.

Introduction

Fungi are known to colonise, multiply and survive in diverse habitats and are cosmopolitan in distribution covering tropics to pole and mountains to deep oceans (Manoharachary *et al.*, 2005). Several fungal species inhabit the soil and exist in both the mycelial and spore stage. The mycelia of the fungi penetrate through soil forming a network which entangles the small soil particles (Powar and Daginawala, 1986). The quality and quantity of organic matter present in the soil have a direct bearing on fungal members in soil since most fungi are heterotrophic in nutrition.

Rhizosphere is the region of the soil subject to to the influence of the plant roots and is characterized by the greater microbial activity determined by the distance to which root exudations from the root system can migrate (Subba Rao, 1999). The area of the soil influenced by root varies with the type of plant, age of the plant, soil conditions, pH, moisture content and other environmental conditions of the soil. The work particularly on soil fungi is sporadic than surface fungi. Among the important contributors, include the work of Marschner *et al.* (2003), Dong *et al.* (2004), and Yu *et al.* (2007). Within the Indian reference, the recent researches particularly related to soil fungi include the work of Ananda and Sridhar (2002), Babu and Manoharachary (2003), Kushwaha and Gupta (2004), Raviraja *et al.* (2006), Satish *et al.* (2007) and Saravanakumar and Kaviyarasan (2010). An intensive survey of literature unrevealed that there is virtually no information available on the soil mycoflora of Uttarakhand in general and Doon Valley in particular. Recently attempt has been made by Guleri *et al.* (2010) with some new and interesting records. Therefore, the present investigation is aimed at providing some preliminary information on rhizosphere mycoflora of some Pea (*Pisum sativum*) crops of Doon Valley of Uttarakhand.

Materials and Methods

Study Area

The study was conduct in the Doon Valley during winter season (2008). The area lies between 30°00′ to 30°35′ N latitude and 77°45′ to 78°15′ E longitude. Pea crops of four sites *viz.*, Bhauwala, Bahadarpur, Selaqui and Doiwala were analyzed to explore the rhizosphere mycoflora of the Doon Valley.

Soil Sampling

Rhizosphere soil of pea plants was collected from different agricultural areas and there were 3 replicates of each rhizosphere soil. In the collection of soil samples, first a soil profile was extracted and then the surface of the profile was cleaned (Brown, 1958). Vertical samples were taken from 0-5cm, 5-10cm, 10-15cm and 15-20cm depths with a disinfected spatula. The spatula was applied perpendicular to the vertical surface of the profile. Samples were kept in sterile polythene bags in a refrigerator.

Soil Analysis

The soil texture was determined by wet sieving technique (Barbour *et al.*, 1980). Soil reaction (pH) was determined using a pH meter with a glass electrode in mixed soil water 1:2 ratio and by electrometric method, following Brady (1990). Moisture content of soil samples was calculated by oven drying the soil and determining the weight loss (Garrett, 1963).

Isolation of Fungal Flora

With the direct plate method, soil samples (0.01g) were dispersed in 1 ml of sterile distilled water in a sterilized petri dish and then approximately 10 ml of molten, cooled sterile agar was added and mixed. The soil particles were distributed throughout the medium by rotating the petri dish. The petri dishes were incubated for 5-7 days at 25 ± 2°C (Warcup, 1950). For serial dilution, soil samples (2.0g) were suspended in 18 ml of sterilized water, which gave a dilution of 1:10. Serial dilutions of 1:100, 1:1000, and 1:10,000 were prepared, and then a 1-ml aliquot from the 1:1000 dilutions was added to a petri dish containing penicillin (20,000 units/l) and streptomycin (200 μg/l). Then, approximately 10 ml of sterile Czapek's Dox agar medium was added to each dish. Each dilution was replicated 3 times and the dishes were incubated for 5-7 days at 25 ± 2 C. The number of colonies produced by a fungus was multiplied by the dilution factor to obtain the total number of propagules/g of soil (Waksman and Fred. 1922). The fungi growing on plates were identified using standard literature (Gilman, 1966; Moubasher, 1993; Barnet, 1967; Domsch *et al.*, 1980, 1991; Ellis, 1971; Raper and Fennell, 1965).

Results and Discussion

Physical and Chemical Properties

Physically, the texture of all rhizosphere soil samples was sandy loam except that of Doiwala, which was sandy. Sandy soils have low porosity. Water moves into and drains out of sandy soil with much greater ease than with a fine textured soil, which means that sandy soils are more permeable than fine textured soils. Such as nutritionally rich, but can be agriculturally problematic due to low permeability and aeration (Barbour *et al.*, 1980). The pH of the sites ranged from 7 to 10 and the water content of the soil samples ranged between 8 and 9 per cent (Table 16.1).

Table 16.1: Physical and Chemical Properties of Rhizosphere Soil Samples

Samples Station	*pH*	*Water Content (per cent)*	*Texture*
Bhauwala	7.89- 10.02	8.22-9.01	Sandy-sandy loam
Bahadarpur	9.82	8.21	Sandy loam
Selaqui	7.89	9.11	Sandy loam
Doiwala	7.85	9.7	Sandy

Isolation of Fungi from Rhizosphere Soil of Pea Plants by Different Techniques

By Direct plating, 18 species of fungi belonging to 10 genera were isolated from the rhizosphere soil of pea plant at 4 different localities (Table 16.2). Bhauwala had 13 species and 8 genera; Bahadarpur had 12 species and 9 genera, Selaqui had 14 species and 9 genera, and Doiwala had 13 species and 9 genera (Figure 16.1) *viz. Curvularia lunata* (Wakk.) Boed.; *Fusarium solani* (Mart.) App. and Woll; *F. poae*

Table 16.2: Rhizosphere Soil Fungi from Pea Plants by Direct Plating

Names of Fungi	*Locations*			
	Bhauwala	*Bahadarpur*	*Selaqui*	*Doiwala*
	Colony No.	*Colony No.*	*Colony No.*	*Colony No.*
Curvularia lunata (Wakk.) Boed.	2	0	3	8
Fusarium solani (Mart.) App. and Woll	25	9	8	4
F. poae (Peck) Wollenweber	3	2	6	0
F. oxysporum Schl.	0	4	0	1
F. sp.	1	0	1	0
Cladosporium sp.	4	3	2	5
Alternaria fasciculata Cooke and Ellis	0	2	0	1
A. sp1	1	0	1	0

Contd...

Table 16.2–Contd...

Names of Fungi	*Locations*			
	Bhauwala	*Bahadarpur*	*Selaqui*	*Doiwala*
	Colony No.	*Colony No.*	*Colony No.*	*Colony No.*
A. sp2	0	1	0	1
Rhizopus arrhizus Fischer	26	30	55	36
*R.*sp	15	0	8	20
Helminthosporium nodulosum (Berkeley and Curtis) Saccardo	3	12	45	0
H.anomalum Gilman and Abbott	2	0	1	3
Aspergillus ustus (Bainier) Thom and Church	1	4	8	1
A. Niger van Tieghem	8	0	4	4
Humicola sp.	0	6	0	13
Mucor sp.	15	44	8	0
Pythium sp.	0	3	20	7

Indications: sp: Species.

Table 16.3: Number of Propagules/g of Soil Based on the Serial Dilution Technique

Names of Fungi	*Locations*			
	Bhauwala	*Bahadarpur*	*Selaqui*	*Doiwala*
Alternaria alternata (fr.) Keissler	2000	0	3000	8000
A.fasciculata Cooke and Ellis	250	90	80	40
A.tenuissima (Nees ex Fr.)	3000	2000	6000	0
A. sp-1	0	400	0	100
A.sp-2	300	0	200	0
Sclerotium rolfsii Saccardo	400	300	200	500
Curvularia tetramera (Mc Kinney) Boedijn ex Gilman	0	2000	0	1000
Macrosporium cladosporioides Desmazieres	1500	0	2300	0
Aspergillus niger van Tieghem	0	1100	0	300
Fusarium oxysporum Schl.	260	300	550	360
Rhizoctonia solani Kuhn	150	0	860	200
Rhizoctonia sp	300	120	450	0
Mucor racemosus Fresenius	2000	0	0	1000
Mucor sp	1,500	4,400	800	0

Indications: sp: Species.

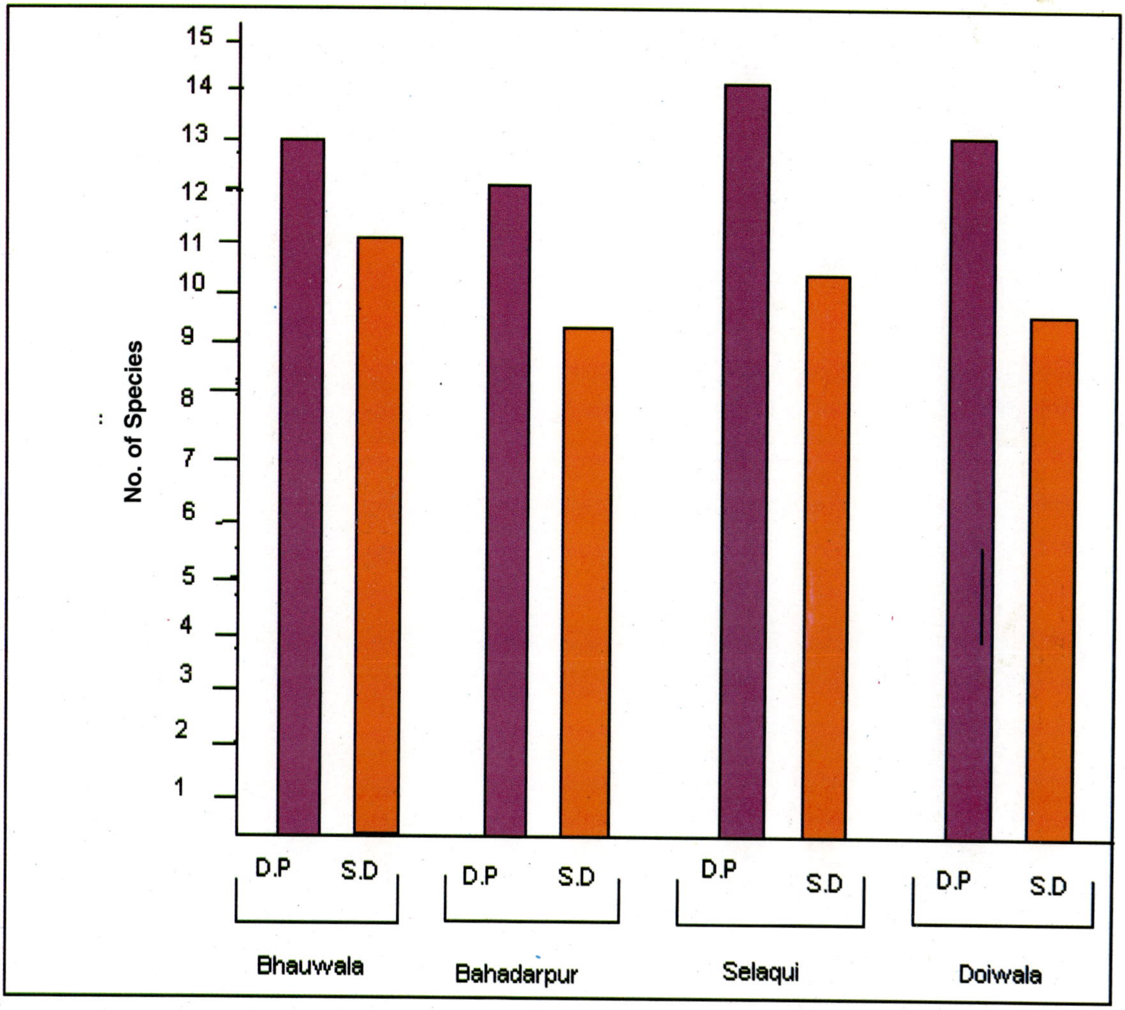

Figure 16.1: Number of Fungal Species Found in Different Areas of Rhizosphere Soil of Pea Plant Indications: D.P: Direct Plating; S.d: Serial dilution

(Peck) Wollenweber; *F. oxysporum* Schl.; *F. sp.*; *Cladosporium* sp.; *Alternaria fasciculata* Cooke and Ellis; A.sp1; A.sp2; *Rhizopus arrhizus* Fischer; R. sp.; *Helminthosporium nodulosum* (Berkeley and Curtis) Saccardo; *H.anomalum* Gilman and Abbott; *Aspergillus ustus* (Bainier) Thom and Church; *A.niger* van Tieghem; *Humicola* sp.; *Pythium* sp. Among these, *F.solani, Cladosporium* sp., *R. arrhizus* and *A. ustus* were observed in all the localities of pea plant. Selaqui yielded the highest number of fungi, followed by Doiwala and Bhauwala, whereas the least number of fungi was observed in Bahadarpur.

With the serial dilution method, 14 fungal species belonging to 8 genera were isolated (Table 16.3). Among these *Alternaria alternata* (fr.) Keissler; *A.fasciculata* Cooke and Ellis; *A.tenuissima* (Nees ex Fr.); A.sp-1; A.sp2; *Sclerotium rolfsii* Saccardo; *Curvularia tetramera* (Mc Kinney) Boedijn ex Gilman;

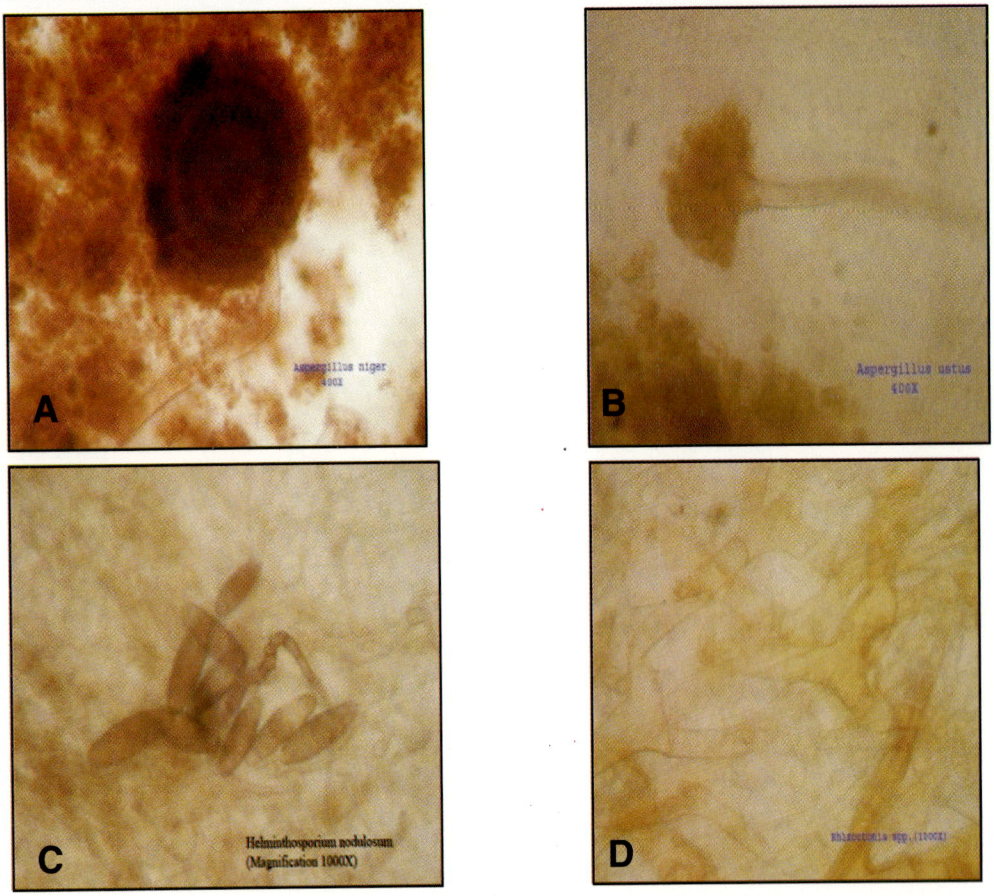

Plate 16.1: (A) *Aspergillus niger* (400x magnification); (B) *Aspergillus ustus* (400x); (C) *Helminthosporium nodulosum* (1000x); (D) *Rhizoctonia* sp. (1000x)

Macrosporium cladosporioides Desmazieres; *Aspergillus niger* van Tieghem; *Fusarium oxysporum* Schl.; *Rhizoctonia solani* Kuhn; *Rhizoctonia* sp.; *Mucor racemosus* Fresenius; *Mucor* sp. were observed (Table 16.3). *A. fasciculata*, *S. rolfsii* and *F. oxysporum* were common in the rhizosphere soil of all the 4 localities of pea plants sampled. The authors observed that Deuteromycota was the most common and dominant group, although members of Zygomycota were also recorded. The present results showed that the greatest number of fungi was isolated by direct plating.

Acknowledgements

The authors (SS and SG) are grateful to the University Grants Commission New Delhi for providing financial support. Thanks are also due to Principal and Head Department of Botany S.G.R.R. (P.G) College for providing the necessary Laboratory facilities.

References

Ananda, K. and Sridhar, K. R. 2002. Diversity of endophytic fungi in the roots of mangrove species on west coast of India. *Can. J. Microbiol*. 48: 871-878.

Babu, K.S. and Manoharachary, C. 2003. Occurrence of Arbuscular mycorrhizal fungi in rhizosphere soils of some medicinal plants. *Indian Phytopath.* 56(2): 223-227.

Barbour, M. G., Bark, J. H. and Pitts, W. D. 1980. Terrestrial Plant Ecology Meulo Park, California.

Barnet, H. L. 1967. Illustrated Genera of Imperfect Fungi, Princeton University Press, Princeton, N J.

Brady, N. D. 1990. The Nature and Properties of Soils. 10th ed. Macmillan pub. Company. New York, USA.

Brown, J. C. 1958. Soil fungi of some British sand dunes in relation to soil type and succession. Ecology, 46: 641-664.

Domsch, K. A., Gams, W. and Anderson, T. H. 1980. Compendium of Soil Fungi (vol.1), Academic Press, London.

Domsch, K. A and Gams, W. 1991. Fungi in Agricultural soils, Longman, New York. Vol.1. pp. 75-125.

Dong, A.R., Lv GZ, Wu QY, Song, R.Q., Song, F.Q. 2004. Diversity of soil fungi in Liangshui natural reserve, Xiaoxinganling forest region. *J. Northeast Forestry University.* 32(1): 8- 10.

Ellis, M. B. 1971. Dematiaceous Hyphomycetes. Kew Surrey UK: Commonwealth Mycological Institute 608p.

Garrett, S. D. 1963. Soil Fungi and Soil Fertility. Oxford, Pergamon Press.

Gilman, J. C. 1966. A Manual of Soil Fungi. The Iowa State University Press, Ames, Iowa.

Guleri, S., Bhandari, B.S. and Saxena, S. 2010. Ecology of Rhizosphere and Non- Rhizosphere soil mycoflora of forest soils of Dehradun district, Uttarakhand. *Int. Trans. Appl. Sci.* 2(1): 69-77.

Kushwaha, R.K.S. and Gupta, M. 2004. Diversity of Keratinophilic fungi in soil and on birds. In: Jaio, P.C. (eds.) *Micobiology and Biotechnology for Sustainable Development* CBS Publishers, New Delhi, 59-70.

Manoharachary, C., Sridhar, K., Singh, R., Adholeya, A., Suryanarayanan, T.S., Rawat, S. and Johri, B.N. 2005. Fungal biodiversity: Distribution, Conservation and prospecting of fungi from India. *Current Science* 89(1): 58-71.

Marschner, P., Kandeler, E., Marschner, B. 2003. Structure and function of the soil microbial community in a long-term fertilizer experiment. *Soil Biol. Biochem.* 35: 453-461.

Moubasher, A. H. 1993. Soil Fungi in Qatar and Other Arab Countries, Doha University of Qatar, Center for Scientific and Applied Research, 566p.

Powar, C.B. and Daginawala, H.F. 1986. *General Microbiology.* Vol II. 2nd Ed. Himalaya Publishing House, Bombay.

Raper, K. B and Fennell, D. L 1965. The Genus Aspergillus. Baltimore: Williams and Wilkins Company. 685p.

Raviraja, N.S., Maria, G.L. and Sridhar, K.R. 2006. Antimicrobial evaluation of endophytic fungi inhabiting medicinal plants of the Western Ghats of India. *Eng. Life Sci.* 6: 515-520.

Saravanakumar, K. and Kaviyarasan, V. 2010. Seasonal distribution of soil fungi and chemical properties of montane wet temperate forest types of Tamil Nadu. *African Journal of Plant Sciences.* 4(6): 190-196.

Satish, N., Sultanas, S., Nanjundiah, V. 2007. Diversity of soil fungi in a tropical deciduous forest in Madumalai, Southern India, *Curr. Sci.* 93: 669-677.

Subba Rao, N.S. 1999. *Soil Microbiology*. 4th Ed., of Soil microorganisms and Plant Growth, Oxford & IBH Publ., Co., Pvt. Ltd., New Delhi, pp. 407.

Waksman, S.A and Fred, E. B. 1922. A tentative outline of plates method for determining the number of microorganisms in soil. Soil Sci., 14: 27-28.

Warcup, J. H. 1950. The soil plate method for isolation of fungi from soil. Nature. Lond. 178:1477.

Yu, C., Lv, D.G., Qin, S.J., Du, G.D., Liu, G.C. 2007. Microbial flora in *Cerasus sachalinensis* rhizosphere. *Chinese. J. Appl. Ecol.*, 18(10): 2277-2281.

Microbes: Diversity and Biotechnology (2012) *Pages* **277–297**
Editors: **Prof. S.C. Sati & Dr. M. Belwal**
Published by: **DAYA PUBLISHING HOUSE, NEW DELHI**

Chapter 17

Dynamics of Leaf Surface Mycoflora of *Ginkgo biloba* and *Taxus baccata* in Kumaun Himalaya

Manju Lata Upadhyaya and R.C. Gupta
Department of Botany, Kumaun University, S.S.J. Campus, Almora - 263 601, Uttarakhand

ABSTRACT

The dynamics of leaf surface mycoflora of *Ginkgo biloba* and *Taxus baccata* in Kumaun Himalaya was studied. The result that of leaf surface mycoflora during plant growth senescence and decay is of great interest. The importance for possible antagonistic action and in decomposition studies on this aspect contribute a lot about the interaction between the foliilcolous pathogens and non-parasitic pathogens onto the leaf surface of *Ginkgo biloba* and *Taxus baccata*.

Keywords: *Mycoflora, Ginkgo biloba, Taxus baccata, Foliilcolous pathogens*

Introduction

The study of leaf surface mycoflora during plant-growth, senescence and decay is of great importance because of its importance for possible antagonistic action against plant pathogens, source for air-borne spores, toxic properties for cattle, role in biological nitrogen fixation and decomposition. The leaf surface mycoflora is under the great influence of host, environment, age of leaf, its location and subjected to its self stimulatory influences also. The investigation in the field of leaf surface mycoflora dates back to the researches of Potter (1910), however, Last (1955) suggested the term 'phyllosphere' to describe leaf surface environment, while Kerling (1958) restricted the use of the term 'phyllosphere' to leaf surface environment and phylloplane to actual leaf surface.

The leaf surface is an important substrate for the growth of micro-organisms as it easily provides essential nutrients required for their life and growth. The nutrients of micro-organisms are available on the surface of living green leaves in the form of organic as well as inorganic substances. Phylloplane supports a rich but varied population of pathogenic as well as non-pathogenic micro-organisms. The colonization of leaf surface micro fungi occurs through landing of air mycoflora and it is quite significant in substrate relationships and disease development (Gregory, 1961).

Plant pathologists are aware that susceptibility or resistance of plants to disease may in part be governed by the conditions prevailing on the leaf surface prior to infection. A definite proportion of mycoflora harbouring the leaf surface of plants and their interference with disease severity have been well documented (Sharma and Mukherji, 1973 and Wildman and Parkinson 1979). The surface of living green leaves, the phylloplane is known to be colonized by a colorful complex array of parasitic and non-parasitic micro-organisms chiefly being fungi and bacteria (Preece and Dickinson, 1971; Dickinson and Preece, 1976, Blakeman, 1972). Increase in total number of fungi in ageing leaves may possibly be attributed to increase in surface area for colonization with a corresponding increase in amount of nutrients on the leaf surface of ageing leaves (Garg and Sharma, 1985; Singh, 1978).

Studies on this aspect have made significant contribution to our knowledge of ecology of fungi colonizing above ground plant parts which are able to grow and multiply even on green living leaves and such colonizers are mostly amongst the residents of the phylloplane. Air-spora studies over crop fields are useful in understanding dissemination of air-borne plant pathogens because air is the most important medium through which many micro-organisms come in contact with various parts of plants and may be able to cause disease (Singh, 1978). Gregory (1961) also concluded that the number of spores deposited on the leaf surface was directly proportional to the number of spores present in the air.

Ecologists now recognize plants detritus as a major component of ecosystem. As we know that in forest ecosystem litter decomposition is a major pathway for the supply of plant nutrients to the soil, fungi play an important role in decomposition of litter amongst various groups of organisms because they are better equipped with enzymes responsible for efficient decomposition. Besides their importance in decomposition, studies on phylloplane micro-organisms have made notable contribution to our understanding of interaction between the foliicolous pathogens and non-parasitic micro-organisms onto the leaf surface.

However, a large number of workers have studied leaf surface mycoflora of a number of plants but almost no work has been done so far on the leaf surface mycoflora in general in particular of *Taxus* and *Ginkgo*. Although a lot of work has been done on different aspects of both the plants but almost no work has been done so far on leaf surface mycoflora in general of the test plants.

Materials and Methods

Collection of Plant Material

The leaves of *G. biloba* were collected from 'Kalika' situated near Ranikhet and leaves of *T. baccata* were collected from Jageshwar forest situated near Almora, Uttarakhand, India.

Isolation of Leaf Surface Mycoflora

Leaf surface mycoflora of *G. biloba* and *T. baccata* was studied at monthly intervals from September, 05 to August, 06. Leaves were removed with sterilized forceps and scissors, placed in clean sterilized polythene bags and brought to the laboratory. Mycoflora was analyzed by the following methods:

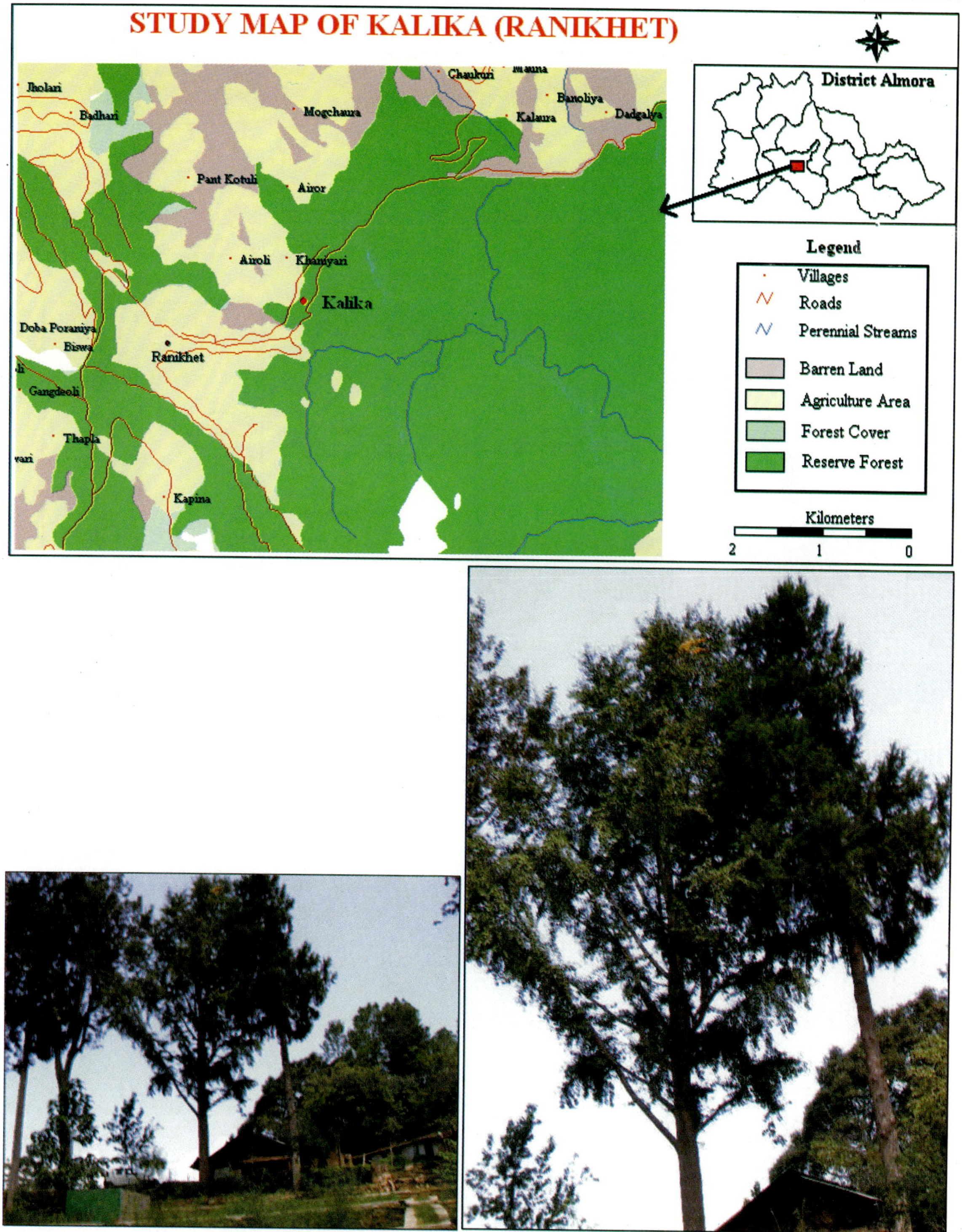

Plate 17.1: The Study Site and Trees of *Ginkgo biloba* and *Taxus baccata*

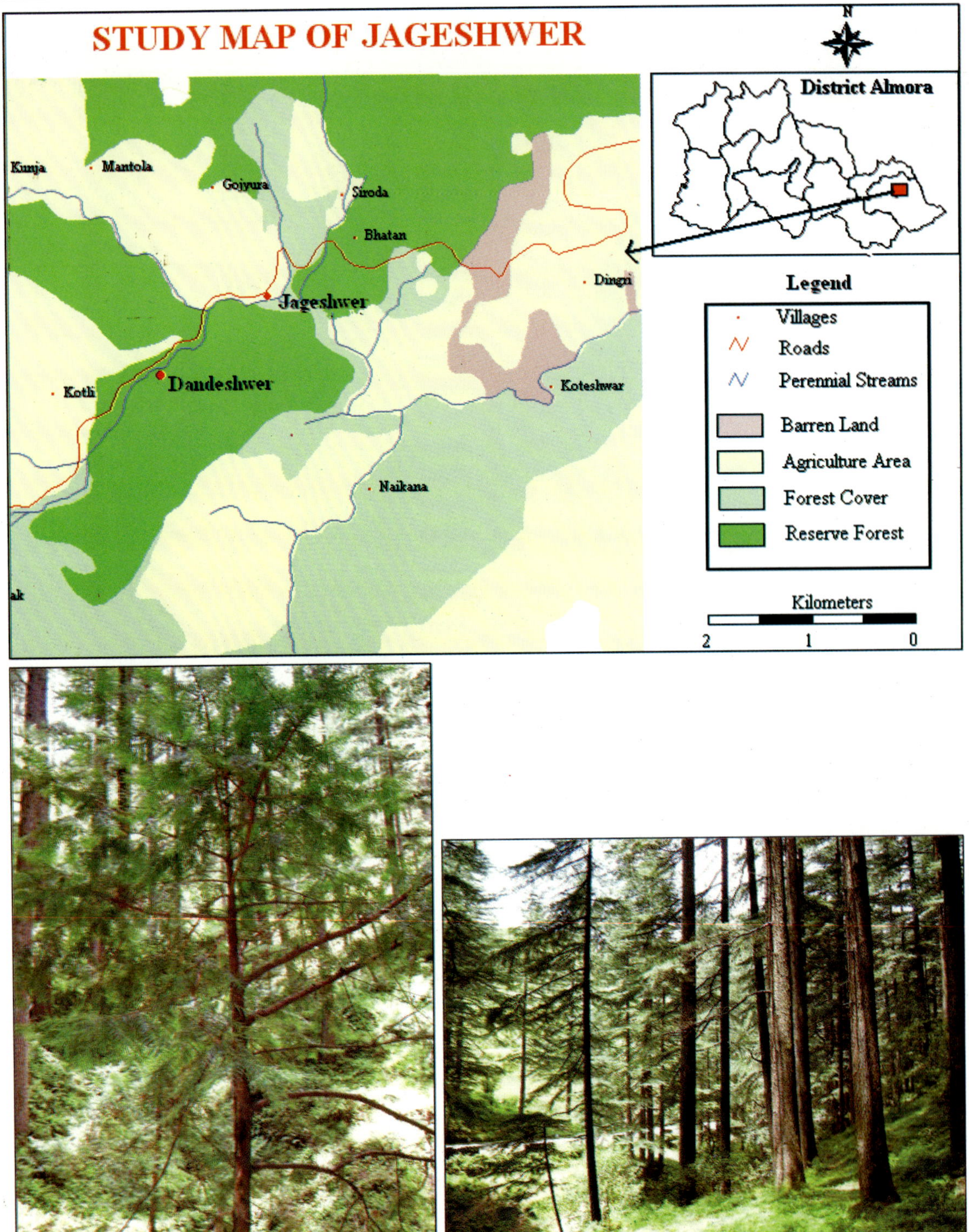

Plate 17.2: The Study Site and Trees of *Taxus baccata*

(A) Dilution Plate Technique (Waksman, 1922)

At each sampling 100 leaf disks (5mm diam.) were cut at random with sterilized cork borer from a number of leaves. The disks were shaken in 100 ml. Erlenmey sterilized water in 250 ml flask for 15-20 minutes with the help of mechanical shaker. The suspension was further diluted by mixing 10 ml original solutions with 90 ml of sterilized distilled water. One ml. suspension was added to each of the 9 cm. sterilized Petri-dish to which approximately 20 ml sterilized and cooled Czapek's agar medium + streptomycin was added aseptically. ($MgSO_4$-0.50g; KH_2Po_4-1.00; $NaNO_3$-2.00g; distilled water-1 liter; pH nearly 5.5; these all are autoclaved at 15 lb pressure/sq. inch for 15 minutes. Streptomycin 30 mg/ml added after autoclaving when temperature of the medium becomes 35°C). Petri-dishes were incubated at 25±1°C for a week. Colonies of individual fungi were identified, counted and results were expressed as fungi/cm^2. The fungi/cm^2 was calculated by applying the following formula:

$$\text{Fungi/cm}^2 = \frac{\text{Total number of fungi in 100 ml}}{\text{Total area of leaf disks}}$$

(*Total area of leaf disks = number of disks × area of both the surfaces of disks*)

(B) Washed Disks Method (Harley and Waid, 1955)

After treating for dilution plate technique the disks were serially washed ten times in successive changes of sterilized distilled water and dried over sterilized filter paper. Five disks were inoculated into each of the ten Petri-dishes containing approximately 20 ml sterilized and cooled Czapek's agar medium, then the Petri-dishes were incubated at 25±1°C for a week and after which fungal colonies were identified and their percentage of occurrence was calculated by applying the following formula:

$$\text{Percent occurrence} = \frac{\text{Occurrence of a species on no. of disks}}{\text{Total number of disks}} \times 100$$

(C) Moist Chamber Method (Keyworth, 1951)

Leaf disks of 5 mm diam were cut from different portions of different leaves with the help of sterilized cork borer. Fifty disks were inoculated in moist chambers by using 10 Petri-dishes containing three layered sterilized blotting paper. Five disks were inoculated into each Petri-dish and then incubated at 25±1°C for 10 days. After which fungi grown were identified and their percentage of occurrence was calculated.

Results

Leaf Surface Mycoflora of *Ginkgo biloba*

Quantitative Variation in Fungi

The results are summarized in Tables 17.1 to 17.4, Figures 17.1a to 17.1e. A total of forty one species were isolated in all samplings by applying a combination of methods amongst which one was Zygomycotina, one was Ascomycotina, Basidiomycotina members were altogether absent and rest 95.12 per cent of total fungal population constituted Deuteromycotina.

Twenty four fungal species were isolated by applying dilution plate technique from which one was Zygomycotina, while Ascomycotina and Basidiomycotina members were absent. One isolate was a bacterium and the rest 95.83 per cent of the total fungal population constituted of Deuteromycotina (Table 17.1). In all fifteen fungal species were isolated by applying washed disks method in all samplings. All isolated fungal species were the members of Deuteromycotina.

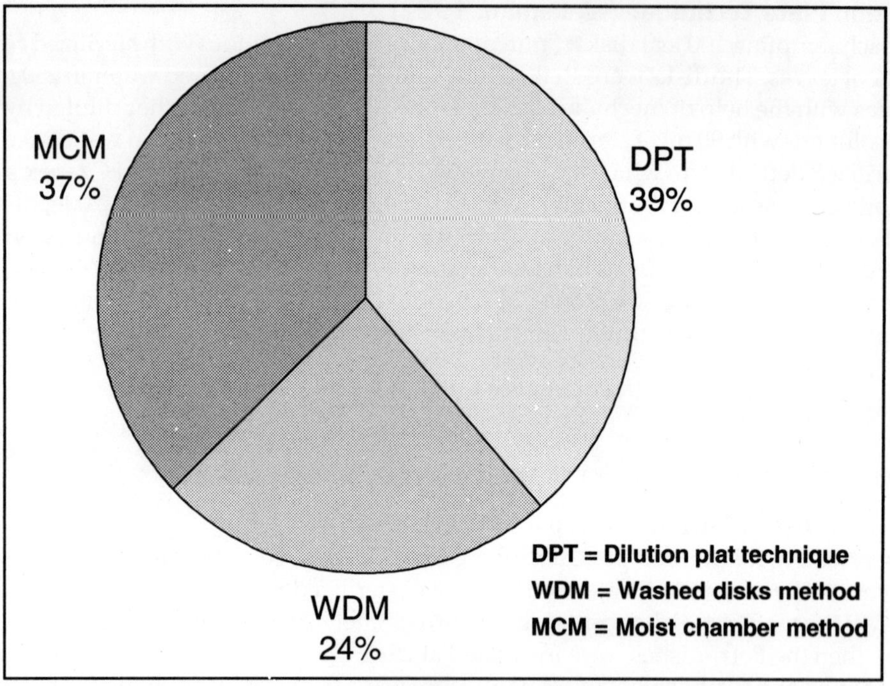

Figure 17.1a: Showing Total Number of Fungal Species Isolated from Leaf Surface of *G. biloba* by Using Various Methods

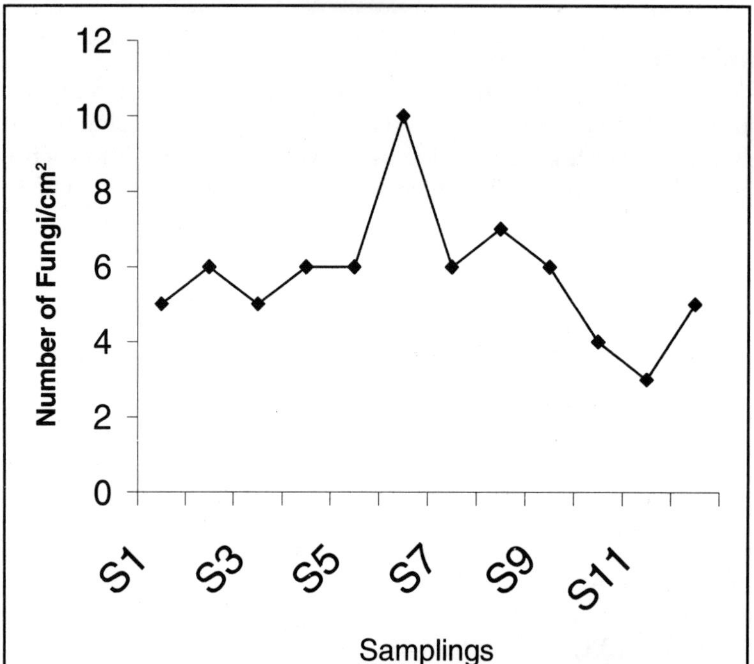

Figure 17.1b: Showing Total Number of Fungi/cm^2 Isolated from Leaf Surface of *G. biloba* at Different Samplings

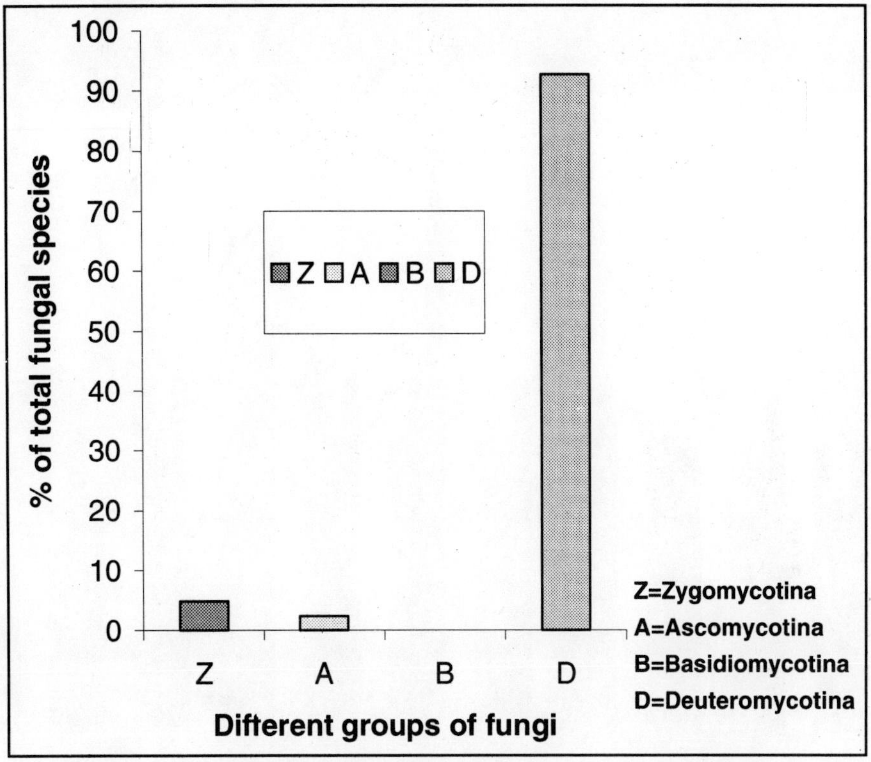

Figure 17.1c: Representation of Different Group of Fungi (*G.biloba*)

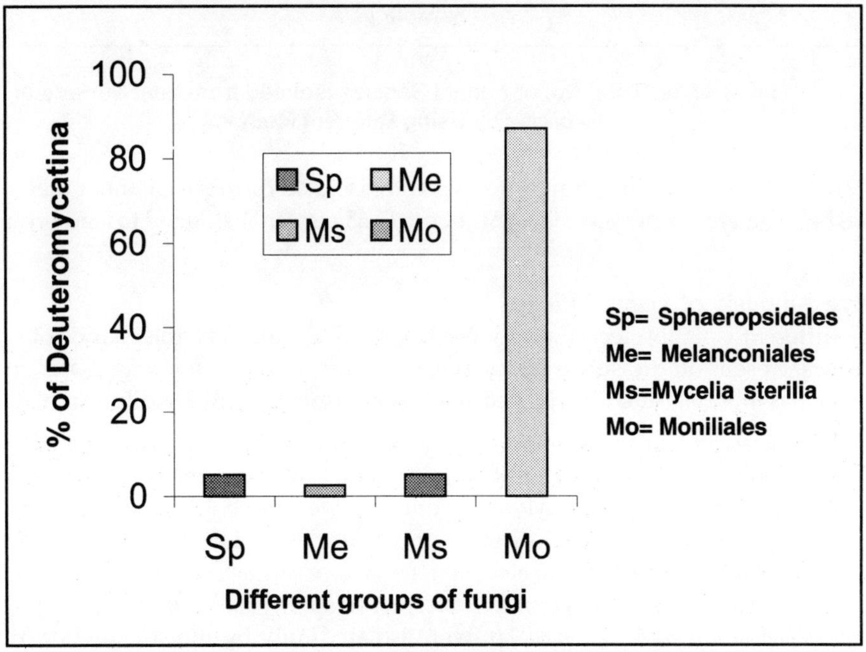

Figure 17.1d: Representation of Different Groups of Fungi (*G. biloba*)

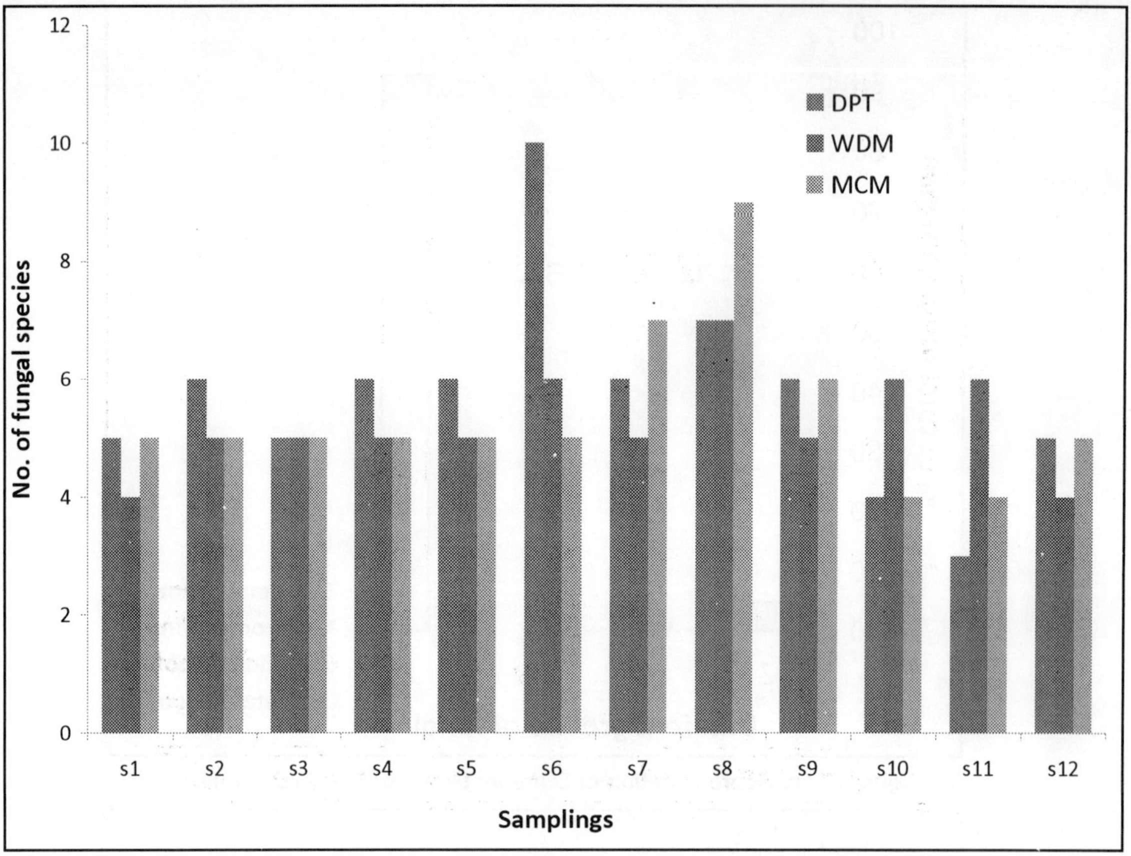

Figure 17.1e: Total No. of Fungal Species Isolated from Leaf Surface of *G. biloba* by Using Different Methods

(Table 17.2) whereas twenty three fungal species were isolated by moist chamber method out of which one species belonged to Ascomycotina and the rest 95.65 per cent belonged to Deuteromycotina (Table 17.3).

Qualitative Analysis of Fungal Flora

Three different isolation techniques were followed to minimize the chance of omission of any fungal species present on the substrate. *Aspergillus niger, Acremonium vitis, Isaria* sp., *Myrothecium roridum, Rhizopus nigricans* and *Trichoderma viride* were isolated only by dilution plate technique.

One bacterial isolate which was tentatively identified as *Bacillus sp.* was isolated by dilution plate technique, was found to have strong antifungal activity towards some fungal species *viz; Alternaria alternata, Fusarium oxysporum* and *Myrothecium roridum* (Plate 17.4). This study needs further investigation. *Gilmaniella* sp. and yellow sterile mycelium were isolated by washed disks methods, whereas *Alternaria cichorii, Chaetomium globosum, Drechslera hawiiensis, Curvularia protuberata, Lacellina* sp., *Memnoniella echinata, Nigrospora oryzae, Pestalotiopsis* sp. *Phoma humicola, Phomopsis* sp., *Septonema* sp. *Pithomyces chartarum* and *Tetraploa* sp. were isolated only by moist chamber method. *Botrytis squamossa, Curvularia protuberata, Lacellina* sp, *Phomopsis* sp., *Penicillium citrinum, Phoma humicola and Ulocladium* sp. were isolated only in one sampling by applying moist chamber method.

Table 171.1: Record of Leaf Surface Mycoflora (Fungi/ cm²) of *Ginkgo biloba*

Fungal Species	*Samplings*											
	S1	*S2*	*S3*	*S4*	*S5*	*S6*	*S7*	*S8*	*S9*	*S10*	*S11*	*S12*
Alternaria alternata	31	20	–	15	–	–	10	15	15	–	–	–
A. solani	–	10	–	–	15	–	–	25	–	–	–	–
A. longipes	–	–	–	25	–	–	–	–	–	–	–	–
A. tenuissima	–	–	–	–	–	–	–	–	–	–	–	–
Aspergillus niger	–	25	41	25	15	5	25	–	–	5	25	15
A. flavus	–	10	20	10	10	31	15	31	31	5	–	–
A. fumigates	–	–	–	–	–	66	–	–	–	–	10	10
Acremonium vitis	–	–	–	–	–	–	–	10	5	–	–	–
Bacillus sp.	5	–	–	–	–	–	–	–	–	–	–	–
Cladosporium sp.	–	–	–	–	–	–	–	–	–	–	–	–
C. herbarum	–	–	–	10	–	–	–	41	31	–	–	–
C. cladosporioides	–	–	20	25	10	10	20	51	41	–	–	–
Fusarium oxysporum	10	–	–	–	–	–	5	–	–	–	–	–
Isaria sp.	–	–	–	–	–	51	–	–	–	–	–	–
Myrothecium roridum	5	–	–	–	–	–	–	–	–	–	–	–
Penicillum citrinum	10	–	41	–	–	–	–	10	15	–	–	15
Penicillum solitum.	–	–	–	–	–	10	–	–	–	–	–	51
Periconia felina	–	10	–	–	10	10	–	–	–	–	–	–
Rhizopus nigricans	–	20	–	–	–	15	–	–	–	81	5	5
Trichoderma harzianum	–	–	15	–	–	–	–	–	–	10	–	–
T. viride	–	–	–	–	–	25	–	–	–	–	–	–
Drechslera holodes	–	–	–	–	–	–	–	–	–	–	–	–
Stachybotrys atra	–	–	–	–	20	20	–	–	–	–	–	–
White sterile mycelium	–	–	–	–	–	–	5	–	–	78	–	52
Total no. of fungal species	5	6	5	6	6	10	6	7	6	4	3	5

Table 17.2: Mycoflora Associated with Leaf Surface of *G. biloba* by Using Washed Disks Method

Fungal Species	*Samplings*											
	S1	*S2*	*S3*	*S4*	*S5*	*S6*	*S7*	*S8*	*S9*	*S10*	*S11*	*S12*
Alternaria alternata	20	30	20	40	30	50	20	50	50	30	30	10
A. tenuissima	–	–	–	20	20	30	–	–	–	–	–	–
A. solani	–	–	10	–	10	–	–	–	–	–	–	–
Aspergillus flavus	–	–	10	–	–	–	–	10	–	–	–	–
A. fumigalus	–	–	–	–	–	–	–	–	–	–	60	50
Cladosporium herbarum	–	20	–	–	–	–	–	–	–	–	–	–
C. cladosporioides	–	10	20	10	–	20	–	20	–	10	10	–

Contd...

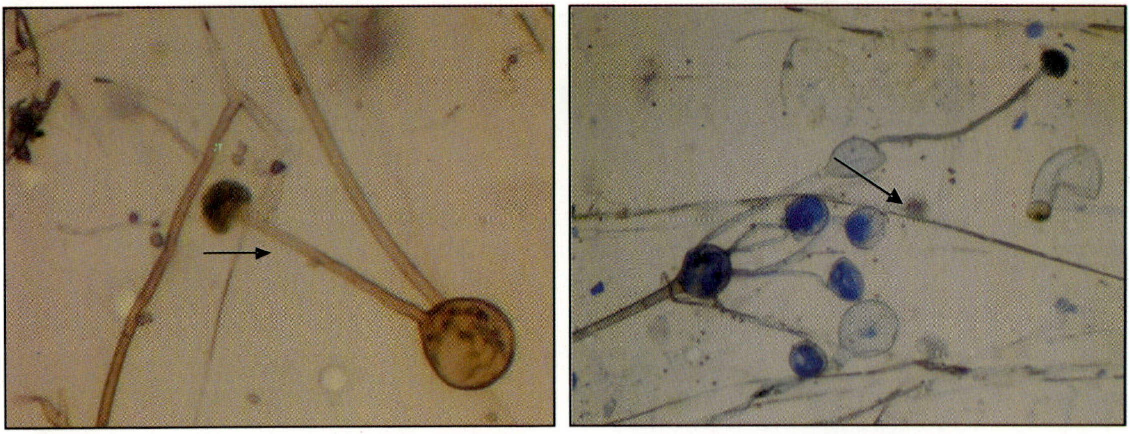

Plate 17.3: *Rhizomucor* Species Showing Proliferation in Sporangium

Plate 17.4: *Ginkgo biloba* Leaf Infected with *Alternaria alternata* and Culture Plates Showing Inhibition Zones

Table 17.2–Contd...

Fungal Species	*Samplings*											
	S1	*S2*	*S3*	*S4*	*S5*	*S6*	*S7*	*S8*	*S9*	*S10*	*S11*	*S12*
Fusarium. oxysporum	–	–	–	–	–	–	30	–	50	–	–	–
Gilmaniella humicola	5	–	–	–	–	20	–	20	–	–	–	–
Epicoccum nigrum	–	–	–	–	10	–	–	–	–	–	–	–
Penicillium citrinum	–	–	–	10	20	20	20	20	20	10	10	10
Penicillium. solitum	–	–	–	–	–	–	–	–	–	20	20	–
Trichoderma harzianum	–	–	–	–	–	–	–	–	20	–	–	10
white sterile mycelium	20	10	10	10	–	–	50	–	20	30	30	30
Yellow sterile mycelium	–	–	–	–	–	10	20	10	20	–	–	–
Total no. of fungal species	4	5	5	5	5	6	5	7	5	6	6	4

Table 17.3: Mycoflora Associated with Leaf Surface of *G. biloba* by Using Moist Chamber Method

Fungal Species	*Samplings*											
	S1	*S2*	*S3*	*S4*	*S5*	*S6*	*S7*	*S8*	*S9*	*S10*	*S11*	*S12*
Alternaria alternata	40	50	30	30	40	30	50	10	70	20	20	30
A. soloni	–	–	20	20	–	–	–	–	–	–	–	20
A. longipes	–	–	–	–	–	–	–	–	–	10	20	–
A. cichori	–	–	–	–	–	–	–	–	–	10	–	–
Botritis squamossa	–	–	–	–	–	–	–	–	–	40	–	–
Cladosporium herbarum	20	30	30	–	30	20	–	–	–	–	–	–
C. cladosporioides	10	40	20	–	40	40	–	60	–	–	–	–
Chaetomium globosum	10	20	–	10	–	–	–	10	–	–	–	–
Drechslera hawiiensis	–	–	–	–	–	–	10	–	10	–	–	–
Curvularia protuberata	–	–	–	10	–	–	30	–	–	–	–	–
Lacellina sp.	–	–	–	–	–	–	40	–	–	–	–	–
Epicoccum purpurescens	–	–	–	–	–	–	–	–	10	–	–	20
Memnoniella echinata	–	–	–	–	–	–	–	30	–	–	10	60
Nigrospora sp.	–	–	–	–	30	–	20	–	–	–	–	–
Periconia felina	–	–	–	–	20	20	50	10	10	–	–	–
Phoma sp.	10	–	–	10	–	10	10	10	–	–	–	–
Phomopsis sp.	–	–	–	–	–	–	–	–	10	–	–	–
Penicillium citrinum	–	–	–	–	–	–	–	–	–	–	20	–
Stachybotrys atra	–	–	–	–	–	–	–	30	–	–	–	30
Septonema sp.	–	–	–	–	–	–	–	10	–	–	–	–
Pithomyces chartarum	–	–	–	–	–	–	–	–	–	–	–	–
Tetraploa sp.	–	–	–	–	–	–	–	–	–	–	–	–
Pestalotiopsis sp.	–	10	20	–	–	–	–	–	40	–	–	–
Ulocladium sp.	–	–	–	–	–	–	–	20	–	–	–	–
Total no. of fungal species	5	5	5	5	5	5	7	9	6	4	4	5

Table 17.4: ANOVA showing effect of different methods and months on isolation of total species of leaf surface fungi of Ginkgo biloba

ANOVA						
Source of Variation	*SS*	*df*	*MS*	*F*	*P-value*	*F crit*
Rows	26.22222	11	2.383838	1.459042	0.21662	2.258518
Columns	2.055556	2	1.027778	0.629057	0.542413	3.443357
Error	35.94444	22	1.633838			
Total	64.22222	35				

Alternaria and *Cladosporium* were the only fungal species which were isolated in all samplings by applying all the tree methods as well as they were common inhabitants of air mycoflora (Singh, 1978). *Aspergillus* spp. was also found to be common on all samplings by applying all the three methods as well as they were common inhabitant of air mycoflora. *Aspergillus* spp. was also found to be common in all samplings by applying dilution plate technique and washed disks method.

Distribution of Different Groups of Fungi

Different groups of fungi recorded by various methods in all samplings are presented in Tables 17.1 to 17.3, and Figures 17.1c and 17.1d. Zygomycotina were represented by two genera with two species *viz., Mucor hiemalis and Rhizopus nigricans,* Ascomycotina was represented by only one genus with one species *i.e. Chaetomium globosum.* Deutromycotina constituted the major portion of mycoflora constituting about twenty eight genera with thirty nine species, they were recorded as dominant member of fungal species isolated in all samplings by applying all the three methods. Zygomycotina were only 4.87 per cent and Ascomycotina were represented by 2.43 per cent, whereas Basidiomycotina were altogether absent. Deuteromycotina were represented by 92.68 per cent of total fungal flora, thus constituting the major part of the mycoflora at each sampling by applying all the three methods. Among Deuteromycotina: Moniliales represented 87.18 per cent, sphaeropsidales 5.12 per cent, Mycelia sterila 5.12 per cent, Melanconiles were constituted about 2.56 per cent of total fungal population. Among Moniliales, Dematiaceous hyphomyecetes were the most common inhabitants of fungal flora. *Ginkgo biloba* leaves were found to be infected with *Alternaria alternata* (Plate 17.5). The identification was confirmed by Indian Agricultural Research Institute, New Delhi. It is a new record.

Leaf Surface Mycoflora of *Taxus baccata*

Quantitative Analysis of Fungal Flora

The results are summarized in Tables 17.4 to 17.7, Figures 17.2a to 17.2e and Plates 17.1 and 17.2). A total of thirty fungal species were isolated in all samplings by applying a combination of methods amongst which three were belonged to Zygomycotina, Ascomycotina and Basidiomycotina were altogether absent and rest 91.89 per cent of total fungal population constituted Deuteromycotina members.

Twenty three fungal species were isolated by applying dilution plate technique, out of which two were Zygomycotina, Ascomycotina and Basidiomycotina members were altogether absent and rest 91.30 per cent of total fungal population to Deuteromycotina (Table 17.4). In all seventeen fungal species were isolated by applying washed disks method (Table 17.5) out of which one was the member of Zygomycotina and rest sixteen fungal species were belonged to Deuteromycotina constituted about 94.12 per cent of total fungal population, whereas sixteen fungal species were isolated by moist chamber method (Table 17.6). All isolated species were belonged to Deuteromycotina.

Qualitative Analysis of Fungal Flora

Three different isolation methods were followed to detect fungal flora present on leaf surface of *T. baccata.* Most of the fungal forms were common in all the three methods but some fungal forms were restricted to one method only. *Chloridium* sp., *Scytalidium lignicola* and *Stemphylium* sp. were isolated only by dilution plate technique, *Botrytis cinerea, Papularia* sp. and *Pellicularia* sp. were restricted only to washed disks method, whereas in moist chamber method *Myrothecium roridum, Periconia felina, Pestalotiopsis* sp., *Diplodina* sp., *Pithomyces Chartarum Phoma humicola, Ulocladium* and an unidentified species were observed (Table 17.4 to 17.6).

Table 17.4: Record of Leaf Surface Mycoflora (fungi/ cm^2) of *Taxus baccata*

Fungal Species	*Samplings*											
	S1	*S2*	*S3*	*S4*	*S5*	*S6*	*S7*	*S8*	*S9*	*S10*	*S11*	*S12*
Aspergillus niger	104	–	–	–	–	–	–	–	–	78	234	26
A. fumigatus	–	–	–	–	–	–	78	–	–	78	104	–
Alternaria alternata	78	–	–	–	52	–	52	–	52	–	–	–
A.tenuis	–	–	–	–	78	78	–	78	–	–	–	–
Cladosporium herbarum	–	130	–	–	–	–	–	–	–	–	–	–
C.cladosporioides	259	519	–	–	–	–	52	260	130	–	–	–
Chloridium sp.	–	–	–	–	–	–	–	–	52	–	–	–
Drechslera australiensis	–	–	–	–	–	–	–	–	52	–	–	–
Epicoccum nigrum	–	–	–	–	–	–	26	–	–	–	–	–
Fusarium oxysporum	–	–	104	–	–	–	–	–	–	–	–	–
Aspergillus flavus	–	–	–	–	–	–	–	–	–	–	–	–
Mucor hiemalis	26	–	26	26	–	–	–	–	–	–	–	–
Mortierella subtilissima	–	–	–	–	–	–	–	–	–	26	–	–
Penicillium sp.	–	–	–	–	–	–	–	78	390	–	–	–
P. citrinum	–	–	–	260	–	–	–	390	–	–	156	52
P. chrysogenum	–	–	–	–	–	–	–	–	779	–	–	–
Scytalidium lignicola	–	–	–	–	–	–	–	–	–	26	–	–
Stemphylium sp.	–	–	52	–	–	–	–	–	–	–	–	–
Trichoderma viride	–	–	–	–	130	208	104	–	–	78	–	52
T. harzianum	–	–	–	–	–	–	–	–	–	78	26	–
Verticillium sp.	–	–	–	–	–	–	–	–	78	–	–	–
white sterile mycelium	52	78	78	104	26	104	–	–	52	–	78	26
yellow sterile mycelium	–	–	–	–	26	103	–	52	–	–	–	–
Total no. of fungal species	5	3	4	3	5	4	5	5	8	6	5	4

Table 17.5 : Frequency of Fungal Species on Leaf Surface of *T. baccata* by Using Washed Disks Method

Fungal Species	*Samplings*											
	S1	*S2*	*S3*	*S4*	*S5*	*S6*	*S7*	*S8*	*S9*	*S10*	*S11*	*S12*
Aspergillus niger	–	–	–	–	–	–	–	–	–	30	30	30
A. flavus	–	–	–	–	–	–	–	–	–	20	–	–
Alternaria alternata	50	30	30	20	30	20	10	50	20	–	–	–
A. tenuis	–	–	–	–	–	–	–	–	–	–	–	–
Botrytis cinerea	–	–	–	30	–	–	–	–	10	–	–	–
Cladosporium herbarum	–	–	70	–	–	–	–	–	–	–	–	–

Contd...

Table 17.5–Contd...

Fungal Species	Samplings											
	S1	S2	S3	S4	S5	S6	S7	S8	S9	S10	S11	S12
C.cladosporioides	100	50	50	50	50	40	20	10	55	–	20	–
Fusarium oxysporum	–	–	–	–	50	20	–	40	–	–	–	–
Mertierella subtillissima	–	–	–	–	–	–	–	–	–	10	–	–
Penicillum citrinum.	–	–	–	–	20	–	80	20	–	–	10	50
P. solitum	–	–	–	–	–	–	20	–	–	–	10	–
Papularia sp.	–	10	–	10	–	–	–	–	–	–	–	–
Pellicularia sp.	–	10	–	10	–	–	–	–	–	–	–	–
Trichoderma viride	–	–	–	–	–	50	–	–	20	–	–	30
T. harzianum	–	–	–	–	–	–	–	–	10	40	–	20
White sterile mycelium	30	10	20	10	20	–	70	20	–	–	10	–
Yellow sterile mycelium.	–	10	30	20	10	–	–	10	–	–	–	–
Rhizomucor sp.	–	–	–	–	–	–	–	–	–	–	10	–
Total no. of fungal species	3	6	5	7	6	4	5	6	5	4	6	4

Table 17.6: Frequency of Fungal Species on Leaf Surface of *T. baccata* by Using Moist Chamber Method

Fungal Species	Samplings											
	S1	S2	S3	S4	S5	S6	S7	S8	S9	S10	S11	S12
A. alternata	50	50	40	10	40	30	10	20	20	30	20	30
A.tenuissima	–	–	–	–	–	–	–	–	–	–	–	–
Cladosporium herbarium	–	30	–	–	–	–	–	–	–	–	30	–
C.cladosporioides	60	20	40	10	30	60	40	30	30	30	20	40
Drechslera australiensis	–	–	–	20	–	–	–	10	10	–	–	–
Diplodina sp.	–	–	–	10	–	–	–	–	–	–	–	–
Myrothecium roridun	–	–	–	10	–	–	–	–	–	–	–	–
Pullularia pullulans	–	–	–	–	–	–	–	–	–	10	–	–
Penicillum solitum.	–	–	–	–	–	–	–	–	–	10	–	–
Periconia sp.	–	–	–	–	–	–	–	10	20	–	–	–
Pestolotiopsis sp.	–	–	–	–	–	–	–	20	20	–	–	–
Pithomyces chartarum.	–	–	–	–	–	–	–	–	10	–	10	–
Stachybotrys atra	40	20	30	10	50	40	40	–	–	20	20	10
Phoma Sp.	–	–	–	10	–	–	–	–	–	–	–	–
Ulocladium sp.	–	–	–	–	–	–	–	10	–	–	–	–
unidentified fungal sp.	–	40	–	10	40	–	–	–	–	–	–	–
Total no. of fungal species	3	5	3	8	4	3	3	6	6	5	4	4

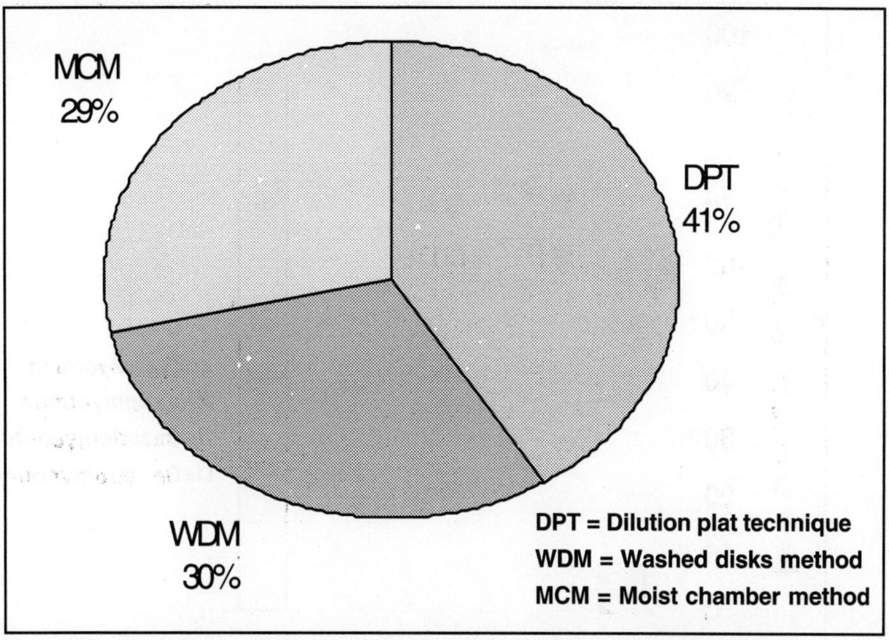

Figure 17.2a: Showing Total Number of Fungal Species Isolated from Leaf Surface of *T. baccata* by Using Various Methods

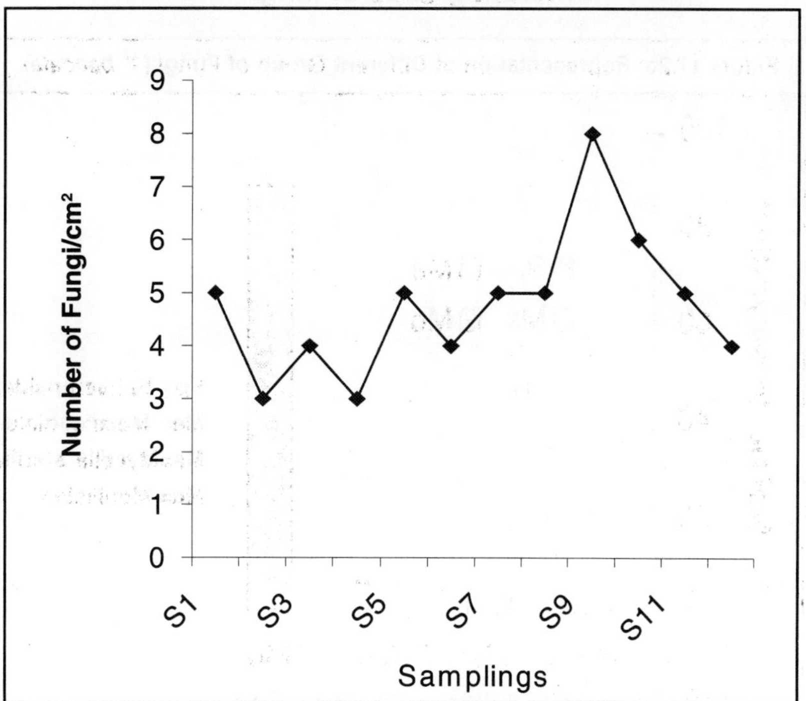

Figure 17.2b: Showing Total Number of Fungi/cm² Isolated from Leaf Surface of *T. baccata* at Different Samplings

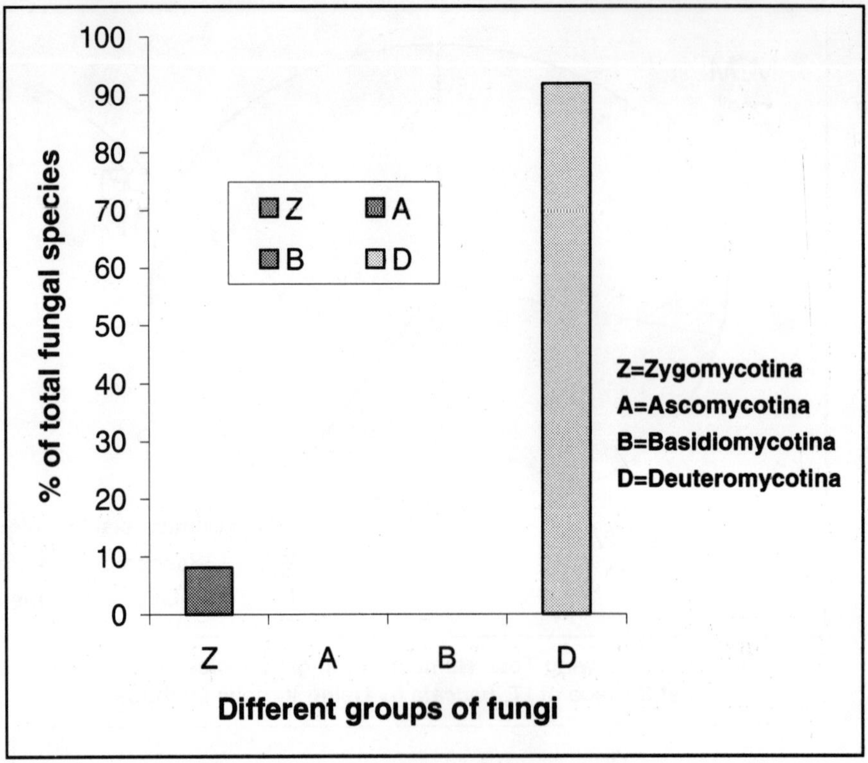

Figure 17.2c: Representation of Different Group of Fungi (*T. baccata*)

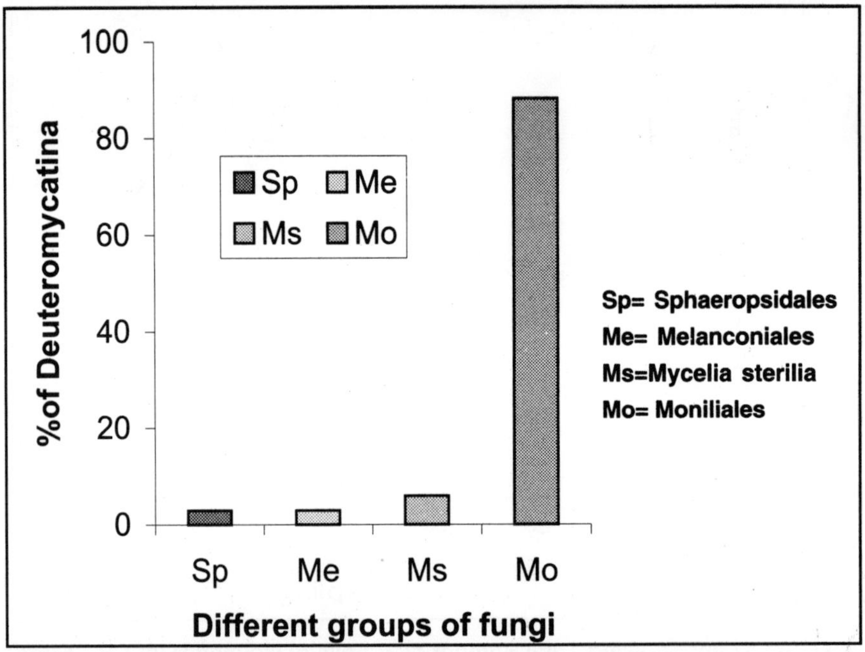

Figure 17.2d: Representation of Different Groups of Fungi (*T. baccata*)

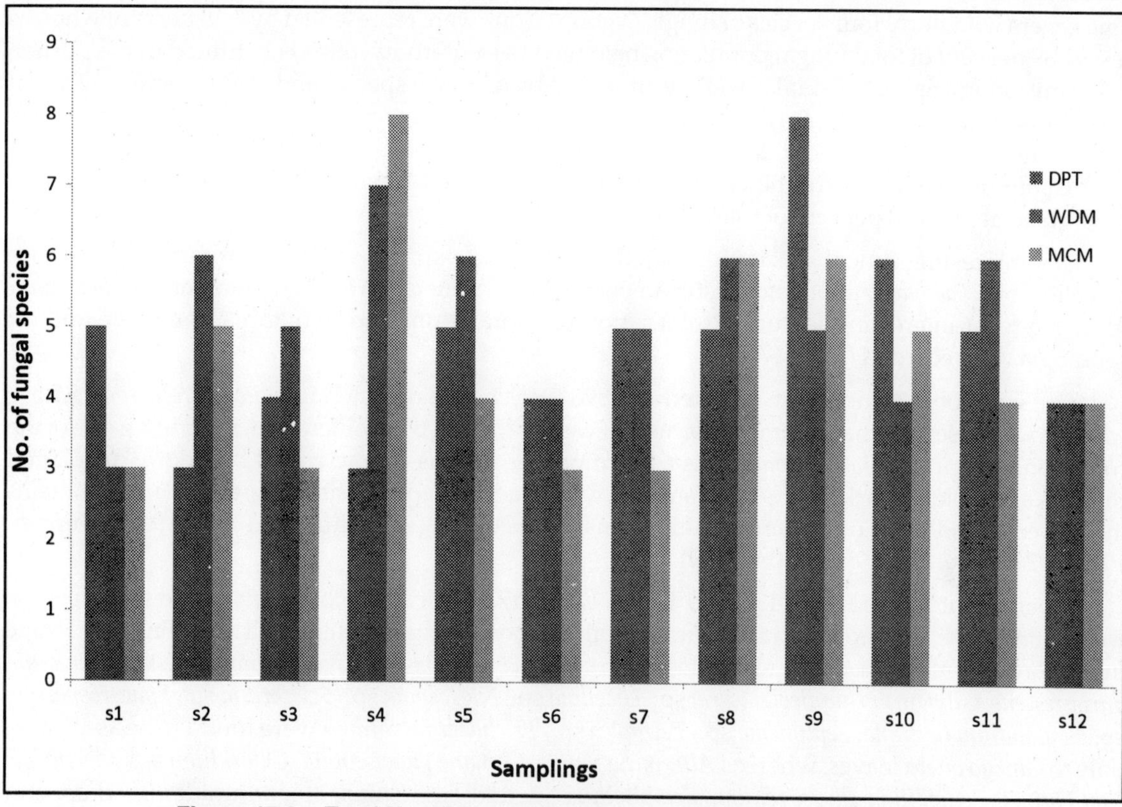

Figure 17.2e: Total No. of Fungal Species Isolated from Leaf Surface of *T. baccata* by Using Different Methods

Table 17.7: ANOVA Showing Effect of Different Methods and Months on Isolation of Total Species of Leaf Surface Fungi of *Taxus baccata*

ANOVA						
Source of Variation	*SS*	*df*	*MS*	*F*	*P-value*	*F crit*
Rows	32.97222	11	2.997475	2.318359	0.04487	2.258518
Columns	1.555556	2	0.777778	0.601563	0.556722	3.443357
Error	28.44444	22	1.292929			
Total	62.97222	35				

Distribution of Different Groups of Fungi in Mycoflora

A total of thirty seven species of fungi were isolated from leaf surface mycoflora *T. baccata* by applying a combination of methods (Table 17.4 to 17.6) and represented by different groups of fungi as illustrated in Figures 17.1c, 17.1d and 17.2c, 17.2d). Zygomycotina were represented by three genera with three species *viz; Mucor hiemalis, Mortierella subtilissima* and *Rhizomucor* sp., thus constituting about 8.10 per cent of total fungal population. Ascomycotina and Deuteromycotina were altogether absent. Deuteromycotina were the most common members of fungal flora representing about twenty nine genera with thirty four species of fungi. Zygomycotina were represented by 8.10 per cent whereas rest 91.89 per cent of total fungal population belonged to Deuteromycotina constituted major portion of the mycoflora. Sphaeropsidales were represented by only one species and thus constituting about 2.94 per cent of total Deuteromycotina members. Melanconials were also represented by only one species thus constituting about 2.94 per cent, while Moniliales constituting about 88.23 per cent of total Deuteromycotina, were the chief component of isolated fungal population. Mycelia sterilia constituted about 5.88 per cent of total Deuteromycotina.

One interesting fungal species identified as *Rhizomucor* sp., was isolated from the leaf surface mycoflora of *T. baccata* by applying moist chamber method. Proliferation in the sporangium of *Rhizomucor* sp. is the most interesting feature. Proliferation of sporangium into two, three, four, five and six sporangia was recorded (Plate 17.1).

The same observation was reported in 1896 by Schostakowitstch in *Mucor proliferous* schost, where it was observed that columella forms a new sporangium by proliferation, reminding one of the condition in Saprolegnia. *Rhizomucor* is a thermophilic fungus characterized by poorly developed rhizoids and stolon a dark colored sporangium which lacks an apophysis and grey colony. It contains three species and is the only thermophilic genus in mucorales (Hasseltine and Ellis, 1973; Schipper, 1970, 1978; Lucet and Constantin, 1899).

A perusal of Tables 17.1 to 17.3 and Tables 17.4 to 17.6 indicates that *Alternaria* and *Cladosporium* were found to be common fungal species to both the host plants isolated in all samplings and some fungal forms were confined only to one host. *Alternaria solani, A. longipes, Chaetomium globosum, Epicoccum purpurescens, Gilmaniella humicola, Isaria* sp., *Lacellina* sp., *Nigrospora* sp., *Sclerotium* sp., *Sphaerosporium* sp., *Scytalidium lignicola, Septonema* sp., *Tetraplea* sp., *Trichoderma koningi* were found to be associated only to *Ginkgo biloba* leaves, whereas *Alternaria tenuis, Cordana pauciseptata, Chloridium* sp., *Mortierella subtilissima, Penicillium chrysogenum, stemphylium* sp. and *Sepedonium* sp., an unidentified sp. and *Rhizomucor* sp. were observed only from the leaves of *Taxus baccata*. Some fungal species were identified by Agharkar Research Institute, Pune.

Discussion

As is evident from the results (Table 17.1–17.6) the number of fungal species isolated by dilution plate technique was higher than that of washed disks method. This difference might be due to the fact that after washing the leaf disk, most of the forms adhering to the leaf surface were detached and only a few forms remains closely attached to the leaf surface or inside the tissue, while in moist chamber competition was least and spores were produced only at the expense of nutrients already present in the mycelium. This method was supposed to be used to detect the presence of potential fungal colonists which were present on the leaf surface in spore form or in vegetative state (Singh, 1978; Pandey, 1988).

The data also reveals that some of the fungal species isolated form leaf surfaces of *G. biloba* and *T. baccata* were common, this similarity might be possible due to their common source of origin through soil and air (Mishra and Srivastava, 1971). Most of the fungal forms are specific to both the host plants, it may be possible due to specificity of some fungal species towards the leaf exudation and physiological conditions of the tissues of both the test plants.

As we know that *Taxus* is an evergreen tree but in case of *Ginkgo* it was observed that old leaves harbour more fungal species than that of young leaves. It may be attributed to an increase in surface area for colonization, with a corresponding increase in the amount of nutrients onto leaf surface of ageing leaves (Singh, 1978: Garg and Sharma, 1985).

There is a variation in the presence of fungal species at various stages of plant growth. Various fungal species appears or present at various stages of leaf growth, thus a definite trend of succession of fungi could not be seen.

Pullularia pullulans was observed only from air mycoflora, the absence of this fungus on leaf surface may be due to the inability of this fungus to form recognizable colony on the leaf disk (Dickinson, 1967; Singh, 1978; Pandey and Dwivedi, 1984, Pandey, 1988). The taxonomic composition of the phylloplane mycoflora and their frequency of occurrence showed much variation in both the test plants, which might be due to host specificity and characteristic variations in meteorological conditions.

Alternaria was found to be dominant form along with *Cladosporium* species in both the host plants. The majority of the phylloplane studies have been carried out in temperate regions and on agricultural crops. The few studies which have been published on tropical plants suggests that many phylloplane fungi are cosmopolitan in distribution (Dickinson, 1976) have described the fungi on wheat and/or barley leaves in England, France and Gorakhpur (India) respectively. In all three countries the leaf surface mycoflora of both the plants were dominated by *Cladosporium* and *Alternaria* was also common, through the percentage of occurrence is more in older leaves than on young leaves. In the present study *Aspergillus* spp. were also found to be common inhabitants of fungal flora, whereas *Aspergillus* and *Curvularia* were not found at all in England, while Diem (1974) lists the common species and Tewari (1973) reported *Aspergillus* and *Curvlaria* on leaf surface mycoflora in Gorakhpur. This might be due to the change in the climatic conditions of India, and England. *Aureobaridium* was common on barley leaves in England (Europe) but Tewari (1973) failed to find this fungus on either plant though he did isolate it from the air above the crop, the same results were also reported by the author during the course of this investigation. Norse (1972) was also reported that *Cladosporium oxysporum*, together with *Alternaria alternata, Epicoccum purpurescens* and *Phoma eupyrena* were the most abundant filamentous saprophytes, the present study also have the same observations.

Table 17.7 and 17.8 shows statistical analysis of data subjected to two way analysis of variance (ANOVA). Table 17.8 shows statistically significant variation in total number of fungal species isolated from leaf surface of *Taxus baccata* in relation to different methods and months.

Alternaria, Epicoccum and *Stemphylium* have all been recorded from the phylloplane (Dickinson, 1976), but proof of their continued growth on healthy leaves is lacking. In an extensive direct microscopy by Dickinson and Wallace (1976) traced to one of these spores. Sprouting colonies of these fungi are, however, common on old leaves of many different plants and Dickinson (1967) has emphasized how quickly they may take advantage of changes on the leaf surface environment at the onset of senescence. Other studies have suggested that these fungi may actually hasten senescence, thus enhance the decomposition process, than would have been the case if phylloplane invaders were absent. The main evidence for this hypothesis comes from a series of field experiments in which the development of cereal leaf surface micro-organisms has been inhibited by applying foliar fungicides.

References

Anonymous. 1996. Current Index of Medicinal Specialities. 19:114.

Barnett, B., Kiefer, D. and Rabago, D. 1999. Assessing the risks and benefits of herbal medicines:an overview of scientific evidence. Altern. Ther. Health Med., 5: 40-49.

Blakeman, J. P. 1972. Effect of plant age on inhibition of *Botrytis cinerea* spores by bacteria on beet root leaves. Physiol. Plant Pathol., 2:143-152.

Champion, H. G. and Seth, S. K. 1968. A revised survey of the forest types of India. Manager of Publications Delhi.

Dickinson, C. H. 1967. Fungal colonization of *Pisum* leaves. Can. J. Bot., 45: 915-927

Dickinson, C. H. 1976. Fungi on aerial surfaces of higher plants. In: Microbiology of aerial plant surface. (Eds. Dickinson, C.H. and Preece, T.F.). Academic Press, London. pp. 293-324.

Dickinson, C. H. and Preece, T. F. 1976. Microbiology of Aerial Plant Surfaces. Academic Press. London.

Dickinson, C. H. and Wallace, B. 1976. Effect of late application of foliar fungicides on activity of micro-organisms on winter wheat flag leaves. Trans. Br. Mycol. Soc., 76: 103-112.

Diem, H. G. 1974. Micro-organisms of the leaf surface: Estimation of the mycoflora of the barley phyllosphere. J. Gen. Microbial., 80: 77-78.

Garg, A. P. and Sharma, P. D. 1985. Ecology of phyllosphere and leaf fungi of *Cyamopsis tetragonoloba* (L.) Taub. Rev. Ecol. Biol. Sci., 22:35-55.

Gregory, P. H. 1961. The leaf as a spore trap. Trans. Br. Mycol. Soc., 44:298-299.

Hasseltine, C. W. and Ellis, J. J. 1973. Mucorales, The Fungi Vol. IV B (Eds. Ainsworth, G.C.; Sparrow, F.K. and Sussman, A.F.)., Academic Press, New York. Pp. 187-217.

Harley, J. L. and Waid, J. S., 1955. A method for studying active mycelia on living roots and other surfaces in the soil. Trans Br. Mycol. Soc., 38: 104-118.

HPPCL. 1998. Herbs Production and Processing Co. Ltd.. His Majesty's Government of Nepal.

Itokava, H., Totsuka, N., Nakehar, K., Takeya, K., Lepoittevlin, J. P. and Asakawa, Y. 1987. Chemical and Pharmaceutical Bull. 35: 3016-3020.

Kerling, L. C. P. 1958. De microflora op het blad van *Beta vulgaris*. Tijdschr. Plziek., 64: 402-410.

Keyworth, W. G. 1951. A Petri-dish in moist chamber. Trans. Br. Mycol. Soc., 34:291-292.

Kim, Y. S., Pyo, M. K. and Park, K. M. 1998. Antiplatlet and antithrombotic effects of a combination of ticlopidine and *Ginkgo biloba* ext (EGb 761). Thromb. Res. 91:33-38.

Last, F. T. 1955. Seasonal incidence of *Sporobolomyces* on cereal leaves. Trans. Br. Mycol. Soc., 38: 221-139.

Mishra, R. R. and Srivastava, V. B.1971. Leaf surface fungi of *Oryza sativa* L. Mycopath. et mycol. Appl., 44:289-294.

Norse, D. 1972. Fungal populations of tobacco leaves and their effect on the growth of *Alternaria longipes*. Trans. Br. Mycol. Soc., 59:261-271.

Pande, P. C. 1991. Gymnospermous Flora of Almora District of Kumaun Himalaya. Vegetos. 4: 33-35.

Pande, P. C., Joshi, G. C. and Kandpal, M. M.1989. Ethnobotany of Kumaun Himalaya. In: Himalayas Environment, Resource and Development. (Eds. N. K. Sah, S. D. Bhatt and R. K. Pande) Shree Almora Book Depot., Almora. pp. 285-294.

Pandey, M. L. 1988. Studies on leaf surface microfungi of *Setaria italica* with reference to antagonism and disease development in Almora hills. Ph.D. Thesis, Kumaun University, Nainital.

Pandey, R. R. and Dwivedi, R. S. 1984. Seasonal incidence of phylloplane mycoflora of guava with reference to fungal pathogens. Acta Bot.Indica, 12: 1-8.

Pillai, S. K. and Sinha, H. C. 1968. Statistical Methods for Biological Workers. Ram Prasad and Sons, Agra.

Potter, M. C. 1910. Bacteria in their relation to Plant Pathology. Trans. Br. Mycol. Soc., 3: 150-168.

Preece, T. F. and Dickinson, C. H. 1971. Ecology of leaf surface micro-organisms. Academic Press, London.

Raizada, M. B. and Sahni, K. C. 1960. Living Indian Gymnosperms. Pt. I (Cycadales, Ginkgoales and Coniferales) Ind. For. Rec. 5:73-150.

Sharma, K. R. and Mukherjee, K. G. 1973. Microbiol colonization of aerial parts of plants- a review. Acta Phytopathologica Acad. Sci. Hung., 8:425-461.

Sikora, R., Sohn, M. and Deutz, F. J. 1989. *Ginkgo biloba* extract in the therapy of erectile dysfunction. J. Urol. 141: 188A.

Singh, D. B. 1978. Studies on leaf surface mycoflora of mustard and barley. Ph.D. Thesis. Banaras Hindu University, India.

Sporne, K. R. 1965. The morphology of Gymnosperms (2nd ed). Hutchinson University Lib. London.

Tewari, R. P. 1973. Investigation into phyllosphere micro-flora of certain crop plants. Ph.D. Thesis, Gorakhpur, India.

Waksman, S. A. 1922. A method for counting the number of fungi in the soil. J. Bact., 7: 339-341.

Wani, M. C., Taylor, H. L., Wall, M. E., Coggon, P. and Mophail, A. T. 1971. Plant anti-tumor agents VI. The isolation and structure of taxol, a novel anti-leukemic and anti-tumor agent from *Taxus brevifolia*. J. Am. Chem. Sopc., 93: 2325-2327.

Wildman, H. G. and Parkinson, D. 1979. Microfungal succession on living leaves of *Populus trimuloides*. Can. J. Bot., 51: 2800-2811.

Microbes: Diversity and Biotechnology (2012) *Pages* **299–324**
Editors: **Prof. S.C. Sati & Dr. M. Belwal**
Published by: **DAYA PUBLISHING HOUSE, NEW DELHI**

Chapter 18

The Aquatic Hyphomycetes: Aspects and Prospects

Pratibha Arya and S.C. Sati
Department of Botany, D.S.B. Campus,
Kumaun University, Nainital - 263 001

ABSTRACT

Fungi, especially aquatic Hyphomycetes are generally considered as the prime microbial decomposers of leaves in streams. A considerable amount of information on aquatic Hyphomycetes and their role in the break down of leaf litter has been produced in recent years. Fungi associated with decomposing leaf litter in streams, especially aquatic Hyphomycetes are affected by a number of factors, which in turn determine their abundance, development and activity. Ecology of aquatic Hyphomycetes and their effects on litter decomposition, including interactions with bacteria and fungi is available in this review.

Introduction

The aquatic Hyphomycetes are a specialized group of anamorphic fungi, usually occurring on dead plant materials, such as leaves and twigs in the beds of streams and rivers. They play an important role in the decomposition of leaf litter or other submerged substrata in aquatic habitats (Barlocher, 1992). Members of this group are intermediaries in energy and food webs in aquatic ecosystems and are distributed worldwide. These fungi are characterized by their magnificent spore types and are anamorphs of Ascomycota and Basidiomycota occurring mainly in lotic water (Pascoal *et al.*, 2005). In traditional systems, these fungi belong to the order Moniliales. Some time these are able to survive in unfavourable environmental conditions in the teleomorphic (sexual) state or as sexual resistant spores, in spite of their conidia (asexual spores) and mycelia.

Aquatic Hyphomycetes are delimited from other fungi by their branched, septate mycelia with normally thin walled hyaline (non-pigmented) asexual (mitotic) spores (conidia) having characteristic

shapes – stauroform (tri or tetraradiate), scolecoform (needle shaped or sigmoid) and helicoform (helical or spiral). According to Descals (2005), aquatic Hyphomycetes are amongst the few groups of fungi where species can in many cases be safely identified by observing conidia from nature, without the need for pure cultures.

De Wildeman (1893, 1894, 1895) was first to record the aquatic Hyphomycetes and described 4 species named as *Tetracladium marchalianum, Fusarium elongatum* (now known as *Anguillospora longissima*), *Clavariopsis aquatica* and *Lemonniera aquatica* from aquatic habitat. In the mid-twentieth century Ingold (1942) described these fungi in detail and named as "Aquatic Hyphomycetes" for the first time as they complete their life cycle on submerged substrate including their vegetative growth, reproduction and spore liberation and spore dispersal in well aerated water. He described 16 species of aquatic Hyphomycetes growing in decaying alder leaves and their conidia in accumulated water foam from fast flowing streams of U.K. Later on, Ingold explored a number of these fungi from various parts of the world (Ingold, 1943 a, b, c; 1944, 1949, 1952, 1956, 1958 a, b; 1960, 1965, 1968, 1974 a, b; 1975 a, b, c; 1976, 1979). Ingold and his co-workers also contributed many new taxa of aquatic Hyphomycetes as well as details of conidial development and morphology (Ingold and Ellis, 1952; Ingold *et al.*, 1968 a, b).

These fungi were also known as 'Fresh water hyphomycetes' (Nilsson, 1964), 'Ingoldian hyphomycetes' (Descals *et al.*, 1978), 'Ingoldian fungi' (Barlocher, 1982; Webster and Descals, 1981; Wood-Eggenschwiler and Barlocher, 1985) and 'Amphibious hyphomycetes' (Michaelides and Kendrick, 1978; Akridge and Koehn, 1987). However, in modern time all conidial fungi, which sporulate under submerged conditions have also been included in this group and named as "water-borne conidial fungi" (Webster and Descals, 1981).

They abundantly sporulate under water and conidia germinate, grow and reproduce preferably in low temperature (Ingold, 1975 a; Webster and Descals, 1981; Barlocher, 1992). A single submerged decaying leaf may harbor more than one species of aquatic Hyphomycetes. Ingold (1943 a) wrote, "One species does not grow on a particular leaf to the exclusion of others. On a single decaying leaf it is usual to find four or five species and not infrequently as many as ten different species may occur".

The anamorph-teleomorph relationships of aquatic Hyphomycetes are known in a very small percentage and hitherto discovered anamorph-teleomorph connections show links mainly to Ascomycota lesser to Basidiomycota (Marvanova, 1997, 2002). This has been supported by some recent studies of phylogenetic relationship based on homologies in rDNA sequences (Campbell *et al.*, 2002; Nikolcheva and Barlocher, 2002; Baschin, 2003; Bauer *et al.*, 2003).

Occurrence and Distribution

Aquatic Hyphomycetes grow well on submerged leaf litter in streams, releasing large numbers of conidia (Barlocher, 1982). They are often trapped by air bubbles and accumulate in persistent foam below water falls (Iqbal and Webster, 1973 a; Ingold, 1975 a; Hyde and Goh, 2003). Foam is indeed the best place to observe the spores of aquatic Hyphomycetes. In some studies occurrence of aquatic Hyphomycetes communities on wood have also been reported (Willoughby and Archer, 1973; Sanders and Anderson, 1979; Shearer and Webster, 1991). Shearer (1992) had reviewed the role of woody debris in the life cycles of aquatic Hyphomycetes.

The majority of studies on the ecology of aquatic Hyphomycetes have been carried out in the aquatic habitat. However, their occurrences on terrestrial habitat have now also been reported. Bandoni (1972, 1974) was the first to find these fungi in terrestrial litter in Canada followed by Koske and

Duncan (1974). Later on, some other workers have also recorded their terrestrial occurrence (Park, 1974; Thakur, 1977; Sridhar and Kaveriappa, 1987, 1988 a; Webster and Descals, 1979; Ando and Tubaki, 1984). Spores of these fungi were also found in unexpected terrestrial environments such as snow (Czeczuga and Orlowska, 1999) or rain water (Czeczuga and Orlowska, 1997, 1999). The occurrence of aquatic Hyphomycetes as endophytes of the plants has also been reported by many workers (Fisher and Petrini, 1989; Clay, 1990; Marvanova and Fisher, 1991; Sridhar and Barlocher, 1992 a; Sati and Belwal, 2005).

The survey of available literature indicates that occurrence and distribution of aquatic Hyphomycetes have been studied all over the world by a number of workers (Ingold and Ellis, 1952; Ranzoni, 1953; Ingold, 1956, 1958 a, b; 1965, 1974 a, b; 1976; Tubaki, 1957; Willen, 1958; Hudson and Ingold, 1960; Cowling and Waid, 1963; Nilsson, 1964; Alasoadura, 1968; Conway, 1970; Park, 1974; Iqbal *et al.*, 1973, Gonczol and Toth, 1974; Marvanova, 1975, 1977, 1986, 1997; Nawawi, 1975, Descals and Webster, 1982 a, b; Chamier, 1985; 1980; Khan, 1986; Kuthubutheen, 1987; Kuthubutheen *et al.*, 1992; Marvanova and Bandoni, 1987; Marvanova and Barlocher, 1988; 1985 a, b, c, d; 1987 a, b; Nawawi and Kuthubutheen, 1987; 1988 a, b, c; Peterson, 1962, 1963 a, b; Sridhar and Kaveriappa, 1984, 1988 b; Webster and Descals, 1979; Webster, 1959, 1961, 1965; Webster *et al.*, 1976, 1994; Willoughby and Archer, 1973; Willoughby and Minshall, 1975; Wolf, 1972). In India, these fungi have also been reported for the first time by Ingold and Webster (1973) followed by Subramanian and Bhat (1981).

The southern part of India has been explored by some mycologists for these fungi (Manoharachary and Murthy, 1981; Sridhar, 1984; Manoharachary and Bhairav Nath, 1985; Chandrashekar *et al.*, 1986, 1991; Manoharachary, 1989; Sridhar and Kaveriappa, 1982, 1984; Sridhar *et al.*, 1992; Soosamma *et al.*, 2001; Sreekala and Bhat, 2002).

In the Kumaun Himalayan region, Mer and Khulbe (1981) reported 5 species of aquatic Hyphomycetes. However, the detailed study of aquatic Hyphomycetes has been carried out by Sati along with his co-workers and explored many species of these fungi from various habitats (Sati *et al.*, 1989, 1992, 2002 a, b; Sati and Tiwari, 1990 a, b, c; 1992 a, b; 1993 a, b; 1997; Tiwari and Sati, 1991 a, b; 1992 a, b; 1993). Nearly 75 species of aquatic Hyphomycetes have been recorded so far from Kumaun Himalaya. Of these, 4 species namely, *Tricladium indicum*, *Pestalotiopsis submerses*, *Pleurophragmium sonam* and *Tetracladium nainitalense* are new to the Science (Sati and Tiwari, 1992 a, 1993 a, 2003; Sati *et al.*, 2009 a). And 36 species are recorded new to India from Kumaun Himalaya (Table 18.1).

The distribution of aquatic Hyphomycetes is world wide. The species of aquatic Hyphomycetes are ubiquitous in nature and the distribution pattern changes with geographical latitude from North to equator and from the equator to the South. There are some species occurring only in the tropics and others found only in cold climates. Temperature appears to be an important factor affecting the occurrence and distribution of these fungi (Suberkropp, 1984). Some of the species of aquatic Hyphomycetes are more in temperate climates, and others are more common in the tropics (Nilsson, 1964; Barlocher, 1992). Altitude is also an important factor in describing the community structure of aquatic Hyphomycetes within a stream (Chauvet, 1991; Raviraja *et al.*, 1998 a). The variation in species composition may be attributed to the altitudinal differences was suggested by Chauvet (1991) and Fabre (1996).

There are several reports indicating the greater species richness of aquatic Hyphomycetes in soft water streams than hard water streams (Wood-Eggenschwiler and Barlocher, 1983; Harrington, 1997; Raviraja *et al.*, 1998 a; Gonczol and Revay, 2003). The abundance and species richness of aquatic

Table 18.1: Aquatic Hyphomycetes New to India Recorded from Kumaun Himalaya

Sl.No.	Fungal Species	Habitat	Locality
1.	*Acaulopage dichotoma* Drechsler	LL, F, S	N, G, SV
2.	*Alatospora pulchella* Marvanova	F	N
3.	*Alatospora flagellata* (Gonczol) Marvanova	LL	SV
4.	*Anguillospora filliformis* Greathead	LL, F	JG
5.	*Anguillospora furtiva* Descals	LL, CN	G, SV
6.	*Dimorphospora foliicola* Tubaki	LL, F, TW	N, G
7.	*Diplocladiella appendiculata* Nawawi	LL, F	JG
8.	*Diplocladiella longibrachiata* Nawawi and Kuthubutheen	LL, F	JE, JG
9.	*Dwayaangam dichotoma* Nawawi	LL, F	K
10.	*Flabellocladia gigantea* Nawawi	F, S	N
11.	*Flabellocladia tetracladia* Nawawi	LL, F	JG
12.	*Flabellospora acuminata* Descals and Webster	LL, F	JE, JG
13.	*Helicosporium lumbricoides* Saccardo	LL, F, TW	N, ST
14.	*Heliscina campanulata* Marvanova	LL, F	JG
15.	*Heliscella stellatacula* Marvanova	CN	N
16.	*Jaculispora submerse* Hudson and Ingold	LL	N, G
17.	*Lemonniera alabamensis* Sinclair and Morgan-Jones	LL	JG
18.	*Lemonniera centrosphaera* Marvanova	LL, F	JE
19.	*Lemonniera cornuta* Ranzoni	LL, CN, F, TW	N, SV
20.	*Lemonniera pseudofloscula* Dyko	LL, CN	N, SV
21.	*Lemonniera terrestris* Tubaki	LL, CN	N
22.	*Magdalaenaea monogramma* Arnaud	LL, F	N
23.	*Naidella fluitans* Marvanova and Bandoni	LL	JG
24.	*Phalangispora nawawii* Kuthubutheen	LL, TW	G
25.	*Setosynnema isthmosporum* Shaw and Sutton	LL, CN, F	N, G, BM
26.	*Speiropsis scopiformis* Kuth. and Nawawi	LL, F	N, SV
27.	*Tetracaldium apiense* Sinclair and Eicker	LL, F	JE, JG
28.	*Tetracaldium maxilliforme* (Rostrup) Ingold	LL, F	SV
29.	*Tetracladium breve Roldan*	RE	G
30.	*Tricladiopsis flagelliformis* Descals	LL, F	N
31.	*Tricladium aciculum* Nawawi	LL, F	JG
32.	*Tricladium anomalum* Ingold	F	N
33.	*Tricladium chaetocladium* Ingold	LL, CN, F, S	N, SV
34.	*Tricladium terrestre* Park	CN	N, SV
35.	*Trinacrium incurvum* Matsushima	LL, F	SV
36.	*Volucrispora ornithomorphora* (Trotter) Haskins	F	JG

LL: Leaf litter; F: Foam; S: Scum; CN: Conifer needle; RE: Root endophyte; TW: Twig; N: Niglat; G: Gufa Mahadev; SV: Snowview; JG: Jageshwar; JE: Jeolikote; K: Khurpatal; ST: SatTal; BM: Bhimtal

Hyphomycetes fungi are also affected by the structure and composition of the riparian plant canopy (Barlocher, 1992). Although highest diversity of aquatic Hyphomycetes has been reported from clear fast flowing streams (Suberkropp, 1992; Garnett *et al.*, 2000), these fungi are also reported colonizing leaf litters in large rivers (Chergui and Pattee, 1988; Baldy *et al.*, 1995). Some of the aquatic Hyphomycetes are also reported from brackish and sea water from different parts of the world (Muller-Haeckel and Marvanova, 1979; Kohlmeyer *et al.*, 1967; Shearer, 1972; Kirk and Brandt, 1980; Raghu-Kumar, 1973; Sridhar and Kaveriappa, 1988 c).

Seasonal Variation in Aquatic Hyphomycetes and Physico-chemical Relation

Species assemblage of aquatic hyphomycetes varies seasonally and site to site and these changes are thought to occur mainly due to differences in water temperature (Barlocher and Kendrick, 1974; Suberkropp, 1992). Temperature affects their growth and sporulation (Koske and Duncan, 1974; Singh and Musa, 1977) and light acts as stimulus for their sporulation on leaf litter (Fisher and Webster, 1978; Thomas *et al.*, 1991, 1992). Suberkropp and Klug (1981) and Suberkropp (1984) distinguished warm season and cold season aquatic hyphomycetous fungi, based on their temperature requirements for growth. Fisher and Webster (1978) concluded that light promotes sporulation in aquatic Hyphomycetes.

A maximum occurrence of aquatic Hyphomycetes has been noted during November and lasting upto December-January in streams of temperate zones (Alasoadura, 1968; Gonczol, 1975; Barlocher *et al.*, 1977; Sanders and Anderson, 1979; Barlocher and Rosset, 1981; Shearer and Webster, 1985). In Australia, Thomas *et al.* (1989) found the maximum number of these fungi from late summer to early autumn and in Pakistan, Iqbal and Bhatty (1979) reported the highest concentration of the conidia of these fungi in April but during heavy rainfall. Similarly, Sridhar and Kaveriappa (1989 a) observed more aquatic Hyphomycetes during monsoon and post monsoon periods than the dry periods in Indian streams at Western Ghats. In a subarctic stream, Muller-Haeckel and Marvanova (1979) found their occurrence maximum in autumn and spring. In the Central Himalayan stream Mer and Sati (1989) also found a maximum frequency of occurrence in autumn and spring.

Sati and Arya (2009) observed a maximum number of the fungal species during spring to early summer and winter seasons, while decline in species number in summer to early autumn seasons (May–Oct). Sridhar and Kaveriappa (1989 b) also observed that the total number of aquatic hyphomycetes was lowest during summer season. Many investigators have observed maximum species during post monsoon periods (Sridhar and Kaveriappa, 1984).

Temperature range of 10-25°C favors the growth and multiplication of conidial fungi in water bodies. In temperate streams sufficient availability of substrate and low temperature results in higher productions of aquatic Hyphomycetes (Ingold and Webster, 1973). The optimum temperature for the growth of *Tricladium chaetocladium* and *Lunulospora curvula* was found within the range of 15-20°C (Iqbal and Webster, 1973 b). Ingold (1975 a) and others from temperate countries have found abundant conidia in autumn, winter and early spring and abundance decreased in the late spring and summer. Mer and Sati (1989) and Raviraja *et al.* (1998 a) found negative correlation with species number and temperature. Manoharachary (1985) have explored *Lunulospora curvula* and *Tetracladium marchalianum* as temperature tolerant species. Similarly, Belwal *et al.* (2008) have also explored *Clavariopsis aquatica*, *Lunulospora curvula*, *Tetracladium marchalianum* and *Triscelophorus monosporus* to be temperature tolerant species.

Aquatic Hyphomycetes are also sensitive to pollution and are generally associated with clean and well aerated freshwater (Barlocher, 1992). A decline in the diversity of these fungi in a stream might be an indicator of organic pollution (Raviraja *et al.*, 1998 b) or heavy metals (Bermingham *et al.*, 1996; Niyogi *et al.*, 2002).

Hydrogen ion concentration (pH) greatly affects the decompositional activities of aquatic hyphomycetes in running fresh water bodies. The occurrence and degradative ability of aquatic Hyphomycetes colonizing on submerged leaf litter is influenced by the pH of water (McKinley and Vestal, 1982). While working in an arctic lake they found a progressive decline of fungi with increasing acidity and their almost complete absence at pH 4.0-3.0. Barlocher and Rosset (1981) suggested that pH close to 7.0 favour higher numbers of fungal species. While some studies concluded that species richness was higher in circumneutral than in alkaline streams (Wood-Eggenschwiler and Barlocher, 1983; Barlocher, 1987; Regelsberger *et al.*, 1987).

These fungi require a fresh oxygenated environment for their occurrence (Webster and Towfic, 1972). Increase in fungal species number is related with increasing dissolve oxygen and dissolve organic matter of the stream (Kaushik and Hynes, 1971).

Several studies indicate that in lotic ecosystems, leaf litter decomposition and fungal activity can be affected by the concentration of nutrients (*e.g.* nitrogen and phosphorus) in the water (Suberkropp and Chauvet, 1995; Sridhar and Barlocher, 2000; Grattan and Suberkropp, 2001; Rosemond *et al.*, 2002). Aquatic hyphomycetes might obtain inorganic nutrients (nitrogen and phosphorus) not only from their organic substrate (leaf litter, wood debris etc.) but also directly from water passing by ravine areas (Suberkropp, 1995; Suberkropp and Chauvet, 1995). In the study of diversity of aquatic Hyphomycetes Pascoal *et al.* (2005) reported that the richness in aquatic Hyphomycetes species was negatively correlated with concentration of phosphate, ammonia and nitrate.

The presence of these fungi in the sulphide containing water bodies indicates the importance of these salts for the growth of aquatic hyphomycetes (Field and Webster, 1985). Recently (Gulis and Suberkropp, 2003, 2004; and Gulis *et al.*, 2004) have initiated some studies on the effect of nutrients on aquatic Hyphomycetes in the American streams. Aquatic Hyphomycetes obtain phosphate and sulphate not only from the leaf litter, wood debris but also directly from water passing by ravine areas (Suberkropp, 1995; Suberkropp and Chauvet, 1995). During the observation Sati and Arya (2009) found a positive correlation between the occurrence of aquatic Hyphomycetes and the sulphate concentration of water. However, the previous work of Field and Webster (1985) shows reduction in the survival of Ingoldian aquatic hyphomycetes with increasing concentration of sulphide. Krauss *et al.* (2001) and Sati and Arya (2009) reported the stimulation of fungus activity (aquatic Hyphomycetes) at high P concentrations.

Sridhar *et al.* (2000) recorded *Tetracladium marchalianum* and *Heliscus lugdunensis* in severely polluted streams of Germany and Pascoal *et al.* (2005) reported *Anguillospora filliformis, Clavariopsis aquatica, Dimorphospora foliicola, Lemonniera aquatica* and *Lunulospora curvula* at the polluted sites but also found these species occurring in unpolluted sites. It was speculated that the tolerance to pollution evolves gradually in some species of aquatic Hyphomycetes living in the stream, along with the proceeding pollution.

The physico-chemical characters of the water were found much influenced to the occurrence and distribution of aquatic Hyphomycetes in a fast flowing stream (Field and Webster, 1985; Sridhar and Kaveriappa, 1989 a; Sridhar *et al.*, 2001; Webster and Descals, 1981).

Colonization Pattern and Litter Decomposition

Among all the fungi which colonize the submerged leaves, aquatic Hyphomycetes are known to be the most active group. More than ten species can commonly be observed colonizing a single leaf.

Fungal colonization of aquatic Hyphomycetes increases the nutritive value of leaves and the detritivores prefer such colonized leaves (Kaushik and Hynes, 1971; Barlocher and Kendrick, 1973 a, b; Kostalos and Seymour, 1976; Suberkropp and Klug, 1981; Graca *et al.*, 1993; Gessner and Chauvet, 1994). Several workers have reported an increment in the nitrogen content of the decaying leaf material in water, due to colonization of aquatic Hyphomycetes (Barlocher and Kendrick, 1973 a, b, 1974; Suberkropp *et al.*, 1983; Suberkropp and Arsuffi, 1984). In the pioneer work of Kaushik and Hynes (1968, 1971) they showed a clear preference of stream feeders for the leaves colonized by fungi then the non-colonized leaves.

The pattern of colonization by aquatic Hyphomycetes on leaf litter has been studied by various workers (Nilsson, 1964; Willoughby and Archer, 1973; Barlocher and Kendrick, 1974; Suberkropp and Klug, 1976; Barlocher *et al.*, 1978; Barlocher and Oertli, 1978 a; Sanders and Anderson, 1979; Chamier and Dixon, 1982). Some of the Indian workers have also studied the colonization of leaf litter by these fungi (Sridhar and Kaveriappa, 1988 a, b, 1989 b; Chandrashekar *et al.*, 1989; Sati and Pant, 2006).

Leaf litter decomposition in streams is an important process of ecosystem and aquatic Hyphomycetes are dominant group of fungi during the first stage of leaf litter decomposition in fresh water streams (Kaushik and Hynes, 1971; Barlocher and Kendrick, 1976; Suberkropp and Klug, 1976; Findlay and Arsuffi, 1989). It has also been known that due to the superior enzyme system, aquatic Hyphomycetous fungi have definite edge over bacteria in initiating the breakdown of solid plant material (Suberkropp *et al.*, 1976; Manoharachary, 1984). Aquatic Hyphomycetes macerate leaf tissues with pectinolytic enzymes (Suberkropp and Klug, 1980; Chamier and Dixon, 1982) and facilitate release of fine particulate organic matter which is an important food resource for aquatic invertebrates.

Duarte *et al.* (2006) observed 14.8 to 59.6 per cent of mass loss of alder leaves after 27 days, inoculated with aquatic Hyphomycetes. Gessner and Chauvet (1997) also reported 58 per cent of initial dry mass loss of ash (*Fraxinus excelsior*) by fungal growth (aquatic Hyphomycetes) after 23 days of inoculation. In the Kumaun Himalaya region Sati and Pant (2000) observed 33 per cent of weight loss within one month of submergence of pine, oak leaf litter in a freshwater stream.

Chandrashekar *et al.* (1989) observed that fishes prefer to eat the rubber leaves which were colonized by aquatic Hyphomycetes *viz.*, *Flagellospora penicillioides*, *Lunulospora curvula*, *Phalangispora constricta*, *Triscelophorus acuminatus* and *Wiesneriomyces javanicus*.

Sridhar (1984) observed *Lunulospora curvula*, *Flagellospora penicillioides* and *Wiesneriomyces javanicus* as fast colonizing species than others. Arshad and Bareen (2007) observed *Sporidesmium ensiforme* and *Lemonniera aquatica* as the early colonizers and *Articulospora proliferate* and *Flagellospora fusarioides* to be as late colonizers.

Substrate Preference

It is a general belief that most species of aquatic Hyphomycetes colonize a wide range of substrates and there is no strong preference for particular tree species. Nevertheless, it seems that some substrate preferences do occur. Leaves and twigs of riparian trees, shrubs and grass blades represent major substrates for aquatic Hyphomycetes in forest stream (Gulis, 2001). Some aquatic Hyphomycetes are known to have substrate selectivity by observation that changes in species composition, frequency of occurrence or conidia concentration in water occur in response to difference in riparian vegetation (Thomas *et al.*, 1991; Gonczol *et al.*, 1999). Thomas *et al.* (1992) concluded that for most of cases it is possible to recognize nature of substrate by their associated fungi. They also noted remarkable differences in fungal assemblages of leaves, bark and wood of the same plant species.

In Spanish stream Chauvet *et al.* (1997) demonstrated, *Lunulospora curvula* having the clear preference for eucalyptus leaves than the alder leaves. Strong preference of *Tetracladium marchalianum* for alder leaves was noted in Hungarian and Swedish streams (Gonczol, 1989; Bengtsson, 1983). Many workers have reported alder (*Alnus glutinosa*) leaves to be better substrate for the abundant colonization of aquatic Hyphomycetes in the stream in temperate regions (Ingold, 1942; Nilsson, 1964; Iqbal and Webster, 1973 b). Sridhar and Kaveriappa (1988 b) in their study in a tropical stream found the leaves of *Ficus bengalensis* harboured more number of Hyphomycetes followed by *Mangifera indica, Tectona grandis* and *Gleichenia pectinata*. There are some reports showing occurrence of aquatic Hyphomycetes in Pines, conifer needles (Sati *et al.*, 1989; Sati and Tiwari, 1990 b). However, it was reported earlier that conifer needles are less intensively colonized due to the presence of thick waxy cuticles and phenolics (Barlocher and Oertli 1978 a, b; Michaelides and Kendrick, 1978).

Oak leaves were found to support lower frequencies of aquatic Hyphomycetes and are processed at slower rates than many other types of deciduous leaf litter in streams (Chamier and Dixon, 1982; Peterson and Cummins, 1974; Suberkropp and Klug, 1976; Triska, 1970). In south Indian waters Banyan, coffee and the rubber leaves are known the best substrates for recovering a maximum number of fungal species (Sridhar and Kaveriappa, 1988 a,b, 1989 a, b; Sridhar *et al.*, 1992).

The work of Suberkropp and Klug (1981) shows that the substrate preference of aquatic Hyphomycetes depends on the changes in chemical composition of leaf materials during processing in stream and the enzymatic capabilities of fungi involved in colonization. Revay and Gonczol (1990) pointed out high percentage of scolecosporous species of aquatic Hyphomycetes from wood. Species with filiform conidia (*Anguillospora* sp., *Filosporella* sp., *F. cavispermum* and *Sporidesmium fuscum*) were also found frequently on wood and grass blades (Gulis, 2001). According to Ummul-Banin (2006), aquatic Hyphomycetes show better survival on hard and persistent substrata. This indicates that harder and persistent debris plays a good role in the survival of aquatic Hyphomycetes during harsh dry periods.

Some substrate support relatively specific aquatic Hyphomycetes assemblages and it is possible to ordinate plant litter types of their fungal complexes. This also supports substrate preferences of aquatic Hyphomycetes, at least in some species.

Endophytic Life and its Significance

Though these fungi are reported from aquatic and terrestrial habitat, Waid (1954) isolated *Varicosporium elodeae* from the root surfaces of beech seedlings and Nemec (1969) isolated *Anguillospora longissima* and *T. marchalianum* from the roots of strawberry plants. Unexpectedly, several lines of evidence now suggest that some aquatic Hyphomycetes are also plant endophytes, as they grow in plants without producing symptoms. It is suspected that the selection pressure might have been induced by scarcity of substrates (Fisher *et al.*, 1991; Sridhar and Barlocher, 1992 a,b). Sridhar and Raviraja (1995) observed that several fungi adopt the endophytic life cycle possibly because of competition of different fungi on suitable substrates.

Fisher and Petrini (1989) isolated 2 endophytic species of fresh water Hyphomycetes *viz.*, *Tricladium splendens* and *Campylospora parvula* from the bark of terrestrial roots of *Alnus glutinosa* for the first time. Later, Fisher *et al.* (1991) reported 12 species of these fungi from the bark of same host as well as root xylem.

Occurrence of aquatic Hyphomycetes as root endophytes of riparian as well as terrestrial plants including grasses, mosses, ferns, conifers and angiosperms has also been reported by many workers

(Bernstein and Carroll, 1977; Carroll and Carroll, 1978; Petrini, 1986; Clay, 1989, 1990; Marvanova and Fisher, 1991; Marvanova *et al.*, 1992; Petrini *et al.*, 1992 a, b; Fisher and Petrini, 1992; Sridhar and Barlocher 1992 a, b; Wilson and Carroll, 1994; Barlocher, 2006). Fernando and Currah (1995) isolated a few dematiaceous aquatic Hyphomycetes from the roots of terrestrial orchids and other plants growing in alpine and sub-alpine habitats. Sridhar and Barlocher (1992 b) reported 12, 5, and 7 species of aquatic Hyphomycetes as root endophytes of spruce, birch and maple respectively. Raviraja *et al.* (1996) encountered 7 species of aquatic Hyphomycetes as endophytes of roots of plantation crops and ferns of which, *Tetracladium furcatum*, *Triscelophorus acuminatus*, *T. konajensis* and *T. monosporus* were recorded for the first time as root endophytes. Ananda and Sridhar (2002) recorded these fungi from the roots of mangrove species on the west coast of India. In the Kumaun Himalayan regions, Sati and Belwal (2005) and Sati *et al.* (2006; 2009 a, b) isolated different aquatic Hyphomycetes from riparian forest plants. Many of them were found with new host records.

Endophytes occupy a unique ecological niche and have major influence on plant distribution, ecology, physiology and biochemistry (Sridhar and Raviraja, 1995). Many pharmaceutical companies and agrobased industries now undertake a large scale screening of these fungi as they provide useful products required in biotechnology and agriculture (Dreyfuss and Chapela, 1992; Bills and Polishook, 1992; Petrini *et al.*, 1992 a, b). Evidences exists that these species have specialization for their endophytic role (Fisher *et al.*, 1991; Marvanova *et al.*, 1992). Endophytic microbes produce a wide variety of bioactive metabolites (Sridhar and Raviraja, 1995).

Fungal endophytes are systemic and often mutualistic (Clay, 1996; Bacon and Hill, 1997) and produce mycotoxins, alkaloids that can enhances resistance to herbivory (Clay, 1996; Clay and Schardl, 2002). Endophytic fungi also increase drought resistance (Elmi and West, 1995), enhance nutrient uptake of their host plant (Malinowski *et al.*, 2000). Clay (1988) observed that endophytic infection enhances the resistance of host plants against insects.

Fisher *et al.* (1991) and Sridhar and Barlocher (1992 b) observed that the roots of riparian vegetation colonized by aquatic Hyphomycetes might have been induced by scarcity of substrates. It seems that aquatic roots provide a stationary refuge for some species of aquatic Hyphomycetes. However, the beneficiary activity of endophytic aquatic Hyphomycetes has been suggested by a few workers as they provide products useful in biotechnology and agriculture (Petrini, 1991; Bills and Polishook, 1992; Dreyfuss and Chapela, 1992).

Recently, the role of root endophytic aquatic Hyphomycetous fungi, *Heliscus lugdunensis* and *Tetrachaetum elegans* in plant health is studied by employing pot experiments (Sati and Arya, 2010 a). Aquatic hyphomycetous fungi isolated as root endophytes of riparian plants were artificially inoculated into two test plants *viz.*, *Solanum melongena* and *Hibiscus esculentus*. *T. elegans* and *H. lugdunensis* showed significant effects (fresh weight, dry weight and length of shoots and roots) on both test plants.

According to Fisher *et al.* (1986), endophytic fungi of healthy plants develop both the asexual and sexual reproductive state but after the death of the plant tissue, it represents usually asexual stage. Since, aquatic Hyphomycetes are established as endophytes in aquatic roots, these may also serve as a helpful tool to establish teleomorph-anamorph connections. Sridhar and Barlocher (1992 b) observed the endophyte, *Heliscus lugdunensis* which was isolated from the aquatic roots of *Picea glauca*, *Acer spicatum* and *Bitula papyrifera*, producing the teleomorph stage after 40 days when subcultured colonies were exposed to continuous light.

Antimicrobial Activity

Antimicrobial activity of some fungi distributed among various taxonomic groups against pathogenic bacteria and fungi are well documented time to time. The intra and interspecific interaction of aquatic Hyphomycetes in relation to aquatic ascomycetes and release of diffusible inhibitory substances has also been reported by a few workers (Khan, 1987; Shearer and Zare-Maivan, 1988; Barlocher, 1991).

The reports on antimicrobial compounds produced by endophytic fungi are now not less known (Pelaez *et al.*, 1998; Rodrigues *et al.*, 2000; Liu *et al.*, 2001; Liu and Zhang, 2003; Corrado and Rodrigues, 2004 and Liu *et al.*, 2004). Many of the aquatic Hyphomycetes have now been reported as root endophytes. These endophytic fungi influence the metabolism of the plants in which they live in different way (Singh and Waingankar, 2005). The endophytic fungi are recognized as being of great importance for their host plants, protecting them against pests, insects, nematodes and plant pathogenic fungi and bacteria (Azevedo *et al.*, 2000).

Some endophytic fungi provide antibiotic compounds in culture that are active against human and plant pathogenic bacteria (Fisher *et al.*, 1984 a, b, 1986; Dreyfuss, 1986). Endophytes are also well known to constitute a valuable source of secondary metabolites for the discovery of new potential therapeutic drugs (Miller, 1995).

Though some workers (Fisher *et al.*, 1986; Clay, 1988; Shearer and Zare-Maivan, 1988; Gloer, 1995; Bills, 1996) focused on fungal secondary metabolites and their importance against pathogens but the experimental report on metabolites and antimicrobial properties of these fungi is almost nil, except the work of Gulis and Stephanovich (1999) and Singh and Waingankar (2005).

Gulis and Stephanovich (1999) studied the antimicrobial effects of culture filtrates of aquatic Hyphomycetes against Gram-positive bacteria, yeast and hyphomycetes, involving in the production of antibiotics. *Fusarium cavispermum*, *Sporidesmium fuscum* and *Anguillospora longissima* have been reported as having antimicrobial effects (Harrigan *et al.*, 1995; Gulis and Stephanovich, 1999). Singh and Waingankar (2005) also studied the endophytic hyphomycetes isolated from medicinal plants for antimicrobial activity against *Bacillus subtilis*, *B. cereus*, *B. mycoides* and *Staphylococcus aureus*.

Interaction of aquatic Hyphomycetes with bacteria and terrestrial fungi has been suggested by some workers (Kaushik and Hynes, 1968; Chamier *et al.*, 1984; Gulis and Suberkropp, 2003). A new antimicrobial compound 'Quinapathin' has been described from the aero-aquatic Hyphomycetes (*Helicoon richonis* (Boud.) Linder (Fisher *et al.*, 1988; Adriaenssens *et al.*, 1994). Like wise the study on isolation and structural determination of antimicrobial compound, 'Anguillosporal', from *Anguillospora longissima* has resulted in the discovery of new metabolite (Harrigan *et al.*, 1995). 'Clavariopsins', an antibiotic produced by an aquatic Hyphomycetes *Clavariopsis aquatica* (Kaida *et al.*, 2001) showed in vitro antifungal activity against *Aspergillus fumigatus*, *A. niger* and *Candida albicans*.

Previously, Platas *et al.* (1998) and Gulis and Stephanovich (1999) demonstrated the antagonistic activity of aquatic Hyphomycetes due to release of diffusible inhibitory substances. According to Gloer (1995) the secondary metabolites of aquatic Hyphomycetes could result in the discovery of new natural bioactive products of medicinal and agricultural importance. In a study on antibiotic effects of aquatic Hyphomycetes Gulis and Stephanovich (1999) suggested that due to their specific habitat, they may have good biosynthetic capabilities different from those of terrestrial fungi.

Endophytes are gaining importance, as they are producing bioactive compounds of agricultural and medicinal importance. Fisher *et al.* (1984 a, b) observed the antibiotic activity of some endophytic fungi. Recently, Singh and Waingankar (2005) have also observed that a majority of endophytes isolated by them showed significant antibiotic activity.

Gulis and Suberkropp (2003) screened 28 isolates of aquatic Hyphomycetes belonging to *Alatospora acuminata* (6 isolates), *Anguillospora filliformis* (5 isolates), *Articulospora tetracladia* (10 isolates), *Tetrachaetum elegans* (4 isolates) and *Tricladium chaetocladium* (3 isolates) against 16 bacterial isolates. These aquatic Hyphomycetes inhibited the bacterial growth by forming the clear zones in the bacterial lawns around wells containing the fungal culture broth.

Recently Antagonistic activity of *Heliscus lugdunensis, Tetrachaetum elegans,* against 7 plant pathogenic fungi was studied by Sati and Arya (2010 b) using a dual culture technique. They found that *Tetrachaetum elegans* had a very significant antagonistic activity against *Colletotrichum falcatum, Fusarium oxysporum, Pyricularia oryzae, Sclerotium sclerotiorum* and *Tilletia indica.* Whereas, *Heliscus lugdunensis* showed its antagonism against only 2 plant pathogenic fungi, *Rhizoctonia solani* and *Colletotrichum falcatum.*

Other Aspects

As the reports indicates that aquatic Hyphomycetes mainly occur in the freshwater streams and remain almost absent in stagnant water, these fungi may well be used as indicator of water pollution. These fungi are dominant group of microbes during the first stage of leaf litter decomposition in fresh water streams. They can be used to enhance the fast litter decomposition in freshwater ecosystem and nutrients release. The root endophytic aquatic Hyphomycetes are known to helpful for plant health and can be used to enhance the host plant biomass. Further study on the mechanism for the health of plants by these fungi can be initiated in this direction. Aquatic Hyphomycetes occurring as endophytes not only help in developmental and physiological activity of plants but also antagonize to their fungal and bacterial pathogens. The antifungal and antibacterial property of these fungi may also be used as naturally occurring compounds, which can compete with synthetic fungicides and bactericides in the modern time.

It can be concluded from the above review that identification and cultivation of many species of aquatic Hyphomycetes is needed. As the results the roles of aquatic Hyphomycetes in the leaf decomposition, in the nutrition of the stream detritivores, their endophytic and antimicrobial activity can be firmly established.

Acknowledgements

Authors are thankful to Head, Department of Botany, Kumaun University, Nainital for providing necessary facilities to carry out present work. Thanks are also due to the DST, New Delhi for financial support.

References

Adriaenssens, P., Anson, E. Begley, M. J. Fisher, P. J., Orrel, K. G., Webster, J. and Whitehurst, J. S. 1994. Quinaphthin, a metabolite produced by *Helicoon richonis*. Journal of Chemical Society. Perkin Transactions 1. No. 14, 2007-2010.

Akridge, R. E. and R. D. Koehn. 1987. Amphibious hyphomycetes from the San Marcos River in Texas. Mycologia, 79: 228-233.

Alasoadura, S. O. 1968. Some aquatic Hyphomycetes from Nigeria. Trans. Brit. Mycol. Soc, 51: 535-540.

Ananda, K. and Sridhar, K. R. 2002. Diversity of endophytic fungi in the roots of mangrove species on the west coast of India. Can. J. of Microb., 48: 871-878.

Ando, K. and Tubaki, K. 1984. Some undescribed hyphomycetes in the rain drops from intact leaf surface. Trans. Mycol. Soc. Japan, 25: 21-37.

Arshad, M. and Bareen, F. 2007. Colonization pattern of aquatic hyphomycetes on persistent substrata in some irrigation channels of Lahore. Mycopath., 5: 67-70.

Azevedo, J. L., Jr. Maccheroni, W., Pereira, J. O. and Araujo, W. L. 2000. Endophytic microorganism: a review on insect control and recent advances on tropical plants. EJB Elect. J. Biotech. 3: 40-65.

Bacon, C. W. and Hills, N. S. 1997. *Neotyphodium / Grass Interactions*. Plenum Press, New York, NY.

Baldy, V., Gessner, M.O. and Chauvet, E. 1995. Bacteria, fungi and the breakdown of leaf litter in a large river. Oikos, 74: 93-102.

Bandoni, R. J. 1972. Terrestrial occurrence of some aquatic Hyphomycetes. Cand. J. Bot., 50: 2283-2288.

Bandoni, R. J. 1974. Mycological observations on aqueous films covering decaying leaves and other litter. Trans. Brit. Mycol. Soc., 15: 309-315.

Barlocher, F and Kendrick, B. 1974. Dynamics of fungal population on leaves in a stream. J. Ecol., 62: 761- 791.

Barlocher, F and Kendrick, B. 1976. Hyphomycetes as intermediaries of energy flow in streams. In; "Recent Advances in Aquatic Mycology", (E.B. Gareth Jones Ed.) 435-446.

Barlocher, F and Oertli, J. J. 1978 a. Colonization of conifer needles by aquatic hyphomycetes. Can. J. Bot. 56: 57-62.

Barlocher, F and Oertli, J. J. 1978 b. Inhibitors of aquatic hyphomycetes in dead conifer needles. Mycologia, 70: 964-974.

Barlocher, F. 1982. On the ecology of Ingoldian fungi. Bioscience, 32:581-586.

Barlocher, F. 1987. Aquatic hyphomycetes spora in 10 streams in New Brunswick and Nova Scotia. Can. J. Bot., 65: 76-79.

Barlocher, F. 1991. Intraspecific Hyphal Interactions among Aquatic Hyphomycetes. Mycologia, 83: 82-88.

Barlocher, F. 1992. The ecology of aquatic hyphomycetes. Berlin:Springer-Verlag. 225p.

Barlocher, F. 2006. Fungal endophytes in submerged roots. In: Schulz BJZ, Boyle CJC, Sieber, TN, eds. Microbial roots endophytes. Berlin Germany, Springer Verlag, 179-190.

Barlocher, F. and Kendrick, B. 1973 a. Fungi and food preferences of *Gammarus pseudolimnaeus*. Arch. Hydrobiol., 72: 501-516.

Barlocher, F. and Kendrick, B. 1973 b. Fungi in the diet of *Gammarus pseudolimnaeus* (Amphipoda) Oikos, 24: 295-300.

Barlocher, F. and Rosset. J. 1981. Aquatic Hyphomycetes Spora of two Black Forest and two Swiss Jura streams. Trans. Brit. Mycol. Soc., 76: 479-483.

Barlocher, F., Kendrick, B. and Michaelides, J. 1977. Colonization of rosin coated slides by Aquatic Hyphomycetes. Can. J. Bot., 55: 1163-1166.

Barlocher, F., Kendrick, B. and Michaelides, J. 1978. Colonization and conditioning of *Prinus resinosa* needles by aquatic Hyphomycetes. Arch. Hydrobiol., 81: 462-674.

Baschien, C. 2003. Development and Evaluation of rDNA Targeted *in situ* Probes and Phylogenetic Relationships of Freshwater fungi. Ph. D. Thesis, Technical University, Berlin.

Bauer, R., Begerow, D. Oberwinkler, F. and Marvanova, L. 2003. Classicula: the teleomorphs of *Naiadella fluitans*. Mycologia, 95: 756-764.

Belwal, M., Sati, S. C. and Arya, P. 2008. Temperature tolerance of water-borne conidial fungi in freshwater streams of Central Himalaya. Nat. Acad. Sci. Lett. 31: 5-6.

Bengtsson, G. 1983. Habitat selection in two species of aquatic Hyphomycetes. Microbial Ecology, 9: 15-26.

Bermingham, S., Maltby, L. and Cooke, R. C. 1996. Effects of a coal mine effluent on aquatic hyphomycetes. I. Field study. J. App. Ecol., 33: 1311-1321.

Bernstein, M. E. and Carroll, G. C. 1977. Internal fungi in old growth Douglas fir foliage. Can. J. Bot., 55: 644-653.

Bills, G. F. 1996. Isolation and analysis of endophytic fungal communities from woody plants. In Endophytic fungi in Grasses and woody plants (eds) Redlin S. S. and Carris L.M.A. PS Press, Saint Paul. Pp. 121-132.

Bills, G. F. and Polishook, J. D. 1992. Recovery of endophytic fungi from *Chamaecykaris thyoides*. Sydowia, 44: 1-12.

Campbell, J., Shearer, C. A. and Marvanova, L. 2002. Evolutionary relationships between aquatic anamorphs and teleomorphs: *Lemonniera*. 7th International Mycological Congress, 11-17 August 2002, Oslo: 193-194. (Abstract).

Carroll, G. C. and Carroll, F. E. 1978. Studies on incidence of coniferous needle endophyte in the Pacific Northwest. Can. J. Bot., 56: 3034-3043.

Chamier, A. C., Dixon, P. A. and Archer, S. A. 1984. The spatial distribution of fungi on decomposing alder leaves in a freshwater stream. Oecologia, 64: 92-103.

Chamier, A. C. 1985. Cell wall degrading enzymes of aquatic hyphomycetes: a rivew. Bot. J. Linn. Soc., 91: 67-81.

Chamier, A. C. and Dixon, P. A. 1982. Pectinases in leaf degradion by aquatic hyphomycetes: The enzymes and leaf maceration. Jour. Gen. Microb. 128: 2469-2483.

Chandrashekar, K. R., Sridhar, K. R. and Kaveriappa, K. M. 1989. Palatibility of Rubber leaves colonized by aquatic hyphomycetes, Arch. Hydrobiol, 115: 361-370.

Chandrashekhar, K. R., Sridhar K. R. and Kaveriappa, K. M. 1991. Aquatic hyphomycetes of a sulphur spring. Hydrobiologia, 218: 151-156.

Chandrashekhar, K. R., Sridhar, K. R. and Kaveriappa. K. M, 1986. Aquatic hyphomycetes of Kempu Hole in Western Ghat forests of Karnataka. Indian Phytopathol., 39: 249-255.

Chauvet, E. 1991. Aquatic hyphomycetes distribution in South-Western France. Journal of Biogeography, 18: 699-706.

Chauvet, E., Fabre, E., Elosegui, A. and Pozo, J. 1997. The impact of eucalypt on the leaf-associated aquatic Hyphomycetes in Spanish streams. Canadian Journal of Botany, 75: 880-887.

Chergui, H. and Pattee, E. 1988.The dynamics of Hyphomycetes on decaying leaves in the network of the River Rhone (France). Arch. Hydrob. 114: 3-20.

Clay, K. 1988. Fungal endophyte of grasses a defensive mutualism between plants and fungi. Ecology, 69: 10-16.

Clay, K. 1989. Clavicipitaceous endophytes of grasses, their potential as biocontrol agents. Mycol. Res., 92: 1-12.

Clay, K. 1990. Fungal endophytes of grasses. Ann Rev Ecol Syst, 21: 275-297.

Clay, K. 1996. Interactions among fungal endophytes, grasses and herbivores. Researches on Population Ecology, 38: 191-201.

Clay, K. and Schardl, C. 2002. Evolutionary origins and ecological consequences of endophyte symbiosis with grasses. Am. Nat., 160: S99-S127.

Conway, K. E. 1970. The aquatic Hyphomycetes of central New York. Mycologia, 62: 516 -530.

Corrado, M. and Rodrigues, K. F. 2004. Antimicrobial evaluation of fungal extracts produced by endophytic strains of *Phomopsis* sp., J. Basic Microbiol., 44: 157-160.

Cowling, S. W. and Waid, J. S. 1963. Aquatic Hyphomycetes in Australia. Australian Jour. *Sci.*, 26:122-124.

Czeczuga, B. and Orlowska, M. 1997. Hyphomycetes fungi in rainwater falling from building roofs. Mycoscience, 38: 447-450.

Czeczuga, B. and Orlowska, M. 1999. Hyphomycetes fungi in rainwater, melting snow and ice. Acta Mycologica, 34: 181-200.

De Wideman, E. 1894. Notes Mycologiques. Fascicle 3 Ann. Soc. Belge. Microsc., 18: 135-161.

De Wildeman, E. 1893. Notes Mycologiques. *Annales de la Societe Belge de Microscopie*, 17: 35-68.

De Wildeman. E. 1895. Notes Mycologiques. Fascicle 6 Ann Soc. Belge. Microsc, 19:191-232.

Descals, E. 2005. Diagnostic characters of propagules of Ingoldian fungi. Mycol. Res., 101: 545-555.

Descals, E. and Webster, J. 1982 a. Taxonomic studies on "aquatic hyphomycetes" III some new species and a new combination. Tran. Brit. Mycol. Soc., 78: 405-437.

Descals, E. and Webster, J. 1982 b. Taxonomic studies on "aquatic Hyphomycetes" IV. Pure culture and typification of various species Trans. Brit. Mycol. Soc. 79: 45-64.

Descals, E., Sander, P. F. and Ugalde, U. 1978. Hifomycetos ingoldianos del pais vasco. Munibe, 29: 237-260.

Dreyfus, M. M. and Chapela, I. H. 1992. The potential of fungi in discovery of novel, low molecular weight Pharmaceuticals. In the Discovery of Novel Natural Products with Therapeutic Potential, Butterworth Publications, Biotechnology Series, London.

Dreyfuss, M. M. 1986. Neue Exkenntnisse aus einem pharmakologischen Pilzscreening. Sydowia, 39: 22- 36.

Dreyfuss, M. M. and Chapela, I.H. 1994. In: The discovery of natural products with therapeutic potential (ed.) V.P. Gullo. Butterworth-Heinemanu, Lond, pp. 44-80.

Duarte, S. Pascoal, C. Cassio, F. and Barlocher, F. 2006. Aquatic hyphomycetes diversity and identity affect leaf litter decomposition in microcosms. Oecologia, 147: 658-666.

Elmi, A. A. and West, C. P. 1995. Endophytic infection effects on stomatal conductance, osmotic adjustment and drought recovery of tall fescue. New Phytol., 131: 61-67.

Fabre, E. 1996. Relationships between aquatic hyphomycetes communities and riparian vegetation in 3 Pyrenean streams. Comptes Rendus Hebdomadaires des Seances des I' Academie des Sciences, Paris 319: 107-111.

Fernando, A. A. and Currah, R. S. 1995. *Leptodontidium orchidicola* (*Mycelium radicis atrovirens* complex): aspects of its conidiogenesis and ecology. Mycotaxon, 54: 287-294.

Field, J. I. and Webster, J. 1985. Effects of sulphide on survival of aero aquatic hyphomycetes. Trans. Brit. Mycol. Soc. 85: 193-198.

Findley, S.*E*.*G*. and Arsuffi, T.L. 1989. Microbial growth and detritus transformations during decomposition of leaf litter in a stream. Fresh water Biol., 21: 261-269.

Fisher P. J., Anson, A.E. and Petrini, O. 1984 a. Antibiotic activity of some endophytic fungi from ericaceous plants. Bot. Helv., 94: 249-253.

Fisher P. J., Anson, A. E and Petrini, O. 1986. Fungal endophytes in *Ulex europaeus* and *Ulex gallii*. Trans. Brit. Mycol. Soc., 86: 153-156.

Fisher P. J., Anson, A.E. and Petrini, O. 1984 b. Novel antibiotic activity of some endophytic *Cryptospriopsis* sp. isolated from *Vaccinium myrytillus*. Trans. Brit. Mycol. Soc., 83: 145- 148.

Fisher, P. J. and Petrini, O. 1989. Two aquatic hyphomycetes as endophytes in *Alnus glutinosa* roots. Mycol. Res., 92: 367-368.

Fisher, P. J. and Petrini, O. 1992. Fungal saprobes and pathogens as endophyte of Rice (*Oryzae sativa*). New Phytol, 120: 137-143.

Fisher, P. J. Petrini, O. and Webster, J. 1991. Aquatic hyphomycetes and other fungi living aquatic and terrestrial roots of *Alnus glutinosa*. Mycol. Res., 95: 543-547.

Fisher, P. J., Anson, A. E. and Webster, J. 1988. Quinaphthin, a new antibiotic, produced by *Helicoon richonis*,. Trans. Brit. Mycol. Soc., 90: 499-502.

Fisher, P. J. and Webster, J. 1978. Sporulation of aero-aquatic fungi under different gas regimes in light and darkness. Trans. Brit. Mycol. Soc., 71: 465-468.

Garnett, H., Barlocher, F. and Giberson, D. 2000. Aquatic hyphomycetes in Catamaran Brook: Colonization dynamics, seasonal patterns and logging effects. Mycologia, 92: 29-41.

Gessner, M. O. and Chauvet, E. 1997. Growth and production of aquatic hyphomycetes in decomposing leaf litter. Limnol. Oceanogr., 42: 496-505.

Gessner, M. O., and Chauvet, E. 1994. Importance of stream microfungi in controlling breakdown rates of leaf litter. Ecology, 75: 1807-1817.

Gloer, J. B. 1995. Bioactive metabolite from aquatic fungi. In The VI International Marine Mycological Symposium (Incorporating Society, 8-15, July 1995, Programme) Abstracts, p. 65, University of Portsmouth: England.

Gonczol, J. 1975. Ecological observations on the aquatic Hyphomycetes of Hungary I. Acta Botanica Academiae Scientiarm Hungaricae, 21:243-264.

Gonczol, J. 1989. Longitudinal distribution pattern of aquatic Hyphomycetes in a mountain stream in Hungary, Experiment with leaf packs. Nova Hedwigia, 48: 391-404.

Gonczol, J. and Revay, A. 2003. Aquatic hyphomycetes in the Morgo stream system. Hungary tributary communities. Nova Hedwigia, 76: 173-189.

Gonczol, J. and Toth, S. 1974. Rare or interesting conidia from stream of Hungary. Bot. Koztem 61. Kotet, 1-fuzet.

Gonczol, J., Revay, A. and Csontos, P. 1999. Studies on the aquatic hyphomycetes of the Morgo stream, Hungary. J. Longitudinal changes of species diversity and conidial concentration. Archiv fir Hydrobiologia, 144: 473-493.

Graca, M. A. S., Maltby, L. and Calow, P. 1993. Importance of fungi in the diet of *Gammarus pulex* and *Asellus aquaticus*. I. Feeding Strategies. Oecologia, 93: 139-144.

Grattan, R. M. II and Suberkropp, K. 2001. Effects of nutrient enrichment on yellow poplar leaf decomposition and fungal activity in streams. Journal of the North American Benthological Society, 20: 33-43.

Gulis, V. 2001. Are there any substrate preferences in aquatic hyphomycetes? Mycol. Res., 105: 1088-1093.

Gulis, V. and Suberkropp, K. 2003. Effect of inorganic nutrients on relative contributions of fungi and bacteria to carbon flow from submerged decomposing leaf litter. Microbial Ecology. 45: 11-19

Gulis, V. and Suberkropp, K. 2004. Effects of whole – stream nutrient enrichment on the concentration and abundance of aquatic hyphomycetes conidia in transport. Mycologia, 96: 57-65.

Gulis, V. I. and Stephanovich, A. I. 1999. Antibiotic effects of some aquatic hyphomycetes. Mycol. Res., 103: 111-115.

Gulis, V., Rosemond, A. D., Suberkropp, K., Weyers, H. S. and Benstead, J. P., 2004. Effects of nutrient enrichment on the decomposition of wood and associated microbial activity in streams. Freshwater Biology, 49: 1437-1447.

Harrigan, G. G., Armentrout, B. L., Gloer, J. B. and Shearer, C. A. 1995. New bioactive natural products from two *Anguillopsora* species. In The VI International Marine Symposium (Incorporating Freshwater Mycology). A Meeting of the British Mycological Society, 8-15 July, 1995. Programme and Abstracts, p. 135, University of Portsmouth., England.

Harrington, T. J. 1997. Aquatic hyphomycetes of 21 rivers in southern Ireland. Biology and Environment: Proceedings of the Royal Irish Academy, 97b: 139-148.

Hudson, H. J. and Ingold, C. T. 1960. Aquatic Hyphomycetes from Jamaica. Trans. Brit. Mycol. Soc., 73: 109-116.

Hyde, K. D. and Goh, T. K. 2003. Adaptations for dispersal in filamentous freshwater fungi. Fungal Diversity Research Series, 10: 231-258.

Ingold, C. T. 1942. Aquatic Hyphomycetes of decaying alder leaves. Trans. Brit. Mycol. Soc., 25: 339-417.

Ingold, C. T. 1943 a. Further observation on aquatic hyphomycetes of decaying leaves. Trans. Brit. Mycol. Soc., 26: 104-115.

Ingold, C. T. 1943 b. *Triscelophorus monosporus* n. gen., n. sp., an aquatic hyphomycetes. Trans. Brit. Mycol. Soc., 26: 148-152.

Ingold, C. T. 1943 c. On the distribution of aquatic hyphomycetes saprophytic on submerged decaying leaves. New Phytol., 42: 139-143.

Ingold, C. T. 1944. Some new aquatic Hyphomycete. Trans. Brit. Mycol. Soc., 27: 35-47.

Ingold, C. T. 1949. Aquatic Hyphomycetes from Switzerland. Trans. Brit. Mycol. Soc., 32: 341-345.

Ingold, C. T. 1952. *Actinispora megalospora* n. sp. an aquatic hyphomycetes. Trans. Brit. Mycol. Soc., 35: 66-70.

Ingold, C. T. 1956. Stream spora in Nigeria. Trans. Brit. Mycol. Soc., 39: 108-110.

Ingold, C. T. 1958 a. Aquatic hyphomycetes from Uganda and Rhodesia. Trans. Brit. Mycol. Soc., 41: 109-114.

Ingold, C. T. 1958 b. New aquatic Hyphomycetes: *Lemonniera brachycladia, Anguillospora crassa* and *Fluminispora ovalis*. Trans. Brit. Mycol. Soc., 41: 365-372.

Ingold, C. T. 1960. Aquatic Hyphomycetes from Canada. Can. J. Bot., 38: 803-806.

Ingold, C. T. 1965. Hyphomycete spores from mountain torrents. Trans. Brit. Mycol. Soc., 48: 453-458.

Ingold, C. T. 1968. More spores from rivers and streams. Trans. Brit. Mycol. Soc., 51: 137-143.

Ingold, C. T. 1974 a. Foam spora from Britain. Trans. Brit. Mycol. Soc., 63: 487-497.

Ingold, C. T. 1974 b. *Tricladium chaetocladium* sp. nov., an Aquatic hyphomycetes from Britain. Trans. Brit. Mycol. Soc., 63: 624-626.

Ingold, C. T. 1975 a. An illustrated guide to aquatic and water-borne hyphomycetes (Fungi Imperfecti) with notes on their biology. Freshwater Biol. Assoc. Scient. Publ. No. 30, England, 96 pp.

Ingold, C. T. 1975 b. Conidia in foam of two English streams. Trans. Brit. Mycol. Soc., 65: 522-527.

Ingold, C. T. 1975 c. Convergent evolution in aquatic fungi. The tetraradiate spore. Biol. J. Linn. Soc., 7: 1-25.

Ingold, C. T. 1979. Advances in the study of so called aquatic Hyphomycetes. Am. J. of Botany, 66: 218-226.

Ingold, C. T., Dann, V. and McDougall, P. J. 1968 a. *Tripospermum camelopardus* sp. nov., Trans. Brit. Mycol. Soc., 51: 51-56.

Ingold, C. T., McDougall, P. J. and Dann, V. 1968 b. *Volucrispora graminea*. sp. nov., Trans. Brit. Mycol. Soc., 51: 325-328.

Ingold, C. T. 1976. The morphology and biology of fresh water fungi excluding Phycomycetes. In "Recent advances in aquatic Mycology". (Ed. E.B. Gareth Jones) Publ. Paul Elek. Ltd. 335-357.

Ingold, C. T. and Ellis, E. A. 1952. On some hyphomycetes spores, including those of *Tetracladium maxilliformis*, from Wheatfen. Trans. Brit. Mycol. Soc., 35: 158-161.

Ingold, C. T. and Webster, J. 1973. Some aquatic Hyphomycetes from India. Kavaka, 1: 5-9.

Iqbal, S. H. and Bhatty, S. F. 1979. Conidia from stream foam. Trans. Mycol. Soc. Japan, 20: 83-91.

Iqbal, S. H. and Webster, J. 1973 a. The trapping of aquatic hyphomycetes spores by air bubbles. Trans. Brit. Mycol. Soc., 60: 37-48.

Iqbal, S. H. and Webster, J. 1973 b. Aquatic Hyphomycetes spora of the River Exe and its tributaries. Trans. Brit. Mycol. Soc., 61: 331-346

Iqbal, S. H., Bhatty, S. F. and Malik, K. S. 1980. Fresh water hyphopmycetes of Pakistan. Bulletin of Mycology, 1: 1-25.

Iqbal, S. H., Sultana, K. and Farzana, S. 1973. Stream spora of Pakistan. Biologia, 19: 79-84.

Kaida, K., Fudou, R., Kameyama, T., Tubaki, K., Suzuki, Y., Ojika, M. and Sakagami, Y. 2001. New Cyclic Depsipeptide Antibiotics, Clavariopsins A and B, Produced by an Aquatic Hyphomycetes *Clavariopsis aquatica*. J. Antibiot, 54: 17-21.

Kaushik, N. K. and Hynes, H. B. N. 1968. Experimental study on the role of autumn shed leaves in aquatic environment. J. Ecol., 56: 229-243.

Kaushik, N. K. and Hynes, H. B. N. 1971. The fate of dead leaves that fall into streams. Arch. Hydrobiol., 68: 465 -515.

Khan, M. A. 1986. Secondary conidial production in *Articulospora tetracladia*. Pakistan J. Bot., 18: 175-177.

Khan, M. A. 1987. Interspecies interactions in aquatic hyphomycetes. Bot. Mag. Tokyo, 100: 295-303.

Kirk, P. W. and Brandt, J. M. 1980. Seasonal distribution of linicolous marine fungi in the lower Chesapeake Bay.-Botanica Marina, 13: 657-668.

Kohlmeyer, J., Schemidt,I. and Nair, N. B. 1967. Eine neue Corollospora (Ascomycetes) aus dem Indischen Ozean und der Ostee.-Ber. Deutsch. Bot. Ges. 80: 98-102.

Koske, R. E. and Duncan, I. W. 1974. Temperature effects on growth, sporulation and germination of some 'aquatic' hyphomycetes. Can. J. Bot., 52: 1387-1391.

Kostalos, M. and Seymour, R. L. 1976. Role of microbial enriched detritus in the nutrition of *Gammarus minus*. Oikos, 27: 512-516.

Krauss, G., Barlocher, F., Schreck, P., Wennrich, R., Glasser, W. and Krauss, G. J. 2001. Aquatic hyphomycetes occurrence in hyper polluted water in Central Germany. Nova Hedwigia, 72: 419-428.

Kuthubutheen, A. J. 1987. A new species of *Phalangispora* and further observation on *P. constricta* from Malaysia. Trans. Brit. Mycol. Soc., 89: 414-420.

Kuthubutheen, A. J., Liew, G. M. and Nawawi A. 1992. *Nawawia nitida* anam. Sp. Nov (Hyphomycetes) and further records of *Nawawia filliformis* from Malaysia. Can. J. Bot., 70: 96-100.

Liu, C. H., Zou, W. X., LU, H. and Tan, R. X. 2001. Antimicrobial activity of Artemisia annua endophyte cultures against phytopathogenic fungi. *J. Biotechnol*, 88: 277-282.

Liu, J. Y., Song, Y. C., Thang, Z. Wang, L., Guo, Z. J. and Zou, W. X. 2004. Aspergillus fumigatus CY018, an endophytic fungus in *Cynodon dactylon*, as a versatile producer of new and bioactive metabolites. J. Biotechnol., 114: 279-289.

Liu, Y. and Zhang, K. 2003. Antimicrobial activities of selected *Cythus* species. Mycopath., 157: 185-189.

Malinowski, D. P., Alloush, G.A. and Belesky, D. P. 2000. Leaf endophyte *Neotyphodium coenophialum* modifies mineral uptake in tall fescue. Plant Soil, 227: 115-126.

Manoharachary, C. 1984. Conidiogenesis in water-borne moulds. Key note lecture annual meeting of microbiologists of India. Pantnagar, India.

Manoharachary, C. 1985. Aspects and Prospects of water-borne conidial fungi from India. Ad. Biol. Res., 4: 160-183.

Manoharachary, C. 1989. Glimpses on water-borne conidial fungi from India- In"*Perspectives in Aquatic Biology*". (Eds. R. D. Khulbe). Papyrus Publishing House, New Delhi.

Manoharachary, C. and Murthy, A. B. S. 1981. Foam spora from Andhra Pradesh. Curr. Sci., 50: 378-379.

Manoharachary, M. and Bhairav Nath, D. 1985. Conidial fungi from fresh water foam. Indian J. Bot., 8: 67-68.

Marvanova, L. 1975. Concerning *Gyoerffyella* Kol. Trans. Brit. Mycol. Soc., 65: 555-565.

Marvanova, L. 1977. Two new *Alatospora* species, Archiv fur Protistenkunde, 119: 68-74.

Marvanova, L. 1986. Three new hyphomycetes from foam. Trans. Brit. Mycol. Soc., 87: 617-625.

Marvanova, L. 1997. Freshwater hyphomycetes: A survey with remarks on tropical taxa *In* Tropical Mycology (eds. K. K. Janardhanan, C. Rajendran, K. Natrajan and D. L. Hawksworth) Science Publishers Inc., 169-226.

Marvanova, L. 2002. Aquatic Hyphomycetes-emerging outlines of their classification in the fungal system. 3rd International Meeting on "Plant Litter Processing in Freshwater", September 7-12, 2002, Szentendre, Hungary: 11. (Abstract).

Marvanova, L. and Bandoni, R. I. 1987. *Naidella fluitans* gen. et. sp. nov,: a conidial basidiomycetes. Mycologia, 79: 573-886.

Marvanova, L. and Barlocher, F. 1988. Hyphomycetes from Canadian streams. I. Basidiomycetous anamorphs. Mycotaxon, 32: 339-351.

Marvanova, L. and Fisher, P. J. 1991. A new endophytic hyphomycetes from alder roots. Nova Hedwigia, 52: 33-37.

Marvanova, L. Fisher, P. J., aimer, R. and segedin, B. C. 1992. A new *Filosporella* from alder roots and foam water. Nova Hedwigia, 54: 151-158.

Mckinley, V. L. and Vestal, J. R. 1982. Effect of acid on plant litter decomposition in an arctic lake. Appl. Eviron. Microb., 43: 1188-1195.

Mer, G. S. and Khulbe, R. D. 1981. Aquatic Hyphomycetes of Kumaun Himalaya, India. Sydowia, Annales Mycologic., 34: 118-124.

Mer, G. S. and Sati, S. C. 1989. Seasonal fluctuation in species composition of aquatic Hyphomycetous flora in a temperate freshwater stream of Central Himalaya, India. Int. Rev. Ges. Hydrobiol., 74: 433-437.

Michaelides, J. and Kendrick. B. 1978. An investigation of factors retarding colonization of conifer needles by amphibious hyphomycetes in streams. Mycologia, 70: 419-430.

Miller, S. L. 1995. Functional diversity in fungi. Can. J. Bot., 73 (Suppl. 1), S50-S57.

Muller-Haeckel, A. and Marvanova, L. 1979. Periodicity of aquatic hyphomycetes in the Sub arctic. Trans. Brit. Mycol. Soc., 73: 109-116.

Nawawi, A. 1975. *Triscelophorous acuminatus* sp. nov. Trans. Brit. Mycol. Soc., 64: 345-348.

Nawawi, A. 1985 a. Aquatic Hyphomycetes and other water-borne fungi from Malaysia. Malay. Nat.

J., 39: 75-134.

Nawawi, A. 1985 b. Another aquatic Hyphomycetes genus from foam. Trans. Brit. Mycol. Soc., 85: 174-177.

Nawawi, A. 1985 c. More *Tricladium* species from Malaysia. Trans. Brit. Mycol. Soc., 85: 177-182.

Nawawi, A. 1985 d. Some interesting hyphomycetes from water. Mycotaxon, 24:217-226.

Nawawi, A. 1987 a. *Clavariopsis azlanii* sp. nov. a new aquatic hyphomycete from Malaysia. Trans. Brit. Mycol. Soc., 88: 428-432.

Nawawi, A. 1987 b. *Diplocladiella appendiculata* sp. nov. a new aquatic hyphomycete. Mycotaxon, 28: 297-302.

Nawawi, A. and Kuthubutheen, A. J. 1987. *Triscelosporium verrucosum* gen. et sp. nov., a dematiaceous aeroaquatic hyphomycetes with tetraradiate conidia. Trans. Brit. Mycol. Soc., 29: 285-290.

Nawawi, A. and Kuthubutheen, A. J. 1988 a. *Tricladiospora*, a new genus of dematiaceous hyphomycetes with staurosporous conidia from submerged decaying leaves. Trans. Brit. Mycol. Soc., 90: 482-487.

Nawawi, A. and Kuthubutheen, A. J. 1988 b. Another new hyphomycetes from leaf litter. Mycotaxon, 31: 339-343.

Nawawi, A. and Kuthubutheen, A. J. 1988 c. *Tricladiomyces geniculatus* sp. nov. a conidial basdidiomycete. Trans. Brit. Mycol. Soc., 90: 670- 673.

Nemec, S. 1969. Sporulation and identification of fungi isolated from root rot diseased Strawberry plants. Phytopathology, 59: 1552- 1553.

Nikolcheva, L. G. and Barlocher, F. 2002. Phylogeny of *Tetracladium* based on 18S rDNA. Czech Mycology, 53: 285-295.

Nilsson, S. 1964. Fresh water Hyphomycetes. Taxonomy, Morphology and Ecology. Symb. Bot. Upsal., 18:1-30.

Niyogi, D. K., Mcknight, D. M., Lewis Jr., W. M. 2002. Fungal communities and biomass in mountain streams affected by mine drainage. Archiv fur Hydrobiologie, 155: 255-271.

Park, D. 1974. Aquatic hyphomycetes on non-aquatic habitats. Trans. Brit. Mycol. Soc., 63: 183-187.

Pascoal, C., Marvanova, L. and Cassio, F. 2005. Aquatic Hyphomycetes diversity in streams of Northwest Portugal. Fungal Diversity, 19: 109-128.

Pelaez, F., Collado, J., Arenal, F., Basilio, A., Cabello, A. and Diez Matas, M. T. 1998. Endophytic fungi from plants living on gypsum soils as a source of secondary metabolites with antimicrobial activity, Mycol. Res., 102: 755-761.

Peterson, R. C. and Cummins, K. W. 1974. Leaf processing in woodland streams. Freshwater Biol. 4: 343-368.

Peterson, R. H. 1962. Aquatic Hyphomycetes from North America. I Aleuriosporae (Part I) and key to genera. Mycologia, 54: 117-151.

Peterson, R. H. 1963 a. Aquatic Hyphomycetes from North America II Aleuriosporae (Part II) and Blastosporae. Mycologia, 55: 18-29.

Peterson, R. H. 1963 b. Aquatic Hyphomycetes from North America III Phialosporae and Miscellaneous species. Mycologia, 55: 570-581.

Petrini, O. 1986. Taxonomy of endophytic fungi of aerial plant tissues. In Microbiology of the phyllosphere (eds.) N. J. Fokkema and J. Van den Henvel. Cambridge University, Cambridge pp. 175-187.

Petrini, O. 1991. Fungal endophytes of tree leaves. In: Microbial Ecology of Leaves. Andrews, J. A. and Hirano, S. S. (ed.) Springer Verlag, New York. Pp. 179-197.

Petrini, O., Sieber, T. N., Toti, L., and Viret, O. 1992 a. Ecology, metabolite production and substrate utilization in endophytic fungi. Natural Toxins, 1: 185-196.

Petrini, O., Fisher, P. J. and Petrini, L. F. 1992 b. Fungal endophyte of bracken (*Pteridium aquilinum*), with some reflections on their use in biological control. Sydowia, 44: 282-293.

Platas, G., Pelaez, F. Collado, J. Villuendas, G., Diez, M. T. 1998. Screening of antimicrobial activities by aquatic hyphomycetes cultivated on various nutrient sources. Cryptogam Mycol., 19: 33-43.

Raghu-Kumar, S. 1973. Marine lignicolous fungi from India. Kavaka, 1: 73-85.

Ranzoni, F. V. 1953. The aquatic Hyphomycetes of California. Farlowia, 4: 353-398.

Raviraja, N. S., Sridhar, K. R. and Barlocher, F. 1996. Endophytic aquatic hyphomycetes of roots of plantation crops and ferns from India. Sydowia, 48: 152-160.

Raviraja, N. S., Sridhar, K. R. and Barlocher, F. 1998 a. Fungal species richness in Western Ghats streams (southern India): is it related to pH, temperature or altitude? Fungal Diversity, 1: 179-191.

Raviraja, N. S., Sridhar, K. R. and Barlocher, F. 1998 b. Breakdown of *Ficus* and *Eucalyptus* leaves in an organically polluted river in India: fungal diversity and ecological functions. Freshwater Biol., 39: 537-545.

Regelsberger, B. Messner, K and Descals, E. 1987. Species diversity of aquatic hyphomycetes in four Austrian streams. Mycotaxon, 30: 439-454.

Revay, A. and Gonczol, J. 1990. Longitudinal distribution and colonization pattern of wood-inhabiting fungi in a mountain stream in Hungary. Nova Hedwigia, 51: 505-520.

Rodrigues, K. F., Hesse, M. and Wernes, C. 2000. Antimicrobial activities of secondary metabolites produced by endophytic fungi from *Spondias mombin*. J. Basic Microbiol., 40: 261-267.

Rosemond, A. D., Pringle, C. M., Ramirez, A., Paul, M. J. and Meyer, J. L., 2002. Landscape variation in phosphorus concentration and effects on detritus based tropical streams. Limnology and Oceanography, 47: 278-289.

Sanders, P. F and Anderson, J.M. 1979. Colonization of wood blocks by aquatic Hyphomycetes. Trans. Brit. Mycol. Soc., 73: 103-107.

Sati, S. C. and (Tiwari) Pant, N. 2006. Colonization and Successional Pattern of Aquatic Hyphomycetes in freshwater steams of Kumaun Himalaya. In "Recent Mycological Researches" (ed. S.C. Sati). I. K. International Publishing House, New Delhi, pp. 145-157.

Sati, S. C. and Arya, P. 2009. Occurrence of Water-borne Conidial Fungi in Relation to Some Physico-chemical Parameters in a Fresh Water Stream. Nature and Science, 7: 20-28.

Sati, S. C. and Belwal, M. 2005. Aquatic hyphomycetes as endophyte of riparian plant roots. Mycologia, 97: 45-49.

Sati, S. C. and Pant, N. 2000. Microbial decomposition of leaf litter in a fast flowing fresh water stream of Kumaun Himalaya. In: Microbes, Agriculture, Industry and Environment. (ed: D. K. Maheshwari, R.C. Dubey, G. Prasad and Navneet) pp. 197-207.

Sati, S. C. and Tiwari, N. 1990 a. Fresh water Hyphomycetes from Jageshwar stream, Kumaun Himalaya, India. Nat. Acad. Sci. Letters, 13: 7-9.

Sati, S. C. and Tiwari, N. 1990 b. Some aquatic hyphomycetes of Kumaun Himalaya, India. Mycotaxon, 39: 407-414.

Sati, S. C. and Tiwari, N. 1990 c. Some aquatic Hyphomycetes from high altitude stream of Kumaun Himalaya. Abs. In Frontiers in Bot. Research. National Symp., Chandigarh. pp 35.

Sati, S. C. and Tiwari, N. 1992 a. A new species of *Tricladium* from Kumaun Himalaya, India. Mycol. Res., 96: 229-232.

Sati, S. C. and Tiwari, N. 1992 b. Colonization, species composition and conidial production of aquatic hyphomycetes on chir pine needle litter in fresh water Kumaun Himalayan stream. Int. Revue. Ges. Hydrobiol., 77: 445-453.

Sati, S. C. and Tiwari, N. 1993 a. A new species of *Pestalotiopsis* on submerged leaf litter. Nova Hedwigia, 56: 543-547.

Sati, S. C. and Tiwari, N. 1993 b. Ingoldian aquatic hyphomycetes from two temperate fresh water streams of Nainital, in Kumaun Himalayas, India. Abstract in XV International Botanical Congress, Tokyo (Japan), 2382p. 324.

Sati, S. C. and Tiwari, N. 2003. A new species of *Pleurophragmium* from Nainital, Kumaun Himalaya, India. Nat. Acad. Sci. Lett., 26: 208-209.

Sati, S. C., Arya, P. and Belwal, M. 2009 a. *Tetracladium nainitalense* sp. nov. a root endophyte from Kumaun Himalaya, India. Mycologia. 101: 692-695.

Sati, S. C., Pargaien, N. and Belwal, M. 2009 b. Diversity of Aquatic Hyphomycetes as Root Endophytes on Pteridophytic Plants in Kumaun Himalaya. Journal of American Science, 5:179-182.

Sati, S. C., Mer, G. S. and Tiwari, N. 1989. Occurrence of water-borne conidial fungi on *Pinus roxburghii* needles, Curr. Science, 58: 918-919.

Sati, S. C., Pargaien, N. and Belwal, M. 2006. Three species of aquatic hyphomycetes as new root endophytes of temperate forest plants. Nat. Acad. Sci. Lett., 29: 9-10.

Sati, S. C., Tiwari, N. and Belwal, M. 2002 a. Species Diversity of water-borne conidial fungi in running freshwater bodies of Kumaun Himalaya, Uttaranchal. In: Microbial Diversity, Status and Potential Applications. Eds. S. C. Tiwari and G. D. Sharma. Pp. 26-35.

Sati, S. C., Tiwari, N. and Belwal, M. 2002 b. Conidial aquatic fungi of Nainital, Kumaun Himalaya, India. Mycotaxon, 81: 445-455.

Sati, S. C., Tiwari, N., and Mer, G. S. 1992. Colonization of aquatic hyphomycetes on pine needles. Indian Phytopath., 45: 106-107.

Sati, S. C. and Arya, P. 2010 a. Assessment of root endophytic aquatic Hyphomycetes fungi on plant growth. Symbiosis, 50: 143-149.

Sati, S. C. and Arya, P. 2010 b. Antagonism of some Aquatic Hyphomycetes against Plant Pathogenic Fungi. The Scientific World Journal, 10: 760-765.

Sati, S. C. and Tiwari, N. 1997. Glimpses of conidial aquatic fungi in Kumaun Himalaya. In "Recent Researches in Ecology, Environment and Pollution" X: 17-37. Eds. S.C.Sati, J. Saxena and R.C. Dubey. Today and Tomorrow printers and Publishers, New Delhi.

Sheare, C. A. 1972. Fungi of the Chesapeake Bay and its tributaries III, The distribution of wood-inhabiting ascomycetes and fungi imperfecti of the Patuxent River. Amer. J. Bot., 59: 961-969.

Shearer, C. A. 1992. The role of woody debris. In Barlocher, F. (ed). The ecology of aquatic hyphomycetes Springer-Verlag, Berlin, p 77-98.

Shearer, C. A. and Zare-Maivan, H. 1988. *In vitro* hyphal interactions among wood and leaf-inhabiting Ascomycetes and Fungi Imperfecti from freshwater habitats. Mycologia, 80: 31-37.

Shearer, C. A., and Webster, J. 1991. Aquatic hyphomycetes in the river Teign. IV. Twig colonization. Mycol. Res., 95: 413-420.

Shearer, C. A. and Webster, J. 1985. Aquatic hyphomycete communities in the river Teign. 1. Longitudinal distribution patterns. Tans. Brit. Mycol. Soc., 84: 509-518.

Singh, N. and Musa, T. M. 1977. Terrestrial occurrence and the effect of temperature on growth, sporulation and spore germination of some tropical aquatic hyphomycetes. Trans. Brit. Mycol. Soc., 63: 103-106.

Singh, S. K. and Waingankar, V. 2005. Endophytic fungi from medicinal plant *Holarrhena antidysenterica* and their potential antibacterial activity in Microbial diversity (Opporunities and Challenges) (Eds. S. P. Gutum *et al.*) from 168-178.

Soosamma, M., Lekha, G., Sreekala, K. N. and Bhat, D. J. 2001. A new species of *Trinacrium* from submerged leaves from India. Mycologia, 93: 1200-1202.

Sreekala, K. N. and Bhat, D. J. 2002. *Dendrospora yessemredda* sp. nov. from fresh water foam. In " Frontiers in Microbial Biotechnology and Plant Pathology". (Eds. C. Manoharachary, *et al.*) Scientific Publishers, Jodhpur. Pp.295-298.

Sridhar, K. R. and Barlocher, F. 1992 a. Aquatic Hyphomycetes in spruce roots. Mycologia, 84: 580-584.

Sridhar, K. R. and Barlocher, F. 1992 b. Endophytic aquatic hyphomycetes of roots of Spruce, Birch and Maple. Mycol. Res., 96: 305-308.

Sridhar, K. R. and Barlocher, F. 2000. Initial colonization, nutrient supply and fungal activity on leaves decaying in streams. App. and Environ. Microbiol., 66: 1114-1119.

Sridhar, K. R. and Kaveriappa, K. M. 1988 c. Occurrence and survival of aquatic hyphomycetes in brackish and sea water. Arch. Hydrobiol., 1: 153-160.

Sridhar, K. R. and Raviraja, N. S. 1995. Endophytes- a crucial issue. Curr. Sci., 69: 570-571.

Sridhar, K. R., Krauss, G., Barlocher, F., Raviraja, N. S., Wennrich, R., Baumbach, R. and Krauss, G. J. 2001. Decomposition of alder leaves in two heavy metal-polluted streams in Central Germany. Aquat. Microb. Ecol., 26: 73-80.

Sridhar, K. R., Krauss, G., Barlocher, F., Wennrich, R., Krauss, G. H. 2000. Fungal diversity in heavy metal polluted waters in Central Germany. Fungal Diversity, 5: 119-129.

Sridhar, K. R. and Kaveriappa, K. M. 1984. Seasonal occurrence of water-borne fungi in Konaje stream (Mangalore), India. Hydrobiologia, 119:101-105.

Sridhar, K. R. and Kaveriappa, K. M. 1987. A new species of *Triscelophorus.* Indian Phytopath., 40: 102-105.

Sridhar, K. R. and Kaveriappa, K. M. 1988 a. New host records of aquatic hyphomycetes. Indian Phytopath., 41: 160-161.

Sridhar, K. R. and Kaveriappa, K. M. 1988 b. Colonization of leaf litter by aquatic hyphomycetes in western ghat stream. Proc. Indian Natl. Sci. Acad. B 54: 199-200.

Sridhar, K. R. and Kaveriappa, K. M. 1989 a. Notes on aquatic hyphomycetes of mountain streams in western ghat region, India. Feddes Report., 100:187-189.

Sridhar, K. R. and Kaveriappa, K. M. 1989 b. Colonization of leaves by water-borne hyphomycetes in a tropical stream. Mycol. Res., 92:392-396.

Sridhar, K. R., Chandrasekhar, K. R. and Kaveriappa, K. M. 1992. Research on the Indian Subcontinent. In "The Ecology of Aquatic Hyphomycetes (ed. Bärlocher, F.) Springer-Verlag, Heidelberg. pp.182-211.

Sridhar, K. R.1984. Studies on water-borne fungi of Dakshina Kannada and Kodagu regions. Ph.D. Thesis, Mangalore University.

Suberkropp, K. 1992. Interactions with Invertebrates. In: The Ecology of Aquatic Hyphomycetes (Ed. F. Barlocher). Springer-Verlag, Berlin. pp. 118-134.

Suberkropp, K. 1995. The influence of nutrients on fungal growth, productivity and sporulation during leaf breakdown in streams. Canadian Journal of Botany, 73: 361-369.

Suberkropp, K. and Arsuffi, T. L. 1984. Degradation, growth and changes in palatability of leaves colonized by six aquatic hyphomycetes species. Mycologia, 76: 398-407.

Suberkropp, K. and Arsuffi, T. L. and Anderson, J. P. 1983. Comparision of degradation ability, enzymatic activity and palatability of aquatic hyphomycetes grown on leaf litter. Appl. Environ. Microbial., 46: 237-244.

Suberkropp, K. and Chauvet, E 1995. Regulation of leaf breakdown by fungi in streams influences of water chemistry. Ecology, 76: 1433-1145.

Suberkropp, K. and Godshalk, G. L. Klug, M. J. 1976. Changes in the chemical composition of leaves during processing in a wood land stream. Ecology, 57: 720-727.

Suberkropp, K. and Klug, M. J. 1980. The maceration of deciduous leaf litter by aquatic hyphomycetes. Can. J. Bot., 58: 1025-1031.

Suberkropp, K. and Klug, M. J. 1981. Degradation of leaf liter by aquatic hyphomycetes. In Wicklow, D.T. and G. C. Carroll (eds). The Fungal Community, its organization and Role in the Ecosystem. Marcel Dekker Inc., New York, 761-776.

Suberkropp, K., and Klug, M. J. 1976. Fungi and bacteria associated with leaves during processing in a woodland stream. Ecology, 57: 707-719.

Suberkropp. K. 1984. Effect of temperature on seasonal occurrence of aquatic Hyphomycetes. Trans. Brit. Mycol. Soc., 82: 53-62.

Subramanian, C. V. and Bhat, D. J. 1981. Conidia from fresh water foam samples from the Western Ghats, Southern India. Kavaka, 9: 45-62.

Thakur, S. B. 1977. Survival of some aquatic hyphomycetes under dry conditions. Mycologia, 69: 843-845.

Thomas, K., Chilvers, G. A. and Morris, R. H. 1989. Seasonal occurrence of conidia of aquatic hyphomycetes (Fungi) in Less Creek. Australian capital territory. Aus. J. Marine and Freshwater Research, 40: 11-23.

Thomas, K., Chilvers, G. A. and Morris, R. H. 1991. Changes in concentration of aquatic hyphomycetes spores in Lees Creek, ACT, Australia. Mycol. Res., 95: 178-183.

Thomas, K., Chilvers, G. A. and Morris, R. H. 1992. Aquatic Hyphomycetes from different substrates, substrate preference and seasonal occurrence. Australian Journal of Marine and Freshwater Research, 43: 491-509.

Tiwari, N. and Sati, S. C. 1991 a. *Dimorphospora foliicola* Tubaki, from pine needle litter- a new addition to India. Nat. Acad. Sci. Lett., 14: 1-2.

Tiwari, N and Sati, S. C. 1991 b. Contribution to Indian aquatic hyphomycetes. Two new additions. Acta Botanica Indica, 19: 256 -258.

Tiwari, N. and Sati, S. C. 1992 a. Two notable additions to Indian aquatic hyphomycetes. Nat. Acad. Sci. Lett., 15: 31-32.

Tiwari, N. and Sati, S. C. 1992 b. Aquatic hyphomycetes, from temperate fresh water streams of Nainital, Kumaun Hills, Abst 79th Indian Sci. Cong. pp. 9.

Tiwari, N. and Sati, S. C. 1993. Some Ingoldian fungi from a freshwater stream of NainiTal, Kumaun Himalaya. Abst. In 80th Indian Science Congress, p 29.

Triska, F. J. 1970. Seasonal distribution of aquatic Hyphomycetes in relation to the disappearance of leaf litter from a wood-land stream. Ph.D. Thesis, University of Pittsburgh. Pittsburgh. PA.

Tubaki, K. 1957. Studies on Japanese Hyphomycetes III Aquatic group. Bull. Nat. Sci. Mus. Tokyo, 41: 294-268.

Umm-ul Banin, 2006. Survival of freshwater Hyphomycetes in different substrata in the Lahore Branch of the BRB Canal under normal conditions and during canal closure. M.Sc Thesis, University of the Punjab, Lahore.

Waid, J. S. 1954. Occurrence of aquatic hyphomycetes upon the roots surface of beech grown in woodland soil. Trans. Brit. Mycol. Soc., 37: 420-421.

Webster, J. 1959. Experiments with spores of aquatic hyphomycetes. I. Sedimentation and impaction on smooth surfaces. Ann. Bot. N.S., 23:595-611.

Webster, J. 1961. The *Mollisia* perfect stage of *Anguillospora crassa*. Trans. Brit. Mycol. Soc., 44: 559-564.

Webster, J. 1965. The perfect stage of *Pyricularia aquatica*. Trans. Brit. Mycol. Soc., 49: 339-343.

Webster, J. and Descals, E. 1979. The teleomorph of water-borne Hyphomycetes from fresh water. In "The Whole Fungus" Vol. 2 (B. Kendrick, Ed.) Ottawa, National Museum of Canada and Kananaskis Foundation. 419-451.

Webster, J. and Descals, E. 1981. Morphology, distribution and ecology of conidial fungi in freshwater habitats. In the Biology of Conidial Fungi (Cole. G. T. and Kendrick, B. eds) N. Y. Academic Press, 295-355.

Webster, J. and Towfic, F. H. 1972. Sporulation of aquatic hyphomycetes in relation to aeration. Trans. Brit. Mycol. Soc., 59: 353-364.

Webster, J. Moran, S. T. and Davey, R. A. 1976. Growth and sporulation of *Tricladium chaetocladium* and *Lunulospora curvula* in relation to temperature. Trans. Brit. Mycol. Soc., 67: 491-499.

Webster, J., Marvanova, L. and Eicker, A. 1994. Spores from foam from South African rivers. Nova Hedwigia., 57: 379-398.

Willen, T. 1958. Conidia of aquatic Hyphomycetes amongst plankton algae. Botaniska Notiser, 111: 431-435.

Willoughby, L. G. and Archer, J. F. 1973. The fungal spora of a fresh water stream and its colonization pattern on wood. Fresh Wat. Biol., 3: 219-239.

Willoughby, L. G. and Minshall, G. W. 1975. Further observation on *Tricladium giganteum.* Trans. Brit. Mycol. Soc., 65: 77-82.

Wilson, D. and Carroll, G. C. 1994. Infection studies of *Discula quercina*, an endophyte of *Quercus garryana*. Mycologia, 86: 635-647.

Wolf, C. C. 1972. Taxonomic considerations in fresh water Hyphomycetes. Virginia. Polytechnic Institute and State University, Ph.D. 1971.

Wood-Eggenschwiler, S. and Barlocher, F. 1983. Aquatic Hyphomycetes in 16 streams in France, Germany and Switzerland. Trans. Brit. Mycol. Soc., 81:371-379.

Wood-Eggenschwiler, S. and Barlocher, F. 1985. Geographical distribution of Ingoldian fungi. Vreh. Internat. Verein. Limnol., 22: 2780-2785.

Microbes: Diversity and Biotechnology (2012) *Pages* **325–332**
Editors: **Prof. S.C. Sati & Dr. M. Belwal**
Published by: **DAYA PUBLISHING HOUSE, NEW DELHI**

Chapter 19

Studies on Antifungal Activity of the Plant *Zanthoxylem armatum* (Rutaceae)

Kapil Khulbe and S.C. Sati

Department of Botany, D.S.B. Campus, Kumaun University, Nainital – 263 001

ABSTRACT

Various organic (hexane, chloroform and methanol) and aqueous extracts of leaves, stem bark and fruit of *Zanthoxylum armatum* DC (Rutaceae) obtained by infusion and maceration were tested for their antimicrobial activity against six fungal strains (*Alternaria alternata, Fusarium solani, Colletotrichum falcatum, Rhizoctonia solani, Aspergillus* sp., *Penicillium* sp.) using food poison method. The methanol extract of stem-bark and fruit showed broad spectrum antifungal activity, with 100 per cent inhibition of test fungi in three cases each. Chloroform extract of fruit, methanol extract of leaves and aqueous extract of bark and fruit also demonstrated complete inhibition at tested concentration of 1000ppm in one case each. *R. solani* was the most sensitive fungi and *Penicillium* sp. was the most resistant fungus to different extracts tested. The activity shown by different extracts was found comparable with standard antibiotic used.

Keywords: *Zanthoxylum armatum, Rutaceae, Antimicrobial, Antifungal.*

Introduction

Diseases of cultivated crops remain the principal limitation to increased agricultural production. The continuous and irrational use of chemicals has resulted into several problems like residues in edible plant parts, resistant strains, and environmental pollution, etc. In the present day agriculture, a lot of emphasis is being given on using eco-friendly means for controlling diseases and pest problems affecting various crops (Thind *et al.*, 2005). Therefore, protection of plants from pathogens remains a primary concern of agricultural scientists.

There is general growing trend among consumers for more natural rather than synthetic products in a whole range of industries, including food and drink, cosmetic, agricultural, and pharmaceuticals

(Bauer *et al.*, 1997; Svoboda and Deans, 1998; Glaser, 1999; Traffic International, 1999; Walton and Brown, 1999; Bansal and Gupta, 2000; Singh *et al.*, 2000; Kamanzi Atindehou, 2002). The use of botanicals for the management of the phytopathogens is gaining ground. Plant extracts and volatile oils may have an important role to play in the preservation of foodstuffs against fungi, in fungicidal application against plant diseases, and in the fight against various human fungal infections. Recent literature has shown the biological activities of plant-extracts, essential oils and their individual pure components, and has documented the inhibitory activity of these substances against the growth of various fungi (Maruzzella and Balter, 1959; Maruzzella, 1963; Awuah, 1994; Abraham and Prakasam, 2000; Pistelli *et al.*, 2002; Kelemu *et al.*, 2004; Romagnoli *et al.*, 2005; Talas-Ogras *et al.*, 2005).

Himalaya has an extraordinarily rich flora and wide knowledge of indigenous medicinal plants. Accordingly, we are investigating the potential of Himalayan medicinal plants as a resource of new biofungicides. The present work focuses mainly on antifungal potential of one of the Himalayan medicinal plants – *Zanthoxylum armatum* DC (Syn. *Z. alatum*) of family Rutaceae. In fact, a little work has been done in the field of antimicrobial potential of this plants against pathogens.

Material and Methods

Plant Material

The plant material was collected from a natural population in Nainital, Kumaun Himalaya of India and authenticated by plant taxonomist Y.P.S. Pangtey from the Department of Botany, Kumaun University, Nainital. A vaucher specimen number MP/84 has been deposited at the herbarium of the above-cited department.

Preparation of Extracts

Different parts of the plant (leaves, stem-bark and fruits) were individually crushed using pestle and mortar and powdered in an electric grinder. Fine powdered plant materials were subjected serially to hexane, chloroform, methanol and water. After extraction, each extract was passed through Whatman filter paper No.1. The filtrate was concentrated on a rotary evaporator under vacuum at 20° C and stored at 4° C for further use.

Microorganism Used

Six filamentous fungi, *Alternaria alternata, Fusarium solani, Colletotrichum falcatum, Rhizoctonia solani, Aspergillus* sp., and *Penicillium* sp. were used as test fungi which were obtained from Plant Pathology Department, Pantnagar University, U.S. Nagar, India.)

Antifungal Testing

The fungitoxic activity of different extracts was tested against six fungal strains employing the poisoned food technique of Grover and Moore (1962) and Perrucii *et al.* (1994). Potato Dextrose Agar (Hi Media, 39 gm of medium dissolved in 1000 ml of distilled water) was used. The medium was autoclaved at 120 °C for 30 minutes. After cooling the medium up to 40°C, 10mg of streptomycin was added to it and mixed thoroughly to prevent bacterial contamination as suggested by Gupta and Banerjee (1970). Requisite amount of the oils were dissolved separately in 0.5 ml of 0.01 per cent of aqueous solution of Tween 80 in Pre-sterilized Petri plates. 9.5 ml of PDA medium was pipetted to each Petri plate and mixed thoroughly to obtain the requisite (1,000 mg/ml) concentration. For control sets, requisite amount of respective solvents in place of extract were added to the medium for negative control and clotrimazole was added for positive control. Each set of experiment was done in triplicate.

Disc of test fungi (5mm diameter), cut with the help of sterilized cork borer from the periphery of a seven day old culture was inoculated aseptically to the center of each Petri plate of treatment and control sets. The Petri plates were incubated at 27° C for six days in incubation chamber. Diameter of fungal colony of treatment and control sets were measured in mutually perpendicular direction on seventh day. The percentage inhibition of growth was calculated by the mean value of colony diameter by the following expression:

$$\text{Percentage of mycelial inhibition} = \frac{\text{Dc} - \text{Dt}}{\text{Dc}} \times 100$$

where,

Dc = Mean colony diameter of control sets

Dt = Mean colony diameter of treatment sets

Result and Discussion

Altogether 72 tests (four extracts of three plants *i.e.* leaf, stem-bark and fruit against six fungi) were performed using four different organic and aqueous extracts of leaves, stem-bark and fruits. The results are summarized in Tables 19.1–19.4.

Hexane Extracts (Table 19.1)

For leaves fraction a low inhibition percent was observed against three fungal strains (*A. alternata*-39 per cent, *F. solani*–45 per cent and *C. falcatum*–17 per cent), whereas no activity was recorded against *R. solani*, *Aspergillus* sp. and *Penicillium* sp. Stem-bark extract showed the maximum inhibition (69 per cent) against *C. falcatum* followed by *A. alternata* (64 per cent). For other tested fungi, a low inhibition activity was recorded (*Penicillium* sp.–47 per cent, *R. solani*–46 per cent, *F. solani* –45 per cent, and *Aspergillus* sp.- 40 per cent). fruit extract showed maximum inhibition of 78 per cent for *C. falcatum*, whereas a comparatively low level of activity was recorded against other test fungi (*A. alternata*–58 per cent, *Penicillium* sp.–52 per cent, *F. solani*–47 per cent, *Aspergillus* sp.–43 per cent and *R. solani*–40 per cent).

Table 19.1: Per cent Inhibition of Leaf, Bark and Fruit Obtained by Infusion in Hexane Against Pathogenic Fungi

Microorganism Used	*Hexane*			*Control*
	Leaf	*Bark*	*Fruit*	*Clt*
A. alternata	39	64	58	33
F. solani	45	45	47	79
C. falcatum	17	69	78	70
R. solani	–	46	40	62
Aspergillus sp.	–	40	43	47
Penicillium sp.	–	47	52	64

Chloroform Extracts (Table 19.2)

Highest percent inhibition for leaves extract was recorded against *R. solani* (73 per cent), followed by *A. alternata* (72 per cent) and *F. solani* (64 per cent). A low inhibition of 47 per cent was observed against *Aspergillus* sp. whereas *C. falcatum* and *Penicillium* sp. were found completely resistant at the tested concentration. Stem-bark extract showed maximum inhibition for *C. falcatum*–93 per cent, followed by *R. solani*–84 per cent, *A. alternata*–74 per cent and *Aspergillus* sp.–72.1 per cent, whereas the extract was found less active against two fungi (*F. solani*–30 per cent and *Penicillium* sp.–29 per cent). Fruit extract exhibited an absolute inhibition (100 per cent) against *C. falcatum*, however *F. solani*, *A. alternata*, *R. solani*, and *Aspergillus* sp. were inhibited 82, 81, 78 and 74 per cent by the extract respectively. The minimum inhibition was recorded against *Penicillium* sp. (20 per cent).

Table 19.2: Per cent Inhibition of Leaf, Bark and Fruit Obtained by Infusion in Chloroform Against Pathogenic Fungi

Microorganism Used	Chloroform			Control
	Leaf	Bark	Fruit	Clt
A. alternata	72	74	81	33
F. solani	64	30	82	79
C. falcatum	–	92	100	70
R. solani	73	84	78	62
Aspergillus sp.	47	72	74	47
Penicillium sp.	–	29	20	64

Methanol Extract (Table 19.3)

Leaves extract exhibited absolute inhibition (100 per cent) against *R. solani*, whereas a significant inhibition was also recorded for *A. alternata* (88 per cent). A comparatively low level of activity was observed against *Aspergillus* sp., *Penicillium* sp., *C. falcatum* and *F. solani* with 52, 47, 46 and 27.8 percent inhibition. Stem-bark fraction exhibited an absolute inhibition (100 per cent) against *F. solani*, *C. falcatum* and *R. solani* whereas for other tested strains a significant inhibition was also recorded (*Aspergillus* sp.–93 per cent, *A. alternata*–80 per cent and *Penicillum* sp.–69 per cent). Fruit extract showed an absolute inhibition (100 per cent) against three test fungi *i.e.* *F. solani*, *R. solani*, and *Aspergillus* sp. whereas a significant inhibition was also recorded against *A. alternata* and *C. falcatum* with percent inhibition of 86 per cent and 84 per cent respectively. A low activity was observed against *Penicillium* sp. (49 per cent).

Table 19.3: Per cent Inhibition of Leaf, Bark and Fruit Obtained by Infusion in Methanol Against Pathogenic Fungi

Microorganism Used	Methanol			Control
	Leaf	Bark	Fruit	Clt
A. alternata	88	80	86	33
F. solani	28	100	100	79
C. falcatum	46	100	85	70
R. solani	100	100	100	62
Aspergillus sp.	52	93	100	47
Penicillium sp.	47	69	49	64

Aqueous Extract (Table 19.4)

The leaves extract was found less active against all the fungi tested whereas in the case of *Aspergillus* sp. it was found completely inactive. Among the other tested fungi the percent inhibition was 42, 41, 39, 18 and 15 per cent in *A. alternata*, *F. solani*, *R. solani*, *Penicillium* sp. and *C. falcatum* respectively. Stem-bark extract showed 100 per cent inhibition against *C. falcatum*. A higher level of activity was also recorded against *F. solani*, *A. alternata*, *Penicillium* sp. and *C. falcatum* with inhibition of 97, 91, 62 and 61 per cent respectively whereas a comparatively low level of activity was recorded against *Aspergillus* sp. (47 per cent). As recorded in the table, the fruit extract completely checked the growth of *F. solani* (100 per cent inhibition), whereas a significant antifungal activity was a low observed for other fungi (*R. solani*–89 per cent, *A. alternata*–84 per cent, *C. falcatum*–80 per cent, *Penicillium* sp.- 70 per cent and *Aspergillus* sp. 58 per cent).

Table 19.4: Per cent Inhibition of Leaf, Bark and Fruit Obtained by Infusion in Water Against Pathogenic Fungi

Microorganism Used	*Water*			*Control*
	Leaf	*Bark*	*Fruit*	*Clt*
A. alternata	42	91	84	33
F. solani	41	97	100	79
C. falcatum	15	61	80	70
R. solani	39	100	89	62
Aspergillus sp.	-	47	58	47
Penicillium sp.	14	62	70	64

In the present antifungal analysis the methanol fraction showed more activity followed by aqueous extract. This might possibly be due to that the substances active against fungi are of high polarity or active substances are more soluble in organic solvents and therefore, not present in water extract.

Out of twelve extracts tested against six test fungi, six extracts showed absolute toxicity in ten cases at 1000ppm, methanol extract of stem-bark and fruit exhibiting complete inhibition of three test fungi each. This finding is important in view of that the effectiveness of particular concentration of plant extracts depends on the metabolism of secondary metabolites by fungal enzymes (Jeffrey *et al.*, 1980). This finding indicates for better exploitation chances of this plant against particular fungal disease management.

The antimicrobial activity of fruit oil of the plant has been screened against *C. falcatum* by Rao and Singh (1994), and the leaf lipophilic extract has also been screened by Gularia and Kumar (2006). However, the stem bark and leaf metabolites have not previously been studied against such a wide variety of test pathogens.

This appears to be for the first attempt to establish the importance of whole plant as a possible source of antifungal agent at one place. The present study was planned to test the whole plant as antifungal agents, which supports the contention that traditional medicine remains a valuable resource in the potential discovery of natural product pharmaceuticals. This paper highlights the evolution of the important research and the development of the technology and methods that are necessary today to prove the essential oils are effective in the war on pathogens.

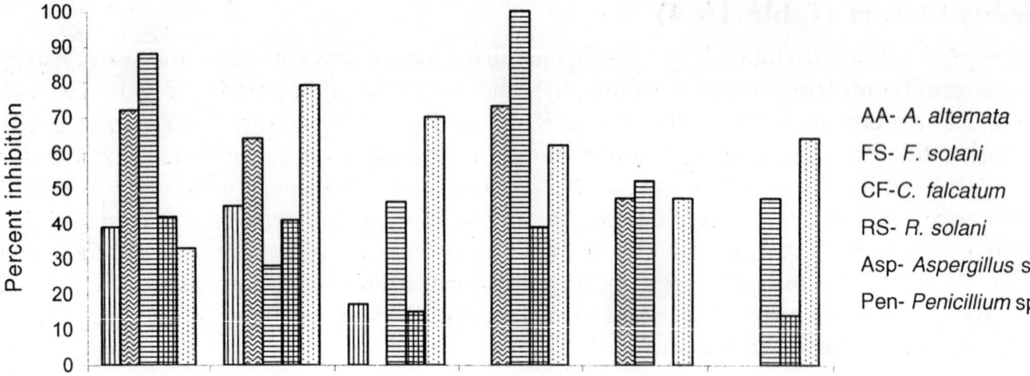

Figure 19.1

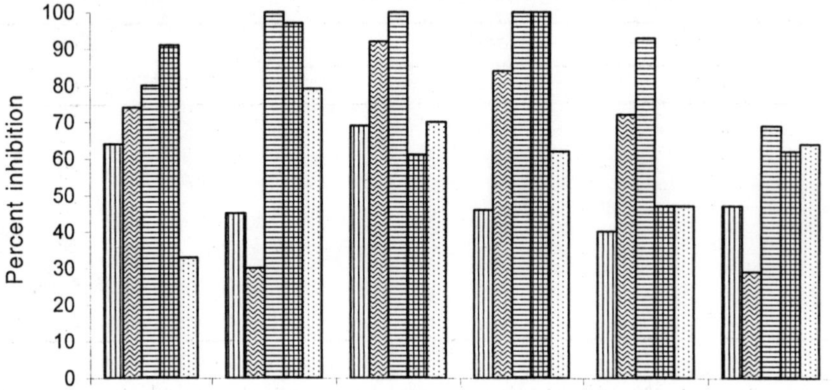

Figure 19.2

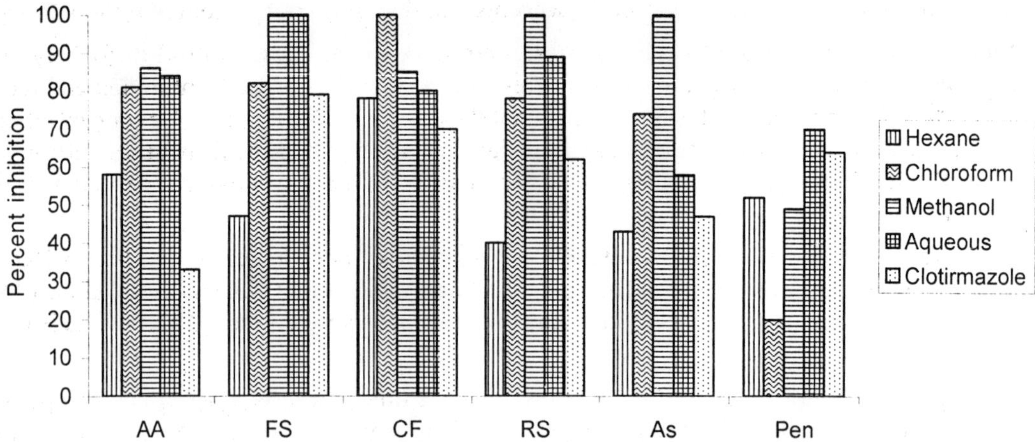

Figure 19.3

Figures 19.1–3: Effect of Hexane, Chloroform, Methanol and Aqueous Extracts of Leaf (Figure 19.1), Stem-bark (Figure 19.2), and Fruit (Figure 19.3) on the Growth of Tested Fungi. Growth inhibition observed 7 days after inoculation is expressed as a percentage of the growth inhibition observed for the control with the correspondent solvent.

Conclusion

The antifungal activity exhibited by the studied plant may be attributed to the various active constituents present in them either individually or in combination. It may also help in the discovery of new chemical classes of antibiotics that could serve as selective agents for the maintenance of human health and may provide biochemical tools for the study of infectious diseases. Therefore, *Z. armatum* can be selected for further studies. The discovery of a potent remedy from plant origin will be a great advancement in fungal and bacterial infection therapies.

Acknowledgements

The authors wish to thank Dr. Y.P.S. Pangtey for plant identification, Department of Plant Pathology, GBPUA&T, Pantnagar, for providing fungal cultures, and Department of Botany, D.S.B. Campus, Kumaun University, Nainital for providing necessary facilities to carry out present investigation.

References

Abraham, S. and Prakasam, V. 2000. Efficacy of botanicals on post harvest pathogens of carrot. J. Mycol Plant Pathol, 30: 257.

Awuah, R. T. 1994. In vivo use of extracts from *Ocimum gratissium* and *Cymbopogon citrates* against *Phytophthora palmivora* causing black pod disease of cocoa. Ann Appl Biol., 124: 173-178.

Bansal, R. K. and Gupta, R. K. 2000. Evaluation of plant extracts against *Fusarium oxysporum* wilt pathogen of fenugreek. Indian Phytopathol., 53: 107-108.

Bauer, K., Garbe, D. and Surburg, H. 1997. Preparation, properties and uses. Third completely revised edition. Common fragrance and flavour materials. Weihiem: Wiley-VCH, p. 269.

Glaser, V. 1999. Billion-dollar market blossoms as botanicals as botanicals take root. Nat Biotechnol, 17: 17.

Grover, R. K. and Moore, J. D. 1962. Toximatic study of fungicides againsts brown rot organism *Sclerotinia facticola* and *S. laxa*. Phytopathol, 52: 876-880.

Guleria S. and Kumar, A. 2006. Antifungal activity of some Himalayan medicinal plants using direct bioautography. J. Cell and Mol. Bio. 5: 95-98.

Gupta, S. and Banerjee, A. B. 1970. A rapid method of screening antifungal antibiotics producing plants. Indian J. Exp. Biol., 8: 148-149.

Jaffery, J. Anderson, and Stanley, Dagley. 1980. Catabolism of Aromatic Aids in *Tricosporum cutaneum*. Journal of Bacteriology, 534-543.

Kamanzi Atindehou, K., Kone, M., Terreaux, C., Traore, D., Hostettmann, K. and Dosso, M. 2002. Evaluation of the antimicrobial potential of medicinal plants from the Ivory Coast. Phytotherapy Res., 16(5): 497-502.

Maruzzella, J. C. and Balter, F. 1959. The action of essential oils on phytopathogenic fungi. Plant Disease Res., 43: 1143.

Maruzzella, J. C. 1963. The effect of perfume oils on the growth of phytopathogenic fungi. Plant Dis Rep., 47: 756-757.

Olalde Rangel, J. A. 2005. The systemic theory of living systems and relevance to CAM. Part I: the theory. Evid Based Complement Alternat Med. 2: 13-18.

Perrucci, S., Mancianti, F., Ciont, P.L., Glamini, G., Morelli, I. and Macchioni, G. 1994. *In vitro* antifungal activity of essential oils against some isolates of *Microsporum cants* and *M. gypseum*. Plant. Med., 60 (20): 184-187.

Pistelli, L., Bertoli, A., Lepori, E., Morelli, I. and Panizzi, L. 2002. Antimicrobial and antifungal activity of crude extracts and isolated saponins from *Astragalus verrucosus*. Fitoterapia, 73 (4): 336-339.

Rao, G. P. and Sing, S. B. 1994. Efficacy of geramiol extracted from the essential oil of *Z. almatum* as a fungitoxicant and insect repellent. Sugar cane, 4: 16-20.

Romagnoli, C., Bruni, R., Andreotti, E., Rai, M. K., Vicentini, C. B. and Mares, D. 2005. Chemical characterization and antifungal activity of essential oil of capitula from wild Indian *Tagetes patula* L. Protoplasma, 225 (1-2): 57-65.

Singh, U. P., Sarma, B. K., Mishra, P. K. and Ray, A. B. 2000. Antifungal activity of venanatine an indole alkaloid isolated from *Alstonia venenata*. Folia- Microbiologia, 45: 173-176.

Svoboda, K. P. and Deans, S. G. 1998. Presentation of the research and development database concerted action air 3 ct 94 2076 final report. 3, 48-60. In: Towards a model of technical and economic optimization of specialist minor crops. Brussels: Commission European, DG VI F, 11.3.

Talas-Ogras, T., Ipekci, Z., Bajrovic, K. and Gozukirmizi, N. 2005. Antibacterial activity of seed proteins of *Robinia pseudoacacia*. Fitoterapia, 76 (1): 67-72.

Tirtha, S. S. 1998. The Ayurvedic Encyclopedia. Sri Satguru Pub., New Delhi.

Traffic International. 1999. Europe's medicinal and aromatic plants. Their use trade and conservation. Goldaming, Surrey.

Walton, N. J., Brown, D. E. editors. 1999. Chemicals from plants: perspectives on plant secondary product. London: Imperial College Press and World Scientific.

Microbes: Diversity and Biotechnology (2012) *Pages* **333–345**
Editors: **Prof. S.C. Sati & Dr. M. Belwal**
Published by: **DAYA PUBLISHING HOUSE, NEW DELHI**

Chapter 20

Antifungal Potential of Gymnosperms: A Review

Savita Joshi* and S.C. Sati

Department of Botany, D.S.B. Campus,
Kumaun University, Nainital – 263 001

ABSTRACT

The trend of using natural products has increased now a days and active plant extract are frequently being screened for new drug discoveries. The gymnosperms plants, which sometimes form dominant trees of forest areas, have also been used in traditional medicines for centuries. The stem, bark, young twigs, leaves, barriers, fruits etc., of these plants have been exploited in various folk medicines. The recent literature has shown the biological activities of gymnosperms plant extracts, essential oils and their pure compounds against the growth of some fungi. The present paper is an attempt to review the work done in the field of antifungal activities of various gymnosperms plant extracts, oils as well as chemical compounds.

Keywords: *Gymnosperms, Antifungal activity, Plant extracts, Essential oil, Kumaun Himalaya.*

Introduction

Throughout the ages, human have relied on nature for their basic needs such as food, shelter, clothing, transportation, fertilizers, flavours, fragrances, and medicines (Cragg and Newman, 2005). Plants have formed the basis of sophisticated traditional medicine systems that have been in existence for thousands of years and continue to provide mankind with new remedies. Although some of the therapeutic properties attributed to plants have proven to be erroneous, medicinal plant therapy is based on the empirical findings of hundreds and probably thousands of years of use.

---

* Corresponding Author: E-mail: savijoshi@ymail.com

The trend of using natural products has increased now a days and the active plant extracts are frequently screened for new drug discoveries (Das *et al.*, 1999). Drug discovery from medicinal plants has evolved to numerous fields of enquiry and various methods of analysis. The process typically begins collection and identification of plant(s) of interest. Collection may involve species with known biological activity from which active compounds have not been isolated (e. g. traditionally used herbal remedies) or may involve taxa collected randomly for a large screening program (Balunas and Kinghorn, 2005).

The interest in nature as a source of potential chemotherapeutic agents continues. Natural products and their derivatives represent more than 50 per cent of all the drugs in clinical use in the world today. Plants have been utilized as medicines for thousands of years (Samuelsson, 2004). These medicines initially took the form of crude drugs such as tinctures, teas, powders, and other herbal formulations (Balick and Cox, 1997; Samuelsson, 2004). The specific plants to be used and the methods of application for particular ailments were passed through oral tradition. Eventually information regarding medicinal plants was recorded in herbal pharmacopeias (Belunas and kinghorn, 2005).

Many efforts have been made to discover new antimicrobial compounds from various kinds of plants. One of such resources is folk medicines. The increasing prevalence of multidrug resistant strains of fungi and the recent appearance of strains with reduced susceptibility to antibiotics raises the presence of untreatable fungal infections and adds urgency to the search for new infection-fighting strategies. This situation forced scientist for searching new antimicrobial substances from various sources like medicinal plants, which are the major source of novel antimicrobial agents.

The fungi are eukaryotic, achlorophyllous and heterotrophic comprises about 1.5 million species of which 74,000 species are described (Hawksworth, 2001). More than 300 species are potentially pathogenic or cause allergy symptoms in man (Gupta *et al.*, 2001). The parasitic fungi cause serious diseases of various plants. Many useful plants attacked by a variety of fungi and are known as the chief causative agents of plant disease (Campbell *et al.*, 2000; Amorim *et al.*, 2004). Similarly, man and other mammalians, fishes, amphibians, reptiles are also susceptible to fungal infections (Dube, 1990).

New antimycotics are badly needed as the world market for antifungal is expanding dramatically (Steinbach and Stevens, 2003). Microorganisms have developed resistance to the synthetic drugs, antifungal and antibiotics. Scientists, therefore, are working on the extraction of anti-infectional compounds from natural source like plants.

Among the use of plants higher plants have more frequently been used for the purpose. Higher plants include angiosperms and gymnosperms. The former group of plants has regularly been explored for the bioactivity than the latter ones.

Gymnosperms are the group of naked seeded vascular plants and sometimes forming dominant trees of forest areas. These plants have also been used in traditional medicines for centuries (Kirtikar and Basu., 1935). The recent finding reports that these plants contains various secondary metabolites such as tannis, terpenoids, alkaloids, glycosides, ligans, phenols, steroids and sugar derivatives (Niemann, 1988: Ster-Mitz *et al.*, 1999: Harborne and Baxtex, 2001).

In folk medicine some gymnospermic plants are being used as an antimalarial, antifungal,antibacterial, antirheumatic, abortifacient and antibronchitis (Bryan- Brown, 1932: Appendino, 1993) as well as antiasthemic (Singh, 1995). Erdemoglu and Sener, (2001) found that gymnosperms possess biological activities such as anticancer, anti inflamentory, antimicrobial and antioxidant.

As evident from the available literature gymnospermic medicinal plants have not been adequately studied as compared to angiospermic plants. Therefore, there is an urgent need to work out for their bioactive potential of these plants which provides a base for the drug development in modern pharmacology.

Gymnosperms are almost a neglected group of plants in Indian subcontinent. However, Himalaya deserves special attention in many respects for gymnosperms as Himalayan region not only dominate forests, but are source of diverse economic and medicinal products. The stem, bark, young twigs, leaves, berries, fruits, etc. of gymnosperms are exploited to obtain medicines and other useful products. Many gymnosperm species, such as *Taxus wallichiana, Ephedra gerardiana* and some *Juniperus* species if used sustainably can provide chief and effective medicines.

Recent literature has shown the biological activities of gymnosperms plant extracts, essential oils and their individual pure compounds (Table 20.1). It has been well documented for the inhibitory activity of these substances against the growth of various fungi.

As compared to other plants, a little work has been done on the antimicrobial activity of gymnosperms, yet ethanobotanical importance of these plants has been documented by some workers. They reported that these plants are of great medicinal importance and are used by the tribal and local people for remedy against various ailments (Dhar *et al.*, 1968). Therefore, this article is an attempt to review the work done in the field of antifungal activities of various gymnosperms plant extracts, essential oils and compounds which will facilitate to unravel the potentiality of gymnosperms plants and plants products for the workers who are engaged in the line of bioactivities of natural products and new drug discoveries.

Pinus

From the chemotaxonomic point of view, regarding the composition of essential oils the genus *Pinus* is divided into two groups (Chalchat and Gorunovic, 1995). One group comprises the species rich in monoterpene hydrocarbons (α-pinene, limonene,_-caryophyllene, germacrene D, Δ-3-carene) and the other is rich in the oxygenated monoterpenes (borneol, bornyl acetate) (Chalchat *et al.*, 1994; Chalchat and Gorunovic, 1995; Yatagai and Hong, 1997). It has several groups of natural compounds mainly flavanoids (Niemann, 1988), piperidine alkaloids (Stermitz *et al.*, 1994) and tannins (Harborne and Baxter, 2001).

The antimicrobial activity of essential oils from the family Pinaceae was investigated by many authors (Bagci and Digrak, 1996a, 1996b; Canillac and Monrey, 1996; Lis-Balchin *et al.*, 1998). It was reported, that essential oils with high monoterpene hydrocarbon levels, such as pine oil, were very active against fungi especially *Fusarium culmorum* (Lis-Balchin *et al.*, 1998).

Digrak *et al.* (1999) investigated antifungal activity of chloroform, methanol and acetone extracts of leaves, resins, bark, cones and fruits of two species of Pinus namely *P. brutia* and *P. nigra.* They found that the methanol and acetone extract were more effective against *Candida tropicalis, C. albicans* and *Penicillium italicum.*

Baranowska *et al.* (2002) investigated antifungal activity of the essential oil from some species of genus *Pinus* towards *Fusarium culmorum, F. solani* and *F. poae*. The strongest activity was observed for the essential oil from *P. ponderosa* which fully inhibited the growth of fungi at the following concentrations *F. culmorum, F. solani* at 2 per cent and *F. poae* at 5 per cent.

Table 20.1: Antifungal Activity of Gymnospermous Plants Extracts/Compounds

Plants	Family	Plant Part	plant extracts/compounds
Abies cilicia	Pinaceae	Leaves, resin, bark, cones, fruit	Chloroform, methanol and acetone (Digrak *et al.*, 1999)
Abies webbiana	Pinaceae	Leaves	Crude extract (Kumar *et al.*, 2006)
Araucaria araucana	Araucariaceae	Heartwood	Ligans (Carlos *et al.*, 2006)
C. libani	Pinaceae	Leaves, resin, bark, cones, fruit	Chloroform, acetone, methanol (Digrak *et al.*, 1999)
Cedrus deodara	Pinaceae	Leaves	Essential oil (Dikshit *et al.*, 1983)
Chamaecypasis obtuse	Taxodiaceae	Leaves	Essential oil (Geong-Ho *et al.*, 2009)
Cryptomeria japonica	Taxodiaceae	Leaves	Essential oil (Geong-Ho *et al.*, 2009)
Cupresses lusitanica	Cupressaceae	Leaves	Hexane extract(Kuiate *et al.*, 2006)
Ginkgo biloba	Ginkgoaceae	Leaves	Bioflavones (Baranowska and Wiwart, 2003)
G. biloba	Ginkgoaceae	Leaves	GBE (Xie *et al.*, 2006)
G. biloba	Ginkgoaceae	Leaves	GAFP protein (Haung *et al.*, 2000)
G. biloba	Ginkgoaceae	Leaves	Methanol, ethyle acetate, n-butanol, aqueous (mazzanti *et al.*, 1999)
Juniperus califirnica	Cupressaceae	Heartwood, bark, leaves	Hexane and Methanol (Clark *et al.*, 1990)
J. communis	Cupressaceae	Leaves	Essential oil (Cavalerio *et al.*, 2006)
J. lucayne	Cupressaceae	Wood	Sesquiterpenes (Nunez *et al.*, 2007)
J. oxycedrus	Cupressaceae	Leaves	Aqueous and Methanol (karaman *et al.*, 2003; Cavalerio *et al.*, 2006)
J. saltuaria	Cupressaceae	Leaves	Essential oil (Wedge *et al.*, 2009)
J. squamata	Cupressaceae	Leaves	Essential oil (Wedge *et al.*, 2009)
J. osteosperma	Cupressaceae	Heartwood, bark, leaves	Hexane and Methanol (Clark *et al.*, 1990)
Juniperus sps.	Cupressaceae	Berry	Essential oil (Filipouioz *et al.*, 2006)
Pinus densiflora	Pinaceae	Leaves	Pinosylvin (Lee *et al.*, 2005)
P. densiflora	Pinaceae	Leaves	Essential oil (Geong- Ho *et al.*, 2009)
P. girardiana	Pinaceae	Nut	Crude extracts (Kumar *et al.*, 2005)
P. nigra	Pinaceae	Leaves, resin, bark, cones and fruit	Chloroform, methanol, acetone extract (Digrak *et al.*, 1999)
P. pinaster	Pinaceae	Leaves	Pycnogenol extract (Torras *et al.*, 2005)
P. ponderosa	Pinaceae	Leaves	Essential oil (Baranowaska *et al.*, 2002)
P. sylvestris	Pinaceae	Heartwood	Pinosylvin (Lee *et al.*, 2005)
P. brutia	Pinaceae	Leaves, resin, bark, cones and fruit	Chloroform, methanol, acetone extract (Digrak *et al.*, 1999)
Taxus baccata	Taxaceae	Needles	Bioflavones (baranowska *et al.*, 2004)
T. baccata	Taxaceae	Leaves	Crude extract (Kumar *et al.*, 2006)
T. orientalis	Cupressaceae	Leaves	Essential oil and compound b1,b2 (Gularia *et al.*, 2008)
T. wallichiana	Taxaceae	Leaves	Methanol extract (Nisar *et al.*, 2008)
T. baccata	Taxaceae	Heartwood	Ethanol extract (Erdemoglu *et al.*, 2004)
Thuja orientalis	Cupressaceae	Leaves	Essential oil (Ezzat *et al.*, 2001)

Pycnogenol, a standardised extract of *Pinus pinaster* Ait. (Pinaceae), was tested for its antifungal activity towards twenty three different yeast and fungi microorganisms (Torras *et al.*, 2005). Pycnogenol inhibited the growth of all the tested microorganisms in minimum concentrations ranging from 20 to 250μg/ml. These results conform with clinical oral healthcare studies describing the prevention of plaque formation and the clearance of candidiasis by pycnogenol.

Lee *et al.* (2005) isolated "Pinosylvin" from *Pinus densiflora* and heartwood of *Pinus sylvestris* were tested for their antifungal activity against *Candida albicans* and *Saccharomyces cerevisiae*. They found that Pinoosylvin inhibited the growth of both tested organisms.

Crude nut extract of *Pinus girardiana* was analyzed for their *in vitro* antifungal activity against a panel of pathogenic fungi (Kumar *et al.*, 2005). They found that the extract was active at 100 μg/ml and 500 μg/ml concentrations against *Candida albicans* and *Aspergillus niger* respectively.

Recently,Geong-Ho *et al.* (2009) investigated the antifungal activity of essential oil of *Pinus densiflora* against 8 pathogenic fungal strains. They found that the essential oil of *P. densiflora* has highest activity against *Cryptococcus neoformans* and *Candida glabrata*.

Juniperus

This genus is well documented for its medicinal value for diarrohea, abdominal pain, tumors, piles, bronchitis and indigestion in traditional system of medicine (Kirtikar and Basu, 1935). Juniper berries (mature female cones) have long been used as flavouring agents in foods and alcoholic beverages such as gin (Clutton, 1972). It contains about 2 per cent volatile oil, juniperin, resins (about 10 per cent), proteins and formic, acetic and malic acids. The dried ripe fruits contain oil of juniper, pinene, cadinenes, camphene and a number of other diterpene acids. Dried berries of juniper and juniper decoction have been evaluated into recent animal studies (Swanston *et al.*, 1990). In herbal medicine, juniper oil has been used as a carminative, diuretic and as a steam inhalant in the management of bronchitis. It has also been used in arthritis as well as antioxidant (Takacsova *et al.*, 1995). Berries are also recommended in cough, infantile tuberculosis and diabetes (Zaman *et al.*, 1970), whereas, ash of the bark is used for certain skin diseases (Baqu:ar, 1989).

Clark *et al.* (1990) assayed hexane and methanol extract of heartwood, bark/sapwood and leaves of 12 taxa of *Juniperus* from the US for their antifungal activity. They found that methanol extract from the leaves of *J. osteosperma* and *J. califirnica* had highest inhibitory activity against *Trichophyton mentagrophytes* and hexane extract showed highest activity against *Cryptococcus neoformans*.

Digrak *et al.* (1999) investigated antifungal activity of chloroform, methanol and acetone extracts of leaves, resins, bark, cones and fruits of *J. oxycedrus*. The methanol and acetone extract were found more effective against *Candida tropicalis*, *C. albicans* and *Penicillium italicum*.

In vitro antifungal activity of aqueous and methanol extracts of *J. oxycedrus* leaves were investigated by Karaman *et al.* (2003) and found that methanol extract had a significant inhibitory effect on the growth of *Candida albicans*, *Alternaria alternata*, *Aspergillus flavus*, *Fusarium oxysporum* and *penicillium* sps. at a concentration 31.25-250 μg/ml.

Filipouioz *et al.* (2003) studied antifungal effects of *Juniperus* berry oil and found that the oil exhibited fungitoxic behaviour against two strains of *Candida albicans* having inhibition zone of 37 and 29mm.

The antifungal activity of essential oil of leaves of *J. oxycedrus* was investigated in order to evaluate its efficacy against dermatophytic fungus *Candida* and *Aspergillus* (Cavaleiro *et al.*, 2006). These results these demonstrated that the essential oil show a good fungicidal activity with from 0.08-0.016 μg/ml for *Candida* and 0.08-0.32 μg/ml for *Aspergillus* sp.

Examples of other antifungal essential oils includes *J. comunis* essential oil which was reported active against the dermatophyte *Aspergillus* and *Candida* strains (Cavaleiro *et al.*, 2006).

Nunez *et al.* (2007) isolated three sesquiterpenes from the ethanolic extract of the wood of *J. lucayna* for their antifungal activity against *Botrytis cineria* and found 71 per cent inhibition on mycelia growth after six days using food poisoned technique.

Wedge *et al.* (2009) investigated chemical composition and antifungal activity of the essential oil of leaves of two Tibetan species of *Juniperus saltuaria* and *J. squamata*. The essential oil tested was found effective on the inactivation of *Colletotrichum aculatum*, *C. fragariae* and *C. gloeosporioides*.

Taxus

The genus *Taxus* has attracted many researchers since the discovery of the anticancer agent paclitaxel (TaxolTM), a diterpenoid alkaloid originally isolated from the bark of the Pacific yew, *T. brevifolia* (Wani *et al.*, 1971). This drug is the first natural product describing that stabilized microtubules and has been approved by the FDA for the treatment of ovarian, breast and stem cell lung carcinomas (Rowinsky, 1997). So far, several hundred different taxoids, lignans, flavonoids, steroids and sugar derivatives have been isolated from different parts of various *Taxus* species (Parmar *et al.*, 1999). Of these compounds, Lignans are known to possess various biological activities such as antibacterial, antifungal, antiviral, antioxidant, anticancer and anti-inflammatory effects (MacRae and Towers, 1984).

In a study ethanol extract of heartwood of *Taxus baccata* showed significant activity aganist five tested fungi *Negrospora oryzae, Microsporum canis, Epidermophyton floceasum, Curvularia lunata* and *Pleuralus astreatu* (Erdemoglu and Sener, 2001). They found that the extract of *T. baccata* showed complete inhibition of mycelium at 400 μg/ml.

Baranowska and Wiwart, (2003) studied antifungal activity of bioflavones isolated from needles of *Taxus baccata*. They found it active towards *Cladosporium oxysporum* and *Fusarium culmorum* at 100 μg/ml and towards *Alternaria alternata* at 200 μg/ml concentration.

Ligans, isolated from *T. baccata* inhibited the growth of *Aspergillus flavus* in minimum concentration 10 μg/ml (Erdemoglu *et al.*, 2004). Similarly, Kumar *et al.* (2006) investigated antifungal activity of crude extracts of leaves of *T. baccata*. and reported that the 100 μg/ml concentration was found active against *Candida albicans* and 500 μg/ml for *Aspergillus niger*.

In a study on the antifungal activity Nisar *et al.* (2008) investigated methanol extract of aerial part of *Taxus wallichiana* against 6 fungal strains using macro-dilution method. They found *Trichophyton longifusus, Microsporum canis* and *Fusarium solani* susceptible to the extract (MIC 0.08 to 200 μg/ml).

Cupresses

This genus consists of nearly a dozen species (Ambasta., 1986). The volatile oil obtained from the cones of *C. torulosa* has been reported anti-inflammatory activity. Whereas, essential oil of the leaves is used to treat rheumatism and whooping cough as well as an astringent (Dhanabal *et al.*, 2000). Biflavones, *viz.* amentoflavone, cupressuflavone, hinokiflavone, and apigenin, are present in the leaves of *C. torulosa* (Natarajan *et al.*, 2001). Chemical analysis of the essential oil present in the foliage of *C.torulosa* indicated presence of mono-, sesqui- and di-terpenes (Cool *et al.*, 1998).

The hexane leaf extract *Cupressus lusitanica* was found to posses good antifungal activity (Kuiate *et al.*, 2006). They found that the extract showed highest activity against *Microsporum audouinii, M. canis, M. langeronis, Trichophyton rubrum* and *T. tonsurans*.

Some *in vitro* antifungal studies on Cupresses extract were carried out by Sellappan *et al.* (2007) using, *Candida albicans, Aspergillus flavus, Trichoderma lignorum* and *Cryptococcus neoformans*. The standard drug griseofulvin (100 ìg/mL) used was for fungi as a positive control. The oil displayed significant antifungal activity against all the pathogen tested.

Ginkgo

The extract of *Ginkgo biloba* and has widely been used in traditional Chinese medicine for treatment of cerebrovascular or cardiovascular diseases for many years (Cheung *et al.*, 1999). In literature the extract of this plant has commanly been decribed as *Gb*E which contains approximately 30 kinds of flavonoids besides these it contains glycosides, diterpenes (including terpene compounds called ginkgolides), bioflavones, quercitin, isorhamnetine kaempferol, proanthocyanidins, sitosterols, lactones, anthocyanin (Cheung *et al.*, 1999; Jacobs and Browner, 2000). It was reported that *Gb*E had antioxidant effects (Calapai *et al.*, 2000). The flavo glycosides of *Ginkgo* are its most active compounds which exhibited remarkable pharmacological capabilities. These chemical constituents have free radical properties and function as antioxidants. The flavonoids include quercitin, kaempferol and isorhamnetine. Similarly, the terpene content of *Ginkgo*, includes the ginkgolides and bilobalides which help to lessen inflammation by inhibiting PAF (Platelet Acti vating Factor) in the blood responsible for diseases such as, asthma, heart attacks and strokes (Yang XF *et al.*, 2003).

Mazzanti *et al.* (1999) investigated antifungal activity of methanolic, ethyl acetate, n-butanol and aqueous extracts of *Ginkgo biloba* and found that these extracts inhibite the growth of *Candida albicans, C. krusei, C. tropicalis* and *C. parapsilosa*. Haung *et al.* (2000) isolated an antifungal protein termed as GAFP (ginkgo antifungal protein) with molecular mass of 4244 Da, from the leaves of *G. biloba*. This protein exhibited potent antifungal activity against *Pellicularia sasakii, Alternaria alternata, Fusarium graminarium and F. moniliforme.*

In a study on antifungal activity of *Ginkgo biloba* Baranowska and Wiwart, (2003) studied antifungal activity of bioflavones isolated from leaves and found it is active against *Alternaria alternata* at 200 µg/ml *Cladosporium oxysporum* and *Fusarium culmorum* at 100 µg/ml.

Antifungal activity of *G. biloba* leaf extract (GBE) and combined effect of GBE+sodium EDTA (sodium Ethylenediaminetetraacetic acid) against *Listeria monocytogens* were also determined by Xie *et al.* (2006). They observed that the GBE was effective in inhibiting microbial growth and addition of EDTA enhances the antimicrobial activity of GBE.

Abies

The genus *Abies* consists of 51 species predominantly occurring in mountainous, temperate and boreal regions of the northern hemisphere (Kim *et al.*, 2004) The available literature on the phytochemical and biological investigations of this genus revealed that nearly 270 compounds have been isolated from 19 species of *Abies*. The chemical constituents are mostly terpenoids, flavonoids, and lignans, together with minor constituents of phenols, steroids, and others. The crude extracts and metabolites have been found to possess various bioactivities including insect juvenile hormone, antitumor, antimicrobial, anti-ulcerogenic, antiinflammatory, antihypertensive, antitussive, and CNS (central nervous system) activities (Yang *et al.*, 2008).

Abies pindrow is regarded as carminative, stomachic, astringent, expectorant, tonic, antispasmodic and antiperiodic (Burdi *et al.*, 2007). Phytochemical studies of the species resulted in isolation of glucopyranoside, hydroxyl-flavanone, chalcone glycoside, bioflavonoids, flavonoids and pindrolactone (Singh *et al.*, 2000) as well as pentacyclic triterpenoids (Manral *et al.*, 1987).

Digrak *et al.* (1999) studied antifungal activity of chloroform, methanol and acetone extracts of leaves, resins, bark, cones and fruits of *Abies Cilicia* and they found that the methanol and acetone extract were more effective against *Candida tropicalis, C. Albicans* and *Penicillium italicum.*

Crude leaf extract of *Abies webbiana* was also analyzed for its antifungal activity against a panel of pathogenic fungi by Kumar *et al.* (2006). They found that the extract the extract was active at 100 μg/ml and 500 μg/ml concentration against *Candida albicans* and *Aspergillus niger* respectively.

Cedrus

Cedrus deodara is an evergreen conifer tree of Himalaya. All parts of its are bitter, slightly pungent, oleaginous. It is very useful in inflammations, dyspepsia, insomnia, cough, fever, urinary discharges, ozoena, bronchitis, itching, elephantiasis, tuberculous glands, leucoderma, opthalmia, plies, disorders of the mind, and diseases of the skin and of the blood (Trease and Evans., 1999). The leaves are used to lessen inflammation when applied in tuberculous glands. The wood is bitter, diuretic, carminative, expectorant, and useful in rheumatism, palsy, epilepsy, stones in kidney and bladder, useful in fever, costiveness and pulmonary complaints (Mukherjee, 2001 and Thirunarayanan, 1994). The oil is analgestic and alexipharmic, useful for bruises and injuries to joints, boils, tubercular glands and skin (Shinde *et al.*, 1999; Jingwen *et al.*, 2006). It is considered to possess diuretic and carminative properties and to be useful in fevers, flatulence and urinary disorders. The bark of *C. deodara* is powerfully astringent and febrifuge (Shinde *et al.*, 1999).

Dikshit *et al.* (1983) determined the fungicidal activity of *Cedrus deodara* oil against 10 strains of fungi. In their this study seeds of two spices *Coriandrum sativum* and *Foeniculum vulgare*, were dressed separately with essential oil of *Cedrus deodara* as well as with five synthetic fungicides *viz.*, phenylmercury acetate, 2-methoxyethyl mercury chloride, copper oxychloride, mancozeb and wettable sulphur. Treated seeds were stored in polythene bags for 12 months. On mycofloral analysis, the oil had checked the appearance of ten fungi, *Absidia* sp., *Alternaria alternata, Aspergillus flavus, A. fumigatus, A. niger, A. ruber, A. versicolor, Cladosporium cladosporioides, Curvularia lunata,* and *Paecilomyces variotii* on the seeds of *C. sativum*, and seven fungi, *Absidia* sp., *A. flavus, A. fumigatus, A. niger, A. ruber, A. versicolor,* and *Rhizopus* spp., on the seeds of *F. vulgare.* The oil proved to be more effective than the synthetic fungicides without show any adverse effect on seed germination and seedling growth of both spices.

In a study on antifungal activity of the leaves, resins, bark, cones and fruit of *Cedrus libani* chloroform, acetone and methanol extracts against various pathogenic fungi (Digrak *et al.* (1999). They found that the methanol extract had the highest activity against *Candida albicans, C. tropicalis* and *Penicillium italicum.*

Thuja

Thuja now known as *Biota* is a small genus belonging to the Cupressaceae family comprising five extant species. In folk medicine, *Thuja occidentalis* has been used to treat bronchial catarrh, enuresis, cystitis, psoriasis, uterine carcinomas, amenorrhea and rheumatism (Peng and Wang, 2008). Leaf oil can be obtained by steam distillation or hydrodistillation of the foliage and is used for the production of perfumes, insecticides, soaps and deodorants (Duke, 1985; Kamden and Hanover, 1993). The major constituent of the oil, the monoterpene thujone, is used pharmacologically as an active ingredient in the production of nasal decongestants and cough suppressants, perfumes etc., while many cultivars are grown for ornamental purposes (FAO, 1995). The work on antifungal activity of *Thuja* is relatively quite meagre.

In vitro antifungal activity of crude methanol extract and essential oil of *Thuja orientalis* were carried by Ezzat, (2001). They found that oil inhibited the growth of *Candida albicans* at 10-80 µg/ml concentration.

Recently, Gularia *et al.* (2008) investigated fungitoxic activity of *Thuja orientalis* leaves essential oil against *Alternaria alternata* in a direct bioautography assay. Two main bioactive compounds named, as b1 and b2 were found active and which produce inhibition zone of 5 and 10mm in diam respectively.

Araucaria

Araucaria is one of the ornamental tree of gymnosperms.Different parts of its are also employed in folk medicine. Tinctures of nodes are traditionally used orally or topically for the treatment of rheumatism. Infusions of nodes are used to treat renal and sexually transmitted diseases. Infusions of barks are used topically to treat muscle strains and varices. The syrup produced with the resin of *Araucaria* is used for the treatment of respiratory tract infections. Extract of leaves are used to treat scrofula, fatigue, and anaemia. Tinctures of leaves are also used to treat the dried skin, wounds and shingles (Freitas *et al.*, 2009). Now a days in spite of its ornamental value and other traditional uses, few phytochemical and pharmacological studies of this plant have been performed (Santi-Gadelha *et al.*, 2006; Vasconcelos *et al.*, 2009). Most of them have focused on biological activity and chemical analysis of *Araucaria angustifolia* seeds.

The antifungal activity of ligans isolated from methanol extract of heartwood of *Araucaria araucana* was studied by Carlos *et al.* (2006) and found that it had a highest activity against *Aspergillus niger, Candida albicans, Fusarium moniliforme, F. sporotrichum and Trichophyton mentagrophytes* at 400 µg/disc.

Conclusion

The aforesaid account on bioactivity of the neglected plant gymnosperms clearly indicates that they have also a significant antifungal potentiality. In addition, these species contains a high tannis, terpenoids, alkaloids, glycosides, ligans, phenols, steroids and sugar derivatives. The different experimental methods and studies indicated various antifungal strength of these gymnosperms plants. In the present contribution more than nine genera of gymnosperms from around the world, are described for their antifungal activities. Fortunately the representatives of all the genera are found either naturally occurring or cultivated in an around Nainital too. Among the known gymnosperms plants *Pinus, Taxus, Cedrus* are well documented for their antifungal activities. Other gymnosperms plants *Cupresses, Thuja* and *Araucaria* are inadequately studied. Whereas *Cryptomeria, Picea* and *Cephalotaxus* are relatively untouched for their bioactivities.

It is also to be underlined that most of the species claimed for potent antifungal activity have not been studied in vivo because *in vivo* trials are more significant than others. Therefore, to exploit the antifungal potentiality of gymnosperms plants more attention to be required for their biotechnological cultivation, conservation and employment of standard methods for their experimentation.

References

Amorim, A., M. Sucena, G. Fernanandes and A. Magalhaes. (2004). Pleural disease and acquired immunodeficiency syndrome. Rev. Port. Pneunol. 10:217-225.

Appendino G. (1993), Taxol (paclitaxel): Historical and ecological aspects. *Fitoterapia* 64, 5-25.

Bagei, E. and M. Digrak (1997). *In vitro* antimicrobial activities of some fir essential oils.*Turkish J. Biol.* 21(3): 273-281.

Balick, M.J. and Cox, P.A. (1997). Plants, people and culture: the secience of ethnobotany. Scientific American Library, New York, NY.

Baquar SR (1989). Medicinal and Poisonous plants of Pakistan. Printas Karachi, Pakistan., pp. 506.

Baranowska, M. K. and Wiwart M. (2003). Antifungal Activity of Biflavones from *Taxus baccata* and *Ginkgo biloba*. Z. Naturforsch. 58c, 65-69.

Baranowska, M.K., Mardarowicz M., Wiwart M., Pob³ocka L. and Dynowska M. (2002). Antifungal Activity of the Essential Oils from Some Species of the Genus *Pinus*. Z. Naturforsch. 57c, 478-482.

Bryan–Brown T. (1932), The pharmacological action of Taxine. *Quat. J. pharma. Pharmacol.* 5, 205-219.

Bulanas, M. and Kinghorn, D.A. (2005). Drug discovery from medicinal plants. Review article. Life Sci. 78(5): 431-441.

Burdi DK, Samejo MQ, Bhanger MI, Khan K M (2007). Fatty acid composition of *Abies pindrow* (West Himalayan fir). Pak J Pharm Sci. 20: 15-9.

Campbell, N.A., Mitchell, G.L. and J.B. Reece. (2000). Biology concepts and connections. 3rd ed. Addison Wesley Longman, Inc. NewYork. Pp: 672-674.

Carlos L. Ce´spedesa, J. G. Avilab, Ana M. Garcý´ab, Jose´ Becerrad, Cristian Floresd, P. Aquevequed, M. Bittnerd, M. Hoeneisend, M. Martinezc, and M. Silvad. (2005). Antifungal and Antibacterial Activities of *Araucaria araucana* (Mol.) K. Koch Heartwood Lignans. Z. Naturforsch. 61c, 35-43.

Carlos L. Ce´spedesa,, J. Guillermo Avilab, Ana M. Garcý´ab, Jose´ Becerrad, Cristian Floresd, Pedro Aquevequed, Magalis Bittnerd, Maritza Hoeneisend, Miguel Martinez, and Mario Silva. (2006). Antifungal and Antibacterial Activities of *Araucaria araucana* (Mol.) K. Koch Heartwood Lignans. Z. Naturforsch. 61c, 35-43.

Cavaleiro, C., Pinto, E., Gonçalves, M.J. and Salgueiro, L. (2006). Antifungal activity of Juniperus essential oils against dermatophyte, *Aspergillus* and *Candida* strains. Journal of Applied Microbiology. 100 (6): 1333-1338.

Clark, M.A. and McChemcr, J.D. Adams R.P. 1999. Antimicrobial Properties of Heartwood, Bark/ Sapwood and Leaves *of Juniperus* Species. Phytotherepy Research. 4(1):15-19.

Clutton DM (1972). History of Gin. Flavour. Ind. 3: 454.

Cragg, G.M. and Newman, D.J. (2005). Biodiversity: A continuing source of novel drug leads. Pure Appl. Chem. 77(1):

Das, S. Das, S. Pal, A. Mujib and S. Dey, (1999)., *Biotechnology of Medicinal Plants- Recent Advances and Potentials.* 1st edition. Vol 11(UK 992 publications, Hyderabad, pp 126-139.

Dhar L.M., M.M., Dhar., B.N. Dhawan., C. Ray (1968). Screening of Indian plants for biological activity. Part I. Indian. J. Exp. Biol., 6: 232-247.

Digrak, M., M. Iicim and N. H. Alwa (1999). Antimicrobial activities of several parts of *Pinus brutia, Juniperus onycedrus, Abies cilicia* and *Pinus nigra. Phytotherapy Res.* 13(7): 584-587.

Dikshit, A., Dubey, N.K., Tripathi, N.N. and Dixit, S.N. (1983). Cedrus oil- A promising storage fungitoxicant. Journal of Stored Products Research. 19(4): 159-162.

Dube, H.C. (1990). An introduction to fungi. 2nd Rev. ed. Vikas Pub. House Pvt. Ltd; 141-176.

Erdemoglu N. and Sener B. (2001), Antimicrobial activity of heartwood of *Taxus baccata. Fitoterapia* 72,59-61.

Erdemoglu N., Sener B., and Choudhary, I.M. (2004). Bioactivity of Lignans from *Taxus baccata*. Z. Naturforsch. 59c, 494-498.

Filipowiez N, Kaminski M, Kurlenda J, Asztemborska M. (2003). Antibacterial and antifungal activity of Juniper Berry oil and its selected components. Phytother Res. 17: 227-231.

Guleria, S., Kumar, A. and Tiku, A. K. (2008). Chemical Composition and Fungitoxic Activity of Essential Oil of *Thuja orientalis* L. Grown in the North-Western Himalaya. Z. Naturforsch. 63c, 211-214.

Gupta., A.K., I. Ahmad and R.C. Summerbell. (2001). Comparative efficacies of commonly used disinfectants and antifungal pharmaceutical spray preparations against dermatophytic fungi. Med. Mycol. 39:321-328.

Harborne J.B. and Baxter W. (2001), the chemical dictionary of economic plants. Wiley and Sons, Chichester, p.582.

Haung, X., W. Xie and Z. Gong. (2000). Characterization and antifungal activity of a chitin binding protein from Ginkgo biloba. FEBS letters. 478: 123-126.

Hawksworth, D.L. (2001). The magnitude of fungal diversity: the 1.5 million species estimate revisited. Mycol. Res. 105: 1422-1432.

Izzat, S.M. (2001). In vitro inhibition of *Candida albicans* growth by plant extracts and essential oils. World Journal of Microbiology and Biotechnology. 17: 757-759.

Jeong-Ho, L., Lee, Byung-Kyu, Jong-Hee Kim, Sang Hee Lee4, and Soon-Kwang Hong. (2009). Comparison of chemical compositions and antimicrobial activities of essential oils from three Conifer Trees; *Pinus densiflora, Cryptomeria japonica*, and *Chamaecyparis obtusa*. J. Microbiol. Biotechnol. (2009), 19(4), 391–396.

Kim H J, Choi EH, Lee IS (2004). Two lanostane triterpenoids from *Abies koreana*. Phytochemistry 65: 2545.

Kirtikar K. R. and Basu B. D. (1935). Indian Medicinal Plants. Vol 3. Periodical experts, Delhi, India.

Kuiate, Jules-Roger, Jean Marie Bessi'ere, Paul Henri Amvam Zollo, Serge Philibert Kuate (2006). Chemical composition and antidermatophytic properties of volatile fractions of hexanic extract from leaves of *Cupressus lusitanica* Mill. from Cameroon. Journal of Ethnopharmacology 103:160–165.

Kumar, R.S., T. Sivakumar, R.S. Sunderam, M. Gupta and N.K. Mazumdar *et al.* (2005). Antioxidant and antimicrobial activities of *Bauhinia racemosa* L. stem bark. Braz. J. Med.Biol. Res., 38, 1015.

Kumar, V. P., N.S. Chauhan., P. Harish., Rajani, M. (2006). Search for antibacterial and antifungal agents from selected Indian medicinal plants. Journal of Ethnopharmacology 107:182–188.

Lee, S.K., H.J. Lee., H.Y. Min, E.J. Park and K.M. Lee (2005). Antifungal and antibacterial activity of pinosylvin, a constitute of pine. Fitoterapia. 76: 258-260.

Manral K, Pathak RP, Khetwal KS (1987). Pentacyclic Triterpenoids from Heart-Wood of *Abies pindrow* Wall. Indian Drugs. 24: 232

Mazzanti, G., M.T. Mascellino, L. Battinelli, D. Coluccia, M. Manganaro, L. Saso. (1999). Antimicrobial investigation of semipurified fractions of *Ginkgo biloba* leaves. Journal of Ethnopharmacology 71: 83–88.

Niemann G.J. (1988), Distribution and evolution of the flavonoids in gymnosperms. In:The Flavonoids. Advances in research since 1980(Harborne J.B., ed). Chapman and hall, London, pp 469-478.

Nisar M., Inamullah Khan, Bashir Ahmad, Ihsan Ali, Waqar Ahmad and Muhammad Iqbal Choudhary. (2008). Antifungal and antibacterial activities of *Taxus wallichiana* Zucc. Journal of Enzyme Inhibition and Medicinal Chemistry. 23(2): 256-260.

Parmar V. S., Jha A., Bisht K. S., Taneja P., Singh S. K., Kumar A., Raijni Jain P., and Olsen C. E. (1999), Constituents of Yew trees. Phytochemistry 50, 1267-1304.

Rowinsky E. K. (1997), The development and clinical utility of the taxane class of antimicrotubule chemotherapy agents. Annu. Rev. Med. 48, 353-374.

Samuelsson, G. (2004). Drugs of natural origin: a textbook of pharmacognosy, 5th Swedish Pharmaceutical press, Stockholm.

Santi-Gadelha, T., Gadelha, C.A.A., Aragão, K.S., Oliveira, C.C., Mota, M.R.L., Gomes, R.C., Pires, A.F., Toyama, M.H., Toyama, D.O., Alencar, N.M., Criddle, D.N., Assreuy, A.M., Cavada, B.S., 2006. Purification and biological effects of *Araucaria angustifolia* (Araucariaceae) seed lectin. Biochemical and Biophysical Research Communications 350, 1050–1055.

Sellappan M., Palanisamy D., Joghee N., Bhojraj S. (2007). Chemical Composition and Antimicrobial Activity of the Volatile Oil of the Cones of *Cupressus torulosa* D. DON from Nilgiris, India. Asian Journal of Traditional Medicines, 2007, 2 (6).

Singh V. (1995), Traditional remedies to treat asthma in northwest and trans- Himalayan region in Jammu and Kashmir state. *Fitoterapia* 66,507-509.

Singh, R.K., S.K. Bhattacharya., S.B. Acharya. (2000). Pharmacological activity of *Abies pindrow.* Journal of Ethnopharmacology 73: 47–51.

Sternitz F.R., Tawara J.N., Boeckl M., Pomeroy M., Foderaro T.A. and Todd F.G. (1994), piperidine alkaloid content of *Picea* (spruce) and *Pinus* (pine). *Phytochemistry* 35,951-953.

Swanston-Flatt DSK, Bailey C, Flatt PR (1990). Traditional plant treatments for diabetes. Studies in normal and streptozotocin diabetic mice. Diabetologia., 33: 462.

Takacsova M, Pribela A, Faktorova M, Nahrung (1995). Study of the antioxidative effects of thyme, sage, juniper and oregano. Nah rung. Food, 39: 241-243.

Torras, M. A., Faura, C. A., Schonlau, F., Rohdewald, P. 2005. Antimicrobial activity of Pycnogenol. Phytother. Res. 19, 647.

Vasconcelos, S.M., Lima, S.R., Soares, P.M., Assreuy, A.M., de Sousa, F.C., Lobato Rde, F., Vasconcelos, G.S., Santi-Gadelha, T., Bezerra, E.H., Cavada, B.S., Patrocinio, M.C., 2009. Central action of *Araucaria angustifolia* seed lectin in mice. Epilepsy Behavior 15, 291–293.

Wani M. C., Taylor H. L., Wall M. E., Coggon P., and McPhail A. T. (1971), Plant antitumor agents. VI. The isolation and structure of taxol, a novel antileukemic and antitumor agent from *Taxus brevifolia.* J. Am. Chem. Soc. 93, 2325-2327.

Wedge DE, Tabanca N, Sampson BJ, Werle C, Demirci B, Baser KH, Nan P, Duan J, Liu Z. (2009). Antifungal and insecticidal activity of two *Juniperus* essential oils. Nat Prod Commun. 2009 Jan;4(1):123-7.

Xie, L., N.S. Hettiarachchy, M.E. Jane, M.G. Johnson. (2006). Antimicrobial activity of *Ginkgo biloba* leaf extract on *Listeria monocytogenes*. Journal of Food Science. 68(1): 268-270.

Yang, Xian-Wen, Su-Mei Li, Yun-Heng Shen, Wei-Dong Zhang. (2008). Phytochemical and Biological Studies of *Abies* Species. Chemistry and Biodiversity. 5(1): 56-81.

Zaman MB, Khan MS (1970). Hundred Drug Plants of West Pakistan. Pakistan Forest Institute Peshawar, Pakistan.

Microbes: Diversity and Biotechnology (2012) *Pages* **347–357**
Editors: **Prof. S.C. Sati & Dr. M. Belwal**
Published by: **DAYA PUBLISHING HOUSE, NEW DELHI**

Chapter 21

Microorganisms: A Source of Bio-pharmaceuticals

Archana Mehta, Jinu John and Pradeep Mehta
Biotechnology Lab, Department of Botany, School of Biological and Chemical Sciences, Dr. H.S.G. University, Sagar – 470 003

ABSTRACT

Microbes possess a wealth of metabolic equipment that brings about diverse chemical transformations. This characteristic of microbes could be exploited in obtaining some valuable products of daily use. The cheap raw materials available in nature as a waste may be converted into useful commercial products by the activity of microbes. Antibiotics are commonly used drugs that are diverse in their actions and side effects. The large number of antibiotics available makes it challenging to understand the characteristics of each of these antimicrobial agents, including important information such as toxicities and indications. Some antibiotics have been employed clinically for their anti-tumour and immuno-suppressive properties, while others for their effectiveness in infections of bacterial or viral origin. Although, certain antibiotics work only on prokaryotes, while others are toxic only to eukaryotes or to both types of organisms. All such studies including antibiotic resistance and emergence of new pathogenic organism pointing towards the need of development of new antibiotics and search for new sources.

Keywords: Microbes, Antibiotics, Antimicrobial agents.

Introduction

Biotechnology utilizes various biological entities for the manufacturing of useful products. Microorganisms are highly exploited in this regard. From time immortal man started utilizing microorganisms for various purposes (food processing, seasoning to wine production, etc). An easy handling and high growth rate made them one of the efficient tools in industrial microbiology. The age old experiences and experiments have made these microorganisms one of the inevitable tools of the day. They have proved their efficiency in various fields such as secondary metabolites, processing of metal ores and even in the production of various biopharmaceuticals, which improved the morbidity considerably like a wide range of antibiotics to different hormones like insulin using genetic engineering. Currently, the complete genetic sequences of more than 580 microbes are known to date. It is possible to identify pathways that produce new compounds by analyzing the DNA sequences and many gene clusters likely to encode natural products. Nature has provided man with exquisitely potent and specific agents such as antibiotics that have proved useful as drugs in the treatment of diseases and as tools in the study of biological processes.

The discovery of antibiotics greatly improved the quality of human life in the last century. The word antibiotic comes from *anti-* "against," and *bios,* meaning "life." Antibiotic drugs are made from living organisms such as fungi, molds, and certain soil bacteria that are harmful to disease-causing bacteria. Antibiotics can also be produced synthetically or combined with natural substances to form semi-synthetic antibiotics. Antibiotics are drugs used to fight infections and infectious diseases caused by bacteria and other organisms. It is believed that antibiotics accomplish these actions by damaging bacterial cell walls or by otherwise interfering with the function of the cells.

Microorganisms have the ability to synthesize a wide range of complex chemicals ranging from bio-plastics to enzymes that are used in laundry detergents. Since 1980s, bacteria have gained importance in the production of many bulk chemicals, including ethanol, from fermented corn. Ethanol is an ingredient of gasohol, a fuel that burns more cleanly than gasoline and uses less petroleum. Chemical's production using bacteria and other microorganisms results in less pollution to the environment than standard chemical production. The development of genetic engineering has opened the way to even greater use of bacteria in large-scale industrial manufacturing and environmentally friendly processes. Many of the microbial products are proved to be potent and efficient in inhibiting the growth or in killing wide range of pathogenic organisms even there are reports for the anticancer properties of microbial products and many are in application also. The main concern of the age is increasing antimicrobial resistance of pathogens; therefore it is necessary to explore new sources having efficiency and lesser or no side effects on human body. The present article mainly deals with the microbial metabolites such as antibiotics and their mode of action.

Fungi: Source of Several Value Added Secondary Metabolites

Microorganisms had now proven not only to be capable producers of antibacterial but also of compounds for other therapeutic applications. The recent and growing understanding of microbial abundance and diversity in particular that of fungi (Hawksworth, 1991), shows that the fungal world has until now been, at best, superficially scratched. Additionally, new technological developments in areas such as robotics, ligand analysis, informatics, chemical analysis, and synthesis have paved the road to a dynamic, discovery-led growth in the pharmaceutical industry. The general scheme for the industrial screening of microorganisms for value added product is given in the Figure 21.1.

The total number of characterized fungal secondary metabolites, including antibiotics, mycotoxins, pharmacologically active compounds, and those without known biological activities, are not easy to find. Highly active areas correspond to fungal groups potentially able to produce a high diversity of secondary metabolites, independent of their commercial usefulness. Genera such as *Aspergillus, Penicillium, Acremonium, and Fusarium* (teleomorphs in the Eurotiales and Sphaeriales) clearly belong to these areas and account for over 40 per cent of the antibiotics known in 1984. Few secondary metabolites are produced by constitutive metabolic machineries, so that specific taxa will always have the potential to produce such "market" metabolites. Pigments, toxins, and other secondary metabolites from mushrooms (basidiomycetes), lichens, and other large, fleshy, or sclerotial fungi are widely accepted as constant characteristics and therefore reliable markers for identification. Although, the estimated 3000 to 4000 known fungal secondary metabolites have been isolated after screening thousands of fungal cultures (Hawksworth and Kirsop 1988; Berdy 1989). The untapped pool of fungal diversity is, however, tremendous. The Table 21.1 depicts the various value added products from different fungi.

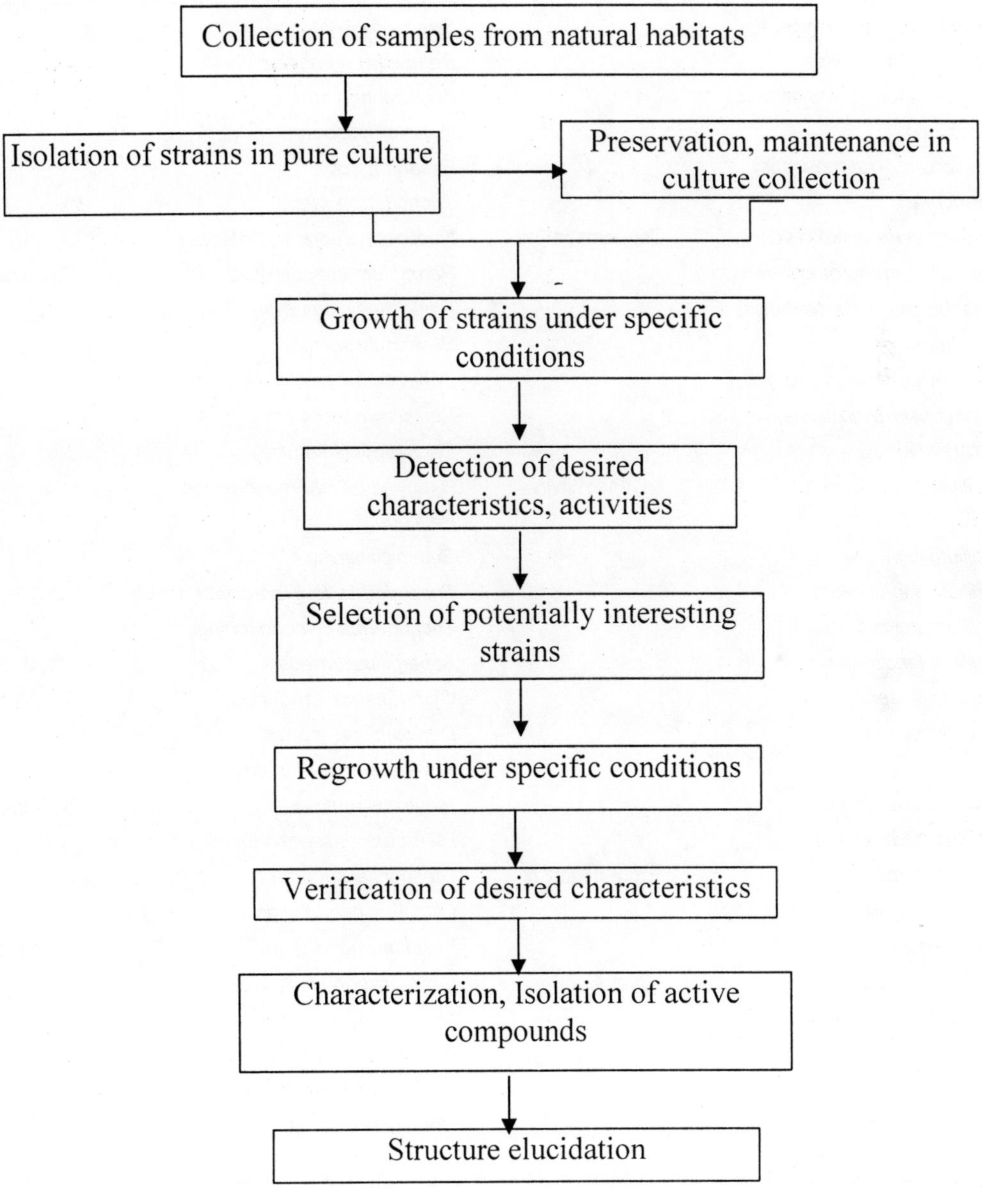

Figure 21.1: General Scheme for Industrial Microbiological Screening (Berdy, 1989)

Table 21.1: Fungal Taxa able to Produce Secondary Metabolites of Past and Present Commercial Relevance, as Pharmaceuticals or for Other Uses (Dreyfuss and Chapela, 1994)

Fungal Taxa	Metabolite/Products	Fungal Taxa	Metabolite/Products
Penicillum chrysogenum	Penicillins	*Chaunopycins alba*	Cyclosporins
Penicillium spp.		*Aphanocladium* sp.	
(Eupenicillum, Talaromyces)		*Beauveria bassiana*	
Aspergillus (Emericella)		*Beauveria brongniarti*	
Epidermophyton (Gymnoascus)		*Acremonium* spp.	
Trichophyton		*Paecilomyces* spp.	
Polypaecilum (Thermoascus)		*Fusarium* spp.	
Malbranchea		*Trichoderma viride*	
Cephalosporium acremonium	Cephalosporins	*Neocosmospora vasinfecta*	
Cephalosporium/Acremonium spp.		*Penicillium brevicompactum*	Compactin
(Nigrosabulum, Emericellopsis)		*Penicillium citrinum*	
Pleurophomopsis		*Penicillium solitum*	
Spiroidium (Arachnomyces)		*Penicillium canescens*	
Scopulariopsis (Anixiopsis)		*Penicillium lanosum*	
Paecilomyces		*Penicillium hirsutum*	
Penicillium griseofulvum	Griseofulvin	*Hypomyces chrysospermus*	
Penicillium patulum		*Paecilomyces* sp.	
Penicillium nigricans		*Eupenicillium* sp.	
Penicillium coprophilum		*Trichoderma longibrachiatum*	
Penicillium canescens		*Trichoderma pseudokoningii*	
Penicillium janczewskii		*Aspergillus terreus*	Mevinolin
Penicillium jensenii		*Gymnoascus umbrinus*	
Penicillium lanosum		*Doratomyces nanus*	
Penicillium raistricki		*Several sterile endophytic fungi*	
Penicillium sclerotigenum		*Pleurotus mutilis*	Pleuromutilin
Penicillium aethiopicum		*Pleurotus passeckerianus*	
Claviceps purpurea	Ergot alkaloids	*Drosophila subatrata*	
Claviceps paspali		*Clitopilus pseudopinsitus*	
"Grass endophytes"		*Fusidium coccineum*	Fusidic acid
Tolypocladium inflatum	Cyclosporins	*Cephalosporium floccosum*	
Tolypocladium geodes		*Microsporum gypseum*	
Tolypocladium nubicola		*Keratinomyces longifuscus*	
Tolypocladium terricola		*Calcarisporium antibioticum*	
Tolypocladium cylindrosporum		*Chrysosporium* sp.	
Tolypocladium tundrense			

Brief History of Antibiotics

The use of microbial products as antibiotics dates back to thousands of years. The ancient Egyptians, the Chinese, and Indians of Central America all used molds to treat infected wounds.

However, they did not understand the connection of the antibacterial properties of mold and the treatment of diseases. The word antibiotic came from antibiosis a term coined in 1889 by Paul Vuillemin that means a process by which life could be used to destroy life. Pasteur observed that "if we could intervene in the antagonism observed between some bacteria, it would offer 'perhaps the greatest hopes for therapeutics'" (Kingston, 2008). Antibiosis was first described in 1877 in bacteria when Louis Pasteur and Robert Koch observed that an airborne bacillus could inhibit the growth of *Bacillus anthracis* (Landsberg, 1949; Waksman, 1947). Technically, an antibiotic is a substance that is produced by one microorganism and it is capable of killing or inhibiting the growth of another microorganism which can be a virus, bacteria, fungi, yeast, protozoan or any other pathogen.

The development of synthetic antibiotic chemotherapy began in Germany with Paul Ehrlich, a German medical scientist in the late 1880s. He noted that certain dyes would bind to and colour human, animal or bacterial cells, while others did not. This observation led to the idea of that it might be possible to make certain dyes or chemicals that would act as selective drug that would bind to and kill bacteria while not harming the human host and series of such research resulted in the discovery of Salvarsan, the first medicinally useful man-made antibiotic drug. However, the adverse side-effect of salvarsan, coupled with the later discovery of the antibiotic penicillin, superseded its use (Bosch and Rosich, 2008). The work of Ehrlich followed by the discovery of sulfa drug resulted in the birth of the antibiotic revolution. In 1928 Alexander Fleming discovered penicillin, which was subsequently developed into a medicine by Florey and Chain in the 1940s. Rene Dubos (1939) isolated gramicidin, one of the first commercially manufactured antibiotics, which was used during World War II and proved highly effective in treating wounds and ulcers (Van, 2006). The antibiotic was hailed as a 'miracle cure' and a golden age of drug discovery followed. All such major events led to the development of antibiotics are summarized in Table 21.2. There are several antibiotics which have been isolated from various microbes till now having potent activity against pathogens (Table 21.3).

Table 21.2: Time Line Mile Stones in the Development of Antibiotics

Late 1800s	With the growing acceptance of the germ theory of disease, scientists began to devote time to searching for drugs that would kill these disease-causing bacteria.
1871	Joseph Lister started studying the phenomenon of antibacterial effect shown by urine contaminated with moid.
1890s	Rudolf Emmerich and Oscar Low, German doctors made an effective medication that they called pyocyanase from microbes, the first antibiotic to be used in hospitals. However, the drug often did not work.
1928	Sir Alexander Fleming observed the antibacterial properties of mold, *Penicillium notatum* that destroyed the colonies of the bacterium *Staphylococcus aureus*.
1935	German chemist Gerhard Domagk discovered the first sulfa drug, Prontosil and he received the Nobel Prize for Medicine in 1939. It got relatively broad effect against Gram-positive *cocci* but not against *enterobacteria*.
1942	Howard Florey, Ernst Chain and A. Fleming invented the manufacturing process for Penicillin G Procaine, and they shared the 1945 Nobel Prize for medicine for their work on penicillin.
1943	American microbiologist Selman Waksman made the drug streptomycin from soil bacteria, the first of a new class of drugs called aminoglycosides. Streptomycin could treat diseases like tuberculosis; however, the side effects were often too severe.
1955	Tetracycline was patented by Lloyd Conover, which became the most prescribed broad spectrum antibiotic in the United States.
1957	Nystatin was patented and used to cure many disfiguring and disabling fungal infections.
1981	SmithKline Beecham patented Amoxicillin or amoxicillin-clavulanate potassium tablets, and first sold the antibiotic in 1998 under the trade names of Amoxicillin, Amoxil, and Trimox. Amoxicillin is a semisynthetic antibiotic.

Table 21.3: Some of the Antibiotics and their Source Organisms

Antibiotic	Source	Active Against
Amphotericin B	*Streptomyces nodosus*	Yeast, fungi
Aterrimin	*Bacillus subtilis*	Gram-positive bacteria
Bacitracin	*Bacillus subtilis*	Gram-positive bacteria
Candicidin B	*Streptomyces griseus*	Yeast, fungi
Cycloheximide	*Streptomyces griseus*	Fungi
Cycloserine	*Streptomyces orchidaceus*	Gram-positive and TB bacteria
Dactinomycin (Actinomycin D)	*Streptomyces antibioticus*	Gram-positive bacteria; anti-tumor principle
Erythromycin	*Streptomuces erythreus*	Gram-positive bacteria
Fusidic acid	*Fusidium coccineum*	Gram-positive bacteria
Gentamycin	*Micromonospora purpurea*	Gram-positive bacteria
Gramicidin	*Bacillus brevis*	Gram-positive bacteria
Griseofulvin	*Penicillium griseofulvum*	Fungi
Hygromycin	*Streptomyces hygroscopicus*	Gram-positive and Gram-negative and TB bacteria
Kanamycin	*Streptomuces kanamyceticus*	Gram-positive, Gram-negative and TB bacteria
Leucomycin	*Streptomyces kitasoensis*	Gram-positive bacteria
Lincomycin	*Streptomuces lincolnensis*	Gram-positive, Gram-negative and TB bacteria
Neomycins	*Streptomyces fradiae*	Gram-positive, Gram-negative and TB bacteria
Novobiocin	*Streptomyces niveus*	Gram-positive bacteria
Nystatin	*Streptomyces noursei*	Fungi and yeast
Penicillin and its Chemical derivatives	*Penicilliium chrysogenum*	Gram-positive bacteria
Polymyxin B	*Aerobacillus polymyxa*	Gram-negative bacteria
Pristinamycin	*Streptomyces sp.*	Gram-positive bacteria
Rifomycin SV	*Streptomyces mediterranei*	Gram-positive and TB bacteria
Ristocetin	*Nocardia lurida*	Gram-positive bacteria
Spiramycin	*Streptomyces ambofaciens*	Gram-positive and Gram-negative bacteria; Rickettsiae
Streptomycin and chemical derivatives	*Streptomyces griseus*	Gram-positive, Gram-negative and TB bacteria
Tetracycline and chemical derivatives	*Streptomyces aureofaciens*	Gram-positive and Gram negative bacteria; Rickettsiae
Tyrothricin	*Bacillus brevis*	Gram-positive and Gram-negative bacteria
Vancomycin	*Streptomyces orientalis*	Gram-positive and TB bacteria

Mechanisms of Action

During infection the causative bacteria invade and multiply in our body. The white blood cells of the immune system multiply to defend the body against these bacteria. The increased number of WBCs indicates the infection in the body. The bacteria usually are able to disturb the immune system and this is where the antibiotics need to be given.

It is the structure of the antimicrobial that determines the mechanism of action, toxicity, and metabolism. Some desirable properties in a antimicrobial are: selective toxicity, bactericidal rather than bacterostatic action, low rate of resistance development, low toxicity, low rate of hypersensitivity reactions, water soluble, activity in a wide variety of chemical environments, wide distribution, and low cost.

Antibiotics targeting process is achieved by exploiting the biochemical difference between microbial and animal cells, which can be used to treat infections without harming host. This is known as selective toxicity. Selective toxicity is the corner stone of effective antimicrobial therapy. Penicillin's are a classic example of excellent selective toxicity. This class of antibiotic interferes with bacterial cell wall synthesis and repair. Animal cells do not have a cell wall and penicillin's have remarkably low direct toxicity to humans, while being highly deleterious to bacteria. However, antibiotic therapy is not risk free, it can illicit serious allergic reactions. Therapeutic index relates to the dose of a drug required to produce the desired effect to the dose associated with its toxicity. In order to be effective the drug must reach the microbe and survive in the chemical environment at the site of infection such as the central nervous system, prostate, joints and eyes. Infection in these sites must be treated with antibiotics that can penetrate them. Abscess are encapsulated and have harsh chemical environments (*e.g.*, pH, redux potentials etc.) limiting the penetration and actions of antibiotics. Often an antibiotic is given to a patient before the organism infecting the patient is identified. This is known as empiric therapy. The choice of drug is guided by knowing which organisms are likely to be causing the specific infection. Organism identification relies on a combination of macroscopic (*e.g.*, colony morphology), microscopic, biochemical, and immunologic characteristics. More recently molecular genetics methods (*e.g.*, PCR), etc. are also contributing to the identification of the organism. With the organism isolated, the sensitivity to different antibiotics can be established, which will guide to proper therapy.

Antibiotics mainly have two types of mechanism of actions. Antibiotics either function as bacteriostatic or as bactericidal. As bacteriostatic they stop the bacterium from multiplying further by interfering with their DNA, but they do not kill the bacteria, while as bactericidal, they kill the bacteria. Penicillin is a bactericidal antibiotic. Some antibiotics do not kill bacteria at normal serum concentrations and these are called bacterostatic. These drugs inhibit bacterial growth, and the growth is resumed upon drug withdrawal. Bactericidal antibiotic are preferable to treat those patients with immunologic compromise or life-threatening infection. Bactericidal drugs are also highly desirable in infections characterized by poor regional host defenses, such as endocarditis and meningitis. Bacterostatic drugs are appropriate when the host's immune system is able to finish the job. Penicillins and the other beta-lactam are examples of bactericidal drugs, while tetracyclines and sulfonamides are bacterostatic. When one drug's mechanism of action makes another drug more effective, they are said to work in synergy. For instance, penicillin's by making the bacterial cell wall more permeable facilitate the action of aminoglycosides that work intra-cellularly and must gain entry into the cell.

The principal goal in the use of antimicrobial agents for the treatment of infections is eradication of the pathogen as quickly as possible with minimal adverse effects on the recipient. For these three basic conditions must be satisfied. First, the antibiotic must bind to a specific target-binding site or 'active site' on the microorganism. Although, the active sites are different for different classes of antibiotics, the principle is the same, namely to disrupt a point of biochemical reaction that the bacterium must undergo as part of its life cycle. If the biochemical reaction is critical to the life of the bacteria, then the antibiotic will have a deleterious effect on the life of the microorganism. The second condition is that the concentration of the antimicrobial is sufficient to occupy a critical number of these specific active sites on the microorganism. Finally, it is important that the agent occupies a sufficient number of active sites for an adequate period of time, is important to the life and death of the bacteria (Capitano and Nightingale, 2001). In addition to understand the drug's mechanism of action and spectrum of activity, the clinical practitioner should know about drug absorption, distribution, metabolism, and excretion. These characteristics are essential in determining aspects of therapy like route of administration and dosing interval (Nightingale *et al.*, 2001).

Apart from the life-saving benefits of antibiotics there are some significant side effects. Common side effects which can be ignored include feeling of uneasiness and minor diarrhea. There is occasional nausea. Infection may appear in vagina, digestive tract and mouth. Apart from these there are some rare side effects which may lead to death if ignored. Consumption of sulphonamides may lead to formation of kidney stones, while cephalosporin may cause abnormal clotting of blood. Tetracycline can make the body more sensitive towards sun. The severity of antibiotics (amino glycosides and erythromycin) may cause deafness in some patients (Nathwani and Wood 1993).

The mode of action of different antibiotics varies according to the site of action. Some of the modes of action by different antibiotics are mentioned below:

Inhibition of Cell Wall Synthesis

Murein or peptidoglycan is one of the major structural components of bacterial cell wall. Cell wall synthesis inhibitors such as beta-lactams, cephalosporins and glycopeptides block the ability of microorganisms to synthesize their cell wall by inhibiting the synthesis of peptidoglycan, leading to lysis and death of the bacteria. Petptidoglycan is made up of polysaccharide chains containing alternating residues of N-acetylmuramic acid and N-acetylglucosamine. Peptide units cross-link extending from the N-acetylmuramic acid and during bacterial wall synthesis these linkages are caused by precursors which are catalyzed by specific enzymes (*e.g.* transpepidases and carboxy-peptidases). These enzymes are regulatory proteins which bind to penicillin and hence are known as penicillin binding proteins (PBPs). The bacteria when exposed to penicillin, the antibiotic binds to the PBPs in the cell membrane followed by the release of autolytic enzymes that degrade the preformed cell wall and arrest further cell wall synthesis leading to death of bacteria.

Penicillin is the first antibiotic developed, which is still in use today, despite the development of antibiotic resistance during the last 70 years. Penicillin is now in a class of antibiotics known as the penicillin, which also includes the more recently developed antibiotics ampicillin, flucloxacillin and methicillin. After that, several other types of antibiotics produced by different species of bacteria and fungi were isolated. Some of these include cephalosporins; aminoglycosides such as streptomycin, gentamicin and kanamycin; ansamycins such as geldanamycin and carabecephems; glycopeptides like vancomycin; macrolides like erythromycin and azithromycin, penicillin, quinolone, polypeptide and sulfonamide.

The earliest cephalosporin was isolated by an Italian scientist in 1948, from bacteria growing in a sewer on the island of Sardinia. The compound was found to be active against *Salmonella typhi*, the pathogen that was responsible for typhoid fever. The earliest cephalosporins were active against just a few gram negative bacteria but chemists worked on the molecule, creating second, third, fourth and fifth generation cephalosporin antibiotics. They generally increased their spectrum of activity, and the most recent ones can be used to treat infections caused by gram positive as well as gram negative bacteria. Glygopeptide antibiotics such as vancomycin, telavancin and decaplanin prevent the adding of new units to the bacterial cell wall. Although they are similar in action to the penicillins and the aminoglycosides, the glycopeptides are extremely toxic and are used in few patients. They are reserved for people whose infections will not respond to other antibiotics because of antibiotic resistance. Until the development of newer antibiotics, glycopeptides were the standard treatment for MRSA (Methicillin resistant *Staphylococcus aureus*).

Interfering with Protein Synthesis

These classes of antibiotics inhibit the protein synthesis machinery in the cell. Some examples include tetracyclines, chloramphenicol, aminoglycosides and macrolides. Macrolids are the class of drug whose activity stems from the presence of a *macrolide ring*, a large macrocyclic lactone ring to which one or more de-oxy sugars, usually cladinose and desosamine, are attached.

Streptomycin, developed in the 1940s, is now in the amino-glycoside class of antibiotics, which also has gentamycin, kanamycin and netromycin as members. All of these antibiotics bind to the smaller subunit of the bacterial ribosome, preventing the bacterium making new proteins. They work well to treat infections by gram negative bacteria such as *Pseudomonas aeruginosa*, *Escherichia coli* and *Klebsiella* bacteria that cause pneumonia. The tetracyclins also act on the bacterial ribosome to prevent the bacterial cell making vital proteins. Tetracyclins have quite significant side effects in addition to the usual digestive disturbances. They increase sensitivity of the skin to sunlight, can affect the enamel on teeth, causing browning, and can slow down the bone growth.

Cell Membrane Inhibitors

Antibiotics such as polymyxins disrupt the integrity and structure of cell membranes, thereby killing them. Such antibiotics are mostly effective on gram negative bacteria because they possess a definite cell membrane.

Effect on Nucleic Acids

Antibiotics such as quinolones and rifamycins bind to the proteins that are required for the processing of DNA and RNA, thus blocking their synthesis and thereby affecting the growth of the cells. Quinolone antibiotics interfere with the bacterial enzymes bacterial DNA gyrase or the bacterial topoisomerase enzyme that help bacterial DNA to replicate during cell division. Commonly used quinolones include ciprofloxacin, lecofloxacin and norfloxacin. They are particularly useful antibiotics for treating infections of the urinary tract, pneumonia that is acquired in the community and diarrhea, which caused by pathogenic bacteria such as *Shigella*. The quinolones cause very few side effects.

Competitive Inhibitors

These are also referred to as anti-metabolites or growth factor analogs. They competitively inhibit the important metabolic pathways occurring inside the bacterial cell. Important ones in this class are sulfonamides such as gantrisin and trimethoprim.

Antibiotic Resistance

Antibiotic resistance is one of the major problems faced during the antibiotic therapy. Resistance to antibiotics existed in the wild before the widespread use of antibiotics in medical practice. These antibiotic resistance genes were designed to combat naturally occurring antibiotics commonly found in nature. However, the tremendous selective pressure of the prevalent antibiotic use and the various mutational and recombinant capabilities of bacteria, has transformed multidrug resistance into a real threat in our time. In the 1940s all isolates of *S. aureus* were found sensitive to penicillin G, but by 1951 approximately 75 per cent of isolates become resistant. Penicillin resistant *Streptococcus pneumoniae*

(PRSP) accounted for approximately 2 per cent of all *S. pneumoniae* isolates in the late 1980s. This number rose to 11.7 per cent in 1995. Some of the most troublesome multidrug resistant bacteria today are Vancomycin resistant *Enterococcus* and Methicillin resistant *Staphylococcus aureus* in the hospital setting; Penicillin resistant *Streptococcus pneumoniae* and the multi-drug resistant *Tubercles Bacillus*, especially in the immune-compromised population. Bacteria develop resistance to antibiotics in a variety of ways, including methods that may decrease the intracellular concentrations of the antibiotic, deactivate the antibiotic, change the binding sites for the antibiotic, and develop adaptations that bypass the need for the binding site targeted by the antibiotic (Kaye *et al.*, 2000).

In general, the biochemical mechanisms of bacterial resistance to antibiotics can be classified into 4 types- (a) alteration of the target site of the antibioitics causing decreased affinity of the antibiotic for the binding site or the inability of the antimicrobial agent to bind, (b) inactivation of the antibiotic by the production of enzymes that are able to convert an active drug into an inactive derivative, (c) decreased permeability of the membranes and entry or accumulation of the antibiotic and (d) a combination of the above three mechanisms.

Antibiotic resistance can be of genetic origin or of non-genetic change. Non-genetic antibiotic resistance is often due to natural resistance, such as with *E. coli* resistance to vancomycin, or mycoplasma resistance to penicillins.

A form of acquired non-genetic drug resistance is metabolic inactivity. As most antibiotics interfere with bacterial metabolic processes, metabolic inactive bacteria are not harmed. With resumption of normal growth these bacteria become susceptible again.

Genetic origin of resistance implies genetic changes and subsequent environmental selection for the mutated bacteria. These resistances may be acquired through mutations in the genes that encode for the target or affected transport proteins. As the bacterial cells without the adaptive mutations are killed as a result of an antibiotic, the cells that have the mutation continue to replicate, replacing the original population with a resistant one. These resistances may also be acquired as a result of the transfer of plasmids or transposons and similar agents. These agents are small segments of DNA that are readily exchanged between bacteria. A plasmid that contains a gene for an adaptive mutation can be shared with a large number of nearby bacteria, which may or may not be the same species. In this manner, resistance can quickly spread from species to species (Normark and Normark, 2002).

Many strategies have been used in an attempt to circumvent the multiple mechanisms of resistance that have developed in bacteria. One way to circumvent the origin of antibiotic resistance is to treat an infection with multiple drugs. Adding beta-lactamase inhibitors to penicillin drugs, chemically altering cephalosporins to create the additional generations of the drugs, and combining sulfa drugs with pyrimethamine, trimethoprim, and erythromycin are examples of these strategies. In addition, new categories of antibiotics are being created in an attempt to stay ahead of the rapid evolution of bacterial resistance. Linezolid, the first oxazolidinone, is an example of this. It is a unique drug that prevents formation of the 70S protein synthesis complex in bacteria, and it may be useful in the treatment of vancomycin-resistant *Enterococci* and methicillin-resistant *Staphylococcus aureus*. Nonetheless, the development of resistance in bacteria is relentless.

In light of the efficient means by which bacteria develop resistances, it is important to avoid practices that contribute to the process. The best way to combat antibiotic resistance is to prevent the infection and spreading of antibiotic resistant microbes by the maintenance of hygiene. Also, it is necessary to find out new sources of antibiotics specific activity without any toxicity.

References

Berdy, J. 1989. The discovery of new bioactive microbial metabolites: screening and identification. Prog. Indust. Microbiol., 27: 3–25.

Bosch, F. and Rosich, L. 2008. The contributions of Paul Ehrlich to pharmacology: a tribute on the occasion of the centenary of his Nobel Prize. Pharmacology, 82(3): 171–9.

Capitano, B. and Nightingale, C. H. 2001. Optimizing antimicrobial therapy through use of pharmacokinetic/pharmacodynamic principles. Mediguide to Infectious Diseases, 21: 1–8.

Dreyfuss, M. M. and Chapela, I. H. 1994. Potential of Fungi in the Discovery of Novel, Low Molecular Weight Pharmaceuticals. The Discovery of Natural Products with Therapeutic Potential. Vincent P Gullo. Eds. 49-80.

Hawksworth, D. L. and Kirsop, B. E. 1988. Living Resources for Biotechnology, Filamentous Fungi. Cambridge University Press, Cambridge M A. 209-214.

Hawkswoth, D. L. 1991. The fungal dimension of biodiversity: Magnitude, significance, and conservation. Mycol. Res., 95(6):641-655.

Kaye, K. S, Fraimow, H. S. and Abrutyn, E. 2000. Pathogens resistant to antimicrobial agents: epidemiology, molecular mechanisms, and clinical management. Infect Dis Clinics North Am., 14(2): 293-319.

Kingston, W. 2008. "Irish contributions to the origins of antibiotics". Irish Journal Medical Science, 177 (2): 87–92.

Landsberg, H. 1949. Prelude to the discovery of penicillin. Isis, 40 (3): 225–227.

Nathwani, D. and Wood, M. J. 1993. Penicillins. A current review of their clinical pharmacology and therapeutic use. Drugs, 45(6):866-894.

Nightingale, C. H., Murakawa, T. and Ambrose, P. G. 2001. Antimicrobial Pharmacodynamics in Theory and Clinical Practice. Marcel Dekker Inc., New York, NY, USA. 1-22

Normark, B. H. and Normark, S. 2002. Evolution and spread of antibiotic resistance. J Intern Med., 252(2): 91-106.

Van Epps, H. L. 2006. Rene Dubos: unearthing antibiotics. J. Exp. Med. 203 (2): 259.

Waksman, S. A. 1947. What is an Antibiotic or an Antibiotic Substance? Mycologia, 39 (5): 565–569.

Microbes: Diversity and Biotechnology (2012) *Pages* **359–374**
Editors: **Prof. S.C. Sati & Dr. M. Belwal**
Published by: **DAYA PUBLISHING HOUSE, NEW DELHI**

Chapter 22

Use of AM Fungi Inoculated Plants in Alleviation of Stress Effects Existing in Mine Wastes

Arun Arya* and Shirali K. Choksi
Department of Botany, Faculty of Science,
The Maharaja Sayajirao University of Baroda, Vadodara – 390 002

ABSTRACT

Minerals are important natural resources which help in economic growth of the nation. Mining areas are classified under the category of disturbed, unproductive land with little or no biological activity. The disturbed soil is characterized by no beneficial microflora. Severe disturbance alters the composition and activity of mycorrhizal fungi as well as the host plants. (Allen *et al.*, 2005). The land reclamation is the process of reconverting disturbed land to its capacity equivalent to the pre-disturbed conditions or any other productive uses. The main objective of reclamation is to reclaim disturbed areas creating a landscape with productive capabilities. The reinstallation of microbiological activities in mining sites is known to help in revegetation and is useful in reclamation process. Growth of plants in any soil depends on its nutrient status and presence of microbes related to geobiochemical cycle that enables the nutrients available for the plant growth. Introduction of beneficial organisms like *Rhizobium* and mycorrhizal fungi etc., as biofertilizer has been found extremely beneficial for a variety of plants.

Keywords: *Biofertilizer, Microflora, AM fungi, Mine wastes.*

* Corresponding Author: E-mail: aryaarunarya@rediffmail.com

Introduction

Reclamation in true sense is the art and science of measures employed to return mined land to a level of productivity and ecological stability that meets the approved post mining land use. Rehabitation of fertility and ecological balance on overburden spoil requires restoration of physical and biological conditions through environmentally sound and sustainable strategies. One such strategy is the use of Arbuscular Mycorrhizae (AM) for alleviating nutrient deficiencies encountered by pioneering plants on mine spoils (Dueck *et al.*, 1986; Diaz and Honrubia, 1994; Jasper *et al.*, 1988). The root system of most of the plants harbour diverse communities of mycorrhizal fungi and are found in wide range of habitats (Read, 1991; Smith and Read, 1997). The paper reports role of AM fungi in restoration of disturbed soils of iron, chromite, coal, lignite, limestone and fluorspar mines.

Role of Endomycorrhizae in Revegetation

Mycorrhizal fungi are essential component of a self sustaining ecosystem. There are abundant examples of the ability of mycorrhizal fungi to enhance growth and nutrition of tree seedlings, both in nursery conditions and in the field after plantation (Danielson and Visser, 1989; Villeneuve *et al.*, 1991; Browning and Whitney, 1993; LeTacon *et al.*, 1994; Gagne *et al.*, 2006 Quoreshi *et al.*, 2008).

The AM fungi are known as bio- ameliorators of saline soils, potential agents in plant protection and pest management, reducing plant mortality, which helps in improving plant establishment, and plant growth (Gould *et al.*, 1996; Sharma and Dohroo, 1996; Al- Karaki *et al.*, 2001; Sylvia and Williams, 1992; Azcon-Aguilar and Barea, 1996). Soil inoculation with *Glomus mosseae* has significantly increased plant growth and biomass production in limestone mine spoil soils (Rao and Tak, 2002). The ectendomycorrhizas are often found in conifers and mostly confined to the genera *Pinus* and *Larix* formed by a small group on the roots of plants growing on disturbed lands.

Although the increased nutrients uptake is the most significant single benefit of mycorrhizae, this fascinating symbiotic relationship offers numerous benefits to their host plants, which can be summarized as follows:

1. Enhance plant efficiency in absorbing water from soil
2. Reduce fertilizer and irrigation requirements
3. Increase drought resistance
4. Increase pathogen resistance
5. Protect against damage from heavy metals and other pollutants
6. Minimize various plant stresses
7. Improve seedling growth and survival
8. Improve soil structure and contribute to nutrient cycling processes
9. Contribute toward carbon sequestration

The significance of AM fungi in disturbed soil remediation has lately been recognized by Gaur and Adholeya (2004), Khan (2006), and Quoreshi (2008). AM fungi and their hyphal network provide an excellent system for plant based environmental clean up and have the potential to take up heavy metals from an enlarged soil volume (Gohre and Paszkowski, 2006). Gaur and Adholeya (2004) have suggested that indigenous AM fungi found naturally in heavy metal – polluted soils are more tolerant than isolates from non-polluted soils, and are reported to colonize plant roots effectively in heavy – metal contaminated environments. According to Oliveira *et al.* (2005) AM strain native to highly alkaline anthropogenic sediment is generally more effective than the non native fungi in improving plant establishment and growth under stressed sediments. Several other studies demonstrated that native AM could perform better in soils from which they are isolated (Enkhtuya *et al.*, 2000; Caravaca *et al.*, 2003; Gohre and Pazkowski, 2006). Many studies have suggested that for restoration of native plant communities, the source of inoculum is an important factor (Klironomos, 2003; Moora *et al.*, 2004). It has been found that locally collected field inoculum is more effective than commercial inocula for establishing late – successional species (Rowe *et al.*, 2007).

Reclamation of Mining Sites

Mycorrhizal inoculation is beneficial for reclamation of a variety of disturbed sites (Danielson and Visser, 1989, Marx, 1991) and had a great potential in the restoration of natural ecosystems (Miller and Jastrow, 1992). Complete absence of mycorrhizal associations on plant root systems is one of the major reasons for failure of plantation establishment and growth in various forests with low inoculum potential, mined sites, and restoration of disturbed areas. It is interesting to note that intensive fertilizer and fungicide use in nursery stock culture to increase seedling growth in single growing season may inhibit mycorrhizal development (Kropp and Langlois, 1990; Quoreshi, 2003). Nevertheless, nursery cultural practices often create cultural conditions that encourage certain mycorrhizal fungi, such as *Telephora terrestris* (Ursic *et al.*, 1997). However, these nursery – adapted fungal strains are often ecologically different from those prevailing in the field, particularly if the seedlings are aimed to be planted in disturbed mined sites.

Three types of inoculum are currently being used in forest nurseries to inoculate seedlings (1) vermiculite–peat based solid – substrate inoculum, (2) liquid/mycelial slurry inoculum, and (3) spore inoculum. There are now many examples in using excised and transformed root organ as a tool for producing inoculum of various AM species (Fortin *et al.*2002). Transformation of roots by *Agrobacterium rhizogenes* has provided a novel way to obtain mass production of exenic roots on artificial media in a very short period. A group of research scientists in Canada have been working on developing techniques for producing improved ectomycorrhizal (ECM) inoculum by using (i) *Agrobacterium rhizogenes* – transformed root culture as a tool for the production ECM inoculum (Coughlan and Piche, 2005); (ii) Chitosan beads for encapsulating fungal mycelia for the production of ECM and *Frankia* sp. inocula (Quoreshi *et al.*, 2008).

The AM fungi are obligates biotrophs and require living host plant for the completion of their life cycle (Fortin *et al.*, 2002; Dalphe and Monreal, 2004). AM fungal propagation can take place either by spore germination or by mycelial extension through soil and roots (Dalphe and Monreal, 2004). Generally AM fungi are propagated through pot culture. In this system, fungal spores, and colonized roots fragment are used as starter inoculum, and mixed with growing substrate for inoculated seedling production subsequently, colonized substrate and root can then serve as AM fungal inoculum. Root organ culture system showed an effective means of production of AM inoculum that can be used either directly as inoculum or as starter inoculum for large scale production. (Fortin *et al.*, 2002). Artificially inoculated mycorrhizal fungi also enhanced growth and nutrition of tree seedling, both in nursery condition and in the field after out planting (Marx *et al.*, 1988, Gagne *et al.*, 2006, Quoreshi *et al.*, 2008). The primary purpose for inoculating seedling with mycorrhizae is to provide planting stock with adequate mycorrhizas to improve their survival and growth after planting. Such approach is particularly essential in revegetation of disturbed sites. The soils of degraded sites are frequently low in available nutrients, mycorrhizal fungi and other beneficial microorganisms (Cooke and Lefor, 1990). Similar to ECM, AM fungi also play a significant role in establishment of plant in disturbed and stressed ecosystem (Gould *et al.*, 1996).

Responses of AM inoculation in Iron Mine Waste Soil

Application of AM fungi generally increased plant growth parameters such as a shoot length and plant dry weight in both crop and tree legumes (Table 22.1). In case of crop legume percentage of increase in shoot length was found to be 55.59 in *Vigna unguiculate* and 101.15 in *Macrotyloma uniflorum*, where as plant dry weight increase was 14.13 per cent and 34.59 per cent respectively (Misra *et al.*, 1996). Among the two crop plants the growth response due to the AM inoculation is found to be more in *M. uniflorum*. However, the degree of stimulation in growth due to AM inoculation was found to vary with plant species as well as the AM fungi inoculated. AM inoculation was seen to enhance nodulation capacity of plants in terms of nodule number as well as dry weight. However, study on nodulation response due to AM inoculation indicated varied responses with plants species. Maximum stimulation in nodulation was observed in *V. unguiculata* among the crop legume and *Leucaena leucocephala* among the tree legumes. Nodules showed improvement in shape, size as well as on increased rhizobial colonization capacity under AM inoculation.

Table 22.1: Effect of AM Inoculation on Growth, Nodulation total N_2 Content and Rhizosphere Rhizobial Population of Tree and Crop Legumes Grown in Iron Mire Waste Soils

	Crop Legume						*Tree Legume*					
	V. unguiculata			*M. uniflorum*			*A. nilotica*			*L. leucocephala*		
	Control	*Treated*	*%*	*Control*	*Treated*	*%*	*Control*	*Treated*	*%*	*Control*	*Treated*	*%*
Shoot Length (cm)	14.93± 0.52	23.25± 2.43	55.59± 0.35	15.66± 1.6	31.50± 0.69	101.15± 1.62	21.5± 0.7	31.5± 0.49	46.51	18.6	29.7	59.67
Plant dry wt (g)	0.679± 0.06	0.775± 0.07	14.13± 0.003	0.311± 0.004	0.4186± 0.16	34.59± 0.29	1.8± +0.32	2.92± 0.11	62.22	2.6	3.38	30.0
Nodule number	13.33± 1.76	27.66± 2.66	107.5	20± 2.08	25± +1.2	25.0	5.67± 0.08	6.3± 0.88	11.11	6.3± 0.88	10.3± 1.2	63.49
Total N_2 Content of plant per cent	1.05± 0	1.47± 0.07	40.0	1.86± 0.04	2.31± 0.04	24.19	0.175± 0.04	1.05± 0.0	5.00	2.45± 0.02	2.87± 0.07	17.14

Per cent increase over control.

Total nitrogen content also showed differential responses in tree and crop legumes. Maximum increase in total N_2 content was found in *V. unguiculata* in crop legume and *A. nilotica* for tree legume (Misra *et al.*1996). AM fungi inoculation greatly stimulated the rhizospheric rhizobial population in both crop as well as tree legumes showing more than 100 percent increase in case in *M. uniflorum* and *leucocephala*. It was interesting to note that in case of tree legumes; increase in plant dry weight was same as was the increase in total N_2 content.

Response of AM inoculation on Chromite Mine Waste Soil

Leguminous plants inoculated with AM fungi showed in general better growth vigour and nodulation capacity along with higher N_2 content. Stimulation activity of AM inoculation however, greatly varied in both crop and tree legumes in chromite mine waste soil.

Improvement in growth of legumes depend upon the effective nodulation and nitrogen fixation which is brought about by successful legume – *Rhizobium* symbiosis (Nutman, 1967). There is report that leguminous plant respond well to mycorrhizal infection, which indirectly increases the capacity of atmospheric nitrogen fixation through P uptake (Bagyaraj *et al.*, 1979). Enhancement of Rhizobial population in rhizospheric soil due to AM inoculation in crop legumes grown in chromite mine soil has been observed by Patnaik *et al.* (1992) This may be responsible for enhanced nodulation in legumes due to AM inoculation.

The poor growth responses of plant is attributed not only to poor nutrient condition of mine soil, but also to other stress situation such as pH imbalance condition, accumulation of heavy metals in higher concentration, poor water holding capacity of the soil and other deterring factors. Study of nodule anatomy of *L. leucocephala* in chromite mine waste soil by Thatoi (1992) showed distorted and poor nodule structure. Disrupted tissue organization along with low rhizobial colonization showing clear evidence of stress effects on nodulation. Inoculation with AM fungi was seen not only increased to nodule number but also improved the shape and size bringing to smooth elongated shape, along with increased colonization area. The bacteriod also appeared more compact per cell. The invaded cells appeared enlarge and healthy, indicating removal of stress effects existing in mines soils. This supports the assumption on the positive role of AM fungi in improving stress tolerant capacity of the host plant as well. They report that mycorrhizal plants improve water uptake and improve drought tolerance (Nelson and Safir, 1982). AM inoculated plant also improve uptake of metals like Cu, Zn, etc. (Cooper and Tinker 1978) and accumulated in the roots thereby developed heavy metal tolerance.

Table2: Effect of VA M inoculation on growth, nodulation, total N_2 content and rhizopheric rhizobial population of tree and crop legumes grown in Chromite mine waste soils

	Crop Legume						*Tree Legume*					
	V. unguiculata			*M. uniflorum*			*A. nilotica*			*L. leucocephala*		
	Control	*Treated*	*%*	*Control*	*Treated*	*%*	*Control*	*Treated*	*%*	*Control*	*Treated*	*%*
Shoot length (cm)	26.83± 2.32	36.26± 3.23	25.14	33.6± 1.48	58.13± 2.19	73.00	21.3± 1.65	28.0± 1.15	31.45	20.36± 1.98	29.3± 1.83	43.90
Plant dry wt (g)	0.855± 0.008	1.859± 0.114	117.42± 0.045	0.890± 0.045	1.8636± 0.089	130.35	1.9± 0.57	3.14± 0.03	65.26	2.75± 0.02	3.35± 0.01	21.81
Nodule mumber	14.33± 1.85	32.0± 1.0	123.30	9.66± 0.66	17.66± 0.33	82.81	3.7± 0.088	5.0± 0.57	35.13	4.3± 0.88	13.0± 1.52	202.32
Total N_2 content of plant %	1.68± 0	1.82± 0	8.33	1.56± 0.06	2.1± 0.04	34.61	0.21± 0.0	0.56± 0.04	166.6	0.175± 0.04	0.945± 0.06	440.0

Per cent increase over control.

The improvement of growth nodulation and rhizobial population of both crop and tree legumes in iron and chromite and mine waste soil under AM inoculation indicates a very special role of AM fungi not only for improvement of plant growth under normal soil conditions but also improving the growth under environmental stress condition through elevating as well as imparting stress tolerating capacity of the host plants. This aspect of AM fungi has top be utilized for better management for mine wastelands where multiple stress factors exist.

Association of AM Fungi with Plants in Sites of Coal Mines

Coal is the dominant energy source in India, accounting for about more than half of the country's energy requirement. In the Indian coal scenario Jharia coalfield occupies a special status as this is the only storehouse of prime coking coal and has been meeting the coking coal needs of the country for over a century. Coal mining in the coalfield was started in the last decade of the 19th century. The coalfield having an area of about 450 sq km belongs to Gondwana group of Permian age and has Talchir, Barakar, Barren and Ranigunj measures. It is a sickle shaped coalfield occurring in the form of a basin truncated with a major boundary fault on the southern flank. Due to coal mining and associated activities the land use in the coalfield has undergone radical changes as seen in Table 22.3, which gives a comparison of the land use in the years 1925, 1974, 1987 and 1993.

Table 22.3: Land Use Pattern in Jharia Coalfield (Area as percentage)

Sl.No	Land Use	1925	1974	1987	1993
1.	Villages, settlements, townships, etc	8.6	16.0	32.3	33.10
2.	Land in mining use including open pits	4.7	17.4	12.5	19.42
3.	Water bodies	7.3	6.7	3.1	2.90
4.	Forests (plantation)	4.9	0.7	0.7	2.45
5.	Agriculture and natural vegetation	65.4	56.8	49.4	39.02
6.	Fallow land and pasture	9.1	2.4	2.0	3.11

Source: Center of Mining Environment, ISM, Dhanbad.

The coalfield has more than 40 workable coal seams out of which the upper seams have coals of superior quality and this quality deteriorates with the seams at deeper horizons. It is generally understood that the seams below VIII-Seam have medium to non-coking coals. The percentage of coal in Barakar measures is about 20, *i.e.*, the overall coal-overburden ratio is 1:5.

Bisen *et al.* (1996) studied plantation of *Eucalyptus*, *Tectona* and *Datura* species with *G. mosseae* in coal mines of Sohagpur in Madhya Pradesh. *Eucalyptus* species showed maximum AM colonization (68 per cent) followed by *Datura* species (55 per cent) and *T. grandis* (49 per cent). Jamaluddin and Chandra (1999) observed an increase in the growth of bamboo species in the nursery soil as well as in coal mine overburden soil. It is well established that many plants can not grow luxuriantly without the presence of AM fungi, the technology of introduction of AM in the roots before planting them in mine soil may prove effective.

In a study undertaken by Swapna and Ammani (2009) at coal mines of Mancheriala, Andhra Pradesh the percentage of root colonization among 20 medicinal plants was found to be highest in *Andrographis paniculata* (70 per cent), followed by *Aloe vera* (36 per cent) and *Ocimum basilicum* (36 per cent). The lowest root colonization was seen in *Acorus calamus* (2 per cent). Vesicles were of different shapes. The occurrence of AM Fungi was also reported from overburden mine spoils of Texas. In the soil samples the spore count was highest in *A. paniculata* (45 per cent) and lowest in *A. calamus* (10 per cent). *Glomus* was found to be most predominant species (Mott and Koske, 1988). Srinivas *et al.* (2005) observed the AM Fungi of coal mines of the Godavari basin and their role in revegetation by *Albizzia lebeck*. The percentage of root infection, and the spore population increased significantly. Srinivas *et al.* (2006) further studied dependency of two multipurpose tree species *Acacia melanoxylon* and *A. nilotica* in eleven soils of coal mining areas at Godavarikhani. Mycorrhizal dependency of these two plants varied with physico-chemical properties. Mycorrhizal colonization of *A. nilotica* ranged between 27.7 and 64.5 per cent and it supported more resting spore population than *A. melanoxylon*.

Responses of AM Inoculation in Flourspar Mine Waste Soil

The area of study is in and around Flourspar mines, situated in the Amabadungar area of Kadipani, in Vadodara district. The Ambadungar hills are situated at the eastern termination of the Vindhyan range and are surrounded by Deccan trap basalt. The mines are located at 744'N and 22 E in Chhotaudepur taluka in Vadodara district. The mines' lease area is about 620 ha and is situated at the elevation of 580 m above mean sea level.

The soils of the area are loamy with some proportions of silt and clay. The depth varies from shallow (0-25cm) and medium (26-50cm) to moderately deep (51-75cm). About 60 per cent of the area has moderate-to-good water holding capacity.

It was only recently that the GMDC got a nod from the Ministry of Forests and Environment to carry out mining in 31.2 hectare land. Under the Project set up nearly 20 years ago, 3 million tonnes flourspar has been extracted from the site while 8 million tonnes deposits still lie to taken out. These deposits include the ones in the additional 52 hectare land which the GMDC has sought from the Forest Department for lease. The mineral is used in steel, aluminum and refrigeration units across the country. The Project expects to extract 20,000 tonnes fluorspar every year whereas the demand in India is 40,000 tonnes. Up gradation is based on French technology. The Department has agreed to spare the officers shortly and their scope of work will be to advise the Corporation in afforestation and remedial measures for preserving the environment. It would be viable for the Corporation to dump the waste in non-forest area which is about 500 metre or one km away. In a Memorandum of Understanding (MoU) signed between the Corporation and the Forest Department, the latter had asked to dump the waste in non-forest area. Till now, the GMDC dumped the waste in the mining site itself.

The climate of the area is sub-humid. Summer temperature reaches a maximum of 43° C and in winter (November-February) the temperature is as low as 6° C. Average annual rainfall is about 1000 mm, mostly during July to September with an average of 35 rainy days per year. The remaining 8-9 months are dry.

Experimental Details

The AM spores were isolated from rhizospheric soil of Flourspar mines by Wet Sieving and decantation technique (Gerdemann and Nicolson, 1963). Mass multiplication of isolated AM spores was done in pots containing maize plants. For single spore culture of AM Fungi Funnel technique was used. The funnel forces the roots to grow near the spores and assures infection even when few spores are used. In our experiments the plastic funnels were used they were lined by thin plastic film and sterilized soil was filled in the center a single AM spore was placed and some soil was filled then seed was placed.This funnel was wet with sterile distilled water and placed in seed germinator. The plants were later on transferred to pots.

Fifty plants each of control, *Leuceana leucocephala* (Lam.) deWit and *Pongamia pinnata* (L.) Pierre were raised in black coloured polybags after incorporating AM spores and soil obtained from maize grown pots. These plants were planted in mining area and performance of control and AM inoculated saplings was analyzed by recording biomass (at 15 days intervals) of these plants. The physicochemical analysis like pH, electric conductivity, percentage moisture, water holding capacity were also done from the soil samples obtained from mining site.

Studies in Flourspar Mining Areas

The analysis of 100g rhizospheric soil samples revealed more number of AM spores (Table 22.4) in *Alysicarpus vaginalis* (143), *Tephrosia senticosa* (162) and *Indigofera linifolia* (118) and least number was recorded in *Anogeissus latifolia* (5).

Table 22.4: Presence of AM Spores per 100g Rhizospheric Soil and Other Characteristics of Soil Obtained from Kadipani, Gujarat

Sl.No.	Plant	Family	No. of Spores/ 100g of Soil	pH	EC	Moisture Per cent
1.	*Acacia nilotica* ssp. *indica* (Benth.) Brenan	Mimosaceae	89	7.90	01	1.90
2.	*Aerva lanata* (L.) Juss.	Amaranthaceae	21	7.82	01	3.11
3.	*Ailanthus excelsa* Roxb.	Simaroubiaceae	30	7.90	01	1.77
4.	*Alysicarpus tetragonolobus* Edgew.	Fabaceae	61	8.33	01	20.48
5.	*Alysicarpus vaginalis* DC Prodr.	Fabaceae	143	7.70	01	20.48
6.	*Anogeissus latifolia* Wall.	Combretaceae	5	7.85	01	1.82
7.	*Argemone maxicana* L.	Papaveraceae	56	7.95	01	1.66
8.	*Bothrichloa pertusa* (L) A. Camus	Poaceae	60	7.90	01	2.90
9.	*Blumea alata* DC.	Asteraceae	32	7.91	01	1.80
10.	*Boreria alata* Meyer.	Rubiaceae	6	7.92	01	0.90
11.	*Butea monosperma* (Lamk).Kuntze.	Fabaceae	52	7.89	01	20.11
12.	*Calotropis procera* (Ait.)Ait.	Asclepediaceae	28	8.28	00	4.44
13.	*Cassia auriculata* Linn.	Caesalpiniaceae	50	7.80	01	13.32
14.	*C. tora* Linn.	"	62	7.46	02	11.45
15.	*Echinops echinatus* R.	Ebenaceae	39	7.95	01	14.33
16.	*Echinochloa colonum* (L.)Link	Poaceae	20	7.55	01	16.76
17.	*Euphorbia hirta* L.	Euphorbiaceae	36	7.44	01	18.20
18.	*Evolvulus alsinoides* L.	"	32	7.76	01	12.32
19.	*Flacourtia montana* Graham	Flacourtiaceae	05	7.91	02	11.11
20.	*Indigofera linifolia* Retz.	Fabaceae	118	7.32	01	13.22
21.	*Lindenbergia urticifolia* Lehm.	Scrophulariaceae	45	8.10	01	10.43
22.	*Ocimum bascilicum* L.	Lamiaceae	28	7.95	01	10.68
23.	*Oldenlandia umbellata* L.	Rubiaceae	39	7.23	01	10.34
24.	*Polygala erioptera* DC.	Polygalaceae	25	7.89	01	11.22
25.	*Prosopis chilensis*(Molina) Stuntz	Mimosaceae	78	8.50	01	10.88
26.	*Ruellia prostrata* Poir.	Acanthaceae	48	8.12	01	10.45
27.	*Salvia aegyptiaca* Linn.	Lamiaceae	22	7.61	01	10.33
28.	*Solanum xanthocarpum* Schl.	Solanaceae	92	7.61	02	11.23
29.	*Tephrosia senticosa* (L.)Pers.	Fabaceae	162	7.20	01	10.22
30.	*T. purpurea* (L.)Pers.	"	40	7.45	01	13.21
31.	*Tridax procumbens* L.	Asteraceae	45	7.54	01	10.23
32.	*Vernonia cinerea* Less.	"	69	7.90	01	12.67
33.	*Viocia auriculata* Cass.	"	15	7.78	02	13.21
34.	*Xanthium strumarium* L.	"	20	7.78	01	10.33
35.	*Ziziphus jujuba* Mill.	Rhamnaceae	89	6.98	01	12.45

The rhizospheric soil of thirty five plants of Kadipani showed dominance of *Glomus fasciculatum*, *G. aggregatum* and *G. mosseae*, *G. claroides* (Plate 22.1A–E) and their population was increased by pot culture method. Successesful inoculation of AM spores was further obtained by Funnel method (Plate 22.1F). The pH of soil ranged from 7.2–8.5 and electric conductivity ranged from 1 – 3 (Table 22.4).

Biomass studies were undertaken in *P. pinnata* (Table 22.5, Figure 22.1) AM inoculated plants, showed maximum 7.5 g dry wt of shoot after 75 days while in control it was 6.0 g. After 75 days root length was 32.6cm as compared to 18.5 in control.

Table 22.5: Showing Number of Leaves and Increase in Biomass of *Pongamia pinnata* After Inoculation of AM Fungi

Period Days	Height of Plant (cm)	No. of Leaves	Shoot Length (cm)	Root Length (cm)	Dry Wt. Shoot (g)	Dry Wt. Root (g)
			Control			
30	20.0±0.0005	10.0±1.15	15.2±0.11	10.0±1.15	4.35±0.17	0.95±0.115
45	32.0±1.15	18.0±1.15	18.5±0.05	12.0±0.577	5.85±0.0005	1.20±0.011
60	33.0±0.577	19.0±0.577	25.30±0.011	15.5±0.0005	5.89±0.0005	2.10±0.0005
75	35.0±0.011	20.0±1.15	33.33±0.005	18.5±0.0005	6.0±0.577	2.20±0.0005
			AM inoculated			
30	47.6±0.0005	15.6±0.057	27.3±0.057	20.3±0.17	6.03±0.001	1.28±0.0005
45	52.6±0.011	22.0±0.577	31.6±0.17	21.0±0.577	5.020±0.0005	1.43±0.0005
60	61.3±0.011	25.6±0.115	33.33±0.011	28.0±0.577	6.16±0.0005	2.23±0.0005
75	68.0±0.02	30.6±0.115	37.33±0.0005	32.6±0.115	7.545±0.0005	2.63±1.115

Table 22.6 Plant Biomass Studies in Control and AM Inoculated *Leucaena leucocephala* Saplings

Period Days	Height of Plant (cm)	No. of Leaves	Shoot Length (cm)	Root Length (cm)	Dry Wt. Shoot (g)	Dry Wt. Root (g)
			Control			
30	58.0±0.577	70.0±1.15	55.5±0.005	15.0±1.73	2.8±0.057	0.5±0.115
45	80.0±0.005	90.0±0.577	68.25±0.005	20.0±1.15	3.5±0.057	0.60±0.017
60	85.0±0.005	91.0±1.15	71.00±0.577	20.0±1.15	3.95±0.057	0.70±0.001
75	91.0±1.15	93.0±0.577	76.55±0.115	20.2±0.11	4.95±0.057	0.80±0.017
			AM inoculated			
30	99.0±0.0115	93.3±0.173	76.23±0.577	22.6±0.115	4.67±0.0057	0.63±0.0115
45	117.0±0.0057	104±0.577	94.00±1.15	23.0±0.577	6.92±0.230	1.20±0.230
60	119.6±1.15	109±0.288	95.00±0.577	32.6±0.115	7.78±0.011	2.79±0.005
75	119.0±0.577	120.0±1.73	95.06±0.017	33.6±0.173	7.90±0.011	2.85±0.005

AM fungus inoculated *Leuceana leucocephala* showed 7.9 g dry wt of shoot while in control it was 4.95 g. These AM inoculated plants further showed better response in Flourspar mining site than control.

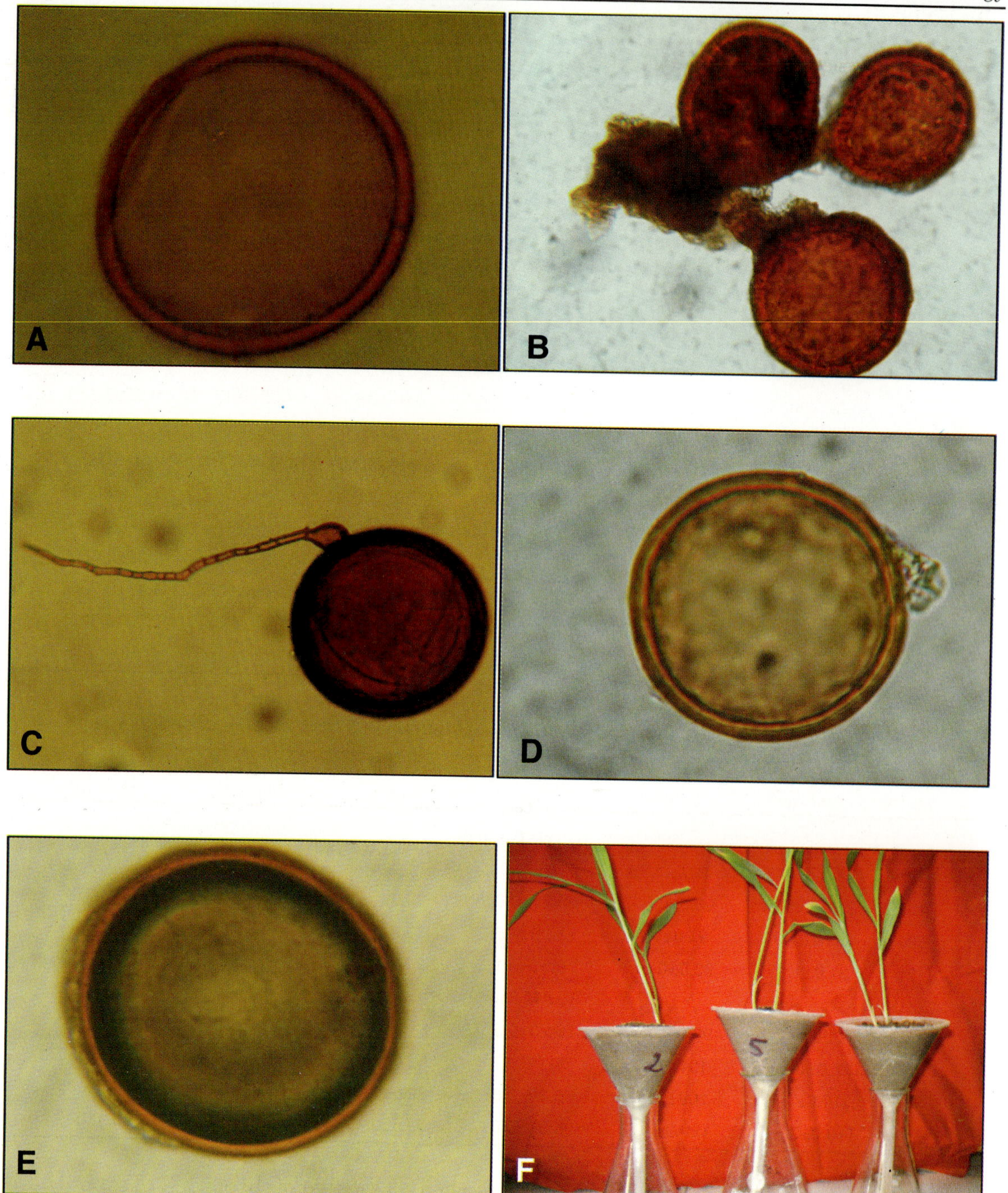

Plate 22.1

A: *Glomus glomerulatum;* B: *G. aggregatum;* C: *Gigaspora* sp.; D: *G. Fasciculatum;* E: *G. claroides;* F: Funnel technique to develop pure culture of *Glomus fasciculatum* in maize roots

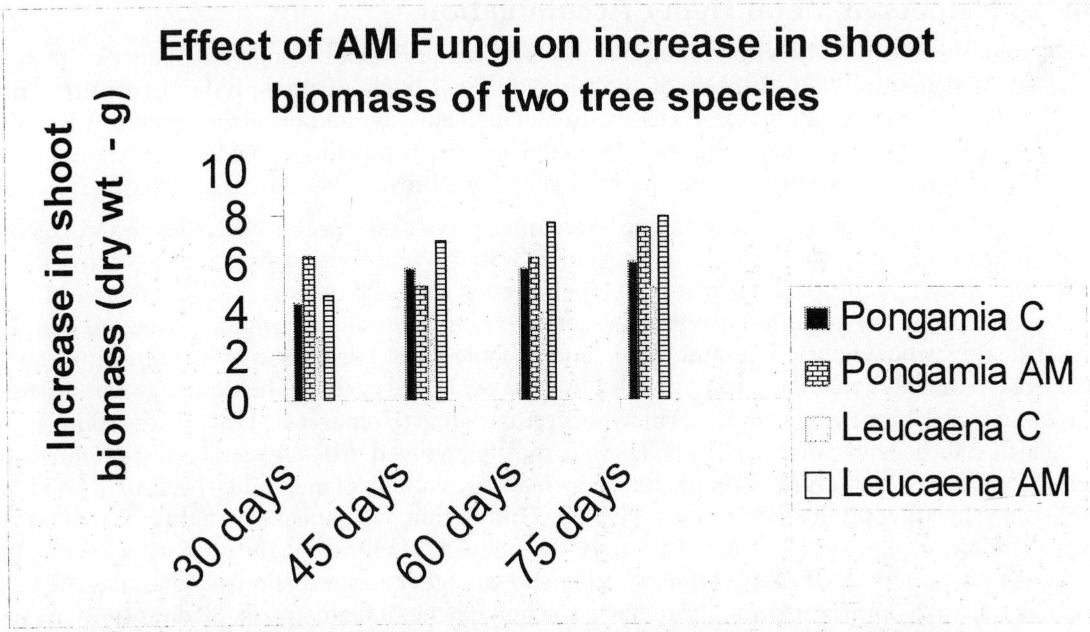

Figure 22.1: Effect of AM Fungi on Incresae in Shoot Biomass to Two Tree Species

Association of AM Fungi with Plants Growing in Sites of Lignite Mines

The association of AM Fungi in soil profiles of Panandhro lignite mines in Kachch, Gujarat was studied by Chatterjee (2006). Native vegetation with *Prosopis juliflora* showed highest mycorrhization. The species isolated and identified were *G. fasciculatum*, *G. albidum*, *G. caledonium*, *G. monosporum* and *G. macrocarpum*. She concluded that mycorrhization is greatly influenced by the host and edaphic factors. AM fungi were present in top 15 cm soil, in the dump soil only two samples showed mycorrrhizal sporulation.

Response of AM inoculation in Soils of Limestone Quarries

Singh and Jamaluddin (2006) found that the population of AM spores increased with increasing age of the dumps. The soils of limestone mined overburden dumps had significant difference in the spore population and further root colonization. Soil from ten year old overburden dumps has the highest spore population. While soils from five-year old and fresh overburden dumps was found to be least populated.

Rao *et al.*(2006) studied the effect of *G. fasciculatum* inoculated seeds of 9 different tree species and one grass on lime stone mined overburden soil. Plants grown on mine spoil showed reduction in concentration of Nand P while, K,Ca and Mg increased as compared to control. Initial activities of all the enzymes in mine spoil as compared to that of normal soil were significantly lower, but the rhizospheric effect was greater. Calcium uptake from mine spoils is significantly higher in all the plant species. Maximum uptake was observed with *S. oleoides* followed by *C. mopane* and *Pithecelobium dulce*. It may be due to higher intensity of AM fungi infection in plants growing in mine spoils where AM fungi acted as scavengers of Ca, which was then passed to shoots. Overall 17 per cent greater conc. of Zn and 13 per cent conc. of Cu was observed.

AMF as Helpers in Metal Hyper Accumulation

The positive role of AMF is due to better exploitation of labile pool of soil phosphate (Smith and Gianinazzi – Pearson 1988). Studies of the physiological basis of phosphate absorption and translocation indicate that all these processes are metabolically dependant. Amongst enzymes, the alkaline phosphotases are known to be involved in the phosphate nutrition of AMF and these enzyme activities also indicates the existence of a functional AM symbiosis (Tisserant *et al.*, 1993).

In nature, some plants hyper accumulate heavy metals. For example, *Viola calaminaria* and *Thlaspi calaminare* grow over calamine deposits in Aachen, in Germany and contain over 1 per cent Zn of their dry weight. Also *Alyssun bertolinni* grow on serpentine soil in Tuscany, Italy and contain over 1 per cent Nickel of their dry weight. Heavy metal complexes in hyper accumulator plants are mainly associated with carboxylic acid like citric, malic, and melonic acids. These organic acids are implicated in the storage of heavy metals in leaf vacuoles. Amino acids like cysteine, histidine, glutamic acid, glycine also form heavy metal complexes in hyper accumulators (Homer *et al.*, 1987). These complexes are stable than those with carboxylic acids. They are mostly involved in heavy metal transport through xylem. Moreover, hyper accumulator plants can increase availability of metals like Fe, Zn, Cu and Mn by releasing chelating phytosiderophores. Hyper accumulation mechanisms may then be related to rhizosphere processes such as release of chelating agents and differences in the number or affinity of metal root transporters. AMF like *G. intraradices* has shown higher accumulation of metals like Zn, Cd, Se (Giasson *et al.*, 2006). They found the metal extraction reaches a plateau after 80 days showing no further phytoaccumulation. Lasat (2002) observed that the effect of AMF on metal root uptake appears to be metal and plant specific. Greater root length densities and presumably more hypahe enable plants to explore a larger soil volumes thus increasing access to cations (metals) not available to known mycorrhizal plants (Mohammed *et al.*, 1995). AM Fungi help in hyper accumulation and this may be one of the factors due to which plants can survive well in higher concentration of metals present in soil.

AMF and Plant Stress Alleviation on Mining Sites

One of the main objectives in mine site reclamation is revegetation. This mining environment is characterized by poor physical and chemical conditions. Mycorrhizal colonization could improve vegetation establishment and survival particularly in such adverse conditions. Young seedling have to be protected from extremely high surface temperatures to prevent heat girdling of stems (Danielson, 1985) By colonizing the roots the fungus enhances plant growth by making soil elements more accessible (George *et al.*, 1992, Gregory 2006) and by improving water absorption (Sweat and Davis 1984). Accordingly mycorrhizal colonization improves vegetation establishment particularly in adverse conditions such as low fertility and arid soils (Smith *et al.*, 1988).

Mine spoils may be extremely acidic or alkaline. Acid mine drainage is very frequent, especially in sulphide metal ore tailing, where rain water reacts with sulphide to form sulphuric acid. Leachate pH existing from tailing could be as low as one. Plant roots can be colonized with myorrhizae with pH value as low as 2.7. Hyphae of AM fungi may extend 8 cm from the root surface or more but rhizomorphs of *Pisolithus* may extend 4 m in to the soil, a result that suggest Ectomycorrhizae are better adapted to long distance transport than AMF (Danielson, 1985). To determine the degree of fungal symbionts adaptation to mine waste conditions, infection levels of each species must be quantified (Danielson, 1985). Chain *et al.* (2007) provided evidences for the potential use of local plant species in combination of AMF for ecological restoration of metalliferrous mine tailing.

Acknowledgements

Authors wish to thank Ministry of Environment and Forests, Govt. of India for financial support to undertake the studies on fluorspar mines in KadiPani area of Gujarat and to GMDC officials for all possible help to carry out the investigations and allowing to do plantation there.

References

Allen, M. F., Allen, E. B. and Gomez-Pompa, A. 2005. Effects of mycorrhizae and non target organisms on restoration of seasonal tropical forest in Quintano Roo, Mexico: factors limiting tree establish mat. Restor. Ecol., 13: 325-333.

Al-Karaki, G. N., Hammad, R. and Rusan, M. 2001. Response of two tomato cultivars differing in salt tolerance to inoculation with mycorrhizal fungi under salt stress. Mycorrhiza, 11: 43 – 47

Azcon-Aguilar, C. and Barea, J. M. 1996. Arbuscular mycorrhizas and biological control of soil – borne plant pathogens; an overview of the mechanisms involved. Mycorrhiza, 6: 457 – 464.

Bagyaraj, D. J., Manjunath, A. and Patil, R. B.1979. Interaction betweenVA mycorrhizal and Rhizobium and their effect on soybean in field. New Phytol., 82: 141-145.

Bisen, P. S., Gour, R. K., Jain, R. K., Dev, A. and Sengupta L. K. 1996. VAM colonization in tree species planted in Cu, Al, and coal mines of Madhya Pradesh with special reference to *Glomus mosseae.* Mycorrhiza News, 8(1): 9-11.

Browning, M. H. R. and Whitney, R. D. 1992. Field performance of black spruce and jack pine inoculated with selected species of ectomycorrhizal fungi. Can. J. For. Res., 22: 1974-1982.

Caravaca, F., Barea, J. M., Palenzuela, J., Figureso, D., Alguacil, M. M. and Roldan, A. 2003. Establishment of shrub species in a degraded semiarid site after inoculation with native or allochthonous arbuscular mycorrhizal fungi. Appl. Soil Ecol., 22: 103 – 111.

Chaterjee, T. 2006. Occurrence of vesicular arbuscular mycorrhizal fungi in Panandhro lignite mines. In: Mycorrhiza, A. Prakash and V.S. Mehrotra (eds.) Pub. by Sci. Pub.(India) Jodhpur pp 241-248.

Chen B. D., Zhu Y. G., Duan J., Xiao X.Y. and Smith S. E. 2007. Effects of the arbuscular mycorrhizal fungus *Glomus mosseae* on growth and metal uptake by four plant species in copper mine tailings. Environ. Pollut., 147: 374-380.

Cooke J. C. and Lefor, M. W. 1990. Comparison of vesicular – arbuscular mycorrhizae in plants from disturbed and adjacent undisturbed regions of a coastal salt marsh in Clinto, Connecticut, USA. Environ. Mange, 14: 212 – 237.

Cooper K. M. and Tinker P. B. H. 1978. Translocation and transfer of nutrients in vesicular mycorrhizae. New Phytol., 81: 43-53.

Coughlan, A. P. and Piche, Y. 2005. *Cistus incanus* root organ cultures: A valuable tool for studying mycorrhizal association. In: *In vitro* culture of mycorrhizas, S. Declerck, D.G. Strullu and A. Fortin (eds.) Springer Berlin. pp. 235-252.

Danielson, R. M. 1985. Mycorrhizae and reclamation of stressed terrestrial environments, In: Soil reclamation processes-micro-biological analyses and applications. R.L. Tate and D.A Klein (eds). Marcel Dekker, New York. Pp173-201.

Danielson, R. M. and Visser, S. 1989 Host response to inoculation and behaviour of induced and indigenous ectomycorrhizal fungi of jack pine grown on oil sands tailings. Can. J. For. Res., 19:1412-1421.

Diaz, G., and Honrubia, M 1994. A mycorrhizal survey of plants growing on mine wastes in southeast Spain. Arid soil Research and Rehabilitation, 8: 59-68.

Dueck, T. A., Visser, P., Ernst, W. H. O. and Schat, H. 1986. Vesicular arbuscular mycorrhizae decrease zinc toxicity to grasses growing in zinc polluted soil. Soil Biol Biochem., 18: 331- 333.

Enkhtuya, B., Rydolva, J., and Vosatka, M. 2000. Effectiveness of indigenous and non –indigenous isolates of arbuscular mycorrhizal fungi in soils from degraded ecosystems and man – made habitats. Appl. Soil Ecol., 14: 201 – 211.

Fortin, J. A., Becard, G., Declerk, S., Dalphe, Y., St–Arnaud, M., Coughan, A. P. and Piche, Y. 2002. Arbuscular mycorrhiza on root–organ cultures. Can. J. Bot., 80: 1 – 20.

Gagne, A., Jany, J. L., Bousquet, J., and Khasa, D. P. 2006. Ectomycorrhizal fungal communities of nursery–inoculated seedlings out planted on clear–cut sites in northern Alberta Can. J. For. Res., 36: 1684 – 1694.

Gaur, A. and Adholeya, A. 2004. Prospects of arbuscular mycorrhizal fungi in phytoremediation of heavy metal contaminated soils. Curr. Sci., 86: 528-534.

George, E., Haussler, K. U., Vetterlein, K. U., Gorgus, E. and Marschner, H. 1992. Water and nutrient translocation by hyphae of *Glomus mosseae*. Can J. Bot., 70: 2130-2137.

Gerdemann, J. W and Nicholson, T. H. 1963. Spores of mycorrhizal endogone species extracted from soil by wet sieving and decanting. Trans. Brit. Mycol. Soc., 46: 235 – 244.

Giasson, P., Jaouich, A., Gagne, S., Massicotte, L., Cayer, P. and Moutoglis, P. 2006. Enhanced phytoremediation: A case study of mycorrhizoremediation of heavy metal contaminated soil. Remediation, 17: 97-110.

Gohre, V. and Paszkowski, U. 2006. Contribution of the arbuscular mycorrhizal symbiosis to heavy metal phytoremediation. Planta, 223: 1115-1122.

Gould, A. B., Hendrix, J. W. and Richard, S. F. 1996. Relationship of mycorrhizal activity to time following reclamation of surface mine land in Western Kentucky. 1. Propagule and spore population densities. Can. J. Bot., 74: 247 – 261.

Gregory, P. J. 2006. Plant root, growth, activity and interaction with soils. Blackwell, Oxford.

Homer, F. A., Reeves, R. D. and Brooks, R. R. 1997. The possible involvement of amino acids in nickel chelation in some nickel-accumulating plants. Curr. Top. Phytochem., 14: 31-33.

Jasper, D. A., Robson, A. D. and Abbott, L. K. 1988. Revegetation in an iron ore mine nutrient requirements for plant growth and the potential role of Vesicular arbuscular (VA) mycorrhizal fungi. Aust J.soil Res., 26: 497-507.

Jamaluddin and Chandra, K. K. 1995. Development of VA-mycorrhiza in tree species planted in coal mine dumps in Maharashtra. Mycorrhiza News, 7: 8-10.

Khan A. G. 2006. Mycorrhizoremediation- an enhanced form of phytoremediation. J. Zhejang Uni. Science, B 7: 503-514.

Kropp, B. R., and Langlois, C. G. 1990. Ectomycorrhizae in reforestation. Can. J. For. Res., 20: 438 – 451.

Klironomos, J. N. 2003. Variation in plant response to native and exotic arbuscular mycorrhizal fungi. Ecology, 84: 2292-2301.

Lasat, M. M. 2002. Phytoextraction of toxic metals: A review of biological mechanisms. J. Environ. Qual. 31: 109-120.

Le Tacon, F., Alvarez, I. F., Bouchard, D., Henrion, B., Jackson, M. R. Luff, S., Parlede, I. J., Pera, J., Stenstrom, E., Volleneuve, N. and Walker, C. 1994. Variations in field response of forest tress to nursery ectomycorrhizal inoculation in Europe. In: Mycorrhizas in eco-systems, D.J., Read *et al.* (eds.) CAB, Wallingford, pp. 119 – 134.

Marx, D. H.1991. The practical significance of ectomycorrhizae in forest establishment. In: Ecophysiology of ectomycorrhizae of forest trees, The Marcus Wallenberg Foundation ed., Stockholm, Sweden, Symposium and Proceedings, pp. 54 – 90.

Marx, D. H., Cordell, C. E., and Clark, III A. 1988. Eight year performance of loblolly pine with *Pisolithus* ectomycorrhiza on good quality forest site. South J. Am. For. 12: 275 – 280.

Miller, R. M. and JAstrow, R. D. 1992. The application of VA mycorrhizae to ecosystem restoration and reclamation. In: Mycorrhizal functioning, M., Allen ed., Chapman and Hall, New York.

Misra, A. K.,Patnaik, R., Thatoin, H. N. and Padhi, G. S. 1996. A new role of VAM fungi for improvement of plant growth through alleviation of stress effects existing in mine waste soil. In: Perspectives in Biological Sciences (eds.) Rai, V. Naik, M.L. and Manoharachary C. Pub. By School of Life Science. Pt. Ravi shankar Shukla Univ. Raipur. pp 131-140.

Mohammad, M. J., Pan, W. L. and Kennedy, A. C. 1995. Wheat responses to vesicular-arbuscular mycorrhizal fungal inoculation of soils from eroded toposequence. Soil Sci. Soc. Am. J., 59: 1080-1090.

Moora, M. Opik, M. and Zobel, M. 2004. Performance of two *Centaurea* species in response to different root-associated microbial communities and to alterations in nutrient availability. Ann. Brot. Fennici, 41: 263-271.

Mott, J. B. and Koske, R. E. 1988. Occurrence of AM in mixed overburden mine spoils of Taxas. Reclamation and Revegetation Research, 6(2): 145-156.

Nelson, C. E. and Safir, G. R.1982. Increased drought tolerance of mycorrhizal onion plants caused by improved Phosphorus nutrition. Planta, 154: 407-413.

Nutman, P. S. 1967. Varietal differences in the nodulation of subterranean clover. Austr. Agric. Res. 18: 381-425.

Oliveira, R. S., Vosatka, M. and Dodd, J. C. 2005. Studies on the diversity of arbuscular mycorrhizal fungi and the efficacy of two native isolates in a high alkaline anthropogenic sediment. Mycorhiza, 16: 23 – 31.

Patnaik, P., Thatoi, H. N., Padhi, G. S. and Mishra, A. K. 1992. Impact of *Glomus fasciculatum* on rhizosphere of *Cajanus cajan* L. grown in chromite mine waste soil. Plant Sci. Res., 14 (1 and 2): 1-3.

Quoreshi, A. M. 2003. Nutritional pre conditioning of and ectomycorrhizal formation of *Picea marina* (Mill) B.S.P. seedling. Eurasia J. For. Res., 6(1): 1 – 63.

Quoreshi, A. M., Piche, Y. and Khasa, D. P. 2008. Field performance of conifer and hardwood species five years after nursery inoculation in the Canadian Prairie Provinces. New For., 35: 235 – 253.

Row, H. I., Brown, C. S. and Classen, V. P. 2007. Comparisons of mycorrhizal responsiveness with field soil and commercial inoculum for six native Montane species and *Bromus tectorum.* Resor. Ecol., 15: 44 – 52.

Rao, A. V. and Tak, R. 2002. Growth of different tree species and their nutrient uptake in limestone mine spoil as influenced by arbuscular mycorrhizal (AM) fungi in Indian Arid Zone. J. Arid. Environ., 51: 113-119.

Rao, A. V., Tarafdar, J. C. and Tak, R. 2006 Effecet of arbuscular mycorrhizal fungus from mine spoils on growth of tree species and grass. In: Mycorrhiza, A. Prakash and V.S. Mehrotra (eds.) Pub. by Sci. Pub.(India) Jodhpur pp 249-253.

Read, D. J. 1999. Mycorrhizas in ecosystems. Experimen., 47: 74-84.

Sharma, S. and Dohroo, N. P. 1996. Vesicular arbuscular mycorrhizae in plant health management. Int. J. Trop. Plant. Dis., 14: 147 – 155.

Singh, A. and Jamaluddin. 2006. Multiplication and trapping of vesicular arbuscular mycorrhiza fungi in soil of dumps of limestone quarries. Mycorrhiza News, 17(4): 17-19.

Smith, S. M. and Read, D. 1997. Mycorrhizal Symbiosis 2nded. Academic Press, London.

Smith M. R., Charvat, I. and Jacobson, R. L. 1998. Arbuscular Mycorrhizae promote establishment of prairie species in a tall grass prairie restoration. Can. J. Bot. 76: 1947-1954.

Srinivas, P., Ram Reddy, S. and Reddy, S. M. 2005. AM fungi of coal mine soils of Godavari basin and their role in revegetation of *Albizzia lebbeck.* J. Mycol. Pl. Pathol, 35 (2).

Srinivas, P., Reddy, G. L., Rao, M. S. and Reddy, S. M. 2006. Mycorrhizal dependency of two multipurpose tree species in coal mine soils. In: Mycorrhiza, A. Prakash and V.S. Mehrotra (eds.) Pub. by Sci. Pub.(India) Jodhpur pp 15-20.

Swapna, V. L. and Ammani, K. 2009. Association of AM fungi on medicinal plants at coal mines of Mancheriala, Andhra Pradesh. Mycorrhiza News, 21 (2): 5-6.

Sweatt, M. R. and Davis, F. T. Jr. 1984. Mycorrhizae, water relations, growth and nutrient uptake of geranium grown under moderately high phosphorus regimes. J. Am. Soc. Hort. Sci., 109:210-213.

Sylvia, D. M., and Williams, S. E. 1992. Vesicular arbuscular mycorrhizae and environmental stress. In: Mycorrhizae in Sustainable Agriculature, R.G. Linderman and G.J., Bethlenfalvay (eds.) Special publication No. 54, American Society of Agronomy Madison, WI, pp. 101 – 124.

Thatoi, H. N. 1992. Growth and nodulation responses of *Leucaena leucocephala* (Lam) De wit. Under Rhizobium and VAM fungi inoculation in chromite mine waste soil. M.Phil Thesis, Utkal Univ.

Tisserant, B., Gianinazzi-Pearson, V., Gianinazzi, S. and Gollet, A. 1993. In plants histochemical staining of fungal alkaline phosphatase activity for analysis of efficient arbuscular mycorrhizal infections. Mycological Research 97(2): 245-250.

Ursic, M., Peterson, R. L. and Husband, B. 1997. Relative abundance of mycorrhizal fungi and frequency of root rot on *Pinus strobes* seedlings in southern Ontario nursery. Can. J. For. Res., 27: 54–62

Villeneuve, N., LeTacon, F. and Bouchard, D. 1991. Survival of inoculated *Laccaria bicolor* in competition with native ectomycorrhizal fungi and effects of the growth of out planted Douglas-fir seedlings. Plant Soil, 135:95-107.

Microbes: Diversity and Biotechnology (2012) *Pages* **375–395**
Editors: **Prof. S.C. Sati & Dr. M. Belwal**
Published by: **DAYA PUBLISHING HOUSE, NEW DELHI**

Chapter 23

Aspergillus spp. Association with Major Crops and its Pathogenic Effects

M.R. Swain and H.N. Thatoi

Department of Biotechnology, College of Engineering and Technology, Techno Campus, Ghatkia, Bhubaneswar – 751 003

ABSTRACT

Aspergillus are widespread in nature due to their capability to grow and develop in a wide range of environmental conditions *i.e.* under dry conditions (9.5 per cent of moisture), in temperature between 4 to 45°C and 65-100 per cent relative humidity. More than 200 species of *Aspergillus* spp. have been documented as per the literature. Due to their saprophytic and opportunistic nature, *Aspergillus* spp. are found to be responsible for several disorders in agricultural crops and commodities such as pulses, cereals, oilseeds, vegetables, fruits, spices, condiments etc. *Aspergillus* spp. have ability to produce various toxic metabolites known as mycotoxins in a wide range of agricultural crops. Mycotoxins produced by *Aspergillus* group have several negative effects on human health such as carcinogenic, immunosuppressive, tremorgenic, teratogenic, hepatotoxic and many more. Besides this there are other manifestations by *Aspergillus* spp. inside seeds which also cause damage to crops.

Keywords: *Aspergillus spp., Mycotoxin, Agricultural crops.*

Introduction

Fungi in the genus *Aspergillus* are comparatively more widespread than others due to their capability to grow and develop in a wide range of environmental conditions *i.e.* under dry conditions, in temperature between 4 to 45°C and 65-100 per cent relative humidity (Doijode, 2001). Although this group of fungus is not considered to be a major cause of plant diseases, due to their saprophytic and opportunistic nature, *Aspergillus* species are found to be responsible for several disorders in various plants, plant products and in agricultural commodities. Although. Micheli described this genus *Aspergillus* as long ago as 1729, it was Link in 1809 who clearly defined it. Raper and Fennel in late

1980s described 132 species of *Aspergillus* in 18 groups which is still used for diagnostic purpose in applied mycology. Now more than 200 species of *Aspergillus* have been documented (Smith and Ross, 1991).

Aspergillus occurs in and on a variety of substrates, including grains, plant parts, decaying vegetation in the field and cattle dung, particularly abundant in soils in the tropics and subtropics. (Zeng *et al.*, 2001 and Winn *et al.*, 2006). Most common species occurring in agricultural crops include *A. niger*, *A. flavus*, *A. parasiticus*, *A. ochraceus*, *A. carbonarius*, and *A. alliaceus* (Perrone *et al.*, 2007). They can contaminate agricultural products at different stages including pre-harvest, harvest and post harvest stages such as during processing and handling (Christensen, 1971and Mycock *et al.*, 1988). Although there are numerous reports about field infection and proliferation of *Aspergillus* on agricultural crops, basically this group is considered as storage fungi due to its virulent growth, sporulation and metabolite production in storage environment (Agrios, 2005 and Christensen, 1991). Aspergilli mainly infect crop seeds, harbor inside, proliferate and produce secondary metabolites either in field or during storage. They can invade crops having moisture content as low as 9.5 percent (Sunflower seeds), 13 percent (Soybean seeds) and 13.5percent (Wheat and Corn).

The primary importance of genus *Aspergillus* is their ability to produce various toxic metabolites known as mycotoxins in a wide range of agricultural crops and commodities. *Aspergillus flavus* was the first member of this group to be recognized as a producer of aflatoxins (a group of mycotoxins) in the 1960 when 100000 Turkeys died by consuming feed contaminated with this fungi and its metabolite (Tajkarimi, 2007). Mycotoxins produced by *Aspergillus* group and other fungi have deleterious effect on agriculture and human health which includes its carcinogenic, immunosuppressive, tremorgenic, teratogenic, hepatotoxic and many more harmful impacts. Mycotoxication by Aspergilli causes great crop, agriculture and economic loss worldwide (Bennett and Klich, 2003; Klich, 2007).

Besides mycotoxication, there are other manifestations by *Aspergillus* species inside seeds which also cause damage to crops. The various secondary metabolites produced by Aspergilli inside seeds leads to numerous disorders in seeds such as seed discolouration, seed rot, ear rot, smuts, kernel death, seedling death, lower germability and many other seed health problems (Zeng *et al.*, 2001). Though the pathogenic effect of members of *Aspergillus* group in agriculture is not wide spread but in some particular crops such as groundnut (Peanut), maize, Soybeans, Barley, Wheat, Rice, Grapes etc, these fungi cause a great crop damage and economic loss. *Aspergillus* species can infect wounded onion, garlic, carrot, pods of peas and beans, ripening tomato fruit, squash, potato tubers, ears of corns, apples, peaches and their produce (Sherf and MacNab, 1986). Here in this review we tried to summarize the deleterious effects caused by various species of genus *Aspergillus*, other than mycotoxication, on agricultural crops. This discussion may draw the attention of *Aspergillus* workers from mycotoxigenic study to the other dark side of crop damage by this group of fungi.

Description of the Genus

Genus *Aspergillus* belongs to Kingdom: Fungi, Phylum: Deuteromycota, Class: Eurotiomycetes, Order: Eurotiales, Family: Trichocomaceae, members are common molds. Colonies effuse generally green and yellowish, sometimes brown and black, mycelium partly immerged and partly superficial. stroma, setae and hyphodia are absent. Conidiophores macronematous, mononematous often with a foot cell, straight or flexuos, colourless or with the upper part mid to dark brown, usually smooth, swollen at the apex in to spherical or clavate vesicle, the surface of which is covered by short branches or phialides. Conidiogenus cell arise at the end of terminal branches, usually determinate, rarely percurrent, ampulliform or lageniform, collarettes sometimes present. Conidia catenate, dry, semiendogenus or acrogenus, spherical, variously coloured, smooth, rugose, verruculose or echinulate, sometimes with spines arranged spirally (Mehrotra and Prasad, 1969; Holliday *et al.*, 1980; Diba *et al.*, 2007; Winn *et al.*, 2006).

Cereals Crops Affected by *Aspergillus* spp.

Maize Seed Damage by *Aspergillus* spp.

Corn or Maize is major source of starch (74.26g Carbohydrate per 100g seed as per (USDA National Nutrient Database for Standard Reference) and used as a staple food in many countries of the world. Worldwide production of maize was over 600 million tonnes in 2003, just slightly more than rice and wheat. In 2004, close to 33 million hectares of maize were planted worldwide, with a production value of more than $23 billion, United States being the highest producer (Commodity online). It is also used as Cornmeal or corn flour popcorn, corn flakes, canjica (in Brazil), salads, garnishes, cooking oil (corn oil), corn syrup and many fermented products (Nebraska Corn Board and Wikipedia).

Different *Aspergillus* species at different seed moisture content cause various diseases in maize such as ear rot (Source: Maize doctor), kernel rot, storage rots, loss of germination, discoloration, caking, heating, and mustiness, resulting in heat-damaged and bin-burned corn. *e.g. A. restrictus* causes blue eye in corn at 14 to 14.5 per cent seed moisture content, *A. glaucus* cause heating and germ discolouration at 14.5 to 15 per cent, *A. candidus* causes kernel discolouration and heating at 15 to 15.5 per cent and *A. flavus* causes rapid heating of seed mass and kernel and germ discolouration at 18 to 18.5 per cent (Brooker *et al.*, 1992). When corn ears infected in the field by insect and mechanical damage and subsequently stored at high moisture contents, common species of *Aspergillus* such as *Aspergillus niger* causes black rot disease, which is it produces black, powdery masses of spores that cover both kernels and cobs. In contrast, *A. glaucus, A. flavus* and *A. ochraceus* normally form yellow-green masses of spores. *Aspergillus parasiticus* is ivy green and less common in maize ((International Maize and Wheat Improvement Center (CIMMYT); Rane and Ruhl, 2007). Insect damage to corn before harvest enhances *Aspergillus* infection and mycotoxins production (Tseng *et al.*, 1995).

A. niger is generally observed in dry years hence suspected that this fungus enters through drought stressed cracks. At moisture level of 32 per cent *A. flavus* readily infects kernels those have been injured by growth cracks, insect damage from corn ear worms, senescing silks and cracks caused by drought stress. Normally it doesn't infect uninjured kernels, however at 32°C to 38°C it can grow on senescent silk to ear in the absence of injuries. Its growth then progresses towards adjacent pericarp at the silk attachment site on the kernel, then over the kernel surface then eventually penetrates the tip-cap region. Spread to other kernels by mycelial growth on pericarps or glumes. (Nyvall, 1999).

At least one species of *Aspergillus* can grow in corn at a moisture content of 12.5 per cent but usually the range is 14 to 22 per cent. *Aspergillus flavus* causes an ear rot known as yellow mold. Like *Penicillium* spp., it is a storage fungus. It is extensively seed-borne, with infection levels up to 79 per cent (McGee, 1987), and can reduce germination. Systemic infection of seedlings from inoculated seed has been reported. *A. niger* causes black mold, an ear rot that can occur worldwide, particularly under moist conditions. It has been detected up to 62 per cent on seed, causing discoloration or rotting. Seed transmission has not been demonstrated (McGee, 1987).

Aspergillus can infect maize seeds from ears irrespective of variety, year, stage of harvest and drying method. Owolade *et al.*, 2005 found highest percentage incidence of *Aspergillus* spp (68.5 per cent) associated with seeds of variety DMRLSR-Y harvested 40 days after tasselling (DAT) and dried in the sun in year 2002 in South West Nigeria (Owolade *et al.*, 2005). In a study by Ghiasian *et al.*, 2004,

8.7 per cent *Aspergillus* infection was recorded in maize seeds collected from four provinces in Iran. *A. flavus* (7.5 per cent incidence) was dominant species followed by *A. niger* (0.9 per cent incidence). (Ghiasian *et al.*, 2004).

Mycological study of twenty maize samples, collected during 2002-2004 from outlets and bazaars of Balikesir, Turkey, showed an interesting pattern of fungal infection between non-disinfected and disinfected seeds. Based on isolation frequency and relative density, *Aspergillus* was the second most

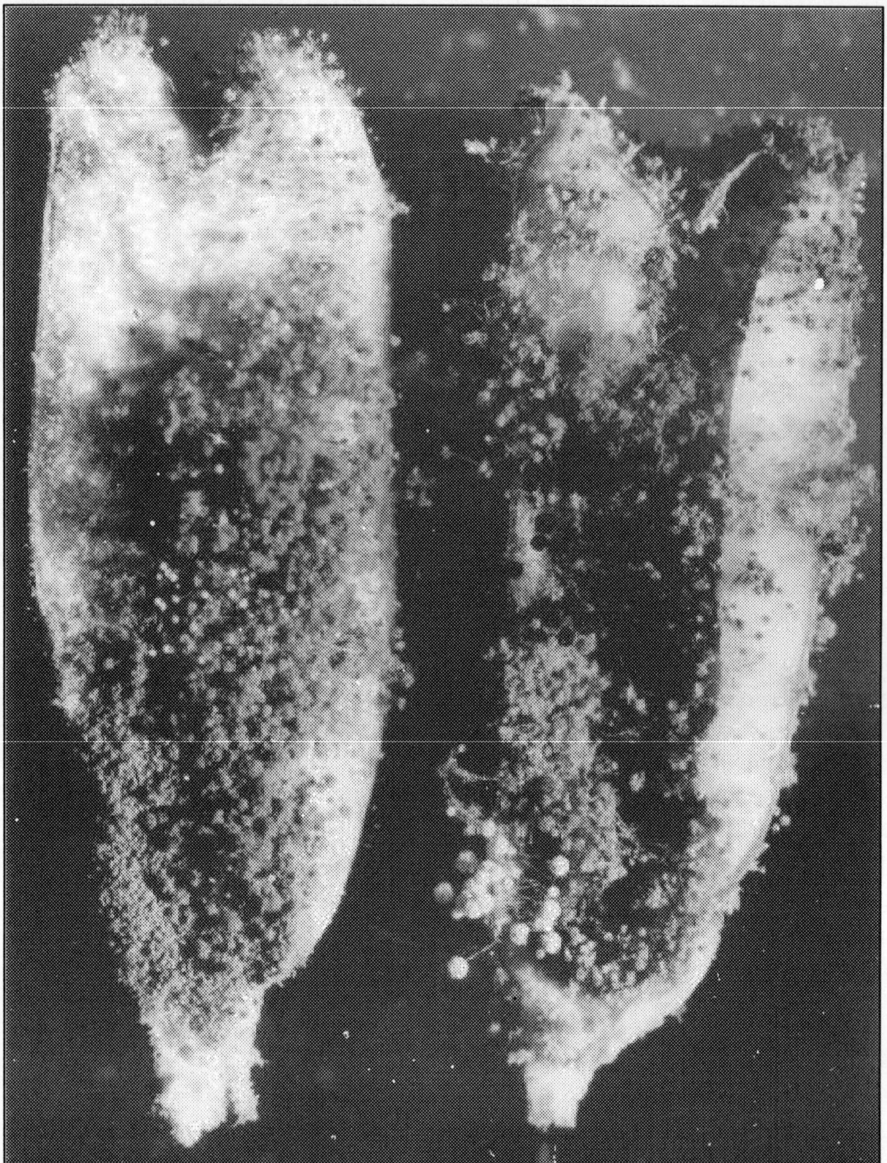

Figure 23.1: *Aspergillus*, a Common Storage Mold, Growing from the Germ of Two Split Corn Kernels (Report on Plant Disease, 1992)

abundant genus in non-disinfected seeds having a relative density (RD) of 19 per cent next to *Rhizopus* (49 per cent RD). But in seed samples, which has been surface disinfected by Chlorine bleach before plating, genus *Aspergillus* was the most abundant that is about 25 per cent (Askun, 2006). This observation clearly depicts the high infectivity of *Aspergillus* species in crop seeds. In non-disinfected seeds, *Rhizopus* suppresses the growth of other fungi during isolation in agar media as its spores remain on the surface of the seeds. Surface sterilization by chlorine reduces spores of *Rhizopus* so that internal mycoflora such as *Aspergillus* comes out on agar media. Among *Aspergillus* species which appeared in both types of seeds are *A. awamori* (RD1.1 per cent and 2.5 per cent in non-disinfected and disinfected seeds respectively), *A. flavus* (2.3 per cent and 4.5 per cent), *A. flavus* var *columnaris* (1.1 per cent and 1.9 per cent), *A. foetidus* var *acidus* (2.3 per cent and 5 per cent), *A. foetidus* var *pallidus* (3.5 per cent and 3.2 per cent), *A. tubingensis* (4.6 per cent and 5 per cent) and *A. wentii* (1.1 per cent and 0.6 per cent). Appearance of *A. niger* (RD 2.3 per cent) was recorded only in non-disinfected seeds. Some species of *Aspergillus* were recovered only from disinfected seeds those are: *A. flavor-furcatus* (06 per cent), *A. foetidus* (0.6 per cent), *A. parasiticus* (0.6 per cent) and *A. terreus* var. *americanus* (0.6 per cent) (Askun, 2006).

Maize is the third most important crop and an important staple food in Burkina Faso. Seed samples of twenty two maize varieties and landraces, collected from 14 locations, were tested by deep freeze blotter method and found be hundred percent infected by *A. niger*, followed by *A. flavus* (86 per cent infection) (Somda *et al.*, 2008).

After infection of corn seeds by storage-rotting molds such as *Aspergillus* species, heat and moisture given off due to their metabolic activity are utilized by their successors to accelerate rotting of the stored grain. The storage rots is manifested by the "cakeing" together of kernels which form a crust, usually at the center and top of a bin. Mold growth is often extensive, and infested bins have a musty odor. Spoilage of the surface grain is often intensified by migration of moisture to the upper layers in bins which lack adequate aeration. (Report on Plant disease, 1992).

Rice Grain Damage by *Aspergillus* spp.

World rice production in 2007 was approximately 645 million tones and considered as third most important cereal crop. At least 114 countries grow rice and more than 50 have an annual production of 100,000 tonnes or more. Asian farmers produce about 90 per cent of the total, with two countries, China and India, growing more than half the total crop (IRRI Statistics Portal, 2008). Hundred gram of rice contains 79.95 g carbohydrates, Protein 7.13 g, dietary fiber 1.3 g and vitamins like thiamine, riboflavin, niacin, pantothenic acid etc.

Mycoflora is a major cause of deterioration in rough rice and results in a loss in quality, economic value and perhaps in quantity. Infection of rice kernels takes place in the field but growth and the resulting deterioration may continue after the harvest until the moisture and temperature of the grain are reduced to inconsistent levels by the fungi. With few exceptions, spoilage of stored dry rice is attributable to storage fungi which such as *A. niger*, *A. candidus*, *A. glaucus* and *A. ochraceus* which were introduced during the post harvest handling process (Carreres *et al.*, 1995). This fact is quit supported by the study of (Haque *et al.*, 2007) where incidence of *Aspergillus* species and other fungi like *Fusarium* was higher in rice samples collected from untrained farmers (seed purpose) than in samples of trained farmers (seed purpose + grain purpose). But the incidence of the fungus *Aspergillus* sp. was significantly higher in the seed sample of untrained farmer's (seed purpose) to trained farmer's (seed purpose). (Haque *et al.*, 2007).

During storage of Rice *Aspergillus restrictus* is often first to appear at a moisture content of 14.0 to 14.5 per cent giving grains a purplish-black cast (Table 23.1). A high incidence of granary and rice weevils may indicate that the grain is heavily inoculated with *A. restrictus*. *A. glaucus*, most common of all storage fungi kills and discolors the germ. An increase in surface disinfected rice grains yielding this fungus when cultured between sampling periods is an indication that spoilage is underway. *A. candidus* discolors the entire kernel, turning the germ black. The presence of this fungus should be an indication that serious grain deterioration is underway as its growth results in heating of the total grain mass up to 130° F at which point the grain is a total loss. *A. flavus* develops at moisture contents between 16 and 18 per cent. It kills the germ, discolors the seed, and causes rapid heating. If *A. flavus* is detected, major spoilage has already occurred. (Rice Quality Workshop 2003) (Mutters, 2003).

Table 23.1: Equilibrium Moisture Content of Rice Grains at Relative Humidities of 65 to 85 per cent and Fungi Likely to be Present (Mutters, 2003)

Relative Humidity (per cent)	*Grain Moisture (per cent)*	*Fungi*
70 – 75	14.0–15.0	*A. glaucus, A. restrictus*
75 – 80	14.5–15.0	*A. candidus*
80 – 85	16.0–18.0	*A. flavus*

13.7 per cent infection by *Aspergillus* sp. was recorded in rice samples stored for six months in Babugonj area of Bangladesh. The incidence of the fungi *Fusarium* sp. and *Aspergillus* sp. differed significantly between the seed samples of trained and untrained farmers. (Haque *et al.*, 2007).

Aspergillus Diseases of Wheat

Wheat (*Triticum* species) is the second most important crop and production up to 593 million metric ton has been achieved worldwide in 1990 and it is growing 100million metric ton per decade. The bread wheat (*Triticum aestivum*) accounts for 80 per cent wheat consumption in the world. (Bajaj, 1990 and Oleson, 1994). 100 g of wheat contains 51.8 g Carbohydrate, 23.15 g protein, 9.72 g of fats, 13.2 g dietary fiber and essential vitamins like thiamine, niacin etc. (USDA National Nutrient Database for Standard Reference). Wheat grain is a staple food used to make flour for leavened, flat and steamed breads; cookies, cakes, pasta, juice, noodles and couscous, and for fermentation to make beer, alcohol, vodka and biofuel (Katina, 2003).

Aspergillus species mainly cause rust, rot and smut diseases in wheat. *A. flavus* and *A. niger* cause coleoptyle and root lesions in wheat. In Pakistan wheat is reported to be infected heavily by species of *Aspergillus* like *A. candidus*, *A. flavus*, *A. fumigatus*, *A. niger* and *A. sulphureus*. It has also been reported that in commercial wheat storage, the invasion of the wheat germ by common species of *Aspergillus* was probably the cause of "sick wheat" (seed with obviously dark germs). The major fungi presenting the "sick" seeds were *Aspergillus restrictus* Link, *A. repens* deBary, *A. candidus* Link, and *A. flavus* Link. (Robertson *et al.*, 1985).

The most common disease of bread wheat in all wheat growing countries is black point. It is characterized by a dark discolouration of the embryo resulting in discoloured and black ended kernels of seeds and embryos are often shriveled and brown to black in color (Toklu *et al.*, 2008). The disease can affect grain quality since food products made from infected kernels have displeasing odour and color. Germination rate, seedling emergence and seedling establishment under the field conditions were reduced by black pointed seeds. When seed moisture content exceeds 20 per cent, coupled with the relative humidity above 90 per cent, the amount of black point increases dramatically. (Hudec, 2007 and Toklu *et al.*, 2008).

Oilseeds

Association of *Aspergillus* Species is a Problem in Peanut or Groundnut

The Peanut or groundnut plant (*Arachis hypogaea*), commonly called groundnut, earthnut, monkey nut, goobers, is unusual as it flowers above ground and pods containing one to five seeds are produced underground. Its seeds are rich source of edible oils and contain 40 -50 per cent fat, 20–50 per cent protein, and 10 to 20 per cent carbohydrate. The seeds are nutritious and contain vitamin E, niacin, folacin, calcium, phosphorus, magnesium, zinc, iron, riboflavin, thiamine, potassium etc. Peanuts, peanut oil and peanut protein meals constitute an important segment of world trade in oilseeds and products. Peanut is the fifth most important oilseed in the world. Peanut is used for different purposes: food (raw, roasted or boiled, cooking oil), animal feed (pressings, seeds, green material, straw), and industrial raw material. (Source: World geography of the Peanut).

From planting to storage, biotic and abiotic agents affect peanut production process by different types of stresses. Biotic agents include insects, fungi, bacteria, virus, nematodes, weeds and abiotic factors include physiological and environmental stresses. A range of *Aspergillus* species are usually associated with peanut or groundnut seeds, which include *Aspergillus awamori*, *A. candidus*, *A. flavus*, *A. japonicus*, *A. luchuensis*, *A. niger*, *A. panamensis*, *A. parasiticus*, *A. penecilloides*, *A. terricola*, *A. terreus* and *A. wentii*. Most prevalent are *A. flavus* and *A. niger* (Rasheed *et al.*, 2004).

A. niger causes crown rot of peanut and along with *A. flavus* is also responsible for pre-emergence and post-emergence of rot of peanut. The fungus attacks the cotyledons soon after germination, forms a brown spot (in substantial part by oxalic acid production). This discoloured area gradually rots and spreads to stem and hypocotyls (Rasheed *et al.*, 2004). *A. flavus* also causes aflaroot rot of groundnut in which abundant growth of mycelia, spore and sclerotia occurs on cotyledons leading to the formation of reddish-brown lesions due to necrosis of central tissues (Desai and Bagwan, 2005).

In Texas, *Aspergillus niger* was found to be the causal agent of Black mold in peanuts caused by Low quality seeds, late plantings and drought stress for the first few weeks after planting have been associated with a high disease incidence. The fungus attacks the crown or collar area near the soil line and may girdle and kill the plant at any stage from seedling to harvest. The disease is symptomised by the black, slightly fluffy fungal growth at the ground line. (Lee *et al.*, 1995). Abundant sporulation by this fungus in cotyledon of peanut causes necrosis, stunting and wilting. (Burgess *et al.*, 2008).

In storage rot of groundnut, which is mostly caused by soil borne and seed-borne *Aspergillus* species, seeds are covered with yellow and green spores which leads to the disintegration of tissues. (Ihejirika *et al.*, 2005). Groundnut seed samples collected from different storage locations in five sampling zones of South Eastern Nigeria were investigated to isolate mycoflora from rot lesions. Highest rot samples were obtained from Owerri (40.2 per cent infection) and lowest from Okigwe sampling area (7.6 per cent). Different species of *Aspergillus* isolated from infected seeds from each zone were observed to be *Aspergillus flavus* Link. (92 per cent occurrence), *A. niger* van Teigh (60 per cent occurrence), *A. versicolor* (Vuill) Tirab (25 per cent occurrence) and *A. fresen* (15 per cent occurrence). High relative humidity, high rainfall and low temperature in Owerri and Mbaise region supposed to increase microbial respiration, growth and spread resulting in disintegration of seed tissue. Where as seeds in an environment of low relative humidity, low rainfall and high temperature in Okigwe area, remained dry hence microbial growth, development and action were restricted. The predominance of *Aspergillus* species in rot infected samples indicates its colonization of wounds of tropical crops and damaged stored seeds. (Ihejirika *et al.*, 2005).

A. niger and *A. pulverulentus* cause Collar rot or crown rot or seedling blight in groundnuts in which the hypocotyls tissue becomes water soaked and light brown, rapid wilting of entire plant or branches. The hypocotyls and tissues of cotyledonary nodes are rotted. The spreading of disease turns the collar region shredded and dark brown. *A. flavus* attacks seeds and un-immerged seedlings and reduce them to shrivelled, dried, black or brown mass covered by yellow or greenish spores, a condition called Yellow mould or Aflaroot disease. Plantation of infected seeds enhances decaying even of the emerging radicles and hypocotyls. Under field conditions, plants become stunted, chlorotic, short leaflets with pointed tips and no secondary root formation in radicles. (Yellow mould or Aflaroot disease: Source: ikisan.com).

Colonization of *Aspergillus* in Soybeans

Soybean (*Glycine max* (L.) Merril) is an Asiatic leguminous plant cultivated in several parts of the world for its oil and protein, which are extensively used in the manufacture of animal and human foodstuffs. (Pimentel *et al.*, 2006).

Kacaniova, 2003 found about 64 per cent natural infection of *Aspergillus* species (Next to *Penicillium* species of about 92 per cent) in soybean seeds. *A. fumigatus* (20 per cent) being the abundant species followed by *A. flavus, A. candidus, A. niger* (16 per cent each), *A. ochraceus, A. sydowii* and *A. versicolor* (8 per cent each). The percentage of occurrence of *Alternaria* was found to be inversely related to the occurrence of *Aspergillus* in soybean seeds. (Kacaniova, 2003).

In a comparative mycological study by Tseng *et al.*, 1995, diseased dry beans from Taiwan were found to be primarily infected by *Aspergillus* species where as no *Aspergillus* infection was found in diseased dry beans of Ontario, Canada. An average of 54.8 per cent and 58.9 per cent of beans from Ontario and Taiwan, respectively, had visible signs of fungal infection. *Aspergillus niger* being the highest infecting fungus (20.6 per cent) followed by *A. flavus* (11.8 per cent), *A. ostianus* (6.2 per cent), *A. flavo-furcatus* (5.1 per cent), *A. japonicus var. aculeatum* (3.1 per cent) and *A. tamari* (1.7 per cent) (Tseng *et al.*, 1995). Infection by these *Aspergillus* species may cause stress on seeds during germination and resulting in the formation of phytoalexins. Besides predominance by *A. flavus, A. candidus* has also been isolated abundantly from surface sterilized and non-sterilized soybean seeds (Smith and Ross, 1991). In warm, moist tropic and sub- tropic climates *Aspergillus* and *Penicillium* species colonized grain before harvest in greater number than it did it in cooler and temperate regions of the world. Also in the tropics and subtropics insect damage to grains is greater than in temperate regions (Tseng *et al.*, 1995). This *Aspergillus* may sometimes be endophytic in Soybeans as it has been isolated from leaf and stem fragments from field and green house (Pimentel *et al.*, 2006).

Heating and Other Effects by *Aspergillus* in Sunflower Seeds

The sunflower (*Helianthus annuus*) is an annual plant mainly grown for Sunflower oil, extracted from the seeds which is used for cooking, as a carrier oil and to produce margarine and biodiesel. Dry seeds contain about 51 per cent lipids, 21 per cent protein and 20 per cent carbohydrate. (USDA National Nutrient Database for Standard Reference).

A. flavus and *A. fumigatus* which are predominant in sunflower seed surface significantly decreased the oil content and iodine and caused pre-emergence rot. (Raj *et al.*, 2007). Stored sunflower seeds may become discolored from the heat generated by the growth of fungi (hot spots), like *Aspergillus* species when moisture levels are adequate for fungal growth Incidence of *Aspergillus* and other fungi increased with increasing moisture content (Robertson *et al.*, 1985). According to the Minnesota Department of Agriculture, Grain Inspection Division, "Heat damaged sunflower seed means seed and pieces of seed, which, when sliced open, show evidence of meats that have been discolored by heat. In their study Robertson *et al.*, 1985 fond that out of 796 fungal infected grade No. I sunflower seeds *Aspergillus* species infection was found to be about 28.51 per cent and that of sample grade sunflower seeds was about 71 per cent (Robertson *et al.*, 1985).

Table 23.2: Different Diseases by *Aspergillus* spp. and its Impact on Infected Plants

Infected Plant	*Aspergillus Species*	*Disease and its Effect*
Groundnut	*Aspergillus species*	Crown Rot leads to Rotting,Kernel rot, pod and seed rot.
	Aspergillus flavus	seedling blight, damping off, yellow mould of nuts
	Aspergillus niger	Crown rot causes Stunting and wilting. seedling blight, damping off, crown rot, black mould.
	A. niger and *A. pulverulentus*	Collar rot, crown rot and seedling blight.
Soybean	*Aspergillus ruber*	Increase in Ergosterol concentration.
	A. candidus, A. flavus, A. fumigatus, A. niger, A. ochraceus, A sydowii, A. versicolor	
Barley	*Aspergillus repens*	Decline in vigour and viability, lower respiration rate, dehydrogenase and diastase activities of germinating deteriorated seeds,
Cowpea	*Aspergillus flavus* *Aspergillus niger*	Seed rot and weak seedling
Cotton	*Aspergillus flavus*	Reduced quality and viability
Black gram (*Vigna mungo* L.)	*Aspergillus niger*	
Corn	*Aspergillus flavus* and *A. parasiticus*	Ear rots
	Aspergillus glaucus	Blue eye of germ leads to Kernel death
	Aspergillus species	Causes storage rot
Grapes	*Aspergillus niger*	Decaying of berries and Bunch rot.
	Aspergillus aculeatus	Rotting
Rape(*Brassica campestris* L.)	*Aspergillus japonicus*	Inhibition of seedling germination and growth
Wheat	*A. candidus, A. flavus, A. niger, A. sulphureus, A. tamarii, A. terreus*	
	Aspergillus restrictus, A. repens, A. candidus, A. flavus	Sick Wheat
	Aspergillus flavus, A. fumigatus., A.niger	
Rice *Oryza sativa* L.	*A. flavus, A. fumigatus, A. niger, A. terreus*	Yellow mould, kernel discolouration

Contd...

Table 23.2–Contd...

Infected Plant	*Aspergillus Species*	*Disease and its Effect*
	A. restrictus	Kills and discolors the germ, giving a purplish-black cast, musty odor in grain.
	A. glaucus	Kills and discolors the germ, causes mustiness and caking
	A. candidus	Discoloration of kernel, turning the germ black
	A. flavus	kills the germ, discolors the seed, and causes rapid heating
Cassava	*A. aculeatus, A. candidus, A. clavatus, A. flavipes, A. flavus, A. fumigatus, A. niger, A. nomius, A. ochraceous, A. parasiticus, A. tamari, A. terreus, A. versicolor*	
Garlic	*Aspergillus alliaceus*	
	Aspergillus alutaceus	Aspergillus rot
	Aspergillus flavus	Aspergillus storage rot
	Aspergillus niger	
	Aspergillus niger	Black rot
Maize	*A. flavus*	Cob rot, ear rot, storage rot and yellow mould.
	Aspergillus niger	Kernel rot and black mould
Vine	*Aspergillus niger*	Vine canker
Sunflower	*Aspergillus* species	Lowered oil quality and increased levels of free fatty acids
	Aspergillus sp.	Head rot
Cardamom	*A. flavus, A. niger*	
Black mustard	*A. proliferans*	
Sorghum	*Aspergillus flavus* and *A.niger*	Seed discoloration
Pistachio *Pistachio vera* L.	*Aspergillus niger* and other *Aspergillus* spp.	Aspergillus blights
	Aspergillus flavus	Yellow mould of nuts
Onion	*Aspergillus niger* (black mould)	Longer roots and shorter shoots than healthy one.Black mould (bulb rot) leads to Black powdery spore masses (also a storage rot).
	(*Aspergillus* spp.)	*Aspergillus* blight
Pawpaw fruit (*Carica papaya* L.)	*A. flavus, A. niger*	Rot
Dry Bean seeds (*Phaseolus vulgaris* L.)	*A. flavus, A. flavo-furcatus, A. japonicus* var. *aculeatum, A. niger, A. tamari, A. ostianus.*	
Lupine (*Lupinus luteus* L.)	*Aspergillus* species	
Parsley Petroselinum sativum Hoffm	*Aspergillus niger v.Tieghem*	

Contd...

Table 23.2–Contd...

Infected Plant	*Aspergillus Species*	*Disease and its Effect*
Coriandrum sativum	*A. nidulans, A. niveus, A. niger*	
Cuminum cyminum	*A. candidus, Aspergillus* spp.	
Curcuma longa	*Aspergillus* spp.	
Trigonella foenum-graecum	*A. fumigatus, A. niger, A. niveus*	
Peaches	*A. flavus*	Fruit rots
Citrus	*Aspergillus flavus*	Fruit and root rot
	Aspergillus niger	Fruit rot
Coconut (*Cocos nucifera* L.)	*Aspergillus flavus*	Yellow mould
Cassava (*Manihot esculenta* Crantz)	*Aspergillus flavus*	Yellow mould
Cotton (*Gossypium hirsutum* L.)	*Aspergillus flavus* *Aspergillus niger*	Boll rot
Potato (*Solanum tuberosum* L.)	*Aspergillus niger*	Secondary tuber rot
Sugarcane (*Saccharum officinarum* L)	*Aspergillus flavus* *Aspergillus niger*	Yellow mould on canes and sugar Black mould on canes and sugar
Sweet potato [*Ipomoea batatas* (L.) Lam.]	*Aspergillus niger*	Secondary storage rot
Yam *Ipomoea batatas* (L.) Lam.	*Aspergillus niger*	Secondary storage rot
Black and white pepper	*Aspergillus flavus* and *A. niger*	
Almond	*A. candidus, A. carneus, A. clavatus, A. fischeri, A. flavus, A. flavipes, A. tumigatus, A. glaucus, A. janus, A. niger, A. ochraceus, A. restrictus, A. sydowi, A. sulphureus, A. terreus, A. versicolor, A. wentii*	

Aspergillus head rot appears under surface of the heads due to water soaking and forms brown discolouration which extends to the stalk to a distance of about 10 to 15 cm. The discoloured under surface of the head becomes very soft and pulpy and the fungus enters into the head through the holes made by the insect attack (Ranasingh *et al.*, 2008).

Aspergillus in Almonds

Almond (*Amygdalus communis* L.) is consumed as both green fresh fruit and kernel. As it contains 51 per cent fat, mainly used for edible oil. In Southeastern Anatolia Project region in Turkey, 26.34 per cent infection of *Aspergillus* species were found in almond seeds (Cimen and Ertugrul 2007). Numerous species of *Aspergillus* was isolated from forty almond seed samples collected from different locations of Pakistan and from fifteen samples collected from different countries like Dubai, Abudhabi, Saudi Arabia (Mecca and Madinna), Kuwait, Iran, Germany, England, Australia, USA and Egypt. The species include *A. candidus, A. carneus, A. clavatus, A. fischeri, A. flavus, A. flavipes, A. fumigatus, A. glaucus, A. janus, A. niger, A. ochraceus, A. restrictus, A. sydowi, A. sulphureus, A. terreus, A. versicolor and A. wentii. A. ustus* was also reported to cause seed rot of almond seeds (Bilgrami, 1998 and Duncan, 2002).

Fruits Rotting by *Aspergillus* Species

Grapes

Jarvis and Traquair (1984) reported *Aspergillus aculeatus* and *Aspergillus violaceo-fuscus* (A possible synonymous species to cause rotting in grape berries which are detrimental to wine industry as they produce so called "Pourriture vulgaire". They found grape cultivars like Aurore, Seyval Blanc, Duthess and Johannesburg Riesling (varieties with tight clusters) as susceptible and cultivars like Marechal Foch and Black cultivar as resistant to *A. aculeatus.* Fungal growth was concentrated at the pedicel, where pressure within the bunch had partially detached the berry, or at rain induced splits, insect punctures. Mycelial growth was profuse on the exposed parenchyma and black conidial heads were abundant. *Aspergillus niger* in complex with *Acetobacter* bacteria cause sour rot (Jarvis and Traquair, 1984).

Aspergillus rot (Aspergillus bunch rot or Aspergillus Summer Bunch Rot; Source: UC-IPM online), common in warmer grape producing countries, is caused by *Aspergillus niger* v. Tieghem. *A. niger* infect surface of the berry and produce abundant black spores resulting in pale and water soaked tissue (UC-IPM Online). Spores are easily liberated resulting in shoot like deposits in adjacent berries. (Sharma and Kaul, 1999). Similarily *A. carbonarious* causes berry rot or bunch rot in both fresh fruit as well as partially-dried and fully-dried grapes. (Leong *et al.*, 2004 and Kazi *et al.*, 2008)

Pawpaw

Baiyewu *et al.*, 2007 tested 90 random fungal infected samples of Pawpaw fruit collected from market places of Oyo, Ogun and Ondo States of Nigeria. In the humid forest of South Western Nigeria, the high rainfall pattern, high humidity and the temperature of between 19 and 31°C prevailing in the agro ecology favors the development of fungal diseases both in the field and the market. The isolation of fungal pathogens like, *Aspergillus niger, Aspergillus flavus* along with *Rhizopus nigricans, Curvularia lunata, Fusarium moniliforme,* and *Colletotrichum capsici* confirmed the studies of Gupta and Pathak (1986) and Kuthe and Spoerhase (1974) that these fungi associated with rotten pawpaw are highly pathogenic causing appreciable losses in pawpaw fruits at post harvest. Incidence and pathogenicity of *A, flavus, A. niger* and other previously mentioned fungi increased from year 2001 to 2002. (Baiyewu *et al.*, 2007). Among all *A. niger* was reported to cause fruit rot in pawpaw, best at30-35°C and 60-80 per cent relative humidity (Baiyewu and Amusa, 2005).

Peaches

A. flavus and *A. niger* causing ripe fruit rot or black rot (Source: Database of food and related sciences) in peaches have reported by several authors in different countries (Barkai-Golan, 1980), first reported in Greece by Michailides and Thomidis, 2007.

Apples

Aspergillus mainly cause various types of rotting in apples such as: soft rot caused by *Aspergillus niger* Van Tiegh (Bisen and Agarwal, 1980), fruit rot by *A. terreus* (Chandra and Tandon, 1963), brown rot caused by *A. flavus* and *A. niger* (Hasan, 2000) and Aspergillus rot caused by *A. niger* (Wilawan *et al.*, 2003). Healthy apple was infected with *A. niger* van Tieghem (63 per cent) and *A. fumigatus* (25 per cent) Fresenius but in rotten apple *A. niger* van Tieghem (83 per cent) and *A. terreus* Thom (8 per cent) causes infection. (Hasan, 2000).

Tuber Crops

Essono *et al.*, 2007 made a detailed study on infection of Cassava chips by *Aspergillus* species in two different locations of Cameroon, Ebolowa (Mengomo) and Yaoundé (Nkometou III) during a two-month monitoring period. Predominant species of *Aspergillus* which caused more than 50 per cent infection in Cassava chips was found to be *A. clavatus* (Presence index of 86.11 per cent and 80.56 per cent in the above two locations respectively) followed by *A. flavus* (69.44 per cent and 77.78 per cent) and *A niger* (50 per cent and 72.22 per cent). The pattern of occurrence of different fungi during storage of cassava chips was very interesting. *A. aculeatus and A. ochraceous* constantly increased their number during a storage period of one to eight weeks in both of the sampling areas. However their infection percentage is relatively lower than other fungi. The incidence pattern of the three highest infecting fungi *A. clavatus, A. flavus* and *A. niger* showed a sharp increase in their occurrence with a sudden decrease. (Essono *et al.*, 2007).

Aspergillus in Some Common Legume Crops

Lupine is a legume, rich in carbohydrates (40 per cent) and proteins (36 per cent) (USDA National Nutrient Database for Standard Reference) and used for soil improvement, grazing, human consumption and medicinal purposes (Source: Healthy pro Food). Based on the completed studies on fungal colonization of Lupine (*Lupinus albus*), a large percentage of saprophyte fungi (*Penicillium* and *Rhizopus*) were found among the total number of isolates (30 per cent and 20 per cent, respectively). Considerably less frequently, fungi representing the genus of *Aspergillus* were isolated (5 per cent of the total number of colonies). The above mentioned fungi were more commonly colonizing the seeds of the Legat cultivar than those of the Markiz cultivar, cultivated in the crop rotation system with 20 per cent of lupine. Their largest communities were observed on the seeds after a 2.5-year period of storage. In the literature, these fungi are described as "storage" fungi and their population increases along with an increase in storage time. (Cwalina-ambroziak and kurowski, 2004).

Hungarian Vetch (*Vicia* L.) is an annual forage legume cultivated for hay, green manure, pasture and seed and also increases soil nitrogen by fixing atmospheric nitrogen. Three different species of *Aspergillus* infected various parts of the plant in both original and newly harvested seeds. Among them *A. niger* was isolated from all samples and seeds parts and *A. alutaceaus* having less incidence in some samples. The re-establishment of *A. niger* in newly harvested seeds was on seed coat and cotyledon and its incidence increased and decreased depending on the seed lines (Coskuntuna and Ozer, 2004).

Pea (*Pisum sativum* L.) is an important leguminous vegetable crop of the tropics and grown all over the world for its fresh use, preservation, high level of digestibility which is more than most of the legumes. Dried peas contain 23.5 per cent crude protein, 1.7 per cent ether extract and 2.9 per cent ash. Owning to great nutritional importance, cultivation of peas in the world is increasing. The association of *Aspergillus* species with pea seeds reduced their germination. (Begum *et al.*, 2004).

Vegetables

Reports of various species of *Aspergillus* are on record from seeds of vegetables, the surfaces of stored seeds of different plants, castor oilseeds and guar seed, cauliflower seed. Metabolites released from this fungus such as *A. japonicus* inhibited the germination and seedling growth rates of Rape, Radish and cucumber up to 54.2, 91.7, and 8.5 per cent, respectively (Zeng *et al.*, 2001).

Prevalence of *Aspergillus* Species in Tomatoes

Tomato (*Lycopersicum esculentum* M.), a good source of vitamins and minerals, has a worldwide production of 80 million tones from a crop area of 3 million hectares. Kalyoncu *et al.* (2005) collected 350 tomato samples from 10 fields (from 2000-2001) and 30 homemade tomato paste samples of Manisa provines of Turkey to study the associated mycoflora. Infection by *Aspergillus* species constituted about thirty four percent of the total mould incidences in tomato field samples and twenty eight percent that of tomato paste samples. Members of *Aspergillus niger* group represented one sixth of all of the identified species. *A. fumigatus* Fresen, *A. terreus* Thom and *A. parasiticus* Speare weren't found in tomato paste samples where as *A. aculeatus* Lizuka, *A. flavus* Link, *A. foetidus* Thom and Raper, and *A. foetidus* var pallidus (Nakaz) Raper and Fennel were found to be associated both with field homemade paste tomato samples. The relation between moulds and tomato and tomato pastes has been studied worldwide and mainly concerns with increasing productivity and nutritional value. The main source of these mold infection is the soil and the intensity is further enhanced by late harvest, inadequate watering period and method, inclusion of mouldy fruits in healthy fruits, late transport, unhygienic processing and improper storage etc. (Kalyoncu *et al.*, 2005). In Pakistan, Tomato seeds are usually collected from rotten fruits which are left in the field. These seeds which play an important role in crop production are vulnerable to attack by saprophytic fungi like *Aspergillus flavus, A. candidus, A. fumigatus, A. niger, A. sulphureus, A. terreus, A. wentii* other *Aspergillus* species. The fungi may cause seed abortion, shrunken seed, seed rot, seed discolouration and reducing germination capacity. (Perveen, 1996).

Black Mould in Onions by *Aspergillus*

Black mould caused by *Aspergillus niger* v. Tiegh causes significant economic losses in onions (*Allium cepa* L.) produced in inland New South Wales, because of surface blemish and rotting in storage. The development of *A. niger*, a common saprophyte, is favoured by warm conditions (28-33°C), which are usual during the harvest period in the onion growing areas of south western New South Wales. In these areas, onions are sown from May to August and harvested from November to March. Black mould is known to develop anywhere on the bulb surface, as well as on interior leaf scale surfaces to which it gains entry via the neck and roots but more common on the middle and base of the bulbs. 77.5 per cent of Cream gold cultivar of onion bulbs had visible infection in one or more zones, compared to 46.6 per cent of Southport White Globe bulbs. (Sinclair and Letham, 1996).

Spices and Condiments

Condiments like *Coriandrum sativum, Cuminum cyminum, Curcuma longa, Trigonella foenum-graecum* are also not spared by *Aspergillus* species. Lal and Raizada, 1975 reported that the fungi found in association with the condiments may be pathogenic to the plants but if the condiment's are thoroughly dried, exposed to bright sunlight and cleaned before utilization, they can be rendered safe for human consumption. (Lal and Raizada, 1975). Similarly considerable infection by *Aspergillus niger* v.Tieghem was also found in Parsley seedlings (Cultivar Berlinska) in both field conditions as well as after storage (Nawrocki, 2004).

Aspergillus, the Storage Fungi: Mode of Infection in Seeds

FAO estimates that 5 per cent of all stored cereal grains are render inedible by storage fungi (Doijode, 2001). Field fungi remain inactive in seeds during drying. Storage fungi such as *Aspergillus* species which are recognized to be saprophytes and opportunistic invaders of naturally dried plant tissue (seeds) and dead organic matter, can tolerate and remain metabolically active in low moisture(as low as 13 per cent) and high temperature. Such fungi can grow and cause internal infection during favorable conditions (Mycock and Berjak, 1995). The avenues for seed infection include 1-Seed surface through the cuticle, natural openings, cracks or injuries caused during threshing, 2- hilum that is covered by the cuticle but with fissures, 3-micropyle, particularly the open type and 4-accessory structures, *e.g.* hairs, wings, aril and caruncle (Singh and Mathur, 2004). Metabolically active storage fungi, which are external to the seed will be true to their opportunistic saprophytic nature and invade seed tissues via any physically injured area. Alternatively if the seed is intact, the fungi grow within the loose peduncle tissue and penetrate through the micropyle in to the underlying seed tissues. The testa of a matured orthodox seed is naturally discontinuous at only one point that is micropyle. This pore in the testa is subtended by the peduncle. Peduncle to micropyle infection is gradual, commencing with mycelial growth on the seed surface, particularly at the micropylar end. The parenchymatus nature of the parenchymatous tissue facilitates hypal growth towards and through the micropyle and in this way fungi gain access to the underlying tissue. Such a method of infection has been reported for *A. candidus*, *A. chevalieri*, *A. flavus*, *A. flavus* var *columnaris*, *A. oryzae*, *A. paraciticus*, *A. ruber*, *A. sydowi*, *A. versicolor* and *P. pinophilum*. (Mycock and Berjak, 1995)

If crop seeds are infected by these type storage fungi and the storage conditions are humid with high temperatures, those crops may fail to germinate or may grow badly because of the allelopathy (Biochemical interaction between all types of microorganisms and plants) of the fungus (Zeng *et al.*, 2001). While the effects of some of the species are confined to storage tissues, others degrade the embryo, starting with discolouration and ending with total decay (Mycock and Berjak, 1995). Metabolites of many fungi may have adverse or stimulatory effects on plants such as suppression of seed germination, malformation, and retardation of seedling growth. Many crop seeds are infected by fungi before harvest or during storage. If conditions are favorable, then the situation is more serious. Some fungi on the surface of seeds may produce mycotoxins that affect food quality and some may produce phytotoxins that affect seed germination and seedling growth (Zeng *et al.*, 2001).

Seeds are a source of many pathogenic factors and play an important role in the wholesomeness of many crop species. Research into the fungal communities colonizing the seeds of papilionaceous plants has been carried out. Seed damage, particularly damages of crown and germ affect the process of germination, especially at lower temperatures. Seed damage means an undisturbed possibility of attack by pathogenic microorganisms in soil, such as *Aspergillus* spp., *Pythium* spp., and *Penicillium* spp. These microorganisms cause rotting of seed and germ. The consequence of this is a thinned stand, which brings about lower yields or the entire plot has to be resown (Simic *et al.*, 2004).

Storage Condition and the Succession of *Aspergillus* Species

The storage temperature, moisture content, presence of oxygen and gaseous composition are the most important factors influencing the development of fungi during storage. Physiological stages of grains or sensitivity of different hybrids to fungi growth are important as well. (Kacaniova, 2003). Oxygen level showed positive correlation with storage fungi *Aspergillus* species such as: *A. flavus*, *A. glaucus*, *A. niger*, *A. terreus* (Chuansin *et al.*, 2006). Under the storage conditions the activity of field fungi is reduced and they are no longer able to perpetuate in the micro-environment of seeds at this time a first succession of *Aspergillus* and *Penicillium* species can establish. *Aspergillus* and *Penicillium* are considered to be most widespread and destructive agents on earth (Mycock and Berjak, 1995).

Chistensen and Kauffman maintained that depending on the grain moisture content, a succession of *Aspergillus* species is manifested which in turn is succeeded by *Penicillium* species as follows: *A. restrictus* group (Seed Moisture Content 13 to 13.5 per cent), *A. glaucus* group (SMC 14 to 14.5 per cent), *A. versicolor* (SMC 14.2 to 15 per cent), *A. ochraceus* (15 to 15.5 per cent), *A. candidus* (SMC 15 to 15.5 per cent), *A. flavus* (17 to 18.5 per cent), *Penicillium* (>18.5 per cent). When comparing the effect of various *Aspergillus* species on seed germination, Christensen and Kauffman concluded that the more xerotolerant species *A. restrictus* have less effect than do those species that require higher SMC such as *A. flavus*. However the duration of the dominance of a particular species within the seed is an important factor. *A. glaucus* is slightly more vigorous and cause heating and significantly reduce seed viability. *A. ruber* can reduce seed germination up to zero in stored pea seeds in eight months. *A. candidus*, *A. flavus*, *A. amstelodami*, *A. glaucus* increase fatty acid content of seeds. *A. candidus* can discolour and kill seeds very quickly (within 4 days of suitable conditions) and rapid growth can cause heating up to 55°C. It also increases seed moisture content within very short period of time (Mycock and Berjak, 1995).

Conclusion

In conclusion, the genus *Aspergillus* is a successful and extremely widespread group of molds encompassing many species of economic importance. Several *Aspergillus* strains are also used as commercial strains for production of several oriental food products such as Temph, Miso, Tamari, Koji etc. in South East Asia, among *Aspergillus* group *Aspergillus oryzae* is the most common. Only a few species of *Aspergillus* are associated with plant disease and found to be responsible for several disorders in agricultural crops and commodities. *Aspergillus* spp. have ability to produce various toxic metabolites known as mycotoxins in a wide range of agricultural crops which has several negative effects on human health such as carcinogenic, immunosuppressive, tremorgenic, teratogenic, hepatotoxic and many more. Other than this *Aspergillus* spp. also causes crop loss and mycotoxins contamination in agricultural commodities which decreases its market value. This discussion may draw the attention of *Aspergillus* workers from mycotoxigenic study to the other dark side of crop damage by this group of fungi.

References

Agrios, G. N. 2005. Post harvest diseases of plant products caused by *Ascomycetes* and *Deuteromycetes*. In: Plant Pathology. Agrios GN (eds), New York, United States, Academic Press, pp 556-560.

Askun, T. 2006. Investigation of fungal species diversity of Maize kernels. J. boil. Sci., 6 (2): 275-281.

Baiyewu, R. A. and Amusa, N. A. 2005. The Effect of Temperature and Relative Humidity on Pawpaw Fruit Rot in South-Western Nigeria. World J. Agri. Sci., 1(1): 80-83.

Baiyewu, R. A., Amusa, N. A., Ayoola, O. A. and Babalola, O. O. 2007. Survey of the post-harvest diseases and aflatoxin contamination of marketed pawpaw fruit (*Carica papaya* L) in South Western Nigeria. African J. Agri. Res., 2(4): 178-181.

Bajaj, Y. P. S. 1990. In vitro technology, establishment of cultures, somatic embryogenesis and micro propagation. In: *Wheat*. Bajaj YPS (ed), Berlin, Germany, Springer, pp 3.

Barkai-Golan, R. 1980. Species of *Aspergillus* causing post-harvest fruit decay in Israel. Mycopathol., 71(1): 13-16.

Begum, N., Alvi, K. Z., Haque, M. I., Raja, M. U. and Chohan, S. 2004. Evaluation of mycoflora associated with Pea seeds and some control measures. Plant Pathol. J. 3(1): 48-51.

Bennett, J. W. and Klich, M. A. 2003. Mycotoxins. Clinic Microbiol. Rev. 16(3): 497–516.

Bilgrami, Z. 1998. Detection and control of seed-borne mycoflora in Almond. Thesis submitted in fulfillment of the requirements of the degree of Doctor of Philosophy, Faculty of Science, University of Karachi, Pakistan.

Bisen, P. S. and Agarwal, G. P. 1980. *In vitro* Production of Pectolytic Enzymes by *Aspergillus niger* Van Tiegh Causing Soft Rot in Apples. J. Phytopathol., 97(4): 317-326.

Brooker, D. B., Bakker-Arkema, F. W. and Hall, C. W. 1992. Grain Storage management. In: *Drying and Storage of Grains and Oilseeds*. Berlin, Germany, Springer, pp 381-382.

Burgess, L. W., Knight, T. E., Tesoriero, L. and Phan, H. T. 2008. Common diseases of some economically important crops. In: *Diagnostic manual for plant diseases in Vietnam*. Burgess, L.W., Knight, T.E., Tesoriero, L. and Phan HT (eds). Canberra, Australian Centre for International Agricultural Research, (http://www.aciar.gov.au/system/files/node/8613/MN129+part1.pdf), Monograph No. 129, pp 156.

Carreres, R., Ballesteros, R. and Sendra, J. B. 1995. Rice diseases in the region of Valencia and methodologies for testing varieties resistance. Cahiers Options Méditerranéennes15(3):21-2// www.medrice.unito.it/Medoryzae/Medoryzae10.pdf.

Chandra, S. and Tandon, R. N. 1963. Studies on the pectolytic enzymes of *Aspergillus terreus* Thom. Mycopathol., 19(3): 216-224.

Christensen, C. M. 1971. Invasion of sorghum seed by storage fungi at moisture contents of 13.5–15 per cent and condition of samples from commercial bins. Mycopathol., 44(3): 277-282.

Christensen, C. M. 1991. Fungi and seed quality. In: Handbook of applied mycology. Arora, D. K., Ajello, L. and Mukerji, K. G. (eds), United States, CRC Press, Taylor and Francis group, pp 113-119.

Chuansin, S., Vearasilpa, S., Srichuwong, S. and Pawelzikb, E. 2006. Selection of packaging materials for Soybean seed storage. Conference on International Agricultural Research for Development, Tropentag 2006, University of Bonn, Bonn October 11–13, 2006.

Cimen, I. and Ertugrul, B. B. 2007. Determination of mycoflora in Almond plantations under drought conditions in southeastern Anatolia project region, Turkey. Plant Pathol. J., 6(1): 82-86.

Commodity online: http://www.commodityonline.com/commodities/cereal/maize.php.

Coskuntuna, A. and Ozer, N. 2004. Seed-borne fungi in Hungarian Vetch and their transmission to the crop. Plant Pathol. J., 3(1): 5-8.

Cwalina-Ambroziak, B. and Kurowski, T. P. 2004. Fungi colonizing seeds of two cultivars of Yellow Lupine (*Lupinus Luteus* L.) cultivated in two rotations. Acta fytotechnica et zootechnica. 7:pp 57-60. Special Number, Proceedings of the XVI. Slovak and Czech Plant Protection Conference organised at Slovak Agricultural University in Nitra, Slovakia. Database of food and related sciences. Our food: http://www.ourfood.com/Phytopathology_diseases_pla.html.

Desai, S. and Bagwan, N. B. 2005. Fungal and Bacterial diseases of Ground nut. In: Diseases of Oilseed Crops. Saharan, G. S., Mehta, N. and Sangwan, M. S. (eds), India, Indus Publications, pp 121.

Diba, K., Kordbacheh, P., Mirhendi, S. H., Rezaie, S. and Mahmoudi, M. 2007. Identification of *Aspergillus* species using morphological characterization. Pak J. Med. Sci., 23(6): 867-872.

Doijode, S. D. 2001. Seed storage in fruit crops: Tropical and subtropical fruits. In: Seed Storage of Horticultural Crops. Philadelphia, United States, Haworth Press, pp 23.

Duncan, R. 2002. Almond Field Day, June 11, 2002, 8:30-11:00 AM, Location: Sherman Thomas Ranch, 25810 Avenue 11, Madera CA, Tree and Vine notes, Cooperative Extension, University of California (Agriculture and Natural resources). http://cemerced.ucdavis.edu/newsletterfiles/Tree_and_Vine_Notes4334.pdf.

Essono, G., Ayodele1, M., Akoa, A., Foko, J., Olembo, S. and Gockowski, J. 2007. *Aspergillus* species on cassava chips in storage in rural areas of southern Cameroon: their relationship with storage duration, moisture content and processing methods. African J Microbiol Res., pp 1-008.

Ghiasian, S., Kord-Bacheh, P., Rezayat, S. M., Maghsood, A. H. and Taherkhani, H. 2004. Mycoflora of Iranian maize harvested in the main production areas in 2000. Mycopathol., 158: 113–121.

Gupta, A. K., Pathak, V. N. 1986. Survey of fruit market for papaya fruit rot by fungi pathogens. Indian J. Mycol., 16:152-154.

Haque, A. H. M. M., Akon, M. A. H., Islam, M. A., Khalequzzaman, K. M. and Ali, M. A. 2007. Study of seed health, germination and seedling vigor of farmers produced Rice seed. Int. J. Sustain. Crop Prod., 2(5): 34-39.

Hasan, H. A. H. 2000. Patulin and aflatoxin in brown rot lesion of apple fruits and their regulation. World J. Microbiol. Biotechnol., 16: 607-612.

Healthy pro Food, RTD programme funded by the European Commission in the framework of the Quality of Life Research Programme: http://users.unimi.it/healthyp/bulletin/e-Bulletin_EN.pdf.

Holliday, P. 1980. Pathogens. In: Fungus Diseases of Tropical Crops. U.K., Cambridge University press, Cambridge, pp-37.

Hudec, K. 2007. Influence of harvest date and geographical location on kernel symptoms, fungal infestation and embryo viability of malting barley. International J. Food Microbiol., 113: 125-132.

Ihejirika, G. O., Nwufo, M. I., Durugbo, C. I., Ibeawuchi, I. I., Onyia, V. H., Onweremadu, E. U. and Chikere, P. N. 2005. Identification of fungi associated with storage rot of groundnut in Imo state south eastern Nigeria. Plant Pathol. J., 4: 110-112.

International Maize and Wheat Improvement Center (CIMMYT): www.cimmyt.org/english/docs/field_guides/maize/pdf/mzDis_earRots.pdf.

IRRI Statistics Portal. 2008. International Rice Research Institute, Philippines: http://beta.irri.org/statistics/index.php?option=com_frontpage&Itemid=1.

Jarvis, W. R. and Traquair, J. A. 1984. Bunch rot of Grapes by *Aspergillus aculeatus*. Plant Disease, 68: 718-719.

Kacaniova, M. 2003. Feeding Soybean Colonization by Microscopic Fungi. Trakya Univ. J. Sci., 4(2): 165-168.

Kalyoncu, F., Tamer, A. U. and Oskay, M. 2005. Determination of fungi associated with Tomatoes (*Lycopersicum esculentum* M.) and Tomato paste. Plant Pathol J., 4(2): 146-149.

Katina, K. 2003. Improving Wheat quality. In: Bread Making: Improving Quality. Cauvain, S. P. (ed), Cambridge, England, Woodhead Publishing Limited, pp 540.

Kazi, B. A., Emmett, R. W., Nancarrow, N. and Partington, D. L. 2008. Berry infection and the development of bunch rot in grapes caused by *Aspergillus carbonarius.* Plant Pathol., 57(2): 301-307.

Klich, M. A. 2007. *Aspergillus flavus*: the major producer of aflatoxin. Mol. Plant. Pathol. 8(6): 713-722.

Kuthe, G. and Spoerhase, H. 1974. Cultivation and use of pawpaw (*Carica papaya* L). Tropen Land Writ., 75:129-139.

Lal, C. B. and Raizada, B. B. S. 1975. Report of some fungi from some of the condiments of daily use in storage. Def. Soi. J. 25: 159-161.

Lee, T. A. Jr., Philley, G. L., Black, M. C. and Kaufman, H. W. 1995. Peanut Disease and Nematode Control Recommendations. Department of Plant Pathology and Microbiology, Texas A and M University. College Station, TX 77843-2132: http://plantpathology.tamu.edu/Texlab/Fiber/Peanuts/pdncr95.html.

Leong, S. L., Hocking, A. D. and Pitt, J. I. 2004. Occurrence of fruit rot fungi (*Aspergillus* section Nigri) on some drying varieties of irrigated grapes. Australian J. Grape and Wine Res., 10(1): 83-88.

MaizeDoctor:http://maizedoctor.cimmyt.org/index.php?option=com_content&task=view&id=227

McGee, D. C. 1987. Seedborne and seed-transmitted diseases of maize in rice-based cropping systems. In: Rice Seed Health Proceedings of the International Workshop on Rice Seed Health 16–20 March 1987. Sponsored by International Rice Research Institute and United Nations Development Programme.

Mehrotra, B. S. and Prasad, R. 1969. Two new species of *Aspergillus* from soil. Trans. Br. Mycol. Soc., 53 (2): 336-340.

Michailides, T. and Thomidis, T. 2007. First report of *Aspergillus flavus* causing fruit rot of Peaches in Greece. New disease Reports: http://www.bspp.org.uk/ndr/jan2007/2006-61.asp.

Mutters, C. 2003. Managing Rice Decay during Storage. In: Rice Quality Workshop: http://www.agronomy.ucdavis.edu/uccerice/QUALITY/rqw2003/C-8RiceDecay 2003.pdf.

Mycock, D. J. and Berjak, P. 1995. The implications of seed associated mycoflora during storage. In: Seed Development and Germination. Kigel, J. and Galili, G. (eds). United States, CRC Press, Taylor and Francis group, pp 749-760.

Mycock, D. J., Lloyd, H. L. and Berjak, P. 1988. Micropylar infection of post-harvest caryopses of *Zea mays* by *Aspergillus flavus* var. columnaris var. nov. Seed science technol., 16(3): 647-653.

Nawrocki, J. 2004. Occurrence of fungal diseases on Parsley seedlings (*Petroselinum sativum* Hoffm.). Acta fytotechnica et zootechnica, 7: 220-223. Special Number, Proceedings of the XVI. Slovak and Czech Plant Protection Conference organised at Slovak Agricultural University in Nitra, Slovakia.

Nebraska Corn Board; Official Nebraska Government Website: http://www.nebraskacorn.org/usesofcorn/index.htm.

Nyvall, R. F. 1999. Diseases of Maize (*Zea mays*). In: Field Crop Diseases. UK, Blackwell Publishing, pp256-257.

Oleson, B. T. 1994. World Wheat production, utilization and trade. In: Wheat production, properties and quality. Bushuk W and Rasper VF (eds). Berlin, Germany, Springer, pp 3.

Owolade, O. F. L., Alabi, B. S., Enikuomehin, O. A. and Atungwu, J. J. 2005. Effect of harvest stage and drying methods on germination and seed-borne fungi of maize (*Zea mays* L.) in South West Nigeria. African J. Biotechnol., 4 (12): 1384-1389.

Perrone, G., Susca, A., Cozzi1, G., Ehrlich, K., Varga, J., Frisvad, J. C., Meijer, M., Noonim, P., Mahakarnchanakul, W. and Samson, R. A.YEAR——-Biodiversity of *Aspergillus* species in some important agricultural products. Studies in Mycol., 59: 53-66.

Perveen, S. 1996. Studies on some seedborne fungal diseases of Tomato. Thesis submitted in fulfillment of the requirement for the degree of Doctor of Philosophy, Faculty of Science, University of Karachi, Pakistan.

Pimentel, I. C., Glienke-Blanco, C., Gabardo, J., Stuart, R. M. and Azevedo, J. L. 2006. Identification and colonization of endophytic fungi from Soybean (*Glycine max* (L.) Merril) under different environmental conditions. Brazilian Arch. Biol. Tech., 49(50): 705-711.

Raj, M. H., Niranjana, S. R., Nayaka, Sc. and Shetty. H. S. 2007. Health status of farmers' saved Paddy, Sorghum, Sunflower and Cowpea seeds in Karnataka, India. World J. Agri. Sci., 3(2):167-177.

Ranasingh, N., Behera, S. K. and Dalai, S. R. 2008. Diseases of Sun Flower and their management. *Orissa* Review 14 (7): 83-84.orissagov.nic.in/e-magazine/Orissareview/2008/feb-march-2008/engpdf/83-84.pdf.

Rane, K. and Ruhl, G. 2007. Crop Diseases in Corn, Soybean, and Wheat: Department of Botany and Plant Pathology Purdue University West Lafayette, IN47907: http://www.btny.purdue.edu/Extension/Pathology/CropDiseases/Corn/corn3.html.

Rasheed, S., Dawar, S. and Ghaffer, A. 2004. Location of fungi in Groundnut seed. Pak. J. Bot., 36 (3): 663-668.

Rasheed, S., Dawar, S., Ghaffer, A. and Shaukat, S. S. 2004. Seed-borne Mycoflora of groundnut. Pak. J. Bot., 36 (1): 199-202.

Report on Plant Disease (1992). Department of Crop Science, university of Illinois extension, College of Agriculture, Consumer and Environmental Sciences: http://web.aces.uiuc.edu/vista/pdf_pubs/206.PDF.

Robertson, J. A., Roberts, R. G. and Chapman, G. W. Jr. 1985. An evaluation of "heat damage" and fungi in relation to sunflower seed quality. Phytopathol., 75:142-145.

Sharma, R. C. and Kaul, J. L. 1999. Post harvest diseases of temperate fruits and their management. In: Diseases of Horticultural Crops: Fruits. Verma, L. R. and Sharma, R. C. (eds), New Delhi, Indus Publishing Co., India, pp 592 and 595.

Sherf, A. F. and MacNab, A. A. 1986. Onion, Garlic, Leeks and Shallots. In: Vegetable Diseases and Their Control. United States, Wiley-IEEE, pp 460.

Simic, B., Popovic, S. and Tucak, M. 2004. Influence of corn (*Zea mays* L.) inbred lines seed processing on their damage. Plant Soil Env., 50 (4): 157–161.

Sinclair, P. J. and Letham, D. B. 1996. Incidence and sites of visible infection of *Aspergillus niger* on bulbs of two onion (*Allium cepa*) cultivars. Australian Plant Pathology, 25: 8-11.

Singh, D. and Mathur, S. B. 2004. Penetration and establishment of fungi in seed. In: Histopathology of Seed-borne Infections. Singh, D. and Mathur, S. B (eds), United States, CRC Press, Taylor and Francis group, pp 93.

Smith, J. E. and Ross, K. 1991. The Toxigenic Aspergilli. In: Mycotoxins and Animal Foods. Smith, J. E. and Henderson, R. S. (ed), United States, CRC Press, Taylor and Francis group, pp102 and 356.

Somda, I., Sanou, J. and Sanon, P. 2008. Seed-borne infection of farmer saved maize seeds by pathogenic fungi and their transmission to seedlings. Plant pathol. J., 7 (1): 98-103.

Tajkarimi, M. 2007. Fungi and Mycotoxins. California Department of Food and Agriculture, Animal Health and Food Safety Services Division. http://www.cdfa.ca.gov/ahfss/Animal_Health/PHR250/2007/25007Myc.pdf.

Toklu, F., Akgul, D. S., Bicici, M. and Karakoy, T. 2008. The Relationship between Black Point and Fungi Species and Effects of Black Point on Seed Germination Properties in Bread Wheat. Turk J. Agric., 32: 267-272.

Tseng, T. C., Tu, J. C. and Tzean, S. S. 1995. Mycoflora and mycotoxin in dry bean (*Phaseolus vulgaris*) produced in Taiwan and in Ontario, Canada.

UC-IPM online: University of California-Integrated Pest Management, Agriculture and Natural resources: Grape, Summer Bunch Rot (Sour Rot) http://www.ipm.ucdavis.edu/PMG/r302100211.html

USDA National Nutrient Database for Standard Reference, Nutrient Data Laboratory <http://www.nal.usda.gov/fnic/foodcomp/search/>.

Wikipedia, the free encyclopedia: http://en.wikipedia.org/wiki/Maize.

Wilawan, K., Yoshie, M. and Yukio, H. 2003. Inhibition of Aspergillus rot by sorbitol in apple fruit with watercore symptoms. Postharvest Biol. technol., 29(2): 121-127.

Winn, W. C., Koneman, E. W., Allen, S. D., Janda, W. M., Schreckenberger, P. C., Procop, G. W. and Woods, G. L. 2006. Hyaline molds and Hylohyphomycosis. In: Koneman's Color Atlas and Textbook of Diagnostic Microbiology. Winn, W. C., Koneman, E. W., Allen, S. D., Janda, W. M., Schreckenberger, P. C., Procop, G. W. and Woods, G. L. (eds), Philadelphia, Lippincott Williams and Wilkins, Wolters and Kluwer Health, 1174- 1177.

World geography of the Peanut. Supported by the USAID through the Peanut CRSP Sustainable Human Ecosystem Laboratory, Department of Anthropology, UGA. http://www.lanra.uga.edu/peanut/knowledgebase.

Yellow mould or Aflaroot disease. Groundnut disease management. ikisan. http://www.ikisan.com/links/ap_groundnutDisease per cent 20Management.shtml#Rhizopus per cent 20Seed per cent 20And per cent 20Seedling per cent 20Rot.

Zeng, R. S., Luo, S. M., Shi, M. B., Shi, Y. H., Zeng, Q. and Tan, H. F. 2001. Allelopathy of *Aspergillus japonicus* on Crops. Agron. J., 93:60–64.

Microbes: Diversity and Biotechnology (2012) *Pages* 397–416
Editors: Prof. S.C. Sati & Dr. M. Belwal
Published by: DAYA PUBLISHING HOUSE, NEW DELHI

Chapter 24

Deep Mycoses in Fish Caused by Water-moulds and Stress Hypothesis: A Review

S.K. Prabhuji[1], *S.K. Sinha*[2] *and Deepanjali Srivastava*[3]
[1]*Biotechnology and Molecular Biology Centre, M.G. Post Graduate College, Gorakhpur – 273 001, U.P.*
[2]*Fisheries Department, U.P., Gorakhnath, Gorakhpur – 273 005, U.P.*
[3]*Department of Home Science, St. Joseph College for Women, Gorakhpur – 273 009, U.P.*

Out of three basic types of mycoses (fungal infection) in fishes, *viz.*, dermatomycoses (infection of the skin), branchiomycoses (infection of the gills) and deep mycoses (infection of the deep tissues and the internal vital organs), the dermatomycosis is associated with the integument and can cause rapid destruction of the epidermis. Deep mycosis is of two basic types: first one, the deep dermal mycosis and the second, the mycosis of vital organs. In deep dermal mycosis which is just a type of dermatomycosis, the tissues beneath the surface of epidermis, *i.e.*, dermal tissues together with scales as well as the underlying musculature are involved whereas in second type the infection involves the vital organs like brain, liver, kidney, intestines etc.

Oomycetes that infect fishes produce easily recognized cottony mycelia on the surface of the affected animal; they have probably been recognized since antiquity (Arderon, 1748; Spallanzani, 1777; Bennet, 1842; Goodsir, 1842; Arechoug, 1844; Unger, 1844; Robin, 1853 and Berkeley, 1864). In 1877, Oomycetes were reported for the first time in European literature in association with an epizootic known as 'Salmon disease' and the causal organisms were identified and further studies were made (Stirling, 1879-80; Huxley, 1882a, b; Willoughby, 1968, 1969, 1971, 1977, 1978; Neish, 1976, 1977). Generally, as reported earlier, saprolegnian (oomycetous) infections are associated with the integument and can cause destruction of the epidermis, thus depriving the fish of the protection of the mucus. In cases where saprolegnian infections are restricted to the integument, it seems reasonable to suppose that the actual cause of death may be related to impaired osmo-regulation and the inability of the fish to maintain its body-fluid balance (Gardner, 1974; Hargens and Perez, 1975).

An important point to emphasize is that saprolegnian fungi are not tissue specific and are capable of attacking virtually any tissue. This aspect has been most carefully documented in the detailed studies of Nolard-Tintigner, (1971, 1973, 1974), however, several other studies have extended and confirmed the general applicability of these observations (Bootsma, 1973; Dukes, 1975; Wolke, 1975; Neish, 1977; Hatai and Egusa, 1977; Papatheodorou, 1980; Srivastava, 1979,1980 a, b, c; Prabhuji, 1988; Srivastava *et al.*, 1994; Prabhuji and Sinha, 1994; Prabhuji *et al.*, 1998). Therefore, the common designation of saprolegniosis (a generalized term use for fish infection caused by oomycetous fungi) as a dermatomycosis is both incorrect and misleading. This contention has probably arisen as a result of the fact that these infections are usually initiated in the integument followed by the death of the fish before the infecting fungus may proceed beneath the skin or the underlying musculature. Are certain mycotoxins, produced by the fungal pathogen, adversely affect the metabolism of the host, sometimes resulting into a seize?

It does not appear likely that saprolegnian fungi produce any toxins (Rucker, 1944; Nolard-Tintigner, 1973; Peduzzi *et al.*, 1976). The extent of damage caused by these fungi may directly be related to the tissue necrosis done by the hyphae in its surrounding area. Assuming that the fungus is the only pathogen, the time of death will be a function of the growth rate of the fungus, the initial site of infection, the type and quantity of tissue destroyed, and the ability of the individual fish to withstand the stress of the disease. However, production of certain pectolytic or proteolytic enzymes by these fungi can not be ruled out. Peduzzi *et al.* (1976) and Peduzzi and Bizzozero (1977) have suggested a correlation between the production of a chymotrypsin-like proteolytic enzyme and the capacity for an isolate of *Saprolegnia ferax* and four isolates in the *S. diclina, S. parasitica* complex to switch over from a saprotrophic to a necrotrophic mode of nutrition.

Saprolegnian infections of fish are frequently associated with the wounds and lesions and also that handling fish may predispose them to infection. The obvious inference, drawn from these observations is that these fungi act as "wound parasites". The integument (skin) of fish in general and the mucus, in particular, both present a physical and a biochemical barrier to the initiation of infection and that if this barrier can be breached, an infection can proceed unrestrained. This has been discussed in some detail by Wilson (1976), Willoughby and Pickering (1977) and Richards and Pickering (1978).

On the basis of his work on Pacific salmon, Neish (1976, 1977) has emphasized the role of "stress" in initiating saprolegnian infections. Although more empirical evidences are required to substantiate this hypothesis, the available evidence is persuasive and provides a mechanism, which explains how physiological changes, which occur in fish, can predispose them to infection by parasites to which they are normally resistant. The basic manner in which the stress response may increase the susceptibility of fish to saprolegniosis is shown in Figure 24.1. A similar outline has been presented by Roddie and Wallace (1975), which is fundamentally based on the work of Selye (1950). How far the Selye's Stress Theory can be applied to fish has been explained by Wedemeyer (1970) and Wedemeyer *et al.* (1976).

Various stressors, which include crowding, injury, sub-optimal water temperature, handling and presence of noxious chemicals in sub-lethal concentrations etc., acting singly or synergistically, operate through the pituitary-interrenal axis to produce an increase in the level of plasma corticosteroids. An increase in plasma corticosteroid levels can impair the inflammatory responses (Mc Leay, 1975) and lead to an increase in corticosteroid regulated protein catabolism and

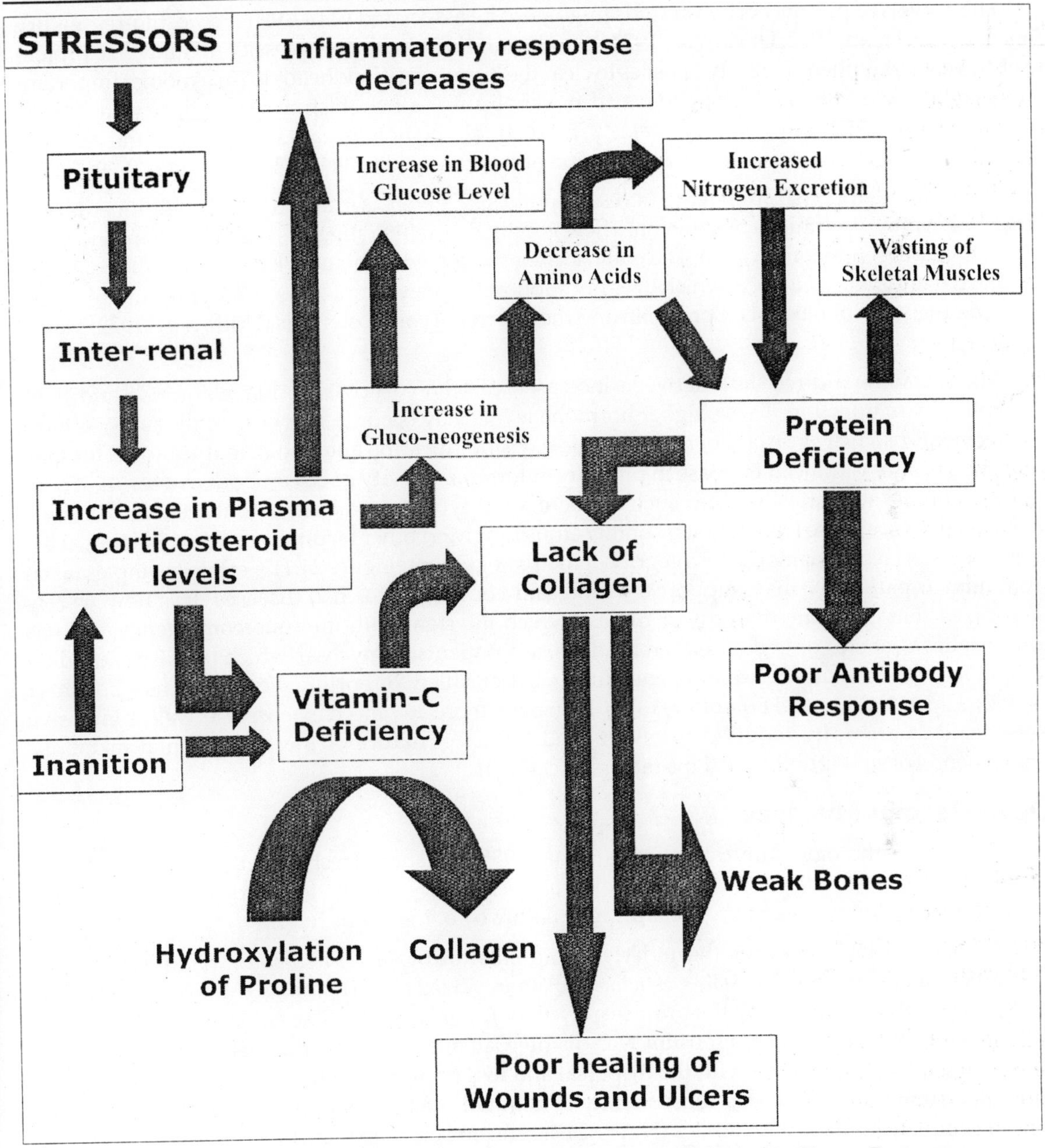

Figure 24.1: Increase in Susceptibility of Fish to Saprolegniosis due to Stress Responses (Selye, 1950; Roddie and Wallace, 1975; and Neish, 1976, 1977)

gluconeogenesis (Woodhead, 1975). This may ultimately result into a protein deficiency, which contributes to the wasting of skeletal muscles and leads to a decrease in antibody production and collagen synthesis. Lack of collagen, in turn, impairs the ability of a fish to heal wounds and ulcers.

High levels of plasma corticosteroids might also be associated with the fishes' osmo-regulatory function (Olivereau, 1962; Utida *et al.*, 1972; Woodhead, 1975), with the necessity to catabolize protein to obtain energy with the inability of fishes to clear the hormone (Woodhead, 1975). Another important factor to be considered is the ascorbic acid metabolism of a fish. Fish, in general, have a dietary requirement for vitamin C (Ashley *et al.*, 1975). It should be noted that an increase in the levels of plasma corticosteroids may also cause depletion of ascorbic acid reserves (Wedemeyer, 1969; 1970).

It should be emphasized that the "stress hypothesis", given, is intended to be complimentary to other observations on mucus production. There is evidence, which suggests that mucus production is controlled by the endocrine system. The main hormone involved in this activity appears to be prolactin; however, there is also evidence, which does not support this view point (Lam, 1972). Presumably there might be an interaction between prolactin and the interrenal corticosteroids (Utida *et al.*, 1972; Meier, 1972).

There may be a direct link between increased plasma corticosteroid levels in fish and their susceptibility to infection. These higher hormone levels may occur in response to the physiological requirements of a fish at certain period of its life like smoltification and sexual maturation. This may be related to stress-induced increase in pituitary-interrenal activity or come about as a result of both factors acting synergistically. With such an increase in typical metabolic activity, the fish become increasingly susceptible to infection by oomycetous fungi and other facultative pathogens and, at the same time less able to maintain the integrity of their integument (Figure 24.1). Certain examples taken from human pathology may support the aforesaid contention for fish diseases. It is now widely recognized that cortisone therapy, or disease which interfere with immuno-competency, such as uncontrolled diabetes or leukemia, can predispose a patients to mycoses which are also caused by ubiquitous and normally non-pathogenic fungi – better called 'opportunistic organisms' (Conant *et al.*, 1971; Chick, Balows and Furcolow, 1975). However, more advanced extensive as well as intensive investigations are aptly required to understand the exact nature of the fundamental metabolic phenomenon of susceptibility and the factors affecting it.

Deep Dermal Mycoses

The histopathological investigations (Sinha, 1985) on some fresh-water fishes, *viz.*, *Anabas testudineus*, *Channa punctatus*, *Chela laubuca* and *Colisa lalius* have shown varying degree of destruction of epidermis, hypodermis and underlying musculature by oomycetous fungi.

The healthy skin tissues and the underlying musculature of *Anabas testudineus* (Figure 24.2) show well-differentiated regions. In *Anabas testudineus* Bl. (parasitized by *Dictyuchus sterile* Coker) the lesions were completely dominated by the profuse growth of fungal mycelia. The epidermis and the scales were missing and stratum spongiosum was disintegrated. Majority of the melanophores showed signs of complete breakdown with the dispersal and loss of pigment granules (Figure 24.3). As the infection progressed the fungus subsequently penetrated the basement membrane (sub cutis) and extended into the underlying muscles (Figure 24.4). The muscle fibre bundles exhibited clear signs of necrosis resulting into the deposition of large number of blood cells (erythrocytes and leucocytes) between the bundles (Figure 24.5). In one individual, epidermis and scales were missing from the affected region and a peculiar inflammatory response of the stratum spongiosum and stratum compactum layers were observed. These two layers could be differentiated by a layer of compressed cells between them (Figure 24.6).

The infected regions in *A. testudineus* (parasitized by *Saprolegnia diclina* Humphrey) showed the presence of dense mycelial tuft, which at first destroyed the epidermis and progressively exhibited

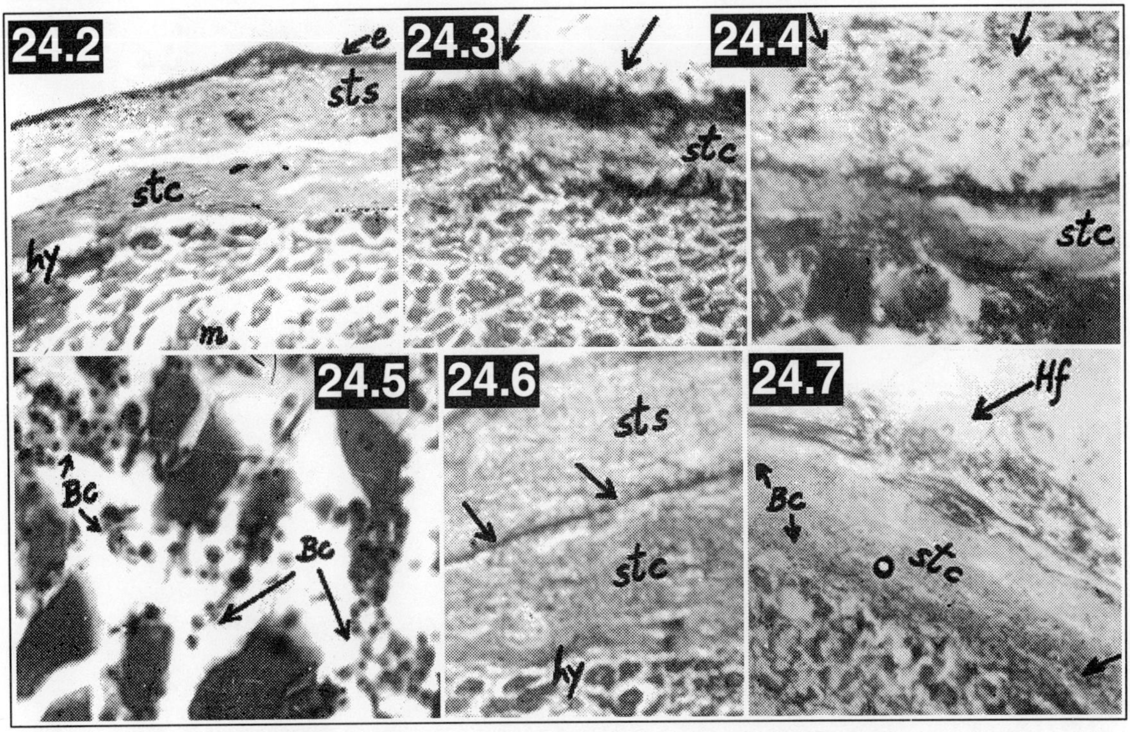

Figures 24.2–24.7: Histopathology of Dermal Tissues of *Anabas testudineus* Bl., Parasitized by Oomycetous Fungi

Figure 24.2: Healthy Dermal Tissues

Figures 24.3–24.6: Fish Parasitized by *Dictyuchus sterile* Coker

Figure 24.3: Complete Loss of Epidermis and Stratum Spongiosum and Accumulation of Melanophores; Arrow shows fungal hyphal fragments.

Figure 24.4: Highly Damaged Upper Tissues; Arrow shows fungal hyphal fragments.

Figure 24.5: Haemorrhage in between Muscles; Arrow shows erythrocytes.

Figure 24.6: Typical Inflammatory Response in Stratum Spongiosum and Stratum Compactum; Arrow shows the compressed layer of cells.

Figure 24.7: Fish Parasitized by *Saprolegnia diclina* Humphrey; Highly Damaged Upper Tissues and Presence of Erythrocytes in Stratum Compactum due to Haemorrhage

(e: Epidermis; sts: Stratum spongiosum; stc: Stratum compactum; hy: Hypodermis; m: Muscle fibre bundles; Bc: Blood cells; Hf: Hyphal fragments)

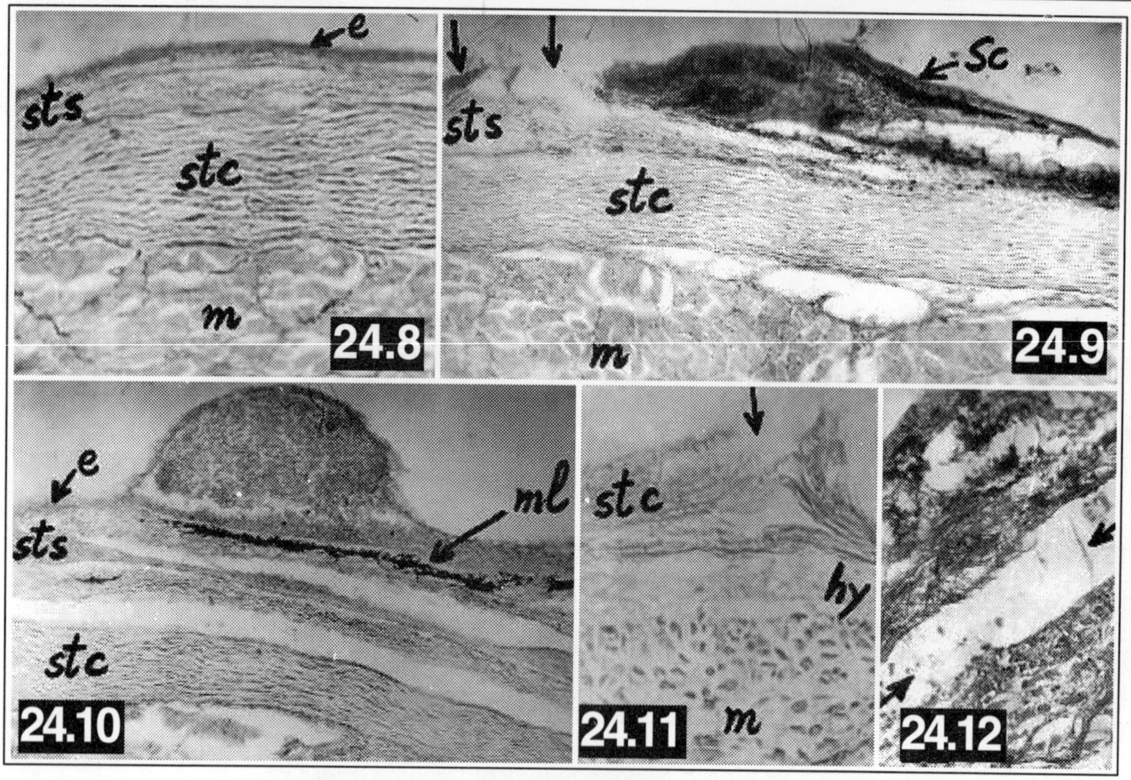

Figures 24.8–24.10: Histopathology of Dermal Tissues of *Channa punctatus* Bl., Parasitized by *Saprolegnia diclina* Humphrey

Figure 24.8: Healthy Tissues

Figure 24.9: Damage and Contraction in Epidermis and Stratum Spongiosum at the Base of Scale (arrows)

Figure 24.10: Development of Hyperplastic Gall-like Tissue Lump and Aggregation of Melanophores at its Base

Figures 24.11–12: Histopathology of Dermal Tissues of *Channa punctatus* Bl., Parasitized by *Saprolegnia ferax* (Gruith.) Thuret

Figure 24.11: Complete Loss of Epidermis and Stratum Spongiosum and Damage in Stratum Compactum

Figure 24.12: Hypertrophied Stratum Spongiosum and Stratum Compactum; Arrows show space occupied by fungal hyphae.

(e: Epidermis; Sc: Scale; sts: Stratum spongiosum; stc: Stratum compactum; hy: Hypodermis; m: Muscle fibre bundles; Bc: Blood cells; Hf: Hyphal fragments; ml: Melanophores)

necrosis of stratum spongiosum. Invasion of erythrocytes and leucocytes into the stratum compactum and the subcutis was distinctly observed (Figure 24.7).

In *Channa punctatus* Bl. (parasitized by *Saprolegnia diclina* Humphrey) the lesions were sparsely covered with fungal hyphae (Figure 24.8 shows the healthy tissues). Epidermis and scales were mostly missing, if present, showed clear signs of disintegration. In majority of the cases stratum spongiosum was disintegrated and the melanophores between the epidermis (if present) and stratum spongiosum were in a contracted state beneath the scale (Figure 24.9). In one of the specimens the epidermis and superficial layers of stratum spongiosum showed abnormal cellular multiplication (hyperplasia) forming a gall like (tumourous) structure. The melanophores between the epidermis and stratum spongiosum were contracted and abnormally aggregated at the base of the gall like structure. However, the stratum compactum and muscle fibre bundles were distinctly intact (Figure 24.10).

The mycelial tufts were present sparsely over the lesions in *C. punctatus* (parasitized by *Saprolegnia ferax* (Gruith.) Thuret.), which destroyed the epidermis and the scales; and these extended through the dermis into the sub cutis (hypodermis). The stratum compactum showed necrosis and the loose connective tissues of hypodermis (sub cutis) exhibited a peculiar inflammatory response (Figure 24.11). On another site, exhibiting inflammatory response, the layers of stratum spongiosum and stratum compactum got hypertrophied developing a space in the sub cutis region occupied by the fungal hyphae (Figure 24.12).

Srivastava *et al.* (1994) have described deep dermal mycoses in *Chela laubuca* Ham. caused by *Achlya orion* Coker and Couch. The lesions were completely dominated by the profuse growth of fungal mycelia. The epidermis and the dermis were completely missing and the sub cutis (hypodermis)

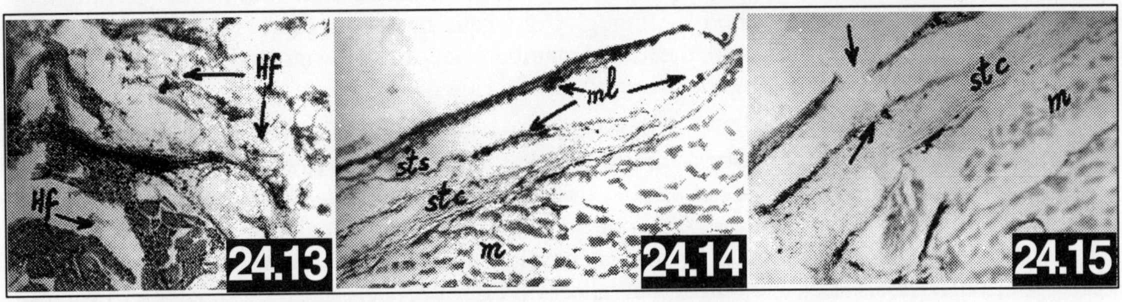

Figure 24.13: Histopathology of Dermal Tissues of *Chela laubuca* Ham., Parasitized by *Achlya orion* Coker and Couch. Tissue damage is deep up to the muscle fibre bundles.

Figyres 24.14–15: Histopathology of Dermal Tissues of *Colisa lalius* Ham., parasitized by *Dictyuchus sterile* Coker

Figure 24.14: Inflammatory Response in the Upper Tissues; Development of Space Between Epidermis and Stratum Spongiosum Occupied by Isolated Patches of Melanophores

Figure 24.15: Epidermis Broken and Damage to Stratum Spongiosum; Arrows indicate hyphal fragments.

(sts: Stratum spongiosum; stc: Stratum compactum; m: Muscle fibre bundles; Hf: Hyphal fragments; ml: Melanophores)

layer also showed disintegration exposing the underlying musculature. Some of the hyphal fragments, penetrating into the muscle tissue, were also observed (Figure 24.13).

The epidermis was disintegrated and the scales were missing in *Colisa lalius* Ham. (parasitized by *Dictyuchus sterile* Coker), however, the melanophores between stratum spongiosum and the epidermal layers were in contracted state. Stratum spongiosum showed degeneration with contracted melanophores at some places (Figure 24.14). At certain points stratum spongiosum was completely disorganized with partial degeneration of stratum compactum (Figure 24.15).

Dermal Ulcerative Syndrome

The salmon disease of 1877 – 1881, on the basis of clinical evidence, has been considered as first well-documented epizootic and designated as Ulcerative Dermal Necrosis (UDN) of salmonids. Various ideas have been advanced on the possible roles of bacteria and even virus, in addition to fungus, in the onset of UDN. According to Roberts *et al.* (1970) there has been an 'initial lesion' on the unscaled dorsal or lateral surface of the fish. It has the white colour and initially has not been occupied by pathogenic microorganisms at all, but, eventually has been found to become ulcerated and raw, later observed to be invaded by fungi and bacteria. Willoughby (1972), however, has serious objection on the use of term – 'initial lesion' (by Roberts *et al.*, 1970) for the site of UDN infection because the skin remains as intact and unbroken. It should have been – 'blanching of the skin', related to contraction or the apparent loss of melanophores, Willoughby (1972) indicated.

During December 1990 to January 1995, Prabhuji and Sinha (unpublished data) had the opportunity to make observations on more than four hundred diseased (in an epizootic form) fish specimens of *Channa punctatus, Puntius sophore, Catla catla, Colisa fasciatus,* and *Carassius carassius.* The affected individuals had shown the symptoms of tail rot, bloated abdomen and the development of several degenerative lesions on the body. Although Oomycetes have been found to be associated with the body lesions, as the study indicated, the fungal members have been designated to be the secondary pathogens; bacteria, probably, being the primary infecting agents.

Ulcerative Dermal Necrosis (UDN), the terminology used by several scientists, is supposed to be erroneous for two fundamental reasons – firstly, the inciting pathogen is not one, it is supposed to be more than one, *i.e.*, bacteria-fungi-sometimes virus too, causing extensive damage and in majority of the cases resulting into an epizootic; and secondly, two words of identical meaning (ulcerative and necrosis) should not be used together to make a complex word. So, the term may either be 'Dermal ulcer' or 'Dermal necrosis', but, it should not be 'Ulcerative dermal necrosis'. It should, therefore, more appropriately be known as "Dermal Ulcerative Syndrome" (DUS). In certain instances even the deeper muscular tissues have been observed to be disorganized. Pickering and Willoughby (1977) have recorded four grades of lesions in Perch (*Perca flaviatilis* L.) having a progressive trend, *i.e.*, the Grade-I lesions were slightly blanched areas of skin; Grade-II with distinct blanched areas of skin having epidermis invariably missing; Grade-III lesions had appeared as an open sore on the flank of the fish, with frequent indications of fungal infection at the centre of the lesions. The Grade-IV lesions were completely dominated by the profuse growth of fungal mycelia, which extended deep into the dermis, damaging the surface scales. Several members of Saprolegniaceae have been found to be associated with the lesions together with several bacteria.

Bucke Maff *et al.* (1979), in their studies on an epizootic of perch (*Perca flaviatilis* L.), have reported the occurrence of large ulcerative lesions on the body and fins of the fish, culminating in deep necrotic areas associated with edema and haemorrhage exposing the underlying musculature and skeletal structures. The fungal and bacterial populations, isolated from the lesions, have been the members of

Saprolegniaceae and bacterial species of *Aeromonas, Pseudomonas* and occasionally myxobacteria. However, Bucke Maff *et al.* (1979) failed to reveal any virus particles, even after using electron-microscopy.

Ichthyophonosis

The fungus, *Ichthyophonus* hoferi Plehn and Mulsow, has been found embedded in the infected fish tissues mostly as thick-walled, spherical, multinucleate structure called a 'resting spore'. The cytoplasm of these resting spores gives a positive PAS and Bauer reaction which indicates that it contains glycogen, a common carbohydrate reserve of fungi. Moreover, strong PAS reaction given by the wall of such spore indicates its polysaccharide nature and thus, confirms the notion that *I. hoferi* is a fungus.

Schaperclaus (1953), Reichenbach-Klinke (1956c) and Reichenbach-Klinke and Elkan (1965) have indicated towards the possibility of the presence of two forms of *Ichthyophonus, i.e.,* salmonoid form and aquarium-fish form, on the basis of two different developmental patterns in this genus (Prabhuji and Sinha, 2009).

Earlier as well as the present day studies, based on the histopathology of tumour tissues and of affected vital organs of fish, have provided sufficient data to indicate the occurrence of a wide range of polymorphism in *I. hoferi.* In this context the most extensive studies, however, are the detailed investigations by Sindermann and Scattergood (1954), Dorier and Degrange (1961), Amlacher (1965), Powles *et al.* (1968), Ruggieri *et al.* (1970), Sinha (1985) and Prabhuji *et al.* (1988).

Infection of the lateral musculature causing the 'sand-paper effect' (roughening of the skin) has been observed by Sindermann and Scattergood (1954) in Atlantic herring. It may be due to the formation of large number of papules caused by proliferation of the fungus and the formation of necrotic areas in the sub-epidermal tissues. However, Srivastava *et al.* (1984), Sinha (1985) and Prabhuji and Sinha (2009) have found roughening of skin, raised spots and irregular tumourous galls on the skin of *Carassius carassius.* Reichenbach-Klinke (1956b) has reported blindness and exophthalmos of serranid fish from the Mediterranean as a result of eye infection. He (Reichenbach-Klinke, 1960) has also reported the cranial and dorsal ulceration.

Generally, no tissue or organ has been found to be immune from infection by *I. hoferi* however, organs with a rich blood supply seem to be more frequently affected. *I. hoferi* elicits a severe focal granulomatous response resulting in cirrhosis and atrophy of the affected organs which can eventually lead to replacement of most of the normal tissue by reticulo-endothelial granulation tissue.

According to Amlacher (1965), the first tissue response of the host consists of an increased activity of the leucocytes, particularly the eosinophilic granulocytes. These leucocytes surround the parasite and many of them are destroyed. During this process, fibrocytes appear and eventually enclose (with one to several layers of long cells) the resting spore of the parasite, the leucocytes and the necrotic debris. The result is a characteristic granuloma consisting of the central, thick-walled, multinucleate resting spore surrounded by necrotic cells enclosed within a connective tissue capsule. In other cases the parasite may be surrounded by long, radially arranged cells or by epithelioid cells surrounded by a connective tissue capsule. Giant cells may also be found, particularly in infected kidneys. Empty resting spores, following germination and release of the plasmodium or spores, may often become infiltrated with connective tissues.

There have been persistent suggestions in the literature that natural *I. hoferi* infections of marine fish may be initiated by ingestion of infected crustaceans, particularly the copepods (Jepps, 1937;

Reichenbach-Klinke, 1956b), Reichenbach-Klinke and Elkan, 1965; Sindermann and Scattergood, 1954; Sindermann, 1958; 1970). However, the definite evidences, in this context, are still lacking. Several experimental studies have shown that the infection is initiated after ingestion of food containing viable *I. hoferi* spores. Gustafson and Rucker (1956) found that feeding fish viscera from infected fish to Rainbow trout, three species of Pacific salmon and a cottid, resulted in infection of these fishes, but, they were unable to establish infections in Gold fish (*Carassius carassius*), Guppies, Squawfish or Brown bullheads.

Sindermann (1958) also carried out infection experiments with immature Atlantic herring and has presented some quantitative data regarding the dosage required to initiate infection. He found that a single exposure of fifty fish to 2x10 spores resulted in 'no infection', but, several successive exposures to the same dose on successive days did result in infection. Using this information, Sindermann produced an experimental epizootic in 2000 immature Atlantic herring which resulted in infection of 23 per cent of this group – 8 per cent acute infections and 15 per cent chronic infections. Mortalities due to acute infections occurred within 2-4 weeks by massive invasion of the heart and degeneration and necrosis of body musculature, and a minimum cellular response from the host. The chronic phase was characterized by a marked host cellular response leading to encapsulation of the parasite by fibrous connective tissue. Such a condition has often been found to exhibit pigment deposition around the spores in the muscles.

Besides the uncoordinated swimming movement of the fish, *i.e.*, the swimming or reeling movement, the affected fish may also exhibit either or both the basic symptoms:

1. Epithelioma or gall or tumour formation on the body surface.
2. Nodulation or necrosis of vital organs like brain, heart, liver, kidney etc.

Significant epithelioma or gall or tumour formation has been reported, for the first time, by Srivastava *et al.* (1984), Sinha (1985) followed by Prabhuji and Sinha (2009). These abnormal hyperplastic structures appear as small or large tissue lumps developing on the body surface (dorsal, lateral, peduncle region or at the base of fins). On dissection the various vital organs of the infected fish, *viz.*, heart, liver, brain, kidney, spleen, intestine and stomach have been observed to exhibit nodulation and necrosis (Pettit, 1913; Rucker and Gustafson, 1953; Sindermann and Scattergood, 1954; Dorier and Degrange, 1961; Erickson, 1965; Srivastava *et al.*, 1984; Sinha, 1985; Prabhuji *et al.*, 1988 and Prabhuji and Sinha, 2009).

Histopathology

Together with the host activities of enclosure of multinucleate resting spore within connective tissue capsule, epitheliod cells may or may not be stimulated for hyperplasia. The hyper plastic growth may result into development of epithlioma or gall or tumour on the body surface or on the surface of affected vital organs. Thus, the histopathological studies have been made on two different aspects, *i.e.*, histopathology of the gall or tumour tissues and the histopathology of different vital organs.

Histopathology of Tumour Tissues

There have been no record of tumourous tissues on the body surface of fish except for the Indian workers (Srivastava *et al.*, 1984; Sinha, 1985) who have reported the occurrence of yellow tissue galls of significant size (5-25 mm in diameter) in *Carassius carassius*. Other report is that of Sindermann and

Scattergood (1954) about the formation of large number of papules in sub epidermal tissues of Atlantic herring (sand-paper effect).

The histopathological studies on the abnormal gall tissues have indicated the presence of different developmental stages of the fungus (*I. hoferi*) embedded within the tumourous tissue. It is still a matter of advanced study whether or not the various developing stages of the fungus cause the cellular stimulation in the host tissues, resulting into the abnormal growth.

All the stages in the development were identifiable and the stages observed during the histopathology of gall tissues (Sinha, 1985; and Prabhuji and Sinha, 2009) were as follows:

Plasmodial Bodies

Galls developed in the pelvic region of the fish, revealed the presence of thin-walled, multinucleated plasmodial bodies within the reticulo-endothelial granulated host tissue.

Multinucleated Cysts

The reticulo-endothelial granulated tissue in the galls on pre-dorsal region of fish exhibited the presence thick-walled, multinucleated cysts.

Resting Spores

The resting spores (cysts) were the prominent stage of the pathogen in almost all the sections of the galls present on different regions of the body of fish. Resting spores (measuring 54-170 mm in diameter) were thick-walled and mostly surrounded by histiocytes and embedded in reticulo-endothelial granulated mass of host tissue. Smear preparations from mesenteric tissue showed the presence of numerous resting spores encapsulated within several layers of histocytes and a massive invasion of blood cells (erythrocytes and leucocytes) were also observed within the tissue.

Germinating Resting Spores

The sections of the gall tissues from the pelvic region of fish revealed the presence of a n interesting stage – 'the germinating resting spore'. Three prominent pseudopodium-like germ-tubes were observed developing from the resting spore embedded within reticulo-endothelial granulation.

Conidial Elements

The mid body region of a fish specimen had a large gall indicating a great deal of necrosis and hyperplasia. Sections of this gall tissue showed the presence of cysts with their inner contents divided into numerous conidial elements. The wall of the cyst was observed to be surrounded by histiocytes embedded within reticulo-endothelial granulation having prominent necrotic areas.

The gall tissues from different regions of fish had different developmental stages of *I. hoferi* (Sinha, 1985; Prabhuji and Sinha, 2009). Basically, the fungus is non-tissue specific as far as the musculature and the gall tissues of different regions are concerned. This characteristic feature has been distinctly depicted by the two developmental stages, *viz.*, the resting spores and the multinucleated cysts, which have shown their presence almost everywhere embedded within reticulo-endothelial granulated tissue. However, certain specific multiplication stages like 'germinating resting spores' and 'the conidial elements' require extra nourishment (derived from disorganized surrounding tissues) and therefore, have been found to be associated with necrotic areas of varying degrees. Earlier report of association of necrotic tissues with the germinating resting spores has been given by Plehn and Mulsow (1911) in deeper body tissues of Rainbow trout (*Salmo gairdneri*).

Histopathology of Vital Organs

Almost all the reports pertaining to Ichthyophonosis (disease caused by *I. hoferi*), including the first record, have been given from deeper tissues of fish, *i.e.*, from various vital organs like brain, heart, liver, spleen, kidney, stomach and intestines together with musculature and mesenteric tissues (Table 24.1).

Table 24.1: Records of *Ichthyophonus* Infections of Fishes (Prabhuji and Sinha, 2009)

Sl.No.	*Fish Species*	*References*
1.	*Alosa pseudoharengus* (Alewife)	Fish (1934), Sindermann (1958).
2.	*Aphanopus carbo* (Black scabbard fish)	Agius (1978) (*Ichthyophonus* – like fungus)
3.	*Clupea harengus harengus* (Atlantic herring)	Cox (1916), Daniel (1933a), Fish (1934), Sindermann and Scattergood (1954), Sindermann (1956, 1958).
4.	*Gadus morhua* (Atlantic cod)	Machado-Cruz (1961), McVicar and McKenzie (1972), Hendricks (1972), Moller (1974)
5.	*Limanda ferruginea* (Yellow-tail flounder)	Powles *et al.* (1968), Ruggieri *et al.* (1970), Hendricks (1972).
6.	*Melanogrammus aeglefinus* (Haddock)	Robertson (1909)
7.	*Myoxocephalus octodecemspinosus* (Longhorn sculpin)	Hendricks (1972)
8.	*Platichthys flasus* (Flounder)	Robertson (1908, 1909)
9.	*Pleuronectes platessa* (Plaice)	Johnstone (1906, 1920)
10.	*Pollachius virens* (Pollock)	Priebe (1973)
11.	*Pseudopleuronectes americanus* (Winter flounder)	Ellis (1928), Fish (1934)
12.	*Salmo trutta* (Brown trout; Sea trout)	Robertson (1909), Neresheimer and Clodi (1914)
13.	*Scomber scombrus* (Atlantic mackerel)	Johnstone (1913), Sproston (1944), Sindermann (1958)
14.	*Oncorhynchus kisutch* (Coho salmon)	Gustafson and Rucker (1956)
15.	*Oncorhynchus nerka* (Sockeye salmon)	Gustafson and Rucker (1956)
16.	*Oncorhynchus tshawytscha* (Chinook salmon)	Gustafson and Rucker (1956)
17.	*Salmo gairdneri* (Rainbow trout)	Laveran and Pettit (1910), Plehn and Mulsow (1911), Pettit (1913), Neresheimer and Clodi (1914), Rucker and Gustafson (1953), Gustafson and Rucker (1956), Ross and Perisot (1958), Bellet (1959), Dorier and Degrange (1961), Erickson (1965), Amlacher (1965)
18.	*Salvelinus fontinalis* (Brook trout)	Pettit (1913), Neresheimer and Clodi (1914), Dorier and Degrange (1961)
19.	*Carassius carassius*	Srivastava *et al.* (1984), Sinha (1985), Prabhuji *et al.* (1988)

The infection takes place by means of spores present in the fish feces, in open ulcers and probably on copepods (Sindermann, 1954). The spores liberate amoeboid bodies (the plasmodial stage) which cross the intestinal mucosa and reach the blood stream. Once in the channel, they are distributed throughout the vital organs by means of blood. They particularly attack the heart, the spleen, the liver and the kidney because the blood flows through these organs slowly and in considerable quantities. The muscles are also attacked.

According to Neresheimer and Clodi (1914) invasion of the tissues produce a chronic inflammation and, in case of long-lasting infection – the granuloma. These lesions are typical in the heart. The numerous granulomas seriously damage the infected organs, which react with an infiltration of round cells and by encapsulating the parasite by connective tissue. This result into cirrhosis and necrotic atrophy within the organs attacked.

Distinct stages in the development of *I. hoferi* have been observed during the histopathology of vital organs of different infected fishes by different aquatic biologists (Prabhuji and Sinha, 2009). These stages have been:

Plasmodium

The sections of brain revealed the presence of thin-walled, multinucleate plasmodial bodies embedded within tissues close to cranial bones in Carassius carassius (Prabhuji *et al.*, 1988). Dorier and Degrange (1961) had reported abundance of the plasmodial stage in liver tissues of Rainbow trout (*Salmo gairdneri*). Ruggieri *et al.* (1970) have reported the presence of thick-walled cysts in liver tissues of Yellow-tail flounder (*Limanda ferruginea*) containing multinucleate 'plasmodium'.

Multinucleated Cysts

The peripheral tissues of the brain, outer membranous tissues of the kidney and the circular muscles of stomach and intestine have been reported to exhibit the presence of thick-walled multinucleate cysts in *Carassius carassius* by Prabhuji *et al.* (1988).

Resting Spores

In *I. hoferi*, the most common and prominent stage during the development is the 'resting spore'. Its abundant occurrence has been recorded in heart and liver tissues of Atlantic herring by Sindermann (1958). Dorier and Degrange (1961) have reported the dominance of resting spore stage in the liver of Rainbow trout (*Salmo gairdneri*). Ruggieri *et al.* (1970) have shown extensive damage and occurrence of numerous 'resting spores' (cysts) in the tissues of liver, kidney, heart, spleen and intestine in Yellow-tail flounder (*Limanda ferruginea*). They have also reported the presence of a resting spore encapsulated within several layers of histiocytes in the testis tissue of the same fish. Resting spores have also been recorded by Reichenbach-Klinke (1964) in the intestines of *Cichlasoma severum* fish. The resting spores (cysts) were the prominent stage of the pathogen in the sections of the brain, kidney, heart, stomach and the intestines of *Carassius carassius* (Prabhuji *et al.*, 1988).

Germinating Resting Spores

As indicated earlier (in histopathology of tumour tissues), the germination of resting spores is almost always associated with the tissue necrosis in the nearby areas. Germination of the resting spores in myocardium, causing necrosis of the heart muscle fibres, has been observed in Atlantic herring (Sindermann, 1958) and in Yellow-tail flounders (Ruggieri *et al.*, 1970). Dorier and Degrange (1961) have recorded the occurrence of germinating resting spores and the necrosis of hepatic (liver) tissues in Rainbow trout. In Gold-fish (*Carassius carassius*) the germinating resting spore has been found within the brain tissues close to cranial bones having two pseudopodia-like germ-tubes (Prabhuji *et al.*, 1988). They have also shown the presence of large necrotic areas in the kidney of the same fish.

Conidial Elements

Ruggieri *et al.* (1970) have reported the conidial elements of *I. hoferi* within the tissues of kidney in Yellow-tail flounders, surrounded by histiocytes. Corresponding with this, later, Prabhuji *et al.* (1988) have also recorded the occurrence of the conidial elements (refer to the histopathology of tumour tissues).

Rucker and Gustafson (1953) have reported the *Ichthyophonus* infection of the brain in Rainbow trout (*Salmo gairdneri*) and have indicated that it was very rare. Erickson (1965), however, recorded brain infection (of *I. hoferi*) in Rainbow trout from southern Idaho and found that it was associated with the spinal curvature caused by atrophic musculature, pulling the spine out of its normal attitude. This has been similar to the sigmoid flexure of the spine observed in Atlantic herring (Sindermann and Scattergood, 1954; Sindermann, 1956) which was thought to be the result of infection of the Central Nervous System (CNS). Such symptoms in Rainbow trout have also been found to be associated with an ascorbic acid or tryptophan deficiency (Ashley *et al.*, 1975).

Sproston (1944) has described *I. hoferi* in Atlantic mackerel (*Scomber scombrus*) and has shown a high degree of polymorphism that include structures like large, elongated, thick 'chlamydospores', dome-shaped 'conidia', 'hyphal bodies' which round up and encyst, branched conidiophores containing endoconidia which are liberated as amoeboid bodies into the blood vessels, simple clavate sporangia having endoconidia (probably endospores), hyphal fusion (probably anastmosis), and thick, resistant 'resting spores' produced following the hyphal fusion. Such polymorphic stages have been completely ignored by the later workers and have been considered to be superficial and cursory observation. Johnson and Sparrow (1961) have attempted to reconcile Sproston's observations with the results of Sindermann and Scattergood (1954), however, her (Sproston's) *I. hoferi* remains an enigmatic organism. Similarly, Agius (1978) could not definitely describe the causal organism in diseased Black scabbard fish (*Aphanopus carbo*) and has identified the organism as *Ichthyophonus*-like fungus.

The polymorphism in *I. hoferi*, as has been described, has probably been related to the problems in distinguishing among the infection of *Ichthyophonus* and infections caused by other organisms, which also exhibit a chronic, proliferative, granulomatous response. Meuron and Burgisser (1973) and Wolke (1975) have discussed the problem of differential diagnosis in brief. Based on histopathological studies, culture studies and infection experiments, Amlacher (1965) has made an extensive study on *Ichthyophonus* infections in Rainbow trout and in a variety of fresh-water aquarium fish and concluded that the 'aquarium fish' infections actually represented symptoms of piscine tuberculosis caused by acid-fast bacteria. He has suggested, in this context, that Ziehl-Neelson method of staining for acid-fast bacteria should be done to confirm *Ichthyophonus* (fungal) infection in fresh-water tropical fish.

There are certain relevant questions regarding the physiology of parasitism that need to be answered as far as the 'Ichthyophonosis' is concerned. Firstly, more intensive researches should be done to have a clear picture of the "hypersensitive reaction" (HSR) between the host and *Ichthyophonus* (the pathogen), *i.e.*, the role of various lytic enzymes as well as the toxin-antitoxin interactions; and secondly, the secretion of certain biochemicals responsible for cellular stimulation like Tumour Inducing Principles (TIPs) causing hypertrophy and hyperplasia in host tissues.

Acknowledgements

Authors are thankful to U.P. State Fisheries Department for the information regarding mass mortality of fishes in some of the ponds for the collection of fish specimens affected with DUS. Thanks are also due to Sri P.N. Srivastava, Secretary and Manager; and to Dr. S.N. Lal, Principal, M.G. Post Graduate College, Gorakhpur; for necessary laboratory facilities and encouragements during course of investigations.

References

Agius, C. 1978. Infection by an *Ichthyophonus*-like fungus in the deep sea Scabbard- Fish, *Aphanopus carbo* (Lowe) (Trichiuridae) in the north east Atlantic, J. Fish Dis., 1: 191-193.

Amlacher, E. 1965. Pathologische und histochemische Befunde bei Ichthyosporidiumbefall der Regenbogenforelle (*Salmo gairdneri*) und am "Aquarienfisch Ichthyophonus", Z. Fisch. (N.F.), 13: 85-112.

Arderon, W. 1748. The substance of a letter from Mr. William Arderon F.R.S. to Mr. Henry Baker F.R.S., Phil. Trans, R. Soc., 45(487): 321323.

Areschoug, J. E. 1844. *Achlya prolifera*, vaxandepa lefvande fisk, Ofvers. K.Vetensk Akad. Forh., 1: 124-126 (summarized in Flora: 28: 59-60, 1845).

Ashley, L. M., Halver, J. E. and Smith, R. R. 1975. Ascorbic acid deficiency in rainbow trout and coho salmon and effects on wound healing, In The Pathology of Fishes (Eds. W.E. Ribelin and G. Migaki), University of Wisconsin Press, Madison, p. 769-786.

Bellet, R. 1959. L'ichthyophonaise des truites d'elevage, Coll. Trav. Path. Comp., Paris, (not seen in original; cited by Amlacher, 1965).

Bennet, J. H. 1842. On the parasitic fungi found growing in living animals, Trans. R. Soc. Edinb., 15(1844): 277-294.

Berkeley, M. J. 1864. Egg parasites and their relatives, Intellectual Observer, 5: 147-153.

Bootsma, R. 1973. Infections with *Saprolegnia* in pike culture (*Esox lucius* L.), Aquaculture, 2: 385-394.

Bucke Maff, D., Cawley Maff, G. D., Craig, J. F., Pickering, A. D. and Willoughby, L. G. 1979. Further studies of an epizootic of perch, *Perca flaviatilis* L., of uncertain etiology. J. Fish Dis., 2: 297-311.

Caullery, M. and Mesnil, F. 1905. Sur les haplosporidies parasites de poisons marins, C.r. Seanc. Soc. Biol., 58: 640-643.

Chick, E. W., Balows, A. and Furcolow, M. L. (1975) Opportunistic Fungal Infections, Charles C. Thomas, Springfield, Illinois, pp. 359.

Conant, N. F., Smith, D. T., Baker, R. D. and Callaway, J. L. 1971. Manual of Clinical Mycology, 3rd.Edition, W.B. Saunders Co., Philadelphia, pp. 755.

Cox, P. 1916. Investigations of a disease of the herring (*Clupea herengus*) in the Gulf of St. Laurence, Contrib. Can. Biol., 1914-1915: 81-85.

Daniel, G. 1933a. Studies on *Ichthyophonus hoferi*, a parasitic fungus of the herring (*Clupea herengus*), I. The parasite as it is found in the herring, Amer. J. Hyg., 17: 267-276.

Daniel, G. 1933b. Studies on *Ichthyophonus hoferi*, a parasitic fungus of the herring (*Clupea herengus*), II. The gross and microscopic lesions produced by the Parasite, Amer. J. Hyg., 17: 491-501.

Dorier, A. and Degrange, C. 1961. L'evolution de l'*Ichthyosporidium* (*Ichthyophonus*) *hoferi* (Plehn et Mulsow) chez les salmonides d'elevage (truite arc en ciel et saumon de fontaine), Trav. Lab. Hydrobiol. Piscic. Univ. Grenoble, 1960/1961: 7-44.

Dukes, T. W. 1975. Ophthalmic pathology of fishes, In The Pathology of Fishes (Eds. W.E. Ribelin and G. Migaki), University of Wisconsin Press, Madison, p. 383-398.

Ellis, M. F. 1928. *Ichthyophonus hoferi* Plehn and Mulsow, a flounder parasite new to North American waters, Proc. Trans. N.S. Inst. Sci., 17: 185-192.

Erickson, J. D. 1965. Report on the problem of *Ichthyosporidium* in a rainbow trout, Progve Fish Cult. 27: 179-184.

Fish, F. F. 1934. A fungus disease in fishes of the Gulf of Maine, Parasitology, 26: 1-16.

Forster, R. P. 1941. The present status of the systemic fungus disease in herring of the Gulf of Maine, Bull. Mt. Desert Isl. Biol. Lab., 1941: 33-35.

Gardner, M. L. G. 1974. Impaired osmoregulation in infected salmon, *Salmo salar*, L. J. mar. biol. Assn., U.K., 54: 635-639.

Goodsir, J. 1842. On the *Conferva* which vegetates on the skin of the Gold fish, Ann. Mag. Nat. Hist., 9: 333-337.

Gustafson, P. V. and Rucker, R. R. 1956. Studies on an *Ichthyosporidium* infection in fish: transmission and host specificity, Spec. Sciet. Rep. U.S. Fish Wildl. Serv., 166:1-8.

Hargens, A. R. and Perez, M. 1975. Edema in spawning salmon, J. Fish Res. Bd. Can., 32: 2538-2541.

Hatai, K. and Egusa, S. 1977. Studies on visceral mycosis of salmonoid fry- Characteristics of fungi isolated from the abdominal cavity of amago salmon fry, Fish Pathol., 11: 187-193.

Hendricks, J. D. 1972. Two new host species for the parasitic fungus *Ichthyophonus Hoferi* in the northwest Atlantic, J. Fish Res. Bd. Can., 29: 1776-1777.

Herkner, H. 1961. Beitrag zur Frag der Art- und Rassenunterschiede beider Fischpathogenen Pilzgattung *Ichthyosporidium* Caullery et Mesnil, 1905, Dissertation, Universitat Munchen, (not seen, cited by Reichenbach-Klike, 973).

Hofer, B. 1893. Eine Salmoniden-Erkrankung. Allg. Fisch Ztg. 18: 168-171.

Huxley, T. H. 1882a. On *Saprolegnia* in relation to the salmon disease, Q.J. microsc. Soc., 22 (N.S.): 311-333.

Huxley, T. H. 1882b. A contribution to the pathology of the epidemic known as the Salmon disease, Proc. R. Soc. 33: 381-389 (also appeared in Nature, London, 25: 437-440.

Jepps, M. W. 1937. On the protozoan parasites of *Calanus finmarchicus* in the Clyde Sea area, Q. J. microsc. Sci. (N.S.), 79: 589-658.

Johnson, T. W., Jr. and Sparrow, F. K., Jr. 1961. Fungi in oceans and estuaries, J. Cramer, Weinheim, 668 pp.

Johnstone, J. 1906. Internal parasites and diseased conditions of fishes, Proc. Trans. Lpool. Biol. Soc., 20: 259-329.

Johnstone, J. 1920. On certain parasites, diseased and abnormal conditions of fishes, Lanc. Sea Fish Lab., Report for 1919, No. 28: 24-33.

Lam, T. J. 1972. Prolactin and hydromineral regulation in fishes, Gen. comp. Endocr. Suppl. 3: 328-338.

Laveran, A. and Pettit, A. 1910. Sur un epizootie des truites, C.r. hebd. Seanc. Acad. Sci., Paris, 151: 421-423.

Leger, L. 1924. Sur un organisme du type Ichthyophone parasite du tube digestif de la Lote d'eau douce, C.r. hebd. Seanc. Acad. Sci., Paris, 179: 785-787.

Leger, L. 1927. Sur la nature et l'evolution des "spherules" decrites chez les Ichthyophones, Phycomycetes parasites de la Truite, C.r. hebd. Seanc. Acad. Sci., Paris, 184: 1268-1271.

Leger, L. and Hesse, E. 1923. Sur un champignon du type *Ichthyophonus* parasite de l'intestin de la Truite, C.r. hebd. Seanc. Acad. Sci., Paris, 167:420-422.

Machado-Cruz, J. A. 1961. Nouveau hote d'*Ichthyosporidium* (*Gadus morhua* L.) Bolm Soc. Port. Cienc. nat., 2 ser., 8: 212-215.

McLeay, D. J. 1975. Variations in the pituitary-interrenal axis and the abundance of circulating blood-cell types in juvenile coho salmon, *Oncorhynchus kisutch*, during stream residence, Can. J. Zool. 53: 1882-1891.

McVicar, A. H. and McKenzie, K. 1972. A fungus disease of fish, Scot. Fish. Bull.,37: 27-28.

Meier, A. H. 1972. Temporal synergism of prolactin and adrenal steroids, Gen. comp. Endocr. Suppl. 3: 499-508.

Meuron, P. A. de and Burgisser, H. 1973. A propos du diagnostic des maladies chez les poisons, Schweizer Arch. Tierheilk, 115: 184-189.

Moller, H. 1974. *Ichthyosporidium hoferi* (Plehn and Mulsow) (Fungi) as parasite in the Baltic cod (*Gadus morhua* L.), Kieler Meeresforsch., 30: 37-41.

Neish, G. A. 1976. Observations on the pathology of saprolegniasis of pacific salmon and on the identity of the fungi associated with this disease, Ph.D. thesis, University of British Columbia, Vancouver, pp. 213, (original not seen; cited in Diseases of Fishes: Book 6: Fungal Diseases of Fishes, by G.A. Neish and G.C. Hughes; T.F.H. Publications Inc. Ltd., Neptune; 1980; pp. 159).

Neish, G. A. 1977. Observations on saprolegniasis of adult sockeye salmon, *Oncorhynchus nerka* (Walbaum), J. Fish Biol., 10: 513-522.

Neresheimer, E. and Clodi, C. 1914. *Ichthyophonus hoferi* Plehn u. Mulsow, der Erreger der Taumelkrankheit der Salmoniden, Arch. Protistenk., 34: 217-248.

Nolard-Tintigner, N. 1971. Cause de la mort dans la saprolegniose experimentale du poisson, Bull. Acad. r. Belg. Cl. Sci., 57: 185-191.

Nolard-Tintigner, N. 1973. Etude experimentale sur l'epidemiologie et la pathogenie de la saprolegniose chez *Lebistes reticulatus* Peters et *Xiphophorus helleri* Heckel, Acta zool. path. Antverp., 57: 1-127.

Nolard-Tintigner, N. 1974. Contribution a l'etude de la Saprolegniose des poisons en region tropicale, Acad. r. Sci. outré-mer, Cl. Sci.nat. med. (NS), 19: 1-58.

Olivereau, M. 1962. Modifications de l'interrenal du smolt (*Salmo salar* L.) au cours du passage d'eau douce en eaude mer, Gen. comp. Endocr. 2: 565-573.

Papatheodorou, V. 1980. Les mycoses chez les poisons: Eltiologie-Pathogenie- Traitement, These le grade de Doctor, L'Inst. Nat. Polytech. De Toulouse pp. 188.

Peduzzi, R., Nolard-Tintigner, N. and Bizzozero, S. 1976. Recherches sur la Saprolegniose, II. Etude du processus de penetration, mise en evidence d'une enzyme proteolytique et aspect histopathologique, Riv. Ital. Piscic. Ittiop., 11: 109-117.

Peduzzi, R. and Bizzozero, S. 1977. Immunological investigation of four *Saprolegnia* species with parasitic activity in fish: serological and kinetic characterization of a chymotrypsin-like activity, Microb. Ecol. 3: 107-118.

Pettit, A. 1911. A propos du microorganisme producteur de la Traumelkrankheit: *Ichthyosporidium* ou *Ichthyophonus*. C.r. Seanc.Soc. Biol., 70: 1045-1047.

Pettit, A. 1913. Observations sur l'*Ichthyosporidium* et sur la maladie qu'il provoque Chez la truite, Annls. Inst. Pasteur, Paris, 27: 986-1008.

Pickering, A. D. and Willoughby, L. G. 1977. Epidermal lesions and fungal infection on the perch, *Perca flaviatilis* L., in Windermere, J. Fish Biol., 11: 349-354.

Plehn, M. and Mulsow, K. 1911. Der Erreger der "Taumelkrankheit" der Salmoniden, Zentbl. Bakt. ParasitKde., Abt. 1, 58: 63-68.

Powles, P. M., Garnett, D. G., Ruggieri, G. D. and Nigrelli, R. F. 1968. *Ichthyophonus* infection in yellow-tail flounder (*Limanda ferruginea*) off Nova Scotia, J. Fish. Res. Bd. Can., 25: 597-598.

Prabhuji, S. K., Srivastava, G. C. and Sinha, S. K. 1988. Observations on the Histopathology of the vital organs of *Carassius carassius* L. parasitized by *Ichthyophonus hoferi* Plehn and Mulsow, Proc. National Seminar on Perspectives in Aquatic Biology, Nainital, (Editor: R.D. Khulbe) pp. 361-368.

Prabhuji, S. K., Sinha, S. K., Srivastava G. C. and Singh, S. B. 1994. Fungal parasites of the eggs of some fresh-water fishes, Proc. National Seminar on Biodiversity and Plant Disease Management, Gorakhpur (Indian Phytopathological Society), pp. 34.

Prabhuji, S. K., Sinha, S. K. and Singh, S. B. 1998. Studies in Aquatic Fungi causing Dermatomycoses and Deep mycoses in fish: *Colisa lalius* Ham. Proc. Xth. Annual Conference of Purvanchal Academy of Sciences, Azamgarh.

Prabhuji, S. K. and Sinha, S. K. 2009. Life Cycle (reproductive) stages of *Ichthyophonus hoferi* Plehn and Mulsow, a parasitic fungus causing deep mycoses, Internat. J. Pl. Repro. Biol., 1 (2): 93–101.

Priebe, K. 1973. Nekrosebenzirk in der Korpermuskulatur eines Kohlers (*Pollachius virens*) mit Befall von *Ichthyosporidium hoferi*, Dt. tierarztl. Wschr., 80: 197-220.

Richards, R. H. and Pickering, A. D. 1978. Frequency and distribution patterns of *Saprolegnia* infection in wild and hatchery-reared brown trout, *Salmo trutta* L. and char, *Salvelinus alpinus* L., J. Fish Dis. 1: 69-82.

Reichenbach-Klinke, H. H. 1954. Untersuchungen uber die bei Fischen durch Parasiten hervorgerufenen Zysten und deren Wirkung auf den Wirtskorper, I. Z. Fisch. (N.S.), 3: 565-636.

Reichenbach-Klinke, H. H. 1955. *in* Pilze in Tumoren bei Fischen. Zool. Anz. Suppl., Verh. dt. Zool.Ges.Tubingen, 1954, Leipzig, pp. 351-357.

Reichenbach-Klinke, H. H. 1956b. Augenschaden bei Meeresfischen durch den Pilz *Ichthyosporidium hoferi* (Plehn et Mulsow) und Bemerkungen zu seiner Verbreitung bei Mittelmeerfischen, Pubbl. Stn. Zool., Napoli, 29: 22-32.

Reichenbach-Klinke, H. H. 1956c. Verbreitung und Bekampfung des Pilzes Ichthyosporidium hoferi (*Plehn et Mulsow*) (*= Ichthyophonus hoferi*)*, Aquar.-u. Terrar.-Z., 9: 70-72.*

Reichenbach-Klinke, H. H. 1960. Die Discus-Krankheit und ihre Ursachen, Aquar.-u. Terrar.-Z., 13: 303-305.

Reichenbach-Klinke, H. H. 1964. Uber den Zusammenhang und die okologische Abhangigkeit zwischen Kiemen- und Lebererkrankungen bei SiiBwasserfischen, Z. Fisch. N.F., 13: 747-767.

Reichenbach-Klinke, H. H. 1973. Reichenbach-Klike's Fish Pathology (in Collaboration by M. Landolt), T.F.H. Publications, Neptune City, New Jersey, pp. 512.

Reichenbach-Klinke, H. H. and Elkan, E. 1965. The principal diseases of lower Vertebrates, Academic Press, New York, pp. 600.

Roberts, R. J., Shearer, W. M., Munro, A. L. S. and Elson, K. G. R. 1970. Studies on Ulcerative Dermal Necrosis of salmonids II. The sequential pathology of the lesions, J. Fish Biol. 2: 373-378.

Robertson, M. 1908. Notes upon a Haplosporidian belonging to the genus: *Ichthyosporidium*. Proc. R. Phys. Soc. Edinb. (1906-1909), 17: 175-187.

Robertson, M. 1909. Notes on an Ichthyosporidian causing a fatal disease in sea Trout, Proc. Zool. Soc., London, 1909: 399-402.

Robin, C. 1853. Histoire naturelle des vegetaux parasites qui croissent sur Phomme Et sur les animaux vivants, J. B. Bailliere, Paris, pp. 702.

Roddie, J. C. and Wallace, W. F. M. 1975. The Physiology of Disease, Lloyd-Luke (Medical Books) Ltd., London, pp. 588.

Ross, A. J. and Perisot, T. J. 1958. Record of the fungus *Ichthyosoporidium* Caullery and Mesnil, 1905, in Idaho, J. Parasit., 44: 453-454.

Rucker, R. R. 1944. A study of *Saprolegnia* infections among fish, Ph.D. thesis, University of Washington, Seattle, pp. 92.

Rucker, R. R. and Gustafson, P. V. 1953. An epizootic among rainbow trout, Progve. Fish Cult. 15: 179-181.

Ruggieri, G. D., Nigrelli, R. F., Powles, P. M. and Garnett, D. G. 1970. Epizootics in Yellow-tail flounder, *Limanda ferruginea* Storer, in the western North Atlantic caused by *Ichthyophonus*, an ubiquitous parasitic fungus, Zoologica, N.Y., 55: 57-62.

Schaperclaus, W. 1953. Fortpflanzung und systematik von *Ichthyophonus*, Aquar.-u Terrar.-Z., 6: 177-182.

Schaperclaus, W. 1954. Fischkrakheiten, Akademie-Verlag, Berlin, pp. 708.

Selye, H. 1950. Stress and the general adaptation syndrome, Br. Med. J., 1: 1383-1392.

Sindermann, C. J. 1956. Diseases of fishes of the western North Atlantic IV. Fungus Disease and resultant mortalities of herring in the Gulf of Saint Lawrence in 1955, Res. Bull. Dept. Sea Shore Fish. Me, 25: 1-23.

Sindermann, C. J. 1958. An epizootic in Gulf of Saint Lawrence fishes, Trans. N. Amer. Wildl. Conf., 23: 349-360.

Sindermann, C. J. and Scattergood, L. W. 1954. Diseases of fishes of the western North Atlantic II. *Ichthysporidium* disease of the sea herring (*Clupea harengus*) Res. Bull. Dep. Sea Shore Fish Me, 19: 1-40.

Sinha, S. K. 1985. Studies in fungi causing fish diseases, Ph. D. thesis, University of Gorakhpur, Gorakhpur, India. pp. 136.

Spallanzani, 1777. Opuscules de physique, Geneve.

Sprague, V. 1965. *Ichthyosporidium* Caullery and Mesnil, 1905, the name of a genus of fungi or a genus of sporozoans?, Syst. Zool., 14: 110-114.

Sprague, V. 1966. *Ichthyosporidium* sp. Schwartz, 1963, parasite of the fish *Leiostomus xanthurus*, is a microsporidian, J. Protozool., 13: 356-358.

Sproston, N. G. 1944. *Ichthyosporidium hoferi* (Plehn and Mulsow, 1911), an internal fungoid parasite of the mackerel, J. mar. Biol. Assn.,U.K., 26: 72-98.

Srivastava, A. K. 1979. Fungal infection of hatchlings of *Labeo rohita*, Mykosen 22(2): 40.

Srivastava, R. C. 1980a. Fungal parasites of certain fresh-water fishes of India, Aquaculture, 21: 387-392.

Srivastava, R.C. 1980b. Studies in Fish mycopathology – a Review, Part I, Mykosen, 23(6): 325-332.

Srivastava, R. C. 1980c. Studies in Fish mycopathology – a Review, Part II, Mykosen, 23(7): 380-391.

Srivastava, G. C., Sinha, S. K. and Prabhuji, S. K. 1984. Water-moulds parasitizing *Carassius carassius* L. causing abnormal growth of fish tissues, J. Ind. bot. Soc., 63 (supple.): 25.

Srivastava, G. C., Sinha, S. K. and Prabhuji, S. K. 1994. Observations on fungal infection of *Chela laubuca* Ham. With special reference to deep mycoses, Current Science, 66(3): 237-239.

Sterling, A. B. 1879. Additional observations on fungus disease of salmon and other fish, Proc. R. Soc. Edinb., 10: 371-378.

Unger, F. 1844. Sur l'*Achlya prolifera*, Annl. Sci. Nat., 3e ser., Bot., 2: 5-20.

Utida, S., Hiranc, T., Oide, H., Ando, M., Johnson, D. W. and Bern, H. A. 1972. Hormonal control of the intestine and urinary bladder in teleost Osmoregulation, Gen. comp. Endocr. Suppl. 3:317-327.

Wedemeyer, G. 1969. Stress induced ascorbic acid depletion and cortisol production in two salmonoid fishes, Comp. Biochem. Physiol. 29: 1247-1251.

Wedemeyer, G. 1970. The role of stress in disease resistance of fishes, Am. Fish. Soc. Spec. Publ. 5: 30-35.

Wedemeyer, G., Meyer, F. P. and Smith, L. 1976. Environmental stress and fish diseases, In Diseases of Fishes, Book 5 (Eds. S.F. Snieszko and H.R. Axelrod), T.F.H. Publications, Neptune, pp. 192.

Willoughby, L. G. 1968. Atlantic Salmon disease fungus, Nature, London, 217: 872- 873.

Willoughby, L. G. 1969. Salmon disease in Windermere and the River Leven; the fungal Aspect, Salmon Trout Mag. No. 186:124-130.

Willoughby, L. G. 1971. Observations on fungal parasites of Lake District salmonids, Salmon Trout Mag. No. 192: 152-158.

Willoughby, L. G. 1972. U.D.N. of Lake District trout and char: outward signs of infection and defense barriers examined further, Salmon Trout Mag. No. 195: 149-158.

Willoughby, L. G. 1977. An abbreviated life cycle of *Saprolegnia* in the salmonid fish, Trans. Br. mycol. Soc., 69: 133-135.

Willoughby, L. G. 1978. Saprolegniasis of salmonid fish in Windermere: a Critical Analysis, J. Fish Dis., 1: 51-67.

Willoughby, L. G. and Pickering, A. D. 1977. Viable Saprolegniaceae spores on the epidermis of the salmonid fish, *Salmo trutta* and *Salvelinus alpinus*, Trans. Br. mycol. Soc., 68: 91-95.

Wilson, J. G. M. 1976. Immunological aspects of fungal disease in fish, In Recent Advances in Aquatic Mycology (Ed. E.B. Gareth Jones), Paul Elek (Scientific Books) Ltd., London, p. 573-601.

Wolke, R. E. 1975. Pathology of bacterial and fungal diseases affecting fish, In The Pathology of Fishes (Eds. W.E. Ribelin and G. Migaki), University of Wisconsin Press, Madison, p. 33-116.

Woodhead, A. D. 1975. Endocrine physiology of fish migration, Oceanogr. Mar. Biol. A. Rev. 13: 287-382.

Microbes: Diversity and Biotechnology (2012) *Pages* **417–427**
Editors: **Prof. S.C. Sati & Dr. M. Belwal**
Published by: **DAYA PUBLISHING HOUSE, NEW DELHI**

Chapter 25

Effect of Temperature, pH and Light on the Growth of Some Aquatic Hyphomycetes

S.C. Sati, Saraswati Bisht and Pratibha Arya
Department of Botany, D.S.B. Campus,
Kumaun University, Nainital – 263 001

ABSTRACT

Six fungal isolates of aquatic hyphomycetes *viz, Tetracheatum elegans, Tetracladium marchalianum, Pestalotiopsis submersus, Flagellospora penicillioides, Beltrania rhombica* and *Heliscus lugdunensis* were examined for the effect of temperature, pH and light on their growth. 25-30°C temperature was found optimum for *P. submersus* and *F. penicillioides.* 20°C for *B. rhombica* and *T. marchalianum.* While for *T. elegans* and *H. lugdunensis* 20-25 °C was found optimum. pH requirement of these species varied species to species. *T. elegans* and *F. penicillioides* had maximum growth in pH 6.5 – 7.0 and *T. marchaliaum* and *P. submersus* in 8.0 –8.5, while *H. lugdunensis* and *B. rhombica* were found to grow in a wide range of pH *i.e.,* 6.5-8.5. White and red light were found the most favorable light sources for these fungi. Where as blue light and darkness were found inhibitory to them.

Keywords: *Aquatic hyphomycetes, Physical factors, Growth.*

Introduction

The freshwater ecosystem has its own mycoflora in which aquatic hyphomycetes occupies an important place. They occur abundantly on submerged substrate in well-aerated waters. The occurrence and distribution of the fungi depend on physical factors like substrate quality, moisture, temperature, light, pH, altitude etc. However, most of the fungi are known to be affected by light, temperature and pH. Though the aquatic hyphomycetes are ubiquitous in occurrence and distributed from arctic to equator (Bhat, 1990, Marvanova, 1997) but majority of them are reported from temperate climate in the running fresh water bodies.

The distribution and variation in species composition is apparently related to temperature (Nilsson, 1964, Ingold, 1975). pH is an another important physical factor which influences the occurrence and degradative ability of aquatic hyphomycetes as well as colonizing capacity on submerged leaf litter (Thompson and Barlocher, 1989).

The evidence from different fungal studies confirmed that different fungal groups might have different pH preferences and it also affects the decompositional activities of aquatic hyphomycetes in running fresh water bodies (Rosset and Barlocher, 1985, Suberkropp and Klug, 1980, Chamier, 1985, Barlocher, 1987, Marvanova 1984).

In general the light does not seem to be an important factor for most of the aquatic fungi but in some instances sporulation are found influenced by the light condition. Though a lot of work has been done on the effect of light on water molds and marine fungi (Alderman and Jones, 1971, Prabhuji, 1979, Mer, 1982) but there is none of the report for aquatic hyphomycetes.

A few reports are available on the studies in the physiology of aquatic hyphomycetes (Thornton, 1963, Suberkropp, 1984, Barlocher 1989, Rosset and Barlocher, 1985) but no information is available on light requirements. Therefore, the present investigation was undertaken as the preliminary but pioneering work in physiology of some aquatic hyphomycetes to determine the effect of temperature, pH and light on their growth. The fungal strains were recovered from a stream in the temperate zone of Nainital, Kumaun Himalaya.

Materials and Methods

Six fungal isolates *viz., Tetracheatum elegans* Ingold, *Tetracladium marchalianum* De Wildeman, *Flagellospora penicillioides* Ingold, *Pestalotiopsis submersus* Sati and Tiwari, *Heliscus lugdunensis* Ingold and *Beltrania rhombica* Penzing were selected for the present investigation. Monohyphal cultures of these isolates were originally obtained by single spore isolation. These strains were previously deposited in the culture collection of the Kumaun University mycological specimen, Department of Botany, Kumaun University, Nainital, India. Pure cultures were maintained on 2 per cent Malt Extract Agar (MEA) slants. The effect of physical factor was examined in a basal medium as recommended by Ranzoni, 1951(4.0 g of Glucose, 1.0 g of KH_2 PO_4, 0.2 g $MgSO_4.7H_2O$, 0.02 g $FeCl_3$, 1.0 g Difco. Yeast extract and 1 litre distilled water). 25 ml of this medium was pipetted into conical flasks (100 ml Cap) and were autoclaved at 15 lbs pressure for 15 minutes, successively for 3 days to reduce the chances of contamination. Agar disks (5 mm diameter) with mycelia mat were cut from 15 days old culture plates and transferred into sterilized conical flasks. The flasks were kept for incubation under different experimental conditions (Plates 25.1A–B).

1. Different temperature treatments ranged 5 to 30 ± 2°C in 5pC intervals, *i.e.*, maintaining 5, 10, 15, 20, 25 and 30°C with an adjusted pH 6.5 of basal media.
2. The range of pH between 6.0 to 8.5 in 0.5 intervals *i.e.*, 6.0, 6.5, 7.0, 7.5, 8.0, 8.5 was adjusted by adding sterile 4 per cent NaOH and 4 per cent HCl in the basal medium.
3. For light, a 60 watt argenta K lamp (emitted 3K-lux intensity of light) manufactured by Pico. Electronics and Electrical limited, Calcutta placed at 50 cm distance was used as the source of light. Red, blue, green and yellow cellophane papers were used as the source of red, blue, green and yellow light. The test flasks were exposed to these light separately for 15 days at 20 ± 2°C with pH 6.5 of basal medium and casually shaken for aeration.

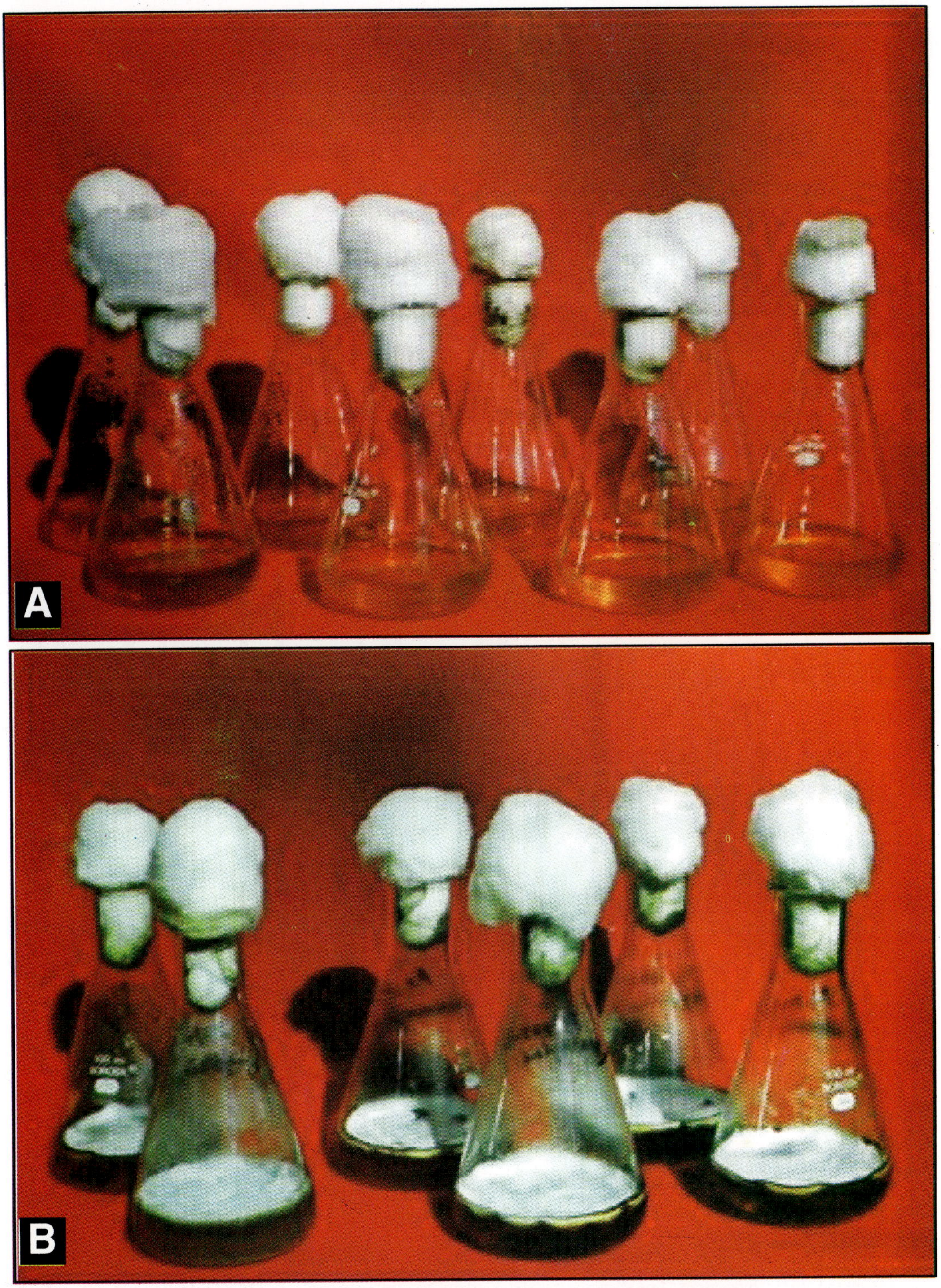

Plate 25.1 A–B: Experiments for Physico-chemical Studies

After 15 days, dry weight biomass was obtained by filtering off the medium through previously weighed filter papers (Whatman No. 1), which were thoroughly rinsed with distilled water to remove residues of medium from the mycelia. Fungal biomass was then dried for 24 h at 80ºC in an oven and cooled in a desiccator prior to reweighing.

Results

Dry weight biomass of each fungal isolated under different experimental conditions are presented in Tables 25.1–25.3.

It is evident from Table 25.3, *P. submersus* and *F. penicillioides* showed their maximum growth in 25-30ºC whereas *B. rhombica* and *T. marchalianum* had their best growth in 20ºC and *T. elegans* and *H. lugdunensis* were recorded their maximum growth in 20-25ºC.A positive correlation was found between the temperature and dry weight of aquatic hyphomycetes (Figure 25.4).

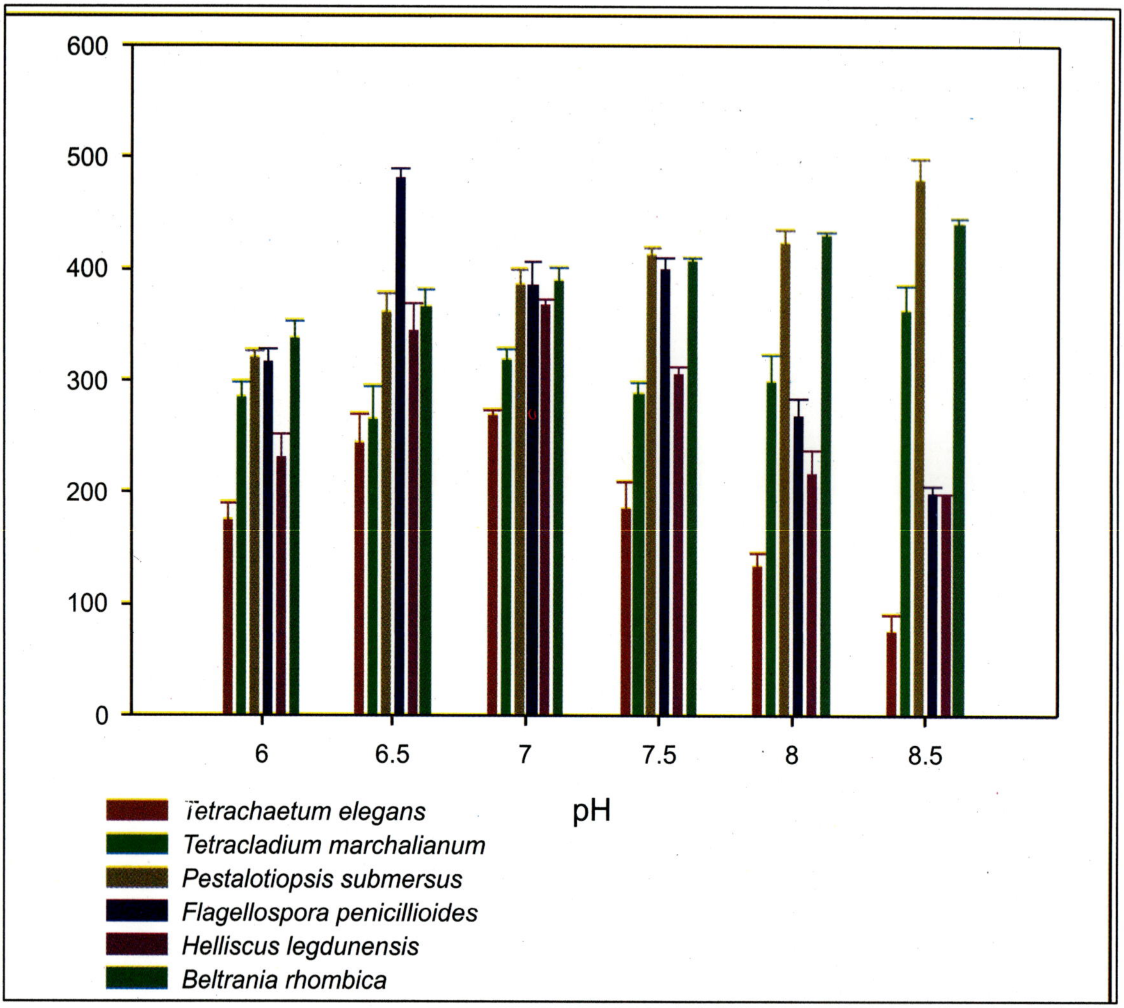

Figure 25.1: Average Dry Weight (±SE) Yields of Six Fungal Isolates in Different pH

The pH requirement of these fungi varied specie to species (Table 25.1). *T. elegans* and *F. penicillioides* were found their best growth in slightly acidic pH 6.5-7.0, while *T. marchalianum* and *P. submersus* were found to grow best in alkaline pH 8.0-8.5 *H. lugdunensis* and *B. rhombica* were found to grow in a wide range of pH *i.e.*, 6.5-8.5 pH (Figure 25.1).

Regression analysis of pH level with average dry weight yields indicates that average dry weights of *T. elegans, F. penicillioides* and *H. lugdunensis* were negatively correlated with pH, where as, *P. submurses, T. marchalianum* and *B. rhombica* were positively correlated with pH levels (Figure 25.5).

In the light requirement experiment the white and red light were found the most favorable light sources. It was interesting to note that the dark and blue lights were found inhibitory for their growth. Green and yellow lights also support good growth of these fungi (Figure 25.2).

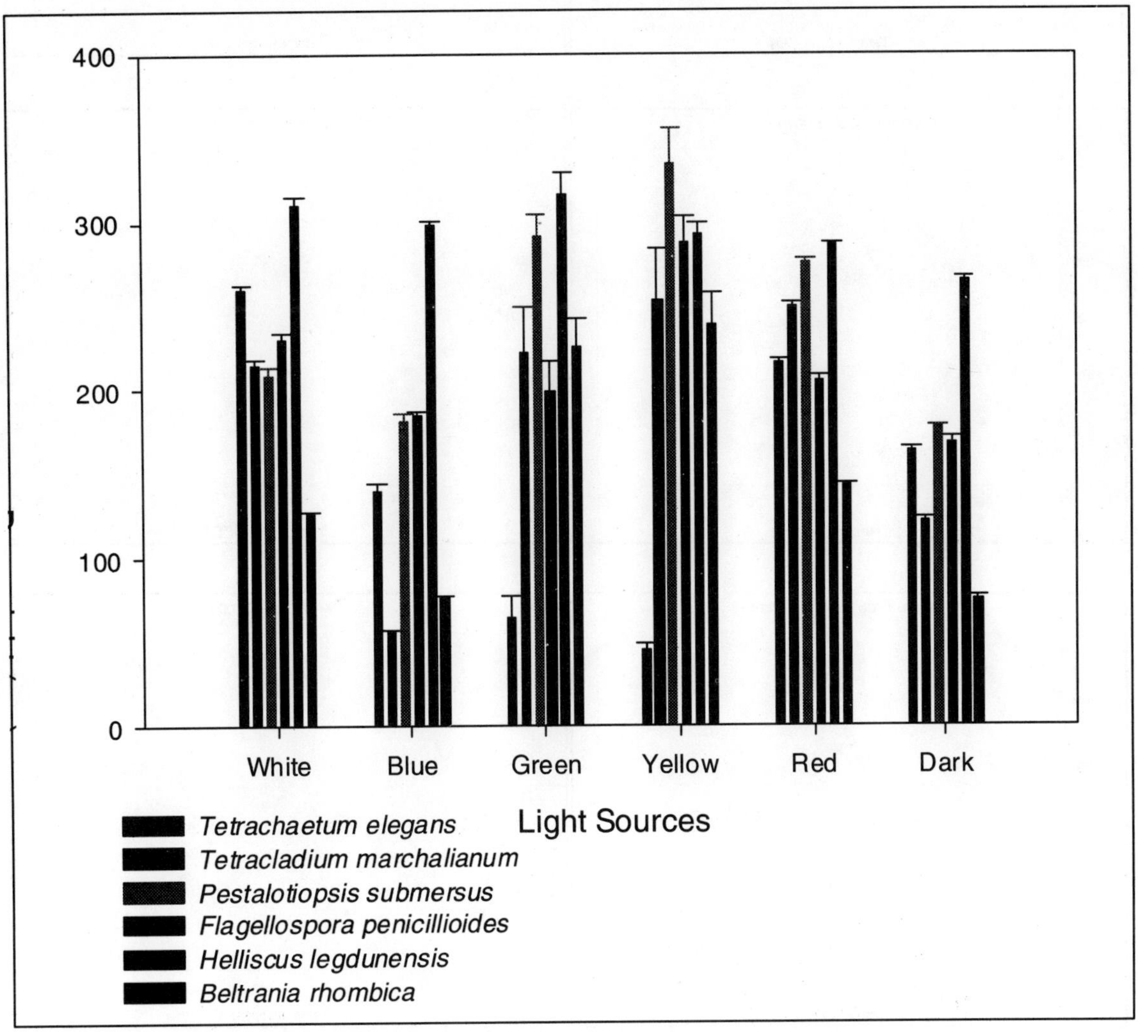

Figure 25.2: Average Dry Weight (±SE) Yields of Six Fungal Isolates in Different Light Sources

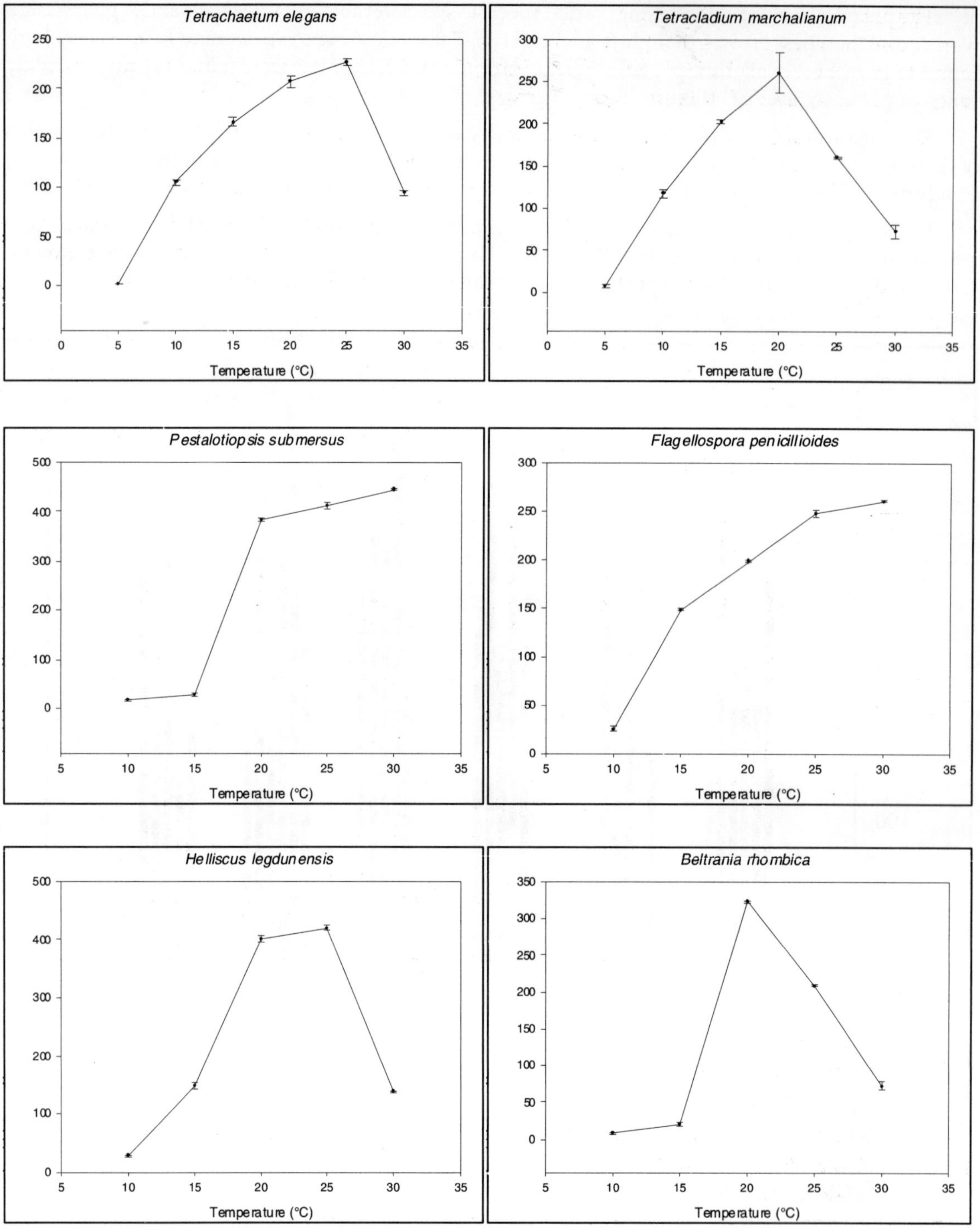

Figure 25.3: Average Dry Weight (±SE) Yields of Six Fungal Isolates at Different Temperatures

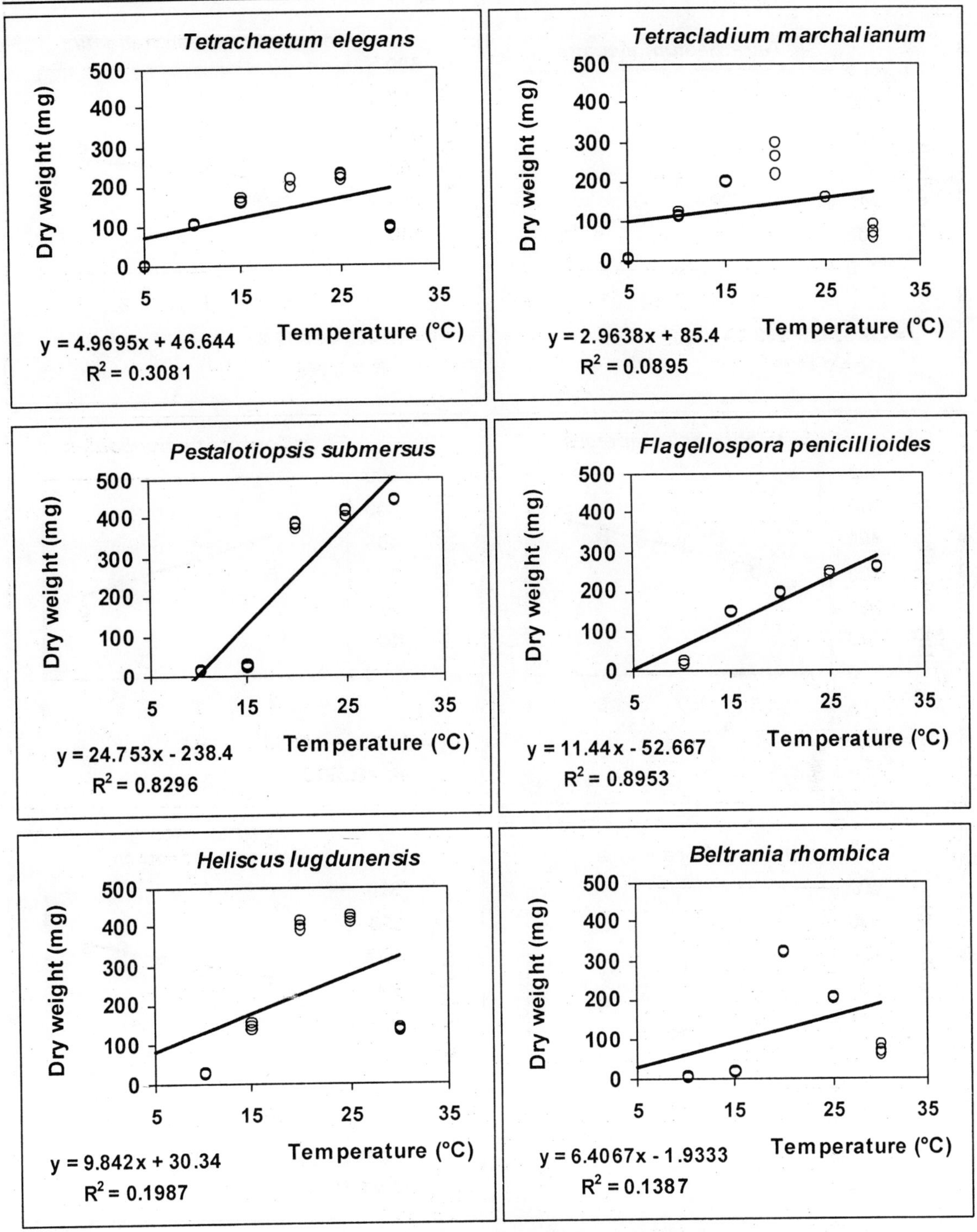

Figure 25.4: Regression Plot of Different Temperatures (°C) with Average Dry Weight (mg) Yields of Six Fungal Species

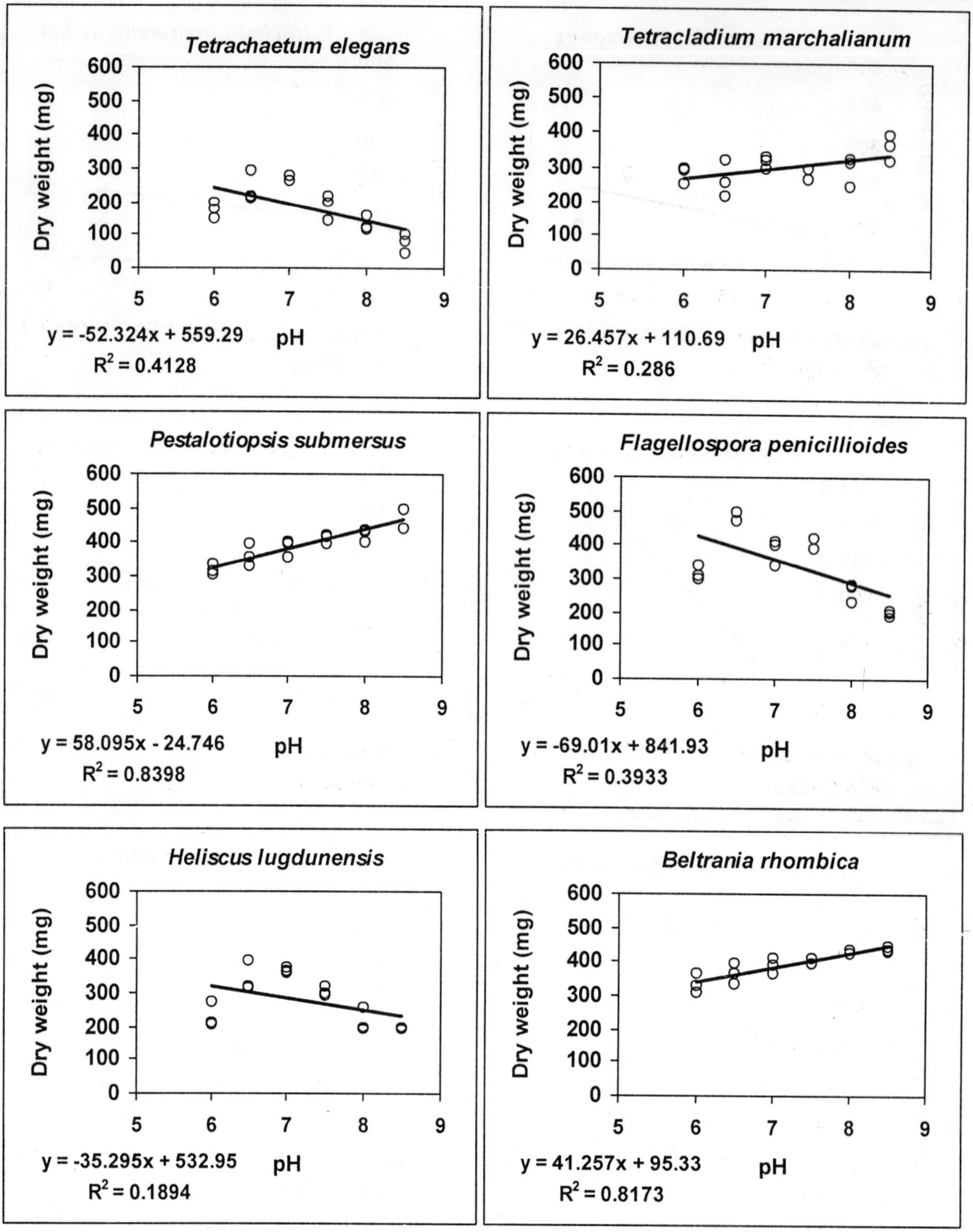

Figure 25.5: Regression Plot of Different pH Levels with Average Dry Weight (mg) Yields of Six Fungal Species

Table 25.1: Average Dry Weight (mg±SE) Yields of Six Fungal Isolates in Different pH

Fungal Species	Average Dry Weight Yield (mg±SE)					
	pH 6	pH 6.5	pH 7	pH 7.5	pH 8	pH 8.5
Tetrachaetum elegans	174.67±15.03	243.67±25.69	268.33±4.91	185.33±23.39	133.33±12.45	74.33±16.33
Tetracladium marchalianum	283.33±14.24	265.00±29.94	318.33±9.91	288.00±10.54	298.33±25.37	362.00±22.50
Pestalotiopsis submersus	319.33±7.88	360.00±18.93	385.33±13.68	411.33±7.22	423.00±11.50	479.67±19.34
Flagellospora penicillioides	316.67±12.44	480.67±8.67	384.67±21.61	400.00±10.00	268.00±16.04	199.67±6.06
Heliscus lugdunensis	231.67±20.67	343.67±25.69	368.33±4.91	304.67±7.06	217.33±20.41	196.67±1.76
Beltrania rhombica	336.67±16.56	366.00±16.46	389.00±12.12	406.00±5.03	429.33±4.33	439.67±5.78

Table 25.2: Average Dry Weight (mg±SE) Yields of Six Fungal Isolates in Different Light Sources

Fungal Species	Average Dry Weight Yield (mg±SE)					
	Blue	Dark	White	Red	Yellow	Green
Tetrachaetum elegans	139.33±5.21	164.00±2.31	260.00±2.89	216.00±2.31	45.00±4.04	64.33±12.99
Tetracladium marchalianum	56.00±1.15	122.33±1.45	215.23±2.77	249.27±2.40	253.00±31.05	222.00±27.51
Pestalotiopsis submersus	181.00±5.20	178.00±1.15	208.67±4.48	276.17±2.32	334.33±21.40	290.67±13.17
Flagellospora penicillioides	185.00±2.89	168.67±3.76	230.67±3.18	205.57±3.04	287.67±15.01	199.33±17.90
Heliscus lugdunensis	298.00±3.21	264.67±2.33	310.17±5.34	286.00±1.73	292.00±7.57	316.00±13.32
Beltrania rhombica	76.00±1.15	75.00±2.65	126.00±2.31	142.67±1.45	239.00±18.68	225.33±17.29

Table 25.3: Average Dry Weight (mg±SE) Yields of Six Fungal Isolates at Different Temperatures

Fungal Species	Average Dry Weight Yield (mg±SE)					
	5°C	10°C	15°C	20°C	25°C	30°C
Tetrachaetum elegans	2.00±0.58	105.00±2.08	166.00±4.16	207.00±6.03	227.33±3.84	94.33±3.18
Tetracladium marchalianum	7.27±1.68	117.67±4.33	203.00±2.08	261.33±23.13	161.00±0.58	73.33±8.82
Pestalotiopsis submersus	–	17.33±1.45	28.00±2.31	382.33±4.06	411.00±5.86	444.67±2.40
Flagellospora penicillioides	–	25.00±2.52	148.00±1.15	198.67±0.88	248.00±3.06	261.00±0.58
Heliscus lugdunensis	–	28.00±1.15	148.00±±4.62	401.67±6.69	420.37±5.23	137.87±2.19
Beltrania rhombica	–	7.33±1.76	19.33±2.40	322.67±1.45	209.00±1.73	72.67±5.93

Discussion

The distribution and occurrence of aquatic hyphomycetes depend on the physical and chemical characteristic of the water body. Nilsson (1964) and Ingold (1975) reported the distribution of certain species and variation in species composition is apparently related to temperature.

Ranzoni, 1951, observed the optimum temperature for the maximum growth of two species of genera *Anguillospora* was 25-28°C. Thornton, (1963) determined optimum temperature requirements for eight hyphomycetous fungus varies from 10-25°C. Similarly the observation of Jones and Stewart (1972) indicated 20-30°C temperature is the best for the growth of *Tricladium varium*. Suberkropp (1984) reported the optimum temperature for few aquatic hyphomycetes ranges between 20-25°C. The results of the present study also support the findings of the above workers.

From the available literature and the findings of present study it is clear that aquatic hyphomycetes require comparatively low temperature. Perhaps it may be a good reason that the most of these fungi have been reported from temperate countries and may have a physiological advantage over their environment competitors. Due to their low temperature need, they colonize more aptly the submerged plant materials and promote the degradation and decompositional biology in fresh water bodies.

As evident from Figure 25.1 the fungal reactions to changing pH are variable and species – specific occurrence, distribution and decompositional activities of aquatic hyphomycetes are highly influenced by water pH. Thompson and Barlocher, (1989), while working with the degradative abilities of three aquatic hyphomycetes, observed *T. marchalianum* reaches its peak degradation activity between pH 8 and 9. The metabolic activity of *C. aquatica* and *A. tetracladia* generally peaked at a pH 6.0 and declined at lower but not at higher pH. Ranzoni (1951) working with the nutrition of two aquatic hyphomycetes found 6.5-6.8 pH was optimum. Marvanova (1984) confirmed that neutral or slightly acidic streams in Czechoslovakia contain more species than hard water streams.

Thus, relying upon the available literature and present finding it could be concluded that aquatic hyphomycetes, belonging to the Duteromycetes distinctly varies from species to species for their pH requirement.

In the light requirement study white and red light were found the most favorable light source (Figure 25.2). It was interesting to note that the dark and blue lights were found inhibitory for the growth of these fungi. Mer (1982) observed that continuous light was found to stimulative for the growth of water moulds. Jones and Ward (1973) have shown that septate conidia are produced in *Asteromyces cruciatus* under dark condition. In a study, Alderman and Jones (1971a) observed that light inhibits the growth of two marine fungi. While working on water molds Prabhuji (1979) reported stimulatory effect on growth and oogonia formation.

The present work is a first report on the growth response of aquatic hyphomycetes to different light sources. As earlier it is well known fact that most of fungi prefer to grow in dark area rather than the lighted areas. But it was note worthy that all the studied fungi were grown well in light than dark or blue light.

Acknowledgements

Authors are thankful to DST, New Delhi, for the financial support, the Head, Department of Botany, Kumaun University, Nainital for providing laboratory facilities and Dr. M. Belwal for his valuable help time to time.

References

Alderman, D. J. and Jones, E. B. G. 1971. Physiological requirement of two marine phycomycetes, *Althornia crouchii* and *Ostracoblabe implexa*. Trans. Br. Mycol. Soc., 57(2): 213-225.

Barlocher, F. 1987. Aquatic hyphomycete spora in 10 streams of New Brunswick and Nova Scotia. Canad. J Bot., 65: 76-79.

Bhat, D. J. and Chien, C. Y. 1990. Water-borne hyphomycetes found in Ethiopia Trans. Mycol. Soc. Jpn., 31:147-157.

Chamier, A. C. 1985. Cell Wall-degrading enzymes of aquatic hyphomycetes a review. Bot. J. Linn. Soc., 91: 67-81.

Ingold, C. T. 1942. Aquatic Hyphomycetes on decaying alder leaves. Trans. Brit. Mycol. Soc., 25: 339 – 417.

Ingold, C. T.1975. An illustrated guide to aquatic and water-borne Hyphomycetes (Fungi imperfecti) with notes on their biology. Freshwater Biol. Assoc. Scient. Publ. No. 30 England 96 pp.

Jones, E. B. G. and Steward, R. J. 1972. *Tricladium varium*, an aquatic Hyphomycetes on wood in water cooling towers. Trans. Brit. Mycol. Soc., 59(1): 163-167.

Jones, E. B. G. and Ward A. W. 1973.. Trans. Br. Mycol. Soc., 61:181-186.

Marvanova, L. 1984. Conidia in water of the protected area Slovensky Raj. Biologia (Bratislava) 39: 821-832.

Marvanova, L. 1997. Fresh water Hyphomycetes: A Survey with Remarks of Tropical Taxa. Tropical Mycology, Sci. Publ. Inc. U.S.A., pp. 169 – 226.

Mer, G. S. 1982. Taxonomic and physiological studies of water molds of Sattal (Nainital). Ph. D. Thesis.

Nilsson, S. 1964. Fresh water hyphomycetes: Taxonomy, Morphology and Ecology. Symb.Bot. Upssal., 18: 1-130.

Prabhuji, S. K. 1979. Studies on some lower fungi occurring in certain soils of Gorakhpur. Ph.D. Thesis, Gorakhpur University, Gorakhpur.

Ranzoni, F. V. 1951. Nutrient requirements for two species of Aquatic Hyphomycetes. Mycologia, 43: 130-141.

Subercropp, K., 1984. Effect of temperature on seasonal occurrence of Aquatic Hyphomycetes. Trans. Br. Mycol. Soc., 82(1): 53-62.

Suberkropp, K. and Klug M. J. 1980. The maceration of deciduous leaf litter by aquatic hyphomycetes. Can J. Bot., 58(9): 1025-1031.

Thompson, P. L. and Barlocher, F. 1989. Effect of pH on leaf breakdown in streams and in the laboratory. J.N. Am. Benthol. Soc., 8(3): 203-210.

Thornton D. R. 1963. The Physiology Nutrition of some Aquatic Hyphomycetes. J. Gen. Mirobiol., 33: 23-31.

Microbes: Diversity and Biotechnology (2012) *Pages* **429–432**
Editors: **Prof. S.C. Sati & Dr. M. Belwal**
Published by: **DAYA PUBLISHING HOUSE, NEW DELHI**

Chapter 26

Endophytic Aquatic Hyphomycetes from Medicinal Plant *Geranium nepalense* Sweet

N. Pargaien, S.C. Sati and M. Belwal
Department of Botany, D.S.B. Campus, Kumaun University, Nainital - 263 001

ABSTRACT

Living roots of a medicinal plant *Geranium nepalense* Sweet. were collected from moist and shady areas of Nainital (1936 m asl) Kumaun Himalaya.The root sample were processed in the lab by using standard techniques. Isolation, culture and identification of root endophytic fungi were made. Six species of aquatic hyphomycetes *viz., Acaulopage tetraceros, Anguillospora longissima, Cylindrocarpon aquaticum, Diplocladiella scalaroides, Speiropsis scopiformis and Tetracladium setigerum* were recorded as root endophyte of *Geranium nepalense*, however, *S. scopiformis* was found as a new root endophyte.

Keywords: *Aquatic Hyphomycetes, Root Endophyte, Geranium.*

Introduction

Aquatic hyphomycetes a peculiar group of fungi imperfectii have been found frequently in decaying leaf litter and foams in fresh water bodies (Ingold, 1942). They play an important role in decomposition of submerged leaf litter (Suberkropp, 1984) and act as intermediaries of energy flow in aquatic ecosystem (Kaushik and Hynes, 1971).

Some recent studies have demonstrated their presence as endophytes in roots of terrestrial as well as riparian vegetations (Fisher and Petrini, 1989; Fisher *et al.*, 1991).Investigation on root endophytic aquatic hyphomycetes was initiated by Fisher and Petrini (1989) who for the first time described two species of aquatic hyphomycetes from the roots of angiosperm plant *Alnus glutinosa*. Later Fisher *et al.*

(1991), Sridhar and Barlocher (1992) also confirmed the occurrence of these fungi from the roots of riparian vegetation. In India Raviraja *et al.* (1996), Ananda and Sridhar (2002) studied root endophytic hyphomycetes from west coast of India. In Kumaun Himalaya, Sati and Belwal (2005) isolated a number of root endophytic hyphomycetes from riparian forest plants.

Kumaun Himalaya is well known for its rich plant diversity consisting of large number of medicinal, aromatic and wood yielding plants. In the present study living roots of a perennial herb *Geranium nepalense* known as medicinally important plant were examined for the isolation of root endophytic fungi.

Methodology

Healthy Roots of *Geranium nepalense* Sweet were collected from moist and shady areas of Nainital, Kumaun Himalaya. Nearly 5-10 cm long roots were cut off with a sharp sterile knife and kept in sterile polythene bags. These root pieces were washed thoroughly to remove extraneous material for 3-4 hours under running tap water and cut into 3-4 cm size segments. These root segments were then rinsed with sterile water, after surface sterilization with 90 per cent alcohol for 2-3 minutes. The segments were incubated at 20±2°C for 5-20 days in sterile Petri dishes containing 30 ml of sterile water. Incubated dishes were observed periodically to detect the conidia of endophytic hyphomycetes under low power of microscope.

Some of the root segments were also used for agar plating following Fisher *et al.* (1991) and Sridhar and Barlocher (1992). The isolated fungi were identified with the help of pertinent literature as well as concern authorieties.

Result and Discussion

Altogether six species of aquatic hyphomycetes *viz., Acaulopage tetraceros, Anguillospora longissima, Cylindrocarpon aquaticum, Diplocladiella scalaroides, Speiropsis scopiformis and Tetracladium setigerum* were isolated in the incubated root segments of *Geranium nepalense* as root endophyte (Figures 26.1A–F). A perusal of available literature indicates that *Speriopsis scopiformis* is being reported for the first time as root endophyte of *Geranium nepalense*. A brief account of these isolates is given here under.

Acaulopage tetraceros Drechsler
Mycologia, 27:195, 1935

Mycelium slender, aseptate, sparingly branched up to 15.6 x 1-2 µm, hyaline. Conidia borne singly on the apices of short branches of the hyphae, inverted bottle shaped with usually 3-4 needle-shaped empty appendages, which are produced from the broad distal ends of conidia. Conidia measured up to 27 x 5-5.5 µm with 13-20.8 x 1-2 µm needle shaped appendages.

Anguillospora longissima (Sacc. and Therry) Ingold
Trans. Brit. Mycol. Soc., 41: 367, 1958.

Root endophytic fungus with branched, septate mycelium. Conidiophore 90-100 x 3-4µm, usually simple bearing conidia singly at its tip. Conidia hyaline, filiform, curved or sigmoid usually spiraled, 5-14 septate, 140-221 x 3-5 µm long and 4-5µm wide at its middle tapering towards the both ends.

Cylindrocarpon aquaticum (Nils.) Marvanova and Descals
Trans. Brit. Mycol. Soc. 89: 499–507, 1987

Root endophytic fungus with septate mycelium. Conidiophore long septate. The distance between two septa of conidium is about 6.6µm. Conidium small, septate and is measured up to 29.7- 42.9 µm long and 4.2- 6.6 µm wide.

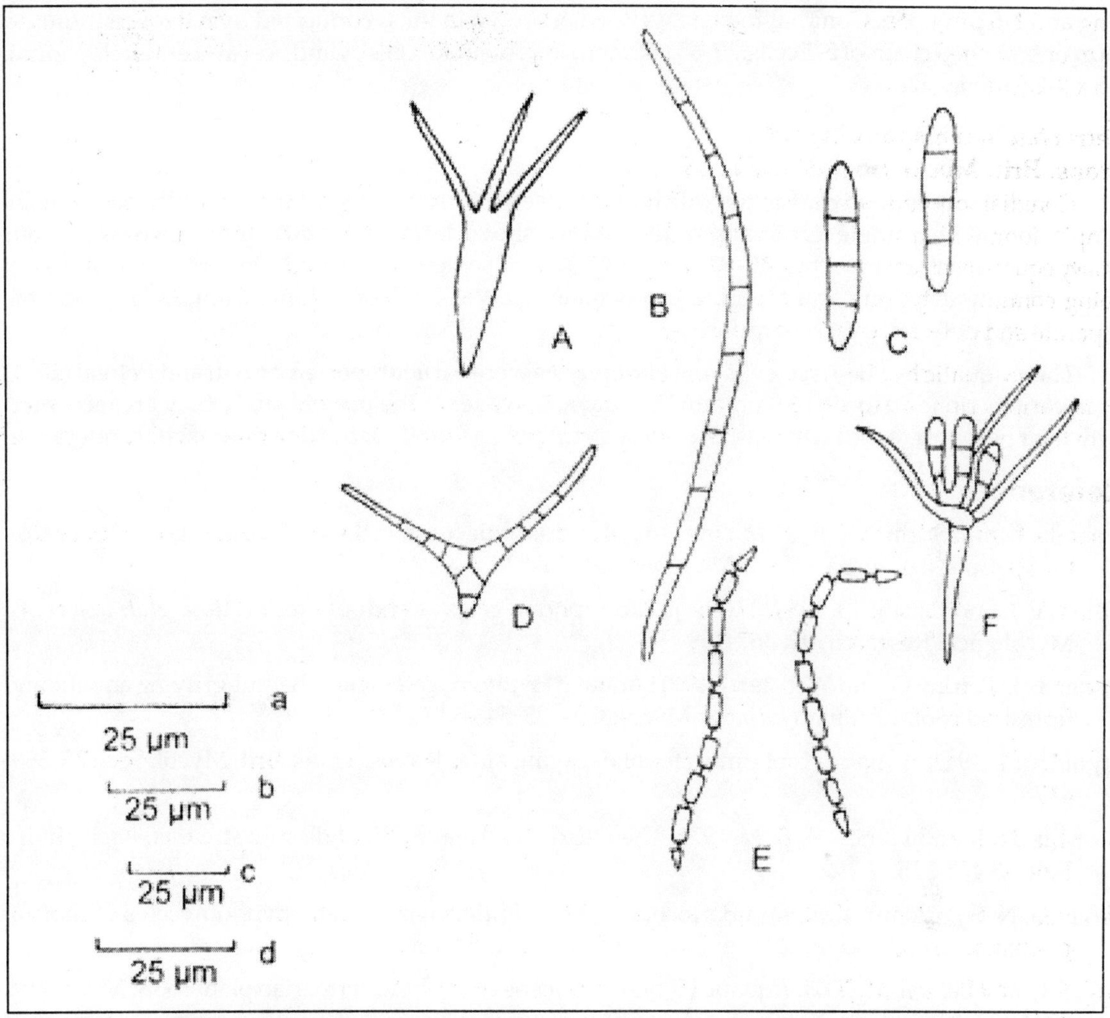

Figures 26.1A–F: Conidia of *Geranium nepalense* as Root Endophytes

A: *Acaulopage tetraceros*, B: *Anguillospora longissima*, C: *Cylindrocarpon aquaticum*, D: *Diplocladiella scalaroides*, E: *Speiropsis scopiformis*, F: *Tetracladium setigerum* (Figure A in scale a; Figures B, D and E in scale b; Figure C in scale c and Figure F in scale d)

Diplocladiella scalaroides Arnaud Bull.
Timest. Soc. Mycol. France, 89: 295, 1954

Root endophytic fungus sub-hyaline to brown triangular, 20-22 μm across (excluding the appendages) 10-12 μm high (wide), 8 celled. The corner or apical cells and the basal cell lighter in colour. Both apical cells with or without an apical, 8-20 μm x 1-2 μm filiform appendages.

Speiropsis scopiformis Kuthu. and Nawawi
Trans. Brit. Mycol. Soc. 89: 584, 1987

Root endophytic conidial fungus with septate, branched partly superficial mycelium. Conidiophores macro-nematus, mono-nematus solitary erect straight simple 2-3 septate, 75-110 μm long and 1-5 µm wide. Conidial hyphae, yellowish brown in mass connected by narrow isthmus to form unbranched chain of 5-7 cells, 37-65 µm long, intermediate cells cylindrical and end cells conical, 6-8 x 2-2.5 µm in size.

Tetrachaetum elegans **Ingold**
Trans. Brit. Mycol. Soc, 25:377, 1942

Conidial root endophytic fungus with branched, septate, hyaline mycelium. Conidiophore usually simple, long with a single terminal conidium which shows tetraradiate structure consisting of four nearly equal divergent branches, 40- 93.5µm x 1.67-3.34µm sometimes tapered. One arm of the conidium being continuously long and tapering forms main axis, 96.4-173.6µm long. Conidia later become separate and detected by a separation cells

These aquatic hyphomycetes as root endophytes were earlier reported by Sati and Belwal (2005) from various riparian forest of Kumaun Himalaya, however in the present study they are recovered with their new host record whereas, *Speiropsis scopiformis* is found for the first time as root endophyte.

References

Ananda, K. and Sridhar K. R. 2002. Diversity of endophytic fungi in the roots of mangrove species on the west coast of India. Can. J. Microbiol.48: 871-878.

Fisher, P. J. and Petrini, O. 1989. Two aquatic hyphomycetes as endophytes in *Alnus glutinosa* roots. Mycological Research, 92: 367-368.

Fisher, P. J., Petrini. O. and Webster. J. 1991. Aquatic Hyphomycetes and other fungi living aquatic and terrestrial roots of *Alnus glutinosa*. Mycol. Res., 95: 543-547.

Ingold, C. T. 1942. Aquatic Hyphomycetes of decaying alder leaves. Trans. Brit. Mycol. Soc., 25: 339-417.

Kaushik, N. K and Hynes, H. B. N. 1971. The fate of dead leaves that falls into streams, Arch Hydro Biol. 68:465-515.

Raviraja, N. S., Sridhar, K. R. and Barlocher, F. 1996. Endophytic aquatic hyphomycetes of roots of plantation crops and ferns from India. Sydowia 48: 152-160.

Sati, S. C. and Belwal, M. 2005. Aquatic Hyphomycetes as endophyte of riparian plant roots. Mycologia, 97 (1), 45-49.

Sridhar, K. R. and Barlocher F. 1992b. Endophytic aquatic hyphomycetes of roots of Spruce, Birch and Maple. Mycol. Res. 96: 305- 308.

Suberkropp, K. 1984. Effect on occurrence of Aquatic Hyphomycetes. Trans. Brit. Mycol. Soc. 82:53-62.

Microbes: Diversity and Biotechnology (2012) *Pages* **433–443**
Editors: **Prof. S.C. Sati & Dr. M. Belwal**
Published by: **DAYA PUBLISHING HOUSE, NEW DELHI**

Chapter 27

Quantitative Estimation of Conidial Production of Water-borne Conidial Fungi on Defined Submerged Substrate

S.C. Sati and M. Belwal
Department of Botany, D.S.B. Campus, Kumaun University, Nainital – 263 002

ABSTRACT

The variation in conidial production of water-borne conidial fungi on submerged leaf litter of three plant species *Betula alanoides, Cedrus deodara* and *Rhododendron arboretum* is studied. Submerged decaying leaf material were sampled and processed. Out of 26 species of water-borne conidial fungi recorded, 12 species were found common to all host substrates. *Cedrus deodara* was colonized by a least number of species while had the maximum conidial production (114102 conidia/cm^2/litre). On the other hand, *Rhododendron arboreum* and *Betula alanoides* were colonized by highest number of water-borne conidial fungi (24 and 18 species respectively) the total average conidial production for both of these two remained in low profile (50084 and 46925 conidia/cm^2/litre respectively). The maximum number of conidial production took place during winter and spring months while the maximum number of species colonization was recorded during autumn and rainy months.

Introduction

Conidial production of water-borne conidial fungi is related to the fertility of fungi particularly, in the case of fungi imperfecti, where the conidia are of paramount importance for their multiplication. These fungi are known as dominant microorganism during the first stages of leaf litter decomposition process in running fresh water bodies (Findlay and Arsuffi, 1989; Barlocher and Kendrick, 1976; Suberkropp and Klug, 1976). For adaptations to stream conditions, these water-borne conidial fungi show variation in their conidial shape and size (Webster 1987). These fungi have tetraradiate appearance, constituting several long appendages diverging from a common point (Bandoni, 1975;

Ingold, 1975); sigmoid appearance in which conidia with curves in one or more planes (Webster, 1981) or with more conventional shapes (rounded, ellipsoid, fusoid, etc.).

These water-borne conidial fungi represent the major microbial element for decay of submerged leaves (Barlocher and Kendrick, 1981; Barlocher *et al.*, 1977; Gareth-Jones, 1976; Subekropp and Klug, 1976). Barlocher and Kendrick (1974) have reported that these fungi act as intermediaries of energy flow in aquatic ecosystem but little is known for their quantitative studies (Willoghby and Archer, 1973; Muller Haeckel and Marvanova, 1979). Iqbal and Webster (1973) were pioneer to understand the rate of conidial concentration in a water body by filtering the water. Sati and Tiwari (1992,1995) developed a simple technique to determine the rate of conidial production by modifying the method of Webster and Towfic (1972).

In this study fallen leaves of three forested plant species were studied for the production of conidia in unit area substrate per litre of water in captivity.

Materials and Methods

To determine the rate of conidial production in per unit area of the substrate in per litre of water, Sati and Tiwari (1995) was followed.

Submerged leaf litter of known three forest trees *i.e. Betula alnoides, Cedrus deodara* and *Rhododendron arboreum* were collected and processed for incubation in the sterile petri dishes containing 20 ml of sterile water at monthly intervals. Prior placing the leaf litter for incubation, the area of each piece of leaf litter was determined with the help of graph paper. After 2-3 days, the incubated dishes containing leaf litter were gently shaken to homogenize the fungal conidia produced in water. The drops of 0.01 ml conidial suspension were pipetted out on glass slides for screening. The counting of conidia was made directly under the low power of microscope and conidial number was recorded individually to each species. Finally the rate of conidial production for each species occurred and total species in unit area (1 cm^2) were calculated using the following formula:

$$RCP = \frac{2000\,n}{a} \text{ Conidia/cm}^2\text{/litre}$$

where,

RCP = Rate of conidial production

n = No. of conidia present in.01 ml of conidial suspension used

a = area of leaf litter substrate incubated (cm^2)

(2000 is used if 20 ml sterile water is supplied to the incubated substrate in dish)

Results

The freshly fallen leaf litter of forest plants *i.e. Betula alanoides, Cedrus deodara* and *Rhododendron arboreum* were studied for a calendar year and production of conidia in per litre of water per unit area of substrate were determined. Data were recorded colonized fungal species wise as well as total conidia produced for each month. Monthly variation in conidial production of aquatic hyphomycetes per litre per unit area of substrate and results are presented in Tables 27.1–27.3. Colonization pattern of 26 species of water-borne conidial fungi on three different host plants as well as the rate of conidial production per cm^2 area in unit volume of water is summarized in Table 27.4.

Betula alanoides Buch-Ham. ex D.Don

The leaves of *Betula alanoides* was colonized by 18 species *viz. Alatospora acuminata, Alatospora pulchella, Anguillospora filliformis, Articulospora tetracladia, Clavariopsis aquatica, Dimorphospora foliicola, Flagellospora penicillioides, Lemonniera cornuta, L. pseudofloscula, L. terrestris, Lunulospora curvula, L.cymbiformis, Speiropsis scopiformis, Tetrachaetum elegans, Tetracladium marchalianum, Triscelophorus acuminatus, T. monosporus* and *Tricladium chaetocladium* (Table 27.1). The total average conidial production was 46925 conidia in unit area of substrate per litre. *Tetrachaetum elegans* accounted the maximum number of conidia (6919 conidia/cm^2/litre) followed by *Alatospora acuminata* (4778), *Lemonniera cornuta* (4623), *Triscelophorus monosporus* (4573) and *Anguillospora filliformis* (4337). On the other hand *Lunulospora cymbiformis* accounted least number of conidia (254 conidia/cm^2/litre).

The maximum number of conidial production on this substrate was reached upto 93550 conidia/cm^2/litre during January while the least number of conidia were observed in the month of August *i.e.* 11620 conidia/cm^2/litre (Figure 27.1a).

Cedrus deodara (Roxb. ex D.Don) G.Don

The incubated leaves of *Cedrus deodara* were colonized by a total of 14 species of water-borne conidial fungi *viz., Alatospora acuminata, Anguillospora longissima, Clavariopsis aquatica, Dimorphospora foliicola, Flagellospora penicillioides, Heliscus lugdunensis, Lemonniera cornuta, Lunulospora curvula, L. cymbiformis, Tetrachaetum elegans, Tricladium chaetocladium. Triscelophorus monosporus, T. acuminatus* and *Tetracladium marchalianum* (Table 27.2). The total average conidial production in *Cedrus deodara* was 114102 conidia/cm^2/litre. The maximum contribution was made by the conidia of *Tetrachaetum elegans* which reached upto an average of 18808 conidia/cm^2/litre while the least number of conidia were contributed by *Triscelophorus acuminatus* (1079 conidia/cm^2/litre).

Maximum conidial production was analyzed during December *i.e.* 229914 conidia/cm^2/litre while the minimum rate of conidial production was occurred in the month of August *i.e.* 26300 conidia/cm^2/litre (Figure 27.1b).

Rhododendron arboreum Smith

There were 24 species *viz Alatospora acuminata, A. flagellata, A. pulchella, Anguillospora longissima, Articulospora tetraciadia, Campylospora chaetocladia, Clavariopsis aquatica, Dimorphospora foliicola, Flagellospora penicillioides, Helicosporium lumbericoides, Lemonniera cornuta, L. pseudofloscula, L. terrestris, Lunulospora curvula, L. cymbiformis, Pestalotiopsis submersus, Speiropsis scopiformis, Tetrachaetum elegans, Tetracladium marchalianum, Tricladium chaetocladium, Triscelophorus acuminatus, T. monosporus, T. konajensis* and *Tricladium splendens* colonizing to incubated leaves of *Rhododendron arboreum* (Table 27.3). This substrate was highly colonized by *Tricladium chaetocladium* and least colonized by *Pestalotiopsis submersus*. Total average conidial production in *Rhododendron arboreum* was 50084 conidia/cm^2/litre. In the total conidial production, maximum contribution was made by conidia of *Tricladium chaetocladium*, which reached upto an average of 5638 conidia/cm^2/litre. The least no. of conidia were contributed by *Pestalotiopsis submersus* as 109 conidia/cm^2/litre.

For the monthly total conidial production maximum conidial production was represented by the month of January during which 96780 conidia in unit area of substrate per litre were observed (Figure 27.1c). On the other hand least number of conidia recorded during April (16710 conidia/cm^2/litre).

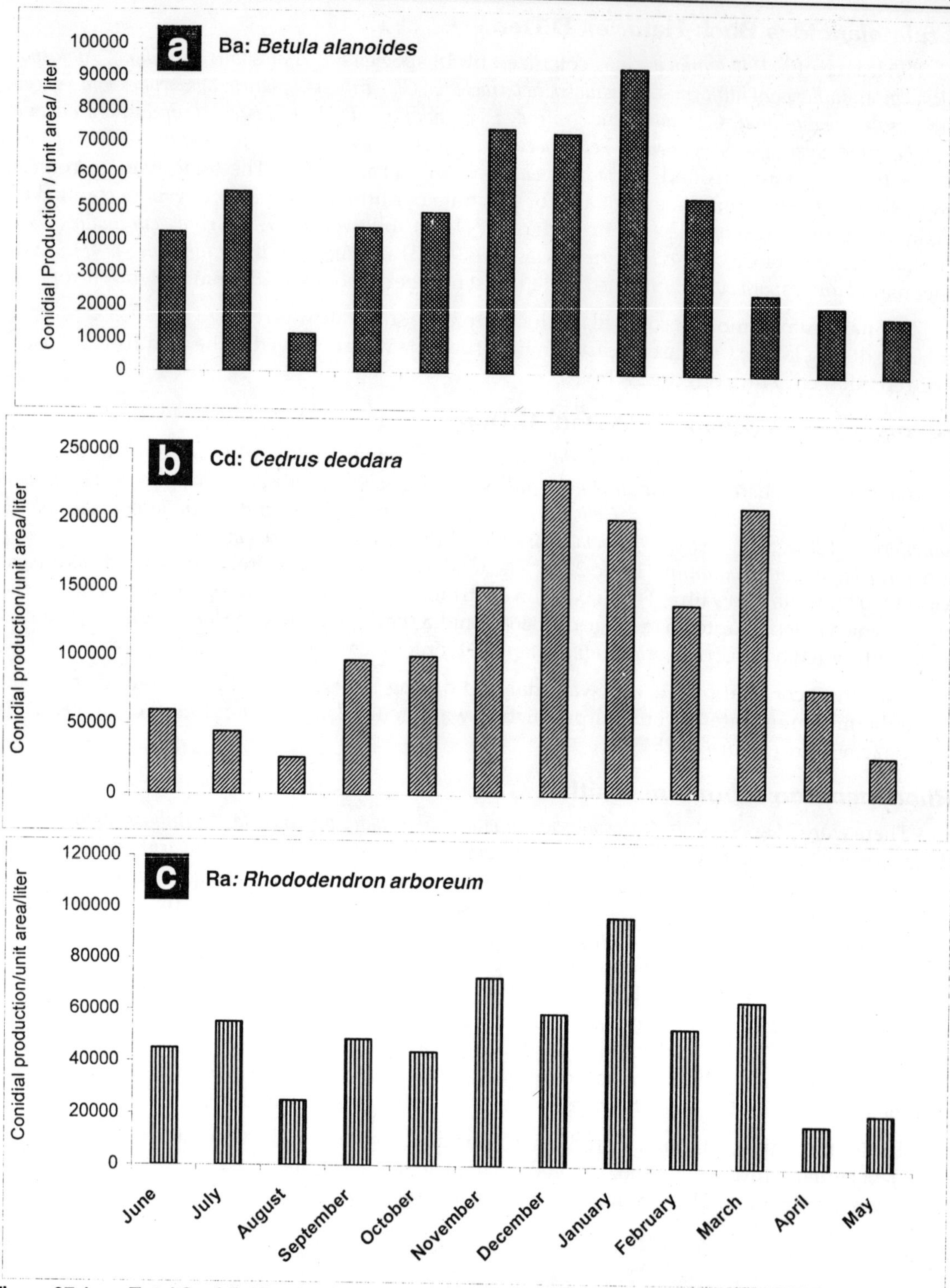

Figure 27.1a-c: Total Conidial Production of Water-borne Conidial Fungi/liter/cm^2 in Different Host Species

Table 27.1: Monthly Variation in Conidial Production of Water-borne Conidial Fungal in per litre/unit Area of *Betula alanoides* Buch-Ham.ex D.Don Leaf Litter in Captivity

Sl.No.	Fungi	Conidia Produced												Average
		June	July	Aug	Sept	Oct	Nov	Dec	Jan	Feb	March	Apl	May	
1.	*Alatospora acuminata*	–	6230	290	13650	10880	8530	17760	–	–	–	–	–	4778
2.	*A. pulchella*	–	–	–	–	5920	–	–	–	–	–	–	–	493
3.	*Anguillospora filliformis*	–	–	–	4660	–	–	3940	22830	20610	–	–	–	4337
4.	*Articulospora tetracladia*	–	–	–	–	–	–	6830	–	–	3530	–	–	863
5.	*Clavariopsis aquatica*	–	–	–	–	1600	–	–	9940	11450	–	8220	–	2600
6.	*Dimorphospora foliicola*	–	–	–	–	–	11250	–	–	–	–	–	–	938
7.	*Flagellospora penicillioides*	6430	6290	–	–	–	12780	–	–	–	630	–	9250	2948
8.	*Lemonniera cornuta*	–	–	–	–	7830	14840	22760	–	–	10040	–	–	4623
9.	*L. pseudofloscula*	–	–	–	–	–	–	–	–	–	5590	–	–	466
10.	*L. terrestris*	–	–	–	–	5770	7050	–	–	–	–	–	–	1068
11.	*Lunulospora curvula*	7330	8270	1370	2810	2700	–	–	15520	5370	–	–	4620	3999
12.	*L. cymbiformis*	–	–	–	–	–	–	–	–	–	–	3050	–	254
13.	*Speiropsis scopiformis*	–	18870	–	–	–	–	–	–	–	–	–	–	1573
14.	*Tetrachaetum elegans*	14594	9900	4100	15830	7630	9700	7200	–	–	5180	4410	4480	6919
15.	*Tetracladium marchalianum*	–	–	–	–	–	–	–	12060	7200	–	–	–	1605
16.	*Tricladium chaetocladium*	–	–	–	–	4970	10470	14878	15490	–	–	–	–	3817
17.	*Triscelophorus acuminatus*	5930	–	1910	5000	–	–	–	–	–	–	–	–	1070
18.	*T. monosporus*	8330	5500	3950	2290	1840	–	–	17710	9670	–	5590	–	4573
	Total	42614	55060	11620	44240	49140	74620	73368	93550	54300	24970	21270	18350	46925

Table 27.2: Monthly Variation in Conidial Production of Water-borne Conidial Fungi in per litre/unit Area of *Cedrus deodara* (Roxb.ex D.Don) G.Don Leaf Litter in Captivity

Sl.No.	Fungi	Conidia Produced												Average
		June	July	Aug	Sept	Oct	Nov	Dec	Jan	Feb	March	Apl	May	
1.	*Alatospora acuminata*	4860	–	–	10860	12730	12650	14320	31170	29550	–	–	12400	10712
2.	*A. longissima*	–	–	–	15150	–	–	–	–	–	–	–	–	1263
3.	*Clavariopsis aquatica*	–	–	–	–	–	–	52260	–	–	14490	–	–	5563
4.	*Dimorphospora foliicola*	–	–	–	–	18240	–	–	–	–	–	–	–	1520
5.	*Flagellospora penicillioides*	–	14420	12910	18270	–	–	2594	–	–	38150	60500	–	12237
6.	*Heliscus lugdunensis*	–	–	–	–	–	–	31450	51940	70580	39510	–	–	16123
7.	*Lemonniera cornuta*	–	–	–	–	–	28230	22010	–	–	–	–	–	4187
8.	*Lunulospora curvula*	7750	10520	10750	–	9730	18360	–	–	–	–	4160	6200	5623
9.	*L. cymbiformis*	14210	5260	2640	–	–	–	52950	48380	–	–	14000	–	11453
10.	*Tetrachaetum elegans*	–	–	–	39330	15870	23000	54330	–	–	93160	–	–	18808
11.	*Tetracladium marchalianum*	–	6410	–	–	14350	–	–	–	–	–	–	–	1730
12.	*Tricladium chaetocladium*	–	–	–	–	13920	23810	–	70380	24870	25580	–	–	13213
13.	*Triscelophorus acuminatus*	12950	–	–	–	–	–	–	–	–	–	–	–	1079
14.	*T. monosporus*	19812	8010	–	13200	15070	44970	–	–	14360	–	–	11690	10593
	Total	59582	44620	26300	96810	99910	151020	229914	201870	139360	210890	78660	30290	114102

Table 27.3: Monthly Variation in Conidial Production of Water-borne Conidial Fungi on per litre/unit Area of *Rhododendron arboreum* Smith Leaf Litter in Captivity

Sl.No.	Fungi	Conidia Produced												Average
		June	July	Aug	Sept	Oct	Nov	Dec	Jan	Feb	March	Apl	May	
1.	*Alatospora acuminata*	–	3800	5990	–	6340	6140	17550	–	–	7210	–	–	3919
2.	*A. flagellata*	–	–	–	–	3060	–	–	–	–	–	–	–	255
3.	*A. pulchella*	–	2950	–	–	3810	–	–	–	–	4490	–	–	938
4.	*A. longissima*	–	–	–	7480	–	–	3020	–	–	–	–	–	875
5.	*Articulospora tetracladia*	–	–	–	–	–	–	5570	–	–	5400	–	–	914
6.	*Campylospora chaetocladia*	–	4620	–	–	–	–	–	–	–	–	–	–	385
7.	*Clavariopsis aquatica*	–	–	–	–	1940	4880	–	12060	–	4720	–	–	1967
8.	*Dimorphospora foliicola*	–	–	–	4200	–	11760	–	46200	–	–	–	–	5180
9.	*Flagellospora penicillioides*	11980	11120	–	1600	–	–	–	–	2200	–	–	–	2242
10.	*Helicosporium lumbericoides*	–	–	2760	–	–	–	–	–	–	–	–	–	230
11.	*Lemonniera cornuta*	–	–	–	–	2350	12190	3160	8770	11970	–	–	–	3203
12.	*L. pseudofloscula*	–	–	–	–	2110	7920	1100	–	–	7980	–	–	1593
13.	*L. terrestris*	–	–	–	–	5640	–	2390	–	12320	–	–	–	1696
14.	*Lunulospora curvula*	10950	3970	6670	14960	7530	2470	–	–	–	–	–	9880	4703
15.	*L. cymbiformis*	–	–	–	7020	–	–	–	–	–	–	–	–	585
16.	*Pestalotiopsis submersus*	–	–	–	–	–	–	–	–	–	–	1310	–	109
17.	*Speiropsis scopiformis*	–	–	5880	–	–	–	–	–	–	–	–	–	490
18.	*Tetrachaetum elegans*	9430	–	3550	4410	–	9500	8760	–	–	11880	–	1670	4100
19.	*Tetracladium marchalianum*	–	12220	–	–	2330	–	–	–	4180	–	–	3790	1877
20.	*Tricladium chaetocladium*	–	–	–	–	2110	13600	8630	16530	4080	22700	–	–	5638
21.	*T. splendens*	–	–	–	–	–	–	8550	–	14290	–	2350	–	2099
22.	*Triscelophorus acuminatus*	6980	–	–	6490	–	–	–	–	–	–	–	–	1123
23.	*T. monosporus*	5590	16290	–	2340	6590	4370	–	13220	4360	–	2300	5780	5070
24.	*T. konajensis*	–	–	–	–	–	–	–	–	–	–	10750		896
	Total	44930	54970	24850	48500	43810	72830	58730	96780	53400	64380	16710	21120	50084

Table 27.4: Occurrence of Fungi on Different Substrates Studied for Rate of Conidial Production in Captivity and their Per cent Frequency

S.No.	*Fungi*	*Average Conidial Production on Different Substrate*			
		Ba	*Ra*	*Cd*	*Per cent Frequency of Occurrence*
1.	*Alatospora acuminata*	4778	3919	10712	100
2.	*A. flagellata*	–	255	–	33.3
3.	*A. pulchella*	493	938	–	66.7
4.	*Anguillospora. filliformis*	4337	–	–	33.3
5.	*A. longissima*	–	875	1263	66.7
6.	*Articulospora tetracladia*	863	914	–	66.7
7.	*Campylospora chaetocladia*	–	385	–	33.3
8.	*Clavariopsis aquatica*	2600	1967	5563	100
9.	*Dimorphospora foliicola*	938	5180	1520	100
10.	*Flagellospora penicillioides*	2948	2242	12237	100
11.	*Helicosporium lumbericoides*	–	230	–	33.3
12.	*Heliscus lugdunensis*	–	–	16123	33.3
13.	*Lemonniera cornuta*	4623	3203	4187	100
14.	*L. pseudofloscula*	466	1593	–	66.7
15.	*L. terrestris*	1068	1696	–	66.7
16.	*Lunulospora curvula*	3999	4703	5623	100
17.	*L. cymbiformis*	254	585	11453	100
18.	*Pestalotiopsis submersus*	–	109	–	33.3
19.	*Speiropsis scopiformis*	1573	490	–	66.7
20.	*Tetrachaetum elegans*	6919	4100	18808	100
21.	*T. marchalianum*	1605	1877	1730	100
22.	*Tricladium chaetocladium*	3817	5638	13213	100
23.	*T. splendens*	–	2099	–	33.3
24.	*Triscelophorus acuminatus*	1070	1123	1079	100
25.	*T. monosporus*	4573	5070	10593	100
26.	*T. konajensis*	–	896	–	33.3
	Total	18	24	14	

Discussion

On reconnaissance of Tables 27.1–27.3 it was interesting to note that a maximum conidial production was accounted during winter months and spring months *i.e.* November to March, when temperature remains low on the other hand less number of species of water-borne conidial fungi were encountered during the summer months. During autumn months like September, October and

November the maximum number of species were observed. Relying upon these results it can be said that the maximum number of conidial production take place during winter and spring months while the maximum number of species variation take place during autumn and rainy months. The findings of some of the previous workers (Iqbal and Webster 1973, Barlocher and Rosset 1981, Mer and Sati 1989) also support the result of present investigation.

The total average conidial production on unit area of different studied tree plant leaf litter in per liter of water is summarized in Table 27.4. As evident from Table 27.4, a total of 26 species of water bone conidial fungi were encountered. A maximum number of species were harbored on the submerged leaves of *Rhododendron arboreum* (24 species). Whereas *Betula alanoides* and *Cedrus deodara* were colonized by 18 and 14 species of water-borne conidial fungi respectively. However, each substrate had different species composition (Table 27.4). It was interesting to note that coniferous substrate allowed less number of species to colonize (14 species), while the conidial production in unit substrate was found maximum (114102 conidia/cm^2/litre). The similar pattern of conidial production for these fungi on coniferous substrate *i.e. Pinus roxburghii* has also been noticed by Sati and Belwal (2009). This suggests, a preferential occurrence of water-borne conidial fungi on the nature of substrates and, it could also be envisaged that the occurrence of water-borne conidial fungi much depend on the availability of submerged leaf litter.

As evident from Table 27.4, *Rhododendron arboreum* and *Betula alanoides* which were colonized by highest number of water-borne conidial fungi (24 and 18 species respectively) the total average conidial production for both of these two remained in low profile (50084 and 46925 conidia/cm^2/litre respectively). Thus, relying up on these observation it could be visualized that number of species colonization and rate of conidial production varies species to species of leaf litter or might be depend on the nature of available substrate. This suggests that the higher rate of production of conidia might be dependent on the nature and nutritive value of substrate. The present observation also confirm the findings of Willoughby and Archer (1973).

On perusing Table 27.4, 12 species *viz Alatospora acuminata,Clavariopsis aqautica, Dimorphospora foliicola, Flagellospora penicillioides, Lemonniera cornuta, Lunulospora curvula, L. cymbiformis, Tetracladium marchalianum, Tetrachaetum elegans, Tricladium chaetocladium, Triscelophorus acuminatus* and *T. monosporus* were found in all the studied plant leaf litter and regarded as most abundant species. As evident from Table 27.4, six species *viz., Alatospora flagellata, Campylospora chaetocladia, Helicosporium lumbericoides, Pestalotiopsis submerses, Tricladium splendens* and *Triscelophorus konajensis* were restricted for their occurrence on Rhododendron arboretum. On the other hand two species of these fungi *viz., Anguillospora filliformis and Heliscus lugdunensis* were restricted to *Betula alanoides* and *Cedrus deodara* respectively, thus showing selective substrate requirement and regarded as least abundant species.

In the present study seven species had their maximum conidial production *i.e.* > 10,000 condia/cm^2/liter being highest for *Tetrachaetum elegans* followed by *Heliscus lugdunensis, Flagellospora penicillioides, Tricladium chaetocladium, Lunulospora cymbiformis, Alatospora acuminata* and *Triscelophorus monosporus*. It was also interesting to note that all the aforesaid species were colonized on the gymnospermus tree *i.e. Cedrus deodara* (Table 27.4).

Acknowledgement

Authors are thankful to the Head, Department of Botany, Kumaun University, Nainital for providing necessary facilities and DST, New Delhi for financial assistance.

References

Bandoni, R. J. 1975. Significance of tetraradiate from in dispersal of terrestrial fungi. Rep. Tottori Mycol. Inst., 12:105-113.

Barlocher, F. and Kendrick, B. 1974. Dynamics of fungal population on leaves in a stream. J. Ecol., 62: 761- 791.

Barlocher, F. and Kendrick, B. 1976 Hyphomycetes as intermediaries of energy flow in streams. In: E.B. Gareth Jones (ed): Recent Advances in Aquatic Mycology, pp 435-446.

Barlocher, F. and Kendrick, B. 1981. Role of aquatic Hyphomycetes in the trophic structure of streams. In: DT Wicklow and GC Carroll (eds) The Fungal community: Its organization and Role in the Ecosystem Marcel Dekker, Newyork, pp. 743-760.

Barlocher, F. and Rosset, J. 1981. Aquatic Hyphomycetes Spora of two Black Forest and two Swiss Jura streams. Trans. Brit. Mycol. Soc., 76: 479-483.

Barlocher, F., Kendrick, B. and Michaelides, J. 1977. Colonization of resin coated slides by Aquatic Hyphomycetes. Can. J. Bot., 55: 1163-1166.

Findley, S.E.G. and Arsuffi. T.L. 1989. Microbial growth and detritus transformations during decomposition of leaf litter in a stream. Fresh water Biol., 21: 261-269.

Gareth Jones, E. B. 1976. Topics of further interest In: EB Gareth Jones(ed)Recent advances in Aquatic mycology Elek Seicnce London pp707-724.

Ingold, C. T. 1975. An illustrated guide to aquatic and water-borne Hyphomycetes (Fungi imperfecti) with notes on their biology. Freshwater Biol. Assoc. Scient. Publ. No. 30 England 96 pp.

Iqbal, S. H. and Webster, J. 1973. The trapping of aquatic Hyphomycetes spores by air bubbles.Trans.Brit.Mycol.Soc., 60: 37-48.

Mer, G. S. and Sati S. C. 1989. Seasonal fluctuation in species composition of aquatic Hyphomycetous flora in a temperate freshwater stream of Central Himalaya. India. Int. Rev. Ges. Hydrobiol., 74: 433-437.

Muller-Haeckel, A. and Marvanova, L. 1979. Periodicity of aquatic hyphomycetes in the Sub arctic. Trans. Brit. Mycol. Soc., 73: 109-116.

Sati, S. C. and Belwal, M. 2009. In vitro conidial production of Aquatic Hyphomycetes on submerged leaf litter. Nature and Science, 7: 78-83.

Sati, S. C. and Tiwari, N. 1992. Colonization, species composition and conidial production of aquatic hyphomycetes on chir pine needle litter in a fresh water Kumaun Himalayan stream. Int. Revue. Ges. Hydrobiol., 77: 445-453.

Sati, S. C. and Tiwari, N. 1995. Counting of conidial production of aquatic hyphomycetes on the substrate-a new method. Nat. Acad. Sci. Letters, 18: 7-8.

Suberkropp, K. and Klug, M. J. 1976. Fungi and bacteria associated with leaves during processing in a woodland stream. Ecology, 57: 707-719.

Webster, J. 1981. Biology and ecology of aquatic hyphomycetes. In: D Wicklow and G Carroll (eds.): The fungal communityM. Dekker, Inc. New York. Pp 691-691.

Webster, J. 1987. Convergent evolution and the functional significance of spore shape in aquatic and semi aquatic fungi. In: ADM Rayner, CM Brasier and D Moore (eds.) Evolutionary Biology of the fungi (Symposium of the British Mycological society, Apr.1986), pp.191-201.

Webster, J. and Towfic, F. H. 1972. Sporualation of aquatic Hyphomycetes in relation to aeration. Trans. Brit. Mycol. Soc., 59: 353-364.

Willoughby, L. G. and Archer, J. F. 1973. The fungal spora of a fresh water stream and its colonization pattern on wood. Fresh Wat. Biol., 3: 219-239.

Microbes: Diversity and Biotechnology (2012) *Pages* **445–467**
Editors: **Prof. S.C. Sati & Dr. M. Belwal**
Published by: **DAYA PUBLISHING HOUSE, NEW DELHI**

Chapter 28

Arbuscular Mycorrhiza: Progression from Womb to World

V.S. Mehrotra*

Agriculture Division, PSS Central Institute of Vocational Education, 131 Zone II, M.P. Nagar, Bhopal – 462 011, M.P.

Introduction

Arbuscular Mycorrhiza (AM) characterized by the formation of arbuscules is an important symbiotic relationship between certain group of fungi and the root. The origin of arbuscular mycorrhizal fungi (AMF) has been traced to 460 million-year-old rocks from the Ordovician of Wisconsin (Redecker *et al.*, 2000). AM colonized fossil roots have been observed in *Aglaophyton major,* which is a non-vascular plant possessing anatomical features intermediate between those of the bryophytes and vascular plants with primitive protostelic rhizomes (Remy *et al.*, 1994). *A. major* is known only from Rhynie chert in Aberdeenshire, Scotland. It lacked roots and like other rootless land plants of the Silurian and early Devonian may have relied on arbuscular mycorrhizal fungi for acquisition of nutrition from the soil. Plants of the Rhynie chert from the Lower Devonian were found to contain structures resembling vesicles and spores of present *Glomus* species. The ubiquitous nature of AM among land plants also suggests that mycorrhizas (or mycorrhizae) were present in the early ancestors of extant land plants.

In the last few years, researchers have suggested that some mycorrhizal fungi do not produce vesicles under all conditions, and so vesicular-arbuscular mycorrhizal fungi (VAMF) should be called arbuscular mycorrhizal fungi (AMF). The presence of arbuscules is normally used to identify AM. Brundrett (2009) opined that it is ironical that as we increasingly tend to drop the V (vesicles) from the name of arbuscular mycorrhizas (from VAM to AM), evidence is accumulating that some of these associations lack arbuscules, that vesicles are used more often than arbuscules in diagnosis of

* E-mail: drvs.mehrotra@gmail.com

associations, and that arbuscules may occur in non-host plants where Glomeromycotan fungi grow as endophytes. After a detailed study on the formation of vesicles in roots of various plants, Mehrotra (1993) concluded that the so called vesicles are chlamydospores, and therefore the name AM is more appropriate than VAM.

Arbuscular mycorrhizal fungi are now known to be associated with the roots of almost all plant species. They are considered as a fundamental part of the plant, as 95 per cent of all plant species could not survive in nature without it. Almost all Angiosperms, Gymnosperms, Pteridophytes and Bryophytes form AM associations (Kendrick and Berch, 1985). In total, mycorrhizologists have presented data on over 10,000 plants species, which equates to about 3 per cent of vascular plants (Brundrett, 2009).

Intensive researches on various aspects of AM throughout the world in the last 10 decades have brought to the fore the importance of mycorrhizal associations in increasing uptake of mineral nutrients, particularly phosphorus (Ryan and Graham, 2002; Mehrotra 2005). Plants also benefit through the association in terms of drought tolerance (Augé and Moore, 2005), greater resistance to attack of certain root pathogens (Vyas and Shukla, 2005), increase in photosynthetic activity (Brown and Bethlenfalvay, 1988), and reproduction in plants (Koide *et al.*, 1988).

The review article examines the progress made in AM with reference to the discoveries in structure, role and functions over the last few years and does not in any way attempt to present the historical development of research on AM. The historical accounts of AM have been given from time to time by several researchers including Rayner (1926-1927), Trappe and Berch (1985), Mosse (1985), Schenck (1985), Harley (1991), Allen (1996), and Koide and Mosse (2004).

Structure of Arbuscular Mycorrhiza

Arbuscular mycorrhizal fungi live in two distinct environments: (*i*) the sheltered place in the root–the womb; and (*ii*) the harsh conditions of the soil-the world (Figure 28.1). The term mycorrhizosphere (Oswald and Ferchau, 1968) is used to refer to the zone of influence of the mycorrhiza in the soil. The mycorrhizosphere has two components. One is the rhizosphere, a thin layer of soil that surrounds the root and is under the direct influence of the root, root hairs, and AM hyphae adjacent to the root. The other, the hyphosphere, is not directly influenced by the root. The hyphosphere is a zone of AM hypha-soil interactions (Marschner, 1995), and may be more or less densely permeated by the AM soil mycelium (8 to 20 km hyphae L^{-1} soil), Schreiner *et al.*, 1997). The growth and development of AM is directly influenced by the changes on root and the mycorrhizosphere. Anatomical features of mycorrhizas must be observed to distinguish them from other fungi in roots (Brundrett, 2004). Spores (40-800 µm in size) containing several hundreds to thousands of nuclei are produced with layers. Spores may be produced singly, in clusters or in morphologically distinct fruit bodies called 'sporocarps'.

The development of AM fungi prior to root colonization, known as pre-symbiosis, consists of three stages: spore germination, hyphal growth, and appressorium formation (Douds and Nagahashi, 2000). AM symbiosis is initiated when fungal hyphae, arising from spores in the soil or adjacent colonized roots, contact the root surface and form appresoria. From the points of root penetration, intraradical hyphae (IRH) grow between epidermal cells through passage cells of the hypodermis, and colonize the intercellular space of the cortex. At the inner cortex IRH penetrate individual cortical cells and differentiate to form highly branched structures known as "arbuscules" (Gallaud, 1905). Following colonization of the root cortex, the fungus develops an extensive network of extraradical

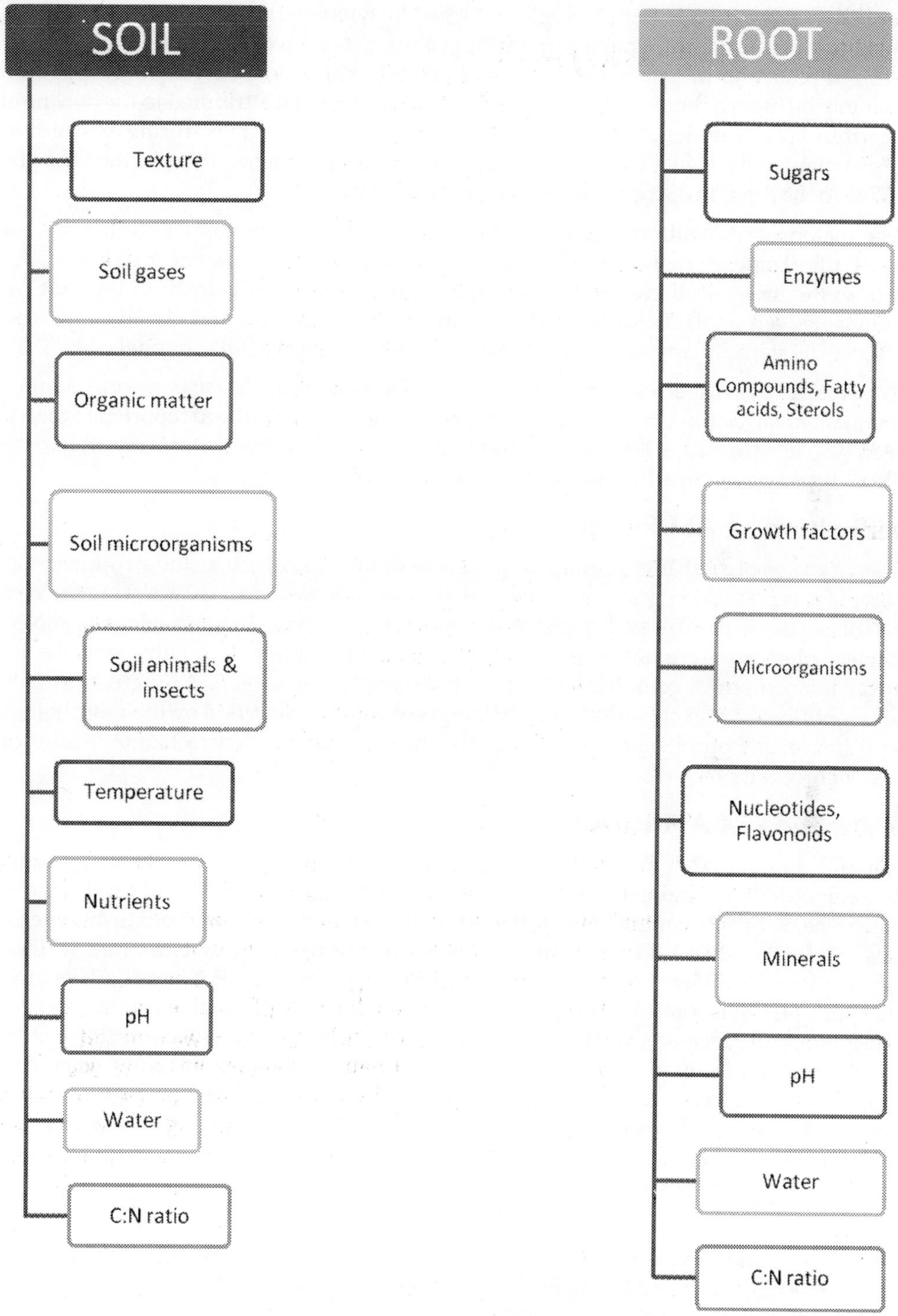

Figure 28.1: Soil and Root Factors Influencing Growth and Development of AM

hyphae (ERH) in the soil (Graham and Miller, 2005). The symbiotic relationship of AM has been found to be mainly mutualistic, but sometime parasitic in nature. It involves exchange of water and nutrients between the plant and the fungus. About 10-12 per cent of the total C fixed by the plant is used for maintaining the fungus (Fitter 1991). The parasitic nature has been attributed to the decline in growth of mycorrhizal plants under certain environmental conditions, especially during early growth stages of the host plant. If the C loss is not compensated by enhanced photosynthesis, then the plant suffer losses due to the drain in nutrients for maintaining AM fungi.

The anatomy of AM differs with the host-endophyte combinations or with certain soil conditions. Two anatomical types of root colonization *Arum* type and *Paris* type have been defined. *Arum-type* is defined on the basis of an extensive intercellular phase of hyphal growth in the root cortex and development of terminal arbuscules on intracellular hyphal branches, and (ii) *Paris-type is* defined by the absence of intercellular phase and presence of extensive intracellular hyphal coils.

Whilst, the fungal species *Glomus intraradices, G.mosseae* and *G.versiforme* form Arum-type, the *Gigaspora margarita, Glomus coronatum* and *Scutellospora calospora* have been reported to form the Paris-type AM (Cavagnaro *et al.*, 2001). Recent studies have shown that the hyphal coils in Paris type have functional homology with arbuscules (Cavagnaro *et al.*, 2003).

Identification of AM Fungi

Identification of AMF is important for a number of reasons, which include: (i)understanding the phenotypic diversity in natural, managed and disturbed ecosystems; (ii)assessing the competitive abilities of isolates in plant growth promotion activity; (iii)studying the physiological, molecular and biochemical phenomenon in mycorrhiza; and (iv)maintaining adequate quality control and purity of commercial inoculum. A complete and accurate description of each AM fungus is essential for its proper identification and classification. AM fungi are usually identified by the morphology of their "spores" but several problems are being faced by the taxonomists in correct identification of the AM fungus (Mehrotra, 1997).

Classification of AM Fungi

Classification of AMF has undergone modifications, changing from Class Zygomycetes to Glomeromycetes, Order Endogonales to Glomales (Morton and Benny, 1990) and Family Endogonaceae to Glomaceae, with new genera being added from time to time to accommodate discoveries of new species (Mehrotra, 2005). Fungi forming AM associations are now classified in the phylum Glomeromycota (Schüßler *et al.*, 2001), which include 11 genera *viz.* (1) Glomus, (2) Sclerocystis, (3) Acaulospora, (4) Gigaspora, (5) Entrophospora, (6) Scutellospora, (7) Archaeospora, (8) Paraglomus, (9) Geosiphon, (10) Pacispora, and (11) Diversispora. The order Glomales was erected by Morton and Benny (1990) to include all AM fungi. The ordinal name Glomales has now been changed to 'Glomerales' (Schüßler *et al.*, 2001) and the family name Glomaceae has been changed to 'Glomeraceae'. Members of the phylum Glomeromycota, once thought to be related to Endogonaceae, are now known to be closer to Dikarya, a sub-kingdom of fungi that includes the phyla Ascomycota and Basidiomycota, both of which in general produce dikaryons, may be filamentous or unicellular, but are always without flagella.

The classification given here is that of Schüßler *et al.* (2001) with emendations of Oehl and Sieverding (2004), Walker and Schüßler (2004), Sieverding and Oehl (2006), Spain *et al.* (2006), Walker *et al.* (2007a, b), and Palenzuela *et al.* (2008).

GLOMEROMYCOTA C. Walker and Schuessler

Glomeromycetes Cavalier-Smith

Archaeosporales C. Walker and Schuessler

Ambisporaceae C. Walker, Vestberg and Schuessler

Ambispora Spain, Oehl and Sieverd.

Archaeosporaceae J.B. Morton and D. Redecker emend. Oehl and Sieverd.

Archaeospora J.B. Morton and D. Redecker

Intraspora Oehl and Sieverd.

Geosiphonaceae Engler. and E. Gilg emend. Schuessler

Geosiphon (Kütz.) F. Wettst.

Diversisporales C. Walker and Schuessler

Acaulosporaceae J.B. Morton and Benny

Acaulospora Gerd. and Trappe emend. S.M. Berch

Kuklospora Oehl and Sieverd.

Acaulospora Gerd. and Trappe emend. S.M. Berch

Diversisporaceae C. Walker and Schuessler

Diversispora C. Walker and Schuessler

Otospora Oehl, J. Palenzuela and N. Ferrol

Entrophosporaceae Oehl and Sieverd.

Entrophospora R.N. Ames and R.W. Schneid. emend. Oehl and Sieverd.

Gigasporaceae J.B. Morton and Benny

Gigaspora Gerd. and Trappe emend. C. Walker and F.E. Sanders

Scutellospora C. Walker and F.E. Sanders

Pacisporaceae C. Walker, Blaszk., Schuessler and Schwarzott

Pacispora Oehl and Sieverd.

Glomerales J.B. Morton and Benny

Glomeraceae Piroz. and Dalpé

Glomus Tul. and C. Tul.

Paraglomerales C. Walker and Schuessler

Paraglomaceae J.B. Morton and D. Redecker

Paraglomus J.B. Morton and D. Redecker

Two new families of Glomales, Archaeosporaceae and Paraglomaceae, with two new genera *Archaeospora* and *Paraglomus*, based on concordant molecular and morphological characters were included in the classification (Morton and Redecker, 2001). In this system of classification, fungi forming or considered to form arbuscular mycorrhizae are placed in four orders, *i.e.*, Archaeosporales,

Diversisporales, Glomerales, and Paraglomerales, comprising ten families and thirteen genera, belonging to the class Glomeromycetes of the phylum Glomeromycota (Oehl and Sieverding, 2004; Palenzuela *et al.*, 2008; Schüßler *et al.*, 2001; Sieverding and Oehl, 2006; Spain *et al.*, 2006; Walker and Schüßler, 2004; Walker *et al.*, 2007a, b). *Geosiphon pyriformis* of the family Geosiphonaceae (Archaeosporales) does not form arbuscular mycorrhizae. It forms endocytosymbioses with cyanobacteria (*Nostocs* sp.) and was included into the Glomeromycota based on only its close molecular relationship.

Members of the order Diversisporales form mycorrhizae with arbuscules, frequently lacking vesicles, with or without auxiliary cells. Spores develop either inside (entrophosporioid spores of the genera *Entrophospora* and *Kuklospora*) or laterally on the neck of a sporiferous saccule (acaulosporioid spores of the genus *Acaulospora* and *Otospora*), from a bulbous base on the sporogenous hypha (gigasporioid spores of the genera *Gigaspora* and *Scutellospora*), or blastically at the tip of a sporogenous hypha (glomoid spores of the genera *Diversispora* and *Pacispora*).

Occurrence and Diversity

Arbuscular mycorrhizal fungi occur globally in most ecosystem including dense rain forests, scrub, savanna, open woodlands, grasslands, heaths, sand dunes, deserts, and more commonly in agricultural lands (Treseder and Cross, 2006; Mehrotra, 2008). Occurrence of AM fungi in diverse environments such as desert soils, degraded grasslands, eroded soils, sewage irrigated soils, reclaimed coal mine spoils, heavy metal polluted soils, have been studied. They are virtually ubiquitous and so have a broad ecological range, being present in temperate, tropical, subtropical, arid, semi-arid and arctic regions of the earth. In a natural ecosystem mixed populations of AMF coexists, with certain fungi becoming dominant in particular patches and subsequently being replaced as environmental and cultural conditions change.

It has been shown that AM fungi occur widely in agricultural farms, vegetable farms, orchards, forest lands, grasslands, tropical rain forests, tropical virgin forests, tropical secondary forest, artificial forests, natural secondary forests, and natural reserves. Tropical plants, alpine plants, halophytes, xerophyte, hydrophyte, geophyte, parasitic plants, are also found to be mycorrhizal. AM occurs in families of Angiosperms including Leguminosae, Poaceae, Rosaceae, Solanaceae, Liliaceae, etc. Plants belonging to families, such as Cyperaceae, Brassicaceae, Caryophyllaceae, Juncaceae, Chenopodiaceae and Amaranthaceae are assumed never to form mycorrhizal associations or do so rarely. The exact reason why some plants do not form mycorrhizas is not fully known, but it may be related to the interaction between the fungus and the plant at the cell wall or middle lamella level (Tester *et al.*, 1987), high concentrations of salicylic acid (Medina *et al.*, 2003) and or presence of fungitoxic compounds in root cortical tissue or in root exudates (Quilambo, 2003).

To date, only fewer than 200 species of AMF have been described (Redecker and Philipp, 2006). Singh and Adholeya (2002) studied the diversity of AMF in 114 samples from 11 wheat growing regions of India. They found 33 species belonging to 5 genera with Glomus as the dominant genus. *Glomus fasciculatum* and *G. albidum* were found to be the most frequently occurring species. In a review on the biodiversity and distribution of AMF in different ecosystems and plant communities in China, Gai *et al.* (2006) concluded that a total of 104 AMF species within nine genera have been reported in the rhizosphere of different plants species in various habitats in China since 1980's and 12 new species were discovered in the surveys. The frequency of occurrence and abundance of sporulation by *Glomus* species are much higher than in other genera. *Acaulospora* is the second most dominant genus in China, ranking after *Glomus* in the north and west and similar to *Glomus* in the south and east.

Gigaspora and *Scutellospora* species are found mostly in the coastal sand dunes and islands, and have been reported from southern and eastern areas of China that are near the sea (Wang *et al.*, 2004). In another study, the mycorrhizal status of plant species in terrestrial ecosystems in China were examined in a total of nearly 800 plant species belonging to 150 families (Wang and Shi, 2008). These plants included food crops, economic crops, vegetables, fruits, ornamental plants, medicinal plants, wild weeds, trees, etc. A total of 122 AM fungal species within 11 genera, including 8 new species, have been reported in various environments. The most common and widely distributed genera is *Glomus* (64 species), followed by *Acaulospora* (26 species) and *Scutellospora* (16 species). Occurrence of certain AMF species have been linked with soil factors: *G. mosseae* with fine textured, fertile and high pH soils; *Acaulospora laevis* with coarse textured soils; and *Gigaspora* species with sand dune soil (Kendrick and Berch 1985). *Glomus intraradices*, the most widely available species, is suitable for soils from about pH 6 to 9. It is known that soil factors such as pH restrict the distribution of some taxa (Abbott and Robson, 1991). It has been interpreted that most species of *Gigaspora* occur in soils with pH values of 5.3 or lower, while most *Glomus* and *Acaulospora* spp. occur at pH 6.1 or higher. *Glomus* appears to dominate in alkaline and neutral soils, while *Acaulopspora* sporulates more abundantly in acid soils (Zhang *et al.*, 1998; Gai and Liu, 2003).

In a study on the AM fungal diversity in forest restoration area in Thailand, Nandakwang (2008) found that AMF belonging to the genera *Glomus* and *Acaulospora* were dominant. Out of 21 AMF species 12 belonged to *Glomus*, 6 to *Acaulospora* and 3 to *Scutellospora*. The most abundant species were *Glomus multicaule*, *Acaulospora elegans* and *Scutellospora pellucida*. Dominance of *Glomus* and *Acaulospora* was attributed to their sporogenous character and production of small spores in a short time, as compared to the large spores produced by *Gigaspora* and *Scutellospora*. The widespread occurrence of *Glomus* may also be attributed to their lower host preference and wide range of pH tolerance. Species of Glomineae tend to have effective spores as well as highly infective extra-radical hyphae but species of Gigasporineae are infective when spores are used as propagules (Klironomos and Hart, 2002).

Different plant species cause the build-up of diverse populations of AMF in the soil. Host plant can be selective in supporting the growth and reproduction of AM fungal species under a particular set of environmental conditions (Mehrotra, 1996). A host variety can switch from compatible to incompatible mycorrhizal associations with a change in environmental variable *e.g.*, phosphorus levels (Li *et al.*, 1991), soil water content, pH, salinity, temperature, intensity and quality of light, etc. As a result, it is doubtful that a single host species will be found that establishes and maintains compatible mycorrhizal associations universally in all environments. Apparent host specificity may, however, occur if host susceptibility does not coincide with the propagules infectivity (Jasper *et al.*, 1987).

Assessment of diversity of AM association has been far more difficult due to rudimentary understanding of the genetics of AM fungi (Kuhn *et al.*, 2001). Intra-specific AM species variability also poses practical problems in using molecular tools for detection of species and strains from field soil (Sanders *et al.*, 1995). Intra-specific diversity has been revealed in *Glomus mosseae* by total protein profiles and ITS-RFLP profiles (Giovannetti *et al.*, 2003) and using rDNA ITS sequence (Antoniolli *et al.*, 2000). Taxon-specific Polymerase Chain Reaction (PCR) primers can be used to amplify fungal DNA from AM fungi mycelial tips (Di Bonito *et al.*, 1995; Reddy *et al.*, 2005) and this approach is now being used to compare fungal diversity from different roots and soil environments. Application of real-time PCR to quantify a single isolate of AM fungus in root segments could offer the opportunity for direct and specific quantification of selected taxa (Alkan *et al.*, 2004), but scaling-up the sampling for root and soil communities will require improved techniques for isolation of DNA and the design of multiple and nested primers procedures (Graham and Miller, 2005).

Seasonal patterns in the formation of AM have been found to vary considerably (Sanders, 1990; Lakshman *et al.*, 2006). Gemma *et al.* (1989) suggested that a combination of abiotic factors such as temperature and light and the biotic factors such as the amount of photosynthates products, quality and quantity of root exudates and fluctuation of root hormone levels occurring during flowering and growth cessation are the primary non-genetic determinants of AMF sporulation. Diaz and Honrubia (1994), however, suggested that increase in spore production with the end of growing season might be related with root senescence; the presence of dead roots perhaps stimulated sporulation. The seasonal breakdown of the fungal network provides the soil with carbon and nitrogen, which have a direct impact on the survival of the soil microflora and microfauna and subsequently on soil biodiversity. The ability of AMF to improve the soil structure by enhancing the stability of soil aggregates (Miller and Jastrow, 2000) through a glycoprotein, termed 'Glomalin' (Wright and Upadhyaya, 1996; Rillig, 2004), has a significant role in changing the dynamics of soil biodiversity. It has been proposed that Glomalin improves the stability of soil by avoiding disaggregation by water (Wright *et al.*, 2007; Bedini *et al.*, 2009) and a strong relationship between Glomalin concentration and the amount of water stable aggregates (WSA) has been demonstrated (Rillig, 2004). Differences in production of Glomalin and the ability to stabilize soil aggregates have been shown among different *Glomus* species (Schreiner *et al.*, 1997) and isolates of the same fungal species (Bedini *et al.*, 2009).

Although considerable research efforts have been made on the factors affecting the occurrence and diversity of AM fungi, but most of these studies are limited to the regional boundaries and specific host plants, as a result conclusive evidence of the effect of the various factors on AM fungi could not be derived.

Interaction of AMF with Other Microorganisms

Mycorrhizas influence soil microbial populations in the 'rhizosphere' and 'hyphosphere'. Rambelli (1973) suggested the term 'mycorrhizosphere' as a substitute to 'rhizosphere', as there is a significant and dynamic microbial interaction in the mycorrhizal roots. Microbial population can either benefit or interfere with the establishment of AM associations (Vosatka and Gryndler, 1999; Gryndler, 2000). The success of AMF is, therefore, dependant upon the mutualistic association and complex interactions between the beneficial microbes and the plants in the dynamic soil system. In the mycorrhizosphere, AMF interact with beneficial rhizosphere microorganisms including free-living nitrogen fixing bacteria and plant growth promoting rhizobacteria (PGPR). Phosphorus solubilizing microorganisms (PSM), which include species of *Pseudomonas*, *Bacillus*, *Flavobacterium*, *Arthrobacter*, *Pencillium* and *Aspergillus* have been found associated with AMF. These microorganisms can produce compounds that increase root cell permeability, thereby increasing the rates of root exudation and in turn stimulate the growth of AMF hyphae in the root and rhizosphere (Jeffries *et al.*, 2003). The quantity of root exudates, which is lost from the living roots affect the microbial population. Physical factors, which can enhance root exudation, include influence of low temperature, water stress, and mechanical contact.

Bacterial communities have been reported to promote germination of AM fungal spores and can increase the rate and extent of root colonisation by AMF. Dual inoculation with AM fungi and PGPRs resulted in enhanced plant biomass compared with inoculation with either one of them (Sumana and Bagyaraj, 2000; Sailo and Bagyaraj, 2006). *Trichoderma harzianum* generally used as a biological control agent is also known to produce plant growth promoting substances, thereby enhancing root growth and anti-microbial compounds deleterious to plant pathogens (Jayanthi *et al.*, 2003). Co-inoculation of *Glomus bagyarajii* with *T. harzianum* has been reported to significantly enhance P uptake by plants compared with AM fungus *G. bagyarajii* inoculation alone (Sailo and Bagyaraj, 2006). Calvet (1992) observed enhanced hyphal development and spore germination of *G. mosseae* in the presence of *Trichoderma* species and attributed the stimulation of plant growth to the production of volatile compounds by the latter. Several researches have demonstrated that AM influence legume-rhizobium symbiosis. Enhanced P nutrition by AM results in increased nodulation and N_2 fixation by the leguminous plants (Vázquez *et al.*, 2002).

Interaction of AMF with Soil Organisms

The AM fungal network in the soil interacts competitively with arthropods (Milleret *et al.*, 2009) and nematodes. On the other hand, AMF are used as food by grazing collembolans (*e.g.*, springtail) (Klironomos *et al.*, 1999), and burrowing earthworms (Fitter and Sanders, 1992). Interaction between AMF and collembola showed that these insects may reduce plant biomass by grazing on hyphae and spores of AMF (Endlweber and Scheu, 2007). Scheu (1994) showed that earthworms may enhance nitrogen mineralization in the soil thus increasing N uptake and therefore plant growth. The presence of earthworms, however, reduced the positive effect of AMF on root biomass.

Effect of Agricultural Practices on AM

Agricultural practices including use of fertilisers and pesticides, tillage, monocultures and the growing of non-mycorrhizal crops have been found to be detrimental to AMF. Application of fertilizers and pesticides has stimulatory, depressive, or no significant effect on the development of AM. Fertilizer application may also affect the composition of AM species such that less efficient mycorrhizal species become dominant (Johnson, 1993). They promote growth of microbial communities either by promoting growth directly by providing nutrients or indirectly by stimulating plant growth and enhancing C flow (Buyanovsky and Wagner, 1987). Pesticides have been found to have both negative and positive effects on the growth of AM. In general, pesticides that enhance root exudation may increase mycorrhizal infection. Systemic fungicides have been found to be more harmful to AM colonization than non-systemic fungicides (Menge, 1982). Organic manures increase the biological activity in the soils and also influence the AMF activity. Hyphae of AMF can grow saprophytically on organic matter (Warner, 1984).

Soil disturbances disrupt the hyphal network resulting in delays in AMF infectivity on plant roots and reduction in spore production (Jasper *et al.*, 1991). Tillage or soil disturbance may disrupt the extra-radical mycelium and thereby reduce nutrient uptake, growth and final yield (Mozafar *et al.*, 2000). Less severe forms of soil disturbance, including agricultural tillage, the rate and method of P application, liming fire and erosion can also influence the mycorrhizal fungus propagules and soil diversity. Propagules of mycorrhizal fungi may be absent from soils, where severe soil disturbance has resulted in topsoil loss or where host plants are limited by adverse soil or site factors such as salinity, aridity, water logging, or climatic extremes. Excessive NaCl levels in soil inhibit mycorrhizal formation and restrict the activity of most AM fungi, but some can tolerate these conditions. The species most commonly found in salinized areas has been *Glomus mosseae,* which suggests that it might be adapted to saline conditions (Maia and Yano-Melo, 2005). Continuous monoculture of certain crops may influence AMF species composition (Johnson *et al.*, 1992). Changes in populations of AMF have been observed when ecosystems are converted to monocultures or severely disturbed, providing indirect evidence for habitat preferences by these fungi.

Quantification

Most information about the dynamics and diversity of AM fungi in field soils is derived from studies of the abundance and types of spores or the total length of mycorrhizal root. There are several methods used to isolate spores of AMF from field soil, which include wet sieving and decanting (Gerdemann and Nicolson, 1963), floatation adhesion (Sutton and Barron, 1972), air stream fractionation (Tommerup, 1982), water-sucrose centrifugation (Ianson and Allen, 1986) and fixing soil slurries to filter paper (Khalil *et al.*, 1994). Identifying AM fungi in roots is made more difficult by a lack of stable morphology and because of the small amount of AM fungal DNA in a rather high background content of DNA of other fungi in roots (Renker *et al.*, 2003; Graham and Miller, 2005). A comparison of visualization techniques for recording AM colonization have been made by Gange *et al.* (1999).

Assessment of AMF species in field soils, however, is not easy because it is mainly based on spore wall characteristics (Walker, 1992; Mehrotra,1997) and spores are present at different stages of development (Morton, 1995). Spore population appears to be governed by a number of interacting factors, among them are initial spore counts, soil nutrients, and texture, moisture, host plant genotype, plant cover, etc. Douds and Millner (1999) have questioned the use of spores for describing AMF community diversity due to the following limitations: (i) the relative abundance of spores of a species may not reflect its functional importance or even its relative biomass contribution to the community as a whole *i.e.*, the number of spores in the soil may not reflect the relative amount of colonization of roots by the fungus or the amount and distribution of hyphae in the soil and (ii) non-sporulating species may be present. The absence of spores of an AMF species does not necessarily indicate its absence in the community. One of the major limitations of direct field assessment is the low level of spores that can be collected. Further, spores of some species can be absent at the time of sampling, even though they may be present within roots, resulting in low species richness.

Succession

Arbuscular mycorrhizal fungi may influence plant communities by affecting species richness (Gange *et al.*, 1990; van der Heijden *et al.*, 1998) or species composition (Klironomos *et al.*, 2000; O'Connor *et al.*, 2002). They also determine the plant biodiversity, ecosystem viability and productivity (Chaudhri, 2005). A change in the species of AM fungi that colonize a host plant has the potential to impact plant carbon exudation (Schwab *et al.*, 1984), rates of decomposition (Hodge *et al.*, 2001), addition of carbon to soil through hyphal turnover (Staddon *et al.*, 2003), retention of carbon and nutrients in recalcitrant fungal tissues (Rillig *et al.*, 2001) and plant nutrient uptake (Cavagnaro *et al.*, 2005). Succession of AM is determined to a large extent by the quality and quantity of mycorrhizal spores present in the soil and by build-up of mycorrhizal networks capable of rapidly colonizing any newly germinating plant. The dominant type of mycorrhiza changes from AM to ecto-and ericoid mycorrhiza as succession progresses from grassland to deciduous forest, boreal forest and heath along the gradient from warm-dry to cold-wet climates (Pankow *et al.*, 1991).

Role of AM in Plant Nutrition

Arbuscular mycorrhiza plays an important role in plant nutrition by making nutrients, especially phosphorus, accessible to the plants beyond the depletion zones (Zhu *et al.*, 2001) and from sources not necessarily otherwise accessible to the roots. However, the effectiveness of AM in enhancing nutrient uptake has been shown to be affected by complex interactions between the nutrients and host-fungus interactions. Differences in performance of AM species have been observed by several workers. For example, Sailo and Bagyaraj (2005) noted that the P content of the root and shoot in plants inoculated with *Glomus bagyarajii* was significantly superior compared to *Scutellopsora calospora, Glomus fasciculatum, G. mosseae, G. monosporum, G. macrocarpum, G. intraradices and Gigaspora margarita.* There was a positive correlation between the intensity of mycorrhizal colonization and growth response.

Increased uptake of one nutrient, but a reduction in another followed by AMF colonisation (Kothari *et al.*, 1990) has been attributed to the effect which may be mediated by the concentration of other soil nutrients (Liu *et al.*, 2000). For example, a reduction in host plant manganese (Mn) absorption following AMF colonisation has been observed, when uptake of other nutrients have increased (Azaizeh *et al.*, 1995). Extraradical hyphae of AM augment the uptake of nutrients from up to 12 cm away from the root (Cui and Caldwell, 1996).

Phosphorus

Several studies have reported that AMF inoculation is likely to be more efficient in extremely P-limited soils than P-fertilized soils (Hu *et al.*, 2009a and b). Sorensen *et al.* (2005) have found that high soil P level or high soil inoculum level was most likely responsible for the limited response of increased mycorrhiza formation on plant growth and nutrient concentrations. AMF species have been reported to differ in hyphal production, hyphal patterns and activity in relation to the host root (Hart and Reader, 2002; Smith *et al.*, 2000). Smith *et al.* (2000) showed that the AM fungus *Scutellospora calospora*, which formed more fungal biomass and thus consumed more carbon was not as efficient in P transfer as *Glomus caledonium*, which produced relatively less biomass in association with *Medicago truncatula*.

Inorganic phosphate (*Pi*) absorbed from soil by the ERH hyphae of AM fungi is transformed into polyphosphate (polyP), which is then translocated into the IRH in mycorrhizal roots and may be supplied to the plant after hydrolysis in arbuscules (Funamoto *et al.*, 2007; Takanishi *et al.*, 2009). PolyP is a linear polymer of *Pi* residues linked by high-energy phosphoanhydride bonds. The chain-length of polyP in AM fungi is variable (Ezawa *et al.*, 1999, 2002), being longer in ERH than in IRH (Solaiman *et al.*, 1999). Both soluble and insoluble (long chain granular) forms of polyP have been found in nature (Clark *et al.*, 1986). Studies with labelled P have revealed that AM increases P in plants (Mehrotra *et al.*, 1995).

Nitrogen

Nitrogen occurs in the soil predominantly in the form of nitrate and ammonia, which are water soluble and readily available for absorption. The availability of N is also affected by the presence of biotrophs, especially *Rhizobium*. AM fungal hyphae have been credited with the uptake and transfer of N from organic sources (Hodge *et al.*, 2001). Studies with labelled N have revealed that AM increase N uptake by plants (Mädder *et al.*, 2000). However, there is little reciprocal transfer of N from the plant to the fungi, which makes uptake and assimilation of N by the symbiont essential for its growth.

Zinc

Zinc uptake by plants is facilitated as diffusion through membranes specific for zinc ion or mediated by specific transporter(s) and about 90.5 per cent of the total zinc required by plants moves towards the roots by diffusion. Zinc is present in the lattice structure of the soil and therefore, unavailable to meet the plant's nutritional requirements. The amount of Zn delivered to the plants via the mycorrhizal pathway has been found to vary significantly between studies (Cooper and Tinker, 1978; Jansa *et al.*, 2003; Mehravaran *et al.*, 2000).

Role in Plant Protection

Different mechanisms have been shown to play a role in plant protection by AMF which include improved plant nutrition, damage compensation, competition for colonization sites and photosynthates, changes in the root system, changes in rhizosphere, microbial population and activation of plant defence mechanisms. Once the symbiosis is established with the host plant AMF are able to confer tolerance against root pathogens (Pozo *et al.*, 2002). AMF could elicit localized defence responses in colonized tissues and systemic responses in non-colonized tissues of mycorrhizal plants (Liu *et al.*, 2007). For example, Aguin-Casal *et al.* (2006) showed that inoculation of grapevine with AMF *Glomus aggregatum* resulted in reduced disease index of *Armillaria mellea* in several vine root stocks. Mycorrhizal root tissues are more lignified than non-mycorrhizal ones, particularly in the vascular region. This restricts the pathogenic fungus to the cortex. AM altered physiology of roots may prevent penetration and retard the development of nematodes. Some workers have suggested that improved nutrition may protect the plant against pathogens (Karagiannidis *et al.*, 2002).

Role of AMF in protection of plants from *Olpidium brassicae* in lettuce and tobacco (Schoenbeck and Dehne, 1979), *Thielaviopsis basicola* in tobacco (Baltruschat and Schoenbeck, 1975), *Pyrenochaeta terrestris* in onion (Becker, 1976), *Fusarium oxysporum* f.sp. *radicis lycopersici* in tomato (Caron *et al.*, 1986), *Macrophomina phaseolina* in mungbean (Jalali *et al.*, 1990) and *Armillaria mellea* in grapevine (Nogales *et al.*, 2009) have been elicited. The variations in levels of polyamines (Pas) detected at the beginning of the pathogenic infection of *Armillaria mellea* causing white root rot disease in grapevines suggested that Pas may have potential role in the signalling mechanism of the tolerance of mycorrhizal plants against *A. mellea* (Nogales *et al.*, 2009). Amino acid content, particularly arginine has been found to be high in mycorrhizal plants. Arginine and root extracts of AM plants reduce chlamydospore production in *Thielaviopsis basicola.* A changed exudation pattern of mycorrhizal plants has been suggested to be at least partially involved in the altered susceptibility of mycorrhizal plants towards soil-borne microorganisms (Vierheilig 2004) such as fungi, bacteria, and nematodes. In *in vitro* studies, root exudates from mycorrhizal strawberry plants reduced the sporulation of *Phytophthora fragariae* (Norman and Hooker 2000), root exudates from mycorrhizal potato plants increased hatching of nematodes (Ryan and Jones 2004) and root exudates from mycorrhizal tomato plants enhanced microconidia germination of the tomato pathogen *Fusarium oxysporum* f. sp. *lycopersici* (Scheffknecht, 2006).

Arbuscular mycorrhiza also provides increased protection against environmental stresses including drought (Augé and Moore, 2005; Abo-Ghalia and Khalafallah, 2008), cold, salinity and pollution. The formation of AM can also protect the plant against the accumulation of Zinc in plant tissues to high concentrations (Timothy, 2008). Mycorrhizal colonization of roots has been shown to increase drought tolerance of wheat, maize, soybean, onion, red clover, lettuce, and other plant species.

Commercial Application

Arbuscular mycorrhizal technology can be successfully employed in situations like reclamation of wastelands and disturbed sites. When total AM fungal biomass is too low in a soil, AMF must be added back into the soil. Low input crop production systems such as organic farming are generally more favourable to AMF as they have the potential to substitute for the fertilizers and biocides which are not permitted in organic systems (Gosling *et al.*, 2006).

An effective and efficient AM fungus should be able to form an extensive network of hyphae in the soil, enhance nutrient uptake by the plants, persists in the soil and should have a broad host range. The selection of an efficient AMF species or a mixture of species should take into account the multiple benefits that the fungus or a consortium of AM fungi will exert on the colonized plant. Trap culture of isolates of AM fungi colonizing roots are to be prepared and used for field inoculation. To establish the trap culture of inoculum from mono-specific cultures, it is necessary to trap the healthy spores of AMF. Spores collected directly from the field soil are not used as they are mixed cultures of different species, contaminated and may not be viable.

Arbuscular mycorrhizal fungi have been applied as (i) a mixture of spores, mycelium, and roots, (ii) peat, sand, vermiculite, perlite and calcined clay based mixes, (iii) gel or liquid based mixes, and (iv) seed pellets. Application of AM inoculum in soil can be done in various ways, which include (i) incorporating a layer of spores, mycelium and pre-colonized roots into the container media or soil, (ii) placing AM inoculum in the dibble hole of the soil before planting the plant material, (iii) pelleting the seed with AM inoculum. The most economical and efficient time of application is during propagation or while transplanting liners. Use of pesticides is to be avoided in the first three weeks of AM inoculation.

The mass production of AM inoculum has been done using various methods, which include soilless methods (*e.g.* aeroponic culture, hydroponic culture, and in vitro dual culture) and soil-based methods (*e.g.*, pot cultures, on-farm production, etc.). Several plant hosts, which include bahia grass (Douds *et al.*, 2006), maize, sorghum, and barley (Chaurasia and Khare, 2005) have been tested for mass production of AM inoculum.

Availability of commercial AM inoculants on a large scale is still a major issue in India. Some companies have adopted procedure for commercial production of AMF inoculants. For example, the technology for mass production of AM fungi suitable for certain crops including plantation crops, developed by The Energy and Resources Institute (TERI), New Delhi has been adopted by two industries namely M/s Cadila Pharamaceutical Ltd., Ahmedabad, and KCP Sugar and Industries (Pvt.) Ltd., Chennai. However, their utilization at a commercial level is still limited because of their incompatibility with the large use of chemical inputs such as fertilizers and pesticides. Different formulated products are now available on the market, which creates the need for the establishment of quality standards for widely accepted quality control. Progress should be made towards registration procedures that stimulate the development of the mycorrhizal industry.

Conclusion

The integration of genetic, molecular, biochemical, physiological, and ecological approaches will lead to a better understanding of AM and their effective use in crop production systems and reclamation of wastelands and disturbed sites. The specific research questions that need to be addressed in future may include but not limited to the following:

1. How to select effective and competitive multifunctional AMF for a crop?
2. What is the amount of AM inoculum or infective propagules in the soil that would be sufficient to infect the root?
3. Which environmental factors are most favourable for AMF to infect the root?
4. Which environmental factors favour AMF the most in the spread of mycelium in the soil and the root?
5. How the host metabolism affects the growth of AM?
6. Which secondary metabolites produced by other microorganisms affect the growth and development of AM?
7. How phytoalexins affect the growth of AM in the root?
8. How AMF persists in different soil environments, especially under stressful conditions?
9. How large scale usage of AM inocula can be made practical and effective in crop production systems?
10. How to assess the number of viable AMF propagules in commercial inocula for quality assurance and regulation?

In recent years, there has been a major shift in cropping patterns due to changes in climatic conditions and rainfall patterns. Mixed cropping, which was practiced in rainfed areas is now limited to hot arid and humid regions. Arbuscular mycorrhiza increases plant efficiency in absorbing water and nutrients from the soil, thus reducing fertility and irrigation requirements. It increases drought resistance, pathogen resistance/protection, seedling growth, rooting of cuttings, and plant transplant establishment. A better understanding of the succession and ecology of AM fungi will be helpful in selection of effective and efficient species of AMF for application in the field.

References

Abbott, L. K. and Robson, A. D. 1991. Factors influencing the occurrence of vesicular-arbuscular mycorrhizas. Agric. Ecosyst. Environ., 35:121-150.

Abo-Ghalia, H. H. and Khalafallah, A. A. 2008. Responses of wheat plants associated with arbuscular mycorrhizal fungi to short-term water stress followed by recovery at three growth stages. Journal of Applied Sciences Research, 4(5): 570-580.

Aguin-Casal, O., Montenegero, D. and Manrilla, J. P. 2006. Proteccion de la vid frente a *Armillaria mellea* mediante la applicacion de hongos micorricicos. Nutri Fitos, 163: 27-33.

Allen, M. F. 1996. The ecology of arbuscular mycorrhizas: a look back into the 20th century and a peek into the 21st. Mycol. Res., 100:769–782

Alkan, N., Gadkar, V., Coburn, J., Yarden, O. and Kapulnik, Y. 2004. Quantification of the arbuscular mycorrhizal fungus *Glomus intraradices* in host tissues using real-time polymerase chain reaction. New Phytol., 161: 877-885.

Antoniolli, Z. I., Schachtman, D. P., Ophel, K. K. and Smith, S. E. 2000. Variation in rDNA, its sequence in *Glomus mosseae* and *Gigaspora margarita* spores from a permanent pasture. Mycol. Res., 104: 708-715.

Augé, R. M. and Moore, J. L. 2005. Arbuscular mycorrhizal symbiosis and plant drought resistance. Pp. 136-162. In: V.S. Mehrotra (ed.) Mycorrhiza: Role and Applications. Allied Publishers Pvt. Ltd., New Delhi.

Azaizeh, H. A., Marschner, H., Roemheld, V. and Wittenmayer, L. 1995. Effects of a vesicular–arbuscular mycorrhizal fungus and other soil microorganisms on growth, mineral nutrient acquisition and root exudation of soil-grown maize plants. Mycorrhiza, 5:321–327.

Baltruschat, H. and Schoenbeck, F. 1975. The influence of endotrophic mycorrhiza on the infection of tobacco by *Thielaviopsis basicola*. Phytopath. Z., 84: 172-188.

Becker, W. N. 1976. Quantification of onion vesicular-arbuscular mycorrhizae and their resistance to *Pyrenochaeta terrestris*. Ph.D. Dissertation, University of Illinois, Urbana.

Bedini, S., Pellegrino, E., Avio, L., Pellegrini, S., Bazzoffi, P., Argese, E. and Giovannetti, M. 2009. Changes in soil aggregation and glomalin-related soil protein content as affected by the arbuscular mycorrhizal fungal species *Glomus mosseae* and *Glomus intraradices*. Soil Biol. Biochem., 41:1491-1496.

Brown, M. S. and Bethlenfalvay, G. J. 1988. The *Glycine-Glomus Rhizobium* symbiosis. VII. Photosynthetic nutrient use efficiency in nodulated mycorrhizal soybeans. Plant Physiol., 86:1292-1297.

Brundrett, M. C. 2004. Diversity and classification of mycorrhizal associations. Biol. Rev. Camb. Philos. Soc., 79: 473-495. DOI 10.1017/s1464793 103006316.

Brundrett, M. C. 2009. Mycorrhizal associations and other means of nutrition of vascular plants: understanding the global diversity of host plants by resolving conflicting information and developing reliable means of disgnosis. Plant and Soil, 320:37-77. DOI 10.1007/s11104-008-9877-9.

Buyanovsky, G. A. and Wagner, G. H. 1987. Carbon transfer in a winter wheat (*Triticum aestivum*) ecosystem. Biol. Fertil. Soils, 5: 76-82.

Calvet, C., Baream, J. M. and Pera, J. 1992. *In vitro* interactions between vesicular-arbuscular fungus *G. mosseae* and some saprophytic fungi isolated from organic substrates. Soil Biol. Biochem., 24: 775-780.

Caron, M., Fortin, J. A. and Richard, C. 1986. Effect of inoculation sequence on the interaction between *Glomus intraradices* and *Fusarium oxysporum* f. sp. *radicis-lycopersici* in tomatoes. Can. J. Plant Pathol., 8:12–16.

Cavagnaro, T. R., Gao, L. L., Smith, F. A. and Smith, S. E. 2001. Morphology of arbuscular mycorrhizas is influenced by fungal identity. New Phytol., 151: 469-475.

Cavagnaro, T. R., Smith, F. A., Ayling, S. M. and Smith, S. E. 2003. Growth and Phosphorus nutrition of a Paris-type arbuscular mycorrhizal symbiosis. New Phytol., 157: 127-134.

Cavagnaro, T. R., Smith, F. A., Smith, S. E. and Jakobsen, I. 2005. Functional diversity in arbuscular mycorrhizas: exploitation of soil patches with different phosphate enrichment differs among fungal species. Plant Cell Environ., 28: 642-650.

Chaudhri, S. 2005. Plant's dependence on arbuscular mycorrhiza. In: V.S. Mehrotra (ed.). Mycorrhiza-Role and Applications. pp. 91-135. Allied Publishers Pvt. Ltd., New Delhi.

Chaurasia, B. and Khare, P. K. 2005. *Hordeum vulgare*: a suitable host for mass production of arbuscular mycorrhizal fungi from natural soil. Applied Ecology and Environmental Research, 4(1): 45-53.

Clark, J. E., Beegen, H., and Wood, H. G. 1986. lsolation of intact chains of polyphosphate from *Propionibacterium sherrnanii* grown on glucose or lactate.*J. Bacteriol.*, 168, 1212-1219.

Cooper, K. M. and Tinker, P. B. 1978. Translocation and transfer of nutrients in vesicular-arbuscular mycorrhizas. II. Uptake and translocation of phosphorus, zinc and sulphur. New Phytol., 81:43-52.

Cui, M. and Caldwell, M. M. 1996a. Facilitation of plant phosphate acquisition by arbuscular mycorrhizae from enriched to soil patches. I. Roots and hyphae exploiting the same volume. New Phytol., 133: 453-460.

Di Bonito, R., Elliot, M. L. and Desjardin, E. A. 1995. Detection of an arbuscular mycorrhizal fungus in roots of different plant species with the PCR. Appl. Environ. Microbiol., 61: 2809-2810.

Diaz, G. and Honrubia, M. 1994. A mycorrhizal survey of plants growing on mine wastes in Southern Spain. Arid Soil Res. Rehabil., 8: 59-68.

Douds, Jr., David, D. and Millner, P. D. 1999. Biodiversity of arbuscular mycorrhizal fungi in agro-ecosystems. Agric. Ecosyst. Environ., 74: 77-93.

Douds, D. D. Jr. and Nagahashi, G. 2000. Signalling and recognition events prior to colonisation of roots by arbuscular mycorrhizal fungi. Pp. 11-18. In: Current Advances in Mycorrhizae Research. GK Podila, DD Douds (eds.) Minnesota: APS Press.

Douds, D. D. Jr., Nagahashi, G., Pfeffer, P. E., Reider, C. and Kayser, W. M. 2006. On-farm production of AM fungus inoculum in mixtures of compost and vermiculite., 97 (6): 809-818.

Endlweber, K. and Scheu, S. 2007. Interactions between mycorrhizal fungi and collembola: effects on root structure of competing plant species. Biol. Fertil. Soils, 43: 741-749. DOI:10.1007/s00374-0060157-7.

Ezawa, T., Kuwahara, S., Sakamoto, K., Yoshida, T. and Saito, M. 1999. Specific inhibitor and substrate specificity of alkaline phosphate expressed in the symbiotic phase of the arbuscular mycorrhizal fungus, *Glomus etunicatum*. Mycologia, 91:636-641.

Ezawa, T., Smith, S. E. and Smith, F. A. 2002. P metabolism and transport in AM fungi. Plant and Soil, 244:221-230.

Fitter, A. H. 1991. Costs and benefits of mycorrhizae: implications for functioning under natural conditions. Experientia, 47: 350-355.

Fitter, A. H. and Sanders, I. 1992. Interactions with the soil fauna. In Allen MF (ed.) Mycorrhizal functioning. Chapman and Hall, New York, pp. 333-354.

Funamoto, R., Saito, K., Oyaizu, H., Saito, M. and Aono, T. 2007. Simultaneous *in situ* detection of alkaline phosphatase activity and polyphosphate in arbuscules within arbuscular mycorrhizal roots. Functional Plant Biology, 34: 803-810.

Gai, J. P., Christie, P., Feng, G. and Li, X. L. 2006. Twenty years of research on community composition and species distribution of arbuscular mycorrhizal fungi in China: a Review. Mycorrhiza, 16: 229–239.

Gai, J. P. and Liu, R. J. 2003. Effect of soil factors on AMF in the rhizosphere of wild plants. Chin. J. Appl. Ecol., 14: 470-472.

Gallaud, I. 1905. Etudes sur les mycorrhizes endophytes. Revue General de Botanique, 17: 5-48, 66-83, 123-136, 223-239, 313-325, 425-433, 479-500.

Gange, A. C., Bower, E. and Farmer, L. M. 1990. A test of mycorrhizal benefit in an early successional plant community. New Phytol., 115: 85-91.

Gange, A. C., Bower, E., Stagg, P. G., Aplin, D. M., Gillam, A. E. and Bracken, M. 1999. A comparison of visualization techniques for recording arbuscular mycorrhizal colonization. New Phytol., 142:123-132.

Gemma, J. N., Koske, R. E. and Carreiro. 1989. Seasonal dynamics of selected species of VA mycorrhizal fungi in a sand dune. Mycol. Res., 92: 317-321.

Gerdemann, J. W. and Nicolson, T. H. 1963. Spores of mycorrhizal Endogone species extracted from soil by wet sieving and decanting. Trans. Br. Mycol. Soc., 46: 235-244.

Giovannetti, M., Sbrana, C., Strani, P., Agnolucci, M., Rinaudo, V. and Avio, L. 2003. Genetic diversity of isolates of *Glomus mosseae* from different geographic areas detected by vegetative compatibility testing and biochemical and molecular analysis. Appl. Environ. Microbiol., 69: 615-624.

Gosling, P., Hodge, A., Goodlass, G. and Bending, G. D. 2006. Arbuscular mycorrhizal fungi and organic farming. Agric. Ecosyst. Environ., 113:17-35.

Graham, J. H. and Miller, R. M. 2005. Mycorrhizas: Gene to function. Plant and Soil., 274:79-100. DOI 10.1007/s11104-004-1419-5.

Gryndler, M. 2000. Interactions of arbuscular mycorrhizal fungi with other soil organisms. In: Kapulnik Y, Douds DD Jr. (Eds). Arbuscular Mycorrhizas: Physiology and Function. Kluwer Academic, Dordrecht, The Netherlands, pp. 239-262.

Harley, J. L. 1991. The history of research on mycorrhiza and the part played by Professor Beniamino Peyronel. In: Estratto da Funghi, Piante e Suolo, Quarat'anni di ricerche del centro di Studio sulla Micologia del Terreno nel centenario della nascita del suo fondatore Beniamino Peyronel. Centro di Studio sulla Micologia del Terreno, CNR, Torino, pp. 31–73.

Hart, M. M. and Reader, R. J. 2002. Taxonomic basis for variation in the colonization strategy of arbuscular mycorrhizal fungi. New Phytologist, 153: 335–344.

Hodge, A., Campbell, C. D. and Fitter, A. H. 2001. An arbuscular mycorrhizal fungus accelerates decomposition and acquires nitrogen directly from organic material. Nature, 413: 297-299.

Hu, J., Lin, X., Wang, J., Chu, H., Yin, R. and Zhang, J. 2009a. Population size and specific potential of P-mineralizing and –solubilising bacteria under long-term P-deficiency fertilization in a sandy loam soil. Pedobiologia. DOI: 10.1016/j.pedobi.2009.02.002.

Hu, J., Lin, X., Wang, J., Dai, J., Cui, X., Chen, R. and Zhang, J. 2009b. Arbuscular mycorrhizal fungus enhances crop yield and P-uptake of maize (*Zea mays* L.): A field case study on a sandy loam soil as affected by long term P-deficiency fertilization. Soil Biol. Biochem. 41: 2460-2465.

Ianson, D. C. and Allen, M. F. 1986. The effects of soil texture on extraction of vesicular-arbuscular mycorrhizal fungal spores from arid sites. Mycologia, 78: 164-168.

Jalali, B. L., Chabra, M. L. and Singh, R. P. 1990. Interaction between vesicular arbuscular mycorrhizal endophyte and *Macrophomina phaseolina* in mung bean. Indian Phytopath., 43:527-530.

Jansa, J., Mozafar, A. and Frossard, E. 2003. Long distance transport of P and Zn through the hyphae of an arbuscular mycorrhizal fungus in symbiosis with maize. Agronomie, 23: 481-488.

Jasper, D. A., Abbott, L. K. and Robson, A. D. 1991. The effect of disturbance on vesicular-arbuscular mycorrhizal fungi in soils from different vegetation types. New Phytol., 118: 471-476.

Jasper, D. A., Robson, A. D. and Abbott, L. K. 1987. Effect of surface mining on the infectivity of vesicular-arbuscular mycorrhizal fungi. Australian Journal of Botany, 35: 641-652.

Jayanthi, S., Bagyaraj, D. J. and Satyanarayana, B. N. 2003. Enhanced growth and nutrition of micropropagated *Ficus benjamina* to *G. mosseae* co-inoculated with *Trichoderma harzianum* and *Bacillus coagulans*. World Journal of Microbiology and Biotechnology, 19: 69-72.

Jeffries, P., Gianinazzi, S., Silvia, P., Turnau, K. and Barea, J. M. 2003. The contribution of arbuscular mycorrhizal fungi in sustainable maintenance of plant health and soil fertility. Biol. Fertil. Soils, 37:1-16.

Johnson, N. C., Copeland, P. J., Crookston, R. K. and Pfleger, F. L. 1992. Mycorrhizae: Possible explanation for yield decline with continuous corn and soybean. Agron J., 84:387-390.

Johnson, N. C. 1993. Can fertilization of soil select less mutualistic mycorrhizae? Ecol. Applic., 3: 749-757.

Karagiannidis, N., Bletsos, F. and Stavropoulos, N. 2002. Effect of Verticillium wilt (*Verticillium dahliae* Kleb.) and mycorrhiza (*Glomus mosseae*) on root colonisation, growth and nutrient uptake in tomato and egg plant seedlings. Sci. Hortic., 94: 145-156.

Kendrick, W. B. and Berch, S. M. 1985. Mycorrhizae: applications in agriculture and forestry. pp. 109-152. In: Comprehensive Biotechnology. Vol.13. (Ed. M Moo-Young). Pergoan Press, Oxford.

Khalil, S., Carpenter-Boggs, L. and Loyanchan, T. E. 1994. Procedure for rapid recovery of VAM fungal spores from soil. Soil Biol. Biochem., 26:1587-1588.

Klironomos, J. N., Bednarczuk, E. M. and Neville, J. 1999. Reproductive significance of feeding on saprobic and arbuscular mycorrhizal fungi by the collembolan, *Folsomia candida*. Functional Ecology, 13: 756–761.

Klironomos, J. N. and Hart, M. M. 2002. Colonization of roots by arbuscular mycorrhizal fungi using different sources of inoculum. Mycorrhiza, 12: 181-184.

Klironomos, J. N., McCune, J., Hart, M. and Neville, J. 2000. The influence of arbuscular mycorrhizae on the relationship between plant diversity and productivity. Ecology Letters, 3: 137–141.

Koide, R., Li, M., Lewis, J. and Irby, C. 1988. Role of mycorrhizal infection in the growth and reproduction of wild vs. cultivated plants. I. Wild vs. Cultivated Oats. Oecologia, 88: 201-208.

Koide, R.T. and Mosse, B. 2004 A history of research on arbuscular mycorrhiza. Mycorrhiza, 14: 145-163.

Kothari, S. K., Marschner, S. and George, E. 1990. Effect of VA mycorrhizal fungi and rhizosphere micro-organisms on root and shoot morphology, growth and water relations in maize. New Phytol., 116:303-311.

Kuhn, G., Hijri, M. and Sanders, I. R. 2001. Evidence for the evolution of multiple genomes in arbuscular mycorrhizal fungi. Nature, 414: 745-748.

Lakshman, H. C., Inchal, R. F. and Mulla, F. I. 2006. Seasonal fluctuations of arbuscular mycorrhizal fungi on some commonly cultivated crops of Dharwad. In: Anil Prakash and V.S.Mehrotra (eds.). Mycorrhiza. pp.11-19. Scientific Publishers (India), Jodhpur.

Li, X. L., George, E. and Marschner, H. 1991. Extension of the phosphorus depletion zone in VA mycorrhizal white clover in a calcareous soil. Plant and Soil, 136: 41-48.

Liu, A., Hamel, C., Hamilton, R. I., Ma, B. L. and Smith, D. L. 2000. Acquisition of Cu, Zn, Mn, and Fe by mycorrhizal maize (*Zea mays* L.) grown in soil at different P and micro-nutrient levels. Mycorrhiza, 9: 331-336.

Liu, J., Maldonado-Mendoza, I., Lopez Meyer, M., Cheung, F., Town, C. D. and Harrison, M. J. 2007. Arbuscular mycorrhizal symbiosis is accompanied by local and systemic alterations in gene expression and increase in disease resistance in the shoots. Plant J., 50: 529-544. DOI: 10.1111/j.1365-313x.2007.03069.x.

Mädder, P., Vierheilig, H., Boller, T., Streitwolf-Engel, Frey, B., Christie, P. and Wiemken, A. 2000. Transport of ^{15}N from a soil compartment separated by a polytetraflouroethylene membrane to plant roots via the hyphae of arbuscular mycorrhizal fungi. New Phytol., 146: 155-161.

Maia, L. C. and Yano-Melo, A. M. 2005. Role of arbuscular mycorrhizal fungi in saline soils. Pp. 282-302. In: V.S. Mehrotra (ed.), Mycorrhiza: Role and Applications. Allied Publishers Pvt. Ltd., New Delhi.

Marschner, H. 1995. Mineral Nutrition of Higher Plants, 2nd Edition. Academic Press, London.

Medina, M. J. H., Gagnon, H., Piche, Y., Ocampo, J. A., Garrido, J. M. G. and Vierheilig, H. 2003. Root colonization by arbuscular mycorrhizal fungi is affected by the salicylic acid content of the plant. Plant Sci., 164:993-998.

Mehravaran, H., Mozafar, A. and Frossard, E. 2000. Uptake and partitioning of ^{32}P and ^{65}Zn by white clover as affected by eleven isolates of mycorrhizal fungi. J. Plant Nutr., 23: 1385-1395.

Mehrotra, V. S. 1993. The so-called vesicles in vesicular arbuscular mycorrhiza are chlamydospores. Philippine Journal of Science, 122: 377-395.

Mehrotra, V. S. 1996. Use of revegetated coal mine spoil as source of arbuscular mycorrhizal inoculum for nursery inoculations. Current Science, 71: 73-77.

Mehrotra, V. S. 1997. Problems associated with morphological taxonomy of AM fungi. Mycorrhiza News, 9: 1-10.

Mehrotra, V. S. 2005. Mycorrhiza: A Premier Biological Tool for Managing Soil Fertility. Pp. 1-65. In: V.S. Mehrotra (ed.), Mycorrhiza: Role and Applications. Allied Publishers Pvt. Ltd., New Delhi.

Mehrotra, V. S. 2008. Diversity of arbuscular mycorrhizal fungi in India. In: M. Tiwari and S.C. Sati (eds). The Mycorrhizae: Diversity, Ecology and Applications. Daya Publishing House, Delhi.

Mehrotra, V. S., Baijal, U., Mishra, S. D., Pandey, D. P. and Mathews, T. 1995. Movement of ^{32}P in sunflower plants inoculated with single and dual inocula of VAM fungi. Current Science, 68: 751-753.

Menge, J. A. 1982. Effect of soil fumigants and fungicides on vesicular-arbuscular mycorrhizal fungi. Phytopath., 72: 1125-1132.

Miller, R. M. and Jastrow, J. D. 2000. Mycorrhizal fungi influence soil structure. In: *Arbuscular Mycorrhizas: physiology and function*, Y. Kapulnik and D. Douds, eds. Kluwer Academic Publishers, Dordrecht, p 4-18.

Milleret, R., Bayon Le Renée-Claire and Gobat Jean-Michel. 2009. Root, Mycorrhiza and Earthworm interactions: their effects on soil structuring processes, plant and soil nutrient concentration and plant biomass. Plant and Soil, 316: 1-2.DOI:10.1007/s11104-008-9753-7.

Morton, J. B. 1995. Taxonomy and phylogenetic divergence among five *Scutellospora* spp. based on comparative developmental sequences. Mycologia, 87: 122-137.

Morton, J. B. and Benny, G. L. 1990. Revised classification of arbuscular mycorrhizal fungi (Zygomycetes): A new order, Glomales, two new suborders, Glomineae and Gigasporineae, and two new families, Acaulosporaceae and Gigasporaceae, with an emendation of Glomaceae. Mycotaxon, 37:471-491.

Morton, J. B and Redecker, D. 2001. Two new families of Glomales, Archaeosporaceae and Paraglomaceae, with two new genera *Archaeospora* and *Paraglomus*, based on concordant molecular and morphological characters. Mycologia, 93: 181-195.

Mosse, B. 1985. Endotrophic mycorrhiza (1885–1950): the dawn and the middle-ages. In: Proceedings of the 6th North American Conference on Mycorrhizae. Forest Research Laboratory, Oregon State University, Corvallis, Ore., pp 48–55.

Mozafar, A., Anken, T., Ruh, R. and Frossard, E. 2000. Tillage intensity, mycorrhizal and non-mycorrhizal fungi and nutrient concentration in maize, wheat and canola. Agron. J., 92:1117-1124.

Nandakwang, P., Elliott, S. and Lumyong, S. 2008. Diversity of arbuscular mycorrhizal fungi in forest restoration area of Doi Suthep-Pui National Park, Northern Thailand. Journal of Microscopy Society of Thailand, 22: 60-64.

Nogales, A., Aguirreolea, J., Marai, E. S., Camprubi, A. and Calvet, C. 2009. Response of mycorrhizal grapevine to *Armillaria mellea* inoculation: disease development and polyamines. Plant and Soil, 317:177-187. DOI:10. 1007/s11104-008-9799-6.

Norman, J. R. and Hooker, J. E. 2000. Sporulation of Phytophthora fragariae shows greater stimulation by exudates of non-mycorrhizal than by mycorrhizal strawberry roots. Mycol. Res., 104:1069–1073.

O'Connor, P. J., Smith, S. E. and Smith, F. A. 2002. Arbuscular mycorrhizas influence plant diversity and community structure in a semiarid herbland. New Phytol., 154: 209-218.

Oehl, F. and Sieverding, E. 2004. Pacispora, a new vesicular-arbuscular mycorrhizal fungal genus in the Glomeromycetes. J. Appl. Bot-Angewandte Bot., 78: 72-82.

Oswald, E. T. and Ferchau, H. A. 1968. Bacterial associations of coniferous mycorrhizae. Plant Soil, 28, 187–92.

Palenzuela, J., Ferrol, N., Boller, T., Azcón-Aguilar, C. and Oehl, F. 2008. *Otospora bareai*, a new fungal species in the Glomeromycetes from a dolomitic shrubland in the Natural Park of Sierra de Baza (Granada, Spain). Mycologia, 100 (2):296-305.

Pankow, W., Boller, T. and Wiemken, A. 1991. The significance of mycorrhizas for protective ecosystems. Experientia, 47: 391-394.

Pozo, M. J., Cordier, C., Dumas, E., Gianinazzi, S., Barea, J. M. and Azcon-Aguilar, C. 2002. Localized versus systemic effect of arbuscular mycorrhizal fungi on defence responses to Phytophthora infection on tomato plants. J. Exp. Bot., 53: 525-534. DOI: 10.1093/jexbot/53.368.525.

Quilambo, O. A. 2003. The vesicular-arbuscular mycorrhizal symbiosis. African Journal of Biotechnology, 2: 539-546.

Rambelli.1973. The Rhizopshere of mycorrhizae. In: GC Marks and TT Kozlowski (eds.). Ectomycorrhizae, their ecology and physiology. New York, USA: Academic Press, 299-349.

Rayner, M. C. 1926–1927. Mycorrhiza. New Phytol, 25:1–50, 65–108, 171–190, 248–263 338–372, 26:22–45, 85–114.

Redecker, D., Kodner, R. and Graham, L. E. 2000. Glomalean fungi from the Ordovician. Science, 289:1920-1921. DOI:10.1126/science.289.5486.1920.

Reddy, S. R., Pindi, P. K. and Reddy, S. M. 2005. Molecular methods for research on arbuscular mycorrhizal fungi in India: problems and prospects. Current Science, 89 (10): 1699-1709.

Redecker, D. and Philipp, R. 2006. Phylogeny of the Glomeromycota (arbuscular mycorrhizal fungi): recent developments and new gene markers. Mycologia, 98(6): 885-895.

Remy, W., Taylor, T., Hass, H. and Kerp, H. 1994. Four hundred million year-old vesicular arbuscular mycorrhizae. Proceedings of the National Academy of Science USA, 91 (25): 11841–11843. DOI: 10.1073/pnas.91.25.11841.

Renker, C., Heinrichs, J., Kaldorf, M. and Buscot, F. 2003. Combining nested PCR and restriction digest of the internal transcribed spacer region to characterize arbuscular mycorrhizal fungi on roots from the field. Mycorrhiza, 13: 191-198.

Rillig, M. C. 2004. Arbuscular mycorrhizae, glomalin, and soil aggregation. Can J. Soil Sci., 84: 355-363.

Rillig, M. C., Wright, S. F., Nichols, K. A., Schmidt, W. F and Torn, M. S. 2001. Large contribution of arbuscular mycorrhizal fungi to soil carbon pools in tropical forest soils. Plant and Soil, 233: 167-177.

Ryan, A. and Jones, P. 2004. The effect of mycorrhization of potato roots on the hatching chemicals active towards the potato cyst nematodes, *Globodera pallida* and *G. rostochiensis*. Nematology, 6:335–342.

Ryan, M. H. and Graham, J. H. 2002. Is there a role for arbuscular mycorrhizal fungi in production agriculture. Plant and Soil, 244: 263-271.

Sailo, G. L. and Bagyaraj, D. J. 2005. Influence of different AM fungi on the growth, nutrition and forskolin content of *Coleus forskohlii*. Mycol. Res., 109: 795-798. DOI:10.1017/s095375605002832.

Sailo, G. L. and Bagyaraj, D. J. 2006. Influence of *Glomus bagyarajii* and PGPRs on the growth and nutrition and forskolin concentration of *Coleus forskohlii*. Biological Agriculture and Horticulture, 23:371-382.

Sanders, I. R. 1990. Seasonal patterns of vesicular-arbuscular mycorrhizal occurrence in grasslands. Symbiosis, 9: 315-320.

Sanders, I. R., Alt, M., Groppe, K., Boller, T. and Wiemken, A. 1995. Identification of ribosomal DNA polymorphisms among and within spores of the Glomales: application to studies on the genetic diversity of arbuscular mycorrhizal fungal communities. New Phytol., 130: 419-427.

Scheffknecht, S., Mammerler, R., Steinkellner, S. and Vierheilig, H. 2006. Root exudates of mycorrhizal tomato plants exhibit a different effect on microconidia germination of *Fusarium oxysporum* f. sp. *lycopersici* than root exudates from non-mycorrhizal tomato plants. Mycorrhiza, (2006) 16: 365–370. DOI 10.1007/s00572-006-0048-7

Schenck, N. C. 1985. VA mycorrhizal fungi, 1950 to the present: the era of enlightenment. In: Molina, R. (ed.) Proceedings of the 6th North American Conference on Mycorrhizae. Forest Research Laboratory, Oregon State University, Corvallis, Ore., pp 56–60.

Scheu, S. 1994. There is an earthworm mobilizable nitrogen pool in soil. Pedobiologia Jena, 38: 243-249.

Schreiner, R. P., Mihara, K. L., McDaniel, H. and Bethlenfalvay, G. J. 1997. Mycorrhizal fungi influence plant and soil functions and interactions. Plant Soil, 188, 199–210.

Schoenbeck, F. and Dehne, H. W. 1979. Unterschungen zum pilzliche spores parasite *Olpidium brassicae*, T.M.V.Z. Pflenzenkrankh. Pfanzenschutz, 86: 103.

Schüßler, A., Schwarzott, D. and Walker, C. 2001. A new fungal phylum, the Glomeromycota: Phylogeny and evolution. Mycol. Res., 105: 1413-1421.

Schwab, S. M., Leonard, R. T. and Menge, J. A. 1984. Quantitative and qualitative comparison of root exudates of mycorrhizal and non-mycorrhizal plant species. Can. J. Bot., 62: 1227-1231.

Sieverding, E. and Oehl, F. 2006. Revision of *Entrophospora* and description of *Kuklospora* and *Intraspora*, two new genera in the arbuscular mycorrhizal Glomeromycetes. J. Appl. Bot. Food Qual., 80: 69-81.

Singh, R. and Adholeya, A. 2002. AMF biodiversity in wheat agro-systems of India. Mycorrhiza News, 14 (3): 21-23.

Smith, F. A., Jakobsen, I. and Smith, S. E. 2000. Spatial differences in acquisition of soil phosphate between two arbuscular mycorrhizal fungi in symbiosis with *Medicago truncatula*. New Phytol., 147: 357-366.

Solaiman, Z. M., Ezawa, T., Kojima, T. and Saito, M. 1999. Polyphosphates in intraradical and extraradical hyphae of arbuscular mycorrhizal fungi. Applied and Environmental Microbiology, 65: 5604-5606.

Sorensen, J. N., Larsen, J. and Jakobsen, I. 2005. Mycorrhiza formation and nutrient concentration in leeks (*Allium porrum*) in relation to previous crop and cover crop management on high P soils. Plant and Soil, 273: 101-114.

Spain, J. L., Sieverding, E. and Oehl, F. 2006. *Appendicispora:* a new genus in the arbuscular mycorrhiza forming Glomeromycetes, with a discussion of the genus *Archaeopsora*. Mycotaxon, 97: 163-182.

Staddon, P. L., Ramsey, C. B., Ostle, N., Ineson, P. and Fitter, A. H. 2003. Rapid turnover of hyphae of mycorrhizal fungi determined by AMS microanalysis of C-14. Science, 300: 1138-1140.

Sumana, D. A and Bagyaraj, D. J. 2000. Interaction between VAM fungus and nitrogen-fixing bacteria and their influence on growth and nutrition of neem (*Azadirachta indica* A Juss). Indian Journal of Microbiology, 42: 295-298.

Sutton, J. C. and Barron, G. L. 1972. Population dynamics of Endogone spores in soil. Canad. J. Bot., 50: 1909-1914.

Takanishi, I., Ohtomo, R., Hayatsu, M. and Saito, M. 2009. Short-chain polyphosphate in arbuscular mycorrhizal roots colonized by *Glomus* spp.: A possible phosphate pool for host plants. Soil Biol. Biochem., 41: 1571-1573. DOI 10.1016/j.soilbio. 2009. 04. 002.

Tester, M., Smith, S. E. and Smith, F. A. 1987. The phenomenon of 'non-mycorrhizal plants'. Can. J. Bot., 65:419-431.

Timothy, R. C. 2008. The role of arbuscular mycorrhizas in improving plant zinc nutrition under low soil zinc concentrations: a review. Plant and Soil, 304:315-325.

Tommerup, I. C. 1982. Airstream fractionation of vesicular-arbuscular mycorrhizal fungi:concentration and enumeration of propagules. Applied and Environmental Microbiology, 44(3) 533-539.

Trappe, J. M. and Berch, S. M. 1985. The prehistory of mycorrhizae: A.B. Frank's predecessors. In Proc. 6th NACOM. R. Molina (editor). Bend, Oreg. pp. 2-11.

Treseder, K. K. and Cross, A. 2006. Global distributions of arbuscular mycorrhizal fungi. Ecosystems, 9: 305-316.

van der Heijden, M. G. A., Klironomos, J. N., Ursic, M., Moutoglis, P., Streitwolf-Engel, R., Boller, T., Wiemken, A. and Sanders, I. R. 1998. Mycorrhizal fungal diversity determines plant biodiversity, ecosystem variability and productivity. Nature, 396: 69–72.

Vázquez, M. M., Barea, J. M. and Azcón, R. 2002. Influence of arbuscular mycorrhizae and a genetically modified strain of *Sinorhizobium* on growth, nitrate reductase activity and protein content in shoots and roots of *Medicago sativa* as affected by nitrogen concentrations. Soil Biol. Biochem., 34: 899-905.

Vierheilig, H. 2004. Regulatory mechanisms during the plant-arbuscular mycorrhizal fungus interaction. Can. J. Bot., 82:1166–1176

Vosatka, M. and Gryndler, M. 1999. Treatment with culture fractions from *Pseudomonas putida* modifies the development of *Glomus fistulosum* mycorrhiza and the response of potato and maize plants to inoculation. Appl. Soil Ecol., 11: 245-251.

Vyas, S. C. and Shukla, B. N. 2005. Practical aspects of arbuscular mycorrhiza in plant disease control. pp. 163-183. In: V.S. Mehrotra (ed.). Mycorrhiza: Role and Applications. Allied Publishers Pvt. Ltd., New Delhi.

Walker, C. 1992. Systematics and taxonomy of the arbuscular endomycorrhizal fungi (Glomales)–a possible way forward. Agronomie, 12: 887-897.

Walker, C. and Schüßler, A. 2004. Nomenclatural clarifications and new taxa in the Glomeromycota. Mycological Research, 108: 981-982.

Walker, C., Vestberg, M., Demircik, F., Stockinger, H., Saito, M., Sawaki, H., Nishmura, I. and Schüßler, A. 2007a. Molecular phylogeny and new taxa in the Archaeosporales (Glomeromycota): *Ambispora fennica* gen. sp. nov., Ambisporaceae fam. nov., and emendation of *Archaeospora* and Archaeosporaceae. Mycol. Res., 111: 137-13.

Walker, C., Vestberg, M. and Schüßler, A. 2007b. Nomenclatural clarifications in Glomeromycota. Mycol. Res., 111: 253-255.

Wang, F. Y., Lin, X. G. and Zhou, J. M. 2004. Biodiversity of AMF in China. Chin. J. Ecol., 23: 149-154.

Wang, F. Y. and Shi, Y. Z. 2008. Biodiversity of arbuscular mycorrhizal fungi in China: a Review, Adv. Environ. Biol., 2 (1): 31-39.

Warner, A. 1984. Colonization of organic matter by vesicular-arbuscular mycorrhizal fungi. Transactions of the British Mycological Society, 82:352-354.

Wright, S. F., Green, V. S. and Cavigelli, M. A. 2007. Glomalin in aggregate size classes from three different farming systems. Soil and Tillage Research, 94:546-549.

Wright, S. F, and Upadhyaya, A. 1996. Extraction of an abundant and unusual protein from soil and comparison with hyphal protein of arbuscular mycorrhizal fungi. Soil Science, 161:575-586.

Zhang, M. Q, Wang, Y. S, and Xing, L. J. 1998. The ecological distribution of AM fungal communities in the south and east coasts of China. Mycosystema, 17: 274-277.

Zhu, Y.G., Cavagnaro, T. R., Smith S. E. and Dickson, D. 2001. Backseat driving? Accessing phosphate beyond the rhizosphere-depletion zone. Trends in Plant Science, 6(5): 194-195.

Microbes: Diversity and Biotechnology (2012) *Pages* **469–476**
Editors: **Prof. S.C. Sati & Dr. M. Belwal**
Published by: **DAYA PUBLISHING HOUSE, NEW DELHI**

Chapter 29

Isolation and Control of Seed Mycoflora of Onion (*Allium cepa* L.) in Nepal

B.K. Chhetri*, U. Budhathoki and C.P. Pokhrel
Central Department of Botany, Tribhuvan University, Kathmandu, Nepal

ABSTRACT

Seed-borne fungi affect the quality and quantity of the crop yield, and these pathogens may either be internal seed-borne or external seed-borne. The present research was carried out on onion (*Allium cepa* L.) to deal with the mycoflora which were invariably seed-borne. Agar plate technique and Standard blotter technique were employed for the study and among these, the Standard blotter technique was found to be the most effective method. Study confirmed *Aspergillus niger, A. flavus, A.* sp., *Chaetomium* sp., *Alternaria tenius, Stemphylium vesicarium, Epicoccum purpurascens, Fusarium* sp., *Curvularia lunata, C. clavata, Penicillium* sp., and an unidentified species. Among them *A. niger* and *C. lunata* were dominant mycoflora, and were selected for testing the antifungal properties of two plant extracts. The assessment of fungitoxicity was carried out by poisoned food technique using three different concentrations (25 per cent, 50 per cent, 100 per cent) against the test fungi in terms of percentage of mycelial growth inhibition. In the present study extracts of *Allium sativum* and *Syzygium aromaticum* were found to be effective against *Aspergillus niger*, with MIC of 50 per cent and 100 per cent respectively. Where as the extract of *Syzygium aromaticum* was found to be effective with MIC of 50 per cent against *Curvularia lunata*.

Keywords: Seed-borne fungi, Allium cepa L., Central Nepal, Control of seed mycoflora.

Introduction

Among many vegetables grown, onion (*Allium cepa* L.) is a popular vegetable grown in Nepal for its pungent bulbs and flavorful leaves. A survey carried out by Seed Entrepreneurs Association of

* Corresponding Author: E-mail: chhetri_vasant217@hotmail.com

Nepal indicates that the seed of radish, carrot and onion are exported from Nepal (FNCCI, 2006). Among the edible *Allium* the onion (*Allium cepa* L.) stands in the first rank, in the warm- temperate hills of eastern Nepal, followed by garlic (*Allium sativum*) and shallot (*Allium cepa* var. *aggregatum*). The bio-physical conditions of the warm-temperate hilly regions of eastern Nepal is congenial for the overall growth and development of onion, garlic and shallot. The onion bulb produced in the hills have a very high potential to supply to the domestic markets of the Terai, and onion seeds have window of opportunity to supply to the south east Asian countries. Similarly, observations on the effect of different sowing dates and methods of production (direct sowing, transplanting in pulverized and transplanting in puddled soil) of these, the direct seeding method produced the highest bulb yields (Gautam *et al.*, 1994). Eight different diseases of onion have been reported from Nepal (HMG/N, 1997). Seed protection by using commercially available chemicals brings a series of environmental problems accompanied with health related problems. The extracts of plants are less phytotoxic, more systemic (Fawcett *et al.*, 1970), easily bio-degradable (Beye, 1978) and environmentally non-pollutive (Costa *et al.*, 2000). Using plant extracts as a means of controlling diseases is also cheaper compared to the commercially available fungicides and thus are economically more feasible to the farmers, and the health and environmental benefits of using such naturally available products are also worth considering as they have no adverse effects on health and environment.

Materials and Methods

Isolation of Fungi

Seed samples were collected from different localities of Kathmandu and from local farmers of Manohara, Thimi. Huge plots of land have been used for the cultivation of vegetables near the manohara river. Among the different vegetables cultivated onion is also one of them. The so collected seeds were incubated for isolation of respective seed-borne fungi. This step was divided into two halves. In the first half of the procedure, fungi were isolated using "Standard blotter technique" and the respective fungi identified. In the second half "Agar plate method" was used for isolation of fungi. In both the above mentioned techniques seeds were divided into two portions *viz;* sterilized and unsterilized seeds. 200 seeds were taken from the bulk, 100 seeds were surface sterilized in 90 per cent ethyl alcohol for 1-3 minutes, followed by thorough rinsing in three changes of sterile distilled water (Kemetz *et al.*, 1979). Remaining 100 seeds were not sterilized. The thus sterilized and unsterilized seeds were plated in moist blotter as for "Standard blotter technique" and on PDA media as for "Agar plate method". The so plated seeds were incubated at 25±2C for about 7 to 10 days in incubator. Fungi were identified by standard criteria on the basis of the macroscopic appearance of the colonies, the dynamics of their growth and the microscopic characteristics of fructification organs.

Control

Two plants i. *Allium sativum* L. and ii. *Syzygium aromaticum* L. were used against the test pathogens. 7 g of required parts of each plant were first surface sterilized with the help of 2-3 per cent sodium hypochlorite for 2-5 minutes and thoroughly washed with sterile distilled water (three changes). The plant parts were then grind finely in sterilized mortar and pistil adding 10 ml of sterile distilled water. The grinded part was filtered using sterile muslin cloth, and the obtained solution was centrifuged at 3000 rpm for 15 minutes, and the upper clear solution was separated. The volume of the obtained clear solution was adjusted to 10 ml by the addition of distilled water. This final volume of extract was used as a "stock solution" ie; 100 per cent concentration. Three different concentrations *viz;* 25 per cent, 50 per cent and 100 per cent of each plant extract were made in separate test tubes.

To study the antifungal mechanism of plant extracts, poisoned food technique (Grover and Moore, 1962) was used. 0.5 ml of each concentration of extract was poured onto sterilized Petri plates, followed by the addition of 9.5 ml of PDA. Temperature of the PDA was constantly monitored, and at the time of pouring the temperature was maintained at 45°C. The Petri dishes were swirled gently to allow thorough mixing of the contents. In the control set no extract was used. After the media solidified, one inoculum disc was made from the pure culture of the respective fungus and inoculated upside down at the centre of each Petri dish and the whole setup was maintained in an incubator at 25±2°C for 7 days. All the experiments were repeated thrice. The average diameter of the fungal colonies was measured on the seventh day of incubation and percentage of mycelial growth inhibition was calculated.

Results

Fungi found to be associated with Onion (*Allium cepa* L.) seeds were *Aspergillus niger, Aspergillus flavus, Aspergillus* sp., *Chaetomium* sp., *Alternaria tenius, Stemphylium vesicariu, Epicoccum purpurascens, Fusarium* sp., *Curvularia lunata, Curvularia clavata, Penicillium* sp., and an unidentified species. Out of these twelve isolates, *Aspergillus niger* and *Curvularia lunata* were found to be dominating fungi, showing the highest occurrence (Table 29.1). In the present investigation, the Standard Blotter Technique was found to be the most ideal method for the isolation of fungi.

Table 29.1: Mycoflora Associated within Seeds of Onion (*Allium cepa* L.)

Sl.No.	Fungal Flora	Methods of Incubation			
		Blotter Method		Agar Plate Method	
		NSS	SS	NSS	SS
1.	*Aspergillus niger*	+	+	+	+
2.	*Aspergillus flavus*	+	–	+	–
3.	*Aspergillus* sp.	+	–	+	+
4.	*Chaetomium* sp.	+	+	+	–
5.	*Alternaria tenius*	+	–	+	–
6.	*Stemphylium vesicarium*	+	+	–	–
7.	*Epicoccum purpurascens*	+	+	–	–
8.	*Fusarium* sp.	+	–	–	–
9.	*Curvularia clavata*	+	–	–	–
10.	*Curvularia lunata*	+	+	+	+
11.	Unidentified species	–	–	+	+
12.	*Penicillium* sp.	+	–	+	–

+: Presence of fungus; –: Absence of fungus; NSS: Non sterilized seeds; SS: Sterilized seeds.

Antifungal activity of each plant extract was assessed in different concentrations against *Curvularia lunata* and *Aspergillus niger* by poisoned food technique method.

The extracts of *Allium sativum* inhibited the growth of *Curvularia lunata* by 26 per cent, 38 per cent and 43 per cent at 25 per cent, 50 per cent and 100 per cent concentrations respectively (Table 29.2).

Table 29.2: Antifungal Activity of Extracts of *Allium sativum* against *Curvularia lunata*

Sl.No.	*Concentration of Extract (per cent)*	*Inoculum Diameter (mm) [B]*	*Mean Colony Diameter (mm) [A]*	*Mycelial Growth Diameter (mm) [A-B]*	*Inhibition (per cent)*
1.	Control	4	80	76	0
2.	25	4	60	56	26
3.	50	4	51	47	38
4.	100	4	47	43	43

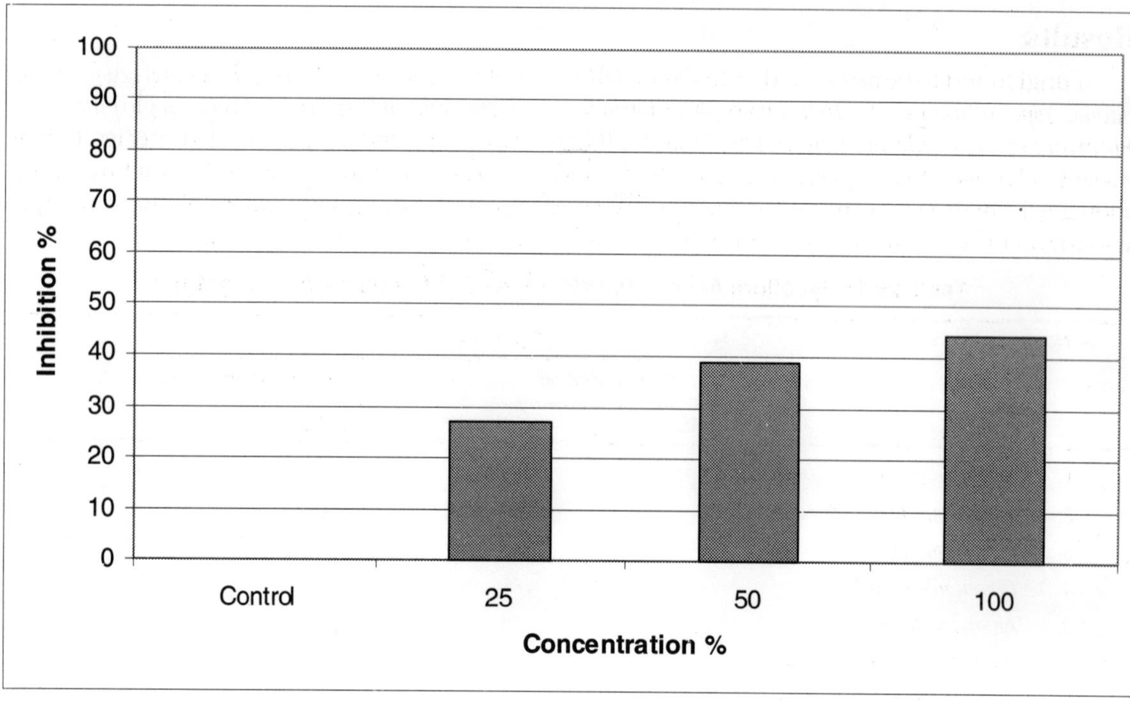

Figure 29.1: Antifungal Activity of Extracts of *Allium sativum* against *Curvularia lunata*

The extracts of *Syzygium aromaticum* inhibited the growth of *Curvularia lunata* by 9 per cent, 100 per cent and 100 per cent at 25 per cent, 50 per cent and 100 per cent concentrations respectively (Table 29.3).

Table 29.3 Antifungal Activity of Extracts of *Syzygium aromaticum* against *Curvularia lunata*

Sl.No.	*Concentration of Extract (per cent)*	*Inoculum Diameter (mm) [B]*	*Mean Colony Diameter (mm) [A]*	*Mycelial Growth Diameter (mm) [A-B]*	*Inhibition (per cent)*
1.	Control	4	50	46	0
2.	25	4	46	42	9
3.	50	4	4	0	100
4.	100	4	4	0	100

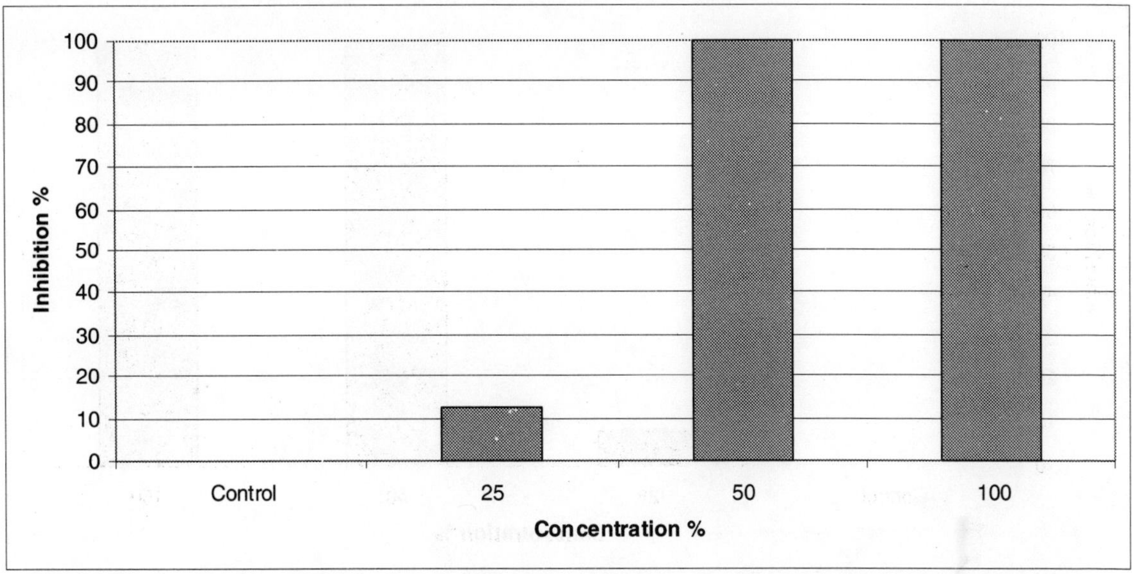

Figure 29.2: Antifungal Activity of Extracts of *Syzygium aromaticum* against *Curvularia lunata*

The extracts of *Allium sativum* inhibited the growth of *Aspergillus niger* by 4 per cent, 100 per cent and 100 per cent at 25 per cent, 50 per cent and 100 per cent concentrations respectively (Table 29.4).

Table 29.4: Antifungal Activity of Extracts of *Allium sativum* against *Aspergillus niger*

Sl.No.	*Concentration of Extract (per cent)*	*Inoculum Diameter (mm) [B]*	*Mean Colony Diameter (mm) [A]*	*Mycelial Growth Diameter (mm) [A-B]*	*Inhibition (per cent)*
1.	Control	4	71	67	0
2.	25	4	68	64	4
3.	50	4	4	0	100
4.	100	4	4	0	100

The extracts of *Syzygium aromaticum* inhibited the growth of *Aspergillus niger* by 4 per cent, 13 per cent and 100 per cent at 25 per cent, 50 per cent and 100 per cent concentrations respectively (Table 29.5).

Table 29.5: Antifungal Activity of Extracts of *Syzygium aromaticum* against *Aspergillus niger*

Sl.No.	*Concentration of Extract (per cent)*	*Inoculum Diameter (mm) [B]*	*Mean Colony Diameter (mm) [A]*	*Mycelial Growth Diameter (mm) [A-B]*	*Inhibition (per cent)*
1	Control	4	60	56	0
2	25	4	58	54	4
3	50	4	53	49	13
4	100	4	4	0	100

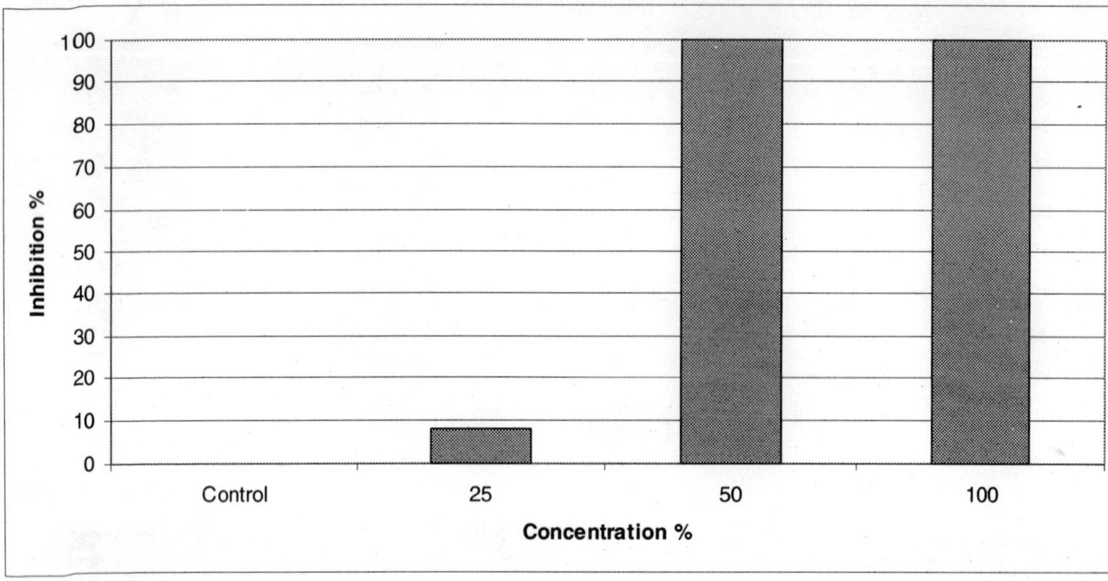

Figure 29.3: Antifungal Activity of Extracts of *Allium sativum* against *Aspergillus niger*

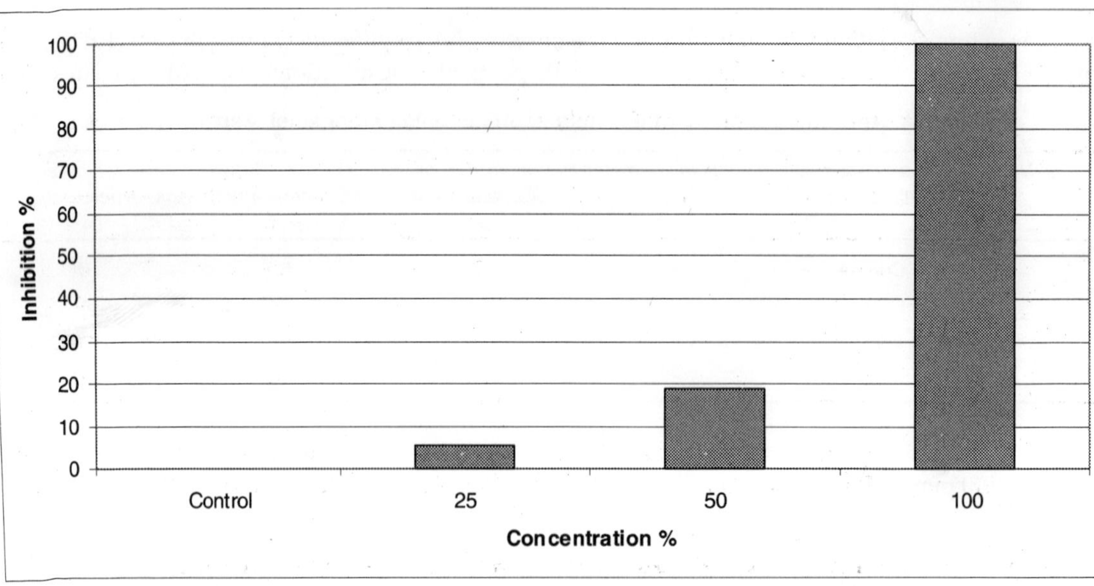

Figure 29.4: Antifungal Activity of Extracts of *Syzygium aromaticum* against *Aspergillus niger*

Discussion

The isolated mycoflora were: *Aspergillus niger, Aspergillus flavus, Aspergillus* sp., *Chaetomium* sp., *Alternaria tenius, Stemphylium vesicarium, Epicoccum purpurascens, Fusarium* sp., *Curvularia clavata, Curvularia lunata* and an Unidentified species. Fungi like; *Aspergillus flavus, Alternaria tenius, Fusarium* sp., *Curvularia clavata,* and *Penicillium* sp. were found mainly as an external contaminant on onion

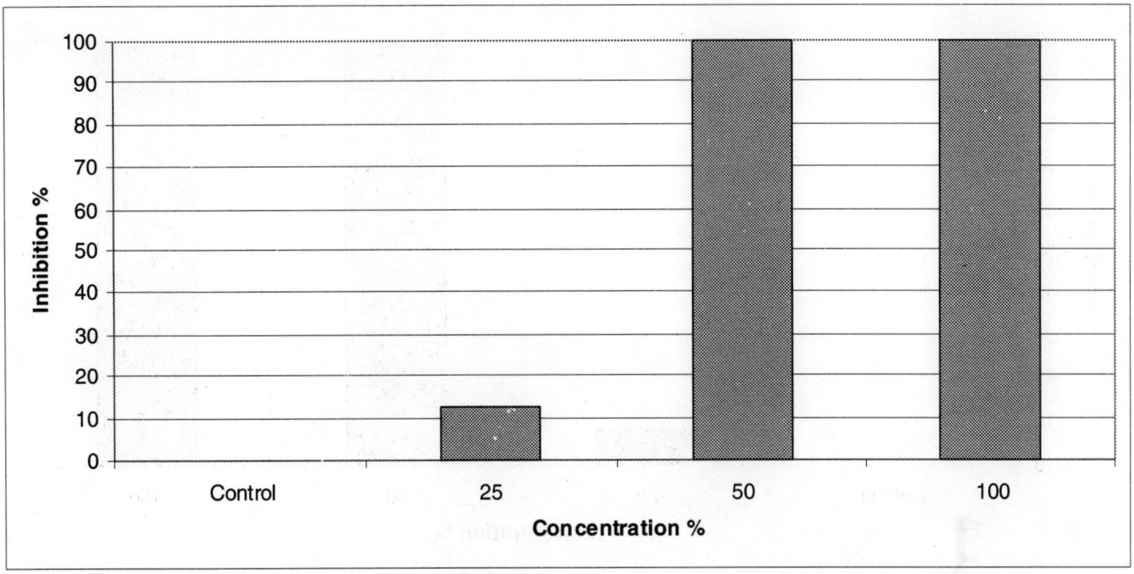

Figure 29.2: Antifungal Activity of Extracts of *Syzygium aromaticum* against *Curvularia lunata*

The extracts of *Allium sativum* inhibited the growth of *Aspergillus niger* by 4 per cent, 100 per cent and 100 per cent at 25 per cent, 50 per cent and 100 per cent concentrations respectively (Table 29.4).

Table 29.4: Antifungal Activity of Extracts of *Allium sativum* against *Aspergillus niger*

Sl.No.	*Concentration of Extract (per cent)*	*Inoculum Diameter (mm) [B]*	*Mean Colony Diameter (mm) [A]*	*Mycelial Growth Diameter (mm) [A-B]*	*Inhibition (per cent)*
1.	Control	4	71	67	0
2.	25	4	68	64	4
3.	50	4	4	0	100
4.	100	4	4	0	100

The extracts of *Syzygium aromaticum* inhibited the growth of *Aspergillus niger* by 4 per cent, 13 per cent and 100 per cent at 25 per cent, 50 per cent and 100 per cent concentrations respectively (Table 29.5).

Table 29.5: Antifungal Activity of Extracts of *Syzygium aromaticum* against *Aspergillus niger*

Sl.No.	*Concentration of Extract (per cent)*	*Inoculum Diameter (mm) [B]*	*Mean Colony Diameter (mm) [A]*	*Mycelial Growth Diameter (mm) [A-B]*	*Inhibition (per cent)*
1	Control	4	60	56	0
2	25	4	58	54	4
3	50	4	53	49	13
4	100	4	4	0	100

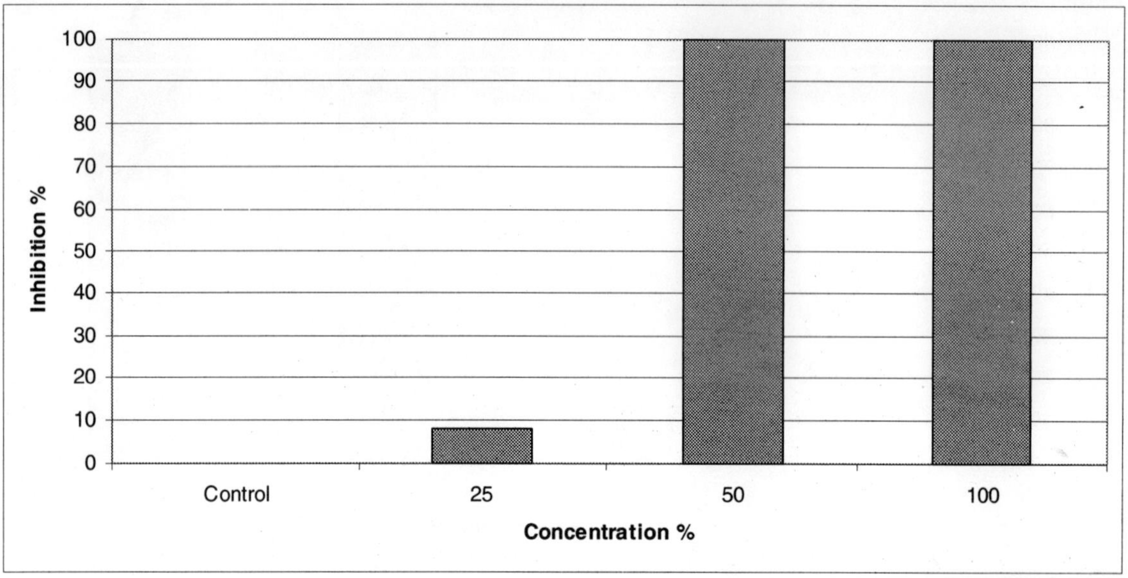

Figure 29.3: Antifungal Activity of Extracts of *Allium sativum* against *Aspergillus niger*

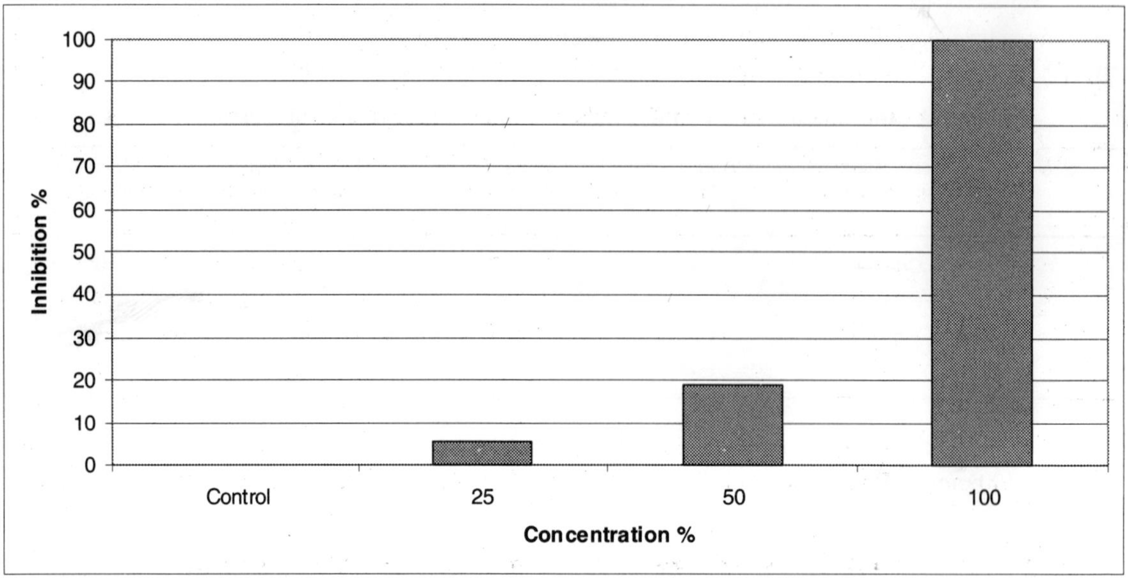

Figure 29.4: Antifungal Activity of Extracts of *Syzygium aromaticum* against *Aspergillus niger*

Discussion

The isolated mycoflora were: *Aspergillus niger*, *Aspergillus flavus*, *Aspergillus* sp., *Chaetomium* sp., *Alternaria tenius*, *Stemphylium vesicarium*, *Epicoccum purpurascens*, *Fusarium* sp., *Curvularia clavata*, *Curvularia lunata* and an Unidentified species. Fungi like; *Aspergillus flavus*, *Alternaria tenius*, *Fusarium* sp., *Curvularia clavata*, and *Penicillium* sp. were found mainly as an external contaminant on onion seeds as these fungi showed only on non-sterilized seeds, whereas their presence was not seen on sterilized seeds. Whereas fungi like *Chaetomium* sp., *Stemphylium vesicarium*, *Epicoccum purpurascens* were found, both on sterilized as well as non-sterilized seeds, indicating their presence as both external as well as internal seed-borne fungi. In this present study *Aspergillus niger* and *Curvularia lunata* were found to be the dominant species among all other species of fungi isolated. El-Nagerabi and Abdalla (2004) studied seed-borne fungi of Sudanese cultivars of onion (*Allium cepa* L.) and they also reported the presence of *Aspergillus niger*, *A. favus*, *Curvularia lunata*, *Penicillium* sp., *Chaetomium elatum*, *Fusarium* sp., *Alternaria alternata*, etc. where *Aspergillus niger* was found to be the dominant species. Similarly, Koycu and Ozer (1997) also reported the presence of fungal species like *Alternaria alternata*, *Fusarium* spp., and *Aspergillus niger* isolated from onion (*Allium cepa* L.) seeds from Turkey. Here also they reported *Aspergillus niger* to be the dominant species. Manandhar and Mathur (2003) recorded species of *Fusarium equiseti*, *F. moniliforme*, *F. pallidoroseum*, *Alternaria alternata*, *Penicillium* sp., and *Stemphylium* sp. on four samples of onion (*Allium cepa* L.) collected from different localities of Nepal. In the current study the Standard blotter technique was found to be most effective method for the isolation of seed mycoflora, compared to Agar plate method. According to Agarawal, Mathur and Neergard (1972), the blotter technique proved better for testing seed-borne fungi in rice, wheat, black gram and soybean.

The effectiveness of the plant extracts used was assessed on the basis of Minimum Inhibitory Concentration (MIC). MIC of plant extract can be expressed as the minimum concentration of the extract required for complete (100 per cent) inhibition of mycelial growth of the test fungus (Rao and Srivastava, 1994). In the present study extracts of *Allium sativum* and *Syzygium aromaticum* were found to be effective against *Aspergillus niger*, with MIC of 50 per cent and 100 per cent respectively. The plant extracts of *Allium sativum* was not found to be effective on the test fungus *Curvularia lunata* on the basis of MIC, where as the extract of *Syzygium aromaticum* was found to be effective with MIC of 50 per cent. Alice and Rao (1986) and Rahman (1992) also reported similar observations on the effectiveness of *Allium sativum* extracts for the control of seed-borne fungus *Drechslera oryzae*. Khan and Tripathi (1994) screened the extracts of different parts of 122 higher plants for their volatile antifungal activity against *Aspergillus niger* and *Curvularia ovoidea*. The flower bud extract of *Syzygium aromaticum* exhibited absolute toxicity against both tested fungi inhibiting the mycelial growth completely. In the present investigation also, the extract of *Syzygium aromaticum* was found to be effective against the test fungi *Aspergillus niger* and *Curvularia lunata*.

In conclusion, it is obvious that onion seeds are vulnerable to contamination with various seed-borne fungi. The seeds also harbor some of the pathogenic fungi together with a considerable number of opportunistic fungi, which cause devastating field diseases and marked post-harvest and storage losses.

Acknowledgements

The authors are grateful to Prof. (Dr.) Krishna Kumar Shrestha, Head of Department, Central Department of Botany, T.U. for providing necessary lab facilities. Thanks are also due to Mr. Shiva Devkota, Lecturer, Central Department of Botany, T.U and Dr. Rosemary Shrestha, Former Lecturer, Central Department of Botany, T.U., for their support. The authors are also grateful to Mr. Gopal Prashad Shrestha, Horticulture Development Research and Training Centre for providing necessary lab facilities.

References

Agarwal, V. K., Mathur, S. B. and Neergard, P. 1972. Some aspects of seed health testing with respect to seed-borne fungi of rice, wheat, black gram, green gram and soybean grown in India. Indian Phytopathol., 25:91-100.

Alice, D. and Rao, A. V. 1986. Management of seed-borne *Drechslera oryzae* of rice with plant extracts. Int Rice Res Newsl., 11: 19-24.

Beye, F. 1978. Insecticide from vegetable kingdom. Plant Res. Devel., 7: 13-31.

Christensen, C. M. and Kaufman, H. M. 1965. Deterioration of storage grain by fungi. Annual Review of Phytopathology, (3): 69-81.

Costa, T., Fernandes, F. L. F., Santas, S. C., Oliveria, C. M. A., Liao, L., Fessi, P. H., Paul, J. R., Ferreira, H. D., Sales, B. H. N. and Silva, M. R. R. 2000. Antifungal activity of volatile constituents of *Eugenia dysentrica* leaf oil. J. Ethophaarm. 72(1-2): 111-117.

Crocker, W. and Barton, L.v. 1957. Physiology of Seeds. Chronica Botanica Company.

El-Nagerabi, S. A. F. and Abdalla, R. M. O. 2004. Survey of seed-borne Fungi of Sudanese Cultivars of Onion, with New Records. Phytoparasitica. 32(4): 413-416.

FAO. 1994. Production Yearbook. Food and Agricultural Organization of the United Nations, Rome, Italy. 48(125).

Fawcett, C. H and Spencer, D. M. 1970. Plant chemotherapy with natural products. Ann. Rev. Phytopath. 8:403.

FNCCI. 2006. Vegetable Seeds. Agro Enterprise Centre, Federation of Nepalese Chambers of Commerce and Industry, Teku. Kathmandu, Nepal.

Gautam, S. R., Neupane, G. Baral, B. H., Rood, P. G. and Pun, L. 1994. Prospects of Onion Cultivation in the warm-temperate hills of Eastern Nepal and its research and development strategies for commercial production. ISHS Acta Horticulturae 433: International Symposium on Edible Alliaceae.

Grover, A and Moore, J. D. 1962. Taximetric studies of fungicides against Brown rot organism *Sclerotina fruticola*. Phytopathology. 52: 876-880.

HMG/N. 1997. Preliminary list of Economic Crop Pests in Nepal. Ministry of Agriculture, Plant Protection Division, Harihar Bhawan, Lalitpur, Nepal.

Kemetz, K. T., Ellett, W. and Schmithenner, A. F. 1979. Soybean Seed Decay: Source of Inoculum and Nature of Infection. Phytopathol., 69(8):798-848.

Khan, S. A. and Tripathi, N. N. 1994. Volatile antifungal activity of some higher plants against storage fungi. Neo-botanica, 2(2):111-119.

Koycu, N. D. and Ozer, N. 1997. Determination of Seed-borne Fungi in Onion and their transmission to Onion Sets. Phytoparasitica. 25(1): 25-31.

Mathur, S. B. and Manandhar, H. K. 2003. Fungi in seeds. The Danish Government Institute of Seed Pathology for Developing Countries, Copenhagen, Denmark.

Rahman, M. 1992. Study on the seed-borne fungi and their control with botanical and chemical fungicides on five local Boro varieties of rice. M.Sc. Thesis. Department of Plant Pathology, Bangladesh Agricultural University, Mymensingh.

Rao, G. P. and Srivastava, A. K. 1994. Toxicity of essential oils of higher plants against fungal pathogens of Sugarcane. Current trends on sugarcane pathology: 347-365.

Microbes: Diversity and Biotechnology (2012) *Pages* **477–493**
Editors: **Prof. S.C. Sati & Dr. M. Belwal**
Published by: **DAYA PUBLISHING HOUSE, NEW DELHI**

Chapter 30
Antimicrobial Potentiality of Bryophytes

Neerja Pande and Prabha Dhondiyal (nee Bisht)
Department of Botany, D.S.B. Campus, Kumaun University, Nainital – 263 001

ABSTRACT

This article is developed around the antimicrobial potential (anti fungal and anti bacterial) of bryophytes with a brief background of these plants, their phyto-chemistry, and ethno medical uses. Bryophytes do possess a variety of secondary metabolites like, monoterpenes, diterpenes, sesquiterpenes, flavoinoids, phenols, and many others as a by product of their metabolism. These secondary metabolites are therefore capable to deter herbivory, insecticidal, piscidal effects. The findings of the researches on the antimicrobial activity of the plant (liverworts and mosses) extracts (crude) isolated in different organic solvents and the chemical compounds exhibiting the inhibitory effect on the growth of microorganisms are also summarized.

Keywords: *Bryophyte, Animicrobial potential, Phytochemistry, Medicinal, Bacteria, Fungi.*

Introduction

The present article reviews principally the antimicrobial potential of liverworts and mosses, which forms an integral component of the ecosystem in wet climates. With the evolution of these plants, they are exposed to different dangers due to this, they are able to adapt themselves to different negative and positive influences by exploiting their biochemical potentialities. For this reason, secondary metabolites of quite numerous chemical compositions are synthesized in their tissues. These metabolites not only confer them resistance towards the infection or external attack but also make them as a potential source of future medicine. This assumption is rested on the ethno-medicinal properties of liverworts and mosses to cure skin diseases, cardiovascular, boils etc. (Mc Cleary and Walkington, 1966; Glime, 2007 and Asakawa, 1990). Moreover, these plants are rarely attacked by the micro organisms, (Ando and Matsuo, 1984) this property attracted the attention of scientist to explore their potential as an antimicrobial agent thus, the aim of this review is to serve as the basis for the exploration of many more bryophyte species containing novel compounds that would serve as stimulus for further researches in antimicrobial potential of bryophytes. In this paper, an attempt will be made to bring together the scattered information available from the studies on the antifungal and antibacterial potential of bryophytes. In this way the gaps in our knowledge will be accentuated by specific allelopathic potentials of bryophytes. Most liverworts are rich in sesquiterpenoids but the compounds responsible for the growth regulatory activities are unknown. Many papers have recently been published on such topics as: biologically active compounds from bryophytes (Benesova and Herout, 1978 and Asakawa, 1998, 2007,Banerjee,2001), the possibilities of using bryophytes in control of diseases and malfunction are exciting, but the exploratory work has yet in its infancy. Almost, three decades ago, virtually very little was known of bryophyte biochemistry, but today it is sure that the variety of chemicals produced by these morphologically simple organism is phenomenal. So far biologically active compounds obtained from bryophytes have not proved economical in practice. Their pharmaceutical worth seems promising as we lack any understanding of potential side effects.

Bryophytes and their Medicinal Uses

Bryophytes are the simplest and primitive group of land plants. Taxonomically, these plants are placed between algae and pteridophytes. They are generally represented by about 21000 species (Schofield, 1985) with a worldwide distribution. As a group, bryophyte is divided into three classes: musci (mosses), hepaticae (liverworts), and anthocerotae (hornworts). Owing to their potential to live on a variety of habitats, they are exposed to differential degree of biotic and environmental hazards. To cope with these dangers, numerous secondary metabolites of several types are synthesized in their tissues as a defense system (Herout, 1990). These chemicals are stored in specialized structures known as oil bodies, oil cells or osmophilic glands, and make the plants non-palatable hence they effectively deter herbivores and pathogens from attacking them.

Bryophytes in natural environment are also linked with the culture, beliefs and ethics of mankind. Since times immemorial these organisms has been used as medicinal plants in North America, China and Europe to cure burns, bruises and external wounds. About 30 to 40 species of bryophytes are used as herbal medicine in China (Ding 1982), for example, Chinese people used *Fissidens* sp. as an antibacterial agent for swollen throats and other symptoms of bacterial infection. Judith Sulliran reported that some of the Chinese herbal medicine that included *Grimmia, Atrichum, Polytrichum* and *Thuidium* were primarily used as antibacterial and anti-inflammatory agents. Germans used *Ceratodon purpureus* and *Bryum argenteum* to cure fungal infections of horses (Glime, 2007). In India, people of Kumaon Himalaya used *Marchantia polymorpha* and *M. palmata* to cure burns, abscesses and to reduce pus formation, while the paste of *Riccia* sp. is applied on the ringworm disease of skin. (Pant and Tewari, 1989).Kumar *et al.* (2000) reported that *Plagiochasma appendiculatum* is used for treating skin diseases by Gaddi tribe in Kangra valley.

Phytochemistry of Bryophyte

With the advancement in the knowledge of the chemistry of medicinal plants preferably angiosperms (flowering plants) and the use of chemicals obtained for therapeutic purposes, or as herbal medicine had attracted the attention of scientists for the search of more and more plants with novel secondary metabolites which can be used as the potential source of herbal medicine. In this context, the tissues of bryophytes were also put to test as the oil bodies of bryophytes are the reservoir of many secondary metabolites and the phytochemical analysis done by various research groups (Asakawa, 1981, 1982; Markham *et al.*, 1969 Markham and Porter,1978; Matsuo *et al.*, 1984) revealed that liverworts and mosses possess thousands of novel metabolites of various orders *viz.*, steroids, flavonoids and phenolic compounds,monoterpenes and sesquiterpenes etc. (Tables 30.1–30.4). In view of the detailed knowledge that has accumulated about bryophyte chemistry there seems a wide possibility that researches with microbial potential of bryophytes can be brought about and the results can be used to recognize the future medicine. Knowledge of the chemical composition of bryophytes has recently expanded at an almost logarithmic rate. Huneck (1983) summarized much of this literature while Surie and Asakawa's review was based more upon usefulness of the compounds in taxonomy. Some unusual forms of lipids and fatty acids have been reported for the mosses, but these may change in response to environmental conditions. Recent investigations on the lipophilic compounds of liverworts resulted in the isolation of numerous new natural products with a high range of biological activity (Asakawa 1982, 1995, 1999, 2007 and Zinsmeister *et al.*, 1991).

Table 30.1: Some Sesquiterpenoids from Liverworts

Sl.No.	*Plant Species*	*Compounds*	*Reference*
1.	*Gymnomitrium obtusum*	gymnomitrial	Connolly,1990
2.	*Bazzania fauriana*	isobazzanene	Matsuo *et al.*, 1977
3.	*Herberta adunca*	Herbertene, alpha- –herbertenol, beta-herbertenol, herbertenediol, herbertenolidie	Matsuo *et al.*, 1982, 1983, 1986
4.	*Bazzania fauriana*	Bazzanenyl caffeate	Toyota and Asakawa,1988
5.	*Marchantia palaecea*	Grimanlodone	Asakawa *et al.*, 1988
6.	*Reboulia hemispherica*	9-hydroxy derivative	Becker *et al.*, 1988
7.	*Ricciocarpus natans*	Furanosesquiterpene lactone	Theoplut Eicher, 1990

Source: Connolly, J.D. (1990).

Table 30.2: Some Monoterpenoids in Liverworts and Mosses

Sl.No.	*Plant Species*	*Compounds*	*Reference*
1.	*Conocephalum conicum*	(-)-thujanol	Connolly,1990
2.	*Targionia hypophylla*	Cis and trans pinocarveyl acetate	*Asakawa *et al.*, 1986 b
3.	*Jungermannia obovata*	Limonene	Connolly,1990
4.	*Porella densifolia*	a-pinene, camphene	Bisht 1998
5.	*Conocephalum conicum*	b-pinene, camphene,a-terpinene, 1-octen-3-yl acetate	Bisht,1998
6.	*Entodon plicatus*	a-thujene,carveol	Joshi,1999
8.	*Wisneralla denudata*	Myrcene,a-pinene,camphene, bornyl acetate,sabinyl acetate	Joshi,1999

* Not seen in original (consulted fromAsakawa,Y. 1990).

Table 30.3: Some Non Phenolic Compounds Isolated from Bryophytes

Sl.No.	Plant Species	Compounds	Reference*
1.	*Conocephalum conicum*	eugenol	Asakawa *et al.*, 1979a
2.	*Bazzania pompeana*	2-hydroxy cuparene	Matsuo *et al.*, 1975
3.	*Radula japonica*	2-hydroxy cuparene	Asakawa *et al.*, 1981 a
4.	*Herberta adunca*	β-herbertenol	Matsuo *et al.*, 1982, 1986
5.	*Pellia endaeviifolia*	Schikmic acid	Kinzel and Walland, 1966
6	*Lophocolea bidentata*	Ellagic acid	Mues and Zinsmeister,1973
7.	*Plagiochila asplenoids*	Ellagic acid	Mues and Zinsmeister,1973
8.	*Anthoceros laevis*	Methyl p-coumarate	Mendez and Sanz Canabinilles, 1979
9.	*Anthoceros punctatus*	Methyl caffeate	Mendez and Sanz Canabinilles,1979

*Not seen in original

Source: Gorham, J. (1990).

Table 30.4: Some Bis (bibenzyls) in Liverworts

Sl.No.	Plant Species	Compounds	Reference*
1.	*Marchantia polymorpha*, *M palacea* var. *diptera*, *M. tosona*	Marchantin A	Asakawa 1982,a; 1983,1984 b
2.	*Marchantia polymorpha*, *M. palacea* var. *diptera*, *M. tosona*	Marchantin B	Asakawa, 1984 b,Asakawa *et al.*, 1983 a
3.	*Plagiochasma intermedium*	Marchantin B	Tori *et al.*, 1985
4.	*Marchantia polymorpha*, *M. palacea* var. *diptera*, *M. tosona*,*M. palmata*	Marchantin C	Asakawa *et al.*, 1983 a; Asakawa, 1984 b
5.	*Plagiochila acanthophylla* var. *Japonica*	Marchantin C	Hashimoto *et al.*, 1987
6	*Marchantia polymorpha*	Marchantin D	Asakawa *et al.*, 1987 d
7.	*Marchantia polymorpha*	Marchantin E	Asakawa *et al.*, 1987 d
8.	*M palacea* var. *diptera*,	Marchantin F	Asakawa *et al.*, 1984 b
9.	*Marchantia polymorpha*	Marchantin G	Asakawa *et al.*, 1987 d
10.	*Plagiochasma intermedium*, *Riccardia multifida*	Marchantin H	Tori *et al.*, 1985
11.	*Plagiochasma intermedium*, *Riccardia multifida*	Marchantin I	Tori *et al.*, 1985

*Not seen in original

Source: Gorham, J. (1990).

Antimicrobial Potential

Antibacterial Potential

The vast knowledge of chemical compounds in bryophytes, presence of characteristic odour (fragrance) in some of the bryophytes, (Schuster 1966) traditional use of many bryophytes as medicine and resistance of bryophytes against micro organisms and insects has led to the pharmaceutical investigations of these plants. Owing to this, scientists put various bryophytes into test against a number of micro organisms (bacteria and fungi), in order to find out their allelopathic potential which can be exploited for the development of herbal medicine and also as a fungicide or herbicide to protect the plants against infection.

It was the study of Hayes (1947) in which the occurrence of antimicrobial substance in *Conocephalum conicum* was reported for the first time. Thereafter, Madsen and Pates (1952) reported that the aqueous extracts of *Sphagnum portoicense,S. strictum, Conocephalum conicum* and *Dumortiera hirsuta* had immense potential to inhibit the growth of bacteria tested (*Pseudomonas aeruginosa* and *Staphylococcus aureus*). These preliminary reports on the antimicrobial potential of these plants had encouraged the researchers dealing with phyto-chemical studies and several other aspects of bryophytes to switch on to observe the effect of bryophytes on growth behaviour of micro organisms. Later in sixties, many bryophytes were tested against the bacteria for their antibiotic properties (Pavelitic and Stilinovic.1963, Wolters1964, Mc Cleary *et al.*, 1960). Some of the bacterial strains sensitive to bryophyte extracts are given in Table 30.5. In the year 1966 a paper of Mc Cleary and Walkington was appeared whereby 18 species of mosses were assessed for anti microbial potential and reported that only 5 genera (*Atrichum, Mnium, Dicranum, Polytrichum, and Sphagnum*) were most active. Out of these, *A. undulatum* seemed to be most effective in inhibiting the growth of all the bacteria tested except *Aerobacter aeurogens* and *Escherchia coli.* These results were based on the differential degree of solubility of active substances to the different solvents and it was presumed that a large number of metabolites (active principle) may likely to be responsible for this antimicrobial potential.

Thereafter, Gupta and Singh (1971) found high antibacterial potential in the extracts of *Barbula* sp. (36.2 per cent) and *Timmiella* sp. (18.8 per cent). Banerjee and Sen (1979) tested 52 species of bryophytes against nine bacteria (both gram positive and gram negative) and three fungi. The water and organic solvents like ethanol, methanol, ether, acetone and chloroform were used for the isolation of active principles. The alcoholic extracts of liverworts exhibited most of the antibacterial activity as compared to other solvents. The aqueous extracts of the selected plants were found inactive in inhibiting the growth of either type of micro organisms. The most notable liverworts and mosses used were: *Asterella angusta, A.sanguinea Lunularia cruciata, Marchantia paleacea M. palmata, Plagiochasma appendiculatum, Reboulia hemispherica, Targionia hypophylla, Hyophila involuta, Mnium sp., Philonotis falcata, Rhodobryum giganteum and Sphagnum sp.* On the basis of solubility and antibiotic spectra, the results indicated the occurrence of a variety of chemicals in the tissues of bryophytes that had potential antibiotic activity. Also, these alcohol soluble chemicals were more frequent in liverworts than in mosses. Interestingly, they also reported that antibiotic activity in a given species may be influenced by the age, season of collection and the habitat of the plant. Matsuo *et al.* (1982a, 1982b, 1983) supported this conclusion by demonstrating that antifungal activity against *Botrytis cinerea, Pythium debaryanum,* and *Rhizoctonia solani* by the liverwort *Herbertus aduncus* was age-dependent. They subsequently isolated three aging substances from it: (-)-α-herbertenol; (-)-β-herbertenol, and (-)-α-formylherbertenol.

Extracts of *Fossombronia pusilla* and *F. himalayensis* had shown bactericidal property and inhibited the growth of *E. coli, Staphylococcus aureus* and *Bacillus subtilis* (Sauerwein and Becker, 1990). Since then researches on antimicrobial potential of bryophytes are being carried out in various laboratories through out the world (Table 30.5). Ichikawa (1982)and Ichikawa *et al.* (1983) examined the antibacterial activity of about 80 mosses and found that all these mosses had this potential to some degree. *Conocephalum conicum, Mnium undulatum* and *Lepidodictyum riparium* were tested against the human pathogenic bacteria by Castaldo Cobianchi *et al.* (1989). The extract of *L. riparium* was found with maximum ability to inhibit the growth of all pathogens including *Pseudomonas aeurogenosa.* Contrary to this, the aqueous and acetonic extracts of other two species had a very weak ability to check the growth of bacteria.

Table 30.5: Bacterial Strains Sensitive to some of the Selected Bryophyte Extracts

Sl.No.	*Bacterial Strains*	*Bryophyte*	*Reference*
1.	*Escherschia coli, Proteus mirabilis, Staphylococcus aureus*	*Marchantiapolymorpha*	Mewari and Kumar, 2008
2	*Micrococcus luteus, Bacillus subtilis, B. cereus, Escherischia coli, Klebsiella pneumoniae, Proteus mirabilis, Staphylococcus aureus, Salmonella typhimurium, Streptococcus pneumoniae, Pseudomonas aeruginosa Enterobacter aerogenus*	*Plagiochasma appendiculatum*	Singh, Govindrajan, Nath, Rawat and Mehrotra, 2006
3.	*Micrococcus luteus, Bacillus subtilis, B. mycoides, Escherischia coli, Klebsiella pneumoniae, Pseudomonas aeruginosa, Enterobacter aerogenus, Yersinia entercolitica*	*Palustriella commutata*	Ilhan *et al.*, 2006
4.	*Micrococcus luteus, Bacillus subtilis, Escheirschia coli, Staphylococcus aureus*	*Bryum argentium*	Sabovljevic *et al.*, 2006
5.	*Escherschia coli, Staphylococcus typhi*	*Conocephalum conicum, Marchantia polymorpha, Plagiochasma appendiculatum*	Vashistha, Dubeyand Pande, 2007
6.	*Escherischia coli, Bacillus subtilis, Streptococcus faecalis*	*Conocephalum conicum, Rebouliahemi spherica, Plagiochasma appendiculatum*	Zehr, 1990
7.	*Escherischia coli, Bacillus subtilis, Streptococcus aureus*	*Fossombronia pusilla, F. himalayensis*	Sauerwein and Becker, 1990
8.	*Escherischia coli, Proteus mirabilis, Staphylococcus aureus, Salmonella typhi, Streptococcus faecalis, Pseudomonas aeruginosa*	*Lunularia cruciata*	Basile *et al.*, 2001
9.	*Salmonella typhi, Bacillus subtilis, M phlei V chlorae, Pseudomonas aeurogenosa*	*Anthoceros erectus, Notothylas indica, Philonotis falcata, Brachythecium procumbens, Rhodobryum giganteum*	Banerjee and Sen, 1979
10.	*Micrococcus luteus, Bacillus subtilis, Staphylococcus aureus*	*Marchantia polymorpha,*	Pavletic and Stilinovic, 1963

Similarly, Latiff *et al.* (1989) screened 14 moss species of Malaysia for antibacterial activity against the growth of *Escherichia coli, Staphylococcus aureus* and *Bacillus subtilis* and found the growth of *S. aureus* was inhibited by all the extracts examined. Joshi (1993,1995 and Joshi *et al.*, 1990) found bacterial growth inhibitory activity of light petroleum ether extract of *Exormotheca tuberifera* and water and DMSO extracts of *Dumortiera hirsuta* and *Lunularia cruciata.* The extract of *Exormotheca tuberifera* showed pronounced activity against *E. coli.* In another paper, Joshi and Desai (1998) described the antibacterial potential in petroleum ether extract of *Fossombronia himalayensis* when tested against *Escherichia coli, Klebsiella pneumoniae, Staphylococcus aureus* and *Bacillus subtilis.* Asakawa (1998) reported the antimicrobial potential of several liverworts (*Bazzania* sp, *C.conicum, D. hirsuta, M. polymorpha, Metzgeria furcata, P. enclvaeutifolia, Plagiochila* sp. *Porella vernicosa* complex, *P. platyphylla, Radula* sp etc.) and mosses (*Atrichum angustatum, A.undulatum, D. scoparium, Isothecium stoloniferum, Mnium punctatum, Orthotrichum rupestre, Polytrichum commune, P. juniperinum, Thuidium recognitum var. delicatulum, Tortula ruralis, Sphagnum fimbriatum, S. palustre and S. strictum*).

Basile *et al.* (1998) did comparison between the antibacterial activities of the extract (liverwort and moss) with the antibiotic drugs. They observed the antibacterial effect of the *Lunularia cruciata* extract against the growth of nine bacterial strains, showing an activity equal to one third that of the antibiotic Nacephotaxime. Basile *et al.* (1998a) also evaluated the action of acetone extract of *Lunularia cruciata* against 13 other bacterial and 2 fungal pathogens and found substantial antibacterial activities for the mature gametophyte of *Lunularia cruciata.* Also, the moss *Rhynchostegium riparioides* was screened against 2 Gram-positive (*Staphylococcus aureus* and *S. faecalis*) and 8 Gram–negative (*Proteus mirabilis, P. vulgaris, Pseudomonas aeruginosa, Klebsiella pneumonia and Bacillus subtilis*) bacteria. The most sensitive strains found were, *E. coli, P. aeruginosa, Enterobacter chlorae* (Basile *et al.*, 1998b).

Ilhan *et al.* (2006) studied antimicrobial activity against 11 bacteria, 1 yeast, and 8 moulds in acetone and methanol extracts of *Palustriella commutata* (Hedw.) by using the disc diffusion method. The antimicrobial test results revealed that acetone extract had a potential activity against 9 test bacteria. Some gram-negative bacteria tested (*Bacillus mycoides, B. subtilis,* and *Micrococcus luteus*) were sensitive to methanol extract while, all gram- negative bacteria tested (*Klebsiella pneumoniae, Yersinia enterocolitica, Pseudomonas areuginosa, Escherichia coli,* and *Enterobacter aerogenes*) were sensitive to the acetone extract but both extracts were inactive against yeast and mould.

Mewari and Kumar (2008) observed antimicrobial activity of crude methanol and flavonoid (free and bound) extracts of *Marchanita polymorpha* L. against three bacterial strains, *viz., Escherichia coli, Proteus mirabilis* (Gram negative), and *Staphylococcus aureus* (Gram positive), and four fungal strains, *viz., Aspergillus flavus, A. niger, Candida albicans,* and *Trychophyton mentagrophytes.* The extracts showed best activity against *Staphylococcus aureus* (IZ 20.6 and 19.6 mm, MIC 0.281 and 0.312 mg/ml, MBC 1.125 and 0.312mg/ml, respectively), however, all the microorganisms were found to be sensitive against the extracts tested. Total activity for *Proteus mirabilis* and *S. aureus* for methanol extract was found to be the same (124 ml) and maximum for free flavonoid against *Candida albicans* (199ml). Bodade *et al.* (2008) screened different bryophytes *viz., Plagiochasma appendiculatum, Thuidium delicatulum, Thuidium cymbifolium, Bryum cellulare, Bryum argenteum* and *Racomitrium crispulum* against 10 bacterial and reported that bryophyte extracts are found to be active against at least one of the test organisms except *Racomitrium crispulum.*

Antifungal Potential

Compared to the antibacterial potential of bryophytes extracts, the information on anti fungal potential of these plants is quite meagre (Table 30.6). Jennings (1926) first of all reported mosses

Table 30.6: Fungal Strains Sensitive to some of the Selected Bryophyte Extracts

Sl.No.	Fungal Strains	Bryophyte	Reference
1.	*Microsporum gypseum, Trichophyton equinum, T. mentagrophytes*	*Ptycanthus striatus*	Dixit *et al.*, 1982
2.	*Candida albicans*	*Polytrichum juniperinum*	Clipa and Pascal, 1986
3.	*Botrytis cinerea, Alternaria solani*	*Bazzania trilobata, Diplophyllum albicans, Sphagnum quinquefarium, Dicranum denudatum, Hylocomium splendens*	Mekuria *et al.*, 1999
4.	*Macrophomina phaseolina*	*Conocephalum conicum, Marchantia polymorpha, Plagiochasma appendiculatum*	Dubey *et al.*, 2001
5.	*Alternaria alternata, Macrophomina phaseolina, Fusarium solani, F.oxysporum, Rhizoctonia solani, Verticillum dahliae, Pythium* sp	*Philonotis marchica, Grimmia pulvinata, Plagiomnium regium, Haplocladium* sp, *Bryum pallens, Drepanocladus aduncus, Pellia, epiphylla*	Shirzadian, *et al.*, 2005
6.	*Aspergillus niger, Penicillum ochrochloron, Candida albicans, Trichophyton*	*Bryum argenteum*	Sabovljevic *et al.*, 2006
7.	*Aspergillus niger, A. flavus, A. spinulosus, A. terreus, A. nidulans, Candida albicans, Trichophyton ruburum, Cryptococcus albidus*	*Plagiochasma appendiculatum*	Singh *et al.*, 2006
8.	*Aspergillus niger, Candida albicans,*	*Conocephalum conicum, Marchantia polymorpha, Plagiochasma appendiculatum*	Vashistha *et al.*, 2007
9.	*Bipolaris sorokiniana, Fusarium solani*	*Rhyncostegium vegans, Entodon plicatus*	Mewari *et al.*, 2007
10.	*Aspergillus niger, A. flavus, Candida albicans, Trichophyton mentagrophytes*	*Marchantia polymorpha,*	Mewari and Kumar, 2008
11.	*Aspergillus niger, A. flavus*	*Bryum argenteum, B. bicolor, B. capillare, Funaria hygrometrica, Hyophila involuta*	Rawat, 2007

immunity towards fungi but never thought that the mosses can be used as remarkable antifungal agents. Wolters (1964) was the pioneer of this field who studied the antifungal activities of 18 species of bryophytes, and found remarkable antifungal activity only of *Diplophyllum albicans, Plagiothecium denticulatum* and *Pogonatum aloides* for the first time. Homans and Fuchs (1970) observed antifungal activity of dichloromethane and methanol extract of the liverwort *Bazzania trilobata* against the plant pathogenic fungus *Cladosporium cucumerinum.* Hoof *et al.* (1981) reported that many moss extracts have a marked activity against *Microsporum canis, Tricophyton rubrum, T. mentagrophytes and Candida albicans.* Dixit *et al.* (1982) investigated the inhibitory effects of the extracts of 6 bryophyte species on the mycelial growth of *Microsporum gypseum, Tricophyton equium* and *T. mentagrophytes.* They reported *Ptycanthus striatus* extract was the most potent, completely inhibiting the growth of *M. gypseum* and *T. equinum.* Raymundo *et al.* (1989) investigated 46 species of pteridophytes and bryophytes for activity against seven test organisms including *Candida utilis, Micrococcus luteus* and *Staphylococcus aureus.* More than 75 per cent plants showed activity against at least one test organism. A study on the toxicity and antimicrobial effect of aqueous extracts of *Polytrichum juniperinum* revealed that they were non-toxic to mice and possessed antimycotinic effect against some strains of *Candida albicans* (Cipla and Pascal, 1986). Asakawa (1998) reported the antifungal potential of several liverworts *viz., Bazzania* sp, *Conocephalum conicum, Lunularia cruciata, Marchantia polymorpha, Plagiochila* sp., and mosses *viz., Atrichum undulatum, Bryum pallens, Dicranella heteromalla, Dicranum scoparium,, Mnium hornum, Plagiothecium denticulatum,, Polytrichum commune, Pogonatum aloides, Sphagnum* sp. Mekuria *et al.* (1999) described bryophyte as a new source of antifungal substance in crop protection. Eighteen bryophyte species were screened for antifungal nature of their ethanolic extracts. *Bazzania trilobata, Diplophyllum albicans, Sphagnum quinquefarium, Dicranum denudatum and Hylocomium spledens* caused the greatest inhibition of mycelial growth of *Botrytis cinerea* and *Alternaria solani.*

Dubey *et al.* (2001) tested acetone and ether-soluble extracts of three bryophytes *Conocephalum conicum, Marchantia polymorpha* and *Plagiochasma appendiculatum* for their antifungal activity against a plant fungus *Macrophomina phaseolina*, which causes charcoal-rot on Soybean. The acetone soluble extract of all three bryophytes had promising antifungal activity.

The effects of water, methanol, ethanol, petroleum ether and acetone extract of 23 bryophyte taxa (21 mosses and two leafy liverworts) were tested against seven different pathogenic fungal species, namely, *Alternaria alternata, Fusarium solani, F.oxysporum, Macrophomina phaseolina, Rhizoctonia solani, Verticillium dahlia* and *Pythium sp.* The broadest spectrum of antifungal activity were shown by the ethanolic extracts of six moss species, namely *Philonotis marchica, Grimmia pulvinata, Plagiomnium regicum, Haplocladium sp., Bryum pallens* and *Drepanocladus aduncus* followed by a liverwort, *Pellia epiphylla.* Ethanol was the most efficient among other experimental solvents (Shirzadian 2005). In an another study, Sabovljevic *et al.* (2006) evaluated the antimicrobial activity of ethanol extracts of *Bryum argenteum* by micro dilution method against four bacterial (*Escherichia coli, Bacillus subtilis, Micrococcus luteus* and *Staphyococcus aureus*) and four fungal species (*Aspergillus niger, Penicillium ochrochloron, Candida albicans* and *Trichophyton mentagrophytes*) and found anti microbial activity of the ethanol extracts against all tested bacteria and fungi.

The traditional knowledge about the use of *Plagiochasma appendiculatum* paste by Gaddi tribe in Kangra valley for treating skin diseases was exploited by Singh *et al.* (2006).These workers performed experiments to assess the antimicrobial potential of this liverwort and found that the alcoholic and aqueous extracts showed significant antibacterial and antifungal activity against thirteen bacterial strains *viz., Micrococcus luteus, Bacillus subtilis, B.cereus, Staphylococcus aureus, Streptococcus pneumoniae, Enterobacter aerogenes, Escherichia coli, Klebsiella pneumoniae, Proteus mirabilis, Pseudomonas areuginosa* and *Salmonella typhimurium Yersinia enterocolitica,,* and eight fungal strains *Aspergillus flavus, A. niger, A. spinulosus, A. terreus, A. nidulans, Candida albicans, Cryptococcus albicans* and *Trychophyton rubrum.* Out of eight fungal strains two are dimorphic fungi. Both type of extracts (alcoholic and aqueous) showed good activity against the dermatophyte and some infectious bacteria with MIC of 2.5 ìg/disc. The result showed that *Plagiochasma appendiculatum* extract has potent wound healing capacity as evident from the wound contraction and increased tensile strength.

Vashistha *et al.* (2007) screened the liverworts, *Conocephalum conicum, Marchantia polymorpha* and *Plagiochasma appendiculatum* against four human pathogens, two fungi (*Aspergillus niger and Candida albicans*) and two Gram-negative bacteria (*Escherichia coli* and *Salmonella typhii*). The acetone soluble extract of all the bryophytes showed inhibitory effect against the pathogens used. However, *S. typhii* was found to be more sensitive than *E. coli.* Singh *et al.* (2007) studied antimicrobial activity of ethanolic extracts of 15 Indian mosses against five Gram (+) and six Gram (-) bacterial strains and eight fungi. The bryophytes, *Sphagnum junghuhnianum, Barbula javanica, B. arcuata, Brachythecium populeum, B. rutabulum, Mnium marginatum* and *Entodon rubicundus* were found to be active against all the organisms.

Mewari *et al.* (2007) reported potent antifungal activities in ethanol and petroleum ether extracts of mosses, (*Rhynchostegium vegans* and *Entodon plicatus*) against two fungal pathogen, (*Bipolaris sorokiniana* and *Fusarium solani*) and found both the extracts of mosses were effective for the tested fungi. Mewari and Kumar (2008) screened antimicrobial activity of crude methanol and flavonoid (free and bound) extracts of *Marchanita polymorpha* L. against four fungal strains, *viz., Aspergillus flavus, A. niger, Candida albicans,* and *Trychophyton mentagrophytes.* All the microorganisms were found to be sensitive against the extracts tested. Total activity was found to be maximal for free flavonoid against *Candida albicans* (199ml). Bodade *et al.* (2008) screened different bryophytes *viz. Plagiochasma appendiculatum, Thuidium delicatulum, Thuidium cymbifolium, Bryum cellulare, Bryum argenteum* and *Racomitrium crispulum* against 3 fungal pathogens and found that *Aspergillus niger* was most sensitive to the ethanol extract of *P. appendiculatum* and *B. argenteum.*

So far, almost all the researches on the bio activity of bryophytes were *in vitro* studies. The first *in vivo* experiment was performed at Bonn University by Frahm, (2004) who tested the alcoholic extract of various bryophytes in field and found active on a variety of crops infected with different fungi. Alcoholic extract of two liverworts showed systemic activity. Green house tests showed that the plants treated with liverwort extract were distinctly less affected by fungus infects *e.g. Phytopthora infestans* than untreated plants.

Table 30.7: Antimicrobial Potential of Compounds Isolated from the Tissues of Bryophytes

Sl.No.	Bryophyte	Chemical Compounds	Antibiotic Effect	Reference
1.	*Porella* sp, *Makonia* sp	cinnamolide	Antifungal	Canonica *et al.*, 1969, Subramonium Subisha, 2005
2.	*Lunularia cruciata*	Lunularin and lunularic acid	Antifungal fabae	Subramonium and Subisha, 2005
3.	*Radula* sp.	3 prenyl bibenzyl	Antibacterial	Asakawa *et al.*, 1982
4.	*Plagiochila stevensoniana*	bibenzyl	Antifungal, antibacterial	Lorimer *et al.*, 1993
5.	*Plagiochila faciculata*	Hydroxyl acetophenones	antifungal	Lorimeret al., 1993
6.	*Lunularia cruciata*	Alpha-D-oligogalacturonides	antibacterial	Basile *et al.*, ,2001
7.	*Bazzania trilobata*	Sesquiterpenes	Antifungal	Scher *et al.*, 2004
8.	*Porella sp*	polygodial	Weak antibiotic activity	Isoe 1983
9.	*Gymnomitrium obtusum*	Gymnomitriol	antifungal	Connolly *et al.*, 1972
10.	*Marchantia polymorpha, M.paleacea*	Marchantin A	Antibacterial, antifungal	Asakawa,1998

Almost, all the researches on the antimicrobial potential of bryophytes were done in crude extracts, prepared in different organic solvents. These aforesaid results provide only the baseline data which recognizes the bryophytes as a group of plants with differential degree of antimicrobial potential but without specific pharmacological studies. On account of this, these plants could not be used as a source of future medicine. In all these studies, target inhibitory compounds are not known, whether the sesquiterpenes were extracted in the solvents or some enzymes were leached out from the tissues or the by products of microbes present in the tissues are responsible for antibiosis. Therefore, there is an immediate need to extract and identify the effective compounds from these plants and to test these chemicals for their antimicrobial potential together with bactericidal and fungicidal capacity, if any. In this respect, a few researches on the antimicrobial potential of the specific compounds extracted from the tissues of liverworts and mosses are summarized in Tables 30.7 and 30.8.

Table 30.8: Compounds of Different Genera of Bryophytes Wffective on Various Microorganisms

Sl.No.	*Plant species*	*Compounds*	*Microorganisms*	*Reference*
1.	*Porella vernicosa* complex	Polygodiol	*Aspergillus fumigatus, A. niger, Candida albicans, Trichophyton mentagrophytes, Cryptococcus neofomans, Staphylococcus aureus*	Asakawa, 1998
2.	*Pellia endvaeviifolia, Tricholeopses sacculata*	Sacculatal	*Artenaria brassicicola, Aspergillus fumigatus, Candida albicans, Cryptococcus neofomans, Sacchromyces cerevisae*	Asakawa, 1998
3.	*Radula complanata, R. kojana*	Prenyl bibenzyl	*Staphylococcus aureus*	Asakawa, 1998
4.	*Lunularia cruciata* and otherhepatics	Lunularic acid	*Artenaria brassicicola, Botrytis cinerea, Septoria nodorum. Uromyces fabae*	Pryce, 1972 a
5.	*Radula complanata*	Prenyl bibenzyl	*Staphylococcus aureus*	Asakawa, 1998
6.	*Porella vernicosa*	Non pinguisone	*Aspergillus. niger*	Canonica *et al.*, 1969
7.	*Odontoschisma denudata*	Acetoxy odonto-schismenol	*Botrytis cinerea*	Matsuo *et al.*, 1985, 1988
8.	*Dicranum scoparium, D. japonicum*	Cyclopentyl fatty acids	*Botrytis cinerea*	Ando and Matsuo, 1984; Ichikawa *et al.*, 1983, 1984
9.	*Trichocolea mollissima, T. tomentella*	Tomentellin, dimethyl tomentellin	*Trichophyton mentagrophytes*	*Perry *et al.*, 1996
10.	*Trichocolea lanata*	trichocolein	*Trichophyton mentagrophytes*	*Perry *et al.*, 1996
11.	*Marchantia polymorpha, M. paleacea* var. *diptera, M.plicata, M. tosana*	Marchantin A	*Bacillus subtilis, B.cereus, B. megaterium, E. coli, Proteus mirabilis, Staphylococcus aureus, Salmonella typhimurium, Pseudomonas aeruginosa, Acinetobacter calcoaceticus, Alcaligenes, faecalis, Cryptococcus neofomans, Enterobacter cloacae, Aspergillus niger, A.fumigatus, Candida albicans, Trichophyton ruburum, Microsporum gypseum, Penicillium chrysogenum, Piricularia oryzae, Rhizoctonia solani, Sacchromyces cerevisiae, Sporothrix schenckii, Alternaria kikuchiana*	Asakawa, 1998

* Not seen in original S (Asakawa,1998).

Conclusions

Studies on the bioactivity of bryophytes suggest that these Lilliputians of plant kingdom are remarkable reservoir of chemical compounds and they have a vast potential as a botanical pesticides as well as modern day drugs in near future. They can serve as a new source of fungicides and antibiotics. Therefore, bryophytes deserve a detailed examination as an effective drug producer prior to their pharmaceutical use which seems to be promising. In the present scenario,lack of clinical trials and little understanding of their harmful side effects, the bryophytes especially liverworts containing biologically active compounds can not be used as herbal medicine or as fungicidal compounds (alternative medicine). Unfortunately, most biologically active substances so far obtained have not proved economical for use, at least in part due to the slow-growing nature, availability of pure plant species and difficulty of culturing bryophytes. Researches indicated that *in vitro* culture of spore of *Fossombronia pusilla* produces the same terpenoids constituents as is produced by the plants in wild and exerted the similar antibacterial activity.

Most of the antimicrobial activity of bryophytes is concerned with the crude extracts prepared in different organic solvents and only a few reports are cited in literature which depicted the antimicrobial potential of a particular chemical compound isolated from bryophytes. Therefore, there is also need to extract more such compounds from these plants and assess their antimicrobial potential individually. Moreover, efforts are required to increase the biomass production of bryophytes in wild as well as in culture in order to obtain large quantity of plants for experimentation and for better yield of biologically active compounds.

References

Ando, H. Matsuo, A.1984 Applied Bryology.In: W.Schultze-Motel(ed.) Advances in Bryology, 2: 133-224, J. Cramer, Vanduz(Germany).

Asakawa Y., Toyota, M. and Takemoto, T. 1980a. Mono- and Sesquiterpenoids from *Wiesnerella denudata*.Phyochemistry.19: 567-569.

Asakawa, Y. 1981. Biologically active substances obtained from bryophytes. *J. Hattori Bot. Lab. 50*: 123-142.

Asakawa, Y. 1981. Chemical constituents of the Hepaticae. Academic Press, London.

Asakawa, Y. 1984b. Some biologically active substances isolated from hepaticae: terpenoids and lipophilic aromatic compounds. *J. Hattori Bot. Lab. 56*: 215-219.

Asakawa, Y. 1988. Biologically active substances found in Hepaticae. In *Studies in Natural Products Chemistry*; Atta-ur-Rahman, (ed), Elsevier: Amsterdam, Vol. 2, 277-292.

Asakawa, Y. 1990a.Terpenoids and aromatic compounds with pharmacological activity from bryophytes In: H. D. Zinsmeister and R. Mues (eds.) Bryophytes: Their Chemistry and Chemical Taxonomy 369-410. Oxford University Press, Oxford.

Asakawa, Y. 1990b. Biologically active substances from bryophytes. In: R. N. Chopra and S. C. Bhatla (eds.) Bryophyte Development: Physiology and Biochemistry: 259-287. CRC Press, Boca Raton.

Asakawa, Y. 1995. Chemical constituents of the bryophytes. *In* W. Herz, G.W. Kirby, R. E., Moore, W. teglich., C. Tamm, (eds.), *Progress in the Chemistry of Organic Natural Products.*; 65: 1-562, Spinger: Vienna.

Asakawa, Y. 1998 Biologically active compounds from bryophytes, *J. Hattori Bot. Lab No. 84*. 91-104.

Asakawa, Y. 1999. Phytochemistry of Bryophytes. Biologically Active Terpenoids and Aromatic Compounds from Liverworts. In *Phytochemicals in Human Health Protection, Nutrition, andDefense;* Romeo, J.T., Ed.; Kluwer Academic/Plenum: New York, 319-342.

Asakawa, Y. 2007. Biologically active compounds from bryophytes. *Pure Appl. Chem.* 79(4): 557-580.

Asakawa, Y. and Aratani, T.1976. Sesquiterpenes of *Porella vernicosa* (Hepaticae). Bull. Soc.Chim.Fr. 1469-1470.

Asakawa, Y. Toyota, K. and Takemoto, T. 1982. Novel bibenzyl derivatives and Ent cup Arena Type sesquiterpenoids from Radula species. *Phytochemistry* 21: 2481.

Asakawa, Y., Tokunaga, N., Toyota, M. and Takemoto, T., Hattori, S., Mizutani, M., and Suire, C.1979 a. Chemosystematics of bryophytes.II the distribution of terpenoids in Hepaticeae and Anthocerotae. *J. Hattori Bot. Lab. 46*: 67.

Asakawa, Y.1982a Chemical constituents of the Hepaticae.Prog. Chem. Org. Natur. Prod.42: 1-285.

Banerjee, R.D. Sen, S.P. 1979 Antibiotic activity of bryophytes. *The Bryologist* 82 (2): 141-153.

Banerjee, R.D.2001. Antimicrobial activities of bryophyte-a review.In Nath, V. and Asthana, A.K, (eds.) Perspectives in Indian bryology, 55-74; Bishen Singh, Mahendra Pal Singh, Dehradun.

Basile, A. Giordano, S. Lopez-Saez, J. A. and Castaldo Cobianchi, R. 1999. Antibiotic activity of pure flavonoids isolated from mosses. *Phytochemistry* 52: 1479-1482.

Basile, A. Vuotto, M. L. and Ielpo, M. T. L. 1998. Antibacterial activity of *Rhynchostegium ripario* (Hedw.) Card. Extract (Bryophyta). *Phytother. Res.* 12: 146-148.

Basile, A., Giordano, S., Sorbo, S., Vuotto, M. L. and Ielpo, M. T. L.and Castaldo Cobianchi, R. 1998.Antibiotic effect of *Lunularia cruciata*(Bryophyta) extract. IntJ Pharmacognosy.36: 1-4.

Basile, A. Violante, U. Vuotto, M. L. and Ielpo, M. T. L. Giordano, S. and Castaldo Cobianchi, R. 1994. Presanza di attaivita antibiotic nel gametofito adulto dell epatica *Lunularia cruciata* (Briofite). Farmacognosia: Moderni suiluppi di una scienza antica. Atti del VII Congresso Nazionale della Societa Italiana di Farmacognosia (Siphar), 30 june-2 july, Bolonga.

Basile, A., Vuotto, M. L. and Ielpo, M. T. L., Giordano, S., Sorbo, S., Laghi, E. and Castaldo Cobianchi, R. 2001.Induction of antibacterial activity by alpha-D-oligogalacturonides in *Lunularia cruciata*(L.)Dum.(Bryophyta) Could this be an immune response in non vascular plants? Hematologica –supplement/27.1-2.

Benesova, V.and Herout, V.(1978). Components of liverworts. Their chemical structures and biological activity. Bryophytorum Biblotheca 13: 355-364.

Bisht, M.1998. Chemical investigation on some Himalayan bryophytes.PhD(Chemistry) thesis, Kumaon University Nainital.

Bodade, R. G. Borkar, P. S. Arfeen, M. S. and Khoragade C. N.2008 *In vitro* screening of bryophytes for antimicrobial activity. *J. of Med. Plants* 7(4): 23-28.

Canonica, L.A. Corbella, A., Gariboldi, P., Jommi, G., Krepinsky, J., Ferrari, G.and Cassagrande, D.!969. Sesquiterpenoids and cinnamomsa fragrans Structure of cinnamolide cinnamosmolide and cinnamodialTetrahedron Lett.25: 3895.-3902.

Castaldo Cobianchi, R. Giordano, S. Basile, A. and Violante, V. 1989. Occurrence of antibiotic activity of *Conocephalum conicum, Mnium undulatum* and *Leptodictyum riparium* (Bryophytes). *G. Bot. Ital.* 122: 303-312.

Clipa, V. and P. Pascal. 1986. The study of toxicity, antibiotic and antimycotic effects of polytrichum juniperinum extracts. IASI Lvcr. Stiini. Ser. Zooth. Med. Vet. 30: 93-94.

Connolly, J. D. 1990. Monoterpenoids and sesquiterpenoids from the Hepaticae. In: Zenmiester, H.D., and Mues, R. (eds) Bryophytes; Their chemistry and chemical taxonomy. Oxford: Clarendon Press, 40-58.

Connolly, J. D. and Harding, A.E. and Thornton, I.M.S. 1972. Gymnomitrol, a novel tricyclic sesquiterpenoid from *Gymnomitrol obtusum* (Lind b.) Pears (Hepaticae). *J. chem. soc. Chem. Comm.* 1320-1321.

Ding, H. 1982. Medical spore bearing plants of China. 499. Shanghai.

Dixit, A. Pandey D.K., Nath S. 1982 Antifungal activity of some bryophytes against human pathogens. *J. Indian Bot. Soc.* 61: 447-448.

Dubey, R. C. Vashistha, H. Tripathi, P. and Tewari, S. D.2001. Antifungal activities of three hepatics against *Macrophomina phaseolina. Indian Phytopahtology* 54 (2): 264-266.

Figueirdo, A. C. Sin-Sim, M. Barroso, J. G. Pedro, L. G. Santos, P. A. G. Schripsema, J. Deans, S. G. and Scheffer, J. J. C.1999. Composition and antibacterial activity of the essential oil from the liverwort *Marchesinia mackii* (Hook) S. F. Gray. Int. sym. On essential oils, Leipzig, Germany.

Frahm, J.P. 2004 New frontiers in bryology and lichenology: Recent developments of commercial products from bryophytes. *Bryologist* 107: 277-283.

Glime, J. M. 2007. Economic and ethnic uses of bryophytes. *In*: Flora of North America Editorial Committee. (eds.). Flora of North America North of Mexico. Vol. 27. Bryophyta, part 1. Oxford University Press, New York. pp. 14-41.

Gorham, J. 1990.Phenolic compoundsa, other than flavoinoids from bryophytes.In: : Zensmiester, H.D., and Mues, R. (eds) Bryophytes; Their chemistry and chemical taxonomy. Oxford: Clarendon Press, 171-200.

Gupta, K.G. and Singh, B.1971 Occurrenceof antibacterial activity in moss extracts.Res.Bull.Punjab Unlv.Sci.22: 237-239.

Hayes, L. E. 1947. Survey of higher plants for presence of antibacterial substances. Bot.Gaz. 108: 408-414.

Herout, V. 1990. Diterpenes and higher terpenes from bryophytes. In: Zenmiester, H.D., and Mues, R. (eds) Bryophytes; Their chemistry and chemical taxonomy. Oxford: Clarendon Press, 83-102.

Homans, A.L. and Fuchs, A. 1970. Direct bioautography on thin layer chromatograms as a method for detecting fungitoxic substances. *J. Chromatogr.* 51: 327-329.

Hoof, L. V. Berghe, D. A. Petit, E. and Vlientinck, A. K. 1981. Antimicrobial and antiviral screening of bryophytes. *Fitoterapia.* 223-229.

Huneck, S. 1983 Chemistry and biochemistry of bryophytes. In Schuster, R.M. (ed.) New Manual of Bryology. 1. Nichinan: Hattori Botanical Laboratory, 1-116.

Ichikaawa, T.1982. Biologically active substances in mosses. Bryon(Kangawa Koke no kai) 2: 1-2(in Japanese).

Ichikaawa, T., Namikawa, M., Yamada, K., Sakai, K., and Kondo, K.1983.Novel cyclopentyl fatty acids from mosses, *Dicranum scoparium* and *Dicranum japonicum.*TetrahedronLett.24: 3327-3340.

Ilhan, S. Savaroulu, F. Olak, F. C. and Erdemgul, F. Z. 2006. Antimicrobial activity of *Palustriella commutata*(Hedw.) Ochyra Extracts (Bryophyta) *Turk J. Biol* 30: 149-152.

Isoe, S.1983.Terpene dials.Biological activity and synthetic study.48th Annual Meeting of the Chemical Society of Japan.Proceeding PapersII, 849-850(In Japanese).

Jennings, O. E. 1926. Mosses immune to molds. *Bryologist* 29: 75-76.

Joshi, D. Y. 1993. In pursuit of the antibacterial property of *Lunularia cruciata* Dum. and *Dumortiera hirsuta Adv. Plant Sci.* 6 (1): 66-70.

Joshi, D. Y. and Desai, A. C. 1998. Antimicrobial properties of *Fossombronia himalayensis*.Kash. *Phytomorphology* 38 (2, 3): 201-203.

Joshi, D. Y., Wani, D. D. and Deshpande, S. 1990. Antimicrobial properties of *Exormotheca tuberifera* Kash. *Adv. Plant Sci.* 3: 147-151.

Joshi. D. Y. 1995 Antimicrobial activity of *Reboulia hemispherica* (L.)Raddi *J. Indian Bot. Soc.* 74: 321-322.

Joshi. P. K 1999. Chemical investigations of some aromatic Himalayan bryophytes. Ph.D. (Chemistry) thesis, Kumaon University, Nainital.

Kinzel, H.and Walland, A. 1966 Vorkommen von Shikimisacure bei moosen undFarnen. Z.Pflanzenphysiol., 54: 371-374.

Kumar, K., Singh, K.K., Asthana, A.K. and Nath, V. 2000. Ethnotheraupeutics of bryophyte *Plagiochasma appendiculatum* among the Gaddi tribes of Kangra valley, Himachal Pradesh, India.Pharmaceutical Biology.38: 353-356.

Latiff, A., Tumiri, S. Z., and Mohammad, A. D. H. 1989. The effect of moss extract on the three species of bacteria. *Appl. Biol.* 18: 77-84.

Lorimer, S. D. Perry, N. D.and Tangney, R. S. 1993. An antifungal bibenzyl from the New Zealand liverworts *Plagiochila stephensoniana*. Bioactivity directed isolation, synthesis and analysis. *J Natl. Products* (Lloydia) 56: 1444-1450.

Madeson, G.C.and Pates, A, L. 1952.Occurrence of antimicrobial substances in chlorophyllose plants growing in Florida.Bot.Gaz. 113: 293-300.

Markham, K. R. Porter, L. J. 1978. Chemical constituents of the bryophytes. In: Reinhold, l., Harborne, J.B. and Swain, T.(eds.)Progress in Phytochemistry 5: 181-272.

Markham, K. R., Porter, L.J.and Brehm B.G., 1969.Flavonoid C-glycosides in the Hepaticae (liverwort. Phytochemistry 8: 2193-2197.

Matsuo, A. Uto, S., Nozaki, H. and Nakayama, M. 1984.Structure and absolute configuration of(+) vitrenal, a novel carbon skeletal sesquiterpenoids having plant growth inhibitory activity, from the liverwort Leoidozia vitrea, J.Chem.Soc> Perkin Trans> !: 215-221.

Matsuo, A. Yuki, S. and Nakayama, M. 1983. -(-)-Herbertenediol and (-)-herbertenolide, two new sesquiterpenoids of the *ent*-herbertane class from the liverwort *Herberta adunca*. Chem. Lett. pp. 1041-1042.

Matsuo, A., Yuki, S., Higashi, R., Nakayama, M., and Hayashi, S. 1982a. Structure and biological activity of several sesquiterpenoids having a novel herbertane skeleton from the liverwort *Herberta adunca*. 25th Symposium on Chemistry of Natural Products. Symposium papers. pp. 242-249.

Matsuo, A., Yuki, S., Nakayama, M., and Hayashi, S. 1982b. Three new sesquiterpene phenols of the *ent*-herbertane class from the liverwort *Herberta adunca*. Chem. Lett. pp. 463-466.

Matsuo, A., Nakayama, M., Maeda, T., Noda, Y.and Hayashi, S.1975 Enatiomeric cupranene-type sesquiterpenoids from *Bazzania pomoeana*.Phyochemistry, 14: .1037.

McCleary, J.A. and Walkington, D.L. 1966. Mosses and antibiosis.Rev.Bryol.Lichenol.24: 309 -314.

McCleary, J.A., Sypherd, P.S. and Walkington, D.L. 1960. Mosses as possible sources of antibiotics.Science 131: 108.

Mekuria, T. Blaeser, P. Steiner, U. Dehne, H. W. Lyr, H. Russell, P. E. and Sisler, H. D. 1999. Bryophytes as a new source of antifungal substances in crop protection. Modern fungicides and antifungal compounds II. 12th Int. Reinhardsbrunn Symposium, Fridridhorda, Thuringia, Germany. 359-360.

Mendez, J and Sanz Cabanilles 1979. Cinnamic esters in Anthoceros species. *Phytochemistry* 18: 1409.

Mewari, N. and Kumar, P. 2008. Antimicrobial activity of extracts of *Marchantia polymorpha*. Pharma. Boil. informa healthcare 46 (10-11): 819-822.

Mewari, N. Chaturvedi, P. Kumar, P. and Rao P. B. 2007. Antimicrobial activity of moss extracts against plant pathogens. *Jour, mycol pl pathol* 37(2): 359-360.

Mues, R. Zinsmeister, H. D. and Brehm, B. G. 1976. Flavinoid variation in the liverwort *Conocephalum conicum*. Evidence for geographic races. *Phytochemistry* 15: 147-150.

Pant, G. and Tewari, S. D. 1989Various human uses of bryophytes in the Kumaon region of north-west Himalaya.Bryologist.92(1): 120-122.

Pavletic, Z.and Stilinovic, B.1963.Istrazivanja antibiotskog djelovanja ekstrakta mahovina na neke bakterije. Acta Bot. Croatica 22: 133-139.

Pryce, R. J. 1972. Metabolism of lunularic acid to a new plant stilbene by *Lunularia cruciata. Phytochemistry* 11: 1355-1364.

Rawat, S.2007 Species diversity of Bryophytes in Oak forests and their allelopathic effects. D thesis in Botany, Kumaon University, Nainital.

Raymundo, A. K. Tan B. C. and Asuncion, A. C. 1989. Antimicrobial activities of some Philippine cryptogams. *Philippine Journal of science*.118: 59-75.

Sabovljevic, D.A. Sokovic, A., Sabovljevic, M. and Grubisic, M. 2006. Antimicrobial activity of *Bryum argenteum. Filoterapia* 77: 144-145.

Sauerwein, M. and Becker, H. 1990. Growth terpenoids production and antibacterial activity of an in vitro culture of a liverwort *Fossombronia pusilla. Planta Med.* 56: 364-367.

Scher, J. M. Speakman, J. B. Zapp, J. Becker. H. 2004.Bioactivity guided isolation of antifungal compounds from the liverwort *Bazzania trilobata* (L.) S. F. Gray *Phytochemistry* 65: 2583-2588.

Schofield, W.B. 1985.Introduction to Bryology. Mac Millan, NewYork.

Schuster, R.M. 1966. The Hepaticae and Anthocerotae of North America.Vol1. Columbia University Press.New York And London 802.

Shirzadian, S. Afshaari, A. Homayoun, Eskandari, M. and Mirza I. M. 2005. Study of antifungal activity of bryophytes (mosses and liverworts). *Agris:* IR2007000976.

Singh, M. Rawat, A. K. S. and Govindarajan, R. 2007. Antimicrobial activity of some Indian mosses. *Fitoterapia.* 78: 156-158.

Singh. M., Govindrajan R., Nath. V., Rawat, A. K. S., Mehrotra., S., 2006 Antimicrobial, wound healing and antioxidant activity of *Plagiochasma appendiculatum* Lehm. Et Lind. *Journal of Ethnopharmacology* 107: 67-72.

Subramonian, A., Subhisha, S. 2005, Bryophytes of India: A potential source of antimicrobial agents.In: Khan, I.A. and Khanum, A. (eds.)Role of Biotechnology in Medicinal and Aromatic Plants.Vol.II.Hyderabad, India: Ukaaz Publications.

Swanson, E.S., Anderson, W.H., Gellerman, J.L.and Schlenk, H. 1976. Ultrastructure and lipid composition of mossesBryologist, 79: 339-349.

Theophil, E. 1990 Recent results from the synthesis of chemical constituents of bryophytes. 209-223.In: Zinsmeister, H. D.and Mues, R. (eds.). Bryophytes; Their chemistry and chemical taxonomy. Oxford: Clarendon Press.

Vashistha, H. Dubey, R. C. and Pandey, N. 2007. Antimicrobial activity of three bryophytes against human pathogens. 47-59 In Nath, V and Asthana, A.K.(eds.) Current Trends in Bryology. Bishen Singh Mahendra Pal Singh, Dehradun.

Wolters, B. 1964. Die Verbrietung antifungaler Eigenschaften bei Moosen. *Planta.* 62: 88-96.

Zehr, D.R. 1990. A simplified technique for assaying the production of bacteriostatic compounds by hepatics.Lindbergia 16: 128-132.

Zinsmeister, H. D.and Mues, R. 1990. Bryophytes; Their chemistry and chemical taxonomy. Oxford: Clarendon Press.

Zinsmeister, H. D.Becker, H. Eicher, Th. 1991. Bryophytes, a source of biologically active, naturally occurring material, Angew. C.